VOLUME SIX HUNDRED AND THIRTY SEVEN

METHODS IN ENZYMOLOGY

Retinoid Signaling Pathways

METHODS IN ENZYMOLOGY

Editors-in-Chief

ANNA MARIE PYLE
Departments of Molecular, Cellular and Developmental Biology and Department of Chemistry
Investigator, Howard Hughes Medical Institute
Yale University

DAVID W. CHRISTIANSON
Roy and Diana Vagelos Laboratories
Department of Chemistry
University of Pennsylvania
Philadelphia, PA

Founding Editors

SIDNEY P. COLOWICK and NATHAN O. KAPLAN

VOLUME SIX HUNDRED AND THIRTY SEVEN

METHODS IN ENZYMOLOGY

Retinoid Signaling Pathways

Edited by

EHMKE POHL

Department of Chemistry; Department of Bioscience; Biophysical Sciences Institute, Durham University, Durham, United Kingdom

ELSEVIER

ACADEMIC PRESS

An imprint of Elsevier

Academic Press is an imprint of Elsevier
50 Hampshire Street, 5th Floor, Cambridge, MA 02139, United States
525 B Street, Suite 1650, San Diego, CA 92101, United States
The Boulevard, Langford Lane, Kidlington, Oxford OX5 1GB, United Kingdom
125 London Wall, London, EC2Y 5AS, United Kingdom

First edition 2020

ISBN: 978-0-12-820144-2
ISSN: 0076-6879

For information on all Academic Press publications
visit our website at https://www.elsevier.com/books-and-journals

Publisher: Zoe Kruze
Editorial Project Manager: Shellie Bryant
Production Project Manager: James Selvam
Cover Designer: Alan Studholme

Typeset by SPi Global, India

Contents

Contributors

Mark K. Adams
Stowers Institute for Medical Research, Kansas City, MO, United States

Mary Ann Asson-Batres
Tennessee State University, Nashville, TN, United States

Olga V. Belyaeva
Department of Biochemistry and Molecular Genetics, Schools of Medicine and Dentistry, University of Alabama at Birmingham, Birmingham, AL, United States

Danika Bongfeldt
Medical Sciences Division, Northern Ontario School of Medicine; Departments of Chemistry and Biochemistry, and Biology and Biomolecular Sciences Program, Laurentian University, Sudbury, ON, Canada

William Bourguet
Center for Structural Biochemistry (CBS), Univ Montpellier, INSERM, CNRS, Montpellier, France

Matteo Bozzo
Laboratory of Developmental Neurobiology, Department of Earth, Environment and Life Sciences, University of Genoa, Genoa, Italy

Simona Candiani
Laboratory of Developmental Neurobiology, Department of Earth, Environment and Life Sciences, University of Genoa, Genoa, Italy

Yunting Chen
Department of Physiology and Cellular Biophysics, Columbia University Irving Medical Center, New York, NY, United States

David R. Chisholm
LightOx Ltd., Newcastle upon Tyne; Centre for Sustainable Chemical Processes, Department of Chemistry, Science Laboratories, Durham University, Durham, United Kingdom

Oliver B. Clarke
Department of Physiology and Cellular Biophysics; Department of Anesthesiology, Columbia University Irving Medical Center, New York, NY, United States

Brianna K. Costabile
Department of Physiology and Cellular Biophysics, Columbia University Irving Medical Center, New York, NY, United States

Lindsay C. Czuba
Department of Pharmaceutics, School of Pharmacy, University of Washington, Seattle, WA, United States

Angel R. de Lera
Departamento de Química Orgánica, Facultade de Química, CINBIO and IBIV, Universidade de Vigo, Campus As Lagoas-Marcosende, Vigo, Spain

Amy E. Defnet
University of Maryland Baltimore, School of Pharmacy, Department of Pharmaceutical Sciences, Baltimore, MD, United States

Patricia García
Departamento de Química Orgánica, Facultade de Química, CINBIO and IBIV, Universidade de Vigo, Campus As Lagoas-Marcosende, Vigo, Spain

Pierre Germain
Center for Structural Biochemistry (CBS), Univ Montpellier, INSERM, CNRS, Montpellier, France

Kelli R. Goggans
Department of Biochemistry and Molecular Genetics, Schools of Medicine and Dentistry, University of Alabama at Birmingham, Birmingham, AL, United States

Timothy Goodman
School of Pharmacy and Bioengineering, Keele University, Keele, United Kingdom

Nina Isoherranen
Department of Pharmaceutics, School of Pharmacy, University of Washington, Seattle, WA, United States

Maureen A. Kane
University of Maryland Baltimore, School of Pharmacy, Department of Pharmaceutical Sciences, Baltimore, MD, United States

Natalia Y. Kedishvili
Department of Biochemistry and Molecular Genetics, Schools of Medicine and Dentistry, University of Alabama at Birmingham, Birmingham, AL, United States

Youn-Kyung Kim
Department of Food Science and Rutgers Center for Lipid Research and New Jersey Institute of Food Nutrition and Health, Rutgers University, New Brunswick, NJ, United States

Alla V. Klyuyeva
Department of Biochemistry and Molecular Genetics, Schools of Medicine and Dentistry, University of Alabama at Birmingham, Birmingham, AL, United States

Azita Kouchmeshky
Institute of Medical Sciences, University of Aberdeen, Aberdeen, Scotland, United Kingdom

Albane le Maire
Center for Structural Biochemistry (CBS), Univ Montpellier, INSERM, CNRS, Montpellier, France

Yu-Lung Lin
Department of Pharmacology, University of Minnesota Medical School, Minneapolis, MN, United States

Paula Lorenzo
Departamento de Química Orgánica, Facultade de Química, CINBIO and IBIV, Universidade de Vigo, Campus As Lagoas-Marcosende, Vigo, Spain

Filippo Mancia
Department of Physiology and Cellular Biophysics, Columbia University Irving Medical Center, New York, NY, United States

Floyd J. Mattie
Department of Nutritional Sciences, Pennsylvania State University, University Park, PA, United States

Peter McCaffery
Institute of Medical Sciences, University of Aberdeen, Aberdeen, Scotland, United Kingdom

Alexander R. Moise
Medical Sciences Division, Northern Ontario School of Medicine; Departments of Chemistry and Biochemistry, and Biology and Biomolecular Sciences Program, Laurentian University, Sudbury, ON, Canada

Joseph L. Napoli
Graduate Program in Metabolic Biology, Nutritional Sciences and Toxicology, University of California, Berkeley, CA, United States

Jennifer Nhieu
Department of Pharmacology, University of Minnesota Medical School, Minneapolis, MN, United States

Christopher W. Norwood
Tennessee State University, Nashville, TN, United States

Judit Osz
Institut de Génétique et de Biologie Moléculaire et Cellulaire (IGBMC); Centre National de la Recherche Scientifique; Institut National de la Santé et de la Recherche Médicale; Université de Strasbourg, Illkirch, France

Carole Peluso-Iltis
Institut de Génétique et de Biologie Moléculaire et Cellulaire (IGBMC); Centre National de la Recherche Scientifique; Institut National de la Santé et de la Recherche Médicale; Université de Strasbourg, Illkirch, France

Ehmke Pohl
Department of Chemistry; Department of Bioscience; Biophysical Sciences Institute, Durham University, Durham, United Kingdom

Loredana Quadro
Department of Food Science and Rutgers Center for Lipid Research and New Jersey Institute of Food Nutrition and Health, Rutgers University, New Brunswick, NJ, United States

Chris P.F. Redfern
School of Natural & Environmental Sciences, Newcastle University, Newcastle upon Tyne, United Kingdom

Natacha Rochel
Institut de Génétique et de Biologie Moléculaire et Cellulaire (IGBMC); Centre National de la Recherche Scientifique; Institut National de la Santé et de la Recherche Médicale; Université de Strasbourg, Illkirch, France

Atharva Piyush Rohatgi
Department of Pathology and Laboratory Medicine, University of California Davis Health, Sacramento, CA, United States

A. Catharine Ross
Department of Nutritional Sciences; Center for Molecular Immunology and Infectious Disease, Pennsylvania State University, University Park, PA, United States

Michael Schubert
Sorbonne Université, CNRS, Laboratoire de Biologie du Développement de Villefranche-sur-Mer, Villefranche-sur-Mer, France

Stephen R. Shannon
Stowers Institute for Medical Research, Kansas City, MO; University of Kansas Medical Center, Department of Anatomy and Cell Biology, Kansas City, KS, United States

Charles W.E. Tomlinson
Department of Chemistry, Durham University, Lower Mountjoy, Durham, United Kingdom

Paul A. Trainor
Stowers Institute for Medical Research, Kansas City, MO; University of Kansas Medical Center, Department of Anatomy and Cell Biology, Kansas City, KS, United States

Yu-Jui Yvonne Wan
Department of Pathology and Laboratory Medicine, University of California Davis Health, Sacramento, CA, United States

Lijun Wang
Department of Pathology and Laboratory Medicine, University of California Davis Health, Sacramento, CA, United States

Cheng-Hsin Wei
Department of Nutritional Sciences, Pennsylvania State University, University Park, PA, United States

Li-Na Wei
Department of Pharmacology, University of Minnesota Medical School, Minneapolis, MN, United States

Andrew Whiting
Department of Chemistry, Durham University, Lower Mountjoy; Centre for Sustainable Chemical Processes, Department of Chemistry, Science Laboratories, Durham University, Durham, United Kingdom

King C. Yabut
Department of Pharmaceutics, School of Pharmacy, University of Washington, Seattle, WA, United States

Hong Sik Yoo
Graduate Program in Metabolic Biology, Nutritional Sciences and Toxicology, University of California, Berkeley, CA, United States

Jianshi Yu
University of Maryland Baltimore, School of Pharmacy, Department of Pharmaceutical Sciences, Baltimore, MD, United States

Guo Zhong
Department of Pharmaceutics, School of Pharmacy, University of Washington, Seattle, WA, United States

Reza Zolfaghari
Department of Nutritional Sciences, Pennsylvania State University, University Park, PA, United States

Preface

Vitamin A is arguably the most important, multifunctional nutrient required from embryogenesis to adulthood; early vitamin A deficiency leads to neonatal growth defects and a wide range of congenital malformations. Shortage of vitamin A is the leading cause of preventable childhood blindness affecting up to half a million malnourished children in developing countries. In adults, vitamin A is indispensable for vision, the function of the immune system, and neural plasticity, which includes the capacity for neurogenesis, neuronal survival, and neurite outgrowth. Consequently, a shortfall has severe effects including night-blindness and a compromised immune system. Vitamin A is available directly from various animal products, and in the form of precursor beta-carotene from plant-based diets. In the body, vitamin A_1 (or retinol) is converted into a number of metabolites often referred to collectively as retinoids. 11-*cis* retinal, for example, is the chromophore in the photoreceptor rhodopsin and therefore indispensable for vision. All-*trans* retinoic acid, also known as ATRA, is generated by a complex pathway that relies on a range of carrier proteins as well as various enzymes of the dehydrogenase/reductase and aldehyde dehydrogenase families. ATRA is the predominant cellular signaling retinoid responsible for a plethora of cellular processes including proliferation, differentiation, and the activation of tumor suppression genes. As a result, natural retinoids and synthetic retinoid analogues have long been investigated for therapeutic applications, however, due to inherent toxicity and instability, the clinical usage has so far been limited.

While vitamin A and its growth-stimulating effects were discovered in the early 20th century, it took many decades to unravel the underlying metabolic and catabolic pathways and to begin to understand the complex networks that underpin vitamin A signaling. To this day, we are still discovering new functions, pathways, novel regulatory mechanisms, and proteins that govern cellular vitamin A signaling. This volume attempts to give an update on the current state of knowledge in the field and provides a wide range of techniques covering the molecular, cellular, and the full-organism levels.

The first three chapters summarize the biochemistry of natural retinoids; followed by Chapters 4–6, describing methods to investigate key components in the metabolic pathways of vitamin A. Chapters 7–10 are focused on the mechanisms of retinoid binding to their respective molecular targets.

They include detailed protocols for a wide range of molecular biology, biochemical, and biophysical techniques available today. Chapters 11 and 12 concentrate on methods to investigate the noncanonical signaling pathways and the analysis of retinoids in their biological environment (Chapter 13). The next chapters are focused on methods in cellular biology describing studies on stem cells (Chapter 14), gene expression patterns (Chapter 15), and the development in the model organism of *amphioxus*, better known as lancelets (Chapter 16). The remaining chapters of this volume give an overview of the design principles of synthetic retinoids (Chapter 17) and summarize key cellular biology methods to investigate the effects of natural and synthetic retinoids (Chapters 18–21). The final chapters also conclude with an overview of the wide therapeutic potential of synthetic retinoids. Taken together this volume aims to represent an extensive compendium of retinoid signaling that will inform and inspire current and future generations of scientists.

Ehmke Pohl
Department of Chemistry; Department of Bioscience;
Biophysical Sciences Institute, Durham University,
Durham, United Kingdom

CHAPTER ONE

Vitamin A and its natural derivatives

Chris P.F. Redfern*
School of Natural & Environmental Sciences, Newcastle University, Newcastle upon Tyne, United Kingdom
*Corresponding author: e-mail address: chris.redfern@ncl.ac.uk

Contents

Abstract

Vitamin A and derivatives, the natural retinoids, underpin signaling pathways of cellular differentiation, and are key chromophores in vision. These functions depend on transfer across membranes, and carrier proteins to shuttle retinoids to specific cell compartments. Natural retinoids, ultimately derived from plant carotenoids by metabolism to all-*trans* retinol, are lipophilic and consist of a cyclohexenyl (β-ionone) moiety linked to a polyene chain. This structure constrains the orientation of retinoids within lipid membranes. *Cis-trans* isomerization at double bonds of the polyene chain and s-*cis*/s-*trans* rotational isomerization at single bonds define the functional dichotomy of retinoids (signaling/vision) and specificities of interactions with specific carrier proteins and receptors. Metabolism of all-*trans* retinol to 11-*cis* retinal, transfer to photoreceptors, and removal and recycling of all-*trans* retinal generated by photoreceptor irradiation, is the key process underlying vision. All-*trans* retinol transferred into cells is metabolized to all-*trans* retinoic acid and shuttled to the cell nucleus to regulate gene expression controlling organ, tissue and cell differentiation, and cellular homeostasis. Research methods need to address the potential of photoisomerization *in vitro* to confound research results, and data should be interpreted in the context of membrane-association properties of retinoids and physiological concentrations *in vivo*. Despite a century of research, there are many fundamental questions of retinoid cellular biochemistry and molecular biology still to be answered. Computational modeling techniques will have an important role for understanding the nuances of vitamin A signaling and function.

Methods in Enzymology, Volume 637
ISSN 0076-6879
https://doi.org/10.1016/bs.mie.2020.02.002

1. Vitamin A: Molecules and biochemical properties

1.1 Vitamin A brief history and nomenclature

Vitamin A has a long history of research, accelerated by the necessity for good nutrition during economic depression and rationing in the first half of the 20th century (Huxley, Lloyd, Goldacre, & Neil, 2000; Mozaffarian, Rosenberg, & Uauy, 2018; Wolf, 2002). This fat-soluble vitamin was linked to deficiencies in vision and growth (Wolf, 2002). Since then, through combinations of studies on humans and animals, vitamin A or retinol has been linked to the development, maintenance and/or function of most of the organ systems in the body (Sun & Kawaguchi, 2011).

Retinol is derived in the diet either directly, or by cleavage of carotenoids. Natural vitamin A and its derivatives comprise a trimethylcyclohexenyl ring (β-ionone) with a conjugated polyene of four double bonds (Fig. 1). One of the first assays used to define "vitamin A activity" was a growth assay in young rats maintained on a vitamin A-deficient diet; although this may seem unrefined, it is an approach still in current use (Riabroy, Dever, & Tanumihardjo, 2014). Using such assays, functionally important structural elements of vitamin A can be identified. For example, moving the retinol C5—C6 double bond to the C4—C5 position (alpha-retinol, Fig. 1) reduces activity, but modifications elsewhere, such as replacing the polyene

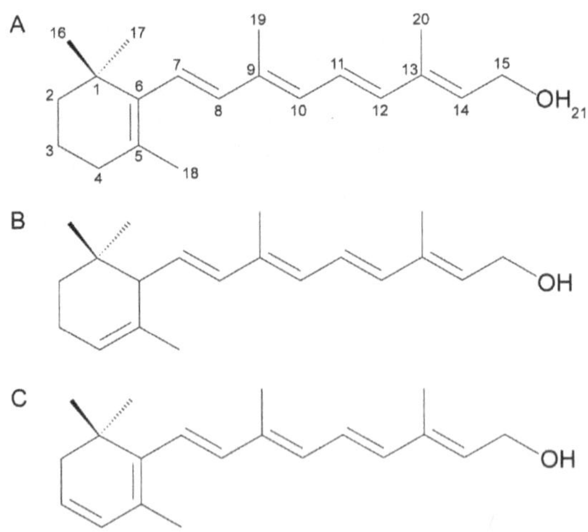

Fig. 1 Structure diagrams for vitamin A (A; retinol); alpha-retinol (B); and 3,4-didehydroretinol (C). The polyene chain is shown in the *s-cis* configuration relative to the cyclohexenyl ring (see also Fig. 2).

alcohol group with an acid, can have little effect on vitamin A activity. Vitamin A may have different chemical forms in different animals: 3,4-didehydroretinol (also referred to as vitamin A2, Fig. 1) is a retinol-equivalent in some fish (La Frano, Cai, Burri, & Thilsted, 2018) and has considerable bioactivity in mammalian cells (Riabroy et al., 2014). This raises the question of nomenclature: vitamin A-active compounds are those, other than carotenoids, having qualitative biological activities similar to retinol; compounds with retinol activity or which are structural analogs of retinol are referred to as "retinoids" (Sporn, Roberts, & Goodman, 1984).

1.2 Natural derivatives of retinol

Despite the epithet "vitamin," retinol is best regarded as a provitamin, along with the carotenoids from which it is derived by central or eccentric cleavage. Retinol or retinyl esters are obtained from dietary animal sources, or derived from plant carotenoids. Plant carotenoids, particularly β-carotene (Fig. 2), are substrates for the cytosolic symmetric cleavage enzyme β-carotene 15,15′-oxygenase (EC 1.13.11.63; BC01) which cleaves the molecule to two molecules of all-*trans* retinaldehyde (retinal) (Devery & Milborrow, 1994); retinal can be converted to all-*trans* retinol *via* aldehyde reductase activity with cellular retinol binding protein (CRBP) and converted to retinyl esters for storage (Ong, 1993) by lecithin retinol acyltransferase (LRAT; phosphatidylcholine-retinol *O*-acyltransferase, EC 2.3.1.135). A related polyene oxygenase, mitochondrial (Raghuvanshi, Reed, Blaner, & Harrison, 2015) β-carotene 9′,10′-oxygenase (EC 1.13.11.71; BCO2) cleaves carotenoids asymmetrically to yield apo-carotenals and β-ionone (Von Lintig, 2012); the latter product may be eliminated (Ide & Toki, 1970), but apo-carotenals may have biological activity as retinoids or retinoid antagonists (Eroglu et al., 2012; Harrison, Dela Sena, Eroglu, & Fleshman, 2012).

In intact biological systems, vitamin A activity is a result of the metabolism of retinol by retinol dehydrogenases to retinal and then to retinoic acid. In broad terms, two retinol derivatives have key biological roles: (1) 11-*cis* retinal is essential for vision as a visual pigment chromophore and (2) all-*trans* retinoic acid (ATRA; Fig. 2) is the key mediator of the role of vitamin A in the development, maintenance and/or function of bodily organ systems (growth regulation) by facilitating the transcriptional regulatory activity of specific receptors within the cell nucleus. Both of these biological roles depend on interactions between the retinoid and receptor

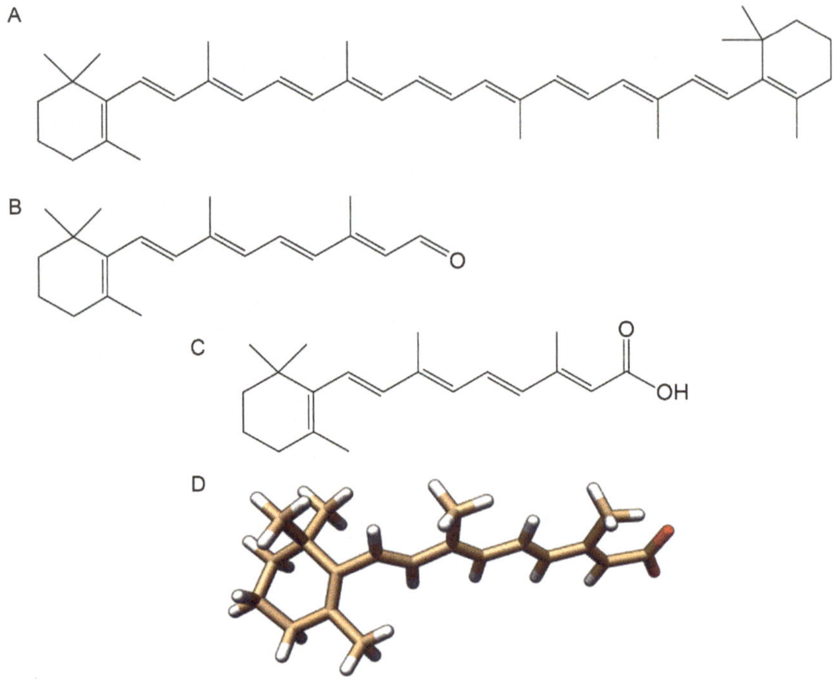

Fig. 2 Structure diagrams for a vitamin A precursor (A; beta-carotene); all-*trans* retinal (B), the product of beta-carotene cleavage by β-carotene 15,15′-oxygenase; all-*trans* retinoic acid (C, ATRA), the oxidation product of retinaldehyde dehydrogenase activity on all-*trans* retinal. (D) ATRA is shown as a 3D stick representation of the lowest-energy conformer calculated with respect to alkene stereochemistry by molecular modeling. *Reproduced from Haffez, H., Chisholm, D. R., Valentine, R., Pohl, E., Redfern, C., & Whiting, A. (2017). The molecular basis of the interactions between synthetic retinoic acid analogues and the retinoic acid receptors.* Medicinal Chemistry Communications, 8, *578–592, by permission of The Royal Society of Chemistry.*

proteins, and these interactions are dependent on the β-ionone-methyl-substituted polyene structure of the natural retinoid molecule.

In the cyclohexenyl ring, carbons 1, 4, 5, and 6 are in one plane with C2 and C3 on opposite sides of the plane but rapidly interconverting in orientation across the plane (Van Aalten, De Groot, Berendsen, & Findlay, 1996). The polyene has rotational freedom relative to the cyclohexenyl ring around C6—C7 with stable and interconverting conformers (Van Aalten et al., 1996). These configurations, together with varying s-*cis*/s-*trans* rotational conformations at other single bonds (Cox, Muccio, & Hamilton, 2013; Haffez et al., 2017, 2018) and endothermic *cis-trans* isomerization at the alternating polyene double bonds (Rando, 1991), determine or allow retinol or its derivatives to fit into different ligand-binding pockets in transport proteins or receptors and influence interactions with lipid membranes.

A key property of the polyene chain of retinol and its derivatives is that this can exist in different isomeric states as a result of *cis-trans* isomerization at double bonds to a mixture of the major isomers all-*trans* (the lowest-energy isomer), 13-*cis*, 9-*cis*, 11-*cis* and minor mono- and di-*cis* isomers (Bull et al., 2019). For vision, *cis-trans* isomerization at C11 generates the chromophore. While ATRA is the main isomer of retinoic acid for transcriptional regulation *via* nuclear receptors, the consequences of isomerization, particularly at C9, for transcriptional regulation is uncertain. The susceptibility of retinoids to *cis-trans* photoisomerization is also an inconvenience for experimental work because it is necessary to ensure that the concentration of the retinoid isomer being studied is as expected (Murayama, Suzuki, & Matsui, 1997).

1.3 Lipid properties of retinoids

Retinol and its derivatives are lipophilic (Herr, Li, Weinberg, Cook, & Storch, 1999), with low solubility in water (Szuts & Harosi, 1991), and will insert into lipid bilayers with the long axis of the retinoid molecule parallel to membrane lipid constituents (Svensson, Lincoln, Nordén, & Esbjörner, 2007). The three main natural retinoid derivatives (retinol, retinal and retinoic acid; all-*trans* isomers) differ in depth and stability within lipid bilayers; this is mainly determined by polarity at the end of the polyene chain with retinoic acid, the most polar, interacting more with lipid bilayer head groups at the membrane-cell lumen interface, whereas retinaldehyde, the least polar, adopts a lower position with greater freedom of movement of the cyclohexenyl moiety within the central hydrophobic domain of the lipid bilayers (Svensson et al., 2007). The *cis-trans* configuration of the polyene chain may also affect the physical properties of retinoid/membrane interactions (Tarapcsák et al., 2017), but this needs to be studied in more detail. However, an important consequence of intercalation of retinoids into cellular membranes will be effects on membrane viscosity, particularly at high levels of micromolar concentrations or above (Meeks, Zaharevitz, & Chen, 1981; Tarapcsák et al., 2017). This is important for *in vitro* experimental work because many experiments investigating the biological properties of retinoids are carried out using μM retinoid concentrations. At these concentrations, physical effects on membrane function will be apparent, and may vary between retinoid isomers (Tarapcsák et al., 2017).

Critically, such effects may have biological consequences, such as the induction of apoptosis, which may not be representative of the signaling function of retinoids at physiological concentrations (Bell et al., 2013).

The lipophilic properties of retinoids necessitate transport mechanisms for delivery to target cells, and for intracellular metabolism and transfer between different biochemical processes and intracellular compartments. These processes are mediated by retinoid binding proteins, but the function of such proteins extends beyond just transport and are essential for isomerization specificity or downstream selectivity of particular retinoid isomers (McBee et al., 2000). In broad terms, there are five classes of retinoid binding proteins necessary for retinoid signaling: (1) eight-stranded β-barrel proteins which bind extracellular retinoids, such as retinol binding protein 4 (RBP4; serum retinol binding protein) and epididymal retinoic acid binding protein (Calderone, Berni, & Zanotti, 2003; Newcomer, 1993); (2) 10-stranded β-barrel proteins of the fatty-acid binding protein family mediating the intracellular trafficking and metabolism of retinoids, such as the cellular retinol binding proteins (CRBPs; represented by CRBP1 and CRBP2), the cellular retinoic acid binding proteins (CRABPs; represented by CRABP1 and CRABP2) (Marcelino, Smock, & Gierasch, 2006); (3) the nuclear receptors for retinoids, represented by retinoic acid receptors (RARs) and retinoid X receptors (RXRs), members of the steroid/thyroid hormone receptor family of proteins; (4) cellular retinaldehyde binding protein (CRALBP), a protein related to α-tocopherol transfer protein (He, Lobsiger, & Stocker, 2009); and (5) the photoreceptor pigments, represented by rhodopsin. Between each group, the mode of retinoid/protein interaction is different (Zhang, Zhao, & Huang, 2012).

Retinoid pathways in the visual cycle deserve wider recognition within the retinoid community in general. Although the endpoints of the two major retinoid signaling pathways—retinoids in growth and in the visual system—represent different problems (nuclear receptor activation and light-dependent neuronal activation), the early stages in these signaling pathways have important commonalities and the lipophilic nature of these molecules means that retinoid-protein interactions are key elements of retinoid signaling pathways at all levels.

1.4 Physiological concentrations of natural retinoids

Vitamin A is present in serum at μM concentrations bound to serum retinol binding protein (RBP4) as a complex with transthyretin (Goodman, 1984).

Retinoic acid is present in serum at nanomolar levels associated with serum albumin (Maiti, Ghosh, Debnath, & Dasgupta, 2006; Smith, Milch, Muto, & Goodman, 1973). Liver is the main site of bodily storage of retinol as retinyl esters (Goodman & Blaner, 1984), but within non-hepatic tissues retinoid concentrations (retinol > retinal > retinoic acid) are of similar orders of magnitude as measured in serum (Napoli, 2012). Tissue and cell concentrations are hard to determine and define with confidence, partly as a result of low concentrations which provide practical limitations of detectability, identification and quantification (Napoli, 2012), but also because of the partitioning of retinoids between lipids and carrier proteins in different cellular compartments. Nevertheless, it is important to stress that cellular physiological concentrations of retinoic acid and retinal are at low-nanomolar levels, much lower than the concentrations routinely used in cell culture experiments.

1.5 Retinoid signaling: Delivery of retinol to tissues and cells

While the endpoints of retinoic acid-mediated pathways of growth regulation (nuclear receptor mediated) and vision (G protein-coupled receptor mediated) differ, earlier stages are mediated by similar biochemical processes: both pathways depend on transfer of all-*trans* retinol into the cell. This can be achieved by various mechanisms (Li, Wongsiriroj, & Blaner, 2014). Retinol circulates in the blood bound to RBP4, and can be complexed with transthyretin. Cell-surface receptors for holo-RBP have been presumed to exist for many years but only recently identified as the large transmembrane protein STRA6 (**ST**imulated by **R**etinoic **A**cid-6) (Berry, Jin, Majumdar, & Noy, 2011). Binding of holo-RBP4 to STRA6 and transfer across the plasma membrane to CRBP1 triggers conformational changes in a cytosolic domain of STRA6. The mechanism of retinol transfer by STRA6 appears to be a mass-action process, with bidirectional transfer into the lipid bilayer (Chen et al., 2016) dependent on intracellular and extracellular concentrations of free- and protein-bound retinoids (Kawaguchi, Zhong, Kassai, Ter-Stepanian, & Sun, 2015). The conformational change of STRA6 results in JAK2-mediated phosphorylation of STRA6, release of holo-CRBP1 and JAK2-mediated phosphorylation of STAT3 or STAT5, facilitating the induction of STAT target genes and the downstream metabolism of retinol (Berry, O'Byrne, Vreeland, Blaner, & Noy, 2012). While STRA6 mediates a role of retinol in regulating insulin responses (Berry et al., 2011), and is essential for the visual system, it does not seem to be necessary for general retinol homeostasis (Berry et al., 2013). Other mechanisms such as diffusion, holo-RBP4 internalization (Matarese & Lodish, 1993), and retinyl ester delivery from chylomicrons (Li et al., 2014) are likely to contribute to retinol transfer into cells. With respect to vision, the demand for retinol in the visual cycle may necessitate high expression of STRA6 (Berry et al., 2013), but in this context STRA6-mediated activation of JAK2/STAT signaling may also be a component of mechanisms to compensate for light-induced photoreceptor stress (Jiang et al., 2014).

2. Retinoid signaling in vision

In the canonical visual cycle, retinol delivered to the retinal pigment epithelium (RPE), either from the general circulation or recycled from photoreceptors, enters the visual cycle after isomerization to 11-*cis* retinol and oxidation to 11-*cis* retinal (Fig. 3). Isomerization of all-*trans* to 11-*cis* retinol is mediated by the isomerohydrolase RPE65 (EC 3.1.1.64; retinoid isomerohydrolase) (Moiseyev, Chen, Takahashi, Wu, & Ma, 2005), a membrane-associated polyene oxygenase related to BC01 and BC02 (Dela Seña et al., 2016). Substrates for RPE65 are retinyl esters such as retinyl palmitate taken up from the plasma membrane or generated from all-*trans* retinol by LRAT (Batten et al., 2004; Jin, Yuan, Li, & Travis, 2007; Kiser & Palczewski, 2010; Saari & Bredberg, 1989; Tsin, Betts-Obregon, & Grigsby, 2018). Protonation after loss of the palmitate anion and charge re-distribution along the polyene backbone reduces energy

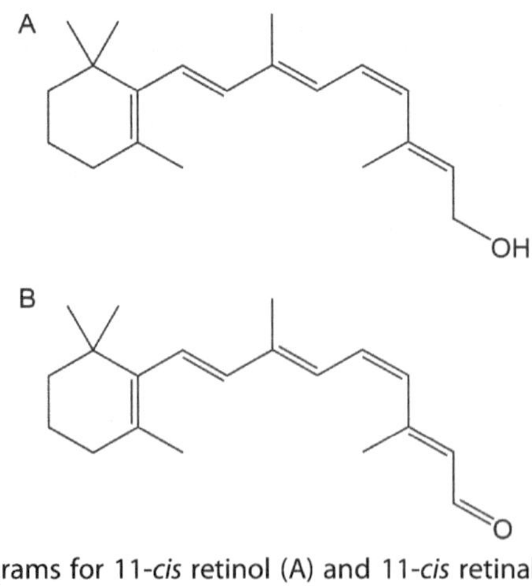

Fig. 3 Structure diagrams for 11-*cis* retinol (A) and 11-*cis* retinal (B).

barriers, facilitating isomerization to 11-*cis* (McBee et al., 2000; Redmond, Poliakov, Kuo, Chander, & Gentleman, 2010). Subsequent oxidation to 11-*cis* retinal is achieved by specific dehydrogenases. The intracellular transfer protein CRALBP has a key function in selection of 11-*cis* isomers in preference to alternative configurations produced from all-*trans* retinol by RPE65 (McBee et al., 2000; Redmond et al., 2010). 11-*cis* retinal can also be generated within the RPE by photoisomerization of all-*trans* retinal bound to peropsin (retinal pigment epithelium-derived rhodopsin homolog) (Hao & Fong, 1999; Nagata, Koyanagi, Lucas, & Terakita, 2018); the extent to which this contributes to the visual cycle is not clear, but these "retinochrome" proteins may also be involved in retinoid transport between RPE and photoreceptor cells (Cook et al., 2017).

Although 11-*cis* retinal is the natural chromophore in visual systems, 9-*cis* retinal, a more-stable isomer, is also an effective chromophore when supplied exogenously (Maeda et al., 2013). 9-*cis* retinal can also be generated endogenously in the absence of RPE65 isomerohydrolase activity (Fan, Rohrer, Moiseyev, Ma, & Crouch, 2003), presumably by isomerization or from dietary 9-*cis*-β-carotenoids (Hebuterne, Wang, Johnson, Krinsky, & Russell, 1995) *via* the 9-*cis* retinol dehydrogenase activities of some retinol dehydrogenase enzymes (Sahu & Maeda, 2016).

Transfer of 11-*cis* retinal, or other chromophores, from the RPE to adjacent photoreceptor cells, and removal of all-*trans* retinol generated by photoreceptor activation, is mediated by interstitial/interphotoreceptor retinoid binding protein (IRBP; RBP3), a large soluble glycoprotein in the interphotoreceptor matrix between the RPE and photoreceptor outer segments (Gonzalez-Fernandez & Ghosh, 2008). IRBP has four retinoid binding sites (Gonzalez-Fernandez et al., 2009). The affinity of all-*trans* retinol for CRBP1 in RPE is higher than it is for IRBP (Edwards & Adler, 1994) and this difference may facilitate the unloading of photoreceptor all-*trans* retinol at the RPE. Although speculative, similar mass-action interactions between membrane lipids with IRBP and with CRALBP might also be involved in facilitating the transfer of 11-*cis* retinal from RPE to photoreceptors (Chen, Houghton, Brenna, & Noy, 1996; Gonzalez-Fernandez et al., 2009; Saari, Nawrot, Stenkamp, Teller, & Garwin, 2009).

11-*cis* retinal transferred to photoreceptors acts as a chromophore, linked by a Schiff base to a lysine residue of the rhodopsin photoreceptor. This stabilizes the protein in an inactive configuration until the chromophore is photoisomerized back to all-*trans* retinal with deprotonation of the Schiff linkage; rapid structural changes of rhodopsin then facilitate G-protein activation leading to membrane hyperpolarization and generation of a nerve impulse (Tsin et al., 2018). Thus, the polyene conformational change and deprotonation of the chromophore lead to conformational changes of the visual pigment resulting in a neuronal signal (Kiser, Golczak, & Palczewski, 2014). The visual cycle is completed by release of all-*trans* retinal from the photoreceptor, a process which may involve condensation with membrane lipid phosphatidylethanolamine (Sun, Molday, & Nathans, 1999), reduction and transfer of all-*trans* retinol away from photoreceptor cells by IRBP to RPE; these processes will be essential for avoiding high concentrations of all-*trans* retinal which would be toxic to photoreceptor cells. Recent evidence also suggests that photoreceptor membranes can facilitate regeneration of the 11-*cis* retinal chromophore directly by all-*trans* to 11-*cis* photoisomerization of the all-*trans* retinal-phosphatidylethanolamine condensate *via* blue light (Kaylor et al., 2017).

A non-canonical visual cycle contributing to cone photoreceptor function (Kiser et al., 2018) involves chromophore regeneration from all-*trans* retinol by Müller cells of the retina (Wang & Kefalov, 2011). These cells lack RPE65, but express an all-*trans* retinol isomerase, DES1 (1.14.19.17: sphingolipid 4-desaturase) which is closely associated with CRALBP (Kaylor et al., 2013).

3. Retinoid signaling in growth and development

In contrast to the visual system, retinoic acid is sufficient to alleviate the effects of vitamin A deficiency on growth and development. The gradual realization that retinoic acid might act *via* specific nuclear receptors (Daly & Redfern, 1987; Wiggert, Russell, Lewis, & Chader, 1977) culminated in the cloning and characterization of RARs (Giguere, Ong, Segui, & Evans, 1987; Petkovich, Brand, Krust, & Chambon, 1987). These have been characterized in detail, and comprise defined domains (A–F) with the A-domain interacting with components of the transcription complex, the C-domain contains zinc-finger DNA binding motifs, and the E-domain binds ATRA as ligand. There are three RAR types (α, β, γ) which are predominantly nuclear proteins, regulating gene expression by binding, as heterodimers with RXRs, to response-element motifs associated with retinoic acid regulated genes (Mangelsdorf & Evans, 1995). As with RARs, three RXR types have been identified (α, β, γ), giving potential for cell-specific patterns of gene regulation by retinoic acid as a result of differential expression of RAR and RXR genes (Al Tanoury et al., 2014;

Samarut et al., 2014). Primary transcripts for RARs and RXRs can be alternatively spliced to give proteins differing in sequence and structure at the N-terminus (A-domain). This is likely to impart additional specificity in gene regulation; for example, alternative splicing of RARβ transcripts can generate mRNA coding for a dominant-negative isoform (Swift, Hays, & Petty, 2008). Otherwise, the functional significance of alternative splicing of RARs and RXRs in relation to patterns of retinoic acid-regulated gene expression during development remains to be explored.

Transcriptional regulation is, therefore, at the sharp end of retinoic acid signaling, and RARs have affinities for ATRA in the sub-nanomolar to low-nanomolar range (Germain et al., 2006). ATRA binds to RARs with the retinoid carboxylate tail buried within a deep pocket in the ligand-binding domain and the cyclohexenyl "head" near the pocket entrance (Fig. 4). Ligand-binding, stabilized by van der Waals interactions, hydrogen bonds with Ser and/or Lys residues, and a salt bridge between the retinoid carboxylate and a conserved Arg residue (Haffez et al., 2017; Renaud et al., 1995), results in a conformational change in the receptor, capping the ligand in the pocket and facilitating ligand-dependent interactions with other transcription complex proteins (Renaud et al., 1995). The retinoid polyene has conformational flexibility with respect to alkene s-*cis* and s-*trans* orientations and, in computational analyses, a bent conformation with a single s-*cis* rotation about the C7—C8 alkene is the preferred polyene conformation over the more linear, s-*trans* conformation (Haffez et al., 2017, 2018) (Fig. 4).

To be an effective signaling molecule for delivering cell and temporal specificity in gene control during development, precise regulation of nuclear concentrations of ATRA is critical. As with the visual cycle, retinol can be taken up from the extracellular milieu and delivered to CRBPs for oxidation to ATRA. Then, CRABP2 plays a key role in mediating transfer of ATRA to nuclear receptors (Budhu & Noy, 2002; Delva et al., 1999; Dong, Ruuska, Levinthal, & Noy, 1999). At a structural level, ATRA binds to a deep pocket in CRABPs with the polar carboxylate interacting with arginine and tyrosine residues at the bottom of the pocket (Chisholm et al., 2019; Kleywegt et al., 1994). Thus, in the CRABP binding pocket ATRA has a similar orientation as in the ligand-binding pocket of RARs. This may be a configuration which facilitates uptake of retinoic acid from cellular membranes. Evidence suggests that the transfer of ATRA from CRABP2 to RAR is *via* a direct but short-lived protein-protein interaction (Budhu, Gillilan, & Noy, 2001; Budhu & Noy, 2002), but conformational

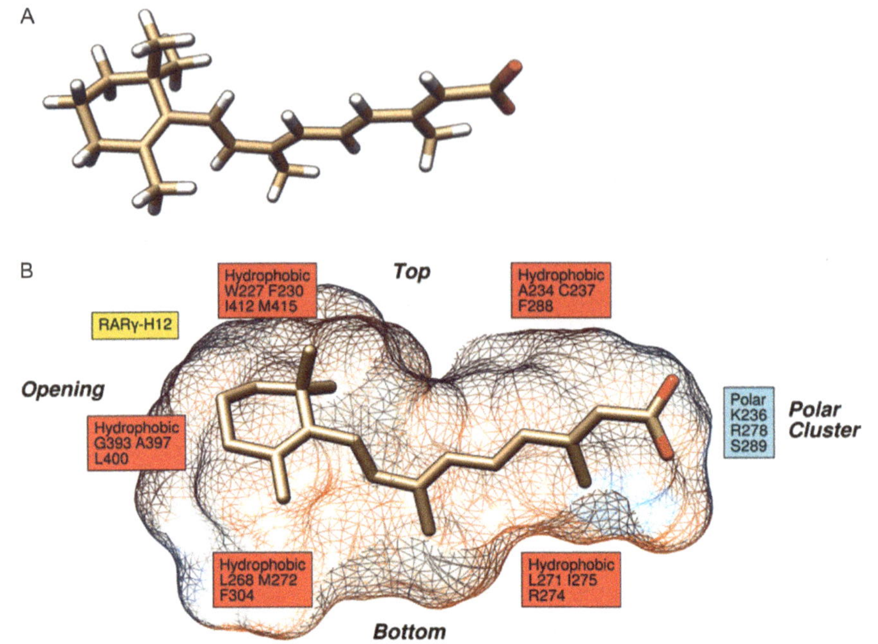

Fig. 4 The binding of ATRA to the ligand-binding domain (LBD) of RARγ. A 3D stick representation (A) of the conformer most-frequently encountered in molecular modeling of ligand/RARγ-LBD pocket interactions, shown in (B) as a ligand interaction diagram with the protein surface of the RARγ-LBD pocket [Protein Data Base: 2LBD (Renaud et al., 1995)] represented as a triangular mesh. ATRA is anchored to the bottom of the pocket by interactions between the carboxylate and a cluster of polar amino acid residues (K236, R278 and S289). The cyclohexenyl of ATRA is positioned near the opening of the pocket. When ligand is bound, a conformational change in the RAR brings helix 12 of the receptor over the pocket entrance. See also Fig. 5. *Reproduced from Haffez, H., Chisholm, D. R., Valentine, R., Pohl, E., Redfern, C., & Whiting, A. (2017). The molecular basis of the interactions between synthetic retinoic acid analogues and the retinoic acid receptors.* Medicinal Chemistry Communications, 8, *578–592, by permission of The Royal Society of Chemistry.*

changes and mechanisms facilitating ligand re-orientation and transfer between proteins remain to be established.

ATRA can be degraded by a range of cytochrome P450 enzymes (Catharine Ross & Zolfaghari, 2011; Thatcher & Isoherranen, 2009), but there is a family of specific retinoic acid degrading enzymes, CYP26, inducible by ATRA (Stevison, Jing, Tripathy, & Isoherranen, 2015; Thatcher & Isoherranen, 2009) and which also show some variation in *cis-trans* specificity (Taimi et al., 2004). CRABP1 can mediate ATRA degradation (Boylan & Gudas, 1992; Ghaffari & Petzold, 2018) and this may be a consequence of a specific CRABP1/CYP26 interaction (Nelson et al., 2016). *In vitro*, CRABP2 is also able to interact with CYP26B1 to channel RA for degradation (Nelson et al., 2016) but whether this is a functional role of CRABP2 *in vivo*, in addition to a role in nuclear ATRA transfer, remains to be established. However, the key message is that retinol dehydrogenases and CYP26 enzymes enable effective regulation of cellular ATRA levels by positive and negative feedback, respectively, providing a high degree of control for developmental regulation. From a therapeutic perspective of maintaining tissue ATRA concentrations, drugs to block CYP26 activity have been developed and may have a future clinical role (Bilip et al., 2020; Gomaa et al., 2012).

Many studies provide evidence for the role of retinoic acid in controlling differentiation and morphogenesis, with perhaps the most striking being the mirror image duplication of chick-wing digits in response to diffusion of ATRA from experimentally-implanted beads (Tickle, 1991, 2006). There is compelling evidence that Hox genes, developmental regulators specifying cell and tissue identity during embryological development, are directly regulated *in vivo* by ATRA (Cunningham & Duester, 2015; Koop et al., 2010). Turing (1953) proposed, from a mathematical analysis, that morphogenesis could result from reaction-diffusion of morphogens during development, predictions since confirmed using simple chemical systems (Tompkins et al., 2014). ATRA would fulfill criteria as a morphogen (Briscoe & Small, 2015) with respect to: (1) endogenous and experimentally induced concentration gradients (Tickle, Lee, & Eichele, 1985) and (2) the ability to interact with other biological response modifiers to control morphogenesis with a high level of specificity in the definition of tissue boundaries (Azambuja et al., 2010; Goldbeter, Gonze, & Pourquié, 2007; Sehring & Weidinger, 2019). Thus, all-*trans* retinoic provides a signaling mechanism to convey positional information, with CYP26 providing sensitive and specific control of ATRA levels (Schilling, Sosnik, & Nie, 2016; White, Nie, Lander, & Schilling, 2007) by negative feedback. However, it is not clear whether gradients of ATRA signaling *in vivo* are established by diffusion of ATRA between cells, or by cell-autonomous changes in ATRA levels as a result of cell-cell interaction signals promoting the intracellular synthesis of ATRA from retinol. The key point is that there are well-defined intracellular mechanisms for precise temporal and cell-specific control of ligand availability to RARs.

When originally discovered, RXRs were thought to represent a novel retinoid response pathway (Mangelsdorf, Ong, Dyck, & Evans, 1990).

However, in a period of rapid discovery, RXRs were identified as heterodimer partners for RARs and other nuclear ligand-dependent receptors (Evans & Mangelsdorf, 2014; Mangelsdorf & Evans, 1995). 9-*cis* retinoic acid was identified as a putative endogenous RXR ligand by chemical analysis of retinoid metabolites or by ligand-trap experiments (Heyman et al., 1992; Levin et al., 1992). Although 9-*cis* retinoic acid induces the transcription of retinoic acid-inducible genes, the dose-response kinetics are very different from ATRA (Redfern, Lovat, Malcolm, & Pearson, 1994). 9-*cis* retinoic acid can bind to RARs, but with higher off-rates in kinetic analyses (Levin et al., 1992); this is in accordance with the positioning of 9-*cis* retinoic acid in the binding pocket which gives reduced contacts between the carboxylate and critical amino acid residues of the ligand-binding pocket (Fig. 5) (Haffez et al., 2017). This emphasizes the role of *cis-trans* isomerization of retinoid polyene double bonds in constraining the molecular interactions in ligand-binding pockets that are possible within the range of alkene s-*cis*/s-*trans* flexibility.

It is, however, questionable whether 9-*cis* retinoic acid is an endogenous retinoid activator of RXR (Dawson & Xia, 2012). Recent studies have identified 9-*cis*-13,14-dihydroretinoic acid (Fig. 5) as an alternative potential endogenous activator of RXR (De Lera, Krezel, & Rühl, 2016; Rühl et al., 2015; Rühl, Krezel, & de Lera, 2018). This retinoid could be generated by saturation of all-*trans* retinol by all-*trans*-retinol 13,14-reductase (EC 1.3.99.23), a membrane-associated enzyme widely expressed, at least in mouse tissues (Moise, Kuksa, Imanishi, & Palczewski, 2004), and *cis-trans* isomerization, perhaps mediated by similar components to the visual cycle such as CRALBP, which is expressed in glial cells (Saari et al., 1997), and RPE65 and DES1 which are expressed in brain and tissues other than the eye (Hinterhuber et al., 2004; Kaylor et al., 2013; Rühl et al., 2018). 9-*cis*-13,14-dihydroretinoic acid also binds to RARs, and therefore could be an activator of RAR-mediated responses, as has been shown *in vitro* (Rühl et al., 2015). Reduced carboxylate polarity as a result of the lower polyene conjugation of 9-*cis*-13,14-dihydroretinoic acid might increase off-rates, making it less effective than 9-*cis* retinoic acid. All-*trans*-13,14-dihydroretinoic acid is an effective RAR ligand *in vitro*, but as an endogenous ligand it may not be relevant because of CRABP2 specificity toward ATRA (Moise et al., 2009). For 9-*cis*-13,14-dihydroretinoic acid as an RXR ligand, it remains to be established what mechanisms might enable the transfer of 9-*cis*-13,14-dihydroretinoic acid or similar retinoids to the nucleus and to RXRs specifically.

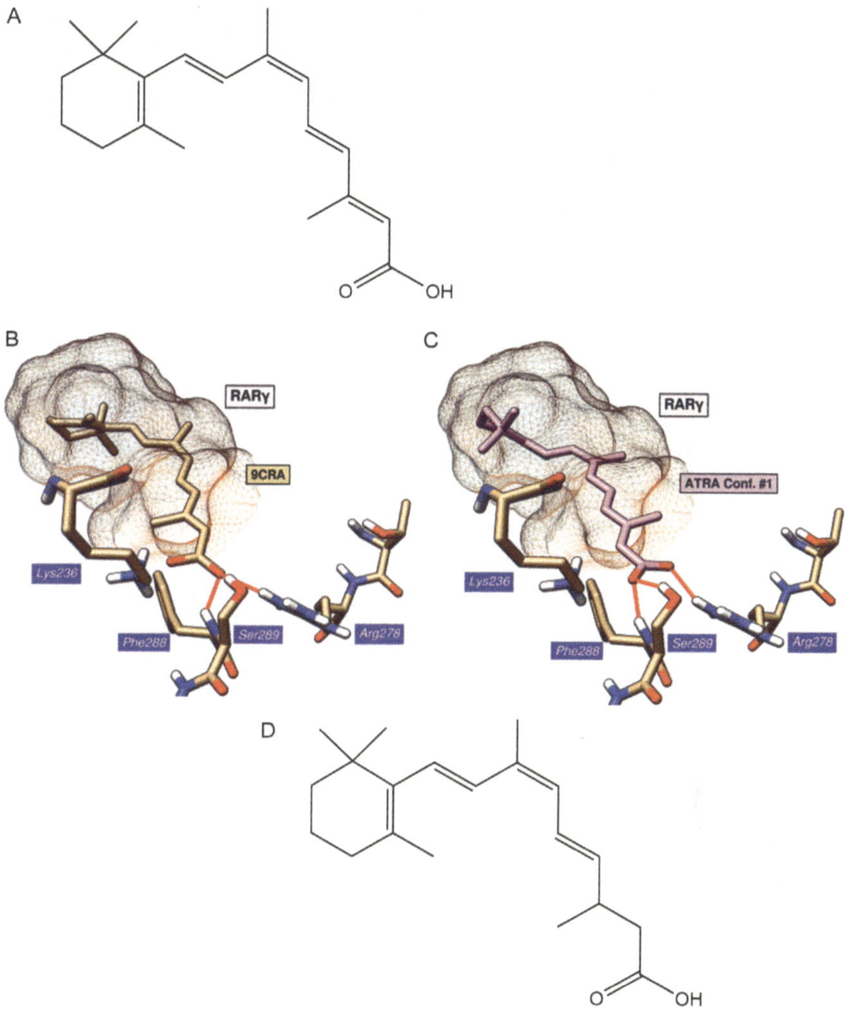

Fig. 5 Hydrogen bonding characteristics of 9-cis retinoic acid (A) binding to the LBD pocket (represented in mesh form) of RARγ (B) in comparison to ATRA (C) from docking studies, showing the key interactions between the carboxylate oxygens and polar amino acid residues at the bottom of the LDB pocket. ATRA (C) can make polar contacts *via* both carboxylate oxygens, but because of positioning only one carboxylate oxygen of 9-*cis* retinoic acid (B) is available for polar interactions. This is likely to increase the off-rates of 9-*cis* retinoic acid compared to ATRA. In (B) and (C), the retinoid is shown in the most frequent ligand conformation from *in silico* molecular modeling. The structure of 9-*cis*-13,14-dihydroretinoic acid, a putative alternative ligand for RXRs and possibly RARs, is shown in (D). *Panels (B) and (C): Reproduced from Haffez, H., Chisholm, D. R., Valentine, R., Pohl, E., Redfern, C., & Whiting, A. (2017). The molecular basis of the interactions between synthetic retinoic acid analogues and the retinoic acid receptors.* Medicinal Chemistry Communications, 8, *578–592, by permission of The Royal Society of Chemistry.*

Interest in RXRs extends beyond conventional retinoid responses because RXRs are heterodimer partners for other ligand-dependent nuclear receptors (NR), such as thyroid hormone receptor, vitamin D receptor, and peroxisome proliferator activated receptors (Dawson & Xia, 2012; Evans & Mangelsdorf, 2014). In such contexts, the NR can be either non-permissive, permissive or conditionally permissive with respect to the effects of ligand binding to the RXR. With a non-permissive NR partner, an RXR ligand is not necessary or has no effect on the activity of the NR-RXR heterodimer; conversely, with a permissive NR partner, ligand binding to either partner facilitates activity, but this can be additively or synergistically enhanced by the RXR ligand. Finally, with a conditionally permissive NR partner, NR-RXR activity is dependent on NR ligand, but can be enhanced by binding of ligand to RXR (Dawson & Xia, 2012; Lammi, Perlmann, & Aarnisalo, 2008; Lefebvre, Benomar, & Staels, 2010). Biochemical modeling studies suggest that ligand-dependent heterodimers provide a mechanism for asymmetry in positive feedback, allowing robust and tuneable responsivity to regulatory signals (Ratushny, Saleem, Sitko, Ramsey, & Aitchison, 2012). For an RXR-RAR heterodimer, positive feedback, which may also be qualitatively distinct with respect to the repertoire of RAR-type-specific patterns of gene expression, can be achieved by the retinoic acid-induced expression of RARβ, a well-characterized retinoic acid-inducible gene (Redfern et al., 1994). Further work is needed to establish mechanisms for the synthesis and delivery of an RXR ligand, and to understand how this might affect RAR-mediated transcriptional regulation and the integration of retinoid responses across different nuclear receptor signaling pathways.

The regulation of gene expression by retinoic acid *via* RARs and RXRs can be considered "genomic" activities. Non-genomic (or non-canonical) activities, independent of retinoic acid-mediated transcriptional regulation, have been described, notably ATRA induction of kinase cascades (Rochette-Egly, 2015) and ATRA-induced translation of mRNA for the GluR1 subunit of the glutaminergic α-amino-3-hydroxy-5-methylisoxazole propionic acid (AMPA) receptor in neuronal cells. The rapid induction of kinase phosphorylation cascades (Cañón, Cosgaya, Scsucova, & Aranda, 2004; Rochette-Egly, 2015) may involve CRABP1 (Persaud, Lin, Wu, Kagechika, & Wei, 2013) and cytosolic RARα (Masiá, Alvarez, De Lera, & Barettino, 2007), and link to the phosphorylation of nuclear proteins, including nuclear retinoid receptors, to control their activity (Rochette-Egly, 2015).

In neuronal cells, ATRA synthesis increases in response to reduced synaptic activity (Aoto, Nam, Poon, Ting, & Chen, 2008). RARα present in the cytoplasm binds *via* the F-domain to the 5′ untranslated region of GluR1 mRNA, inhibiting translation initiation. Translation inhibition is relieved by ATRA binding to RARα, either as a result of adding exogenous ATRA to the cells or by blocking synaptic activity to allow increased retinoic acid synthesis; a conformational change as a result of ATRA binding then releases GluR1 mRNA, allowing translation to proceed (Aoto et al., 2008; Chen & Napoli, 2008; Chen, Onisko, & Napoli, 2008).

4. Future work

Apart from intellectual curiosity, the benefit of a detailed understanding of how vitamin A works is that we may be able to use this knowledge for therapeutic benefit by designing drugs to target particular cellular or tissue-specific responses. Retinoic acid already has an important clinical use in acute promyelocytic leukemia (Schenk, Stengel, & Zelent, 2014; Zhou, Zhang, Wang, Chen, & Chen, 2007), and may have a role for neurological and eye diseases. In these respects, it will be important to be able to deliver a therapeutic retinoid to the appropriate tissue, and, once there, to target the appropriate cellular compartment or transcriptional response, and to control or limit degradation. Simple assays of growth responses, whether at a cellular or multicellular scale, are important as measures of vitamin A activity. However, with RNA sequencing technology it is possible to obtain precise, global measures of transcriptional activity in individual cells and tissues. This offers great potential to refine our understanding of retinoid biology and to test the specificity of responses to synthetic retinoids for future drug development (Chisholm et al., 2019). While there has been substantial progress in understanding the biochemistry of retinoid pathways and how retinoid responses are regulated (Napoli, 2012; Ratushny et al., 2012), we need a better understanding of how individual components of retinoid signaling pathways are integrated into cellular signaling pathways in different tissue contexts. *In silico* biochemical modeling, using resources such as the open-source CoPaSi platform (Mendes, Messiha, Malys, & Hoops, 2009), is important as a research tool and deserves greater prominence for understanding the nuances of retinoid signaling in growth, development and vision.

References

Al Tanoury, Z., Piskunov, A., Andriamoratsiresy, D., Gaouar, S., Lutzing, R., Ye, T., et al. (2014). Genes involved in cell adhesion and signaling: A new repertoire of retinoic acid receptor target genes in mouse embryonic fibroblasts. *Journal of Cell Science*, *127*, 521–533.

Aoto, J., Nam, C. I., Poon, M. M., Ting, P., & Chen, L. (2008). Synaptic signaling by all-trans retinoic acid in homeostatic synaptic plasticity. *Neuron*, *60*, 308–320.

Azambuja, A. P., Portillo-Sánchez, V., Rodrigues, M. V., Omae, S. V., Schechtman, D., Strauss, B. E., et al. (2010). Retinoic acid and VEGF delay smooth muscle relative to endothelial differentiation to coordinate inner and outer coronary vessel wall morphogenesis. *Circulation Research*, *107*, 204–216.

Batten, M. L., Imanishi, Y., Maeda, T., Tu, D. C., Moise, A. R., Bronson, D., et al. (2004). Lecithin-retinol acyltransferase is essential for accumulation of all-trans-retinyl esters in the eye and in the liver. *The Journal of Biological Chemistry*, *279*, 10422–10432.

Bell, E., Ponthan, F., Whitworth, C., Westermann, F., Thomas, H., & Redfern, C. P. F. (2013). Cell survival signalling through PPARδ and arachidonic acid metabolites in neuroblastoma. *PLoS One*, *8*(7), e68859. https://doi.org/10.1371/journal.pone.0068859.

Berry, D. C., Jacobs, H., Marwarha, G., Gely-Pernot, A., O'Byrne, S. M., DeSantis, D., et al. (2013). The STRA6 receptor is essential for retinol-binding protein-induced insulin resistance but not for maintaining vitamin A homeostasis in tissues other than the eye. *The Journal of Biological Chemistry*, *288*, 24528–24539.

Berry, D. C., Jin, H., Majumdar, A., & Noy, N. (2011). Signaling by vitamin A and retinol-binding protein regulates gene expression to inhibit insulin responses. *Proceedings of the National Academy of Sciences of the United States of America*, *108*, 4340–4345.

Berry, D. C., O'Byrne, S. M., Vreeland, A. C., Blaner, W. S., & Noy, N. (2012). Cross talk between signaling and vitamin A transport by the retinol-binding protein receptor STRA6. *Molecular and Cellular Biology*, *32*, 3164–3175.

Bilip, M., Shah, S., Mathiyalakan, M., Tagalakis, A. D., Hart, S. L., Maeshima, R., et al. (2020). Liposomal delivery of hydrophobic RAMBAs provides good bioavailability and significant enhancement of retinoic acid signalling in neuroblastoma tumour cells. *Journal of Drug Targeting*, 1–12. https://doi.org/10.1080/1061186X.2019.1710157.

Boylan, J. F., & Gudas, L. J. (1992). The level of CRABP-I expression influences the amounts and types of all-trans-retinoic acid metabolites in F9 teratocarcinoma stem cells. *The Journal of Biological Chemistry*, *267*, 21486–21491.

Briscoe, J., & Small, S. (2015). Morphogen rules: Design principles of gradient-mediated embryo patterning. *Development*, *142*, 3996–4009.

Budhu, A., Gillilan, R., & Noy, N. (2001). Localization of the RAR interaction domain of cellular retinoic acid binding protein-II. *Journal of Molecular Biology*, *305*, 939–949.

Budhu, A. S., & Noy, N. (2002). Direct channeling of retinoic acid between cellular retinoic acid-binding protein II and retinoic acid receptor sensitizes mammary carcinoma cells to retinoic acid-induced growth arrest. *Molecular Cell Biology*, *22*, 2632–2641.

Bull, J. N., West, C. W., Anstöter, C. S., Da Silva, G., Bieske, E. J., & Verlet, J. R. R. (2019). Ultrafast photoisomerisation of an isolated retinoid. *Physical Chemistry Chemical Physics*, *21*, 10567–10579.

Calderone, V., Berni, R., & Zanotti, G. (2003). High-resolution structures of retinol-binding protein in complex with retinol: pH-induced protein structural changes in the crystal state. *Journal of Molecular Biology*, *329*, 841–850.

Cañón, E., Cosgaya, J. M., Scsucova, S., & Aranda, A. (2004). Rapid effects of retinoic acid on CREB and ERK phosphorylation in neuronal cells. *Molecular Biology of the Cell*, *15*, 5583–5592.

Catharine Ross, A., & Zolfaghari, R. (2011). Cytochrome P450s in the regulation of cellular retinoic acid metabolism. *Annual Review of Nutrition*, *31*, 65–87.

Chen, Y., Clarke, O. B., Kim, J., Stowe, S., Kim, Y. K., Assur, Z., et al. (2016). Structure of the STRA6 receptor for retinol uptake. *Science*, *353*(6302), aad8266. https://doi.org/10.1126/science.aad8266.

Chen, Y., Houghton, L. A., Brenna, J. T., & Noy, N. (1996). Docosahexaenoic acid modulates the interactions of the interphotoreceptor retinoid-binding protein with 11-cis-retinal. *The Journal of Biological Chemistry*, *271*, 20507–20515.

Chen, N., & Napoli, J. L. (2008). All-trans-retinoic acid stimulates translation and induces spine formation in hippocampal neurons through a membrane-associated RARα. *FASEB Journal*, *22*, 236–245.

Chen, N., Onisko, B., & Napoli, J. L. (2008). The nuclear transcription factor RARα associates with neuronal RNA granules and suppresses translation. *The Journal of Biological Chemistry*, *283*, 20841–20847.

Chisholm, D. R., Tomlinson, C. W. E., Zhou, G. L., Holden, C., Affleck, V., Lamb, R., et al. (2019). Fluorescent retinoic acid analogues as probes for biochemical and intracellular characterization of retinoid signaling pathways. *ACS Chemical Biology*, *14*, 369–377.

Cook, J. D., Ng, S. Y., Lloyd, M., Eddington, S., Sun, H., Nathans, J., et al. (2017). Peropsin modulates transit of Vitamin A from retina to retinal pigment epithelium. *The Journal of Biological Chemistry*, *292*, 21407–21416.

Cox, B. D., Muccio, D. D., & Hamilton, T. P. (2013). Conformational analysis of retinoic acids: Effects of steric interactions on nonplanar conjugated polyenes. *Computational & Theoretical Chemistry*, *1011*, 11–20.

Cunningham, T. J., & Duester, G. (2015). Mechanisms of retinoic acid signalling and its roles in organ and limb development. *Nature Reviews Molecular Cell Biology*, *16*, 110–123.

Daly, A. K., & Redfern, C. P. F. (1987). Characterisation of a retinoic-acid-binding component from F9 embryonal-carcinoma-cell nuclei. *European Journal of Biochemistry*, *168*, 133–139.

Dawson, M. I., & Xia, Z. (2012). The retinoid X receptors and their ligands. *Biochimica et Biophysica Acta, Molecular and Cell Biology of Lipids*, *1821*, 21–56.

De Lera, Á. R., Krezel, W., & Rühl, R. (2016). An endogenous mammalian retinoid X receptor ligand, at last!. *ChemMedChem*, *11*, 1027–1037.

Dela Seña, C., Sun, J., Narayanasamy, S., Riedl, K. M., Yuan, Y., Curley, R. W., et al. (2016). Substrate specificity of purified recombinant chicken β-carotene 9′,10′-oxygenase (BCO2). *The Journal of Biological Chemistry*, *291*, 14609–14619.

Delva, L., Bastie, J.-N., Rochette-Egly, C., Kraïba, R., Balitrand, N., Despouy, G., et al. (1999). Physical and functional interactions between cellular retinoic acid binding protein II and the retinoic acid-dependent nuclear complex. *Molecular and Cellular Biology*, *19*, 7158–7167.

Devery, J., & Milborrow, B. V. (1994). β-Carotene-15,15′-dioxygenase (EC 1.13.11.21) isolation reaction mechanism and an improved assay procedure. *The British Journal of Nutrition*, *72*, 397–414.

Dong, D., Ruuska, S. E., Levinthal, D. J., & Noy, N. (1999). Distinct roles for cellular retinoic acid-binding proteins I and II in regulating signaling by retinoic acid. *The Journal of Biological Chemistry*, *274*, 23695–23698.

Edwards, R. B., & Adler, A. J. (1994). Exchange of retinol between IRBP and CRBP. *Experimental Eye Research*, *59*, 161–170.

Eroglu, A., Hruszkewycz, D. P., Dela Sena, C., Narayanasamy, S., Riedl, K. M., Kopec, R. E., et al. (2012). Naturally occurring eccentric cleavage products of provitamin A β-carotene function as antagonists of retinoic acid receptors. *The Journal of Biological Chemistry*, *287*, 15886–15895.

Evans, R. M., & Mangelsdorf, D. J. (2014). Nuclear receptors, RXR, and the big bang. *Cell*, *157*, 255–266.

Fan, J., Rohrer, B., Moiseyev, G., Ma, J. X., & Crouch, R. K. (2003). Isorhodopsin rather than rhodopsin mediates rod function in RPE65 knock-out mice. *Proceedings of the National Academy of Sciences of the United States of America*, *100*, 13662–13667.

Germain, P., Chambon, P., Eichele, G., Evans, R. M., Lazar, M. A., Leid, M., et al. (2006). International union of pharmacology. LXIII. Retinoid X receptors. *Pharmacological Reviews*, *58*, 760–772.

Ghaffari, H., & Petzold, L. R. (2018). Identification of influential proteins in the classical retinoic acid signaling pathway. *Theoretical Biology & Medical Modelling*, *15*(1), 16. https://doi.org/10.1186/s12976-018-0088-7.

Giguere, V., Ong, E. S., Segui, P., & Evans, R. M. (1987). Identification of a receptor for the morphogen retinoic acid. *Nature*, *330*, 624–629.

Goldbeter, A., Gonze, D., & Pourquié, O. (2007). Sharp developmental thresholds defined through bistability by antagonistic gradients of retinoic acid and FGF signaling. *Developmental Dynamics*, *236*, 1495–1508.

Gomaa, M. S., Bridgens, C. E., Illingworth, N. A., Veal, G. J., Redfern, C. P. F., Brancale, A., et al. (2012). Novel retinoic acid 4-hydroxylase (CYP26) inhibitors based on a 3-(1H-imidazol- and triazol-1-yl)-2,2-dimethyl-3-(4-(phenylamino)phenyl)propyl scaffold. *Bioorganic & Medicinal Chemistry*, *20*, 4201–4207.

Gonzalez-Fernandez, F., Bevilacqua, T., Lee, K. I., Chandrashekar, R., Hsu, L., Garlipp, M. A., et al. (2009). Retinol-binding site in interphotoreceptor retinoid-binding protein (IRBP): A novel hydrophobic cavity. *Investigative Ophthalmology and Visual Science*, *50*, 5577–5586.

Gonzalez-Fernandez, F., & Ghosh, D. (2008). Focus on molecules: Interphotoreceptor retinoid-binding protein (IRBP). *Experimental Eye Research*, *86*, 169–170.

Goodman, D. S. (1984). Plasma retinol binding protein. In M. B. Sporn et al., (Eds.), *Vol. 2. The retinoids* (pp. 42–88). Academic Press.

Goodman, D. S., & Blaner, W. S. (1984). Biosynthesis, absorption, and hepatic metabolism of retinol. In M. B. Sporn et al., (Eds.), *Vol. 1. The retinoids* (pp. 2–34). Academic Press.

Haffez, H., Chisholm, D. R., Tatum, N. J., Valentine, R., Redfern, C., Pohl, E., et al. (2018). Probing biological activity through structural modelling of ligand-receptor interactions of 2,4-disubstituted thiazole retinoids. *Bioorganic & Medicinal Chemistry*, *26*, 1560–1572.

Haffez, H., Chisholm, D. R., Valentine, R., Pohl, E., Redfern, C., & Whiting, A. (2017). The molecular basis of the interactions between synthetic retinoic acid analogues and the retinoic acid receptors. *Medicinal Chemistry Communications*, *8*, 578–592.

Hao, W., & Fong, H. K. W. (1999). The endogenous chromophore of retinal G protein-coupled receptor opsin from the pigment epithelium. *The Journal of Biological Chemistry*, *274*, 6085–6090.

Harrison, E. H., Dela Sena, C., Eroglu, A., & Fleshman, M. K. (2012). The formation, occurrence, and function of β-apocarotenoids: β-Carotene metabolites that may modulate nuclear receptor signaling. *The American Journal of Clinical Nutrition*, *96*, 1189S–1192S.

He, X., Lobsiger, J., & Stocker, A. (2009). Bothnia dystrophy is caused by domino-like rearrangements in cellular retinaldehyde-binding protein mutant R234W. *Proceedings of the National Academy of Sciences of the United States of America*, *106*, 18545–18550.

Hebuterne, X., Wang, X. D., Johnson, E. J., Krinsky, N. I., & Russell, R. M. (1995). Intestinal absorption and metabolism of 9-cis-β-carotene in vivo: Biosynthesis of 9-cis-retinoic acid. *Journal of Lipid Research*, *36*, 1264–1273.

Herr, F. M., Li, E., Weinberg, R. B., Cook, V. R., & Storch, J. (1999). Differential mechanisms of retinoid transfer from cellular retinol binding proteins types I and II to phospholipid membranes. *The Journal of Biological Chemistry*, *274*, 9556–9563.

Heyman, R. A., Mangelsdorf, D. J., Dyck, J. A., Stein, R. B., Eichele, G., Evans, R. M., et al. (1992). 9-cis retinoic acid is a high affinity ligand for the retinoid X receptor. *Cell, 68*, 397–406.

Hinterhuber, G., Cauza, K., Brugger, K., Dingelmaier-Hovorka, R., Horvat, R., Wolff, K., et al. (2004). RPE65 of retinal pigment epithelium, a putative receptor molecule for plasma retinol-binding protein, is expressed in human keratinocytes. *The Journal of Investigative Dermatology, 122*, 406–413.

Huxley, R. R., Lloyd, B. B., Goldacre, M., & Neil, H. A. W. (2000). Nutritional research in World War 2: The Oxford Nutrition Survey and its research potential 50 years later. *British Journal of Nutrition, 84*, 247–251.

Ide, H., & Toki, S. (1970). Metabolism of beta-ionone. Isolation, characterization and identification of the metabolitesin the urine of rabbits. *The Biochemical Journal, 119*, 281–287.

Jiang, K., Wright, K. L., Zhu, P., Szego, M. J., Bramall, A. N., Hauswirth, W. W., et al. (2014). STAT3 promotes survival of mutant photoreceptors in inherited photoreceptor degeneration models. *Proceedings of the National Academy of Sciences of the United States of America, 111*, E5716–E5723.

Jin, M., Yuan, Q., Li, S., & Travis, G. H. (2007). Role of LRAT on the retinoid isomerase activity and membrane association of Rpe65. *The Journal of Biological Chemistry, 282*, 20915–20924.

Kawaguchi, R., Zhong, M., Kassai, M., Ter-Stepanian, M., & Sun, H. (2015). Vitamin A transport mechanism of the multitransmembrane cell-surface receptor STRA6. *Membranes (Basel), 5*, 425–453.

Kaylor, J. J., Xu, T., Ingram, N. T., Tsan, A., Hakobyan, H., Fain, G. L., et al. (2017). Blue light regenerates functional visual pigments in mammals through a retinyl-phospholipid intermediate. *Nature Communications, 8*(1), 16. https://doi.org/10.1038/s41467-017-00018-4.

Kaylor, J. J., Yuan, Q., Cook, J., Sarfare, S., Makshanoff, J., Miu, A., et al. (2013). Identification of DES1 as a vitamin A isomerase in Müller glial cells of the retina. *Nature Chemical Biology, 9*, 30–36.

Kiser, P. D., Golczak, M., & Palczewski, K. (2014). Chemistry of the retinoid (visual) cycle. *Chemical Reviews, 114*, 194–232.

Kiser, P. D., & Palczewski, K. (2010). Membrane-binding and enzymatic properties of RPE65. *Progress in Retinal and Eye Research, 29*, 428–442.

Kiser, P. D., Zhang, J., Sharma, A., Angueyra, J. M., Kolesnikov, A. V., Badiee, M., et al. (2018). Retinoid isomerase inhibitors impair but do not block mammalian cone photoreceptor function. *The Journal of General Physiology, 150*, 571–590.

Kleywegt, G. J., Bergfors, T., Senn, H., Le Motte, P., Gsell, B., Shud, K., et al. (1994). Crystal structures of cellular retinoic acid binding proteins I and II in complex with all-trans-retinoic acid and a synthetic retinoid. *Structure, 2*, 1241–1258.

Koop, D., Holland, N. D., Sémon, M., Alvarez, S., de Lera, A. R., Laudet, V., et al. (2010). Retinoic acid signaling targets Hox genes during the amphioxus gastrula stage: Insights into early anterior-posterior patterning of the chordate body plan. *Developmental Biology, 338*, 98–106.

La Frano, M. R., Cai, Y., Burri, B. J., & Thilsted, S. H. (2018). Discovery and biological relevance of 3,4-didehydroretinol (vitamin A2) in small indigenous fish species and its potential as a dietary source for addressing vitamin A deficiency. *International Journal of Food Sciences and Nutrition, 69*, 253–261.

Lammi, J., Perlmann, T., & Aarnisalo, P. (2008). Corepressor interaction differentiates the permissive and non-permissive retinoid X receptor heterodimers. *Archives of Biochemistry and Biophysics, 472*, 105–114.

Lefebvre, P., Benomar, Y., & Staels, B. (2010). Retinoid X receptors: Common heterodimerization partners with distinct functions. *Trends in Endocrinology and Metabolism, 21*, 676–683.

Levin, A. A., Sturzenbecker, L. J., Kazmer, S., Bosakowski, T., Huselton, C., Allenby, G., et al. (1992). 9-cis retinoic acid stereoisomer binds and activates the nuclear receptor RXRα. *Nature, 355*, 359–361.

Li, Y., Wongsiriroj, N., & Blaner, W. S. (2014). The multifaceted nature of retinoid transport and metabolism. *Hepatobiliary Surgery and Nutrition, 3*, 126–139.

Maeda, T., Dong, Z., Jin, H., Sawada, O., Gao, S., Utkhede, D., et al. (2013). QLT091001, a 9-cis-retinal analog, is well-tolerated by retinas of mice with impaired visual cycles. *Investigative Ophthalmology and Visual Science, 54*, 455–466.

Maiti, T. K., Ghosh, K. S., Debnath, J., & Dasgupta, S. (2006). Binding of all-trans retinoic acid to human serum albumin: Fluorescence, FT-IR and circular dichroism studies. *International Journal of Biological Macromolecules, 38*, 197–202.

Mangelsdorf, D. J., & Evans, R. M. (1995). The RXR heterodimers and orphan receptors. *Cell, 83*, 841–850.

Mangelsdorf, D. J., Ong, E. S., Dyck, J. A., & Evans, R. M. (1990). Nuclear receptor that identifies a novel retinoic acid response pathway. *Nature, 345*, 224–229.

Marcelino, A. M. C., Smock, R. G., & Gierasch, L. M. (2006). Evolutionary coupling of structural and functional sequence information in the intracellular lipid-binding protein family. *Proteins: Structure, Function, and Genetics, 63*, 373–384.

Masiá, S., Alvarez, S., De Lera, A. R., & Barettino, D. (2007). Rapid, nongenomic actions of retinoic acid on phosphatidylinositol-3-kinase signaling pathway mediated by the retinoic acid receptor. *Molecular Endocrinology, 21*, 2391–2402.

Matarese, V., & Lodish, H. F. (1993). Specific uptake of retinol-binding protein by variant F9 cell lines. *The Journal of Biological Chemistry, 268*, 18859–18865.

McBee, J. K., Kuksa, V., Alvarez, R., De Lera, A. R., Prezhdo, O., Haeseleer, F., et al. (2000). Isomerization of all-trans-retinol to cis-retinols in bovine retinal pigment epithelial cells: Dependence on the specificity of retinoid-binding proteins. *Biochemistry, 39*, 11370–11380.

Meeks, R. G., Zaharevitz, D., & Chen, R. F. (1981). Membrane effects of retinoids: Possible correlation with toxicity. *Archives of Biochemistry and Biophysics, 207*, 141–147.

Mendes, P., Messiha, H., Malys, N., & Hoops, S. (2009). Chapter 22 enzyme kinetics and computational modeling for systems biology. *Methods in Enzymology, 467*, 583–599.

Moise, A. R., Alvarez, S., Domínguez, M., Alvarez, R., Golczak, M., Lobo, G. P., et al. (2009). Activation of retinoic acid receptors by dihydroretinoids. *Molecular Pharmacology, 76*, 1228–1237.

Moise, A. R., Kuksa, V., Imanishi, Y., & Palczewski, K. (2004). Identification of all-trans-retinol:all-trans-13,14-dihydroretinol saturase. *The Journal of Biological Chemistry, 279*, 50230–50242.

Moiseyev, G., Chen, Y., Takahashi, Y., Wu, B. X., & Ma, J. X. (2005). RPE65 is the isomerohydrolase in the retinoid visual cycle. *Proceedings of the National Academy of Sciences of the United States of America, 102*, 12413–12418.

Mozaffarian, D., Rosenberg, I., & Uauy, R. (2018). History of modern nutrition science-implications for current research, dietary guidelines, and food policy. *BMJ, 361*, k2392. https://doi.org/10.1136/bmj.k2392.

Murayama, A., Suzuki, T., & Matsui, M. (1997). Photoisomerization of retinoic acids in ethanol under room light: A warning for cell biological study of geometrical isomers of retinoids. *Journal of Nutritional Science and Vitaminology (Tokyo), 43*, 167–176.

Nagata, T., Koyanagi, M., Lucas, R., & Terakita, A. (2018). An all-trans-retinal-binding opsin peropsin as a potential dark-active and light-inactivated G protein-coupled receptor. *Scientific Reports, 8*, 3535. https://doi.org/10.1038/s41598-018-21946-1.

Napoli, J. L. (2012). Physiological insights into all-trans-retinoic acid biosynthesis. *Biochimica et Biophysica Acta, Molecular and Cell Biology of Lipids, 1821*, 152–167.

Nelson, C. H., Peng, C. C., Lutz, J. D., Yeung, C. K., Zelter, A., & Isoherranen, N. (2016). Direct protein–protein interactions and substrate channeling between cellular retinoic acid binding proteins and CYP26B1. *FEBS Letters*, *590*, 2527–2535.

Newcomer, M. E. (1993). Structure of the epididymal retinoic acid binding protein at 2.1 Å resolution. *Structure*, *1*, 7–18.

Ong, D. E. (1993). Retinoid metabolism during intestinal absorption. *The Journal of Nutrition*, *123*, 351–355.

Persaud, S. D., Lin, Y. W., Wu, C. Y., Kagechika, H., & Wei, L. N. (2013). Cellular retinoic acid binding protein I mediates rapid non-canonical activation of ERK1/2 by all-trans retinoic acid. *Cellular Signalling*, *25*, 19–25.

Petkovich, M., Brand, N. J., Krust, A., & Chambon, P. (1987). A human retinoic acid receptor which belongs to the family of nuclear receptors. *Nature*, *330*, 444–450.

Raghuvanshi, S., Reed, V., Blaner, W. S., & Harrison, E. H. (2015). Cellular localization of β-carotene 15,15′ oxygenase-1 (BCO1) and β-carotene 9′,10′ oxygenase-2 (BCO2) in rat liver and intestine. *Archives of Biochemistry and Biophysics*, *572*, 19–27.

Rando, R. R. (1991). Membrane phospholipids as an energy source in the operation of the visual cycle. *Biochemistry*, *30*, 595–602.

Ratushny, A. V., Saleem, R. A., Sitko, K., Ramsey, S. A., & Aitchison, J. D. (2012). Asymmetric positive feedback loops reliably control biological responses. *Molecular Systems Biology*, *8*, 577. https://doi.org/10.1038/msb.2012.10.

Redfern, C. P. F., Lovat, P. E., Malcolm, A. J., & Pearson, A. D. J. (1994). Differential effects of 9-cis and all-trans retinoic acid on the induction of retinoic acid receptor-β and cellular retinoic acid-binding protein II in human neuroblastoma cells. *The Biochemical Journal*, *304*, 147–154.

Redmond, T. M., Poliakov, E., Kuo, S., Chander, P., & Gentleman, S. (2010). RPE65, visual cycle retinol isomerase, is not inherently 11-cis-specific: Support for a carbocation mechanism of retinol isomerization. *The Journal of Biological Chemistry*, *285*, 1919–1927.

Renaud, J. P., Rochel, N., Ruff, M., Vivat, V., Chambon, P., Gronemeyer, H., et al. (1995). Crystal structure of the RAR-γ ligand-binding domain bound to all-trans retinoic acid. *Nature*, *378*, 681–689.

Riabroy, N., Dever, J. T., & Tanumihardjo, S. A. (2014). α-Retinol and 3,4-didehydroretinol support growth in rats when fed at equimolar amounts and α-retinol is not toxic after repeated administration of large doses. *The British Journal of Nutrition*, *111*, 1373–1381.

Rochette-Egly, C. (2015). Retinoic acid signaling and mouse embryonic stem cell differentiation: Cross talk between genomic and non-genomic effects of RA. *Biochimica et Biophysica Acta, Molecular and Cell Biology of Lipids*, *1851*, 66–75.

Rühl, R., Krezel, W., & de Lera, A. R. (2018). 9-cis-13,14-dihydroretinoic acid, a new endogenous mammalian ligand of retinoid X receptor and the active ligand of a potential new vitamin A category: Vitamin A5. *Nutrition Reviews*, *76*, 929–941.

Rühl, R., Krzyżosiak, A., Niewiadomska-Cimicka, A., Rochel, N., Szeles, L., Vaz, B., et al. (2015). 9-cis-13,14-dihydroretinoic acid is an endogenous retinoid acting as RXR ligand in mice. *PLoS Genetics*, *11*(6), e1005213. https://doi.org/10.1371/journal.pgen.1005213.

Saari, J. C., & Bredberg, D. L. (1989). Lecithin:retinol acyltransferase in retinal pigment epithelial microsomes. *The Journal of Biological Chemistry*, *264*, 8636–8640.

Saari, J. C., Huang, J., Possin, D. E., Fariss, R. N., Leonard, J., Garwin, G. G., et al. (1997). Cellular retinaldehyde-binding protein is expressed by oligodendrocytes in optic nerve and brain. *Glia*, *21*, 259–268.

Saari, J. C., Nawrot, M., Stenkamp, R. E., Teller, D. C., & Garwin, G. G. (2009). Release of 11-cis-retinal from cellular retinaldehyde-binding protein by acidic lipids. *Molecular Vision*, *15*, 844–854.

Sahu, B., & Maeda, A. (2016). Retinol dehydrogenases regulate vitamin A metabolism for visual function. *Nutrients*, *8*(11), 746. https://doi.org/10.3390/nu8110746.
Samarut, E., Gaudin, C., Hughes, S., Gillet, B., de Bernard, S., Jouve, P. E., et al. (2014). Retinoic acid receptor subtype-specific transcriptotypes in the early zebrafish embryo. *Molecular Endocrinology*, *28*, 260–272.
Schenk, T., Stengel, S., & Zelent, A. (2014). Unlocking the potential of retinoic acid in anticancer therapy. *British Journal of Cancer*, *111*, 2039–2045.
Schilling, T. F., Sosnik, J., & Nie, Q. (2016). Visualizing retinoic acid morphogen gradients. *Methods in Cell Biology*, *133*, 139–163.
Sehring, I. M., & Weidinger, G. (2019). Recent advancements in understanding fin regeneration in zebrafish. *WIREs Developmental Biology*, *e367*, 1–16.
Smith, J. E., Milch, P. O., Muto, Y., & Goodman, D. S. (1973). The plasma transport and metabolism of retinoic acid in the rat. *Biochemical Journal*, *132*, 821–827.
Sporn, M. B., Roberts, A. B., & Goodman, D. S. (1984). Introduction. In M. B. Sporn et al., (Eds.), *Vol. 1. The retinoids* (pp. 1–5). Academic Press.
Stevison, F., Jing, J., Tripathy, S., & Isoherranen, N. (2015). Role of retinoic acid-metabolizing cytochrome P450s, CYP26, in inflammation and cancer. *Advances in Pharmacology*, *74*, 373–412.
Sun, H., & Kawaguchi, R. (2011). The membrane receptor for plasma retinol-binding protein, a new type of cell-surface receptor. *International Review of Cell and Molecular Biology*, *288*, 1–41.
Sun, H., Molday, R. S., & Nathans, J. (1999). Retinal stimulates ATP hydrolysis by purified and reconstituted ABCR, the photoreceptor-specific ATP-binding cassette transporter responsible for Stargardt disease. *The Journal of Biological Chemistry*, *274*, 8269–8281.
Svensson, F. R., Lincoln, P., Nordén, B., & Esbjörner, E. K. (2007). Retinoid chromophores as probes of membrane lipid order. *The Journal of Physical Chemistry B*, *111*, 10839–10848.
Swift, C., Hays, J., & Petty, W. (2008). Distinct functions of retinoic acid receptor beta isoforms: Implications for targeted therapy. *Endocrine, Metabolic & Immune Disorders Drug Targets*, *8*, 47–50.
Szuts, E. Z., & Harosi, F. I. (1991). Solubility of retinoids in water. *Archives of Biochemistry and Biophysics*, *287*, 297–304.
Taimi, M., Helvig, C., Wisniewski, J., Ramshaw, H., White, J., Amad, M., et al. (2004). A novel human cytochrome P450, CYP26C1, involved in metabolism of 9-cis and all-trans isomers of retinoic acid. *The Journal of Biological Chemistry*, *279*, 77–85.
Tarapcsák, S., Szalóki, G., Telbisz, Á., Gyöngy, Z., Matúz, K., Csosz, É., et al. (2017). Interactions of retinoids with the ABC transporters P-glycoprotein and breast cancer resistance protein. *Science Reports*, 7, 41376.
Thatcher, J. E., & Isoherranen, N. (2009). The role of CYP26 enzymes in retinoic acid clearance. *Expert Opinion on Drug Metabolism & Toxicology*, *5*, 875–886.
Tickle, C. (1991). Retinoic acid and chick limb bud development. *Development*, *112*, 113–121.
Tickle, C. (2006). Making digit patterns in the vertebrate limb. *Nature Reviews Molecular Cell Biology*, 7, 45–53.
Tickle, C., Lee, J., & Eichele, G. (1985). A quantitative analysis of the effect of all-trans-retinoic acid on the pattern of chick wing development. *Developmental Biology*, *109*, 82–95.
Tompkins, N., Li, N., Girabawe, C., Heymann, M., Ermentrout, G. B., Epstein, I. R., et al. (2014). Testing Turing's theory of morphogenesis in chemical cells. *Proceedings of the National Academy of Sciences of the United States of America*, *111*, 4397–4402.
Tsin, A., Betts-Obregon, B., & Grigsby, J. (2018). Visual cycle proteins: Structure, function, and roles in human retinal disease. *The Journal of Biological Chemistry*, *293*, 13016–13021.

Turing, A. M. (1953). The chemical basis of morphogenesis. *Philosophical Transactions of the Royal Society B*, *237*, 37–72.
Van Aalten, D. M. F., De Groot, B. L., Berendsen, H. J. C., & Findlay, J. B. C. (1996). Conformational analysis of retinoids and restriction of their dynamics by retinoid-binding proteins. *Biochemical Journal*, *319*, 543–550.
Von Lintig, J. (2012). Metabolism of carotenoids and retinoids related to vision. *The Journal of Biological Chemistry*, *287*, 1627–1634.
Wang, J. S., & Kefalov, V. J. (2011). The cone-specific visual cycle. *Progress in Retinal and Eye Research*, *30*, 115–128.
White, R. J., Nie, Q., Lander, A. D., & Schilling, T. F. (2007). Complex regulation of cyp26a1 creates a robust retinoic acid gradient in the zebrafish embryo. *PLoS Biology*, *5*, 2522–2533.
Wiggert, B., Russell, P., Lewis, M., & Chader, G. (1977). Differential binding to soluble nuclear receptors and effects on cell viability of retinol and retinoic acid in cultured retinoblastoma cells. *Biochemical and Biophysical Research Communications*, *79*, 218–225.
Wolf, G. (2002). The experimental induction of vitamin A deficiency in humans. *The Journal of Nutrition*, *132*, 1805–1811.
Zhang, Y. R., Zhao, Y. Q., & Huang, J. F. (2012). Retinoid-binding proteins: Similar protein architectures bind similar ligands via completely different ways. *PLoS One*, 7(5), e36772. https://doi.org/10.1371/journal.pone.0036772.
Zhou, G. B., Zhang, J., Wang, Z. Y., Chen, S. J., & Chen, Z. (2007). Treatment of acute promyelocytic leukaemia with all-trans retinoic acid and arsenic trioxide: A paradigm of synergistic molecular targeting therapy. *Philosophical Transactions of the Royal Society B: Biological Sciences*, *362*, 959–971.

CHAPTER TWO

Post-natal all-*trans*-retinoic acid biosynthesis

Joseph L. Napoli*
Graduate Program in Metabolic Biology, Nutritional Sciences and Toxicology, University of California, Berkeley, CA, United States
*Corresponding author: e-mail address: jna@berkeley.edu

Contents

Abstract

Generation of the autacoid all-*trans*-retinoic acid (ATRA) from retinol (vitamin A) relies on a complex metabolon that includes retinol binding-proteins and enzymes from the short-chain dehydrogenase/reductase and aldehyde dehydrogenase gene families. Serum retinol binding-protein delivers all-*trans*-retinol (vitamin A) from blood to cells through two membrane receptors, Stra6 and Rbpr2. Stra6 and Rbpr2 convey retinol to cellular retinol binding-protein type 1 (Crbp1). Holo-Crbp1 delivers retinol to lecithin: retinol acyl transferase (Lrat) for esterification and storage. Lrat channels retinol directly

ISSN 0076-6879
https://doi.org/10.1016/bs.mie.2020.02.003

into its active site from holo-Crbp1 by protein-protein interaction. The ratio apo-Crbp1/holo-Crbp1 directs flux of retinol into and out of retinyl esters, through regulating esterification vs ester hydrolysis. Multiple retinol dehydrogenases (Rdh1, Rdh10, Dhrs9, Rdhe2, Rdhe2s) channel retinol from holo-Crbp1 to generate retinal for ATRA biosynthesis. β-Carotene oxidase type 1 generates retinal from carotenoids, delivered by the scavenger receptor-B1. Retinal reductases (Dhrs3, Dhrs4, Rdh11) reduce retinal into retinol, thereby restraining ATRA biosynthesis. Retinal dehydrogenases (Raldh1, 2, 3) dehydrogenate retinal irreversibly into ATRA. ATRA regulates its own concentrations by inducing Lrat and ATRA degradative enzymes. ATRA exhibits hormesis. Its effects relate to its concentration as an inverted J-shaped curve, transitioning from beneficial in the "goldilocks" zone to toxicity, as concentrations increase. Hormesis has distorted understanding physiological effects of ATRA post-nataly using chow-diet fed, ATRA-dosed animal models. Cancer, immune deficiency and metabolic abnormalities result from mutations and/or insufficiency in Crbp1 and retinoid metabolizing enzymes.

1. Introduction

Activation of the nutrient all-*trans*-retinol (vitamin A, retinol) into the autacoid all-*trans*-retinoic acid (ATRA) relies on a complex metabolon, fueled by availability of retinol (Napoli, 2012). Cells access retinol from the serum retinol binding-protein, sRbp (encoded by *rbp4*) (Fig. 1). sRbp interacts with two intermembrane receptors, Stra6 (stimulated by RA6) and Rbpr2 (Rbp receptor 2). Cellular retinol binding-protein type 1 (Crbp1) and lecithin: retinol acyl transferase (Lrat) guarantee that retinol flux into cells occurs one way physiologically. Crbp1 sequesters retinol, preventing its exit from cells, and delivers retinol to Lrat. Lrat, which synthesizes retinyl esters, accesses retinol bound to Crbp1 and channels retinol directly into its active site by protein-protein interaction. A second esterifying enzyme, diacylglycerol acyltransferase type 1 (Dgat1) also can synthesize retinyl esters, especially in skin, but does not make a major contribution in most cells. The ratio apo-Crbp1/holo-Crbp1 directs flux of retinol into and out of retinyl esters, through stimulation of retinyl ester hydrolases (Reh) and inhibition of Lrat by apo-Crbp1. The latter has an inhibitory constant of ~0.2 μM for Lrat. Therefore, with saturated holo-Crbp1 (no apo-Crbp1) both Lrat and retinol dehydrogenases (Rdh) have maximum activity and generate retinyl esters for retinol storage and retinal for ATRA biosynthesis, respectively. As the concentration of apo-Crbp1 increases (deficit of retinol), holo-Crbp1 continues delivering retinol to support RA biosynthesis, but apo-Crbp1 limits storing retinol as retinyl esters.

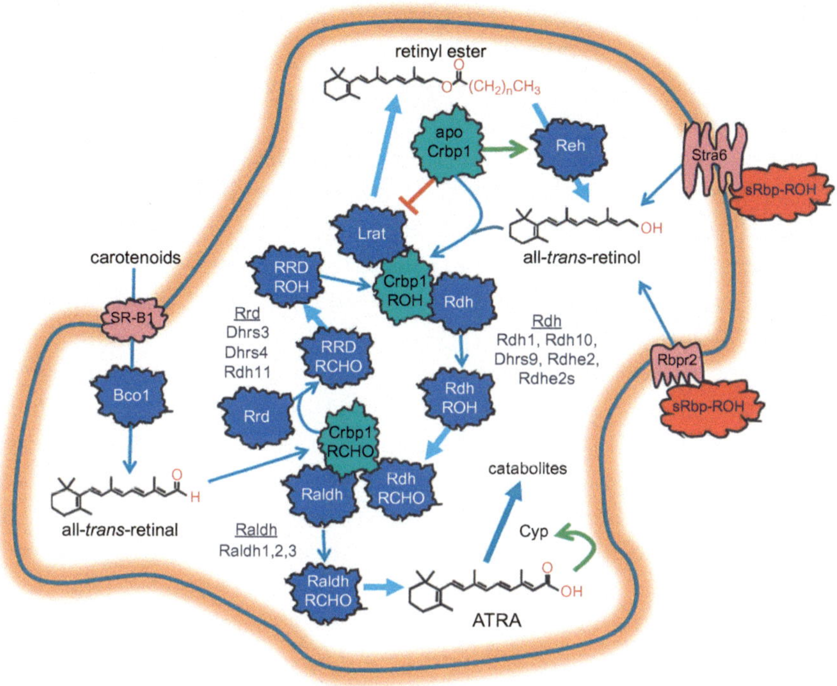

Fig. 1 A model of the retinoid metabolon. Light blue indicates retinoid binding-proteins. Dark blue indicates enzymes. Pink indicates membrane receptors. Red indicates sRbp. The red line indicates inhibition. The green arrows show induction of activity. The wide blue arrows indicate catalytic events. The thin blue arrows indicate retinoid transfer.

Two reactions generate ATRA from all-*trans*-retinol. Rdh of the short-chain dehydrogenase/reductase gene (*sdr*) family catalyze the first reaction, the reversible dehydrogenation of retinol into all-*trans*-retinal (retinal). Rdh occur in microsomes (Rdh1) or associate with lipid droplets (Rdh10). The NAD^+/NADH ratio and the irreversible second reaction that converts retinal into ATRA drive the reversible Rdh-catalyzed reaction forward. Multiple Rdh have been identified (Rdh1, Rdh10, Dhrs9, Rdhe2, Rdhe2s). Those that have been tested access retinol from holo-Crbp1 by channeling through protein-protein interactions, and do not rely on retinol diffusion from Crbp1.

The membrane receptor scavenger receptor-B1 (SR-B1) transports carotenoids into cells, where β-carotene oxidase type 1 (Bco1) converts those that have a β-ionone ring into retinal (Harrison, 2012). Retinal reductases (Rrd), also of the *sdr* gene family, catalyze reduction of retinal into retinol. Outside the retina, these include Dhrs3, Dhrs4 and Rdh11, and are either microsomal or peroxisomal. Where tested, Rrd channel retinal from Crbp1.

Retinal dehydrogenases (Raldh), of the aldehyde dehydrogenase gene family (*aldh*), catalyze irreversible retinal dehydrogenation into ATRA. These cytosolic enzymes include: Raldh1, Raldh2 and Raldh3, and also channel retinal directly from Crbp1.

Retinol and retinal are chaperoned intracellularly by Crbp1 (except for the intestinal enterocyte) (Napoli, 2017; Ross, 1993). Crbp1 has high affinity ($\leq$1 nM) for retinol. Therefore, intracellular retinol occurs as holo-Crbp1 in most tissues. Ablation of Crbp1 disrupts retinol metabolism and causes metabolic disease and cancer (Kane et al., 2011; Pierzchalski, Yu, Norman, & Kane, 2013).

ATRA controls its own intracellular concentrations by inducing Lrat, thereby directing more retinol into retinyl esters, and also by inducing cytochrome P-450's (Cyp26a1, b1 and c1) to initiate its own catabolism.

Two cytosolic enzymes of the alcohol dehydrogenase gene family (*adh*), also known as medium-chain alcohol dehydrogenases, convert retinol into retinal *in vitro*. Adh do not access retinol bound to Crbp1—the form of retinol that occurs in cells. Adh isozymes active with retinol *in vitro* include AdhI and AdhIV in the mouse. AdhIII has little to no activity with retinol. Adh, however, do not contribute to ATRA biosynthesis *in vivo* from physiological amounts of retinol. *In vitro* and *in vivo*, they access retinol concentrations in the toxic range, which exceed capacities of binding-proteins, and overcome restrictions on ATRA biosynthesis. This action contributes to retinol toxicity.

2. ATRA and hormesis

Dosing larger amounts of retinol or ATRA reduces obesity in rodents fed high-fat diets, i.e., ameliorates diet-induced obesity (Berry, DeSantis, Soltanian, Croniger, & Noy, 2012; Mercader et al., 2006). Retinol and ATRA, however, exhibit hormesis; their effects relate to the amount dosed as an inverted J-shaped curve (Fig. 2). As concentrations increase beyond optimum, beneficial effects diminish and toxic effects ensue (Hayes, 2007). Thus, dosing high amounts of retinol or ATRA confounds physiology with toxicology. Chow diets, which contain copious amounts of vitamin A, exacerbate hormesis effects. Chow diets also contain pro-vitamin A carotenoids, which raise ATRA levels. Hormesis has distorted understanding physiological ATRA effects post-nataly, particularly with respect to energy balance, in chow-diet fed, retinol/ATRA-dosed rodents.

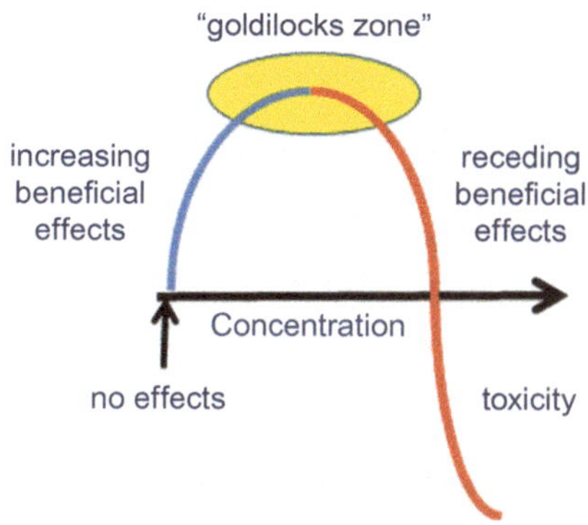

Fig. 2 ATRA has hormesis effects. Only in the "goldilocks" zone does ATRA exert optimum physiological activity. As concentrations increase toxicity sets in.

3. Retinol dehydrogenases

3.1 Background

Initial detailed characterization of ATRA biosynthesis was conducted in an established cell line (the pig kidney epithelial cell line LLC-PK_1) to preserve cell organization, and let the "cell speak for itself," rather than rely on subcellular fractions. This study established retinol dehydrogenation as the rate-limiting reaction, and indicated that the retinol concentration was a primary determinant of the ATRA synthesis rate over a wide range. In other words, substrate availability helps drive retinol dehydrogenation. Retinol esterification/de-esterification and cell uptake control the retinol concentration and therefore substrate availability. In contrast to retinol dehydrogenation, retinal reduction and dehydrogenation occur 4–8-fold and 30–60-fold faster, respectively. Consistent with this, retinal generated ~11-fold more ATRA than retinol. This study also established that ATRA up to 100 nM, 2–20-fold higher than occurs *in vivo*, did not inhibit its own synthesis. The potent aldehyde oxidase inhibitor 4′-(9-acridinylamino)methanesulfon-m-anisidide did not inhibit ATRA biosynthesis, nor did 220 mM ethanol, or higher concentrations (<100 mM) of the potent AdhI and AdhII inhibitor 4-methylpyrazole. These data excluded aldehyde oxidase, AdhI, and AdhII from contributing to ATRA biosynthesis, at least in this cell line. The cytochrome P-450 (Cyp) inhibitor ketoconazole increased the ATRA concentration during biosynthesis, indicating that the ATRA concentration depends on the catabolic rate, in addition to the production rate, suggesting tight control of intracellular ATRA.

The next phase involved identifying enzymes in the physiological pathway of ATRA biosynthesis. Intracellularly, retinol occurs bound to cellular retinol binding-protein type 1 (Crbp1) with high affinity ($k_d \leq 1\,nM$) (Newcomer, Jamison, & Ong, 1998; Ross, 1993). Crbp1 engulfs retinol buried in its binding pocket with the alcohol functional group oriented deep inside. Holo-Crbp1 resists proteolysis by Arg-C during an hour of incubation, indicating a tightly closed structure (Jamison, Newcomer, & Ong, 1994). This presents a conundrum of how metabolic enzymes would access retinol. Access could occur if select enzymes interacted with holo-Crbp1, capturing its retinol via protein-protein interaction, channeling retinol directly from binding-protein to enzyme. Kinetic experiments comparing esterification of free retinol with Crbp1-bound retinol demonstrated occurrence of channeling. Free retinol can be esterified using acyl-CoA as substrate (Arat), but bound retinol undergoes esterification using acyl moieties from the SN1 position of phosphatidylcholine, i.e., lecithin: retinol acyltransferase (Lrat) (Ong, MacDonald, & Gubitosi, 1988). Lrat has identical K_m and V_{max} values in presence of either free or Crbp1-bound retinol. This verifies that Lrat interacts with holo-Crbp1 and directly captures its retinol, because Crbp1 renders the free retinol concentration negligible.

In this context, a physiologically relevant Rdh should have capacity to recognize holo-Crbp1 and capture its retinol through channeling. Kinetics of retinol conversion into retinal in the presence of holo-Crbp1, with added apo-Crbp1 to ensure total retinol sequestration, revealed that rodent liver microsomes express such activity (Posch, Boerman, Burns, & Napoli, 1991). This identification of holo-Crbp1 as the authentic substrate for converting retinol into retinal in the path of ATRA biosynthesis allowed comparison of activities in microsomes vs cytosol and classification of the activities (Boerman & Napoli, 1996). Microsomes from liver, kidney, testis and lung have 4–13-fold higher capacity to generate retinal in a comparison to cytosolic activity, with 5 μM holo-Crbp1 as substrate in the presence of 2 μM apo-Crbp1, a reasonable approximation of their concentrations in liver. Neither ethanol nor the Adh inhibitor 4-methylpyrazole inhibited the microsomal reactions. In contrast, the IC_{50} values of ethanol and 4-methylpyrazole for the cytosolic reaction were 20 μM and 180 mM, respectively. These results exclude Adh2 and Adh3 from consideration for the cytosolic activity, because these two Adh have far different sensitivities to ethanol and 4-methypyrazole than the cytosolic retinol dehydrogenase activity. Carbenoxolone, a potent inhibitor of Sdr, inhibited both the microsomal and cytosolic activities with IC_{50} values of 55 and 70 μM, respectively.

These data, along with previous data, exclude Adh isoforms from contributing to cytosolic retinol dehydrogenase activity and reveal that microsomal Rdh generated most retinal for ATRA biosynthesis.

An Rdh involved in ATRA biosynthesis was identified through cross-linking with holo-Crbp1, using either microsomes or an enzyme purified to two gel bands (Boerman & Napoli, 1995). Both sources of Rdh produced the same results, showing specificity of cross-linking. Cross-linking required pyridine nucleotide cofactor, consistent with an ordered bi-substrate reaction mechanism of a dehydrogenase. The ~36 kDa protein identified had a subunit molecular weight and other attributes typical of an Sdr, including the highly-conserved sequences GXXXGXG (cofactor binding), WXLVNNAG, and YCISK (active site residues), sensitivity to carbenoxolone ($IC_{50} = 55\,\mu M$), and insensitivity to inhibition by 860 mM ethanol and the Adh inhibitor 4-methylpyrazole. The rat, mouse and human orthologs were cloned, characterized and dubbed RodhI and II, Rdh1, and Rodh4/Rodh-E/Rdh-E (Chai, Boerman, Zhai, & Napoli, 1995; Chai, Zhai, Popescu, & Napoli, 1995; Gough, VanOoteghem, Sint, & Kedishvili, 1998; Jurukovski et al., 1999; Zhang, Chen, Smith, & Napoli, 2001). Through the peculiarities of nomenclature revision these are now known as Rdh2, Rdh1 and Rdh16, respectively.

Some concern has been raised about the purine nucleotide preferred cofactor (NAD^+ vs $NADP^+$) by Rdh, based on the perception that dehydrogenases prefer NAD^+, whereas $NADP^+$ preferring enzymes should be considered reductases that really use NADPH. This is not a biochemically sound concern, because most Rdh can use either cofactor, although the K_m value usually is lower with one or the other. The "preference" depends not only on the K_m value, but also on the cellular concentration of the cofactor. The function of the enzyme in an intact cell provides the answer, when presented with retinol vs retinal. All known Rdh generate ATRA in the presence of an Raldh in intact cells, as shown for Rdh1 (Zhang et al., 2001).

3.2 Functions of mRdh1/hRdh16

Rdh1 associates with the microsomal membrane with the body of the enzyme projecting into the cytoplasm (Zhang, Hu, & Napoli, 2004). An N-terminal sequence of 22 residues forms a hydrophobic helix that ends in a net positive charge. This leader anchors Rdh1 in the membrane bilayer. Mutating the leader sequence to place the positive charges at the beginning, and replacing the positively charged residues at the end with neutral residues, completely reverses Rdh1 orientation to face the lumen of the smooth endoplasmic reticulum. Thus, Rdh1 adheres to the "positive inside rule," which defines the orientation of many microsomal and plasma membrane-associated proteins with respect to the cytoplasm. Simply put, the positive charges in the leader sequence occur in the cytoplasm.

The mouse embryo expresses Rdh1 by embryo day 7 (e7), with expression increasing each embryo day until parturition (Zhang et al., 2001). Rdh1 mRNA expression relates inversely to Cyp26a1 and 26b1 expression during embryogenesis. Cyp26 catabolize ATRA. Therefore, diametrically opposed expression between Cyp and Rdh1 suggests coordination between the catabolic and biosynthetic enzymes to maximally reduce or enhance the ATRA concentration during development.

Rdh1/Rdh16 has enzymatic activity with all-*trans*-retinol, 9-*cis*-retinol, 5α-androstane-3α,17β-diol (3-adiol) and androsterone *in vitro* (M Zhang et al., 2001). But also accesses retinol bound with Crbp1 (Jurukovski et al., 1999). Human Rdh16 now has the designation Sdr9c8.

The *rdh1* mouse knockout has a phenotype consistent with a decrease in ATRA, but serum steroids do not change. The knockout generates mice in Mendelian frequency with no obvious phenotypic differences from wild type, when fed a chow diet (copious amount of vitamin A). *Rdh1*-null male mice fed a diet with 4 IU vitamin A/g (as recommended by the National Research Council) have liver retinol levels ~2-fold greater than wild type littermates, consistent with decreased ATRA biosynthesis. Feeding the 4 IU vitamin A/g diet to *rdh1*-null mice revealed that *cyp26a1* mRNA and protein decreased by 2.5-fold. This compensatory decrease in ATRA catabolism confirms importance of Rdh1 to generating ATRA. Knockout mice fed a vitamin A-restricted diet become heavier due to increased adiposity, even when fed a low-fat diet. At 33 weeks old, male *rdh1*-null mice weigh 10.6 g (37%) more than wild type; females weigh 5.8 g (23%) more than WT. Increased adiposity accounts for the increase in weight. Mesentery, femoral and inguinal fat pads increase relative to body weight (% body weight) and per g/mouse. Knockout mice did not consume more food than WT, nor reduce their activity.

The *rdh1*-null mouse fed a low-fat diet with recommended retinol levels offered opportunity to study physiological functions of lowering endogenous ATRA. *Rdh1* has widespread tissue expression, but neither muscle or nor epidydymal white fat pads express *rdh1*. Brown adipose tissue (BAT) expresses *rdh1* more intensely than most other tissues, and *rdh1*-null mice have an average body temperature ~0.8 °C lower than WT under normal vivarium conditions (Krois et al., 2019). These data indicate flawed

BAT function. BAT differentiation seems unaffected, but mitochondrial function was damaged, as indicated by decreased proteins such as cytochrome C and Ucp1, a decreased oxygen consumption rate, and disrupted mitochondrial membrane potential. Knockout BAT had increased fatty acid uptake, but decreased lipolysis, resulting in increased lipid content and larger lipid droplets. RNAseq revealed 424 dysregulated BAT genes in knockout mice, which mostly segregated by differences during fasting (92 genes) or re-feeding (328 genes). The mRNA of only four genes differed from WT during both conditions. Sixty percent of the genes had confirmed RARE. One of the genes whose mRNA had changed during both fasting and re-feeding encoded the serum Rbp (i.e., *rbp4*), which increased 2- to 3-fold, indicating *rdh1* suppression of *rbp4* and *rbp4* promotion of adiposity. The mRNA of the GTPase, *gbp2b*, increased 200–400-fold. Over-expression of *gbp2b* in immortalized brown adipocytes recapitulated accumulation of larger lipids droplets. WT BAT cells showed a transient increase in ATRA 4 h after re-feeding, absent in *rdh1*-null BAT. *Rdh1*-null BAT had increased *dhrs9* and *aldh1a1* (encodes Raldh1) mRNA during fasting, and decreased *cyp26b1* mRNA, as apparent compensation for *rdh1* loss. These data demonstrate a crucial purpose of Rdh1 in generating ATRA to maintain BAT function, not duplicated by other Rdh. These data also demonstrate the pervasive function of ATRA in normalizing BAT function through regulation of multiple genes in an energy status-specific manner. *Rdh1* loss impairs glucose tolerance and causes insulin resistance, due to increased adiposity, reminiscent of metabolic syndrome.

Mouse kidney and skin have the highest *rdh1* mRNA (Krois et al., 2019). Accordingly, the human ortholog *rdh16* contributes to ATRA biosynthesis in human skin (Pavez Loriè, Chamcheu, Vahlquist, & Törmä, 2009; Pavez Loriè, Li, Vahlquist, & Törmä, 2009). Keratinocyte differentiation increased *rdh16* expression, while reducing *cyp26b1* expression (Jurukovski et al., 1999). ATRA treatment of organotypic epidermis reduces *rdh16* mRNA expression by 80%. These data show that *rdh16* contributes to keratinocyte differentiation and skin function.

Nerve growth factor (NGF) up regulates *rdh2* expression in primary hepatocytes isolated from Sprague-Dawley rats and *rdh1* expression in cholestatic mouse livers (Kao et al., 2018). NGF functions via an NF-κB pathway to protect against oxidative damage. These data suggest that NGF intensifies the anti-fibrotic and anticarcinogenic effects of ATRA by inducing its biosynthesis. The authors suggested that *rdh16* could serve as a biomarker of cholestatic liver disease.

Alcohol-associated hepatocarcinogenesis down regulates *rdh16* through promoter DNA methylation (Udali et al., 2015). Most human hepatocarcinoma samples and cell lines have down-regulated *rdh16* mRNA and protein. Down-regulation associates with marginal survival, because *rdh16* generates ATRA to suppress cell growth and division (Zhu et al., 2019).

3.3 Rdh10 functions

Rdh10 was identified as an Rdh/*sdr* that contributes to 11-*cis*-retinal biosynthesis in the retinal pigment epithelium and all-*trans*-retinal biosynthesis in Muller cells of the retina. Rdh10 interacts with the retinoid binding-protein cellular retinal binding-protein (Cralbp) of the retinal pigment epithelium, effecting direct transfer of 11-*cis*-retinal (Farjo, Moiseyev, Takahashi, Crouch, & Ma, 2009; Wu et al., 2002, 2004). Subsequently, a contribution of Rdh10 to generating retinal for ATRA biosynthesis was demonstrated during mid-gestation (Ashique et al., 2012; Cammas, Romand, Fraulob, Mura, & Dollé, 2007; Romand, Kondo, Cammas, Hashino, & Dollé, 2008; Siegenthaler et al., 2009), and during pancreas organogenesis (Arregi et al., 2016). The next stage involved delineating Rdh10 post-natal functions, which involve generating retinal for ATRA biosynthesis in multiple organs (Parker & Crouch, 2010).

3.3.1 Post-natal, non-ocular Rdh10 functions

Spermatogenesis relies on Rdh10 in both Sertoli and germ cells (Tong, Yang, Davis, & Griswold, 2013). Knocking out *rdh10* in Sertoli and germ cells causes testicular atrophy and prevents spermatogenesis in juvenile mice, due to ATRA absence. The effect is temporary. Males less than 7 weeks old have impaired fertility, but males older than 9 weeks are fertile. This phenomenon resembles the Rdh10 contribution to embryogenesis, i.e., its total absence has a devastating effect exerted during a limited time frame. Retinal supplementation during this time (embryo day 7.5 through 11.5) allows production of mice in Mendelian frequency, but with distal limb abnormalities and atypical physical behavior (Rhinn, Schuhbaur, Niederreither, & Dollé, 2011).

A subset of mouse and human dendritic cells express *rdh10* (along with *aldh1a2,* which encodes Raldh2) (Gyöngyösi et al., 2013). These cells generate ATRA to promote T-cell specification. These include human monocyte-derived dendritic cells and mouse CD103(+), granulocyte-macrophage colony-stimulating factor, and interleukin-4-treated bone marrow-derived dendritic cells. Thus, immune cells express both *rdh10* and *dhrs9* (see below).

Both visceral and subcutaneous fat express *rdh10* (Sima, Manolescu, & Bhat, 2011). Mice with a heterozygote whole-body *rdh10* knockout revealed the contribution of Rdh10 to adipose function (Yang et al., 2018). Although born in Mendelian frequency, these mice present with metabolic abnormalities of white fat, liver, muscle and bone, when fed a high-fat diet. By 16 weeks old, male heterozygotes weigh >5.3 g more than controls (~14%) and females weigh >4.7 g more than controls (~18%). Fat accounted for the difference in weight. The increased weight led to decreased glucose tolerance and increased insulin resistance. Inguinal (subcutaneous), perirenal (visceral) and epidydymal (visceral) fat pads were enlarged, inflamed and fibrotic. Males suffer greater hepatic steatosis than controls, but females did not. In contrast, females have increased femur marrow adipocyte formation, clustered near the growth plate, regardless of a high or low-fat diet, but males did not. These data illustrate the importance of Rdh10 to overall metabolic health, and show that modestly decreased ATRA has profound metabolic effects. This model provides opportunity to study physiological actions of moderately decreased endogenous ATRA. Studying decreases in endogenous ATRA avoids complications and artifacts caused by dosing ATRA, because of its hormesis properties.

Rdh10 has been reported as a major Rdh in skin (Lee, Belyaeva, Wu, & Kedishvili, 2011). This conclusion was based on over-expression, rather than on examination of endogenous Rdhs. It is now clear that Rdh16, Dhrs9, RdhE2 and E2S account for physiological generation of retinal for ATRA biosynthesis in skin (see below).

3.3.2 Rdh10 and cancer

Grade 3, stage III non-small-cell lung tumors have mutated *rdh10* (Bankovic et al., 2010). This indicates a contribution of low Rdh10 and low ATRA to cancer promotion and progression. In a different approach, a computational strategy ranked large numbers ($2^{10}-2^{20}$) of genes vis-a-via their contribution to prostate cancer (Nim, Furtado, Ramialison, & Boyd, 2017). The results revealed that patients with an 85% decrease in *rdh10*, among other genes, had a mean survival rate of ~63 months compared to 150 months for patients without alterations.

3.4 Dhrs9 contributions to ATRA biosynthesis

Dhrs9 has been cloned by multiple labs and has been christened 3α-HSD, Rdh15, Rdh-TBE, eRoldh, Rodh-E2, RdhE and RdhL (Chetyrkin, Belyaeva, Gough, & Kedishvili, 2001; Soref et al., 2001; Rexer & Ong, 2002; Nedialka G. Markova, Pinkas-Sarafova, Karaman-Jurukovska, Jurukovski, & Simon, 2003). Human *dhrs9* now has the designation Sdr9C4. Dhrs9 has activity with all-*trans*-retinol and various androgens (Chetyrkin et al., 2001; Soref et al., 2001). Early publications reported it as an androgen-related SDR without activity toward retinol, because of "low" activity with retinol. The low retinol activity, however, resulted from evaluation of *dhrs9* (3α-HSD) expressed in Sf9 cells, which greatly reduced its activity, as Sf9 cells do for Rdh10. Subsequent publications showed activity with free retinol and with Crbp1-bound retinol, identifying Dhrs9 (eRoldh, Rodh-E2) as an Rdh (Markova et al., 2003; Rexer & Ong, 2002).

Dhrs9, as with *rdh1* and *rdh10*, has widespread tissue expression, especially in epithelial tissues. Research has focused on its functions in hair and skin, as a tumor suppressor, and as an immune stimulant. Significantly, Dhrs9 (hRoDH-E2) may also "moonlight" as a transcription repressor (Markova, Pinkas-Sarafova, & Simon, 2006). Dhrs9 interacts with Sp1, converting it from an activator to a repressor, in an ATRA independent manner. Affirmation of this ability was provided by the observation that knockdown of *dhrs9* in astrocytes increased ATRA synthesis ~40%, by increasing *aldh1a1* (encodes Raldh1) expression (Wang, Kane, & Napoli, 2011). This regulatory cross-talk provides a mechanism for controlling ATRA biosynthesis.

3.4.1 Dhrs9 in hair and skin

The hair follicle, sebaceous gland, and interfollicular epidermis of mice each express a complete retinoid metabolon (pathway of ATRA biosynthesis) and signaling apparatus. The components include Dhrs9 (aka eRodh), Raldh1,2,3 (*aldh1a1,2,3*), Crbp1, Crabp2 and Rarα (Everts, Sundberg, King, & Ong, 2007). Expression varies with the hair cycle, indicating spatial and temporal regulation of ATRA biosynthesis. Components were not always expressed in the same cell layer, consistent with epithelial–mesenchymal interactions in ATRA synthesis and actions. Focal alopecia in C57BL/6J mice resembles the human disease central centrifugal cicatricial alopecia. In these mice and humans, an increase in *dhrs9* expression occurred in mild disease, but not in severe disease (Everts et al., 2013). In alopecia areata in mice (C3H/HeJ), rat and human genes of the ATRA biosynthetic metabolon increased, whereas genes of ATRA catabolism decreased (Duncan et al., 2013). *Dhrs9* increased in mice, but not in humans. Overall, however, ATRA concentrations increased in the disease, but restricting vitamin A resulted in more severe disease. This phenomenon is consistent with the hormesis properties of vitamin A/ATRA; precise amounts of ATRA support vitamin A function in a restricted "goldilocks" concentration zone.

3.4.2 Dhrs9 and tumor progression

Human colon adenomas and carcinomas, caused by the adenomatous polyposis coli gene mutation, have reduced *dhrs9* (aka RDHL) (Jette et al., 2004). Multiple colon carcinoma cell lines also have reduced *dhrs9* expression. Reduced *dhrs9* expression associates with reduced ATRA biosynthesis, leading to inhibited colonocyte differentiation. A second study demonstrated that human colorectal cancer samples have propensity to down-regulate *dhrs9* relative to normal tissue (Hu et al., 2016). Decreased *dhrs9* expression associates with increased lymph node metastasis, disease recurrence, and death.

Reduced proliferation of human renal cell carcinoma can be achieved with treatments that increase *dhrs9* expression (Tavares, Nanus, Yang, & Gudas, 2008). Conversely, loss of *dhrs9* expression associates with Burkitt lymphomas, T-cell acute lymphoblastic leukemia, and acute myeloid leukemia (Limongi, Curatolo, Pelliccia, & Rocchi, 2005; Meester-Smoor et al., 2008). Reduced *dhrs9* expression occurs in >69% of patients with oral squamous cell carcinoma, and correlates with relatively poor prognosis (Shimomura, Sasahira, Nakashima, Shimomura-Kurihara, & Kirita, 2018).

Dhrs9-driven differentiation is not always beneficial. Differentiated epithelial cells allow lytic Epstein-Barr virus replication, but undifferentiated cells do not. Gastric carcinomas and Epstein-Barr positive Burkitt lymphomas have increased *dhrs9* expression, increasing ATRA biosynthesis (Jones, Dickerson, Bhende, Delecluse, & Kenney, 2007). Enhanced ATRA-driven differentiation would enhance viral replication.

3.4.2.1 Dhrs9 and immunity

Immune cells express *dhrs9*, although reports disagree as to the specific cells involved. One report concluded that macrophages do not express *dhrs9*, whereas granulocyte-macrophage colony-stimulating factor-derived dendritic cells do (Kim et al., 2019). Dendritic cells would secrete ATRA to resist pathogenesis in *M. tuberculosis*-infected macrophages. Contrary reports find that a subset of monocyte-derived macrophages, human regulatory macrophages, express *dhrs9* specifically (Riquelme et al., 2017; Riquelme & Hutchinson, 2018). The secreted ATRA would contribute to Treg development.

3.5 RdhE2 and E2S (retinol dehydrogenases epidermal 2 and 2 similar)

RdhE2 and RdhE2S function as Rdh when expressed in HEK293 cells; RdhE2S has ~5-fold higher activity (Wu et al., 2019). Retinol and ATRA down-regulate RdhE2 mRNA in a skin raft culture model of stratified human epidermis (Adams, Lee, Belyaeva, Wu, & Kedishvili, 2017). A double knockout lacking both E2 and E2S produces offspring in a Mendelian ratio. Rdh activity of microsomes and mitochondria from double knockout skin decreases by ~10-fold and ~6-fold, respectively, but no phenotype develops when mice are fed a diet with 24 IU vitamin A/G—a high amount. The retinoid metabolon compensated for the dual loss by up-regulating *rdh10*, *dhrs9*, *crabp2*, and *aldh1a2* (encodes Raldh2), and down regulating *cyp26b1*. When fed a vitamin A-deficient diet, the dual knockout has larger sebaceous glands and longer hair shafts, suggesting disrupted growth. These data show that RdhE2 and RdhE2s have a non-critical function in embryogenesis, but moderate the hair growth cycle post-nataly. Thus, like Rdh1, RdhE2 and E2s seem especially important for post-natal ATRA biosynthesis.

3.6 Other Rdh

SRP-35 has been reported as a Rdh of particular importance in muscle and adipose (Treves et al., 2012). The enzymatic characterization, however, does not support this claim unequivocally.

3.7 A cautionary note

The NCBI gene bank confuses Dhrs9 and RdhE2 as the same SDR. Confusion extends to declaring identity between Rdh16 and Rdh5, and Rdh16 and Rdh6 and Crad1. It is likely that there occurs further NCBI gene bank mis-identification of retinoid associated *sdr*.

4. Retinal reductases (Rrd) and retinoid metabolism

4.1 The reductase Dhrs3 (mouse and human retSDR1, hSdr16C1, hRdh17)

Dhrs3 initially was identified as retSDR1 in cone photoreceptors (Haeseleer, Huang, Lebioda, Saari, & Palczewski, 1998). It has widespread tissue expression, however. Knockout mice provide insight into essential extra-retinal function (Billings et al., 2013). *Dhrs3* ablation increases embryonic ATRA amounts by ~40%, while decreasing retinol and retinyl esters by ~60% and 55%, respectively. Embryo death late in gestation of knockout mice, from multiple defects in heart development, illustrate Dhrs3 importance in controlling ATRA concentrations. These data show that Rrd contribute crucially to controlling ATRA concentrations, by controlling retinal produced from β-carotene and/or from retinol.

Even though Dhrs3 contributes essentially to ATRA homeostasis, it does not recognize retinal as its only substrate (Lundová et al., 2015). Dhrs3 also uses as substrates androstenedione, estrone, glyceraldehyde and multiple xenobiotics. In this respect, it resembles activities of other Rdh/*sdr*.

Interestingly, Dhrs3 heterodimerizes with Rdh10, when both are overexpressed in mammalian cells (Adams, Belyaeva, Wu, & Kedishvili, 2014). In contrast, Rdh10 does not form a heterodimer with the Rrd Rdh11. Dimerization of Dhrs3 and Rdh10 enhances both activities. This may seem counterintuitive, because the net effect might provide futile cycling between retinol and retinal. The net outcome, however would be determined by substrate availability. In other words, an influx of retinal, from β-carotene cleavage, would drive the reaction to retinol production. A decrease in ATRA would drive the reaction to retinal production by release of retinol from RE and irreversible retinal conversion into ATRA.

4.1.1 Dhrs3 and cancer

Neuroblastomas that regress spontaneously, rather than progress to fatality, overexpress *dhrs3* (and *cyp26a1*) (Kamei et al., 2009). Ironically, overexpression of these two genes should decrease ATRA. This seems counterintuitive because ATRA generally reduces tumor proliferation and induces differentiation. Perhaps, overexpression represents compensation for enhanced expression of ATRA biosynthetic enzymes, or countermands ATRA biosynthesis in cells in which ATRA prompts proliferation by activating Pparδ. A different study concluded that *dhrs3* activated by p53 and p63 suppressed tumor formation (Kirschner, Rother, Müller, & Engeland, 2010). These contradictory reports illustrate the problem with basing insight into ATRA functions by assessing only one (or two) gene(s) of the retinoid metabolon, and not quantifying ATRA.

4.2 An uncertain contribution of Rdh13 (Pan2)

Rdh13 (*aka pan2*) has widespread tissue expression (Belyaeva & Kedishvili, 2002). Although more active *in vitro* in reduction than in dehydrogenation, this is a common characteristic of dehydrogenases. Cofactor concentrations and enzymes that remove product determine the direction *in vivo*. Moreover, Rdh13 kinetic constants were determined with enzyme expressed in Sf9 cells, which often decreases Sdr activities extensively. Until activity is assessed in intact cells and it is ablated, its contribution to retinoid homeostasis remains uncertain.

4.3 The reductase Dhrs4 (mouse Rrd; human 2,4-dienoyl-CoA reductase)

Dhrs4 has been characterized as a peroxisomal retinal reductase, expressed by multiple tissues (Lei, Chen, Zhang, & Napoli, 2003). In contrast to Rdh1, which only generates retinal from retinol, when expressed in CHO-K1 cells, Dhrs4 only reduces retinal to retinol. This observation was reproduced in CHO cell lysates. Dhrs4 functions in the presence of a 2-fold molar excess of Crbp1/retinal, but holo-Crbp1 and apo-Crbp1 compete with each other. Thus, the ratio of holo-Crbp1/apoCrbp1 controls the amount of retinal available for ATRA biosynthesis. With higher amounts of apo-Crbp1, i.e., lower amounts of cellular retinol, retinal reduction is inhibited to direct flux of retinal into ATRA.

An enzymatic lesson was learned with Dhrs4. Assessment of only one time point or temperature indicated that Dhrs4 *appeared* to function as a dehydrogenase. But the amount of retinal produced did not increase with increasing temperature or time. Indeed, similar amounts of "product" were produced in a zero time control. These data indicate that "product" formation was not enzymatic in the dehydrogenation direction. In contrast, reduction of retinal was protein-, time-, and temperature-dependent. These data illustrate the importance of evaluating enzymes under varying conditions, in accordance with Michaelis-Menten kinetics.

4.4 The reductase Rdh11 (mouse and human RalR1, Psdr1 scald and human Sdr7C1)

Rdh11 initially was identified as a Rrd in the visual cycle (Parker & Crouch, 2010). *Rdh11*, however, has widespread extra-ocular expression in the mouse and human. The knockout mouse has no identified phenotype, but shows lower reduction rates of retinal in liver and testes, and impaired generation of retinol from retinal derived from β-carotene. Reduced dietary vitamin A accentuates differences in tissue retinal. These data suggest Rdh11 contributes to retinoid homeostasis during vitamin A dietary insufficiency, but the lack of an identified phenotype restricts insight into its functional contribution to retinoid homeostasis.

5. Regulation of ATRA tissue concentrations

Disruption of FoxO1 lowers expression of *dhrs9* in hepatocytes (Shin et al., 2012). This observation begged the question whether FoxO1 regulates ATRA concentrations. Based on the contribution of ATRA to controlling energy balance and adiposity, it would make sense if energy status regulated ATRA concentrations. Indeed, fasting increases FoxO1 presence in the nucleus, and FoxO1 induces transcription of *rdh1* and *rdh10* in liver, which leads to increased ATRA biosynthesis (Obrochta, Krois, Campos, & Napoli, 2015). Fasting also enhances glucagon and cortisol secretion, which inhibit the ATRA catabolic enzyme Cyp26a1. Inhibition of Cyp26a1 prevents futile cycling, because ATRA is a potent inducer of *cyp26a1* transcription. *Re*-feeding decreases the liver concentration of ATRA by ~50%. Glucose gavage and insulin recapitulates this phenomenon. Insulin functions by causing phosphorylation of FoxO1, which excludes it from the nucleus. These data reveal a mechanism for regulating the opposing effects of ATRA and insulin on energy use, and suggest a nexus between ATRA and insulin-signaling related diseases (Fig. 3).

Observation of insulin regulation of ATRA biosynthesis has been extended. A tissue bank of diabetic transgenic pigs (INS^{C94Y}) and their WT littermates has been established in Munich: the Munich MIDY (mutant insulin gene-induced diabetes of youth). Livers from the MIDY pigs have a "marked" increase in *rdh16* (pig designation for mouse *rdh1*), which associates with an ~2.5-fold increase in ATRA (Backman et al., 2019).

The importance of the reciprocal relationship between insulin and ATRA lies in their opposing actions. Whereas insulin stimulates lipogenesis,

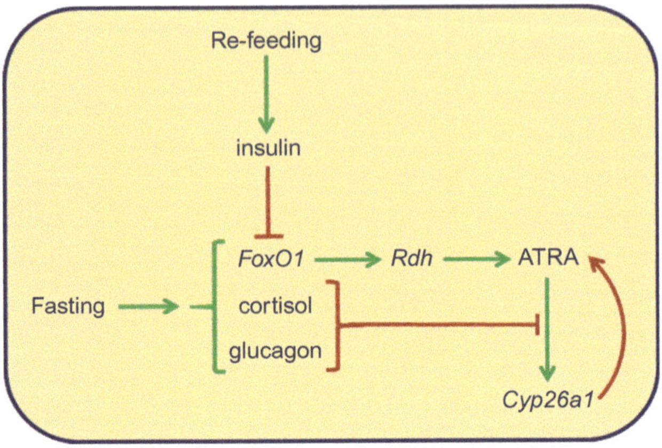

Fig. 3 Regulation of liver ATRA concentrations by fasting and re-feeding. Green arrows indicate stimulation/increased secretion; red arrows indicate inhibition, or in the case of the red arrow extending from Cyp26a1 to ATRA, catabolism. Fasting activates FoxO1, which induces *rdh* transcription to increase ATRA biosynthesis. Concurrently, cortisol and glucagon ameliorate *cyp26a1* transcription, which ATRA induces, to reduce ATRA catabolism. *Re*-feeding stimulates insulin secretion, which excludes FoxO1 from the nucleus. This reduces *rdh* transcription and accelerates *rdh* mRNA degradation. Re-feeding also reduces cortisol and glucagon secretion, allowing maximum *cyp26a1* expression to catabolize ATRA. Thus, during fasting ATRA concentrations exceed those during re-feeding. ATRA concentrations inverse to insulin, optimize the contrary actions of each hormone.

ATRA promotes fatty acid oxidation; insulin retards lipolysis, ATRA induces lipolysis; insulin arrests gluconeogenesis, ATRA stimulates gluconeogenesis. It is reasonable to conclude that the insulin-driven decrease in ATRA augments insulin actions. On the other hand, the FoxO1 increase in ATRA would augment adaptation to energy use caused by fasting.

6. The multiplicity of enzymes in the retinoid metabolon

Tissues, and often the same cells, co-express multiple *rdh*, *rrd* and *aldh1a*; gene ablations reveal different phenotypes for each, consistent with non-redundant functions. Yet, ablation of any one often prompts compensation by others, consistent with redundant functions in some locations. For example, astrocytes express *rdh1*, *dhrs9*, *rdh10* and all three *aldh1a* (Wang et al., 2011). Knocking down each one decreases ATRA biosynthesis, with the exception of *dhrs9* (see above). These data suggest both redundancy and differing metabolons generating ATRA for different functions. The foregoing discussion strongly supports this specificity involved in ATRA biosynthesis.

7. Alcohol dehydrogenases

Adh have been postulated to contribute to ATRA biosynthesis for decades, but the totality of data indicate no physiological contribution of Adh to ATRA biosynthesis (see above). The supposed Adh contribution to ATRA biosynthesis has prompted the suggestion that ethanol inhibits ATRA biosynthesis, thus accounting (in part) for ethanol toxicity and fetal alcohol spectrum disease. As discussed above, ethanol does not inhibit Rdh. In fact, the converse occurs (Kane, Folias, Wang, & Napoli, 2010; McCaffery et al., 2004; Napoli, 2011). Acute ethanol dosing increases ATRA in adult hippocampus, liver, and testis by ~2-fold. Dosing dams with 6.5% ethanol from embryonic day 13–19 increases ATRA in fetal hippocampus up to 20-fold and in fetal cortex up to 50-fold, depending on the dam blood alcohol content. Chronic ethanol dosing increases ATRA in adult hippocampus 20-fold, cortex 2-fold, testis 2-fold, and serum 10-fold. Tissue-specific increases in *rdh* mRNAs and Rdh activities, extrahepatic retinol concentrations, and ATRA catabolism combine to produce site-specific effects. Moreover, ethanol stimulates stem cell differentiation by increasing retinol cell uptake and conversion into ATRA to activate RARγ-dependent transcription (Serio, Laursen, Urvalek, Gross, & Gudas, 2019). Thus, the hypothesis has been disproved that an aspect of fetal alcohol spectrum disease involves inhibition of ATRA biosynthesis or function.

8. Retinal dehydrogenases

The retinal dehydrogenases (Raldh) that catalyze irreversible conversion of retinal into ATRA belong to the *aldh1a* gene family. They are Raldh1 (*aldh1a1*), Raldh2 (*aldh1a2*) and Raldh3 (*aldh1a3*).

Raldh1 was the first retinal dehydrogenase determined unequivocally as a physiological contributor to ATRA biosynthesis (Posch, Burns, & Napoli, 1992). This Raldh was identified and distinguished from other cytosolic Aldh, some of which had been postulated to serve in ATRA biosynthesis based on *in vitro* activity with retinal, along with many other substrates. To clarify which Aldh contributed to ATRA biosynthesis, retinal was generated *in situ* in microsomes (a source of most Rdh) from retinol complexed with Crbp1. Cytosolic fractions (source of Aldh) were added to determine which would generate ATRA from holo-Crbp1-produced retinal. This experiment identified the first Raldh. These data were re-enforced by experiments that showed direct access of Raldh1 to retinal bound with Crbp1 (Ottonello, Scita, Mantovani, Cavazzini, & Rossi, 1993; Penzes, Wang, & Napoli, 1997). In contrast to Raldh2, apo-Crbp1 inhibits Raldh1. This suggests primacy of other Raldh in generating ATRA. In fact, the *aldh1a1* knockout has decreased ATRA and increased retinal in liver, but generally not in parametrical, epidydymal or femoral fat in either males or females (Yang et al., 2017). This can be explained by the nature of Raldh1. As an enzyme it has multiple substrates, and as a protein, it has multiple non-enzymatic functions (Grünblatt & Riederer, 2016). Thus, it is not reasonable to attribute all Raldh actions to ATRA generation.

Aldh1a1-knock out mice, in utter disagreement with ATRA actions and *rdh* knockout phenotypes, *resist* adiposity compared to WT controls. One lab attributed this to Raldh1 function as a retinal dehydrogenase, and concluded that an increase in retinal in the knockout prevented adiposity in females, and low concentrations of ATRA promote adiposity (Ziouzenkova et al., 2007). The data are overwhelming, however, that restriction of vitamin A or ATRA enhances adiposity (Berry et al., 2012; Blaner, 2019). A more reasonable interpretation would have taken into account the multiple Raldh1 functions. A re-examination of the phenotype showed that Raldh1 functions as a retinoid-independent promoter of adiposity only during adolescence in both males and females, through pre-adipocyte cell autonomous actions (Yang et al., 2017). Raldh1, therefore, acts as a retinal dehydrogenase only in select tissues and cells, not universally, and not likely in white adipose.

Upon identification as a testes Raldh, *aldh1a2* (Raldh2) was cloned, expressed in *E. coli* and characterized, and subsequently cloned from embryonic cells (Wang, Penzes, & Napoli, 1996; Zhao et al., 1996). As for Raldh1, the purified recombinant enzyme produces ATRA from retinal generated *in situ* by microsomes from Crbp1-retinol. Equivalent rates with unbound retinal and in the presence of a 2:1 ratio of Crbp1/retinal confirm that Raldh2 can access retinal bound to Crbp1. In primary astrocytes, Raldh2 localizes most intensely in the perinuclear regions, whereas Raldh1 localizes to both cytosol and nuclei. These differences suggest that multiple Raldh exist to generate ATRA in specific areas of the cell, perhaps to support different vitamin A functions. Nuclear localization of Raldh1 also suggests activity in gene regulation.

Raldh3 (aka Aldh6), which also accesses retinal bound with Crbp1, completes the complement of known Aldh that contribute to ATRA biosynthesis (Arnold et al., 2015; Grün, Hirose, Kawauchi, Ogura, & Umesono, 2000).

8.1 Raldh and disease

Complex connections occur among Raldh and diseases. This undoubtedly reflects the multifunctional nature of Raldh1, and perhaps other Raldh. For example, over-expression of Raldh1 in colon macrophages, through generating ATRA, was postulated to increase inflammation in Crohn's disease (Sanders et al., 2014). This work did not quantify ATRA, but relied on estimation of Raldh1 activity by Aldefluor assay. This overlooks the multiple substrates metabolized by Raldh1 and its multiple functions. In other words, it does not provide any evidence of ATRA involvement in exacerbating Crohn's disease. Raldh1 seems to confer multi-drug resistance to cancer (Chefetz et al., 2019; Kozovska et al., 2018). The latter may explain frequent association of Raldh1 with cancer progression, not its ability to generate ATRA in some loci. In contrast, Raldh2 contributes to ATRA biosynthesis to trigger an antimicrobial response in tuberculosis: low expression of Raldh2 in lung characterizes the disease (Kim et al., 2019).

References

Adams, M. K., Belyaeva, O. V., Wu, L., & Kedishvili, N. Y. (2014). The retinaldehyde reductase activity of DHRS3 is reciprocally activated by retinol dehydrogenase 10 to control retinoid homeostasis. *The Journal of Biological Chemistry*, *289*(21), 14868–14880. https://doi.org/10.1074/jbc.M114.552257.

Adams, M. K., Lee, S.-A. , Belyaeva, O. V., Wu, L., & Kedishvili, N. Y. (2017). Characterization of human short chain dehydrogenase/reductase SDR16C family members related to retinol dehydrogenase 10. *Chemico-Biological Interactions*, *276*, 88–94. https://doi.org/10.1016/j.cbi.2016.10.019.

Arnold, S. L. M., Kent, T., Hogarth, C. A., Griswold, M. D., Amory, J. K., & Isoherranen, N. (2015). Pharmacological inhibition of ALDH1A in mice decreases all-trans retinoic acid concentrations in a tissue specific manner. *Biochemical Pharmacology*, *95*(3), 177–192. https://doi.org/10.1016/j.bcp.2015.03.001.

Arregi, I., Climent, M., Iliev, D., Strasser, J., Gouignard, N., Johansson, J. K., et al. (2016). Retinol dehydrogenase-10 regulates pancreas organogenesis and endocrine cell differentiation via paracrine retinoic acid signaling. *Endocrinology*, *157*(12), 4615–4631. https://doi.org/10.1210/en.2016-1745.

Ashique, A. M., May, S. R., Kane, M. A., Folias, A. E., Phamluong, K., Choe, Y., et al. (2012). Morphological defects in a novel Rdh10 mutant that has reduced retinoic acid biosynthesis and signaling. *Genesis (New York, NY: 2000)*, *50*(5), 415–423. https://doi.org/10.1002/dvg.22002.

Backman, M., Flenkenthaler, F., Blutke, A., Dahlhoff, M., Ländström, E., Renner, S., et al. (2019). Multi-omics insights into functional alterations of the liver in insulin-deficient diabetes mellitus. *Molecular Metabolism*, *26*, 30–44. https://doi.org/10.1016/j.molmet.2019.05.011.

Bankovic, J., Stojsic, J., Jovanovic, D., Andjelkovic, T., Milinkovic, V., Ruzdijic, S., et al. (2010). Identification of genes associated with non-small-cell lung cancer promotion and progression. *Lung Cancer*, *67*(2), 151–159. https://doi.org/10.1016/j.lungcan.2009.04.010.

Belyaeva, O. V., & Kedishvili, N. Y. (2002). Human pancreas protein 2 (PAN2) has a retinal reductase activity and is ubiquitously expressed in human tissues. *FEBS Letters*, *531*(3), 489–493. https://doi.org/10.1016/s0014-5793(02)03588-3.

Berry, D. C., DeSantis, D., Soltanian, H., Croniger, C. M., & Noy, N. (2012). Retinoic acid upregulates preadipocyte genes to block adipogenesis and suppress diet-induced obesity. *Diabetes*, *61*(5), 1112–1121. https://doi.org/10.2337/db11-1620.

Billings, S. E., Pierzchalski, K., Butler Tjaden, N. E., Pang, X.-Y. , Trainor, P. A., Kane, M. A., et al. (2013). The retinaldehyde reductase DHRS3 is essential for preventing the formation of excess retinoic acid during embryonic development. *FASEB Journal*, *27*(12), 4877–4889. https://doi.org/10.1096/fj.13-227967.

Blaner, W. S. (2019). Vitamin A signaling and homeostasis in obesity, diabetes, and metabolic disorders. *Pharmacology & Therapeutics*, *197*, 153–178. https://doi.org/10.1016/j.pharmthera.2019.01.006.

Boerman, M. H., & Napoli, J. L. (1995). Characterization of a microsomal retinol dehydrogenase: A short-chain alcohol dehydrogenase with integral and peripheral membrane forms that interacts with holo-CRBP (type I). *Biochemistry*, *34*(21), 7027–7037. https://doi.org/10.1021/bi00021a014.

Boerman, M. H., & Napoli, J. L. (1996). Cellular retinol-binding protein-supported retinoic acid synthesis. Relative roles of microsomes and cytosol. *The Journal of Biological Chemistry*, *271*(10), 5610–5616.

Cammas, L., Romand, R., Fraulob, V., Mura, C., & Dollé, P. (2007). Expression of the murine retinol dehydrogenase 10 (Rdh10) gene correlates with many sites of retinoid signalling during embryogenesis and organ differentiation. *Developmental Dynamics*, *236*(10), 2899–2908. https://doi.org/10.1002/dvdy.21312.

Chai, X., Boerman, M. H., Zhai, Y., & Napoli, J. L. (1995). Cloning of a cDNA for liver microsomal retinol dehydrogenase. A tissue-specific, short-chain alcohol dehydrogenase. *The Journal of Biological Chemistry*, *270*(8), 3900–3904.

Chai, X., Zhai, Y., Popescu, G., & Napoli, J. L. (1995). Cloning of a cDNA for a second retinol dehydrogenase type II. Expression of its mRNA relative to type I. *The Journal of Biological Chemistry*, *270*(47), 28408–28412.

Chefetz, I., Grimley, E., Yang, K., Hong, L., Vinogradova, E. V., Suciu, R., et al. (2019). A pan-ALDH1A inhibitor induces necroptosis in ovarian cancer stem-like cells. *Cell Reports*, *26*(11), 3061–3075.e6. https://doi.org/10.1016/j.celrep.2019.02.032.

Chetyrkin, S. V., Belyaeva, O. V., Gough, W. H., & Kedishvili, N. Y. (2001). Characterization of a novel type of human microsomal 3alpha -hydroxysteroid dehydrogenase: Unique tissue distribution and catalytic properties. *The Journal of Biological Chemistry*, *276*(25), 22278–22286. https://doi.org/10.1074/jbc.M102076200.

Duncan, F. J., Silva, K. A., Johnson, C. J., King, B. L., Szatkiewicz, J. P., Kamdar, S. P., et al. (2013). Endogenous retinoids in the pathogenesis of alopecia areata. *The Journal of Investigative Dermatology*, *133*(2), 334–343. https://doi.org/10.1038/jid.2012.344.

Everts, H. B., Silva, K. A., Montgomery, S., Suo, L., Menser, M., Valet, A. S., et al. (2013). Retinoid metabolism is altered in human and mouse cicatricial alopecia. *The Journal of Investigative Dermatology*, *133*(2), 325–333. https://doi.org/10.1038/jid.2012.393.

Everts, H. B., Sundberg, J. P., King, L. E., & Ong, D. E. (2007). Immunolocalization of enzymes, binding proteins, and receptors sufficient for retinoic acid synthesis and signaling during the hair cycle. *The Journal of Investigative Dermatology*, *127*(7), 1593–1604. https://doi.org/10.1038/sj.jid.5700753.

Farjo, K. M., Moiseyev, G., Takahashi, Y., Crouch, R. K., & Ma, J. (2009). The 11-cis-retinol dehydrogenase activity of RDH10 and its interaction with visual cycle proteins. *Investigative Ophthalmology & Visual Science*, *50*(11), 5089–5097. https://doi.org/10.1167/iovs.09-3797.

Gough, W. H., VanOoteghem, S., Sint, T., & Kedishvili, N. Y. (1998). CDNA cloning and characterization of a new human microsomal NAD+-dependent dehydrogenase that oxidizes all-trans-retinol and 3alpha-hydroxysteroids. *The Journal of Biological Chemistry*, *273*(31), 19778–19785.

Grün, F., Hirose, Y., Kawauchi, S., Ogura, T., & Umesono, K. (2000). Aldehyde dehydrogenase 6, a cytosolic retinaldehyde dehydrogenase prominently expressed in sensory neuroepithelia during development. *The Journal of Biological Chemistry*, *275*(52), 41210–41218. https://doi.org/10.1074/jbc.M007376200.

Grünblatt, E., & Riederer, P. (2016). Aldehyde dehydrogenase (ALDH) in Alzheimer's and Parkinson's disease. *Journal of Neural Transmission*, *123*(2), 83–90. https://doi.org/10.1007/s00702-014-1320-1.

Gyöngyösi, A., Szatmari, I., Pap, A., Dezso, B., Pos, Z., Széles, L., et al. (2013). RDH10, RALDH2, and CRABP2 are required components of PPARγ-directed ATRA synthesis and signaling in human dendritic cells. *Journal of Lipid Research*, *54*(9), 2458–2474. https://doi.org/10.1194/jlr.M038984.

Haeseleer, F., Huang, J., Lebioda, L., Saari, J. C., & Palczewski, K. (1998). Molecular characterization of a novel short-chain dehydrogenase/reductase that reduces all-trans--retinal. *The Journal of Biological Chemistry*, *273*(34), 21790–21799. https://doi.org/10.1074/jbc.273.34.21790.

Harrison, E. H. (2012). Mechanisms involved in the intestinal absorption of dietary vitamin A and provitamin A carotenoids. *Biochimica et Biophysica Acta*, *1821*(1), 70–77. https://doi.org/10.1016/j.bbalip.2011.06.002.

Hayes, D. P. (2007). Nutritional hormesis. *European Journal of Clinical Nutrition*, *61*(2), 147–159. https://doi.org/10.1038/sj.ejcn.1602507.

Hu, L., Chen, H.-Y. , Han, T., Yang, G.-Z. , Feng, D., Qi, C.-Y. , et al. (2016). Downregulation of DHRS9 expression in colorectal cancer tissues and its prognostic significance. *Tumour Biology*, *37*(1), 837–845. https://doi.org/10.1007/s13277-015-3880-6.

Jamison, R. S., Newcomer, M. E., & Ong, D. E. (1994). Cellular retinoid-binding proteins: Limited proteolysis reveals a conformational change upon ligand binding. *Biochemistry*, *33*(10), 2873–2879. https://doi.org/10.1021/bi00176a017.

Jette, C., Peterson, P. W., Sandoval, I. T., Manos, E. J., Hadley, E., Ireland, C. M., et al. (2004). The tumor suppressor adenomatous polyposis coli and caudal related homeodomain protein regulate expression of retinol dehydrogenase L. *The Journal of Biological Chemistry*, *279*(33), 34397–34405. https://doi.org/10.1074/jbc.M314021200.

Jones, R. J., Dickerson, S., Bhende, P. M., Delecluse, H.-J. , & Kenney, S. C. (2007). Epstein-Barr virus lytic infection induces retinoic acid-responsive genes through induction of a retinol-metabolizing enzyme, DHRS9. *The Journal of Biological Chemistry*, *282*(11), 8317–8324. https://doi.org/10.1074/jbc.M608667200.

Jurukovski, V., Markova, N. G., Karaman-Jurukovska, N., Randolph, R. K., Su, J., Napoli, J. L., et al. (1999). Cloning and characterization of retinol dehydrogenase transcripts expressed in human epidermal keratinocytes. *Molecular Genetics and Metabolism*, *67*(1), 62–73. https://doi.org/10.1006/mgme.1999.2840.

Kamei, N., Hiyama, K., Yamaoka, H., Kamimatsuse, A., Onitake, Y., Sueda, T., et al. (2009). Evaluation of genes identified by microarray analysis in favorable neuroblastoma. *Pediatric Surgery International*, *25*(11), 931–937. https://doi.org/10.1007/s00383-009-2448-1.

Kane, M. A., Folias, A. E., Pingitore, A., Perri, M., Krois, C. R., Ryu, J. Y., et al. (2011). CrbpI modulates glucose homeostasis and pancreas 9-cis-retinoic acid concentrations. *Molecular and Cellular Biology, 31*(16), 3277–3285. https://doi.org/10.1128/MCB.05516-11.

Kane, M. A., Folias, A. E., Wang, C., & Napoli, J. L. (2010). Ethanol elevates physiological all-trans-retinoic acid levels in select loci through altering retinoid metabolism in multiple loci: A potential mechanism of ethanol toxicity. *FASEB Journal, 24*(3), 823–832. https://doi.org/10.1096/fj.09-141572.

Kao, Y.-H. , Lee, P.-H. , Chiu, T.-C. , Lin, Y.-C. , Sun, C.-K. , Chen, P.-H. , et al. (2018). Transcriptome analysis reveals a positive role for nerve growth factor in retinol metabolism in primary rat hepatocytes. *Cytokine, 107*, 74–78. https://doi.org/10.1016/j.cyto.2017.11.018.

Kim, E. W., De Leon, A., Jiang, Z., Radu, R. A., Martineau, A. R., Chan, E. D., et al. (2019). Vitamin A metabolism by dendritic cells triggers an antimicrobial response against mycobacterium tuberculosis. *mSphere, 4*(3), pii: e00327-19. https://doi.org/10.1128/mSphere.00327-19.

Kirschner, R. D., Rother, K., Müller, G. A., & Engeland, K. (2010). The retinal dehydrogenase/reductase retSDR1/DHRS3 gene is activated by p53 and p63 but not by mutants derived from tumors or EEC/ADULT malformation syndromes. *Cell Cycle*, 9(11), 2177–2188. https://doi.org/10.4161/cc.9.11.11844.

Kozovska, Z., Patsalias, A., Bajzik, V., Durinikova, E., Demkova, L., Jargasova, S., et al. (2018). ALDH1A inhibition sensitizes colon cancer cells to chemotherapy. *BMC Cancer, 18*(1), 656. https://doi.org/10.1186/s12885-018-4572-6.

Krois, C. R., Vuckovic, M. G., Huang, P., Zaversnik, C., Liu, C. S., Gibson, C. E., et al. (2019). RDH1 suppresses adiposity by promoting brown adipose adaptation to fasting and re-feeding. *Cellular and Molecular Life Sciences: CMLS, 76*(12), 2425–2447. https://doi.org/10.1007/s00018-019-03046-z.

Lee, S.-A. , Belyaeva, O. V., Wu, L., & Kedishvili, N. Y. (2011). Retinol dehydrogenase 10 but not retinol/sterol dehydrogenase(s) regulates the expression of retinoic acid-responsive genes in human transgenic skin raft culture. *The Journal of Biological Chemistry, 286*(15), 13550–13560. https://doi.org/10.1074/jbc.M110.181065.

Lei, Z., Chen, W., Zhang, M., & Napoli, J. L. (2003). Reduction of all-trans-retinal in the mouse liver peroxisome fraction by the short-chain dehydrogenase/reductase RRD: Induction by the PPAR alpha ligand clofibrate. *Biochemistry, 42*(14), 4190–4196. https://doi.org/10.1021/bi026948i.

Limongi, Z. M., Curatolo, A., Pelliccia, F., & Rocchi, A. (2005). Biallelic deletion and loss of expression analysis of genes at FRA2G common fragile site in tumor-derived cell lines. *Cancer Genetics and Cytogenetics, 161*(2), 181–186. https://doi.org/10.1016/j.cancergencyto.2005.01.018.

Lundová, T., Zemanová, L., Malčeková, B., Skarka, A., Štambergová, H., Havránková, J., et al. (2015). Molecular and biochemical characterisation of human short-chain dehydrogenase/reductase member 3 (DHRS3). *Chemico-Biological Interactions, 234*, 178–187. https://doi.org/10.1016/j.cbi.2014.10.018.

Markova, N. G., Pinkas-Sarafova, A., Karaman-Jurukovska, N., Jurukovski, V., & Simon, M. (2003). Expression pattern and biochemical characteristics of a major epidermal retinol dehydrogenase. *Molecular Genetics and Metabolism, 78*(2), 119–135. https://doi.org/10.1016/s1096-7192(02)00226-3.

Markova, N. G., Pinkas-Sarafova, A., & Simon, M. (2006). A metabolic enzyme of the short-chain dehydrogenase/reductase superfamily may moonlight in the nucleus as a repressor of promoter activity. *The Journal of Investigative Dermatology, 126*(9), 2019–2031. https://doi.org/10.1038/sj.jid.5700347.

McCaffery, P., Koul, O., Smith, D., Napoli, J. L., Chen, N., & Ullman, M. D. (2004). Ethanol increases retinoic acid production in cerebellar astrocytes and in cerebellum. *Brain Research Developmental Brain Research*, *153*(2), 233–241. https://doi.org/10.1016/j.devbrainres.2004.09.003.

Meester-Smoor, M. A., Janssen, M. J. F. W., Grosveld, G. C., de Klein, A., van IJcken, W. F. J., Douben, H., et al. (2008). MN1 affects expression of genes involved in hematopoiesis and can enhance as well as inhibit RAR/RXR-induced gene expression. *Carcinogenesis*, *29*(10), 2025–2034. https://doi.org/10.1093/carcin/bgn168.

Mercader, J., Ribot, J., Murano, I., Felipe, F., Cinti, S., Bonet, M. L., et al. (2006). Remodeling of white adipose tissue after retinoic acid administration in mice. *Endocrinology*, *147*(11), 5325–5332. https://doi.org/10.1210/en.2006-0760.

Napoli, J. L. (2011). Effects of ethanol on physiological retinoic acid levels. *IUBMB Life*, *63*(9), 701–706. https://doi.org/10.1002/iub.500.

Napoli, J. L. (2012). Physiological insights into all-trans-retinoic acid biosynthesis. *Biochimica et Biophysica Acta*, *1821*(1), 152–167. https://doi.org/10.1016/j.bbalip.2011.05.004.

Napoli, J. L. (2017). Cellular retinoid binding-proteins, CRBP, CRABP, FABP5: Effects on retinoid metabolism, function and related diseases. *Pharmacology & Therapeutics*, *173*, 19–33. https://doi.org/10.1016/j.pharmthera.2017.01.004.

Newcomer, M. E., Jamison, R. S., & Ong, D. E. (1998). Structure and function of retinoid-binding proteins. *Sub-Cellular Biochemistry*, *30*, 53–80. https://doi.org/10.1007/978-1-4899-1789-8_3.

Nim, H. T., Furtado, M. B., Ramialison, M., & Boyd, S. E. (2017). Combinatorial ranking of gene sets to predict disease relapse: The retinoic acid pathway in early prostate cancer. *Frontiers in Oncology*, *7*, 30. https://doi.org/10.3389/fonc.2017.00030.

Obrochta, K. M., Krois, C. R., Campos, B., & Napoli, J. L. (2015). Insulin regulates retinol dehydrogenase expression and all-trans-retinoic acid biosynthesis through FoxO1. *The Journal of Biological Chemistry*, *290*(11), 7259–7268. https://doi.org/10.1074/jbc.M114.609313.

Ong, D. E., MacDonald, P. N., & Gubitosi, A. M. (1988). Esterification of retinol in rat liver. Possible participation by cellular retinol-binding protein and cellular retinol-binding protein II. *The Journal of Biological Chemistry*, *263*(12), 5789–5796.

Ottonello, S., Scita, G., Mantovani, G., Cavazzini, D., & Rossi, G. L. (1993). Retinol bound to cellular retinol-binding protein is a substrate for cytosolic retinoic acid synthesis. *The Journal of Biological Chemistry*, *268*(36), 27133–27142.

Parker, R. O., & Crouch, R. K. (2010). Retinol dehydrogenases (RDHs) in the visual cycle. *Experimental Eye Research*, *91*(6), 788–792. https://doi.org/10.1016/j.exer.2010.08.013.

Pavez Loriè, E., Chamcheu, J. C., Vahlquist, A., & Törmä, H. (2009). Both all-trans retinoic acid and cytochrome P450 (CYP26) inhibitors affect the expression of vitamin A metabolizing enzymes and retinoid biomarkers in organotypic epidermis. *Archives of Dermatological Research*, *301*(7), 475–485. https://doi.org/10.1007/s00403-009-0937-7.

Pavez Loriè, E., Li, H., Vahlquist, A., & Törmä, H. (2009). The involvement of cytochrome p450 (CYP) 26 in the retinoic acid metabolism of human epidermal keratinocytes. *Biochimica et Biophysica Acta*, *1791*(3), 220–228. https://doi.org/10.1016/j.bbalip.2008.12.004.

Penzes, P., Wang, X., & Napoli, J. L. (1997). Enzymatic characteristics of retinal dehydrogenase type I expressed in Escherichia coli. *Biochimica et Biophysica Acta*, *1342*(2), 175–181. https://doi.org/10.1016/s0167-4838(97)00102-7.

Pierzchalski, K., Yu, J., Norman, V., & Kane, M. A. (2013). CrbpI regulates mammary retinoic acid homeostasis and the mammary microenvironment. *FASEB Journal*, *27*(5), 1904–1916. https://doi.org/10.1096/fj.12-219410.

Posch, K. C., Boerman, M. H., Burns, R. D., & Napoli, J. L. (1991). Holocellular retinol binding protein as a substrate for microsomal retinal synthesis. *Biochemistry*, *30*(25), 6224–6230. https://doi.org/10.1021/bi00239a021.

Posch, K. C., Burns, R. D., & Napoli, J. L. (1992). Biosynthesis of all-trans-retinoic acid from retinal. Recognition of retinal bound to cellular retinol binding protein (type I) as substrate by a purified cytosolic dehydrogenase. *The Journal of Biological Chemistry*, *267*(27), 19676–19682.

Rexer, B. N., & Ong, D. E. (2002). A novel short-chain alcohol dehydrogenase from rats with retinol dehydrogenase activity, cyclically expressed in uterine epithelium. *Biology of Reproduction*, *67*(5), 1555–1564. https://doi.org/10.1095/biolreprod.102.007021.

Rhinn, M., Schuhbaur, B., Niederreither, K., & Dollé, P. (2011). Involvement of retinol dehydrogenase 10 in embryonic patterning and rescue of its loss of function by maternal retinaldehyde treatment. *Proceedings of the National Academy of Sciences of the United States of America*, *108*(40), 16687–16692. https://doi.org/10.1073/pnas.1103877108.

Riquelme, P., Amodio, G., Macedo, C., Moreau, A., Obermajer, N., Brochhausen, C., et al. (2017). DHRS9 is a stable marker of human regulatory macrophages. *Transplantation*, *101*(11), 2731–2738. https://doi.org/10.1097/TP.0000000000001814.

Riquelme, P., & Hutchinson, J. A. (2018). Novel molecules mediate specialized functions of human regulatory macrophages. *Current Opinion in Organ Transplantation*, *23*(5), 533–537. https://doi.org/10.1097/MOT.0000000000000560.

Romand, R., Kondo, T., Cammas, L., Hashino, E., & Dollé, P. (2008). Dynamic expression of the retinoic acid-synthesizing enzyme retinol dehydrogenase 10 (rdh10) in the developing mouse brain and sensory organs. *The Journal of Comparative Neurology*, *508*(6), 879–892. https://doi.org/10.1002/cne.21707.

Ross, A. C. (1993). Cellular metabolism and activation of retinoids: Roles of cellular retinoid-binding proteins. *FASEB Journal*, 7(2), 317–327. https://doi.org/10.1096/fasebj.7.2.8440409.

Sanders, T. J., McCarthy, N. E., Giles, E. M., Davidson, K. L. M., Haltalli, M. L. R., Hazell, S., et al. (2014). Increased production of retinoic acid by intestinal macrophages contributes to their inflammatory phenotype in patients with Crohn's disease. *Gastroenterology*, *146*(5), 1278–1288.e1-2. https://doi.org/10.1053/j.gastro.2014.01.057.

Serio, R. N., Laursen, K. B., Urvalek, A. M., Gross, S. S., & Gudas, L. J. (2019). Ethanol promotes differentiation of embryonic stem cells through retinoic acid receptor-γ. *The Journal of Biological Chemistry*, *294*(14), 5536–5548. https://doi.org/10.1074/jbc.RA118.007153.

Shimomura, H., Sasahira, T., Nakashima, C., Shimomura-Kurihara, M., & Kirita, T. (2018). Downregulation of DHRS9 is associated with poor prognosis in oral squamous cell carcinoma. *Pathology*, *50*(6), 642–647. https://doi.org/10.1016/j.pathol.2018.06.002.

Shin, D.-J. , Joshi, P., Hong, S.-H. , Mosure, K., Shin, D.-G. , & Osborne, T. F. (2012). Genome-wide analysis of FoxO1 binding in hepatic chromatin: Potential involvement of FoxO1 in linking retinoid signaling to hepatic gluconeogenesis. *Nucleic Acids Research*, *40*(22), 11499–11509. https://doi.org/10.1093/nar/gks932.

Siegenthaler, J. A., Ashique, A. M., Zarbalis, K., Patterson, K. P., Hecht, J. H., Kane, M. A., et al. (2009). Retinoic acid from the meninges regulates cortical neuron generation. *Cell*, *139*(3), 597–609. https://doi.org/10.1016/j.cell.2009.10.004.

Sima, A., Manolescu, D.-C. , & Bhat, P. (2011). Retinoids and retinoid-metabolic gene expression in mouse adipose tissues. *Biochemistry and Cell Biology*, *89*(6), 578–584. https://doi.org/10.1139/o11-062.

Soref, C. M., Di, Y. P., Hayden, L., Zhao, Y. H., Satre, M. A., & Wu, R. (2001). Characterization of a novel airway epithelial cell-specific short chain alcohol dehydrogenase/reductase gene whose expression is up-regulated by retinoids and is involved in the metabolism of retinol. *The Journal of Biological Chemistry, 276*(26), 24194–24202. https://doi.org/10.1074/jbc.M100332200.

Tavares, T. S., Nanus, D., Yang, X. J., & Gudas, L. J. (2008). Gene microarray analysis of human renal cell carcinoma: The effects of HDAC inhibition and retinoid treatment. *Cancer Biology & Therapy*, 7(10), 1607–1618. https://doi.org/10.4161/cbt.7.10.6584.

Tong, M.-H. , Yang, Q.-E. , Davis, J. C., & Griswold, M. D. (2013). Retinol dehydrogenase 10 is indispensible for spermatogenesis in juvenile males. *Proceedings of the National Academy of Sciences of the United States of America, 110*(2), 543–548. https://doi.org/10.1073/pnas.1214883110.

Treves, S., Thurnheer, R., Mosca, B., Vukcevic, M., Bergamelli, L., Voltan, R., et al. (2012). SRP-35, a newly identified protein of the skeletal muscle sarcoplasmic reticulum, is a retinol dehydrogenase. *The Biochemical Journal, 441*(2), 731–741. https://doi.org/10.1042/BJ20111457.

Udali, S., Guarini, P., Ruzzenente, A., Ferrarini, A., Guglielmi, A., Lotto, V., et al. (2015). DNA methylation and gene expression profiles show novel regulatory pathways in hepatocellular carcinoma. *Clinical Epigenetics*, 7, 43. https://doi.org/10.1186/s13148-015-0077-1.

Wang, C., Kane, M. A., & Napoli, J. L. (2011). Multiple retinol and retinal dehydrogenases catalyze all-trans-retinoic acid biosynthesis in astrocytes. *The Journal of Biological Chemistry, 286*(8), 6542–6553. https://doi.org/10.1074/jbc.M110.198382.

Wang, X., Penzes, P., & Napoli, J. L. (1996). Cloning of a cDNA encoding an aldehyde dehydrogenase and its expression in Escherichia coli. Recognition of retinal as substrate. *The Journal of Biological Chemistry, 271*(27), 16288–16293. https://doi.org/10.1074/jbc.271.27.16288.

Wu, L., Belyaeva, O. V., Adams, M. K., Klyuyeva, A., Lee, S.-A. , Goggans, K. R., et al. (2019). Mice lacking the epidermal retinol dehydrogenases SDR16C5 and SDR16C6 display accelerated hair growth and enlarged meibomian glands. *The Journal of Biological Chemistry, 294*(45), 17060–17074. https://doi.org/10.1074/jbc.RA119.010835.

Wu, B. X., Chen, Y., Chen, Y., Fan, J., Rohrer, B., Crouch, R. K., et al. (2002). Cloning and characterization of a novel all-trans retinol short-chain dehydrogenase/reductase from the RPE. *Investigative Ophthalmology & Visual Science, 43*(11), 3365–3372.

Wu, B. X., Moiseyev, G., Chen, Y., Rohrer, B., Crouch, R. K., & Ma, J.-X. (2004). Identification of RDH10, an all-trans retinol dehydrogenase, in retinal Muller cells. *Investigative Ophthalmology & Visual Science, 45*(11), 3857–3862. https://doi.org/10.1167/iovs.03-1302.

Yang, D., Krois, C. R., Huang, P., Wang, J., Min, J., Yoo, H. S., et al. (2017). Raldh1 promotes adiposity during adolescence independently of retinal signaling. *PLoS ONE, 12*(11) e0187669 https://doi.org/10.1371/journal.pone.0187669.

Yang, D., Vuckovic, M. G., Smullin, C. P., Kim, M., Lo, C. P.-S. , Devericks, E., et al. (2018). Modest decreases in endogenous all-trans-retinoic acid produced by a mouse Rdh10 heterozygote provoke major abnormalities in adipogenesis and lipid metabolism. *Diabetes, 67*(4), 662–673. https://doi.org/10.2337/db17-0946.

Zhang, M., Chen, W., Smith, S. M., & Napoli, J. L. (2001). Molecular characterization of a mouse short chain dehydrogenase/reductase active with all-trans-retinol in intact cells, mRDH1. *The Journal of Biological Chemistry, 276*(47), 44083–44090. https://doi.org/10.1074/jbc.M105748200.

Zhang, M., Hu, P., & Napoli, J. L. (2004). Elements in the N-terminal signaling sequence that determine cytosolic topology of short-chain dehydrogenases/reductases. Studies with retinol dehydrogenase type 1 and cis-retinol/androgen dehydrogenase type 1. *The Journal of Biological Chemistry, 279*(49), 51482–51489. https://doi.org/10.1074/jbc.M409051200.

Zhao, D., McCaffery, P., Ivins, K. J., Neve, R. L., Hogan, P., Chin, W. W., et al. (1996). Molecular identification of a major retinoic-acid-synthesizing enzyme, a retinaldehyde-specific dehydrogenase. *European Journal of Biochemistry*, *240*(1), 15–22. https://doi.org/10.1111/j.1432-1033.1996.0015h.x.

Zhu, Y.-H. , Li, J.-B. , Wu, R.-Y. , Yu, Y., Li, X., Li, Z.-L. , et al. (2019). Clinical significance and function of RDH16 as a tumor-suppressing gene in hepatocellular carcinoma. *Hepatology Research*, *50*, 110–120. https://doi.org/10.1111/hepr.13432.

Ziouzenkova, O., Orasanu, G., Sharlach, M., Akiyama, T. E., Berger, J. P., Viereck, J., et al. (2007). Retinaldehyde represses adipogenesis and diet-induced obesity. *Nature Medicine*, *13*(6), 695–702. https://doi.org/10.1038/nm1587.

CHAPTER THREE

Retinoid metabolism and functions mediated by retinoid binding-proteins

Joseph L. Napoli*, Hong Sik Yoo
Graduate Program in Metabolic Biology, Nutritional Sciences and Toxicology, University of California, Berkeley, CA, United States
*Corresponding author: e-mail address: jna@berkeley.edu

Contents

Abstract

Cellular retinoid-binding proteins (BP) chaperone retinol through esterification, conversion of retinol into retinal, reduction of retinal, conversion of retinal into all-*trans*-retinoic acid (ATRA), and ATRA to catabolism. They also deliver ATRA to nuclear receptors and mediate non-genomic ATRA actions. These retinoid-specific binding-proteins include: cellular retinol binding-protein, type 1 (Crbp1), cellular retinol binding-protein type 2

Methods in Enzymology, Volume 637
ISSN 0076-6879
https://doi.org/10.1016/bs.mie.2020.02.004

(Crbp2), cellular retinol binding-protein type 3 (Crbp3), cellular retinoic acid binding-protein type 1 (Crabp1); cellular retinoic acid binding-protein type 2 (Crabp2). Retinoid BP bind their ligands specifically and with high-affinity. These BP seemingly evolved to solubilize the lipophilic retinoids in the aqueous cellular medium, and allow retinoid access only to enzymes that recognize both the BP and the retinoid. By chaperoning retinoids through cells, retinoid BP provide specificity to retinoids' metabolism and protect the scarce resource from dispersing into cell membranes and/or undergoing catabolism as xenobiotics. Other functions include non-genomic actions of Crabp1, delivery of ATRA to RAR by holo-Crabp2, and stabilization of HuR by apo-Crabp2. In addition to the retinoid-specific BP, Fabp5 also binds ATRA and delivers it to Pparδ. This article describes these BP and their functions, with a focus on experimental protocols to distinguish protein-protein interactions from diffusion-mediated transfer of ligand from BP to enzymes or receptors, and methods for quantifying retinoids.

1. Nature of the BP

Retinoid BP belong to the intracellular lipid binding-protein gene family and occur as flattened β-clam structures of ∼16 kDa. The Crbp structures enclose retinol and retinal deep within the proteins with the functional groups (alcohols and aldehydes) projecting away from the entry portal (Fig. 1). In Crabp, ATRA also orients with the functional group (a carboxyl) buried within the protein. In this case, however, the protein allows greater access to

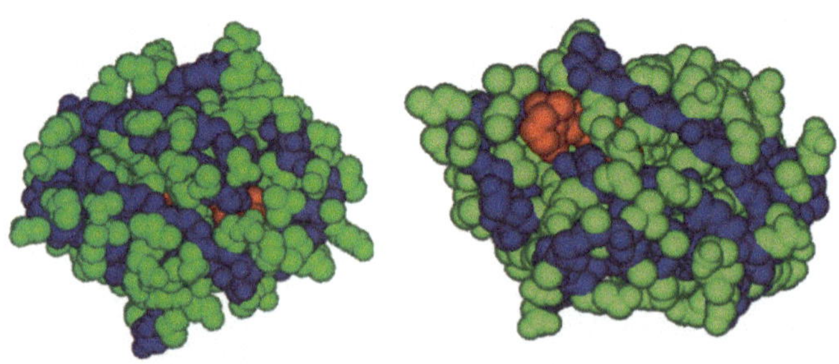

Fig. 1 Space filling models of human cellular retinoid BP. The Crbp1 orientation shows the most visible aspect of retinol, with the red residues showing the β-ionone ring. The hydroxyl group buries itself inside the protein. The red residues in Crabp2 show the ATRA β-ionone ring. The carboxyl group buries itself inside the protein. In both cases, blue shows the protein backbones; green shows the protein side-chains. Structural information was downloaded from https://www.ncbi.nlm.nih.gov/Structure/CN3D/cn3dwin.shtml (Cowan, Newcomer, & Jones, 1990; Menozzi et al., 2017; Vaezeslami, Mathes, Vasileiou, Borhan, & Geiger, 2006).

the β-ionone ring, where catabolism begins at the C4 and C18 carbons. Table 1 provides a brief summary of retinoid BP.

1.1 Crbp2 (encoded by *rbp2*)

Crbp2, which accounts for ~1% of the soluble protein in the small intestinal enterocyte, tightly controls retinoid absorption and metabolism (Kakkad & Ong, 1988; Ong, 1984; Ong, Kakkad, & MacDonald, 1987; Ong, MacDonald, & Gubitosi, 1988). Diet provides pro-vitamin A carotenoids and retinyl esters, with the former prevailing quantitatively. A brush border lipase hydrolyzes luminal retinol esters to allow uptake of the released retinol, which binds with Crbp2 and undergoes lecithin: retinol acyltransferase (Lrat)-catalyzed re-esterification and incorporation into chylomicrons for systemic transport (Fig. 2). Pro-vitamin A carotenoids undergo conversion into retinal catalyzed by the enzyme β-carotene oxidase type 1 (Bco1) (Harrison, 2012). The retinal formed binds to Crbp2. Crbp2 prevents dehydrogenation of retinal into ATRA, and directs retinal to reduction into retinol, which then undergoes esterification. Because carotenoids usually provide the major dietary source of vitamin A, without direction into retinol rather than ATRA, they could cause ATRA toxicity upon metabolism.

Table 1 Cellular retinoid binding-proteins, distributions, ligands and their affinities.

Protein	Post-natal distribution	Ligands	k_d (nM)
Crbp1	Ubiquitous? (low in small intestine)	all-*trans*-retinol	≤1
		3,4-didehydroretinol	–
		9-cis-retinol	11
		all-*trans*-retinal	9
Crbp2	High in small intestine	all-*trans*-retinol	10
		all-*trans*-retinal	11
Crbp3	Heart, muscle, adipose, mammary	all-*trans*-retinol	109
		9-cis-retinol	
Crabp1	Widespread	ATRA	0.06
Crabp2	Limited? (skin, uterus, and ovary)	ATRA	0.13
Fabp5	Widespread	ATRA	35
		Long-chain fatty acids	15–20

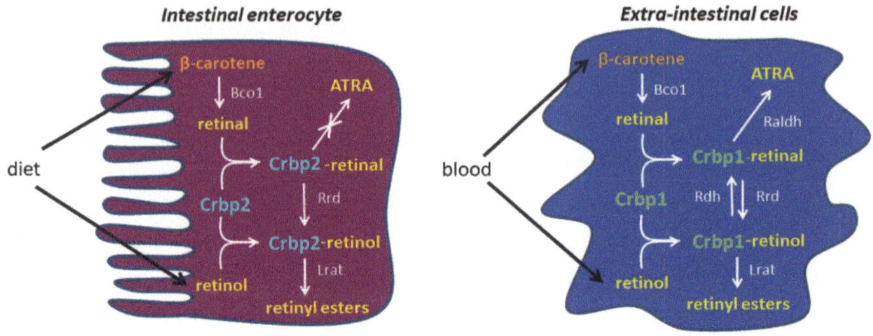

Fig. 2 Effects of Crbp1 and 2 on retinol and retinal metabolism. Crbp2 in the small intestine enterocyte prevents massive ATRA biosynthesis from retinal generated by dietary β-carotene cleavage. Outside the intestine, Crbp1 chaperones retinol and retinal for ATRA biosynthesis. Bco1, β-carotene oxidase type 1; Rrd, retinal reductase; Lrat, lecithin: retinol acyltransferase.

Thus, Crbp2 provides for efficient uptake of retinol by mass action and prevents retinoid toxicity by limiting ATRA biosynthesis.

The phenotype of *rbp2*-null mice revealed another advantage of Crbp2. *Rbp2*-null mice pups die by 24 h after birth when delivered from dams fed a diet marginal in vitamin A (E et al., 2002). This indicates essentiality of Crbp2 in promoting efficient retinoid acquisition from diet and their systemic delivery.

1.2 Crbp1 (encoded by *rbp1*)

Crbp1 promotes retinol uptake, storage as retinyl esters and delivery of retinol for conversion into ATRA (Napoli, 2017). In liver, Crbp1 delivers retinol specifically to Lrat for esterification, and mediates transfer of retinol hepatocytes to stellate cells (MacDonald & Ong, 1988). In extra-hepatic tissues, Crbp1 sequesters retinol transferred into cells by the membrane receptors for the serum Rbp (encoded by *rbp4*), i.e., Stra6 and Rbpr2 (Berry, O'Byrne, Vreeland, Blaner, & Noy, 2012) (Fig. 2). By mass action and delivery of retinol to Lrat, Crbp1 ensures one-way uptake of retinol from serum. Crbp1 also delivers: retinol to retinol dehydrogenases (Rdh) for conversion into retinal; retinal to reductases (Rrd) for conversion into retinol; retinal to retinal dehydrogenases (Raldh) for conversion into ATRA.

Rbp1-null mice seem normal morphologically, but have low concentrations of retinyl esters in liver and eliminate them sixfold faster than wild-type mice (Matt et al., 2005). Disrupted hepatic storage likely relates to the function of Crbp1 in transferring retinol to stellate cells. *Rbp1*-null mice also have impaired glucose tolerance, as a result of pancreatic abnormalities (Kane et al., 2011). These include reduced islet expression of genes involved in glucose sensing and insulin secretion. The null mice also suffer from α-cell infiltration into the β-cell field of pancreatic islets, which contributes to reduced glucose-stimulated insulin secretion, elevated glucagon secretion, increased gluconeogenesis and hyperglycemia.

2. BP contributions to retinoid metabolism

2.1 Crbp contributions to retinol and retinal metabolism

Isolation of retinoids from the cellular milieu within Crbp1 (and Crbp2) with the hydroxyl function buried deeply in the BP and high affinity binding, begs the question of how enzymes can access these substrates. The issue was resolved by establishing that Lrat, retinol dehydrogenases (Rdh), retinal reductases (Rrd), and retinal dehydrogenases (Raldh) interact with BP and "tease" retinoids from the BP into their active sites (Napoli, 2016). Several experiments have confirmed direct-transfer from BP to enzymes, i.e., no requirement for substrate diffusion through the aqueous medium. Michaelis-Menten kinetics (K_m and $V_{\max}$ values) of retinol and Crbp1-retinol with Lrat are similar (Ong et al., 1988). Moreover, apo-Crbp inhibits Lrat, but stimulates retinyl ester hydrolysis (Boerman & Napoli, 1991; Herr & Ong, 1992) (Fig. 3). Likewise, the same kinetics are observed with

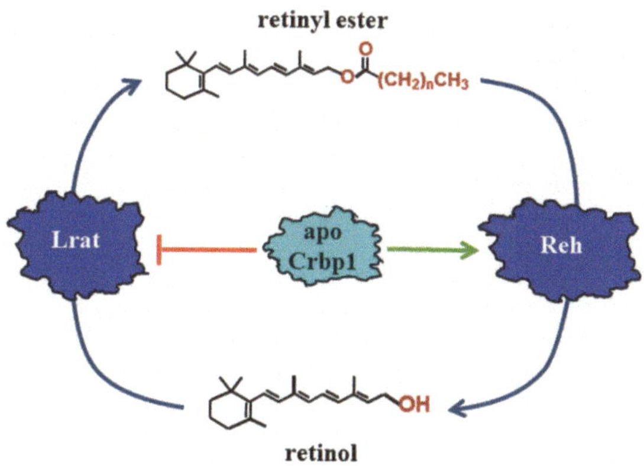

Fig. 3 Function of apo-Crbp1. The relative amounts of holo- and apo-Crbp1 reflect cellular retinol status, and thereby direct flux of retinol into and from retinyl esters. The mechanism involves apo-Crbp1 inhibiting Lrat and stimulating retinyl ester hydrolysis. In contrast, holo-Crbp1 chaperones retinol into retinyl esters and into retinal for ATRA biosynthesis.

Crbp1-retinal and Raldh1 and 2, i.e., unchanging Michaelis-Menten kinetics with either free retinal or Crbp1-retinal as substrate provider (Penzes, Wang, & Napoli, 1997; Wang, Penzes, & Napoli, 1996). Apo-Crbp1 inhibits Raldh1, but not Raldh2. This suggests different contributions of Raldh1 and 2 to ATRA biosynthesis. Crbp1 also chaperones retinol to Rdh, whereas Crbp2 prevents Rdh access. In the case of retinal biosynthesis, V_{max} values of holo-Crbp1-delivered retinol are much lower than that of free retinol, perhaps indicating that the rate-limiting step involves retinol transfer from the BP to Rdh, i.e., that transfer occurs more slowly than the catalytic cycle (Posch, Boerman, Burns, & Napoli, 1991). These data provide a teleological argument for direct BP participation in retinoid metabolism, insomuch as they indicate specific interactions with holo- vs apo-Crbp, and Crbp1 vs Crbp2.

2.2 Crabp contributions to ATRA catabolism

Crabp1 and Crabp2 doubly-null-mice have a nearly fourfold higher death rate compared to wild type mice by 6 weeks after birth from unknown causes (Lampron et al., 1995). Nevertheless, the survivors seem normal morphologically, when fed a diet containing copious vitamin A, as occurs in rodent chow. The single observed exception involves a small outgrowth anomaly, predominantly in forelimbs, for both Crabp2 and the double knockout. Double null-mice do not have increased sensitivity to exogenous ATRA, suggesting that the BP are not required to protect against ATRA toxicity. Thus, ablations of Crabp1 and Crabp2 have not clarified their functions.

The first and rate-limiting reactions of ATRA catabolism occur at C4 and C18, mediated by cytochrome P450 (Cyp)-catalyzed hydroxylation. Microsomal Cyp26a1, b1 and c1 function as the most dominant hydroxylases (Ross & Zolfaghari, 2011; Topletz et al., 2012, 2015). ATRA induces these enzymes with exquisite sensitivity. Holo-Crabp1 increases catalytic efficiency sevenfold (k_{cat}/K_m) by delivering ATRA to catabolism catalyzed by microsomes, driven largely by a lower K_m value (<2 nM for holo-Crabp1 vs ~50 nM for free ATRA) (Fiorella & Napoli, 1991). Both holo-Crabp1 and Crabp2 deliver ATRA to recombinant Cyp26b1 (Nelson et al., 2016). Cyp26a1 and c1 have not been tested. K_m values were lower with holo-Crabp1 (K_m ~22 nM) and holo-Crabp2 (K_m ~24 nM) delivery of ATRA to Cyp26b1, than with delivery of free ATRA (K_m ~65 nM). Holo-Crabp delivery to Cyp26b1 enhances catalytic efficiencies (k_{cat}/K_m) of hydroxylation, generating relative values of 8 and 12 for holo-Crabp1 and Crabp2, respectively, vs 7 for unbound ATRA. With recombinant Cyp2b1, apo-Crabp1 and Crabp2 added in graded amounts in excess of ATRA substantially reduce rates of catabolism. An equimolar amount of apo-Crabp1 relative to total ATRA reduces the V_m by 50%, whereas apo-Crabp2 has less of an impact, reducing the V_m only by ~25%. Note, however, that even a 10-fold excess of apo-Crabp1 does not totally disable ATRA catabolism. Even a threefold excess of total Crabp to total ATRA, given the k_d values, would allow little to no free ATRA in the reaction mixture. Thus, Cyp26b1 must access ATRA bound with Crabp. Inhibition of catabolism by apo-Crabp may allow preservation of limited ATRA, consistent with holo-Crabp2 delivery of ATRA to nuclear RAR, and non-genomic regulatory functions of both Crabp1 and Crabp2.

2.3 Fabp5 and ATRA function

Fabp5 has lower affinity for ATRA (k_d ~42 nM) than that of Crabp. Yet Fabp5 mediates transfer of ATRA to nuclear Pparδ (Noy, 2016). Activation of Rar vs Pparδ distinguishes ATRA actions of inducing differentiation (Rar) from stimulating proliferation (Pparδ).

3. Experimental methods

3.1 Definition

These protocols will address large-scale preparation of BP, evaluation of their integrity and the integrity of retinoids, application of kinetics to distinguish BP-chaperoned delivery through direct transfer of retinoids to acceptor proteins (enzymes and receptors) vs diffusion-controlled transfer, and how to quantify retinoids in tissues and as products of metabolic reactions.

3.2 Rationale

These protocols relate up-to-date methods for handling and analytically rigorous quantification of retinoids with specificity and retinoid BP (Kane & Napoli, 2010). They include use of internal standards (so-called external standards should be avoided), verification of peak identity using HPLC/UV (internal standard needed) or mass fragmentation, reasons for obtaining total scans during UV assessment, and issues for developing sound kinetic data. Application of these techniques avoid spurious results.

HPLC with UV detection has relatively low equipment costs, relative to LC/MS/MS, can be specific for retinoids due to their unique absorbance spectra, but has a lower limit of quantification (LOQ) ~0.4–1 pmol. HPLC/UV can work well with analyzing enzymatic production of ATRA and tissue concentrations of retinol and retinyl esters. GC/MS assays, with a lower limit of detection <250 fmol, are more sensitive than HPLC/UV assays, but require ATRA derivatization. Methyl retinoate eluates only from packed GC columns, not from capillary GC columns. LC/MS/MS assays based on triple-quadrupole instruments offers sensitivity, specificity, no requirement for derivatization, and mass identification. The fragments analyzed and internal standards used must be chosen to avoid interfering peaks.

Substitutes for the analytically validated assays described here, such as "reporter" assays lack specificity and quantification. Expression assays that rely on Rar-driven reporters are not analytically rigorous, are not specific for ATRA, are not especially sensitive, and can produce both false-positive and false-negative results. These assays do not detect retinoids in real time because they reflect the time it takes to activate the receptor and generate the reporter, which can occur well after the retinoid has been catabolized.

3.3 Materials, equipment and reagents

Specialized equipment includes HPLC/UV, LC/MS/MS, shaking incubator (for *E. coli* production of BP), UV spectrophotometer, temperature controlled water bath (for kinetic studies). Specialized regents include vectors to generate recombinant BP and enzymes, available from labs referenced herein. A variety of retinoid internal standards can be used, such as 4,4-dimethyl-ATRA, 3,3-dimethyl-ATRA, and 3,4-didehydroretinal (for retinal quantification). Pentadeuterated ATRA should be used with caution. It served as a valid internal standard with GC/MS, but depending on the mass transitions used it can co-migrate with inferring peaks on LC/MS, rendering it invalid as an internal standard. LC/MS grade solvents should be used routinely. Only glass or stainless steel containers, pipettes, and syringes should be used to handle retinoids, as retinoids adhere to other surfaces, even Teflon. Gasses to evaporate samples should be applied via rubber tubing (not Tygon) to avoid contaminating samples with plasticizers. This is especially important for MS analyses.

4. Protocols

4.1 BP preparation

Recombinant production provides the most efficient method for pure obtaining apo- and holo-cellular retinoid BP in larger amounts (Levin, Locke, Yang, Li, & Gordon, 1988; Napoli, Sperkova, Sperka, & Fiorella, 1998). This approach allows generation of ten's of mg of nearly pure protein.

The BP should be stored at −80 °C as apo-proteins. On the day of use, an aliquot can be divided into two parts. One should remain as apo and the other can be charged with excess ligand to ensure complete saturation, and then "dropped" through a Sephadex G-25 column, with a buffer containing freshly added sulfhydryl group protecting reagent, such as dithiothreitol. The holo-BP elute rapidly, given their sizes, free of unliganded retinoids, which remain on the column. This allows preparing precise mixtures of apo- and holo-BPs.

Retinol bound to holo-Crbp1, even stored at −80 °C, partially isomerizes to 9-*cis*-retinol over a month. When used for assaying ATRA biosynthesis, this gives the appearance of ATRA, because ATRA and 9-*cis*-RA closely migrate on LC columns and/or gives the false impression that holo-Crbp1 somehow generates 9-*cis*-RA. The ratios of the elution time of an internal standard with the elution times of ATRA vs 9-*cis*-RA remain constant, and can be used to distinguish one retinoid from the other, even if only one peak is present.

With BP isolated from tissues, exact apo- to holo-ratios cannot be determined without saturating an aliquot with the cognitive ligand and comparing UV spectra before and after saturation. Some reports have attempted ligand removal by irradiating the BP to destroy retinoids, or by stirring a layer of ether (in the cold) over the BP in aqueous buffer to extract retinoids. In our hands, these approaches destroyed/removed the retinoid, but also denatured Crbp, as indicated by attempting to re-charge an aliquot to determine whether the BP retained capacity to bind retinol. It did not. Without this verification by UV analyses, it is easy to mislead oneself into believing that the absence of retinoid leaves behind an intact apo-BP.

4.2 Retinoid purification

Retinoids should be handled under red or yellow light. Fluorescent white light and sunlight can isomerize and degrade retinoids within minutes. Retinol should be purified by HPLC before use. We find that impurities in retinol inhibit ATRA biosynthesis, especially in cell lines. On the other hand, retinol can be contaminated with retinal, which generates ATRA and gives the impression of Rdh activity.

4.3 UV analyses

UV analyses provides the "gold-standard" of retinoid and BP purities and quantitation. Weighing is inaccurate, given the quantities used and impurities in most commercial retinoids that do not absorb in the UV range.

Retinol has an A_{max} at 325 nm in ethanol with an ε (molar absorptivity) value of 52,770 (Fig. 4). ATRA has an A_{max} at 350 nm and an ε value of 45,300. Retinal has an A_{max} at 383 nm and an ε value of 42,880. The A_{max} values of *trans* retinoids distinguish them from geometric isomers, which have lower A_{max} values (National Research Council, 1972). A PDF with UV spectra information can be downloaded from https://www.nap.edu/catalog/21491/specifications-and-criteria-for-biochemical-compounds-third-edition.

The UV spectra of holo-Crbp1 and Crabp2 indicate quality and quantity by the A_{max} and the ratio A_{max}/A_{min}, which should be 1.7 for a pure protein (Ong & Chytil, 1978a, 1978b). Complete scans from at least 220–420 nm

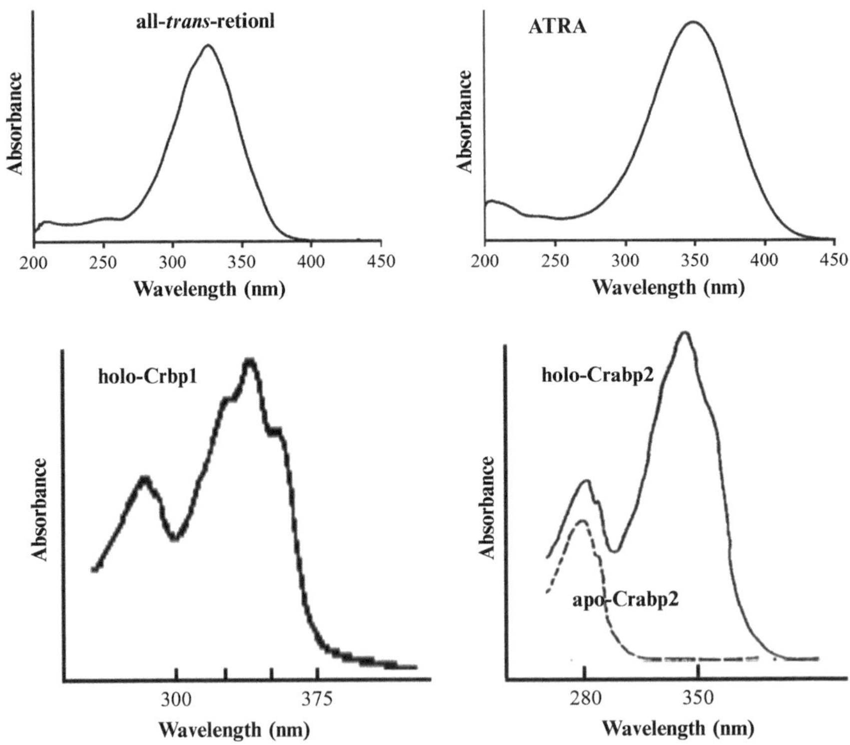

Fig. 4 UV spectra of retinoids and retinoid BPs. Spectra of retinoids were obtained in ethanol. The spectrum of holo-Crbp1 was taken in 20 nM Hepes, 150 mM KCl, 2 mM DTT at pH 7.5. The A_{max} of holo-Crbp1 occurs at 350 nM. A ratio A_{250}/A_{350} of 1.7 indicates a pure protein. The spectra of holo-Crabp1 (solid line) and apo-Crabp1 (dashed line) were obtained in 20 mM bis-Tris-HCl (pH 7.0), 2 mM β-mercaptoethanol, 0.05% sodium azide, 150 mM NaCl at pH 7.0. A molar absorptivity (ε) of 50,000 at A_{350} was used to quantify holo-Crabp1.

should be recorded, rather than just assessing absorbance at the peak. The quality of the complete spectrum reflects the quality of the sample. Spectra with "filled in" absorbance between 220 and 300 nm can indicate an impure sample, or an impure solvent.

4.4 Tissue, cell and homogenate extraction

Retinoid tissue and serum concentrations should be compared intra-strain, among animals fed the same diets, and in either the fasted or fed states (Obrochta, Krois, Campos, & Napoli, 2015; Obrochta & Napoli, 2014). The nature of the diet (chow vs AIN formula), the length of time fed the diet (first generation or subsequent generations), the mouse strain, and the timing of sample collection all affect concentrations of tissue retinoids, especially ATRA.

The extraction procedure has been modified from that originally published to provide a "cleaner" extract (Kane, Folias, Wang, & Napoli, 2008).

1. Add 1.5 mL methanol in a ground glass hand homogenizer (Duall, size 22 or similar, Kontes).
2. Add ~100 mg tissue and homogenize using a ground glass pestle with a motorized homogenizer (Heidolph).
3. Transfer the homogenate into a glass conical centrifuge tube and add 10 μL 1 μM 3,3-dimethyl-ATRA (or other internal standard) and 10 μL 50 μM 3,4-didehydroretinol (internal standard for retinol). Vortex 10 s.
4. Centrifuge 5 min at 1200 × *g*. Transfer the supernatant to a round-bottom glass tube.
5. To recover retinol and retinyl esters, add 100 μL 0.5 M KOH in methanol. Vortex 10 s.
6. Add 10 mL hexane. Vortex 20 s. Centrifuge 3 min at 1200 × *g*.
7. Transfer the upper hexane phase to a clean glass tube, and evaporate the solvent under a stream of N_2 or Ar.
8. Add 40 μL methanol. Vortex 20 s. Centrifuge.
9. Transfer the supernatant to an HPLC vial insert. Inject 20 μL onto the column. HPLC/UV quantifies retinol and retinyl esters as published (Kane, Folias, & Napoli, 2008).
10. To recover RA, add 150 μL 4 M HCl to the remaining bottom phase. Vortex 10 s.
11. Add 10 mL hexane. Vortex 20 s. Centrifuge 3 min at 1200 × *g*.
12. Transfer the upper hexane phase to a clean round-bottom glass tube and evaporate the hexane under a stream of N_2 or Ar.

13. For HPLC/UV add 40 μL methanol to the tube. Vortex 20 s. Centrifuge 3 min at 1200 × *g*. Transfer the supernatant to a HPLC vial insert. Inject ~10 μL onto the LC column.
14. For GC/MS or LC/MS/MS transfer the hexane (upper) phase to a glass tube and evaporate the solvent with a stream of N_2 or Ar with heating at 25–30 °C in a water bath. Suspend the residue in 40 μL acetonitrile. Inject 5 μL onto the LC-MS/MS column.

Typically 15–35 mg tissue or 100–200 μL serum from adult mice generate reliable data (Kane, Chen, Sparks, & Napoli, 2005; Kane, Folias, Wang, et al., 2008; Napoli, Pramanik, Williams, Dawson, & Hobbs, 1985).

4.5 Chromatography

Multiple HPLC systems can resolve retinol and its esters (Kane, Folias, & Napoli, 2008). The choice depends on the goal—quantifying retinol vs distinguishing retinol and ATRA isomers. For example a normal-phase HPLC system with UV detection resolves most common retinoids (Fig. 5A). An ultra-high performance LC reverse-phase system with MS/MS detection can quantify retinal. This system has a 20 fmol lower limit of quantification (Fig. 5B) (Wang, Yoo, Obrochta, Huang, & Napoli, 2015). RA isomers can also be quantified with a LC reverse-phase system with MS/MS detection (Fig. 5C). For the system shown, a 0.75 fmol lower limit of quantification can be achieved when assaying tissues (Kane, Folias, Wang, et al., 2008).

4.6 Kinetic studies

Complete Michaelis-Menten curves should be generated to truly assess enzymes' activity with substrates, especially when assessing whether a BP delivers retinoid through protein-protein contact to an enzyme. This involves a range of concentrations that surround the inflection point; the K_m value for Michaelis-Menten kinetics. Failure to include points that extend beyond the inflection point generates unreliable kinetic constants. Data should be generated in a linear range of time and amount of protein. Because enzyme-catalyzed reactions produce product proportionally to substrate concentration, large decreases in the substrate concentration during assay also will produce unreliable kinetic constants. Therefore, Michaelis-Menten kinetics rely on the steady-state assumption, which postulates little change in substrate concentration during the assay. No change is impractical, so a "rule of thumb" relies on less than a 10% decrease in substrate. This produces data well within the margin of analytical variation. The rate of

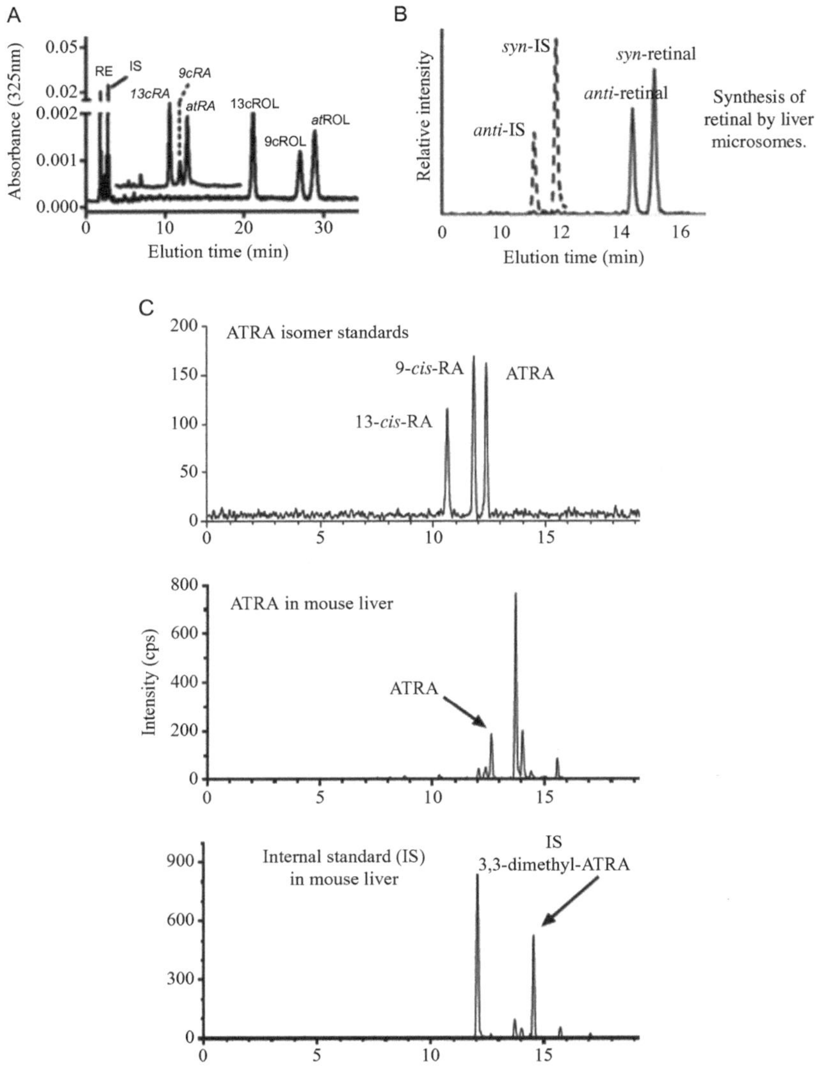

Fig. 5 Chromatographic resolution of retinoids. (A) Normal phase HPLC separation of retinol and ATRA isomers with 0.4% 2-propanol/hexane with an elution of 2 mL/min: retinyl palmitate (RE) (2.0 min); internal standard (retinyl acetate), 3.6 min; 13-*cis*-RA, 10.9 min; 9-*cis*-RA, 12.1; ATRA, 13.1 min; 13-*cis*-retinol, 20.9 min; 9-*cis*-retinol, 27.0 min; all-*trans*-retinol, 28.9 min. Absorbance of retinol isomers was monitored at 325 nm (left y-axis); absorbance in the overlay showing RA isomers was monitored at 340 nm (Kane, Folias, & Napoli, 2008). (B) Resolution of *syn*- and *anti*- all-*trans*-retinal oximes and oximes of the internal standard 3,4-didehydro-all-*trans*-retinal (IS). Ultra-high-performance LC

(Continued)

an especially active enzyme can be reduced by decreasing the assay temperature from 37 to 25 °C, which decreases the rate ~eightfold, as was done for Raldh1 (Penzes et al., 1997). To skip these steps risks generating faulty kinetic values or even completely misleading data. A case in point involved Dhrs4 (aka Rrd) (Lei, Chen, Zhang, & Napoli, 2003). This enzyme appeared to function as a dehydrogenase when assayed at one protein concentration and one time. Time-course and protein concentration experiments, however, showed no increase in retinal production from retinol as time and protein increased, indicating non-enzymatic (i.e., artifactual) retinal production. As much retinal was recovered at time 0 as after a 15 min assay. This illustrates the propensity of retinol to oxidize chemically. When assayed as a retinal reductase, Dhrs4 behaved as a typical enzyme, with increasing product formation as protein concentrations and assay times increased.

Several approaches can reveal whether an enzyme recognizes only free retinoid or also interacts with a BP to tease the retinoid into its active site. The easiest protocol simply uses a slight excess of BP to total retinoid as substrate, to ensure total retinoid (substrate) sequestration by the BP. If a substantial reaction occurs, the enzyme must recognize the "cassette" of holo-BP and remove the retinoid into its active site via protein-protein interaction. This ensues because the off-rate of retinoids from BP occur exceedingly slowly without an "active" receptor, and the affinity of retinoids with BP results in nearly total sequestration. With total BP ~1 to total retinoid ~0.9, virtually no free retinoid occurs in the assay mixture. Significant product formation must result from the enzyme recognizing the BP. Examples of this include interactions of Crbp1-retinol with Lrat and Crbp1-retinal with Raldh2 (Fig. 6A).

Fig. 5—Cont'd used a C18 reverse-phase column (2.1 × 150 mm) at a flow of 0.4 mL/min. The column was eluted with 80% A (0–2 min), 80–85% A (2–2.5 min), 85–95% A (2.5–17.5 min), 95–100% A (17.5–18 min), 100% A (18–20 min), 100–80% A (20–20.5 min), re-equilibrating 3.5 min. Mobile phase A: 0.1% formic acid in acetonitrile. Mobile phase B: 0.1% formic acid in water. Retinal oximes used a *m/z* 328.5 to 236.4 transition; 3,4-didehydroretinal oximes used a *m/z* 326.6 to 234.4 transition (Wang et al., 2015). (C) Resolution of ATRA from its isomers. An Ascentis RP-Amide column (Supelco, 2.1 × 150 mm, 3 μm) was eluted at 0.4 mL/min with a linear gradient: 70% A (0–3 min); 70–95% A (3–15 min); 95% A (15–20 min); 5–70% A (20–21 min); re-equilibrate at 70% A (21–25 min). RA isomers were monitored using a *m/z* 301.1 [M + H] + to *m/z* 205.0 transition; the IS, 3,3-dimethyl-RA was monitored with a *m/z* 329.4 [M + H] + to *m/z* 151.3 transition.

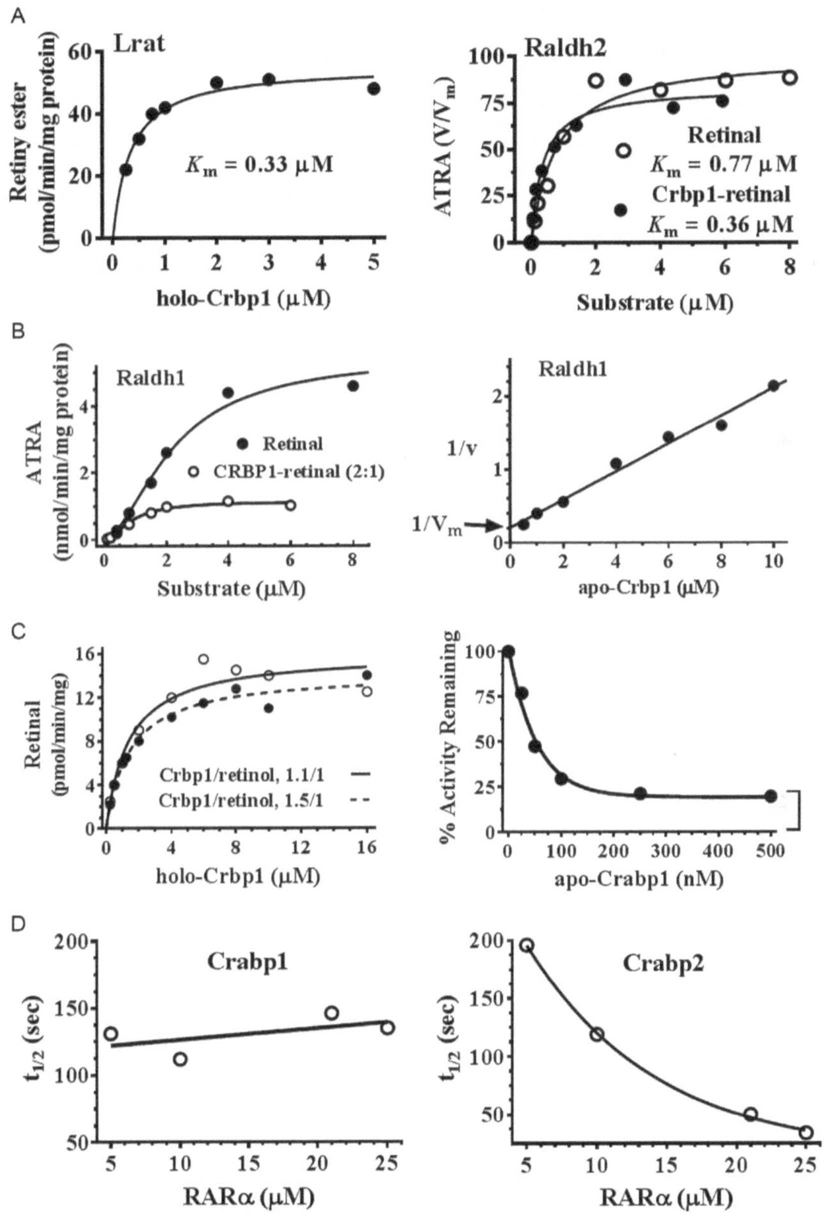

Fig. 6 Effects of BP on reactions/transfers. (A) Crbp1 delivers substrates to Lrat and Raldh2 through channeling. The Lrat assay used a 1:1 ratio of BP/retinol. The Raldh2 assay used a ratio of 2:1 Crbp1/retinal. Both enzymes had similar V_m values for bound and free retinoid, indicating that each accessed retinoid bound with BP. Lrat had similar K_m values for free (not shown) and bound retinol (MacDonald & Ong, 1988).

(Continued)

A complication arises when using apo-Crbp1 in excess of retinoid, if the apo-BP inhibits the enzyme. The impact of apo-Crbp1 on Raldh1 provides an example of this (Fig. 6B). In this case a lower V_m in the presence of excess Crbp1 vs retinal (2:1) gave an impression of a lower V_m with the BP. The Dixon plot (right) allowed extrapolation to a theoretical value of zero apo-Crbp1, and showed that the V_{m} with a precise 1:1 ratio of Crbp1 to retinal was nearly equivalent to that of the reaction with free retinal. These data allow two conclusions: (1) Raldh1 recognizes holo-Crbp1 and accesses its retinal; (2) apo-Crbp1 inhibits Raldh1, in contrast to Raldh2.

A third kinetic display occurs if the rate-limiting event in the reaction becomes transfer of retinoid from the BP to the enzyme. The reaction with Rdh1 provides an example of this (Posch, Burns, & Napoli, 1992). In this case, presentation of holo-Crbp1 lowered the V_{m} of Rdh1 (not shown). Yet two different ratios of total Crbp1/retinol both generated Michaelis-Menten kinetics with nearly the same K_{m} and V_{m} values (Fig. 6C). Because the increase in Crbp1 would substantially lower the amount of free retinol present (if any), these data indicate that Rdh1 recognizes holo-Crbp1 and can tease retinol into its active site; otherwise there would have occurred a substantial decrease in V_{m} with increased apo-Crbp1. These data can be accompanied by a titration of apo-BP into the reaction mixture to determine whether inhibition also occurs. In the case of Rdh1, inhibition by apo-Crbp1 did not occur until higher concentrations of apo-Crbp1 were reached (not shown). Incidentally, using holo-Crbp1 to deliver retinol allows no artifactual generation of retinal, whereas adding retinol to an aqueous homogenate generates retinal in the absence of enzyme.

apo-Crabp1 titration into assessment of ATRA catabolism, catalyzed by recombinant Cyp26B1, illustrates inhibition by BP, as the BP also serves

Fig. 6—Cont'd Crbp1 lowered the K_{m} value for retinal by ~ twofold, indicating a twofold increased k_{cat} value (Wang et al., 1996). (B) Effects of Crbp1 on Raldh1 reaction kinetics. The assay used a ratio of 2:1 Crbp1/retinal. The intersect with the Y-axis in the Dixon plot (right) allowed extrapolation to a theoretical V_{m} value with zero apo-Crbp1. (C) Crbp1 delivers retinol to Rdh1. The lack of difference in the V_{m} values with different amounts of total Crbp1 (increased apo-Crbp1) indicates Rdh1 can access retinol bound to Crbp1. (D) Titration of apo-Crabp1 into ATRA catabolism catalyzed by recombinant Cyp26B1. 50 nM ATRA were used (Nelson et al., 2016). 4-Hydroxy-ATRA was the metabolite monitored. Failure to completely prevent the reaction shows that Cyp26B1 recognizes BP and accesses its ATRA. (E) Rates of ATRA transfer from Crabp to Rarα. The increase in the rate as the Rarα concentration increases indicates protein-protein interaction between the BP and the nuclear receptor (Dong, Ruuska, Levinthal, & Noy, 1999).

as substrate (Nelson et al., 2016). 50 nM ATRA provided the substrate in this reaction (Fig. 6D). These data illustrate several points. At equimolar amounts of Crabp1/ATRA, the reaction rate decreased by 50%. This alone does not necessarily indicate inhibition, but may indicate a shift in the rate-limiting event from catalysis of hydroxylation to ATRA release from holo-Crabp1. A further decrease in reaction rate with added apo-Crabp1 beyond a 1:1 ratio with ATRA indicates inhibition by apo-Crabp1. Note, however, that inhibition never reaches 100%. Even at the ratio Crabp1/ATRA (10/1), 4-hydroxylation proceeds. Given the k_d, virtually no free ATRA should occur even at a 1:1 ratio. Thus, these data show that Cyp26B1 can access ATRA bound to Crabp1 and that apo-Crabp1 inhibits Cyp26B1.

Another kinetic approach to distinguish diffusion-contingent transfer of ligand from BP to acceptor vs chaperoning by direct contact, relies on kinetics generated by changing the acceptor's concentration. The exchange of ATRA between Crabp1 and Rarα vs Crabp2 and Rarα illustrates this approach (Dong et al., 1999). Both Crabp1 and Crabp2 have high affinity and lethargic off-rates, with Crabp1 having about twice the affinity of Crabp2. Crabp1 has a k_{off} value of 0.22 min^{-1} and a K_d value of 0.06 nM, whereas the values for Crabp2 equal 0.42 min^{-1} and 0.13 nM, respectively. Given the slow off-rates, little to no ATRA would occur in solution during the course of the experiment. If diffusion were the mechanism of transfer, increasing the acceptor concentration (Rarα) should not alter the rate of transfer, as observed for Crabp1 (Fig. 6E). In contrast, increasing the concentration of acceptor (Rarα) increases the rate of transfer from Crabp2, indicating that transfer occurs via protein-protein interaction.

4.7 Analysis and statistics

For quantifying retinoids in model animal tissues, a minimum of six animals should be assayed; many more are desirable. Extensive biological variation occurs even with so-called genetically identical strains. Back-crossing is essential in using knockouts, as different strains of mice have different set-points of retinoid concentrations (Obrochta & Napoli, 2014).

4.8 Non-genomic BP functions

HuR binds AU-rich sections in 3′-untranslated regions of mRNAs associated with cell proliferation, apoptosis and differentiation, and thereby protects transcripts from degradation, while regulating translation. Apo-Crabp2 binds to HuR, which enhances HuR affinity for mRNAs >3 orders of magnitude (Vreeland, Yu, Levi, de Barros Rossetto, & Noy, 2014). ATRA binding changes the Crabp2 conformation, causing its release from HuR and exposing a nuclear localization signal (Sessler & Noy, 2005). The holo-BP then undergoes SUMOylation, which prompts nuclear transfer and ATRA delivery to Rar (Majumdar, Petrescu, Xiong, & Noy, 2011). The newly formed apo-BP rapidly returns to cytosol to re-bind HuR.

Rar do not medicate all ATRA functions. During embryonic stem cell stimulation by growth factors, such as EGF, competitive binding of holo-Crabp1 with cRAF dampens cell proliferation through decreasing ERK1/2 activation (Nagpal & Wei, 2019). In contrast, in the absence of growth factor stimulation, holo-Crabp1 activates ERK1/2, which reduces cell growth by activating protein phosphatase 2A. Further, holo-Crabp1 induces differentiation by repressing *Oct4*, a gene essential for maintaining undifferentiated stem cells.

5. Summary

Despite discovery of the first retinoid BP more than 47 years ago, the field still has not developed comprehensive insight into their complex functions (Bashor, Toft, & Chytil, 1973). Clearly, retinoid BP contribute crucially to retinoid metabolism. Apparently, BP chaperone retinol, retinal and ATRA throughout cells, such that retinoids do not exist in their free forms. This has effects of solubilizing the lipophilic retinoids and securing specificity for their metabolism. Retinoid BP also contribute to non-genomic mechanisms of retinoid actions. BP ablations have not produced disastrous morphological consequences (perhaps with exception of Crbp1 on pancreas islets), but such observations do not support conclusions of physiological irrelevance. First, the BP undoubtedly promote efficient retinoid use. Vitamin A has been a scarce but essential nutrient and still remains the third most prevalent nutritional deficiency in many areas of the globe. Second, in general, BP knockouts have not been subject to sufficient stress and observation to reveal other than gross morphological defects. Exceptions have included studies that have associated lack of Crbp1 with increased breast cancer and glucose metabolism abnormalities. Also, *Crabp1*-null mice have reduced basal cardiac function and cardiac hypertrophy, and suffer more damage than WT when challenged chronically with β-adrenergic agents (Park et al., 2018). The null mice also have arrested hippocampal neural stem cell proliferation, impacting memory formation. Moreover, reports continually increase of Crabp1 and Crabp2 associations with incidences of various cancers and their prognoses.

References

Bashor, M. M., Toft, D. O., & Chytil, F. (1973). In vitro binding of retinol to rat-tissue components. *Proceedings of the National Academy of Sciences of the United States of America*, *70*(12), 3483–3487.

Berry, D. C., O'Byrne, S. M., Vreeland, A. C., Blaner, W. S., & Noy, N. (2012). Cross talk between signaling and vitamin A transport by the retinol-binding protein receptor STRA6. *Molecular and Cellular Biology*, *32*(15), 3164–3175.

Boerman, M. H., & Napoli, J. L. (1991). Cholate-independent retinyl ester hydrolysis. Stimulation by Apo-cellular retinol-binding protein. *The Journal of Biological Chemistry*, *266*(33), 22273–22278.

Cowan, S. W., Newcomer, M. E., & Jones, T. A. (1990). Crystallographic refinement of human serum retinol binding protein at 2A resolution. *Proteins*, *8*(1), 44–61.

Dong, D., Ruuska, S. E., Levinthal, D. J., & Noy, N. (1999). Distinct roles for cellular retinoic acid-binding proteins I and II in regulating signaling by retinoic acid. *The Journal of Biological Chemistry*, *274*(34), 23695–23698.

E, X., Zhang, L., Lu, J., Tso, P., Blaner, W. S., Levin, M. S., et al. (2002). Increased neonatal mortality in mice lacking cellular retinol-binding protein II. *The Journal of Biological Chemistry*, *277*(39), 36617–36623.

Fiorella, P. D., & Napoli, J. L. (1991). Expression of cellular retinoic acid binding protein (CRABP) in *Escherichia coli*. Characterization and evidence that holo-CRABP is a substrate in retinoic acid metabolism. *The Journal of Biological Chemistry*, *266*(25), 16572–16579.

Harrison, E. H. (2012). Mechanisms involved in the intestinal absorption of dietary vitamin A and provitamin A carotenoids. *Biochimica et Biophysica Acta*, *1821*(1), 70–77.

Herr, F. M., & Ong, D. E. (1992). Differential interaction of lecithin-retinol acyltransferase with cellular retinol binding proteins. *Biochemistry*, *31*(29), 6748–6755.

Kakkad, B. P., & Ong, D. E. (1988). Reduction of retinaldehyde bound to cellular retinol-binding protein (type II) by microsomes from rat small intestine. *The Journal of Biological Chemistry*, *263*(26), 12916–12919.

Kane, M. A., Chen, N., Sparks, S., & Napoli, J. L. (2005). Quantification of endogenous retinoic acid in limited biological samples by LC/MS/MS. *The Biochemical Journal*, *388*(Pt. 1), 363–369.

Kane, M. A., Folias, A. E., & Napoli, J. L. (2008). HPLC/UV quantitation of retinal, retinol, and retinyl esters in serum and tissues. *Analytical Biochemistry*, *378*(1), 71–79.

Kane, M. A., Folias, A. E., Pingitore, A., Perri, M., Krois, C. R., Ryu, J. Y., et al. (2011). CrbpI modulates glucose homeostasis and pancreas 9-cis-retinoic acid concentrations. *Molecular and Cellular Biology*, *31*(16), 3277–3285.

Kane, M. A., Folias, A. E., Wang, C., & Napoli, J. L. (2008). Quantitative profiling of endogenous retinoic acid in vivo and in vitro by tandem mass spectrometry. *Analytical Chemistry*, *80*(5), 1702–1708.

Kane, M. A., & Napoli, J. L. (2010). Quantification of endogenous retinoids. *Methods in Molecular Biology*, *652*, 1–54.

Lampron, C., Rochette-Egly, C., Gorry, P., Dollé, P., Mark, M., Lufkin, T., et al. (1995). Mice deficient in cellular retinoic acid binding protein II (CRABPII) or in both CRABPI and CRABPII are essentially normal. *Development (Cambridge, England)*, *121*(2), 539–548.

Lei, Z., Chen, W., Zhang, M., & Napoli, J. L. (2003). Reduction of all-trans-retinal in the mouse liver peroxisome fraction by the short-chain dehydrogenase/reductase RRD: Induction by the PPAR alpha ligand clofibrate. *Biochemistry*, *42*(14), 4190–4196.

Levin, M. S., Locke, B., Yang, N. C., Li, E., & Gordon, J. I. (1988). Comparison of the ligand binding properties of two homologous rat apocellular retinol-binding proteins expressed in Escherichia coli. *The Journal of Biological Chemistry*, *263*(33), 17715–17723.

MacDonald, P. N., & Ong, D. E. (1988). A lecithin:Retinol acyltransferase activity in human and rat liver. *Biochemical and Biophysical Research Communications*, *156*(1), 157–163.

Majumdar, A., Petrescu, A. D., Xiong, Y., & Noy, N. (2011). Nuclear translocation of cellular retinoic acid-binding protein II is regulated by retinoic acid-controlled SUMOylation. *The Journal of Biological Chemistry*, *286*(49), 42749–42757.

Matt, N., Schmidt, C. K., Dupé, V., Dennefeld, C., Nau, H., Chambon, P., et al. (2005). Contribution of cellular retinol-binding protein type 1 to retinol metabolism during mouse development. *Developmental Dynamics: An Official Publication of the American Association of Anatomists*, *233*(1), 167–176.

Menozzi, I., Vallese, F., Polverini, E., Folli, C., Berni, R., & Zanotti, G. (2017). Structural and molecular determinants affecting the interaction of retinol with human CRBP1. *Journal of Structural Biology*, *197*(3), 330–339.

Nagpal, I., & Wei, L.-N. (2019). All-trans retinoic acid as a versatile cytosolic signal modulator mediated by CRABP1. *International Journal of Molecular Sciences*, *20*(15), E3610.

Napoli, J. L. (2016). Functions of intracellular retinoid binding-proteins. *Sub-Cellular Biochemistry*, *81*, 21–76.

Napoli, J. L. (2017). Cellular retinoid binding-proteins, CRBP, CRABP, FABP5: Effects on retinoid metabolism, function and related diseases. *Pharmacology & Therapeutics*, *173*, 19–33.

Napoli, J. L., Pramanik, B. C., Williams, J. B., Dawson, M. I., & Hobbs, P. D. (1985). Quantification of retinoic acid by gas-liquid chromatography-mass spectrometry: Total versus all-trans-retinoic acid in human plasma. *Journal of Lipid Research*, *26*(3), 387–392.

Napoli, J. L., Sperkova, Z., Sperka, M., & Fiorella, P. D. (1998). Expression and purification of CRABPs from *E. coli*. *Methods in Molecular Biology*, *89*, 105–110.

National Research Council. (1972). *Specifications and criteria for biochemical compounds* (3rd ed.). The National Academies Press.

Nelson, C. H., Peng, C.-C., Lutz, J. D., Yeung, C. K., Zelter, A., & Isoherranen, N. (2016). Direct protein-protein interactions and substrate channeling between cellular retinoic acid binding proteins and CYP26B1. *FEBS Letters*, *590*(16), 2527–2535.

Noy, N. (2016). Non-classical transcriptional activity of retinoic acid. *Subcellular Biochemistry*, *81*, 179–199.

Obrochta, K. M., Krois, C. R., Campos, B., & Napoli, J. L. (2015). Insulin regulates retinol dehydrogenase expression and all-*trans*-retinoic acid biosynthesis through FoxO1. *The Journal of Biological Chemistry*, *290*(11), 7259–7268.

Obrochta, K. M., & Napoli, J. L. (2014). Effects of diet and strain on mouse serum and tissue retinoid concentrations. *PLoS One*, *9*, e99435.

Ong, D. E. (1984). A novel retinol-binding protein from rat. Purification and partial characterization. *The Journal of Biological Chemistry*, *259*(3), 1476–1482.

Ong, D. E., & Chytil, F. (1978a). Cellular retinol-binding protein from rat liver. Purification and characterization. *The Journal of Biological Chemistry*, *253*(3), 828–832.

Ong, D. E., & Chytil, F. (1978b). Cellular retinoic acid-binding protein from rat testis. Purification and characterization. *The Journal of Biological Chemistry*, *253*(13), 4551–4554.

Ong, D. E., Kakkad, B., & MacDonald, P. N. (1987). Acyl-CoA-independent esterification of retinol bound to cellular retinol-binding protein (type II) by microsomes from rat small intestine. *The Journal of Biological Chemistry*, *262*(6), 2729–2736.

Ong, D. E., MacDonald, P. N., & Gubitosi, A. M. (1988). Esterification of retinol in rat liver. Possible participation by cellular retinol-binding protein and cellular retinol-binding protein II. *The Journal of Biological Chemistry*, *263*(12), 5789–5796.

Park, S. W., Persaud, S. D., Ogokeh, S., Meyers, T. A., Townsend, D., & Wei, L.-N. (2018). CRABP1 protects the heart from isoproterenol-induced acute and chronic remodeling. *The Journal of Endocrinology*, *236*(3), 151–165.

Penzes, P., Wang, X., & Napoli, J. L. (1997). Enzymatic characteristics of retinal dehydrogenase type I expressed in Escherichia coli. *Biochimica et Biophysica Acta, 1342*(2), 175–181.
Posch, K. C., Boerman, M. H., Burns, R. D., & Napoli, J. L. (1991). Holocellular retinol binding protein as a substrate for microsomal retinal synthesis. *Biochemistry, 30*(25), 6224–6230.
Posch, K. C., Burns, R. D., & Napoli, J. L. (1992). Biosynthesis of all-trans-retinoic acid from retinal. Recognition of retinal bound to cellular retinol binding protein (type I) as substrate by a purified cytosolic dehydrogenase. *The Journal of Biological Chemistry, 267*(27), 19676–19682.
Ross, A. C., & Zolfaghari, R. (2011). Cytochrome P450s in the regulation of cellular retinoic acid metabolism. *Annual Review of Nutrition, 31*, 65–87.
Sessler, R. J., & Noy, N. (2005). A ligand-activated nuclear localization signal in cellular retinoic acid binding protein-II. *Molecular Cell, 18*(3), 343–353.
Topletz, A. R., Thatcher, J. E., Zelter, A., Lutz, J. D., Tay, S., Nelson, W. L., et al. (2012). Comparison of the function and expression of CYP26A1 and CYP26B1, the two retinoic acid hydroxylases. *Biochemical Pharmacology, 83*(1), 149–163.
Topletz, A. R., Tripathy, S., Foti, R. S., Shimshoni, J. A., Nelson, W. L., & Isoherranen, N. (2015). Induction of CYP26A1 by metabolites of retinoic acid: Evidence that CYP26A1 is an important enzyme in the elimination of active retinoids. *Molecular Pharmacology, 87*(3), 430–441.
Vaezeslami, S., Mathes, E., Vasileiou, C., Borhan, B., & Geiger, J. H. (2006). The structure of Apo-wild-type cellular retinoic acid binding protein II at 1.4 A and its relationship to ligand binding and nuclear translocation. *Journal of Molecular Biology, 363*(3), 687–701.
Vreeland, A. C., Yu, S., Levi, L., de Barros Rossetto, D., & Noy, N. (2014). Transcript stabilization by the RNA-binding protein HuR is regulated by cellular retinoic acid-binding protein 2. *Molecular and Cellular Biology, 34*(12), 2135–2146.
Wang, X., Penzes, P., & Napoli, J. L. (1996). Cloning of a cDNA encoding an aldehyde dehydrogenase and its expression in Escherichia coli. Recognition of retinal as substrate. *The Journal of Biological Chemistry, 271*(27), 16288–16293.
Wang, J., Yoo, H. S., Obrochta, K. M., Huang, P., & Napoli, J. L. (2015). Quantitation of retinaldehyde in small biological samples using ultrahigh-performance liquid chromatography tandem mass spectrometry. *Analytical Biochemistry, 484*, 162–168.

CHAPTER FOUR

Generation and isolation of recombinant retinoid oxidoreductase complex

Mark K. Adams[a,*], Olga V. Belyaeva[b], Natalia Y. Kedishvili[b]
[a]Stowers Institute for Medical Research, Kansas City, MO, United States
[b]Department of Biochemistry and Molecular Genetics, Schools of Medicine and Dentistry, University of Alabama at Birmingham, Birmingham, AL, United States
*Corresponding author: e-mail address: maa@stowers.org

Contents

Abstract

All-*trans*-retinoic acid (RA) is a bioactive lipid that influences many processes in embryonic and adult tissues. Given its bioactive nature, cellular concentrations of this molecule are highly regulated. The oxidation of all-*trans*-retinol to all-*trans*-retinaldehyde represents the first and rate-limiting step of the RA synthesis pathway. As such, it is the target of mechanisms that fine-tune RA levels within the cell. RDH10 is one enzyme responsible for the oxidation of all-*trans*-retinol to all-*trans*-retinaldehyde, and together with the all-*trans*-retinaldehyde reductase DHRS3 forms an oligomeric protein complex. The resulting retinoid oxidoreductase complex (ROC) is bifunctional and has the capacity to regulate steady-state levels of the direct

Methods in Enzymology, Volume 637
ISSN 0076-6879
https://doi.org/10.1016/bs.mie.2020.02.005

precursor of RA, all-*trans*-retinaldehyde. As ROC represents a major regulatory element within the RA synthesis pathway, it is essential that methods are in place that allow for the study of this complex. Here we describe the production and isolation of recombinant ROC using a baculovirus expression system. Recombinant proteins retain enzymatic activities in intact microsomes and can be affinity purified for analysis. These methods can be used to assist in the assessment of ROC properties and the regulation of this protein complex's functional attributes.

1. Introduction

All-*trans*-retinoic acid (RA) has well established roles in many developmental and adult processes (Huang, Chandra, & Rastinejad, 2014). While RA is a potent activator of nuclear retinoic acid receptors, bioactive RA does not represent a major stored or transported form of retinoids (Kedishvili, 2016). Instead, RA is generated via a two-step process from the alcohol precursor of vitamin A, all-*trans*-retinol. All-*trans*-retinol is first oxidized to all-*trans*-retinaldehyde and all-*trans*-retinaldehyde is further oxidized to RA. As RA binding to nuclear receptors potently activates transcription, it is not surprising that mechanisms are in place that regulate the concentrations of this molecule within the cell.

The rate-limiting step of the RA synthesis pathway, the oxidation of all-*trans*-retinol, is a major control point of RA concentrations within cells and several enzymes have demonstrated capacities to oxidize all-*trans*-retinol *in vitro* (Belyaeva, Adams, Popov, & Kedishvili, 2020). Additionally, retinaldehyde reductases, capable of counteracting retinol dehydrogenases by reducing all-*trans*-retinaldehyde to all-*trans*-retinol, are present within the cell (Kedishvili, 2016). The combined activities of retinol dehydrogenases and retinaldehyde reductases influence the availability of all-*trans*-retinaldehyde for RA synthesis. Among retinol dehydrogenases, retinol dehydrogenase 10 (RDH10, SDR16C4) is perhaps the best characterized and is the main enzyme responsible for retinol oxidation during early- and mid-embryogenesis (Cunningham, Chatzi, Sandell, Trainor, & Duester, 2011; Farjo et al., 2011). Interestingly, RDH10 is capable of forming a protein complex with dehydrogenase reductase 3 (DHRS3, SDR16C1) (Belyaeva, Adams, Wu, & Kedishvili, 2017), an all-*trans*-retinaldehyde reductase (Haeseleer, Huang, Lebioda, Saari, & Palczewski, 1998). The resulting protein complex, retinoid oxidoreductase complex (ROC), possesses both retinol dehydrogenase and retinaldehyde reductase activities (Adams, Belyaeva, Wu, & Kedishvili, 2014). By coupling retinol dehydrogenase and retinaldehyde reductase activities, an elegant system is formed that can fine-tune steady-state levels of all-*trans*-retinaldehyde, and consequently RA, concentrations within the cell.

Here we describe methods for the isolation and characterization of the bifunctional ROC. Using a baculovirus expression system, catalytically active ROC is generated and enriched via subcellular fractionation. Recombinant ROC can be further enriched from microsomes via the described affinity purification procedures. As the regulation of ROC formation and function is poorly understood, this protein complex is a focus of ongoing studies that aim to elucidate complex properties. The provided methods describe a streamlined procedure that enables the assessment of the functional attributes of this protein complex.

2. The short-chain dehydrogenase/reductase 16C (SDR16C) family of enzymes

Members of the short-chain dehydrogenase/reductase (SDR) superfamily are NAD(P)-dependent enzymes present within prokaryotes and eukaryotes (Kavanagh, Jörnvall, Persson, & Oppermann, 2008). A common feature of SDRs is a propensity to form oligomeric structures and most SDRs are known to exist in dimer and tetramer forms (Jörnvall et al., 1995). While available three-dimensional structures suggest that many features, such as the Rossmann fold, are conserved among members of this enzyme superfamily (Shin & Kihara, 2019), the substrates recognized by members of this superfamily are diverse. Notably, several SDRs have demonstrated capacities to oxidize or reduce retinoids (Bhatia et al., 2015).

The short-chain dehydrogenase/reductase 16C family (SDR16C) of SDRs consists of five enzymes within humans and contains many of the enzymes with confirmed or proposed capacities to recognize the RA precursors, retinol and retinaldehyde, as substrates (Belyaeva, Chang, Berlett, & Kedishvili, 2015; Ma et al., 2019; Takahashi, Moiseyev, Farjo, & Ma, 2009). At least three enzymes within this family have reported roles in the oxidation of retinol. Mammalian embryos depend on RDH10 for adequate retinol oxidation (Chatzi, Cunningham, & Duester, 2013; Farjo et al., 2011; Friedl et al., 2019; Kurosaka, Wang, Sandell, Yamashiro, & Trainor, 2017; Metzler et al., 2018; Sandell, Lynn, Inman, McDowell, & Trainor, 2012). RDH10 also has an established role as a retinol dehydrogenase in juvenile (Tong, Yang, Davis, & Griswold, 2013) and adult tissue (Yang et al., 2018). A second member of this family, retinol dehydrogenase epidermal 2 (RDHE2, SDR16C5), likely functions as a retinol dehydrogenase in epidermal tissue (Adams, Lee, Belyaeva, Wu, & Kedishvili, 2017; Wu et al., 2019). Additionally, there is emerging evidence that 17β-hydroxysteroid dehydrogenase 13 (HSD17B13, SDR16C3) may contribute to retinol oxidation in a tissue-specific manner (Ma et al., 2019).

In addition to oxidative enzymes, the SDR16C family also contains the retinaldehyde reductase DHRS3. Results obtained by others suggest that DHRS3 is a critical regulator of RA synthesis (Billings et al., 2013). However, early reports indicated that DHRS3 has a low capacity to function as a retinaldehyde reductase (Haeseleer et al., 1998). These findings suggest that, if DHRS3 functions as a retinaldehyde reductase, additional factors may be needed for the potentiation of this activity.

3. The retinoid oxidoreductase complex (ROC)

Structural descriptions of SDRs indicate that many of these enzymes are capable of forming homo-oligomeric structures (Jörnvall et al., 1995). However, there is evidence that 17β-hydroxysteroid dehydrogenase 8 (HSD17B8, SDR30C1) and carbonyl reductase 4 (CBR4, SDR45C1), two SDRs that share 50% sequence similarity, form hetero-oligomeric structures via associations with one another (Chen et al., 2009; Venkatesan et al., 2014). Interestingly, RDH10 and DHRS3 also share approximately 50% sequence similarity with one another at the protein level. Our findings demonstrate that, like HSD17B8 and CBR4, RDH10 and DHRS3 form stable hetero-oligomeric complexes. Data suggest that this protein complex, ROC, is larger than a dimer (Adams et al., 2014; Belyaeva et al., 2017).

The formation of ROC influences the catalytic properties of both RDH10 and DHRS3 subunits. DHRS3's capacity to reduce retinaldehyde is potentiated by interactions between DHRS3 and RDH10. Likewise, retinol oxidation by RDH10 is stimulated by interactions between RDH10 and DHRS3 (Adams et al., 2014). As expression of the gene encoding DHRS3 is stimulated by RA (Feng, Hernandez, Waxman, Yelon, & Moens, 2010; Zolfaghari, Chen, & Ross, 2012), it is likely that the formation of ROC is influenced by cellular RA concentrations and that this protein complex is a key component in the feedback regulation of RA synthesis. However, mechanisms regulating ROC function and formation remain poorly defined.

4. The baculovirus expression system

Many protein expression systems exist, each with their own benefits and shortcomings. *Escherichia coli* is the most popular host organism within which to produce recombinant proteins (Pandey, Shin, Patterson, Liu, & Rainey, 2016). However, prokaryotic expression hosts, such as *E. coli*, are often poorly suited for the expression of integral membrane proteins like RDH10 and DHRS3. Established mammalian cell lines can also be used to generate functional recombinant integral membrane proteins, though high costs and low levels of recombinant protein are major limitations of systems utilizing these host cells. Additionally, the isolation of recombinant mammalian proteins from these cells for *in vitro* analyses may be complicated by the presence of factors that interact with or influence properties of the recombinant mammalian proteins.

The baculovirus expression system is an alternative expression system that utilizes a eukaryotic host cell. As such, it offers many of the benefits offered by mammalian cells but generally produces higher levels of recombinant protein. Recombinant protein can represent over 50% of total soluble protein content within the cell (Gómez-Sebastián, López-Vidal, & Escribano, 2014). To generate high levels of recombinant protein, the existence of a naturally occurring viral pathogen that targets *Spodoptera frugiperda*, *Autographa californica* multiple nucleopolyhedrovirus (*Ac*MNPV), is exploited (O'Reilly, Miller, & Luckow, 1992). This expression system typically utilizes one of two cell lines derived from *S. frugiperda* ovaries, Sf21 and Sf9 cells, that are easily cultured and maintained in the laboratory. To produce recombinant protein, Sf9 or Sf21 cells are infected with modified *Ac*MNPV baculovirus whose genome contains a gene of interest downstream of the polyhedrin promoter. Infected cells are harvested, and protein is isolated from cells.

Prior to the generation of recombinant protein, the baculoviral genome must be modified so that it contains a gene of interest. Two major systems exist that accomplish this recombination; however, these systems differ in the stage at which the recombinant baculoviral genome is created. The generation of recombinant baculovirus genomes, or bacmids, in *E. coli* is the most recently developed approach to produce baculovirus particles for recombinant protein generation. The second approach, co-transfection of linearized genomic *Ac*MNPV DNA along with a transfer vector containing a gene of interest downstream of the polyhedrin promoter, is an older technique than bacmid generation but is less labor-intensive. Unlike bacmids that are generated in *E. coli*, recombination between genomic *Ac*MNPV DNA and the transfer vector during the co-transfection approach occurs within Sf9 or Sf21 cells. Owing to the capacity of this approach to quickly and cost-effectively produce high yields of properly folded integral membrane proteins, we have chosen it to generate baculovirus particles and recombinant ROC complex.

5. Calcium phosphate precipitation

Several transfection methods have been successfully used to introduce baculoviral and transfer vector DNA into host cells (O'Reilly et al., 1992). Among the available methods, calcium phosphate precipitation is well documented and does not require the use of expensive transfection reagents. However, the use of calcium phosphate precipitation as a transfection method requires that the composition of culture medium be known. Some commercially available media that are widely used, such as Sf-900™ III SFM, do not have published compositions. Thus, calcium phosphate precipitation is not suited for this medium. Instead, it is advisable to perform calcium phosphate precipitation in a medium of known composition, such as Grace's Insect Medium. The co-transfection approach of baculovirus generation using calcium phosphate precipitation in Grace's Insect Medium has been described in the methods provided (Fig. 1).

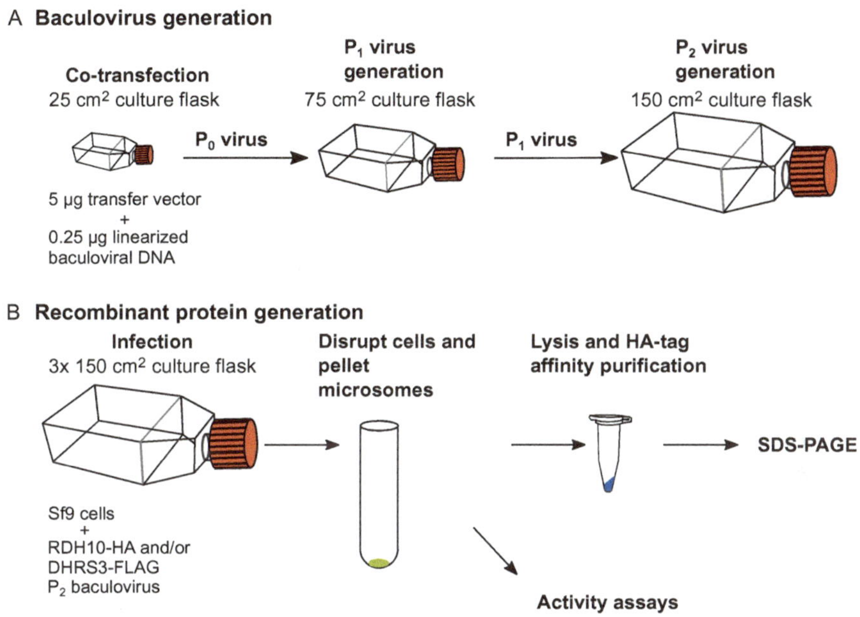

Fig. 1 Diagram of recombinant ROC production. (A) Generation of baculovirus. Sf9 cells in 25 cm^2 culture flasks are co-transfected with baculoviral DNA and transfer vector containing an open reading frame encoding DHRS3-FLAG or RDH10-HA variants via calcium phosphate precipitation. Medium from transfected Sf9 cells is harvested and retained as P_0 virus. P_0 is added to Sf9 cells in a 75 cm^2 culture flask to increase viral titer. Medium from infected Sf9 cells is harvested and retained as P_1 virus. To increase the volume of the baculovirus stock, P_1 virus is added to Sf9 cells in a 150 cm^2 culture flask. Medium from infected Sf9 cells is harvested and retained as P_2 virus. (B) Recombinant protein generation. P_2 virus is added to Sf9 cells in 150 cm^2 culture flasks and cells are harvested for protein isolation. A French pressure cell is used to disrupt cells and samples are centrifuged to pellet microsomes (green). Microsomes are resuspended and used in activity assays or affinity purification assays. For affinity enrichment of ROC, microsomes are lysed and incubated with Pierce™ Anti-HA Agarose (blue). Bound proteins are eluted in SDS-PAGE loading buffer and analyzed via SDS-PAGE.

6. Protocol

6.1 Equipment

Equipment	Source	Catalog number
1. Cell scraper	Corning	3011
2. Hemocytometer	Hausser Scientific	1490
3. 25 cm^2 cell culture flask	Corning	430639
4. 75 cm^2 cell culture flask	Corning	430641U
5. 150 cm^2 cell culture flask	Corning	430825

Continued

—cont'd

Equipment	Source	Catalog number
6. FRENCH® press	Thermo Electron	FA-078A
7. FRENCH pressure cell	Thermo Electron	FA-032
8. JA-17 fixed-angle aluminum rotor	Beckman Coulter	369691
9. Avanti Centrifuge J-20 XPI	Beckman Coulter	368613
10. Optima Ultracentrifuge LE-80K	Beckman Coulter	365668
11. 5 mL, open-top thin wall polypropylene tube	Beckman Coulter	326819
12. SW 55 Ti swinging-bucket rotor	Beckman Coulter	342194
13. Bel-Art® pestle for 1.5 mL microcentrifuge tubes	Sigma-Aldrich	BAF199220000

6.2 Reagents

Reagent	Source	Catalog number
1. Gibco™ Sf9 cells in Grace's	Thermo Fisher Scientific™	B82501
2. pVL1393 transfer vector	Expression systems	91-013
3. BestBac 1.0 Linearized Baculovirus DNA	Expression systems	91-001
4. HEPES	Sigma-Aldrich	H4034
5. Calcium chloride dihydrate	Sigma-Aldrich	C7902
6. NaCl	Sigma-Aldrich	S5886
7. Gibco™ Grace's Insect Medium, supplemented	Thermo Fisher Scientific™	11605094
8. Gibco™ fetal bovine serum, heat inactivated	Thermo Fisher Scientific™	10082139
9. Aprotinin	VWR Life Science	97062
10. Leupeptin	APExBIO	A2570
11. Pepstatin A	APExBIO	A2571
12. PMSF	Sigma-Aldrich	10837091001
13. EDTA	Sigma-Aldrich	EDS

—cont'd

Reagent	Source	Catalog number
14. Benzamidine	Sigma-Aldrich	12072
15. 2-Mercaptoethanol	Sigma-Aldrich	M3148
16. Pierce™ Anti-HA Agarose	Thermo Fisher Scientific™	26181
17. Monoclonal ANTI-FLAG® M2 antibody	Sigma-Aldrich	F1804
18. RDH10 antibody	Proteintech Group	14644-1-AP
19. Pierce™ Protein A Agarose	Thermo Fisher Scientific™	20333
20. Triton™ X-100	Sigma-Aldrich	T8787
21. DL-Dithiothreitol solution	Sigma-Aldrich	646563
22. IGEPAL CA-630	Sigma-Aldrich	I8896
23. Sodium deoxycholate	Sigma-Aldrich	30970
24. SDS	Sigma-Aldrich	L3771
25. KH_2PO_4	Sigma-Aldrich	P5655
26. KCl	Sigma-Aldrich	P9541

6.3 Buffer preparation

6.3.1 Recombinant baculovirus generation buffers

6.3.1.1. Prepare Grace's Complete Medium. Add heat inactivated FBS to Grace's Insect Medium (Supplemented) to a final concentration of 10%. Store at 4 °C. *It is recommended that serum is heat inactivated as baculoviruses can be inactivated by serum complement* (Tani et al., 2003).

6.3.1.2. Prepare Transfection Reagent B (25 mM HEPES pH 7.1, 125 mM $CaCl_2$, 140 mM NaCl). Filter sterilize and store at 4 °C. *Accurate pH is critical for calcium phosphate precipitation.*

6.3.2 Subcellular fractionation buffers

6.3.2.1. Prepare the following protease inhibitor stock solutions: 2.2 mg/mL aprotinin, 10 mg/mL leupeptin, 1 mg/mL pepstatin A, 0.5 M EDTA, and 1 M benzamidine. Store aprotinin and EDTA stock solutions at 4 °C. Store leupeptin, pepstatin A, and benzamidine stock solutions at −20 °C.

6.3.2.2. Prepare Buffer A. To 10 mL of 50 mM Tris–HCl, pH 7.4, add 9 μL aprotinin stock solution, 2 μL leupeptin stock solution, 10 μL pepstatin A stock solution, 20 μL EDTA stock solution, 50 μL benzamidine stock solution, and 3.5 μL β-mercaptoethanol. *After cell lysis and prior to subcellular fractionation, supplement buffer with PMSF to a final concentration of 5 mM.*

6.3.2.3. Prepare Reaction Buffer (90 mM KH_2PO_4, pH 7.4, 40 mM KCl). Adjust pH to 7.4 and supplement with EDTA stock solution to a final concentration of 0.1 mM. Store at 4 °C. *After protein quantitation, add DTT to samples in Reaction Buffer to a final concentration of 1 mM.*

6.3.3 Affinity purification buffers

6.3.3.1. Prepare Solubilization Buffer (25 mM Tris, pH 7.4, 150 mM NaCl, 0.1% SDS, 0.5% sodium deoxycholate, 1% IGEPAL CA-630).

6.4 Generation of recombinant baculovirus particles by co-transfection and calcium phosphate precipitation (P_0 virus production)

Note: perform all steps in a cell culture laminar flow hood

6.4.1. Maintain Sf9 cells in Grace's Complete Medium as a monoculture in 75 cm^2 culture flasks.

6.4.2. Gently resuspend cells with a cell scraper.

6.4.3. Count cells with a hemocytometer. Adjust volume with Grace's Complete Medium so that cell concentration is 1.2×10^6/mL. Aliquot 2 mL (2.4×10^6 cells) into a 25 cm^2 culture flask and allow cells to attach at 27 °C for approximately 30 min.

6.4.4. Add 2.5 μL (0.25 μg) of BestBac 1.0 Linearized Baculovirus DNA to a 1.5 mL microcentrifuge tube.

6.4.5. Add 5 μg of pVL1393 transfer vector containing open reading frames encoding RDH10-HA or DHRS3-FLAG variants to the BestBac 1.0 Linearized Baculovirus DNA. Tap gently to mix and incubate for 5 min at room temperature. *RDH10-HA, RDH10 Y210A-HA, DHRS3-FLAG, and DHRS3 Y188A-FLAG in pVL1393 vector were previously described* (Adams et al., 2014; Belyaeva et al., 2017). *RDH10 G43A/G47A/G49A-HA and DHRS3 G49A/G51A-FLAG were generated using targeted mutagenesis. Oligonucleotides are available upon request. RDH10*

Y210A-HA and DHRS3 Y188A-FLAG contain mutations within the active sites of the enzymes whereas RDH10 G43A/G47A/G49A-HA and DHRS3 G49A/G51A-FLAG contain mutations within the cofactor binding sites of the enzymes.

6.4.6. Remove the 2 mL of medium from attached cells and replace with 1 mL of fresh Grace's Complete Medium.

6.4.7. Add 1 mL of Buffer B to the microcentrifuge tube containing BestBac 1.0 Linearized Baculovirus DNA and pVL1393 transfer vector. Gently invert the microcentrifuge tube to mix.

6.4.8. Add Buffer B + DNA solution dropwise to the flask of cells from step 6.4.6. Gently swirl the flask during the addition of Buffer B + DNA solution.

6.4.9. Incubate flask in the dark at 27 °C for 4–5 h.

6.4.10. Add 7 mL of PBS, pH 7.4, to the flask. Gently swirl the flask to rinse. Aspirate the full 9 mL of solution from the cells.

6.4.11. Add 5 mL of Grace's Complete Medium to the cells. Incubate in the dark at 27 °C for 4–5 days.

6.4.12. Collect medium and centrifuge at 500 × *g*. The supernatant contains baculovirus particles. Retain the supernatant as P_0 virus stock and store in the dark at 4 °C.

6.5 Increasing titer of baculovirus stocks (P_1 and P_2 baculovirus production)

Note: perform all steps in a cell culture laminar flow hood

6.5.1. Seed 5×10^6 fresh Sf9 cells in a 75 cm^2 culture flask containing 12 mL of Grace's Complete Medium.

6.5.2. Add 2 mL of P_0 baculovirus stock to the cells prepared in step 6.5.1. Incubate in the dark at 27 °C for 3–4 days. Cells should display signs of infection by 72 h of culture. *Infected cells will have increased diameters and will begin detaching from the culture flask.*

6.5.3. Collect medium and centrifuge at 500 × *g*. Retain the supernatant as P_1 baculovirus stock and store in the dark at 4 °C.

6.5.4. Seed 2×10^7 fresh Sf9 cells in a 150 cm^2 culture flask containing 30 mL of Grace's Complete Medium.

6.5.5. Add 1 mL of P_1 baculovirus stock to cells prepared in step 6.5.4. Incubate in the dark at 27 °C for 3–4 days. Cells should display signs of infection by 48 h of culture. *While P_1 baculovirus stocks are of sufficient titer to generate recombinant protein, it is often necessary to generate P_2 virus to increase the volume of baculovirus stocks. It is recommended that* *passages beyond P_2 are not generated as deleterious mutations may arise within the baculoviral genomes.*

6.5.6. Collect medium and centrifuge at 500 × *g*. Retain the supernatant as P_2 baculovirus stock and store in the dark at 4 °C.

6.6 Protein production and subcellular fractionation

6.6.1. Seed 2×10^7 fresh Sf9 cells in a 150 cm² culture flask containing 30 mL of Grace's Complete Medium. Prepare at least three flasks of cells. Allow cells to adhere for at least 30 min at 27 °C.

6.6.2. To each flask, add 2 mL of DHRS3-FLAG variant P_2 virus, RDH10-HA variant P_2 virus, or 2 mL of both P_2 viruses.

6.6.3. Culture cells in the dark at 27 °C for approximately 4 days.

6.6.4. Scrape cells and transfer the cell-containing medium from the 150 cm² culture flasks to 50 mL centrifuge tubes. Centrifuge cells at 1000 × *g* for 10 min. Discard supernatant and retain pellets.

6.6.5. Gently resuspend and wash each cell pellet with 10 mL room temperature PBS, pH 7.4. Combine the resuspended cells from all 50 mL centrifuge tubes.

6.6.6. Centrifuge resuspended cells with a JA-17 fixed-angle rotor in an Avanti Centrifuge J-20 XPI at 1000 × *g* for 10 min at 4 °C. Discard the supernatant.

6.6.7. Resuspend the cell pellet in 10 mL of Buffer A and homogenize once with a FRENCH Press and a FA-032 FRENCH pressure cell at 800 psi. Collect cells in a 15 mL centrifuge tube on ice.

6.6.8. Repeat step 6.6.7.

6.6.9. Add PMSF, reconstituted in DMSO, to a final concentration of 5 mM. *DMSO should not account for more than 1% of final volume.*

6.6.10. Place tubes in a JA-17 fixed-angle rotor. Centrifuge in an Avanti Centrifuge J-20 XPI at 3000 × *g* for 10 min at 4 °C. Transfer the supernatant to a new conical centrifuge tube.

6.6.11. Place tubes in a JA-17 fixed-angle rotor. Centrifuge in an Avanti Centrifuge J-20 XPI at 10,000 × *g* for 35 min at 4 °C (the pellet is the mitochondrial fraction). *If mitochondrial protein will be analyzed, the pellet can be resuspended in Reaction Buffer with 20% glycerol and snap frozen in liquid N_2. Store the resuspended mitochondrial fraction at −80 °C.*

6.6.12. Collect supernatant and distribute into 5 mL open-top thin wall polypropylene tubes. Balance tubes and place in a prechilled SW 55 Ti swinging-bucket rotor. Centrifuge the supernatant at 105,000 × *g* for 90 min at 4 °C in a prechilled Optima Ultracentrifuge LE-80K.

6.6.13. Discard the supernatants. Wash the pellets with 1 mL of prechilled Reaction Buffer.

6.6.14. Resuspend pellets in 200 μL of prechilled Reaction Buffer using a microcentrifuge tube pestle.

6.6.15. Add glycerol to final concentration of 20%.

6.6.16. Take aliquots to determine protein concentration. After protein quantitation, add DTT to a final concentration of 2 mM. Freeze in liquid nitrogen and store at −80 °C. *At this point, microsomal fractions can be utilized for activity assays as previously described* (Adams et al., 2014) *or recombinant protein isolation as described in Section 6.7.*

6.7 Isolation of recombinant protein

6.7.1. Solubilize 20–40 μg of isolated microsomes containing recombinant proteins with 300 μL of Solubilization Buffer for 10 min on ice.

6.7.2. Aliquot 15 μL of Pierce™ Anti-HA Agarose slurry into a 1.5 mL microcentrifuge tube. Briefly centrifuge beads at 200 × *g* and discard supernatant. Resuspend beads in 1 mL of Solubilization Buffer and invert tubes to rinse. Briefly centrifuge beads at 200 × *g* and discard supernatant.

6.7.3. Block Pierce™ Anti-HA Agarose in 500 μL TBST with 2% BSA for 2 h at 4 °C. Briefly centrifuge beads at 200 × *g* and discard supernatant. *If performing assays with preblocked beads, this step can be omitted.*

6.7.4. During step 6.7.3, clear the solubilized microsomes. Transfer the solubilized microsomes to a 1.5 mL microcentrifuge tube containing 15 μL of Pierce™ Protein A Agarose or Pierce™ Protein G Agarose. Incubate with rotation for 2 h at 4 °C. Briefly centrifuge beads at 200 × *g* and retain supernatant as cleared microsomes.

6.7.5. Transfer the cleared microsomes to the 1.5 mL microcentrifuge tube containing blocked Pierce™ Anti-HA Agarose from step 6.7.3. Incubate samples with rotation for 4 h at 4 °C. *Alternatively, samples can be incubated overnight at 4 °C with rotation.*

6.7.6. Briefly centrifuge beads at 200 × *g* and discard supernatant.

6.7.7. Wash by resuspending beads in 1 mL of Solubilization Buffer. Incubate samples for 10 min at 27 °C with rotation. Briefly centrifuge beads at 200 × *g* and discard supernatant.

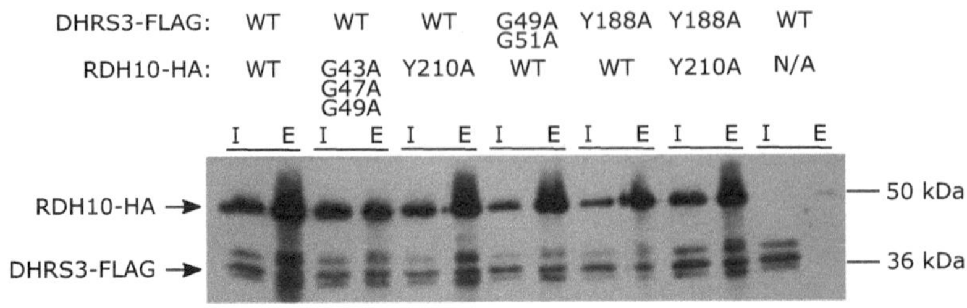

Fig. 2 Western blot of affinity purified proteins. Protein was affinity purified from Sf9 microsomes containing wild-type or mutant forms of DHRS3-FLAG, RDH10-HA, or both proteins. Protein was eluted by boiling beads with immobilized protein in 30 μL SDS-PAGE Loading Buffer. Protein was separated on a 12% polyacrylamide gel. Sf9 microsomes containing recombinant protein were loaded as inputs. To detect recombinant RDH10 and DHRS3, protein was transferred to a PVDF membrane and the membrane was probed with a monoclonal ANTI-FLAG® M2 antibody and an anti-RDH10 antibody. Both antibodies were diluted 1:3000 in TBST containing 1% BSA. The cofactor binding mutants, RDH10 G43A/G47A/G49A-HA and DHRS3 G49A/G51A-FLAG, retain the capacity to form complexes with wild-type protein partners. Similarly, active site mutants, RDH10 Y210A-HA and DHRS3 Y188A-FLAG, retain the capacity to form complexes with wild-type protein partners. Thus, catalytically active proteins are not necessary for complex formation. I, input; E, eluate.

6.7.8. Repeat steps 6.7.6 and 6.7.7 three additional times for a total of four washes.

6.7.9. Elute protein by boiling beads for 20 min in 30 μL of an SDS-PAGE loading buffer supplemented with a reducing agent, such as DTT.

6.7.10. The analysis of complex isolation is achieved via SDS-PAGE (Fig. 2).

7. Summary

As the rate-limiting step of RA synthesis, the conversion of all-*trans*-retinol to all-*trans*-retinaldehyde is a target of mechanisms that regulate RA synthesis. ROC, consisting of the retinol dehydrogenase RDH10 and the retinaldehyde reductase DHRS3, is a critical component of RA synthesis regulation. We describe a procedure for the production and enrichment of recombinant ROC. Additionally, the enzymatic properties of the complex in microsomal fractions can be examined as described previously (Adams et al., 2014; Belyaeva et al., 2017). As ROC is an important regulator of RA synthesis, the production and isolation of recombinant ROC allows for the study of the RA synthesis regulatory mechanisms.

Acknowledgments

Research reported in this publication was supported by the National Institute of General Medical Sciences of the National Institutes of Health under Award Number F32GM122215 (M.K.A.) and the National Institute on Alcohol Abuse and Alcoholism of the National Institutes of Health under Award Number R01AA012153 (N.Y.K.). The content is solely the responsibility of the authors and does not necessarily represent the official views of the National Institutes of Health.

References

Adams, M. K., Belyaeva, O. V., Wu, L., & Kedishvili, N. Y. (2014). The retinaldehyde reductase activity of DHRS3 is reciprocally activated by retinol dehydrogenase 10 to control retinoid homeostasis. *Journal of Biological Chemistry*, *289*(21), 14868–14880. https://doi.org/10.1074/jbc.M114.552257.

Adams, M. K., Lee, S. A., Belyaeva, O. V., Wu, L., & Kedishvili, N. Y. (2017). Characterization of human short chain dehydrogenase/reductase SDR16C family members related to retinol dehydrogenase 10. *Chemico-Biological Interactions*, *276*, 88–94. https://doi.org/10.1016/j.cbi.2016.10.019.

Belyaeva, O. V., Adams, M. K., Popov, K. M., & Kedishvili, N. Y. (2020). Generation of retinaldehyde for retinoic acid biosynthesis. *Biomolecules*, *10*(1). 5, https://doi.org/10.3390/biom10010005.

Belyaeva, O. V., Adams, M. K., Wu, L., & Kedishvili, N. Y. (2017). The antagonistically bifunctional retinoid oxidoreductase complex is required for maintenance of all-*trans*-retinoic acid homeostasis. *Journal of Biological Chemistry*, *292*(14), 5884–5897. https://doi.org/10.1074/jbc.M117.776914.

Belyaeva, O. V., Chang, C., Berlett, M. C., & Kedishvili, N. Y. (2015). Evolutionary origins of retinoid active short-chain dehydrogenases/reductases of SDR16C family. *Chemico-Biological Interactions*, *234*, 135–143. https://doi.org/10.1016/j.cbi.2014.10.026.

Bhatia, C., Oerum, S., Bray, J., Kavanagh, K. L., Shafqat, N., Yue, W., et al. (2015). Towards a systematic analysis of human short-chain dehydrogenases/reductases (SDR): Ligand identification and structure–activity relationships. *Chemico-Biological Interactions*, *234*, 114–125. https://doi.org/10.1016/j.cbi.2014.12.013.

Billings, S. E., Pierzchalski, K., Butler Tjaden, N. E., Pang, X. Y., Trainor, P. A., Kane, M. A., et al. (2013). The retinaldehyde reductase DHRS3 is essential for preventing the formation of excess retinoic acid during embryonic development. *The FASEB Journal*, *27*(12), 4877–4889. https://doi.org/10.1096/fj.13-227967.

Chatzi, C., Cunningham, T. J., & Duester, G. (2013). Investigation of retinoic acid function during embryonic brain development using retinaldehyde-rescued Rdh10 knockout mice. *Developmental Dynamics*, *242*(9), 1056–1065. https://doi.org/10.1002/dvdy.23999.

Chen, Z., Kastaniotis, A. J., Miinalainen, I. J., Rajaram, V., Wierenga, R. K., & Hiltunen, J. K. (2009). 17β-hydroxysteroid dehydrogenase type 8 and carbonyl reductase type 4 assemble as a ketoacyl reductase of human mitochondrial FAS. *The FASEB Journal*, *23*(11), 3682–3691. https://doi.org/10.1096/fj.09-133587.

Cunningham, T. J., Chatzi, C., Sandell, L. L., Trainor, P. A., & Duester, G. (2011). Rdh10 mutants deficient in limb field retinoic acid signaling exhibit normal limb patterning but display interdigital webbing. *Developmental Dynamics*, *240*(5), 1142–1150. https://doi.org/10.1002/dvdy.22583.

Farjo, K. M., Moiseyev, G., Nikolaeva, O., Sandell, L. L., Trainor, P. A., & Ma, J. (2011). RDH10 is the primary enzyme responsible for the first step of embryonic vitamin A metabolism and retinoic acid synthesis. *Developmental Biology*, *357*(2), 347–355. https://doi.org/10.1016/j.ydbio.2011.07.011.

Feng, L., Hernandez, R. E., Waxman, J. S., Yelon, D., & Moens, C. B. (2010). Dhrs3a regulates retinoic acid biosynthesis through a feedback inhibition mechanism. *Developmental Biology*, *338*(1), 1–14. https://doi.org/10.1016/j.ydbio.2009.10.029.

Friedl, R. M., Raja, S., Metzler, M. A., Patel, N. D., Brittian, K. R., Jones, S. P., et al. (2019). RDH10 function is necessary for spontaneous fetal mouth movement that facilitates palate shelf elevation. *Disease Models & Mechanisms*, *12*(7). https://doi.org/10.1242/dmm.039073.

Gómez-Sebastián, S., López-Vidal, J., & Escribano, J. M. (2014). Significant productivity improvement of the baculovirus expression vector system by engineering a novel expression cassette. *PLoS One*, *9*(5), e96562, https://doi.org/10.1371/journal.pone.0096562.

Haeseleer, F., Huang, J., Lebioda, L., Saari, J. C., & Palczewski, K. (1998). Molecular characterization of a novel short-chain dehydrogenase/reductase that reduces all-*trans*-retinal. *Journal of Biological Chemistry*, *273*(34), 21790–21799. https://doi.org/10.1074/jbc.273.34.21790.

Huang, P., Chandra, V., & Rastinejad, F. (2014). Retinoic acid actions through mammalian nuclear receptors. *Chemical Reviews*, *114*(1), 233–254. https://doi.org/10.1021/cr400161b.

Jörnvall, H., Persson, B., Krook, M., Atrian, S., Gonzalez-Duarte, R., Jeffery, J., et al. (1995). Short-chain dehydrogenases/reductases (SDR). *Biochemistry*, *34*(18), 6003–6013. https://doi.org/10.1021/bi00018a001.

Kavanagh, K. L., Jörnvall, H., Persson, B., & Oppermann, U. (2008). Medium- and short-chain dehydrogenase/reductase gene and protein families: The SDR superfamily: Functional and structural diversity within a family of metabolic and regulatory enzymes. *Cellular and Molecular Life Sciences*, *65*(24), 3895–3906. https://doi.org/10.1007/s00018-008-8588-y.

Kedishvili, N. Y. (2016). Retinoic acid synthesis and degradation. In M. Asson-Batres & C. Rochette-Egly (Eds.), *Vol. 81. The biochemistry of retinoid signaling II Subcellular biochemistry*. Dordrecht: Springer. https://doi.org/10.1007/978-94-024-0945-1_5.

Kurosaka, H., Wang, Q., Sandell, L., Yamashiro, T., & Trainor, P. A. (2017). Rdh10 loss-of-function and perturbed retinoid signaling underlies the etiology of choanal atresia. *Human Molecular Genetics*, *26*(7), 1268–1279. https://doi.org/10.1093/hmg/ddx031.

Ma, Y., Belyaeva, O. V., Brown, P. M., Fujita, K., Valles, K., Karki, S., et al. (2019). 17-Beta hydroxysteroid dehydrogenase 13 is a hepatic retinol dehydrogenase associated with histological features of nonalcoholic fatty liver disease. *Hepatology*, *69*(4), 1504–1519. https://doi.org/10.1002/hep.30350.

Metzler, M. A., Raja, S., Elliott, K. H., Friedl, R. M., Tran, N. Q. H., Brugmann, S. A., et al. (2018). RDH10-mediated retinol metabolism and RARα-mediated retinoic acid signaling are required for submandibular salivary gland initiation. *Development*, *145*(15), dev164822. https://doi.org/10.1242/dev.164822.

O'Reilly, D. R., Miller, L. K., & Luckow, V. A. (1992). *Baculovirus expression vectors: A laboratory manual*. Oxford, United Kingdom: Oxford University Press.

Pandey, A., Shin, K., Patterson, R. E., Liu, X. Q., & Rainey, J. K. (2016). Current strategies for protein production and purification enabling membrane protein structural biology. *Biochemistry and Cell Biology*, *94*(6), 507–527. https://doi.org/10.1139/bcb-2015-0143.

Sandell, L. L., Lynn, M. L., Inman, K. E., McDowell, W., & Trainor, P. A. (2012). RDH10 oxidation of vitamin A is a critical control step in synthesis of retinoic acid during mouse embryogenesis. *PLoS One*, 7(2), e30698. https://doi.org/10.1371/journal.pone.0030698.

Shin, W. H., & Kihara, D. (2019). 55 Years of the Rossmann fold. In A. Kister (Ed.), *Vol. 1958. Protein supersecondary structures. Methods in molecular biology*. New York, NY: Humana Press. https://doi.org/10.1007/978-1-4939-9161-7_1.

Takahashi, Y., Moiseyev, G., Farjo, K., & Ma, J. (2009). Characterization of key residues and membrane association domains in retinol dehydrogenase 10. *Biochemical Journal*, *419*(1), 113–123. https://doi.org/10.1042/BJ20080812.

Tani, H., Limn, C. K., Yap, C. C., Onishi, M., Nozaki, M., Nishimune, Y., et al. (2003). In vitro and in vivo gene delivery by recombinant baculoviruses. *Journal of Virology*, 77(18), 9799–9808. https://doi.org/10.1128/JVI.77.18.9799-9808.2003.

Tong, M. H., Yang, Q. E., Davis, J. C., & Griswold, M. D. (2013). Retinol dehydrogenase 10 is indispensible for spermatogenesis in juvenile males. *Proceedings of the National Academy of Sciences of the United States of America*, *110*(2), 543–548. https://doi.org/10.1073/pnas.1214883110.

Venkatesan, R., Sah-Teli, S. K., Awoniyi, L. O., Jiang, G., Prus, P., Kastaniotis, A. J., et al. (2014). Insights into mitochondrial fatty acid synthesis from the structure of heterotetrameric 3-ketoacyl-ACP reductase/3R-hydroxyacyl-CoA dehydrogenase. *Nature Communications*, *5*, 4805. https://doi.org/10.1038/ncomms5805.

Wu, L., Belyaeva, O. V., Adams, M. K., Klyuyeva, A. V., Lee, S. A., Goggans, K. R., et al. (2019). Mice lacking the epidermal retinol dehydrogenases SDR16C5 and SDR16C6 display accelerated hair growth and enlarged meibomian glands. *Journal of Biological Chemistry*, *294*(45), 17060–17074. https://doi.org/10.1074/jbc.RA119.010835.

Yang, D., Vuckovic, M. G., Smullin, C. P., Kim, M., Lo, C. P. S., Devericks, E., et al. (2018). Modest decreases in endogenous all-*trans*-retinoic acid produced by a mouse *Rdh10* heterozygote provoke major abnormalities in adipogenesis and lipid metabolism. *Diabetes*, *67*(4), 662–673. https://doi.org/10.2337/db17-0946.

Zolfaghari, R., Chen, Q., & Ross, A. C. (2012). DHRS3, a retinal reductase, is differentially regulated by retinoic acid and lipopolysaccharide-induced inflammation in THP-1 cells and rat liver. *American Journal of Physiology—Gastrointestinal and Liver Physiology*, *303*(5), G578–G588. https://doi.org/10.1152/ajpgi.00234.2012.

CHAPTER FIVE

Sample preparation for structural and functional analyses of the STRA6 receptor for retinol-binding protein

Brianna K. Costabile[a], Youn-Kyung Kim[b], Yunting Chen[a], Oliver B. Clarke[a,c], Loredana Quadro[b], Filippo Mancia[a,]*

[a]Department of Physiology and Cellular Biophysics, Columbia University Irving Medical Center, New York, NY, United States
[b]Department of Food Science and Rutgers Center for Lipid Research and New Jersey Institute of Food Nutrition and Health, Rutgers University, New Brunswick, NJ, United States
[c]Department of Anesthesiology, Columbia University Irving Medical Center, New York, NY, United States
*Corresponding author: e-mail address: fm123@cumc.columbia.edu

Contents

Methods in Enzymology, Volume 637
ISSN 0076-6879
https://doi.org/10.1016/bs.mie.2020.03.005

Abstract

STRA6 (stimulated by retinoic acid 6) is a 75 kDa polytopic transmembrane protein that facilitates cellular retinol uptake from retinol-binding protein (RBP). Structural characterization of STRA6 from *Danio rerio* purified in detergent and reconstituted in amphipol A8-35 was achieved by single-particle cryo-electron microscopy (cryo-EM). This provided the first high-resolution snapshot of this protein, showing a novel topology of a tightly assembled homodimer, and an unexpected physiological association with calmodulin in addition to insights into its potential mechanism of function. Specifically, a large hydrophobic cavity in the center of STRA6 linked to the known site of interaction with RBP suggested a route of retinol entry into the cell by diffusion into the membrane through a lateral opening of the cavity directly into the bilayer. Moreover, the capability to produce pure and homogeneous protein has allowed previously unattainable functional characterization of STRA6 in a reconstituted system. Here, we describe detailed methods for *Danio rerio* STRA6 expression in insect cells, purification in detergent and reconstitution in amphipol for structural characterization by cryo-EM. Furthermore, we show reconstitution of the protein in liposomes for an *in vitro* proteoliposome-based assay of STRA6-mediated retinol uptake. Finally, we present methods and preliminary cryo-EM data for STRA6 incorporated in lipid-filled nanodiscs, a close to native milieu to study membrane protein structure and function.

1. Introduction

Retinoids, vitamin A derivatives, have diverse roles throughout the body including vision, immune response and embryonic development (Blomhoff & Blomhoff, 2006). Within the cell, each retinoid form has a specific function. Retinoic acid for example, functions as a ligand for the transcription factors retinoic acid receptors (RARs) and retinoid X receptors (RXRs) (Allenby et al., 1993; Giguere, Ong, Segui, & Evans, 1987; Mangelsdorf, Ong, Dyck, & Evans, 1990; Petkovich, Brand, Krust, & Chambon, 1987). Photoisomerization of 11-*cis*-retinaldehyde to all-trans-retinaldehyde meanwhile, is a key step in the vision cycle (Wald, 1935, 1968). During times of vitamin A deficiency, retinol, the alcohol form of vitamin A, can be mobilized from storage tissues, predominantly the liver, for delivery to peripheral tissues (Paik et al., 2004; Quadro et al., 1999). Retinol, which is intrinsically water-insoluble, is secreted from the cell bound to its sole specific transporter retinol-binding protein (RBP, also called RBP4) and circulates in the bloodstream, solubilized by this protein carrier (Blaner & Olson, 1994; Kanai, Raz, & Goodman, 1968). Once it reaches a target cell, the transmembrane (TM) receptor STRA6—the product of a previously reported gene responsive to retinoic acid (Bouillet et al., 1997), and identified as the cellular receptor for RBP in a landmark study (Kawaguchi et al., 2007)—facilitates the release of retinol from RBP and the subsequent translocation of this hydrophobic molecule across the plasma membrane into the cell through an as yet undetermined mechanism (Kawaguchi et al., 2007). STRA6 was found to be expressed to particularly high levels in blood-tissue barriers where a regulated delivery of an adequate supply of retinol is most crucial, including the brain, eye and placenta (Amengual et al., 2014; Bouillet et al., 1997; Kelly, Widjaja-Adhi, Palczewski, & von Lintig, 2016).

To begin to understand how STRA6-mediated translocation of retinol across the membrane occurs at a molecular level, we determined the structure of STRA6 from *Danio rerio* (zebrafish) reconstituted in amphipol by single-particle cryo-EM (Fig. 1) (Chen et al., 2016). We chose to work on the protein from zebrafish as, of the orthologs we screened, this was the highest expressing and most stable one in detergents compatible with

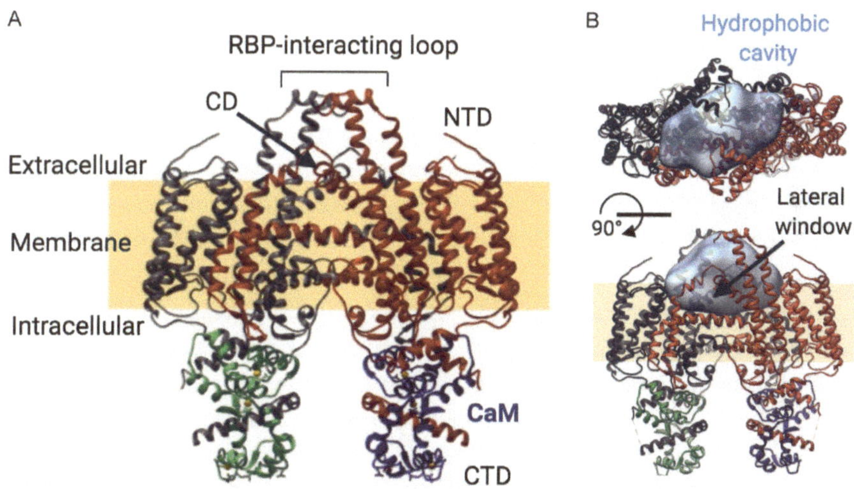

Fig. 1 *Danio rerio* STRA6 structure reconstituted in amphipol and determined by single-particle cryo-EM to 3.9Å resolution. (A) Protomers are colored in gray and red, one molecule of CaM is associated with each protomer in green and blue, respectively. The membrane is indicated by a yellow rectangle. The residues of STRA6 known to interact with RBP protrude above the membrane on the extracellular side and are labeled accordingly. The N- and C-terminal domains are labeled (NTD and CTD) and the central domain is labeled (CD). (B) A central hydrophobic cavity is indicated by the purple spacefill and its direct opening to the bilayer as defined by TM helices 8 and 9, and by the IM helix is labeled as lateral window. The volume model was created using the Voss Volume Voxelator (Voss & Gerstein, 2010).

structural studies. We learned from this structure and from related biochemical work, that (i) the protein is an intrinsic homodimer displaying a previously unobserved fold consisting of nine TM and an intramembrane (IM) helix, comprising an N-terminal penta-helical TM bundle (TM helices 1–5), a central domain along the dimer interface (TM helices 6–9 and IM helix) and a C-terminal cytoplasmic domain; (ii) that each protomer is tightly—albeit unexpectedly—associated with calmodulin (CaM) at the C-terminal domain, posing a potential link between calcium homeostasis and retinol metabolism; (iii) that the central domains of the two protomers form a striking central hydrophobic cavity spanning from ~25 Å above the putative membrane at the previously identified site for interaction with RBP (Fig. 1B) (Kawaguchi, Yu, Wiita, Honda, & Sun, 2008), to approximately mid-bilayer as delineated by the two (one for each protomer) long IM helices that lie essentially parallel to the membrane itself. We hypothesized that interaction of RBP with the loop at the apex of this central cavity, where the residues involved in RBP interaction are located (Kawaguchi et al., 2008) may trigger the release of retinol into the hydrophobic milieu of the cavity itself. Retinol could in turn diffuse into the bilayer through a triangular lateral opening of the cavity delineated by TM helices 8 and 9 on the sides and the IM helix on the bottom. Indeed, in the amphipol structure, the cavity appeared to be lipid filled likely taking lipids from the *Spodoptera frugiperda* (Sf9) cell membrane where it was expressed before purification in detergent and reconstitution in amphipol. We surmised from this result that the membrane itself might play a role in the interaction of STRA6 and retinol, necessitating a lipid bilayer in which to study the purified protein.

Here, we present methods for expression and purification of STRA6 in detergents, reconstitution of this protein in amphipol and nanodisc for structural analysis, as well as reconstitution in liposomes for assays of STRA6-mediated retinol uptake in a proteoliposome-based reconstituted system. These methods may also be useful to the study of other retinoid-interacting transmembrane proteins such as ABC transporters (Tarapcsák et al., 2017) or retinoid-linked G protein-coupled receptors like rhodopsin (Zhao et al., 2019), if not integral membrane proteins, transporters in particular, at large.

2. Purification of the STRA6 protein in detergent and reconstitution in amphipol

STRA6 from zebrafish was chosen as a structural target after screening nine orthologs for expression and monodispersion in detergents using a GFP tag at either the N- or C-terminus. We performed anti-GFP western blot analysis on HEK293 cells transiently transfected with STRA6 GFP fusion DNA constructs to compare expression and subsequent fluorescence-coupled size exclusion chromatography (FSEC) to further probe expression and also measure protein stability—assessed from the height, sharpness and symmetry of the elution peak—in detergents amenable to structural analysis (Kawate & Gouaux, 2006). Initial screening of the orthologs was performed utilizing dodecyl-β-D-maltopyranoside (DDM; Anatrace) with and without the addition of cholesteryl hemisuccinate (CHS; Anatrace) to further stabilize the protein (Weiss & Grisshammer, 2002). The results of these screening experiments are described further in our publication (Chen et al., 2016). After settling on the zebrafish STRA6 ortholog, which was by far the best expressing and most stable of those screened, we also tested lauryl maltose neopentyl glycol (LMNG; Anatrace) as a detergent for solubilization and purification with and without the addition of CHS using size exclusion chromatography (SEC). LMNG in the presence of CHS in a 10:1 ratio (w/w) LMNG:CHS produced the sharpest and most symmetric peak by SEC. We scaled up expression by switching to the baculovirus expression system, following standard protocols. We purified the protein following a two-step purification consisting of metal-affinity chromatography (taking advantage of a genetically-engineered deca-histidine tag at the C-terminus of the protein) followed by SEC. We first attempted crystallization experiments. Crystals were formed from a preparation of STRA6 in DDM/CHS, using standard vapor diffusion conditions. However they were not sufficiently ordered for structure determination by X-ray crystallography. Despite extensive rounds of optimization, the crystals never diffracted past 14 Å.

During purification of STRA6 from Sf9 cells, we noticed that it formed a hard to dissociate association with a low molecular weight protein, which was initially thought to be a recalcitrant contaminant. N-terminal sequencing and mass spectrometry of the proteolytic peptide fragments of the low molecular weight extra band excised from an SDS-PAGE denaturing gel, informed us that the supposed "contaminant" was CaM from the expression host. Through co-immunoprecipitation studies from zebrafish membranes, we later determined that this interaction was physiological (Chen et al., 2016). Dynamic light scattering experiments coupled to SEC showed unambiguously that STRA6 in solution was of approximately 180 kDa. We surmised that the protein was a dimer (75 kDa × 2 = 150 kDa) and that each monomer was associated with one molecule of CaM

(CaM 15 kDa × 2 = 30 kDa). This molecular weight put us in the realm (although on the smaller side) of what was achievable by cryo-EM in terms of atomic resolution (Merk et al., 2016). In initial attempts, the protein structure could not be solved by cryo-EM using detergent as this masked the structural features of the protein, resulting in a lack of fiducial markers with which to align particles. The protein was then reconstituted in amphipol A8-35 from LMNG/CHS and the structure was analyzed by single-particle cryo-EM. Amphipol has become a commonly used molecule for membrane protein reconstitution after detergent purification as it stabilizes membrane proteins while showing minimal background for cryo-EM experiments (Tribet, Audebert, & Popot, 1996). Reconstitution of STRA6 using this amphipathic molecule allowed us to solve the STRA6 structure to 3.9 Å resolution (Chen et al., 2016). We present here a detailed method for detergent purification of STRA6 and reconstitution in amphipol A8-35.

2.1 Materials: Purification of STRA6 protein in detergent

1. Baculovirus expression system of STRA6 in Sf9 cells.
2. Sorvall RC 30 Plus centrifuge, rotor and 800 mL volume collection buckets for large-scale cell collection.
3. 1 × Phosphate-buffered saline (PBS).
4. Beckman Coulter Optima L-90K Ultracentrifuge, 45 Ti rotor and tubes.
5. 100 mL Dounce homogenizer.
6. Low salt buffer: 10 mM Hepes pH 7.5, 10 mM KCl, 10 mM $MgCl_2$, 1 mM phenylmethylsulfonyl fluoride (PMSF), cOmplete mini EDTA-free protease inhibitor cocktail, DNase and RNase.
7. High salt buffer: Same as low salt buffer with the addition of 1 M NaCl.
8. Storage buffer: 20 mM Hepes pH 7.5, 200 mM NaCl, 1 mM PMSF and cOmplete mini EDTA-free protease inhibitor cocktail.
9. Rotator kept at 4 °C.
10. 3% Lauryl maltose neopentyl glycol/0.3% cholesteryl hemisuccinate tris salt detergent solution in water (LMNG and CHS both from: Anatrace).
11. LMNG detergent buffer: 20 mM Hepes pH 7.5, 200 mM NaCl, 1% LMNG/0.1% CHS.
12. Wash buffer: 20 mM Hepes pH 7.5, 200 mM NaCl, 60 mM imidazole, 0.1% LMNG/0.01% CHS.

13. Elution buffer: 20 mM Hepes pH 7.5, 200 mM NaCl, 200 mM imidazole, 0.05% LMNG/0.005% CHS.
14. Ni-NTA agarose (Qiagen).
15. Poly-Prep chromatography column (Bio-Rad).
16. Amicon 100 kDa centrifugal concentrator (Millipore).
17. Ultra-free-MC centrifugal micro-filter, GV 0.22 μm, 0.5 mL volume (Millipore).
18. Tabletop centrifuge (Eppendorf 5415R).
19. AKTA system—Superdex 200 Increase 10/300 GL column.
20. AKTA buffer: 20 mM Hepes pH 7, 150 mM NaCl.

2.2 Protocol: Purification of STRA6 protein in detergent

1. Produce a recombinant baculovirus that expresses zebrafish STRA6 with a C-terminal 10 × Histidine tag (Chen et al., 2016; Goehring et al., 2014). Amplify the virus to P4 following the procedures detailed in Goehring et al. (2014).
2. Use P4 of the baculovirus to scale up expression of STRA6 in Sf9 cells. Virus potency is not always consistent; a titration to determine the optimal infection concentration in small scale is recommended before scaling up expression.
3. Grow cells for 72 h in shaker flasks at 27 °C.
4. Harvest cells:
   a. Centrifuge the cultures at 2400 RPM for 17 min in the Sorvall centrifuge.
   b. Pour off the supernatant and gently wash the centrifuged cell pellet in ice-chilled 1 × PBS.
   c. Transfer each resuspended cell pellet to a 50 mL Falcon tube and centrifuge for 17 min at 2400 RPM.
   d. Pour off the supernatant and store the pellets at −80 °C until use.
5. Isolate the membranes of the collected cells:
   a. Thaw one frozen cell pellet, suspend the pellet in low salt buffer—use 40 mL buffer per cell pellet from one 800 mL SF9 culture. Pellets can range in weight from 12 to 20 g depending on the cell growth.
   b. Lyse the pellet by homogenization in a Dounce homogenizer.
   c. Centrifuge the lysate at 34,000 RPM for 30 min.
   d. Pour off the supernatant and resuspend the pellet in 40 mL high salt buffer.

e. Homogenize the pellet in a 100 mL glass homogenizer.
f. Centrifuge the lysate at 34,000 RPM for 30 min.
g. Repeat steps d–f—the supernatant should be clear after the second high salt wash.
h. Resuspend the final pellet in ice-chilled storage buffer. Store the pellet at −80 °C until ready to use.

6. To solubilize the prepared membranes, add 4 mL LMNG/CHS detergent buffer per 1 g of original cells so that the final membrane protein concentration is around 2 mg/mL.
7. Rotate 1.5 h at 4 °C.
8. Centrifuge 34,000 RPM for 30 min.
9. Add solubilized protein to Ni-NTA resin along with 30 mM imidazole and rotate overnight at 4 °C.
10. The next day, settle the resin on ice.
11. Wash the resin with 10 column volumes of wash buffer.
12. Elute the protein from the resin using 3.5 column volumes of elution buffer.
13. Concentrate the protein using a 100 kDa cutoff centrifugal concentrator.
14. Filter the protein using an ultra-free MC centrifugal filter.
15. Perform size exclusion chromatography on the filtered protein by injecting the sample on an AKTA system with a Superdex 200 Increase 10/300 GL column to analyze the protein.
16. Run an SDS-PAGE gel of the injected protein and collected peak fractions to confirm that the molecular weight of the protein from the chromatogram peak matches that of STRA6.

2.3 Materials: Reconstitution of STRA6 in amphipol

1. Disposable PD-10 Desalting column with Sephadex G-25 resin (GE Healthcare).
2. Desalting buffer: 20 mM Hepes pH 7.5, 200 mM NaCl.
3. NanoDrop Microvolume Spectrophotometer (ThermoFisher).
4. Amphipol A8-35 (Anatrace).
5. Prepared Bio-beads SM-2 Resin (Bio-Rad).

2.4 Protocol: Reconstitution of STRA6 in amphipol

1. Purify the STRA6 protein in detergent following steps 1–15 of the 2.1 Protocol.
2. Desalt the protein to remove imidazole.

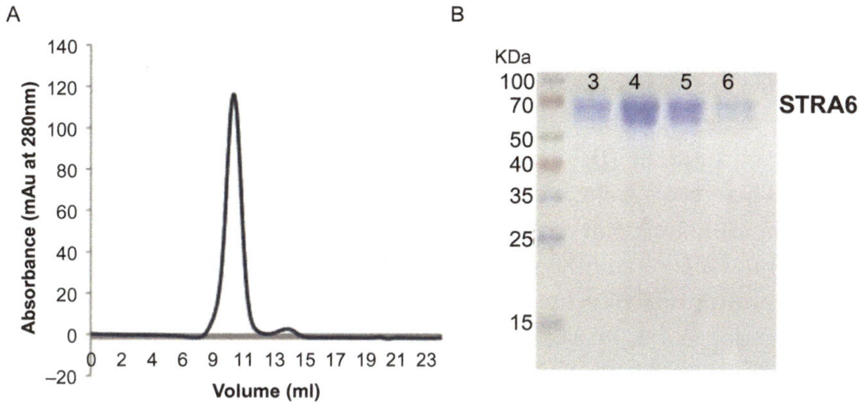

Fig. 2 *Danio rerio* STRA6 purification and reconstitution in amphipol. (A) Representative SEC profile of recombinant zebrafish STRA6 purified from Sf9 cells in LMNG/CHS and reconstituted in amphipol. (B) SDS-PAGE gel of purified STRA6. First lane is a molecular weight marker and lanes 2–5 are eluted fractions from the AKTA purification system showing the protein after a two-step purification.

3. Measure the protein concentration using a Nanodrop.
4. If necessary, concentrate the protein to 0.5 mg/mL or higher using an Amicon centrifugal concentrator with a 100 kDa cutoff.
5. Add Amphipol dissolved in 200 μL desalting buffer in a 1:3 (*w*/w) ratio of STRA6:amphipol to the protein sample.
6. Rotate at 4 °C for 4–5 h protected from light.
7. Equilibrate 0.2 g of Bio-Beads per 1 mL of sample in desalting buffer.
8. Add Bio-Beads to the sample and rotate at 4 °C overnight protected from light.
9. The next day remove the sample from the Bio-Beads.
10. Concentrate to 500 μL or less if necessary using an Amicon 100 kDa cutoff concentrator.
11. Filter the sample using an Ultra-free-MC centrifugal filter.
12. Inject the filtered sample on the AKTA system with a Superdex 200 increase 10/300 column (Fig. 2A).
13. Collect the peak fractions and run an SDS-PAGE gel to confirm correct size of STRA6 protein (Fig. 2B).

3. Reconstitution of STRA6 in nanodisc

Numerous structures of membrane proteins in detergent have been solved to date and have provided great insight into TM protein structures

(Efremov, Gatsogiannis, & Raunser, 2017). However, advances in nanodisc protein reconstitution using membrane scaffolding proteins (MSPs) or styrene maleic acid (SMA) copolymers have made it possible to study membrane proteins in soluble lipid bilayers at the single-particle level, in a close to native environment (Bayburt, Grinkova, & Sligar, 2002; Knowles et al., 2009). These tools have allowed us to study membrane proteins in a more physiologically relevant environment compared to reconstitution using amphipol, which—and the same being true for detergents—is not a particularly good mimic of a lipidic bilayer in a membrane (Bayburt & Sligar, 2003; Boldog, Li, & Hazelbauer, 2007). Setting up a system to reconstitute STRA6 in nanodisc became most relevant here given the hydrophobic nature of retinol, and our mechanistic hypothesis involving its direct interaction with the membrane bilayer. Indeed, as discussed above, we observed in the amphipol structure of STRA6 a large hydrophobic cavity at the interface between the two protomers of the dimer that appeared to be lipid-filled (presumably lipids from the SF9 cells in which the protein was originally expressed). We hypothesized that retinol may be deposited into this cavity from RBP as part of the STRA6-mediated retinol uptake process, for direct diffusion into the membrane through a lateral opening of the cavity directly into the bilayer. Although the cavity appeared to be lipid filled in the amphipol structure, we knew that in order to further study this interaction, we needed to analyze the protein in a membrane environment. We present here methods to reconstitute zebrafish STRA6 in MSP nanodiscs to study this interaction.

Efficient nanodisc reconstitution conditions are specific to the protein target and as such each new protein target must be tested in relation to a number of conditions including choice of detergent, MSP length, lipid species and ratio of the three components (protein solubilized in detergent, MSP and lipid) (Ritchie et al., 2009). In brief, nanodisc reconstitution protocols consist of initial membrane protein purification in detergent followed by addition of lipid and MSP. Nanodisc formation is then initiated by removal of the detergent solubilizing the protein by polystyrene beads (Rigaud, Levy, Mosser, & Lambert, 1998) or dialysis, resulting in embedding of the protein in a lipid bilayer formed by the added lipid and bound by the MSP (Bayburt & Sligar, 2010).

We utilized FSEC to determine the optimal reconstitution parameters and subsequent SDS-PAGE gel to confirm presence of the protein and MSP in the elution peak from SEC (Bayburt & Sligar, 2010). FSEC has been often used in screening of protein purification conditions including temperature and choice of detergent (Kawate & Gouaux, 2006). We decided to take advantage of a C-terminal GFP tagged STRA6 to screen MSP and lipid species for efficient nanodisc reconstitution conditions. To screen nanodisc conditions it may be useful to start with lipids closely related to the membrane composition of the cell type where the protein is naturally expressed. Once a lipid species and MSP length is chosen, relative ratios of protein, MSP and lipid must be tested. Starting ratios should take into account the size of the protein and weight of the lipid species. We screened two MSP lengths and two lipid species by FSEC and SDS-PAGE gel. Once the MSP and lipid parameters were determined, we used SEC and SDS-PAGE gel analysis of the peak fractions to optimize the ratio of the three components for optimal protein reconstitution. The following is a protocol specific to the incorporation of zebrafish STRA6 in POPG filled nanodiscs bound by MSP1E3D1. MSP1E3D1 is a variation of the original MSP1D1 which has three 22-mer helices added in the center of the apolipoprotein A1 sequence (Denisov, Grinkova, Lazarides, & Sligar, 2004). POPG lipid was purchased in powder form from Avanti. In our hands, we have found that the optimal nanodisc incorporation molar ratio is 1:5:300 (STRA6 in DDM/CHS:MSP1E3D1:POPG).

3.1 Materials: Reconstitution of STRA6 in nanodisc

1. Purified MSP1E3D1 protein (Bayburt et al., 2002).
2. Prepared Bio-beads SM-2 Resin (Bio-Rad).
3. POPG (1-palmitoyl-2-oleoyl-*sn*-glycero-3-[phospho-rac-(1-glycerol)]) lipid sodium salt (Avanti).
4. DDM detergent buffer: 20 mM Hepes pH 7.5, 200 mM NaCl, 1% DDM/0.1% CHS
5. Wash buffer: 20 mM Hepes pH 7.5, 200 mM NaCl, 60 mM imidazole, 0.1% DDM/0.01% CHS.
6. Elution buffer: 20 mM Hepes pH 7.5, 200 mM NaCl, 200 mM imidazole, 0.03% DDM/0.003% CHS.
7. Detergent-free wash buffer: Same as wash buffer minus the DDM/CHS.
8. Detergent-free elution buffer: Same as elution buffer minus the DDM/CHS.
9. AKTA buffer: 20 mM Hepes pH 7, 150 mM NaCl.
10. Probe sonicator (Fisher Scientific Sonic Dismembrator F60).

3.2 Protocol: Reconstitution of STRA6 in nanodisc

1. Prepare zebrafish STRA6 protein following the 2.1 protocol steps 1–15 using the modified buffers containing DDM/CHS instead of LMNG/CHS. In our experience, nanodisc incorporation is not easily accomplished from LMNG compared to DDM detergent.
2. Prepare POPG lipid by suspending powder in AKTA Buffer to a final concentration of 20 mM. Sonicate the solution using a probe sonicator until the solution is slightly opaque rather than white. Do not over-sonicate; a completely clear solution indicates that the lipid may have oxidized and is unlikely to form nanodiscs.
3. Add appropriate amount of MSP1E3D1 to the protein (1:5 M ratio STRA6:MSP1E3D1).
4. Add appropriate amount of POPG lipid to the protein and MSP1E3D1 (1:5:300 STRA6:MSP1E3D1:POPG molar ratio).
5. Rotate at 4 °C in dark for 2 h.
6. Equilibrate 0.2 g Bio-Beads per 1 mL of protein sample in AKTA bffer.
7. Add protein/MSP/lipid mix to the equilibrated Bio-Beads.
8. Rotate in dark 4 °C overnight.
9. Equilibrate 300 μL of Ni-NTA resin in AKTA buffer.
10. Carefully pipette the protein off of the Bio-Beads and apply it to the equilibrated Ni-NTA resin.
11. Rotate 4 °C overnight protected from light.
12. Wash the resin using 10 column volumes of detergent-free wash buffer.
13. Elute the protein from the Ni-NTA resin using 3.5 column volumes of detergent-free elution buffer.
14. Concentrate the protein to less than 500 μL.
15. Filter the protein using an Ultra-free-MC centrifugal filter at top speed in a tabletop centrifuge for 1 min.
16. Inject the protein on gel filtration AKTA system with a Superdex 200 Increase 10/300 column attached (Fig. 3A).
17. Collect peak protein fractions and run an SDS-PAGE gel to confirm correct size of STRA6 protein and presence of MSP1E3D1 (Fig. 3B).

4. Single-particle cryo-EM analysis of STRA6 reconstituted in nanodisc

Freezing conditions were identical to those used for structure determination of STRA6 protein reconstituted in amphipol (Chen et al., 2016).

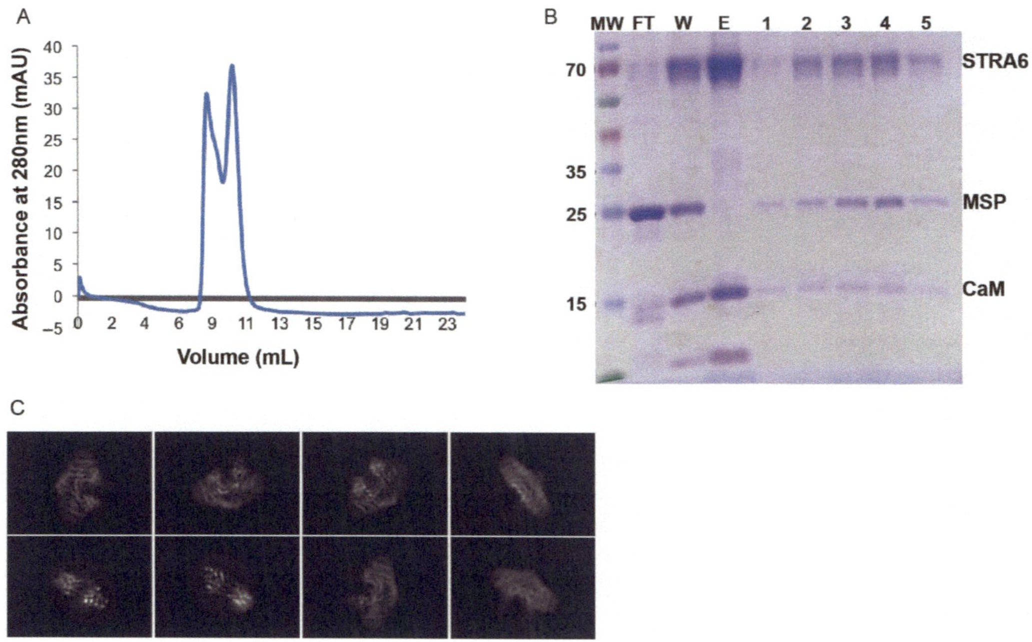

Fig. 3 *Danio rerio* STRA6 reconstitution in nanodiscs. (A) SEC profile of STRA6 purified in DDM/CHS detergent and reconstituted in nanodiscs. The first peak is aggregation and the second peak corresponds to the protein successfully reconstituted in nanodiscs. (B) SDS-PAGE gel analysis of the reconstituted protein. FT, flow-through of the nanodisc reconstitution. E, elution of the nanodisc reconstitution. D, detergent-purified STRA6 prior to nanodisc incorporation. Numbered lanes correspond to the elution fractions from the AKTA purification system. (C) 2D class averages of the STRA6 protein in nanodisc. CaM and RBP-interacting loops can be seen protruding from either side of the lipid bilayer bounded by the MSP.

We were able to obtain 2D class averages of the STRA6 protein reconstituted in nanodisc by single-particle cryo-EM (Fig. 3C). The structure of the protein in nanodisc as judged from the 2D class averages is similar overall to the protein in amphipol (Chen et al., 2016). The TM region is surrounded by the circular nanodisc; the C-terminus of each protomer bound to CaM can be seen protruding from one side of the nanodisc while the RBP binding loop protrudes from the opposite side of the nanodisc lipid bilayer. This resemblance supports the relevance of our amphipol-reconstituted structure, at least to first approximation. Further 3D and high-resolution analysis of the structure of STRA6 in nanodisc is needed to determine whether reconstitution in a lipid bilayer results in subtle changes in the protein conformation compared to in amphipol. Indeed, a number of proteins including bacteriorhodopsin (Etzkorn et al., 2013) and the TRPV1 channel (Gao, Cao, Julius, & Cheng, 2016) showed similar overall structural features in amphipol *versus* nanodisc with slight differences that did not affect the overall protein conformation.

5. Liposome-based assay of STRA6-mediated retinol uptake

To study STRA6 function by probing structure-based mechanistic hypotheses in a tightly-controlled reconstituted system as well as to confirm the functionality of recombinant, detergent-purified STRA6, we developed an assay of liposome-based STRA6 retinol uptake. A number of studies have utilized crude membranes or membrane vesicles from cells expressing STRA6 (or unknown RBP receptor at the time) to demonstrate retinol incorporation into a lipid bilayer (Kawaguchi et al., 2007; Sivaprasadarao & Findlay, 1988a, 1988b). Retinoid incorporated liposomes have also been characterized and used as a retinoid delivery tool for therapeutic purposes (Mehta, Estey, & Lopez-Berestein, 1998; Pedersen et al., 2010; Sinico et al., 2005). Utilizing STRA6 reconstituted in brain polar lipid liposomes (Anderson, Frase, Michaelis, & Hrycyna, 2005), we demonstrated STRA6-dependent retinol uptake from purified RBP refolded with retinol (prepared as described in Chen et al., 2016). Previous studies have demonstrated that RBP can transfer retinol to liposomes composed of egg phosphatidylcholine and that retinol can be translocated between liposomes in the absence of a retinol receptor or other exchange protein (Fex & Johannesson, 1987, 1988; Rando & Bangerter, 1982). Our data show some incorporation of retinol directly into liposomes from RBP, but uptake is highly increased in liposomes with reconstituted STRA6. Our studies were made possible by the availability of purified, structure-determination quality, homogeneous protein.

The liposome-based assay of retinol uptake described here essentially recapitulates the assays using membrane vesicles, but allows us to study function in a controlled system using purified protein, absent endogenous cellular components. Furthermore, detergent purification has often been shown to negatively impact membrane protein function. Given that our solved structures of the STRA6 protein in amphipol requires an initial detergent purification step, we wanted to develop a method of assaying retinol uptake of the protein after this potentially detrimental step (Seddon, Curnow, & Booth, 2004). In brief, STRA6 protein is purified in detergent as described above and reconstituted in liposomes formed from porcine brain polar lipid extract (Avanti) (Anderson et al., 2005) in an 80:1 ratio of lipid to STRA6 (w/w). Purified RBP refolded with retinol (Chen et al., 2016) is incubated with the STRA6-reconstituted liposomes in a 1:10 M ratio of STRA6:RBP for 1 h at room temperature. The liposomes are collected by centrifugation and washed through a series of centrifugation and buffer resuspension steps in order to remove background RBP (Julkowska, Rankenberg, & Testerink, 2013). The retinol content of the liposomes is then measured by reverse phase high-performance liquid chromatography (HPLC) (Kim & Quadro, 2010; Perduca, Nicolis, Mannucci, Galliano, & Monaco, 2018) (Fig. 4). This assay has allowed us to confirm the

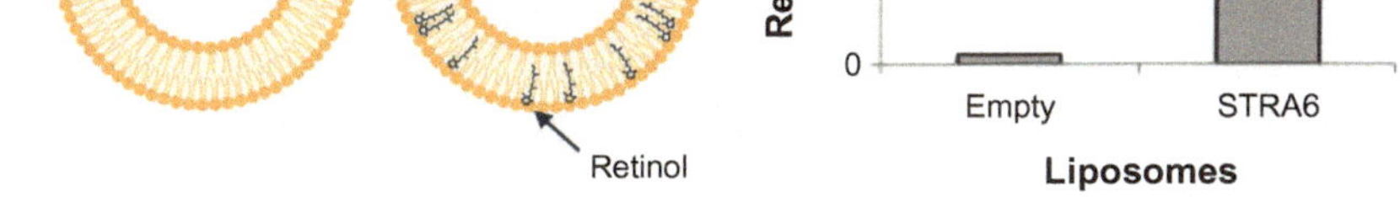

Fig. 4 Proteoliposome-based uptake assay. (A) Schematic of the proteoliposome-based assay of STRA6-mediated retinol uptake. On the left, empty liposomes treated the same as the STRA6-incorporated liposomes on the right. (B) STRA6-incorporated proteoliposomes showed increased retinol uptake from RBP compared to empty liposomes, demonstrating that the assay works, as well as functionality of the detergent-purified protein reconstituted in an artificial bilayer.

functionality of our detergent-purified STRA6 protein and provides us a tool to study factors that may influence STRA6-mediated retinol uptake in a controlled system.

5.1 Materials: Preparation of brain polar lipid liposomes

1. Porcine brain polar lipid extract (Avanti).
2. Chloroform.
3. 50 mL glass beaker.
4. Desiccator.
5. Argon gas.
6. Liposome formation buffer: 20 mM Hepes pH 7.5, 150 mM NaCl, 2 mM β-mercaptoethanol, 1.5% octylglucoside.
7. Slide-a-lyzer dialysis cassette.
8. Dialysis buffer: 20 mM Hepes pH 7.5, 150 mM NaCl, 2 mM β-mercaptoethanol.
9. 1 L glass beaker.
10. Magnetic stirrer and stir bar.
11. Liquid nitrogen for rapid freezing.
12. Avestin LiposoFast basic extruder and 100 nm filters.

5.2 Protocol: Preparation of brain polar lipid liposomes

1. Weigh out 100 mg of brain polar lipid extract into a 50 mL glass beaker.
2. Suspend the lipid extract in 1 mL chloroform, gently swirl until completely dissolved.
3. Dry the lipids under a steady stream of argon gas until the chloroform has completely dried and an even layer of dried lipid is formed on the bottom of the beaker.
4. Dry the lipids overnight by placing the beaker in the desiccator attached to a vacuum.
5. The next day, resuspend the dried lipids in 5 mL of liposome formation buffer by pipetting and swirling the beaker (Mimms, Zampighi, Nozaki, Tanford, & Reynolds, 1981; Woodle & Papahadjopoulos, 1989).
6. Using a needle and syringe, inject the resuspended lipids into a 3–12 mL Slide-a-lyzer dialysis cassette.
7. Dialyze the lipids overnight in a 1 L beaker of dialysis buffer stirring overnight (Alpes et al., 1986).
8. The next day remove the lipids from the dialysis cassette, using needle and syringe and aliquot 500 μL each into 10 1.5 mL tubes.
9. Flash freeze the lipids in liquid nitrogen then thaw at RT.

10. Extrude the liposomes using the Avestin LiposoFast basic extruder with a 100 nm filter, following the manual (Elorza, Elorza, Sainz, & Chantres, 1993; MacDonald et al., 1991).

5.3 Materials: Reconstitution of STRA6 in liposomes and liposome uptake assay

1. Porcine brain polar lipid liposomes prepared as described.
2. LMNG/CHS-purified STRA6 as described above.
3. Triton-X 100.
4. Protein incorporation buffer: 20 mM Hepes pH 7 and 150 mM NaCl.
5. Prepared Bio-beads SM-2 Resin (Bio-Rad).
6. Biofuge Sorvall Centrifuge (Thermo Fisher).
7. Purified zebrafish RBP (Chen et al., 2016).

5.4 Protocol: Reconstitution of STRA6 in liposomes

1. Purify STRA6 in LMNG/CHS detergent as described above.
2. Pipette 160 μL of 20 mg/mL prepared brain polar lipid liposomes into a tube.
3. Add 10% Triton-X 100 so that the final volume is 0.11%, vortex briefly to mix.
4. Add 40 μg of STRA6 and however much extra buffer is needed so that the final volume of the reaction is 320 μL and therefore the liposomes are diluted to a final concentration of 10 mg/mL.
5. Rotate in dark at RT 15 min.
6. Add 60 mg semi wet bio-beads per 1 mL reaction (in this case 19.2 mg) to remove the detergent (Lévy, Bluzat, Seigneuret, & Rigaud, 1990).
7. Rotate another 1 h RT.
8. Add another 19.2 mg semi wet bio-beads.
9. Rotate 1 h RT.
10. Add double the amount of bio-beads (38.4 mg) and rotate overnight at 4 °C.
11. The next day carefully pipette the liposomes away from the bio-beads and into a new tube.
12. Centrifuge the liposomes at 22,000 RPM for 20 min at 4 °C.
13. Resuspend the pellet in buffer

5.4.1 Protocol: Liposome-based assay of retinol uptake

1. Add 100 μL of prepared STRA6 liposomes to an eppendorf tube.
2. Add a 10-fold molar excess of purified holo-hRBP (Chen et al., 2016).
3. Incubate rotating in the dark 1 h RT.

4. Collect the liposomes by hard spin at 4 °C 22,000 RPM in a Biofuge Stratos Centrifuge (44,910 × *g*) for 20 min.
5. Wash the liposomes by resuspending them in buffer.
6. Repeat steps 4 and 5 a second time, then transfer the resuspended liposomes to a new eppendorf tube.
7. Spin the liposomes a final time at 22,000 RPM.
8. Remove the supernatant and store the pellet at −80 °C until HPLC analysis.

5.5 Materials: HPLC analysis of the liposome retinol content

1. Retinoid standards (retinol and retinyl acetate, Sigma) prepared in ethanol (ACS grade).
2. Phosphate-buffered saline (PBS).
3. Probe sonicator (Branson).
4. Extraction solvent: Hexane.
5. Nitrogen gas.
6. HPLC system (Ultimate Thermo Fisher).
7. HPLC column, Denali reverse C18, 250 mm × 4.6 mm, 5 μM (Grace) and HPLC guard column (PerkinElmer).
8. HPLC mobile phase: Acetonitrile 70%: Methylenechloride 15%: Methanol 15% (HPLC grade solvents).

5.6 Protocol: HPLC analysis of the liposome retinol content

1. Add 200 μL of PBS to the liposome pellet.
2. Sonicate and vortex the pellet until no clumps are visible.
3. Transfer the pellet suspension into a borosilicate glass tube.
4. Add 20 μL of retinyl acetate (1 ng/μL) as an internal standard and 180 μL of ethanol—bringing the volume to 200 μL, equivalent to the PBS and vortex well.
5. Add 3 mL of hexane and vortex (30 s × 2) to extract retinol from the liposome.
6. Collect the supernatant containing retinol by centrifugation at 3000 RPM in a tabletop centrifuge for 3 min.
7. Wash the extract by adding 0.4 mL of HPLC grade water and vortex.
8. Repeat step 6, then transfer the supernatant to a new borosilicate glass tube.
9. Completely dry the extract under stream of nitrogen gas.
10. Resuspend the extract in 50 μL HPLC mobile phase and vortex.

11. Inject 20 μL into the HPLC system with the mobile phase flowing at 1.8 mL/min.
12. Analyze the retinol content by comparing retention times and the peak spectra with those of the authentic standards (Kim & Quadro, 2010).

6. Summary

Membrane protein reconstitution tools such as nanodiscs and liposomes have made it possible to study STRA6 structure and function in a purified system while maintaining a close to native lipid bilayer environment. This chapter details our developed methods of STRA6 purification and reconstitution in amphipol and nanodisc for structural studies of the protein. The similarity of the nanodisc to the original amphipol structure of STRA6 as could be judged by 2D class averages, supports the significance of the nanodisc reconstitution technique and suggests the relevance of artificial environments such as those created by amphipols to a physiological setting. We also describe a liposome-based assay of retinol uptake in which STRA6 is reconstituted in liposomes composed of brain polar lipid and treated with purified RBP to assess retinol uptake. This liposome assay has provided us a cell-like environment in which we can specifically study the factors governing STRA6 function. Liposomes in functional study have indeed been invaluable and their use in cryo-EM protein structural studies is progressing rapidly, adding a potentially exciting component to the field (Wang, 2018). With these tools at our disposal, it is now possible to begin answering the longstanding questions as to how STRA6 directly interacts with retinol in a membrane environment, and what role CaM might have in this process.

Acknowledgments

We are grateful to members of the Mancia Lab, in particular Jonathan Kim, for helpful discussions and Khuram Ashraf for help with the figure design. This work was supported by NIH-NEI grant R01 EY027405 (to F.M.) and NIH-NEI grant F31 EY029591 (to B.K.C.).

References

Allenby, G., Bocquel, M. T., Saunders, M., Kazmer, S., Speck, J., Rosenberger, M., et al. (1993). Retinoic acid receptors and retinoid X receptors: Interactions with endogenous retinoic acids. *Proceedings of the National Academy of Sciences of the United States of America*, *90*(1), 30–34. https://doi.org/10.1073/pnas.90.1.30.

Alpes, H., Allmann, K., Plattner, H., Reichert, J., Rick, R., & Schulz, S. (1986). Formation of large unilamellar vesicles using alkyl maltoside detergents. *Biochimica et Biophysica Acta-Biomembranes*, *862*(2), 294–302. https://doi.org/10.1016/0005-2736(86)90231-2.

Amengual, J., Zhang, N., Kemerer, M., Maeda, T., Palczewski, K., & Von Lintig, J. (2014). STRA6 is critical for cellular vitamin A uptake and homeostasis. *Human Molecular Genetics*, *23*(20), 5402–5417. https://doi.org/10.1093/hmg/ddu258.

Anderson, J. L., Frase, H., Michaelis, S., & Hrycyna, C. A. (2005). Purification, functional reconstitution, and characterization of the Saccharomyces cerevisiae isoprenylcysteine carboxylmethyltransferase Ste14p. *The Journal of Biological Chemistry*, *280*(8), 7336–7345. https://doi.org/10.1074/jbc.M410292200.

Bayburt, T. H., Grinkova, Y. V., & Sligar, S. G. (2002). Self-assembly of discoidal phospholipid bilayer nanoparticles with membrane scaffold proteins. *Nano Letters*, *2*(8), 853–856. https://doi.org/10.1021/nl025623k.

Bayburt, T. H., & Sligar, S. G. (2003). Self-assembly of single integral membrane proteins into soluble nanoscale phospholipid bilayers. *Protein Science*, *12*(11), 2476–2481. https://doi.org/10.1110/ps.03267503.

Bayburt, T. H., & Sligar, S. G. (2010). Membrane protein assembly into nanodiscs. *FEBS Letters*, *584*(9), 1721–1727. https://doi.org/10.1016/j.febslet.2009.10.024.

Blaner, W., & Olson, J. (1994). *The retinoids: Biology, chemistry, and medicine*. New York: Raven Press, pp. 229–255.

Blomhoff, R., & Blomhoff, H. K. (2006). Overview of retinoid metabolism and function. *Journal of Neurobiology*, *66*(7), 606–630. https://doi.org/10.1002/neu.20242.

Boldog, T., Li, M., & Hazelbauer, G. L. (2007). [14]—Using Nanodiscs to create water-soluble transmembrane chemoreceptors inserted in lipid bilayers. In *Vol. 423,* M. I. Simon, B. R. Crane, & A. Crane (Eds.), *Methods in enzymology* (pp. 317–335). Academic Press.

Bouillet, P., Sapin, V., Chazaud, C., Messaddeq, N., Decimo, D., Dolle, P., et al. (1997). Developmental expression pattern of Stra6, a retinoic acid-responsive gene encoding a new type of membrane protein. *Mechanisms of Development*, *63*(2), 173–186. https://doi.org/10.1016/s0925-4773(97)00039-7.

Chen, Y., Clarke, O. B., Kim, J., Stowe, S., Kim, Y. K., Assur, Z., et al. (2016). Structure of the STRA6 receptor for retinol uptake. *Science*, *353*(6302). https://doi.org/10.1126/science.aad8266.

Denisov, I. G., Grinkova, Y. V., Lazarides, A. A., & Sligar, S. G. (2004). Directed self-assembly of monodisperse phospholipid bilayer nanodiscs with controlled size. *Journal of the American Chemical Society*, *126*(11), 3477–3487. https://doi.org/10.1021/ja0393574.

Efremov, R. G., Gatsogiannis, C., & Raunser, S. (2017). lipid nanodiscs as a tool for high-resolution structure determination of membrane proteins by single-particle cryo-EM. In *Vol. 594,* C. Ziegler (Ed.), *Methods in enzymology* (pp. 1–30). Academic Press. (chapter one).

Elorza, B., Elorza, M. A., Sainz, M. C., & Chantres, J. R. (1993). Comparison of particle size and encapsulation parameters of three liposomal preparations. *Journal of Microencapsulation*, *10*(2), 237–248. https://doi.org/10.3109/02652049309104390.

Etzkorn, M., Raschle, T., Hagn, F., Gelev, V., Rice, A. J., Walz, T., et al. (2013). Cell-free expressed bacteriorhodopsin in different soluble membrane mimetics: Biophysical properties and NMR accessibility. *Structure*, *21*(3), 394–401. https://doi.org/10.1016/j.str.2013.01.005.

Fex, G., & Johannesson, G. (1987). Studies of the spontaneous transfer of retinol from the retinol: Retinol-binding protein complex to unilamellar liposomes. *Biochimica et Biophysica Acta (BBA)—Biomembranes*, *901*(2), 255–264. https://doi.org/10.1016/0005-2736(87)90122-2.

Fex, G., & Johannesson, G. (1988). Retinol transfer across and between phospholipid bilayer membranes. *Biochimica et Biophysica Acta (BBA)—Biomembranes*, *944*(2), 249–255. https://doi.org/10.1016/0005-2736(88)90438-5.

Gao, Y., Cao, E., Julius, D., & Cheng, Y. (2016). TRPV1 structures in nanodiscs reveal mechanisms of ligand and lipid action. *Nature*, *534*(7607), 347–351. https://doi.org/10.1038/nature17964.

Giguere, V., Ong, E. S., Segui, P., & Evans, R. M. (1987). Identification of a receptor for the morphogen retinoic acid. *Nature*, *330*(6149), 624–629. https://doi.org/10.1038/330624a0.

Goehring, A., Lee, C.-H., Wang, K. H., Michel, J. C., Claxton, D. P., Baconguis, I., et al. (2014). Screening and large-scale expression of membrane proteins in mammalian cells for structural studies. *Nature Protocols*, *9*(11), 2574–2585. https://doi.org/10.1038/nprot.2014.173.

Julkowska, M. M., Rankenberg, J. M., & Testerink, C. (2013). Liposome-binding assays to assess specificity and affinity of phospholipid–protein interactions. In T. Munnik & I. Heilmann (Eds.), *Plant lipid signaling protocols* (pp. 261–271). Totowa, NJ: Humana Press.

Kanai, M., Raz, A., & Goodman, D. S. (1968). Retinol-binding protein: The transport protein for vitamin A in human plasma. *The Journal of Clinical Investigation*, *47*(9), 2025–2044. https://doi.org/10.1172/jci105889.

Kawaguchi, R., Yu, J., Honda, J., Hu, J., Whitelegge, J., Ping, P., et al. (2007). A membrane receptor for retinol binding protein mediates cellular uptake of vitamin A. *Science*, *315*(5813), 820–825. https://doi.org/10.1126/science.1136244.

Kawaguchi, R., Yu, J., Wiita, P., Honda, J., & Sun, H. (2008). An essential ligand-binding domain in the membrane receptor for retinol-binding protein revealed by large-scale mutagenesis and a human polymorphism. *The Journal of Biological Chemistry*, *283*(22), 15160–15168. https://doi.org/10.1074/jbc.M801060200.

Kawate, T., & Gouaux, E. (2006). Fluorescence-detection size-exclusion chromatography for precrystallization screening of integral membrane proteins. *Structure*, *14*(4), 673–681. https://doi.org/10.1016/j.str.2006.01.013.

Kelly, M., Widjaja-Adhi, M. A. K., Palczewski, G., & von Lintig, J. (2016). Transport of vitamin a across blood-tissue barriers is facilitated by STRA6. *FASEB Journal: Official Publication of the Federation of American Societies for Experimental Biology*, *30*(8), 2985–2995. https://doi.org/10.1096/fj.201600446R.

Kim, Y. K., & Quadro, L. (2010). Reverse-phase high-performance liquid chromatography (HPLC) analysis of retinol and retinyl esters in mouse serum and tissues. *Methods in Molecular Biology*, *652*, 263–275. https://doi.org/10.1007/978-1-60327-325-1_15.

Knowles, T. J., Finka, R., Smith, C., Lin, Y.-P., Dafforn, T., & Overduin, M. (2009). Membrane proteins solubilized intact in lipid containing nanoparticles bounded by styrene maleic acid copolymer. *Journal of the American Chemical Society*, *131*(22), 7484–7485. https://doi.org/10.1021/ja810046q.

Lévy, D., Bluzat, A., Seigneuret, M., & Rigaud, J.-L. (1990). A systematic study of liposome and proteoliposome reconstitution involving bio-bead-mediated triton X-100 removal. *Biochimica et Biophysica Acta (BBA)—Biomembranes*, *1025*(2), 179–190. https://doi.org/10.1016/0005-2736(90)90096-7.

MacDonald, R. C., MacDonald, R. I., Menco, B. P., Takeshita, K., Subbarao, N. K., & Hu, L. R. (1991). Small-volume extrusion apparatus for preparation of large, unilamellar vesicles. *Biochimica et Biophysica Acta*, *1061*(2), 297–303. https://doi.org/10.1016/0005-2736(91)90295-j.

Mangelsdorf, D. J., Ong, E. S., Dyck, J. A., & Evans, R. M. (1990). Nuclear receptor that identifies a novel retinoic acid response pathway. *Nature*, *345*(6272), 224–229. https://doi.org/10.1038/345224a0.

Mehta, K., Estey, E., & Lopez-Berestein, G. (1998). Liposome-mediated delivery of retinoids. In G. Gregoriadis & B. McCormack (Eds.), *Targeting of drugs 6: Strategies for stealth therapeutic systems* (pp. 87–94). Boston, MA: Springer US.

Merk, A., Bartesaghi, A., Banerjee, S., Falconieri, V., Rao, P., Davis, M. I., et al. (2016). Breaking cryo-EM resolution barriers to facilitate drug discovery. *Cell*, *165*(7), 1698–1707. https://doi.org/10.1016/j.cell.2016.05.040.

Mimms, L. T., Zampighi, G., Nozaki, Y., Tanford, C., & Reynolds, J. A. (1981). Phospholipid vesicle formation and transmembrane protein incorporation using octyl glucoside. *Biochemistry*, *20*(4), 833–840. https://doi.org/10.1021/bi00507a028.

Paik, J., Vogel, S., Quadro, L., Piantedosi, R., Gottesman, M., Lai, K., et al. (2004). Vitamin A: Overlapping delivery pathways to tissues from the circulation. *The Journal of Nutrition*, *134*(1), 276S–280S. https://doi.org/10.1093/jn/134.1.276S.

Pedersen, P. J., Adolph, S. K., Subramanian, A. K., Arouri, A., Andresen, T. L., Mouritsen, O. G., et al. (2010). Liposomal formulation of retinoids designed for enzyme triggered release. *Journal of Medicinal Chemistry*, *53*(9), 3782–3792. https://doi.org/10.1021/jm100190c.

Perduca, M., Nicolis, S., Mannucci, B., Galliano, M., & Monaco, H. L. (2018). Human plasma retinol-binding protein (RBP4) is also a fatty acid-binding protein. *Biochimica et Biophysica Acta*, *1863*(4), 458–466. https://doi.org/10.1016/j.bbalip.2018.01.010.

Petkovich, M., Brand, N. J., Krust, A., & Chambon, P. (1987). A human retinoic acid receptor which belongs to the family of nuclear receptors. *Nature*, *330*(6147), 444–450. https://doi.org/10.1038/330444a0.

Quadro, L., Blaner, W. S., Salchow, D. J., Vogel, S., Piantedosi, R., Gouras, P., et al. (1999). Impaired retinal function and vitamin A availability in mice lacking retinol-binding protein. *The EMBO Journal*, *18*(17), 4633–4644. https://doi.org/10.1093/emboj/18.17.4633.

Rando, R. R., & Bangerter, F. W. (1982). The rapid intermembraneous transfer of retinoids. *Biochemical and Biophysical Research Communications*, *104*(2), 430–436. https://doi.org/10.1016/0006-291X(82)90655-6.

Rigaud, J. L., Levy, D., Mosser, G., & Lambert, O. (1998). Detergent removal by non-polar polystyrene beads. *European Biophysics Journal*, *27*(4), 305–319. https://doi.org/10.1007/s002490050138.

Ritchie, T. K., Grinkova, Y. V., Bayburt, T. H., Denisov, I. G., Zolnerciks, J. K., Atkins, W. M., et al. (2009). reconstitution of membrane proteins in phospholipid bilayer nanodiscs. *Methods in Enzymology*, *464*, 211–231. https://doi.org/10.1016/S0076-6879(09)64011-8. (chapter 11).

Seddon, A. M., Curnow, P., & Booth, P. J. (2004). Membrane proteins, lipids and detergents: Not just a soap opera. *Biochimica et Biophysica Acta*, *1666*(1–2), 105–117. https://doi.org/10.1016/j.bbamem.2004.04.011.

Sinico, C., Manconi, M., Peppi, M., Lai, F., Valenti, D., & Fadda, A. M. (2005). Liposomes as carriers for dermal delivery of tretinoin: In vitro evaluation of drug permeation and vesicle–skin interaction. *Journal of Controlled Release*, *103*(1), 123–136. https://doi.org/10.1016/j.jconrel.2004.11.020.

Sivaprasadarao, A., & Findlay, J. B. (1988a). The interaction of retinol-binding protein with its plasma-membrane receptor. *The Biochemical Journal*, *255*(2), 561–569.

Sivaprasadarao, A., & Findlay, J. B. (1988b). The mechanism of uptake of retinol by plasma-membrane vesicles. *The Biochemical Journal*, *255*(2), 571–579.

Tarapcsák, S., Szalóki, G., Telbisz, Á., Gyöngy, Z., Matúz, K., Csősz, É., et al. (2017). Interactions of retinoids with the ABC transporters P-glycoprotein and breast cancer resistance protein. *Scientific Reports*, 7(1), 41376. https://doi.org/10.1038/srep41376.

Tribet, C., Audebert, R., & Popot, J. L. (1996). Amphipols: Polymers that keep membrane proteins soluble in aqueous solutions. *Proceedings of the National Academy of*

Sciences of the United States of America, 93(26), 15047–15050. https://doi.org/10.1073/pnas.93.26.15047.
Voss, N. R., & Gerstein, M. (2010). 3V: Cavity, channel and cleft volume calculator and extractor. *Nucleic Acids Research, 38*, W555–W562. https://doi.org/10.1093/nar/gkq395. web server issue.
Wald, G. (1935). Carotenoids and the visual cycle. *The Journal of General Physiology, 19*(2), 351–371. https://doi.org/10.1085/jgp.19.2.351.
Wald, G. (1968). The molecular basis of visual excitation. *Nature, 219*(5156), 800.
Wang, L. (2018). Random spherically constrained single-particle (RSC) method to study voltage-gated ion channels. *Methods in Molecular Biology, 1684*, 265–277. https://doi.org/10.1007/978-1-4939-7362-0_20.
Weiss, H. M., & Grisshammer, R. (2002). Purification and characterization of the human adenosine A(2a) receptor functionally expressed in Escherichia coli. *European Journal of Biochemistry, 269*(1), 82–92. https://doi.org/10.1046/j.0014-2956.2002.02618.x.
Woodle, M. C., & Papahadjopoulos, D. (1989). [9] liposome preparation and size characterization. In *Vol. 171, Methods in enzymology* (pp. 193–217). Academic Press.
Zhao, D. Y., Poge, M., Morizumi, T., Gulati, S., Van Eps, N., Zhang, J., et al. (2019). Cryo-EM structure of the native rhodopsin dimer in nanodiscs. *The Journal of Biological Chemistry, 294*(39), 14215–14230. https://doi.org/10.1074/jbc.RA119.010089.

CHAPTER SIX

Use of fixatives for immunohistochemistry and their application for detection of retinoic acid synthesizing enzymes in the central nervous system

Azita Kouchmeshky, Peter McCaffery*
Institute of Medical Sciences, University of Aberdeen, Aberdeen, Scotland, United Kingdom
*Corresponding author: e-mail address: p.j.mccaffery@abdn.ac.uk

Contents

Methods in Enzymology, Volume 637
ISSN 0076-6879
https://doi.org/10.1016/bs.mie.2020.03.010

Abstract

Retinoic acid (RA) is a lipid signaling molecule that has a crucial role in growth and survival of neurons as well as regulation of neuronal plasticity in the central nervous system. Complete understanding of the distribution of RA is necessary to identify foci of RA signaling. However, RA itself is very difficult to detect by immunohistochemistry as there are few effective antibodies to this lipid and it can be difficult to fix in place in tissue. A set of retinaldehyde dehydrogenases (RALDHs) catalyze the last step of RA synthesis and their distribution can be used as a proxy for RA distribution. This protocol uses the example of RALDH2 expression in the motor neurons of the spinal cord as a demonstration of the approach and describes methods that can improve its effectiveness. Immunodetection is impacted by protein cross linking and protein denaturation as well as antigen/epitope masking by various fixatives. Finding a suitable fixative that preserves morphology and tissue structure by linking cellular component such as proteins and lipids in an indissoluble macromolecular network while keeping functional groups, including antigens, intact is essential. Here, we discuss fixatives in general and also describe a fixation protocol that allows detection of multiple antigens in the same section with a periodate-lysine-paraformaldehyde (PLP) fixative. This keeps tissue structure and antigen well preserved in the adult spinal cord to show RALDH2 expression in motor neurons.

1. Introduction

Immunohistochemistry (IHC) uses antibodies to determine the tissue distribution of a specific target protein or other antigen in tissue sections. The location that the antibody binds to is visualized by microscopy using a marker such as an enzyme or a fluorescent dye, which is directly conjugated to the primary antibody or linked to an appropriate secondary antibody. Tissue preparation is a critical step to maintain morphology of the cell/tissue and retain the antigenicity of the epitope(s) of interest. These are both improved by rapid collection and preservation of the tissue;

perfusion is a common technique that quickly fixes tissue and flushes out blood-derived antigens. After perfusion the tissue is post-fixed by immersion fixation, placing the tissue in fixative to diffuse into the tissue. This method is very much influenced by tissue thickness, the thinner the tissue the better; while if thick there will be a gradient in fixation quality from surface to center. If the tissue is very thick the inner part will undergo various changes such as autolysis, cellular membrane damage and microbial degradation (McFadden et al., 2019). Several other factors including temperature, time, penetration rate, fixative to tissue volume ratio, pH, concentration, and osmolarity may affect the effectiveness and rate of the fixation process (Singh et al., 2019). Room temperature works well for many fixatives to preserve tissue morphology. Higher temperatures may increase the rate of fixation but will increase the tissue autolysis rate. On the other hand, lower temperatures will reduce autolysis but also reduce the fixation rate (Knoblaugh, Randolph-Habecker, & Rath, 2012). Preparing fixatives in phosphate buffered saline (PBS) not only maintains the pH but also reduces the problem of incorrect osmolarity (Knoblaugh et al., 2012). Usually fixatives are buffered, maintaining the fixative pH between 7.2 and 7.4 which, for instance, reduces the formation of fixative pigments (Knoblaugh et al., 2012). Choice of time of fixation is dependent on the fixative type, e.g., over-fixation with a coagulating fixative and mercury salts results in tissue break-down, while over-fixation by formalin masks antigens of interest and under-fixation by formalin can lead to tissue deterioration (Knoblaugh et al., 2012). Specimens need an adequate amount of fixative for suitable fixation; during the fixation process, binding of the fixative to the tissue leads to reduction of available fixative. Therefore, volume of the fixation solution must be at least 15–20 times greater than specimen volume (Knoblaugh et al., 2012).

Various fixation solutions have been used for preserving intact cellular structure and components (including membrane bound and soluble cellular constituents). Appropriate fixatives will inhibit cell degeneration that starts once the blood supply is cut and also prevent autolysis of tissue due to intracellular enzyme digestion after membrane rupture and bacterial decomposition. Fixation can also protect cellular components from damage by chemicals used in the IHC procedure (Ramos-Vara, 2005). Holding the free substances of the cell, such as soluble protein constituents, is an essential task of fixation. These may be bound to the cells insoluble components by precipitation, coagulation, or cross-linking (Eltoum, Fredenburgh, Myers, & Grizzle, 2001). Alternatively, the fixation of subcellular compartments can trap soluble or free-floating components. In the process of fixation tissue/cells can undergo several changes including swelling, shrinkage, hardening, and color variation (Eltoum, Fredenburgh, Myers, et al., 2001).

Fixation can be physical or chemical (Eltoum, Fredenburgh, Myers, et al., 2001) but whereas physical methods (freeze drying, heating, and microwaving) are not common procedures chemical fixative solutions are used regularly in pathology and histology (Eltoum, Fredenburgh, Myers, et al., 2001). Such fixatives can be dissolved in water, alcohol or buffers for a specific pH. Fixative solutions used in histopathology can be either coagulating (precipitating) or non-coagulating (cross-linking) (Eltoum, Fredenburgh, Myers, et al., 2001; Janardhan, Jensen, Clayton, & Herbert, 2018).

1.1 Coagulating fixatives

Coagulating fixatives break hydrogen bonds and denature and coagulate (precipitate) proteins. Key proteins providing structure to cells are lipoproteins (in the plasma membrane and other organelles) and fibrous proteins (nucleoproteins and collagen) and coagulating these proteins by fixation can preserve tissue morphology required for IHC by rendering soluble proteins into insoluble form (Eltoum, Fredenburgh, Myers, et al., 2001; Ramos-Vara, 2005) and quickly dehydrate tissue (Knoblaugh et al., 2012). The typical fixatives in this group are dehydrants such as ethanol and other alcohols or acetone (Eltoum, Fredenburgh, Myers, et al., 2001; Ramos-Vara, 2005). The hydrophobic amino acids of soluble proteins localize to the interior part of the protein to minimize their exposure to the aqueous environment, the water outside forcing these domains into closer contact to stabilize their hydrophobic bonding. In contrast the hydrophilic amino acids of soluble proteins are at the surface and form hydrogen bonds with water. Dehydrants remove and replace free water within tissue/cells and such water removal destabilizes both hydrophobic bonding and hydrophilic hydrogen bonding; when water is removed, in the absence of the repulsion force of water on hydrophobic domains they freely occupy greater space and thus destabilize the protein tertiary structure (Eltoum, Fredenburgh, Myers, et al., 2001; Ramos-Vara, 2005). In addition, the reduction in water molecules is associated with decrease and destabilization of hydrogen bonding in the hydrophilic domains. Thus, dehydrants unfold and disturb the tertiary structure of protein and this conformational change in the protein structure alters the protein solubility, rendering them insoluble (coagulated) that can mask the antigen/epitope of interest since the paratope, the antigen binding site on the antibody, recognizes and interacts with the tertiary structure of the target epitope via non-covalent bonds (Eltoum, Fredenburgh, Myers, et al., 2001; Ramos-Vara, 2005).

Organic acids (such as picric acid and trichloroacetic acid) are also coagulant fixatives with the ability to disturb electrostatic and hydrogen bonding by changing the charge of protein side chains (Eltoum, Fredenburgh, Myers, et al., 2001). They can also disturb tertiary protein structure by inserting lipophilic anions into hydrophilic regions of protein converting basic side chains to the corresponding salt and coagulating proteins. They are usually used in combination with other fixatives, e.g., acetic acid is added to other fixatives to coagulate and preserve nucleic acids (Eltoum, Fredenburgh, Myers, et al., 2001). However, neutralizing the solution that the tissue is in may re-solubilize the coagulated proteins. In addition, the low pH of acid fixatives eventually leads to hydrolysis of nucleic acids.

1.2 Cross-linking fixatives

Non-coagulant or cross-linking fixatives bind proteins and other cell components by generating inter- and intra-molecular cross-links. Aldehydes such as formaldehyde, paraformaldehyde (PFA), and glutaraldehyde are among the best-known fixatives in this group (Fig. 1). Formaldehyde is a small vapor molecule that by dissolving in water make an aqueous solution (formalin) consisting of 37%–40% formaldehyde that rapidly penetrates tissues (more rapidly than, e.g., glutaraldehyde) and forms cross-links within peptides, proteins, nucleic acids as well as between proteins and nucleic acids (Singh et al., 2019). However, although moving quickly into the tissue the rate of cross-linking is slow (Fox, Johnson, Whiting, & Roller, 1985). Tissues can undergo autolysis during this slow fixation and chilling both fixative and tissue during the fixation process can slow autolysis down (Fox et al., 1985). Dissolving formaldehyde in water results in the formation of methylene glycol and the equilibrium between these two is in favor of glycol. During fixation the reduction of formaldehyde in the solution as it forms

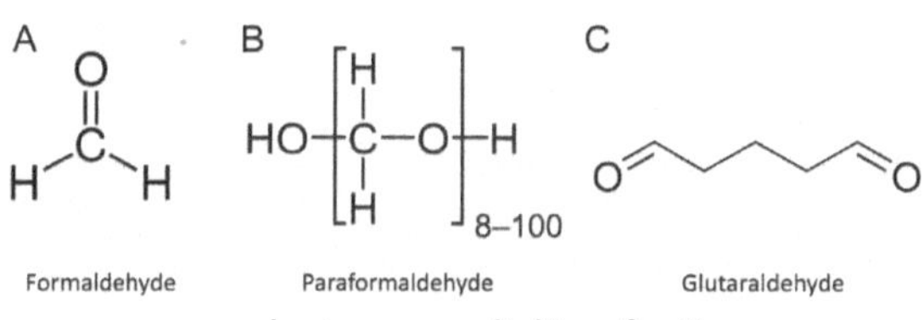

Fig. 1 Aldehydes as non-coagulant or cross-linking fixatives.

cross-links in the tissue leads to formation of more formaldehyde from methylene glycol dissociation.

Formaldehyde in aqueous solution is in a hydrated, carbonium ion form ($+CH_2OH$, Fig. 2), which is a reactive electrophile and reacts with the electron-rich center of several protein side chains including uncharged reactive amino groups ($=NH$ or NH_2), peptide bonds, hydroxyl groups, sulfhydryl group, and phenyl groups, and produces reactive hydroxymethyl compounds. The latter product in turns forms cross-link with a second reactive hydrogen on the protein to generate a methylene bridge or reacts with hydrogen groups (Jones, 1972; Ramos-Vara, 2005) (Fig. 2A). Arginine, asparagine, cysteine, glutamine, histidine, lysine, and tyrosine are amino acids with the highest affinity for formaldehyde (Eltoum, Fredenburgh, Myers, et al., 2001; Ramos-Vara, 2005). The production of reactive carbonium ions increases at lower pH, therefore buffering formaldehyde fixative can reduce the power of formaldehyde containing fixative (Jones, 1972). Formaldehyde can cross-link between nucleic acids and nuclear proteins (Fig. 2B), reacting with free amino groups in nucleotides. In addition, cross-links form at adenine-thymidine (A-T) rich regions in naked DNA. In unsaturated lipids, the $C=C$ and $-SH$ bonds are the site of reaction for formaldehyde when the solution contains calcium (minimizing phospholipid solubility by increasing phospholipid bridging), while formaldehyde does not act at all on carbohydrates (Eltoum, Fredenburgh, Myers, et al., 2001; Jones, 1972; Ramos-Vara, 2005).

A problem though with cross-linking based fixatives is that either over or under fixation results in unreliable or unreproducible IHC outcome, for instance masking the target epitopes bound by the antibody or greatly increasing non-specific binding (Ramos-Vara, 2005). There are conformational changes in macromolecules, mostly in tertiary and quaternary protein structure and less in secondary and primary structure, as a result of formaldehyde fixation (Mclean & Nakane, 1974; Ramos-Vara, 2005). This happens because of the inter- and intra-molecular cross-links that aldehyde makes. The application of cross-linking fixatives are generally progressive, time and temperature sensitive procedures (Ramos-Vara, 2005). For optimum fixation the tissue should have less than 5 mm thickness and fixation time of 24–48 h to obtain optimal formation of cross links in both periphery and core areas of the specimen (Janardhan et al., 2018; Ramos-Vara, 2005).

A way to limit cross-linking with formaldehyde based fixatives is by washing out the excess formaldehyde and the addition of glycine which will terminate the cross-link reaction by quenching the excess formaldehyde

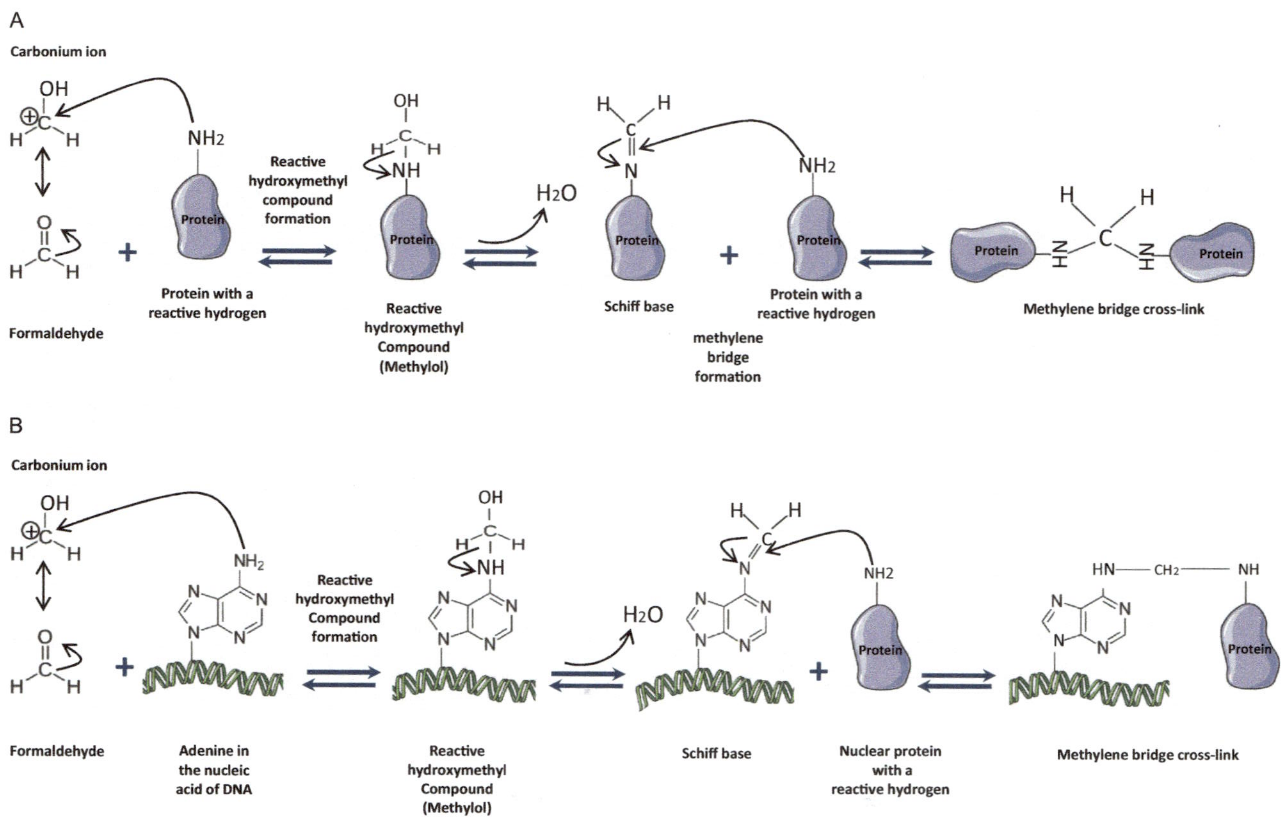

Fig. 2 Cross-linking and formation of methylene bridges by formaldehyde between (A) proteins and (B) proteins and nucleic acids.

(Hoffman, Frey, Smith, & Auble, 2015). When the fixative is then washed out the reactive hydroxymethyl groups (Fig. 2) quickly revert to their original form, but the methylene cross-link bridges already formed stay intact (Eltoum, Fredenburgh, Myers, et al., 2001). With extended washing though the cross-linking bridges can break because they have differing degrees of stability based on the protein side chain group that makes up the bridge (Eltoum, Fredenburgh, Myers, et al., 2001). Several environmental situations such as pH and temperature can also change the stability of the cross-linking bridges which is discussed in Section 1.3. Long term fixation of the tissue in formaldehyde results in oxidation of the reactive hydroxymethyl groups to a stable form (Eltoum, Fredenburgh, Myers, et al., 2001) and stable cross-links can also form but this takes up to 30 days and of course can result in loss of antigens (Eltoum, Fredenburgh, Myers, et al., 2001; Thavarajah, Mudimbaimannar, Rao, Ranganathan, & Elizabeth, 2012).

10% neutral buffered formalin (NBF) is a commonly used aldehyde-based fixative for tissues. NBF contains 10% formalin v/v, an aqueous solution of formaldehyde which is equal to 4% formaldehyde w/v (Eltoum, Fredenburgh, Myers, et al., 2001). The typical recipe for NBF starts with a 37% formaldehyde solution diluted in phosphate buffer, though some commercial NBF solutions might contain stabilizer agents such as 10% methanol (Eltoum, Fredenburgh, Myers, et al., 2001), since storage of concentrated formaldehyde results in spontaneous condensation reactions which can be prevented by addition of a small percentage of alcohol in the solution (Fox et al., 1985). Another common aldehyde containing fixative is 4% PFA. It is the same as 10% NBF based on the formaldehyde percentage, though it does not require stabilizer (Janardhan et al., 2018). Paraformaldehyde is a polymerized formaldehyde that is not itself active. When PFA depolymerizes to formaldehyde by heating and alkalinization it can then cross-link reactive sites such as amino groups, histidine and lysine (Pieri, Sassoli, Romagnoli, & Domenici, 2010; Sasse, 1986). It must be freshly prepared because it will re-polymerize over time and its formaldehyde content and fixation capacity will fall (Janardhan et al., 2018).

Glutaraldehyde is a third fixative in this group. It has two aldehyde groups and thus can act directly as a protein cross-linker. In contrast with formaldehyde it is faster to form cross-links and generates a greater number of them so is poorer for IHC than formaldehyde by increasing the inaccessibility of antigens to a greater degree. Glutaraldehyde fixation however can preserve the activity of some enzymes allowing these enzymes to be localized through this activity, such as beta-galactosidase, unlike several other aldehyde-based fixatives (Ma, Rogers, Zbar, & Schmidt, 2002). Usually glutaraldehyde does not react with lipids and nucleic acids. Like PFA glutaraldehyde can deteriorate over time in its activity, forming cyclic and oligomeric compounds in aqueous solution and can oxidize to glutaric acid. It can be kept stable though by keeping at a temperature of around 4 °C and pH 5 (Eltoum, Fredenburgh, Myers, et al., 2001). It exerts stronger osmotic effects than some other aldehyde fixatives which can alter cell/tissue ultrastructure.

A variety of chemical agents can be added to fixatives, or are used after fixation, to improve the fixation process such as zinc chloride, zinc sulfate and osmium tetroxide (Cohen, 1992; Eltoum, Fredenburgh, Myers, et al., 2001). Zinc will reduce the epitope-masking effect of PFA by directly reacting with negative groups of proteins, precipitating them, and maintaining the protein tertiary structure while osmium tetroxide is used for electron microscopy to make cell membranes insoluble, dark and electron dense by forming bridges between unsaturated fatty acids (Cohen, 1992). A list of major fixatives and their combinations can be found in Table 1.

1.3 Antigen retrieval

As mentioned, antigen/epitope masking due to over-fixation, especially with formaldehyde, is a common problem in IHC. Preserving, or restoring, antigenicity is essential in tissue used for this technique. One way to solve this problem is antigen retrieval which unmasks antigen/epitopes and restores immunoreactivity. Enzyme or protease-induced epitope retrieval (PIER), heat-induced epitope retrieval (HIER), or a combination of both methods can be used (Janardhan et al., 2018; Ramos-Vara, 2005). PIER can use enzymes such as proteinase-K, pronase, pepsin or trypsin controlling time, temperature and pH. HIER involves heating the tissue section in a buffer of specific pH, usually, citrate buffer (pH 6.0), EDTA buffer (pH 8.5), or Tris-EDTA buffer (10 mM Tris base, 1 mM EDTA solution, pH 9) at high temperature (Janardhan et al., 2018; Ramos-Vara, 2005). Overall though there is seldom one fixative or specific antigen retrieval method that fits all antibodies desired. However, a mild cross-linking fixative can provide a way to more easily avoid excessive cross-linking or the loss of many cell epitopes that results from coagulant fixatives. Such a fixative consists of a periodate-lysine-paraformaldehyde (PLP) mix, which is mild and also reduces cross-linking of protein epitopes.

Table 1 Action of popular fixatives and their combinations.

Fixative category	Example	Fixation effect: Proteins	Lipids	Carbohydrates	Nucleotide	Others	Ultrastructural effect	Components	Special consideration	Ref.
Dehydrants (Coagulating fixatives)	Ethanol Methanol Acetone	Coagulates by altering tertiary structure	None	None	Very slightly	Preserves small non-lipid molecules (e.g., glycogen) and some enzyme activity	Destroys many cytoplasmic organelles	Solution of 70%–100% or in combination with other fixative	• Acetone is not preferred as a fixative because it is volatile and flammable and affects some tissues and plastic materials • Must be used cold at 4 °C • Many dehydrants are lipid solvents, and acetone in particular rapidly makes tissue brittle	Eltoum, Fredenburgh, Myers, et al. (2001), Singh et al. (2019)
Aldehyde (cross-link fixative)	Formaldehyde Paraformaldehyde Formalin NBF	Cross-links by adding active hydroxymethyl groups to amine/ amide, sulfhydryl groups	None	Acts only slightly on polysaccharides (and not at all on soluble polysaccharides or acid mucopolysaccharides)	Cross-links	It does not cross-link carbohydrate but can cross-link glycoproteins	Ranging from good preservation (NBF) to excellent preservation (glutaraldehyde)	10% NBF (37%–40% formaldehyde) 4% PFA buffered with phosphate buffer pH 7.2–7.4	• pH (formic acid and formalin pigment formation) • PFA needs to be depolymerised by heating • Cross-links can be reversed • Is considered a histochemical fixative • Unlike alcohol, formaldehyde results in less tissue shrinkage and distortion	Eltoum, Fredenburgh, Myers, et al. (2001), Fox et al. (1985), Janardhan et al. (2018), Ramos-Vara (2005)
	Glutaraldehyde	Has two aldehyde groups directly cross-linking proteins	None (unless free amino groups are available such as some phospholipids)	Unknown	At room temperature, in the absence of nucleohistones is not able to cross-link nucleic acids; however, at >45 °C it can form cross-links	Unlike other aldehydes preserves the activity of some enzymes (beta-galactosidase)	Excellent preservation (suitable for electron microscopy)	Glutaraldehyde 2% diluted in buffer pH 7.4	• Stable at 4 °C and pH 5 • Deteriorates over time, forming cyclic and oligomeric compounds in aqueous solution and can oxidize to glutaric acid • In comparison with formaldehyde, penetrates slowly but fixes faster, generating higher numbers of cross-links but masking more antigens • Exerts some osmotic effects and high concentration leads to cell shrinkage • Makes irreversible cross-links • Good fixative for electron microscopy • High background for IHC due to non-specific binding of antibody to free aldehydes • Is considered a histochemical fixative	Eltoum, Fredenburgh, Myers, et al. (2001), McLean and Nakane (1974), Pieri et al. (2010), Ramos-Vara (2005)
	PLP	Cross-links	Cross-links	Cross-links the carbohydrate moieties of glycoproteins	Unknown	Poor fixation of glycolipids	Preserves ultrastructure	Sodium m-periodate, PFA, and lysine	• Preserves antigenicity as well as ultrastructure	Accart, Sergi, and Rooke (2014), Brenes, Harris, Paz, Petrovic, and Scheuer (1986), Gendelman, Moench, Narayan, and Griffin (1983), Liu, Brown, and Fraser (2003), McLean and Nakane (1974), Pieri et al. (2010), Prasadarao, Tobet, and Jungalwala (1990)

Continued

	Acrolein (acrylic aldehyde) 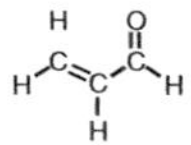	Cross-links macromolecules in a similar way to formaldehyde	Can react with fatty acids and potentially cross-links to proteins	Unknown	Unknown	Used to maintain enzyme activity and for plant fixation	Poor	4%–10% acrolein with 0.1% hydroquinone as a stabilizer buffered to pH 7.2	• Reversible cross-links • More cross-links than formaldehyde • Penetrates rapidly • High reactivity • Potent lachrymatory agent • Easily forms polymers (unstable) • Unstable at alkaline pH • Good fixative for large sized tissue	Eltoum, Fredenburgh, and Grizzle (2001), Jones (1969, 1972), Saito and Keino (1976)
Oxidative agents (Cross-linking fixatives)	Osmium tetroxide (OsO_4)	Crosslinks with amino acid side chains including disulfide, carboxyl, hydroxyl, sulfhydryl, amide groups (poor antigenicity because of extensive denaturation)	Reacts with phospholipids and the unsaturated bonds of lipids	None	Cross-links slightly (reacts with the terminal ribose group at the 2,3-glycol moiety and with thymine residues at 5,6-double bonds)	Unknown	Is used as a post fixative to improve ultrastructure, used for ultrastructural visualization of membrane and for visualization of lipids on frozen sections	1% (w/v) buffered to 7.4	• Highly toxic and volatile • Must be used under the fume hood (It can readily fix the nasal mucosa and the conjunctiva of the eyes) • Can form black insoluble precipitant on membrane due to osmium tetroxide reduction • As a heavy metal it scatters electrons making it suitable for en bloc staining for transmission electron microscopy. For light microscopy, it can be used for myelin tracts, unsaturated lipids, and for Golgi impregnation procedure • Is used to post-fix tissue with (1%–3%) glutaraldehyde or (2%–4%) PFA for electron microscopy • Limited penetration ability (<1 mm) • Slow rate of fixation • It has some osmotic effect (tissue swelling)	Eltoum, Fredenburgh, Myers, et al. (2001), Ganjali and Ganjali (2013), Pieri et al. (2010), Singh et al. (2019)
	Potassium dichromate ($K_2Cr_2O_7$)	Oxidation of proteins to cross-link carboxyl and hydroxyl side chains. Amino groups will remain available for acid dyes	Reacts with unsaturated lipids (becoming insoluble after prolonged fixation >48 h) Good for mitochondrial fixation	Unknown	None (dissolves DNA)	Is used as a co-fixative to fix endocrine tissues (adrenal medulla and tumors of the adrenal medulla)	Unknown	1%–3% (w/v)	• A cross-linking fixative but at lower pH can act as a coagulant fixative • Tissue must be washed in water after fixation to avoid formation of insoluble chromate suboxide	Eltoum, Fredenburgh, Myers, et al.(2001), Singh et al. (2019)

Continued

Table 1 Action of popular fixatives and their combinations.—cont'd

Fixative category	Example	Fixation effect					Ultrastructural effect	Components	Special consideration	Ref.
		Proteins	Lipids	Carbohydrates	Nucleotide	Others				
Miscellaneous	Metallic ions (e.g., Hg^{2+} and Zn^{2+})	Mercuric salts powerfully coagulate proteins (good for acid dyes)	Unknown	Unknown	Zinc and mercury salts enhance fixation of nuclei (by reacting with phosphate residues of nucleic acids and fixing nucleoproteins)	Mercuric salts are very useful in the fixation of hematopoietic tissues	Unknown	Zinc salts 1% Mercury salts 6%	• Mercuric salts (one of the oldest fixatives called corrosive sublimate). It was a major component in some of combinative fixatives such as Helly's fixative and B5 fixatives • Mercuric salts can be combined with either acetic acid plus dichromate or formaldehyde with acetate • Mercuric chloride reacts and forms cross links mostly with cysteine but also reacts with amines, amides, amino acids and sulfhydryl groups • Mercuric salts are now infrequently used as they are highly toxic, can be absorbed through the skin, are corrosive, need specific disposal, have poor penetration ability, cause tissue shrinkage, and harden tissue with prolonged exposure • Zinc salts (zinc chloride and zinc sulfate) and barium are used as a substitute for mercuric salts in many fixatives • Zinc sulfate is often preferred because it is less corrosive than the chloride salt • Zinc salts are used with other fixatives, e.g., increases the penetration of formaldehyde and when used in combination with PFA reduces its epitope-masking effect • Phosphate buffer should not be used with fixative containing zinc salts to avoid precipitation	Eltoum, Fredenburgh, Myers, et al. (2001), Singh et al. (2019)

Continued

Miscellaneous	Organic acids	Picric acid	Coagulates Disrupts electrostatic and hydrogen bonding on proteins by changing the charges of ionizable side chains In addition, it can insert a lipophilic anion into the hydrophilic part of proteins and alter its tertiary structure	Very poor (destroys mitochondria and nuclear membranes)	Very poor	None (hydrolyses nucleic acids because of low pH, around 2)	Recommended for glycogen fixation	It is an acid and needs to be removed completely after fixation. If remains in the tissue it results in distortion of cellular structure	Saturated picric acid 2.1% (w/v) in distilled water 5% (w/v) in absolute ethanol	• Makes salt (picrates) with basic protein groups and coagulates the proteins • Disintegrates calcium deposits • Is used in combination with other fixatives such as formalin plus acetic acid	Eltoum, Fredenburgh, Myers, et al. (2001), Singh et al. (2019)
		Trichloroacetic acid	As above	Unknown	Unknown	Extracts nucleic acids	Unknown	Unknown	Up to 10%	None	
		Acetic acid	None	Unknown	Unknown	Coagulates nucleic acids	Unknown	Unknown	Up to 50% (v/v)	• Often added to other fixatives to fix nucleic acids • Fast penetration • Combined with ethanol to provide an effective cytological fixative	
Combination fixative	Alcoholic formalin		Cross-links and coagulates	Extensively extracts lipid	None	Very slight	Unknown	Poor	Alcohol (90% Ethanol) combined with formalin 10% formaldehyde	None	Eltoum, Fredenburgh, Myers, et al. (2001)
	Bouin's Fixative		Unknown	Unknown	Unknown	Unknown	Unknown	Unknown	Picric acid (main component), acetic acid and formaldehyde	• Is used for delicate and soft tissues to preserve structure	Singh et al. (2019)

1.4 Periodate-lysine-paraformaldehyde (PLP) fixative

PLP was first introduced by Mclean and Nakane (1974) and modified by Gendelman et al. (1983) as a fixative that maintains tissue ultrastructure but improves antigenicity compared to other aldehyde fixatives by, in theory, cross-linking the carbohydrate moieties of glycoproteins. PLP contains sodium m-periodate, PFA and lysine (Mclean & Nakane, 1974). Sodium periodate oxidizes polysaccharide chains, producing aldehyde groups which with the help of lysine, a divalent amine, directly or indirectly forms cross-links between carbohydrate groups (Mclean & Nakane, 1974; Pieri et al., 2010). A low concentration of PFA in PLP further cross-links proteins but this relatively low concentration means there is less masking of protein antigen (Accart et al., 2014; Brenes et al., 1986; Mclean & Nakane, 1974). In addition, studies have suggested that lysine and formaldehyde interact while linked to the oxidized carbohydrate molecules which increases fixative efficacy (Pieri et al., 2010). A later analysis of the mechanism of action of this fixative suggests that it does not cross-link carbohydrates (Hixson, Yep, Glenney, Hayes, & Walborg, 1981), however, despite an evident complex mode of action it can maintain antigenicity of multiple antigens (Liu et al., 2003; Pieri et al., 2010).

1.5 Retinoic acid signaling in the adult central nervous system

Retinoic acid (RA) signaling has been extensively studied in its control of the developing central nervous system (CNS) but it is now being understood to have a critical role in the adult CNS, controlling neuronal plasticity in regions such as the hippocampus, hypothalamus and olfactory bulb (Shearer, Stoney, Morgan, & McCaffery, 2012). The receptors for RA are of the nuclear receptor family, ligand-gated transcription factors regulating gene expression. RA exerts its function through both genomic and non-genomic pathways (Al Tanoury, Piskunov, & Rochette-Egly, 2013; Rochette-Egly, 2015). RA metabolic and catabolic enzymes, receptors, and binding proteins are expressed and distributed in different regions of the adult CNS with varied functions in each area (Shearer et al., 2012). Although RA is itself difficult to immunolocalize in tissue because there are few good antibodies against this lipid and it is difficult to fix in place in tissue, its region of synthesis can be determined through localization of the enzymes catalyzing the last step of RA synthesis, the retinaldehyde dehydrogenases (RALDHs). A major reason why there has been less emphasis in the past on RA in the adult versus embryonic CNS is that

RA levels are lower in the adult as there is a steady decline in RA synthesis as the animal develops from embryo to adult (Rossant, Zirngibl, Cado, Shago, & Giguere, 1991; Shearer et al., 2012). Thus, to localize the RALDHs in the adult CNS by IHC, a highly sensitive and specific technique is needed.

In this study, the expression of RA enzymes in the CNS is shown by the example of RALDH2 in motor neurons of the spinal cord. The spinal cord has been relatively little studied with regard to RA signaling. However, RA is likely to be important in terms of the innervation of the limb muscle by motor neurons of the spinal cord given vitamin A deficiency results in lack of coordination and a dragging of the hindlegs and eventually paralysis in various animals (Corcoran, So, & Maden, 2002) which is reviewed by Maden (2007). Thus, vitamin A deficiency pathophysiology suggests a role for RA in motor neuron function and perhaps motor neuron disease. This study applies IHC on lumbar spinal cord fixed with PLP as a highly effective fixative that maintains the epitopes recognized by many antibodies. RALDH2 positive motor neurons are identified by double-labeling with antibodies to either choline acetyltransferase (ChAT) the enzyme responsible for biosynthesis of the neurotransmitter acetylcholine (ACh), or the vesicular ACh transporter (VAChT), a neurotransmitter transporter that translocates ACh from cytoplasm into synaptic vesicles (Malomouzh, 2012).

2. Rationale

Before starting IHC, the choice of the appropriate tissue fixative is crucial to maintain the correct shape of the antigens to which the antibodies are directed. An incorrect choice can result in loss of the antigen of interest (false negative) or increase in non-specific background staining (false positive). The main fixatives in current use for IHC are of the aldehyde group, which are not always suitable for IHC detection. For instance, glutaraldehyde is good for preservation of ultrastructure but in most cases, does not retain antigenicity. On the other hand, PFA maintains ultrastructure quite well and is better at retaining antigenicity but many antigens can still be lost (Mclean & Nakane, 1974). PLP, however, provides a good compromise between preservation of ultrastructure and maintaining a greater number of antigens (Mclean & Nakane, 1974). Here we show that PLP is a fixative effective for maintaining antigenicity of antibodies to RALDH2 as well as several antibodies used as markers for motor neurons.

3. Materials, equipment and reagents

3.1 Materials

C57/BL6 mice

3.2 Equipment

1. Coverslips (CAT# 631-0167, VWR)
2. Cryostat
3. Filter paper, 3MM, Whatman (0.34 mm thickness) (CAT# 10441412, Fisher Scientific)
4. Falcon tubes (50 mL) (CAT# 430921, Corning)
5. Fluorescent microscope
6. Freezer (−20 °C, −70 °C)
7. Fridge (4 °C)
8. Funnel (Glass)
9. Heat plate stirrer
10. IHC slide box
11. Hydrophobic barrier PAP pen (e.g., ImmEdge CAT# H-4000, Vectorlabs)
12. Needle (20–26 gauge blunted needle)
13. pH meter
14. Perfusion pump (optional)
15. Positively charged adhesion slides, Menzel Gläser, Polysine (CAT# 631-1349, VWR)
16. Syringe (60 mL BD Plastipak syringe, CAT# 300866, Becton Dickinson) (1 mL latex free syringe, CAT# 309625, Becton Dickinson)
17. Pipettes (10 mL, 25 mL) (CAT# 4488 & # 4489, Corning)
18. Water bath

3.3 Chemicals

1. Bisbenzimide H 33258 ($C_{25}H_{24}N_6O{\cdot}3HCl$; CAT# 14530, Sigma-Aldrich)
2. DABCO (CAT# 10214620, Fisher Scientific)
3. Glycine (CAT# 10061073, Fisher Scientific)
4. Glycerol (CAT# 5516, Sigma-Aldrich)
5. Donkey serum (CAT# AS-228/500, Biosera)
6. L-Lysine monohydrochloride ($H_2N(CH_2)_4CH(NH_2)CO_2H{\cdot}HCl$; CAT# L5626, Sigma-Aldrich)

7. Milli-Q water
8. Paraformaldehyde ($OH(CH_2O)$ n; CAT# 50-276-34)
9. Sodium phosphate monobasic anhydrous (NaH_2PO_4; CAT# 10135013, Fisher Scientific)
10. Sodium phosphate dibasic anhydrous (Na_2HPO_4; CAT# 10182863, Fisher Scientific)
11. Sodium phosphate monobasic monohydrate ($NaH_2PO_4 \cdot H_2O$; CAT# 10754534, Fisher Scientific)
12. Sodium phosphate dibasic heptahydrate ($Na_2HPO_4 \cdot 7H_2O$; CAT# 10420101, Fisher Scientific)
13. Sodium chloride (NaCl; CAT# 10112640, Fisher Scientific)
14. Sodium (meta) Periodate ($NaIO_4$; CAT# s1878, Sigma-Aldrich)
15. Triton X-100 (CAT# X100-500ML, Sigma-Aldrich)

3.4 Antibodies

Antibodies are listed in Tables 2 and 3.

4. Protocols

Prepare the following stocks and working solutions:

1. *10× Phosphate Buffered Saline (PBS) stock solution*: Dissolve 2.28 g of sodium phosphate monobasic anhydrous (NaH_2PO_4) and 11.5 g

Table 2 Primary antibodies used in study.

Antibody name	Supplier	Product code	Species	Monoclonal/ polyclonal (primary)	Dilution
Primary					
Anti-RALDH2 (ALDH1A2) Antibody	Millipore	ABN420	Rabbit	Polyclonal	1/300
Anti-Choline Acetyltransferase (ChAT) Antibody	Abcam	ab18736	Sheep	Polyclonal	1/100
Anti-Choline Acetyltransferase (ChAT) Antibody	Abcam	ab6168	Rabbit	Polyclonal	1/100
Anti-Vesicular Acetylcholine Transporter (VAChT)	Millipore	ABN100	Goat	Polyclonal	1/400

Table 3 Secondary antibodies used in study.

Antibody name	Supplier	Product code	Isotope (secondary)	Dilution	Dye/fluorochrome	Wavelength (nm)
Secondary						
Donkey Anti-Rabbit	Jackson Immuno Research	705-545-152-JIR	IgG (H + L)	1/100	Alexa Fluor	488
Donkey Anti-Goat	Abcam	ab150132	IgG (H + L)	1/100	Alexa Fluor	594
Donkey Anti-Sheep	Biotium	20152-1 LOT:10C0702	IgG (H + L) Highly cross-absorbed	10 μg/mL	CF	594

of sodium phosphate dibasic anhydrous (Na_2HPO_4), and 87.68 g of sodium chloride in about 900 mL Milli-Q water. Adjust the pH to 7.4 using hydrochloric acid or sodium hydroxide. After dissolving, add Milli-Q water to reach a final volume of 1 L. Note: to make 1 × PBS, add 100 mL of the 10 × PBS stock solution to 900 mL Milli-Q water and keep at RT.

2. *Phosphate Buffer (PB) 0.4 M*: Dissolve 10.55 g of sodium phosphate monobasic monohydrate ($NaH_2PO_4 \cdot H_2O$) and 86.8 g of sodium phosphate dibasic heptahydrate ($Na_2HPO_4 \cdot 7H_2O$) in about 900 mL Milli-Q water. Adjust the pH to 7.4 and adjust the final volume to 1 L with Milli-Q water.
3. *8% Paraformaldehyde (PFA)*: Heat 800 mL Milli-Q water in a beaker on a heat/stirrer plate to 60 °C. Add 80 g of PFA and stir continuously. Note: do not allow the solution to heat much beyond 60 °C. Add NaOH 1 M, one to two drops at the time until reaching pH 7.4 and the solution clears. Note: the presence of a small amount of flocculent material is unavoidable. Filter the solution using 3MM Whatman filter paper. Adjust the pH to 7.4 and add Milli-Q water to reach a final volume of 1 L. Ideally used fresh but can still be effective when used from 50 mL aliquots kept at −20 °C and reheating to 60 °C.
4. *4% Paraformaldehyde*: Dilute 500 mL of 8% PFA with 500 mL of PB 0.2 M.
5. *Lysine-HCl 0.2 M*: Dissolve 7.32 g Lysine-HCl powder in 200 mL of 0.1 M PB.
6. *Periodate-lysine-paraformaldehyde (PLP)*: Mix 50 mL of PB 0.1 M, 50 mL of 0.2 M Lysine HCl, and 33.3 mL of 8% PFA. Add 0.27 g of Sodium m-Periodate powder to the solution and shake until the powder is completely dissolved. If needed re-adjust the pH to 7.4 with 1 M NaOH or HCl. Final concentration of PLP will be: 2% PFA, [$NaIO_4$]: 0.2 g/100 mL, [lysine]: 0.075 M, and [PB]: 0.04 M. Note: PLP must be made freshly just before use.
7. *30%, 20%, 10% Sucrose*: Prepare sucrose in 1 × PBS by dissolving 30 g, 20 g, and 10 g of sucrose in 100 mL 1 × PBS.
8. *PBS-T*: Prepare by mixing PBS and triton to a final concentration of 0.1% triton X-100.
9. *Blocking solution*: Add normal serum 10% (v/v) to PBS-T. Note: the choice of serum species is based on the species that the secondary antibody was raised and for this study was donkey.

10. *Mounting mix*: Mix 8% PFA with glycerol (1:1). Add bisbenzimide (1 mg/mL) and DABCO (2.5%). Wrap the bottle with foil and keep at room temperature.

4.1 Transcardial perfusion and fixation

1. Warm up PBS and PLP in a water bath to 37 °C.
2. Fill a 50 mL syringe with PBS and another with PLP. Note: if many animals are to be perfused a perfusion pump can be used.
3. Attach the tube connected to the perfusion needle to the syringe containing PBS and flush out the tube and needle with PBS to clean and clear it of air bubbles.
4. Weigh animal and inject pentobarbitone (Euthatal 150 mg/kg) intraperitoneally.
5. When the mouse is deeply anesthetized and unresponsive to sensory stimuli, e.g., toe pinch response, place the animal in a well-ventilated fume hood.
6. Stabilize the animal using needles to hold the paws to the dissection board.
7. Make an incision just beneath the rib cage. Carefully detach the liver from the diaphragm. Carefully cut away diaphragm with curved blunt scissors to reach the chest cavity.
8. Cut a central strip of the rib cage free from the remaining ribs, clamp the sternum tip with a hemostat and pull back to one side of the head and fix with a needle to the dissection board to expose completely the chest cavity.
9. Displace lungs to have a clear view of heart. The heart should still be beating for the best perfusion.
10. Insert an approximately 20–26 gauge blunted needle into the outer wall of the left ventricle (not too deep to pass through the ventricle and penetrate the opposing wall). The needle can be clamped to the apex of the heart with a small hemostat to hold in place.
11. Immediately after placing the needle into the left ventricle make a small incision into the right atrium and start the flow of the perfusion solution (PBS) using the syringe (or pump) running quickly and evenly.
12. When the fluid running out of atrium is clear the first step of perfusion is complete (20–40 mL of PBS is usually enough, and a pale liver is a good indicator of effective perfusion).
13. Change the syringe to the one containing PLP and start the flow of fixative.

14. Animal will show contraction and spasms of muscles at the beginning of the fixation and will be stiff when the fixation is almost complete. Note: if the neck and tail are stiff these are signs of good fixation.
15. The waste PLP must be stored for appropriate disposal.

4.2 Dissection, post fixation, cryoprotection, sectioning and mounting

1. Dissect spinal cord by total laminectomy (Kennedy, Jones, & Caplazi, 2013) or simply cutting the lumbar segment and taking out the spinal cord (Sleigh, Weir, & Schiavo, 2016).
2. Post-fix with fresh fixative solution (PLP) for 24 h on the shaker followed by cryoprotection with 30% sucrose at 4 °C. Note: it is better to increase the sucrose concentration stepwise with constant shaking progressing from 10% sucrose for 2 h to 20% sucrose for 2 h and finally 30% sucrose overnight and kept at 4 °C. Note: the volume for fixative should be at least 15–20 times more than the tissue volume.
3. Make serial transverse sections of 8–10 μm thickness using a cryostat and cutting through the lumbar spinal cord enlargement. The sections should be mounted onto positively charged slides. Keep the slides for half an hour at room temperature and then transfer to −70 °C.

4.3 Immunohistochemistry

1. Transfer the slides to be used for the experiment to room temperature for half an hour.
2. Place the slides in a slide jar and submerge in PLP for 20 min.
3. Wash three times for 5 min with PBS-T.
4. Wet some paper tissue and place in an IHC slide box (a dark box with lid and appropriate place for slides).
5. Use a barrier pen (e.g., ImmEdge pen) to draw a barrier around the tissue on the slides and place the slides in the IHC slide box.
6. Dissolve glycine to a final concentration of 0.3 M in blocking solution (PBS-T + 10% Donkey serum).
7. Add blocking solution on top of the slides and incubate for 2 h at room temperature.
8. Prepare the dilution of primary antibody in blocking solution (Table 2).
9. Discard blocking solution and add primary antibody on top of the slides and keep overnight at 4 °C. Note: keep one slide as a no primary antibody control, adding just blocking solution.

10. The next day the slides are washed three times for 5 min with PBS-T. Note: the time for exposure to primary antibody can be increased to 72 h to increase signal.
11. Prepare the dilution of secondary antibody in blocking solution (Table 3).
12. Add the secondary antibody on the slides for 2 h at room temperature protected from light.
13. Wash the slides three times for 5 min with PBS-T.
14. Discard the PBS-T, add a drop of mounting medium and gently place a cover slip over the tissue on the slide. Wipe off the extra mounting medium and put some transparent nail polish around the edges of the slide to hold the coverslip in place and allow to dry in the dark at room temperature.
15. Based on the wavelength of the labeled secondary antibody view the slides under a fluorescence microscope.

5. Results

5.1 Acetone/ethanol or paraformaldehyde fixation and RALDH2 immunolabeling of cells in the spinal cord

Immunolabeling of RALDH2 (green) in transverse sections (8–10 μm) of lumbar spinal cord fixed in (A) cold 1:1 acetone:ethanol for 20 min and (B) 4% PFA for 20 min are shown in Fig. 3. With acetone/ethanol fixation there was excessive non-specific immunostaining including labeling of blood vessels (A) while with 4% PFA staining there was an absence of immunostaining (B).

5.2 Periodate-lysine-paraformaldehyde (PLP) fixation and RALDH2 immunolabeling of cells in the spinal cord

Immunolabeling of RALDH2 was then studied in PLP fixed spinal cord as shown in Fig. 4 in transverse sections (8–10 μm) of lumbar spinal cord with RALDH2 in green. The results show the presence of RALDH2 in the soma of large cells in the ventral/lateral spinal cord which have the appearance of motor neurons.

5.3 Motor neurons immunolabeled for ChAT in PLP fixed spinal cord

Antibodies that specifically label spinal cord motor neurons were then tested on PLP fixed spinal cord. Immunolabeling of ChAT (green) in transverse

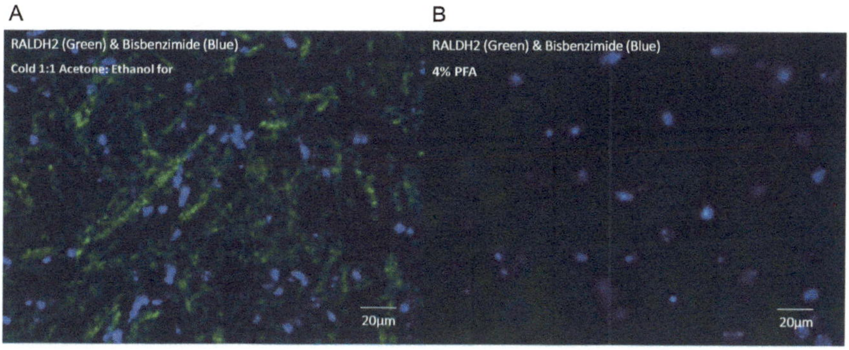

Fig. 3 RALDH2 immunolabeling of transverse ventral lumbar spinal cord is poor using either acetone/ethanol or PFA fixation: Merged immunolabeling image of RALDH2 (green) and bisbenzimide (blue), a DNA fluorescent stain to label nuclei. Tissue was fixed in (A) cold 1:1 acetone:ethanol for 20 min and (B) 4% PFA for 20 min. RALDH2 immunolabeling either showed excessive background staining with acetone:ethanol (A) or complete lack of immunostaining with 4% PFA fixation (B).

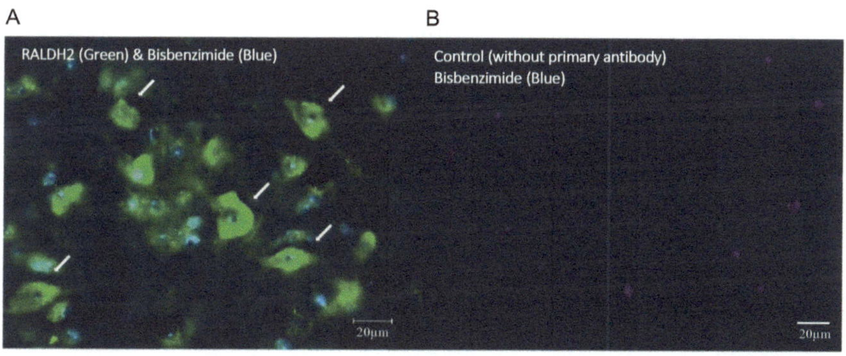

Fig. 4 RALDH2 immunolabeling of transverse ventral lumbar spinal cord is greatly improved with periodate-lysine-paraformaldehyde (PLP) fixation: (A) Merged immunolabeling image of RALDH2 (green) and bisbenzimide (blue); Arrows point to what appear to be motor neurons from their large size and position in spinal cord. (B) Merged image of negative control, no primary antibody followed by incubation with secondary antibody.

sections (8–10 μm) of lumbar spinal cord is shown in Fig. 5. The ChAT antibody effectively labels large cells in the ventral spinal cord and identified as motor neurons.

5.4 Motor neurons double immunolabeling with RALDH2 and ChAT on PLP fixed spinal cord

Double immunolabeling of RALDH2 and ChAT in transverse sections (8–10 μm) of lumbar spinal cord is shown in Fig. 6. Immunolabeling of

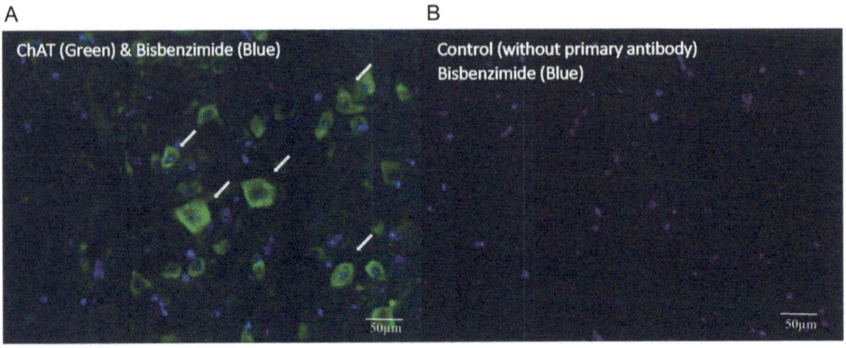

Fig. 5 ChAT Immunolabeling identifies motor neurons in PLP fixed transverse ventral lumbar spinal cord: (A) Merged immunolabeling image of ChAT (green) and bisbenzimide (blue); Arrows point to cells likely to be motor neurons from their expression of ChAT. (B) Merged image of negative control, no primary antibody followed by incubation with secondary antibody.

ChAT showed bright granular labeling in the soma of motor neurons (red) where it is synthesized within the rough endoplasmic reticulum and then transported to the nerve terminal by axoplasmic transport. The RALDH2 antibody labeled the soma of the same cells in green. Double labeling of cells indicates the presence RALDH2 as a synthetic enzyme for RA in motor neurons.

5.5 Double immunolabeling of motor neurons with RALDH2 and VAChT on PLP fixed spinal cord

Double immunolabeling of RALDH2 and VAChT in transverse sections (8–10 μm) of lumbar spinal cord is shown in Fig. 7. The VAChT antibody was used to identify motor neurons and labeled innervating cholinergic interneuron terminals in red. Large puncta were seen in close proximity to the motor neuron cell body and in proximal dendrites and along the axon. This type of labeling is unique for alpha motor neurons specifically in the lumbar region and it is due to cholinergic C-bouton synaptic terminals that makes contact with alpha motor neurons and regulates the action of the motor neuron (Arvidsson, Riedl, Elde, & Meister, 1997; Gilmor et al., 1996). RALDH2 antibody labeled the soma of the same cells in green with high intensity and with lower intensity at axon hillocks and along the axons. This labeling pattern confirms the presence of RALDH2 as a synthetic enzyme for RA in motor neurons.

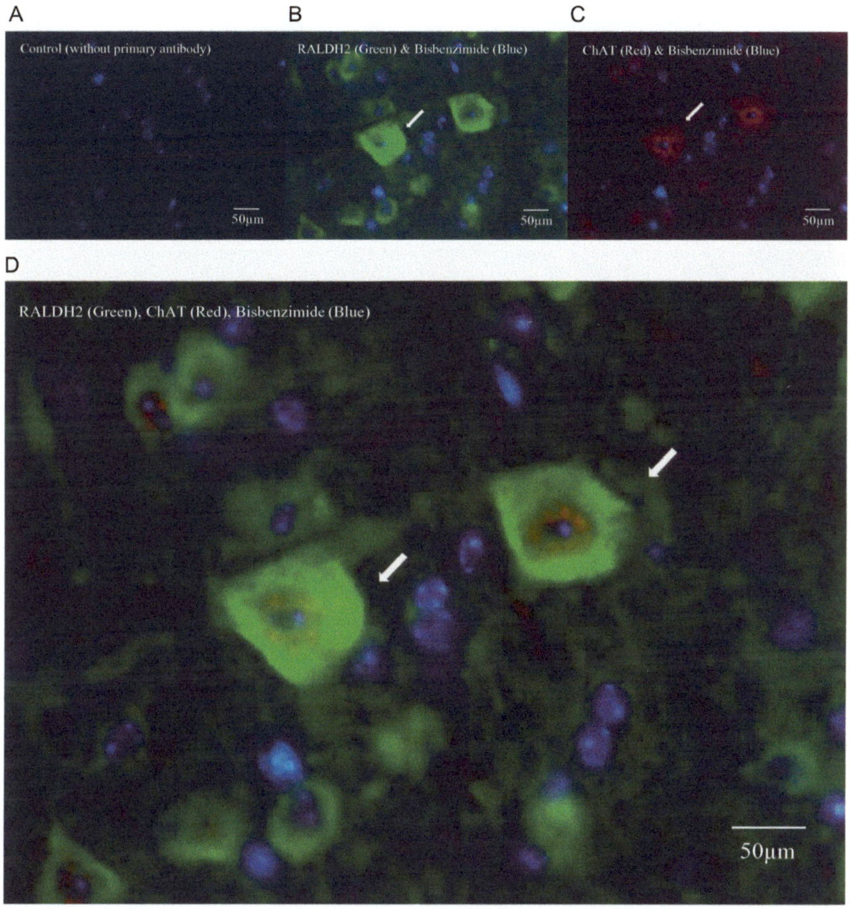

Fig. 6 Double immunolabeling shows the colocalization of RALDH2 and ChAT in cells likely to be motor neurons in PLP fixed transverse ventral lumbar spinal cord: (A) Merged image of negative control, no primary antibody followed by incubation with secondary antibody; (B) Merged immunolabeling image of RALDH2 (green) and bisbenzimide (blue); (C) Merged immunolabeling image of ChAT (red) and bisbenzimide (blue); (D) Double immunolabeling image of RALDH2 (green) and ChAT (red) merged with bisbenzimide (blue); Arrows point to cells likely to be motor neurons from their expression of ChAT with presence of RALDH2 in their cell bodies.

5.6 Distribution of motor neurons expressing RALDH2 in transverse sections of the lumbar region of PLP fixed spinal cord

IHC of transverse lumbar sections of spinal cord revealed expression of RALDH2 in motor neurons that are mostly in the ventral horn of the

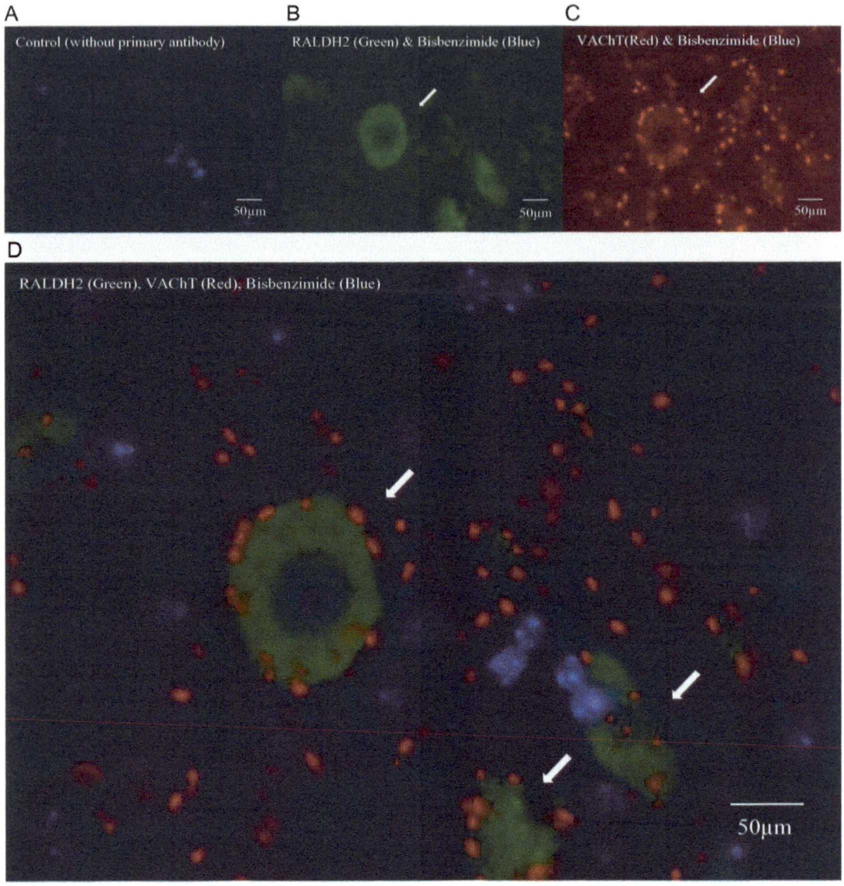

Fig. 7 Double immunolabeling shows the colocalization of RALDH2 and VAChT in cells likely to be motor neurons in PLP fixed transverse ventral lumbar spinal cord: (A) Merged image of negative control, no primary antibody followed by incubation with secondary antibody; (B) Merged immunolabeling image of RALDH2 (green) and bisbenzimide (blue); (C) Merged immunolabeling image of VAChT (red) and bisbenzimide (blue); (D) Double immunolabeling image of RALDH2 (green) and VAChT (red) merged with bisbenzimide (blue); Arrows point to cells likely to be motor neurons from their specific VAChT labeling and with the expression of RALDH2 in their cell bodies.

gray matter (Fig. 8). Yellow arrows point out cells with the appearance of alpha motor neurons, multipolar large-sized cell body. Other small-labeled cells in the spinal cord may be gamma motor neuron, interneurons or glial cells.

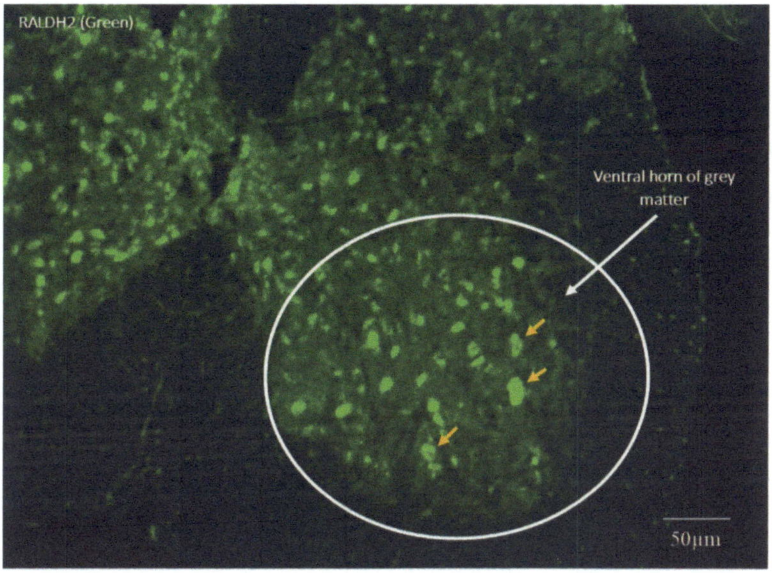

Fig. 8 RALDH2 immunolabeling in the lumbar region of PLP fixed spinal cord shown at low magnification to illustrate the distribution of labeled cells across the spinal cord: RALDH2 was expressed in many cells of the spinal cord as well as in motor neurons of the ventral horn.

6. Safety considerations and standards

Paraformaldehyde must be used under a properly ventilated fume hood. Both bisbenzimide and PFA are toxic, and safety must be considered when working with these.

7. Pros and cons

Pros	Cons
Preserves many antigens	Possibly poor fixative for glycolipids
Maintains tissue ultrastructure	Requires extra steps of preparation compared to other aldehyde-based fixatives
Inexpensive	
No extensive tissue shrinkage	

8. Troubleshooting and optimization

Problem	Solution
Fixative	Formaldehyde naturally undergoes oxidation and produce formic acid as a result of storage conditions (temperature, oxygen, and pH). An excess amount of formic acid in the formaldehyde solution during fixation of a blood-rich tissue results in formalin pigment formation (Fox et al., 1985). At acidic pH (below 6) formaldehyde react with hemoglobin and produces intra- and extra- cellular brownish granular deposit of hematin called formalin pigment. Adding calcium carbonate to the diluted formaldehyde to form a salt with formic acid, or maintaining the pH at a predetermined range with buffer (specifically phosphate buffer) will prevent excess formation of formic acid (Fox et al., 1985). Storage of concentrated formaldehyde results in spontaneous condensation reactions. A low amount of alcohol can prevent this reaction at room temperature and stabilize the solution. Almost all manufactured formaldehyde solution contains 10% methanol, or sometimes butanol, as a stabilizer. It should be noted that methanol can affect the fixation process in some experiments and the addition of stabilizer is not always mentioned on the label of commercial formaldehyde solutions (Fox et al., 1985). PFA, as a polymer of formaldehyde, must be converted with heat to a formaldehyde solution (formalin) in order to fix tissue. When dissolving the temperature must be kept around 60 °C to depolymerize the PFA to formaldehyde. However, avoid higher temperatures (e.g., boiling) to prevent the production of formic acid. If your antigen of interest is lost PLP can be modified to improve for certain antigens changing either sodium m-periodate (0–0.1 M) or PFA (ranging from 0% to 4%) while glutaraldehyde can also be added (Gendelman et al., 1983).

—cont'd

Problem	Solution
Fixation	For successful perfusion and fixation air bubbles should be avoided in the syringe and tubing.
IHC (non-specific labeling and background)	For the blocking solution the serum (10% v/v) should ideally be from the same species as the host species of the labeled secondary antibody to reduce background. (Note: if bovine serum albumin (BSA) is used in the block this can cause background problems if the antibody is raised in goat). For double immunostaining, the primary antibodies should be raised in two different species, and the secondary antibodies that recognize each must be specific for that species. There are many excellent on-line guides to IHC trouble-shooting from companies producing antibodies, e.g., https://www.abcam.com/kits/troubleshooting-ihc-experiments.
Unsuccessful perfusion	When extremities (liver, nose, ear, and paws) become pale this is a primary sign of successful perfusion. The dissected tissue, such as brain and spinal cord, should also be white to pale yellow in appearance showing that blood is replaced with fixative. The needle for perfusion needs to be carefully inserted into the atrium of the heart and not pass through the atrium or reach the aortic arch. Between each perfusion the tube must be washed with 1 × PBS to flush out the remaining PLP.

9. Summary

Mice transcardially perfused and post-fixed with PLP, sequentially cryoprotected with 10%–20%–30% sucrose followed by cryosectioning provides well preserved sections of spinal cord with good morphology and intense immunolabeling. Previously it has been shown that various protein antigens are protected when PLP is used as a fixative including that of cytokeratins; enzyme activity can also be protected such as that of tyrosinase in melanocytes while the antigenicity of bromodeoxyuridine (BrdU) used for cell proliferation studies is maintained (Pieri et al., 2010).

Here we showed that PLP is an effective fixative for tissue from the CNS allowing immunostaining for multiple markers of neurons in the spinal cord.

The IHC results showed the expression of RALDH2 in the cell bodies of motor neurons. Previous research had indicated that RALDH1 and RALDH2 were highly expressed in the meninges and blood vessels (Maden, 2007) in the spinal cord. Here we showed that RALDH2 is also present in neurons in lumbar segments of spinal cord. Double immunolabeling of RALDH2 with ChAT, and the VAChT positive C-boutons surrounding the motor neurons, showed specificity of RALDH2 for alpha motor neurons over other neurons (Witts, Zagoraiou, & Miles, 2014).

Acknowledgments

This work was supported by the Elphinstone Scholarship University of Aberdeen, the Sir Richard Stapley Educational Trust and the "SPRINT-MND/MS PhD programme," which is funded by the Chief Scientist Office (ref MMPP/01) and the Universities of Aberdeen, Dundee, Edinburgh, Glasgow and St Andrews.

References

Accart, N., Sergi, F., & Rooke, R. (2014). Revisiting fixation and embedding techniques for optimal detection of dendritic cell subsets in tissues. *Journal of Histochemistry & Cytochemistry*, *62*(9), 661–671. https://doi.org/10.1369/0022155414539963.

Al Tanoury, Z., Piskunov, A., & Rochette-Egly, C. (2013). Vitamin A and retinoid signaling: Genomic and nongenomic effects. *Journal of Lipid Research*, *54*(7), 1761–1775. https://doi.org/10.1194/jlr.R030833.

Arvidsson, U., Riedl, M., Elde, R., & Meister, B. (1997). Vesicular acetylcholine transporter (VAChT) protein: A novel and unique marker for cholinergic neurons in the central and peripheral nervous systems. *The Journal of Comparative Neurology*, *378*(4), 454–467. https://doi.org/10.1002/(SICI)1096-9861(19970224)378:4<454::AID-CNE2>3.0.CO;2-1.

Brenes, F., Harris, S., Paz, M. O., Petrovic, L. M., & Scheuer, P. J. (1986). PLP fixation for combined routine histology and immunocytochemistry of liver biopsies. *Journal of Clinical Pathology*, *39*(4), 459–463. https://doi.org/10.1136/jcp.39.4.459.

Cohen, A. L. (1992). Electron microscopy: Principles and techniques for biologists. John J. Bozzola, Lonnie D. Russell. *The Quarterly Review of Biology*, *67*(2), 247–249. https://doi.org/10.1086/417649.

Corcoran, J., So, P. L., & Maden, M. (2002). Absence of retinoids can induce motoneuron disease in the adult rat and a retinoid defect is present in motoneuron disease patients. *Journal of Cell Science*, *115*(Pt 24), 4735–4741. https://doi.org/10.1242/jcs.00169.

Eltoum, I., Fredenburgh, J., & Grizzle, W. E. (2001). Advanced concepts in fixation: 1. Effects of fixation on immunohistochemistry, reversibility of fixation and recovery of proteins, nucleic acids, and other molecules from fixed and processed tissues. 2. Developmental methods of fixation. *Journal of Histotechnology*, *24*(3), 201–210. https://doi.org/10.1179/his.2001.24.3.201.

Eltoum, I., Fredenburgh, J., Myers, R. B., & Grizzle, W. E. (2001). Introduction to the theory and practice of fixation of tissues. *Journal of Histotechnology*, *24*(3), 173–190. https://doi.org/10.1179/his.2001.24.3.173.

Fox, C. H., Johnson, F. B., Whiting, J., & Roller, P. P. (1985). Formaldehyde fixation. *Journal of Histochemistry & Cytochemistry*, *33*(8), 845–853. https://doi.org/10.1177/33.8.3894502.

Ganjali, H., & Ganjali, M. (2013). Fixation in tissue processing. *International Journal of Farming and Allied Sciences*, *2*(18), 686–689.

Gendelman, H. E., Moench, T. R., Narayan, O., & Griffin, D. E. (1983). Selection of a fixative for identifying T cell subsets, B cells, and macrophages in paraffin-embedded mouse spleen. *Journal of Immunological Methods*, *65*(1–2), 137–145. https://doi.org/10.1016/0022-1759(83)90310-1.

Gilmor, M., Nash, N., Roghani, A., Edwards, R., Yi, H., Hersch, S., et al. (1996). Expression of the putative vesicular acetylcholine transporter in rat brain and localization in cholinergic synaptic vesicles. *The Journal of Neuroscience*, *16*(7), 2179–2190. https://doi.org/10.1523/JNEUROSCI.16-07-02179.1996.

Hixson, D. C., Yep, J. M., Glenney, J. R., Hayes, T., & Walborg, E. F. (1981). Evaluation of periodate/lysine/paraformaldehyde fixation as a method for cross-linking plasma membrane glycoproteins. *Journal of Histochemistry & Cytochemistry*, *29*(4), 561–566. https://doi.org/10.1177/29.4.6265547.

Hoffman, E. A., Frey, B. L., Smith, L. M., & Auble, D. T. (2015). Formaldehyde crosslinking: A tool for the study of chromatin complexes. *Journal of Biological Chemistry*, *290*(44), 26404–26411. https://doi.org/10.1074/jbc.R115.651679.

Janardhan, K. S., Jensen, H., Clayton, N. P., & Herbert, R. A. (2018). Immunohistochemistry in investigative and toxicologic pathology. *Toxicologic Pathology*, *46*(5), 488–510. https://doi.org/10.1177/0192623318776907.

Jones, D. (1969). Acrolein as a histological fixative. *Journal of Microscopy*, *90*(1), 75–77. https://doi.org/10.1111/j.1365-2818.1969.tb00696.x.

Jones, D. (1972). Reactions of aldehydes with unsaturated fatty acids during histological fixation. *The Histochemical Journal*, *4*, 421–465. https://doi.org/10.1007/BF01012533.

Kennedy, H. S., Jones, C., & Caplazi, P. (2013). Comparison of standard laminectomy with an optimized ejection method for the removal of spinal cords from rats and mice. *Journal of Histotechnology*, *36*(3), 86–91. https://doi.org/10.1179/014788813X13756994210382.

Knoblaugh, S., Randolph-Habecker, J., & Rath, S. (2012). Necropsy and histology. In *Comparative anatomy and histology* (pp. 15–40). Elsevier. https://doi.org/10.1016/B978-0-12-381361-9.00003-2.

Liu, W. G., Brown, D. A., & Fraser, J. R. (2003). Immunohistochemical comparison of anti-prion protein (PrP) antibodies in the CNS of mice infected with Scrapie. *Journal of Histochemistry & Cytochemistry*, *51*(8), 1065–1071. https://doi.org/10.1177/002215540305100810.

Ma, W., Rogers, K., Zbar, B., & Schmidt, L. (2002). Effects of different fixatives on β-galactosidase activity. *Journal of Histochemistry & Cytochemistry*, *50*(10), 1421–1424. https://doi.org/10.1177/002215540205001015.

Maden, M. (2007). Retinoic acid in the development, regeneration and maintenance of the nervous system. *Nature Reviews Neuroscience*, *8*(10), 755–765. https://doi.org/10.1038/nrn2212.

Malomouzh, A. I. (2012). Non-cholinergic Signaling pathways at vertebrate neuromuscular junctions. In *Skeletal muscle—From myogenesis to clinical relations*. InTechhttps://doi.org/10.5772/48337.

McFadden, W. C., Walsh, H., Richter, F., Soudant, C., Bryce, C. H., Hof, P. R., et al. (2019). Perfusion fixation in brain banking: A systematic review. *Acta Neuropathologica Communications*, 7(1), 146. https://doi.org/10.1186/s40478-019-0799-y.

Mclean, I. W., & Nakane, P. K. (1974). Periodate lysine paraformaldehyde fixative. A new fixative for immunoelectron microscopy. *Journal of Histochemistry & Cytochemistry*, *22*(12), 1077–1083. https://doi.org/10.1177/22.12.1077.

Pieri, L., Sassoli, C., Romagnoli, P., & Domenici, L. (2010). Use of periodate-lysine-paraformaldehyde for the fixation of multiple antigens in human skin biopsies. *European Journal of Histochemistry*, *46*(4), 365. https://doi.org/10.4081/1749.

Prasadarao, N., Tobet, S. A., & Jungalwala, F. B. (1990). Effect of different fixatives on immunocytochemical localization of HNK-1-reactive antigens in cerebellum: A method for differentiating the localization of the same carbohydrate epitope on proteins vs lipids. *Journal of Histochemistry & Cytochemistry*, *38*(8), 1193–1200. https://doi.org/10.1177/38.8.1694876.

Ramos-Vara, J. A. (2005). Technical aspects of immunohistochemistry. *Veterinary Pathology*, *42*, 405–426. https://doi.org/10.1354/vp.42-4-405.

Rochette-Egly, C. (2015). Retinoic acid signaling and mouse embryonic stem cell differentiation: Cross talk between genomic and non-genomic effects of RA. *Biochimica et Biophysica Acta (BBA)—Molecular and Cell Biology of Lipids*, *1851*(1), 66–75. https://doi.org/10.1016/j.bbalip.2014.04.003.

Rossant, J., Zirngibl, R., Cado, D., Shago, M., & Giguere, V. (1991). Expression of a retinoic acid response element-hsplacZ transgene defines specific domains of transcriptional activity during mouse embryogenesis. *Genes & Development*, *5*(8), 1333–1344. https://doi.org/10.1101/gad.5.8.1333.

Saito, T., & Keino, H. (1976). Acrolein as a fixative for enzyme cytochemistry. *Journal of Histochemistry & Cytochemistry*, *24*(12), 1258–1269. https://doi.org/10.1177/24.12.187691.

Sasse, D. (1986). Histology and histochemistry. In *Regulation of hepatic metabolism* (pp. 57–86). Boston, MA: Springer US. https://doi.org/10.1007/978-1-4684-5041-5_3.

Shearer, K. D., Stoney, P. N., Morgan, P. J., & McCaffery, P. J. (2012). A vitamin for the brain. *Trends in Neurosciences*, *35*(12), 733–741. https://doi.org/10.1016/j.tins.2012.08.005.

Singh, H., Bishen, K. A., Garg, D., Sukhija, H., Sharma, D., & Tomar, U. (2019). Fixation and fixatives: Roles and functions—A short review. *Dental Journal of Advance Studies*, *07*(02), 051–055. https://doi.org/10.1055/s-0039-1693098.

Sleigh, J. N., Weir, G. A., & Schiavo, G. (2016). A simple, step-by-step dissection protocol for the rapid isolation of mouse dorsal root ganglia. *BMC Research Notes*, *9*(1), 82. https://doi.org/10.1186/s13104-016-1915-8.

Thavarajah, R., Mudimbaimannar, V., Rao, U., Ranganathan, K., & Elizabeth, J. (2012). Chemical and physical basics of routine formaldehyde fixation. *Journal of Oral and Maxillofacial Pathology*, *16*(3), 400. https://doi.org/10.4103/0973-029X.102496.

Witts, E. C., Zagoraiou, L., & Miles, G. B. (2014). Anatomy and function of cholinergic C Bouton inputs to motor neurons. *Journal of Anatomy*, *224*(1), 52–60. https://doi.org/10.1111/joa.12063.

CHAPTER SEVEN

Classical pathways of gene regulation by retinoids

Ehmke Pohl[a,b,c,*], **Charles W.E. Tomlinson**[a]
[a]Department of Chemistry, Durham University, Durham, United Kingdom
[b]Department of Bioscience, Durham University, Durham, United Kingdom
[c]Biophysical Sciences Institute, Durham University, Durham, United Kingdom
[*]Corresponding author: e-mail address: ehmke.pohl@durham.ac.uk

Contents

Abstract

Retinoic acid receptors were discovered during early studies of the actions and mechanisms of essential vitamins. Vitamin A is metabolized in the body to retinoic acid (RA) which is a key compound in the control of many developmental processes in chordates.

Methods in Enzymology, Volume 637
ISSN 0076-6879
https://doi.org/10.1016/bs.mie.2020.03.008

These functions are mediated by a subfamily of nuclear receptors, divided into two classes, the retinoic acid receptors (RAR) and the retinoid X receptors (RXR). Each class is encoded by three closely related genes that are located on different chromosomes. The three proteins in each class are designated α, β and γ, respectively. A wealth of structural studies have shown that they all share the same architecture including a DNA-binding domain connected by a flexible linker to the ligand and co-activator binding domain. Retinoic acid incorporation into the ligand-binding domain leads to a conformational change enabling the formation of RAR homodimers or RAR/RXR heterodimers that in turn bind specifically to target DNA sequences. The consensus sequences located on the promotors of regulated genes are known as retinoic acid response elements (RARE). The activated RAR/RXR homodimers recruit co-activators with histone acetylase activity leading to an opening of the chromatin structure and enabling downstream transcription of regulated genes. These canonical pathways describe the control mechanism for the majority of developmental processes mediated by retinoic acid and its derivatives.

Abbreviations

***13/9-cis*-RA**	*13/9-cis* retinoic acid
ATRA	all-*trans* retinoic acid
CRABP1/2	cellular retinoic acid binding protein 1/2
DHA	docosahexaenoic acid
FPLC	fast protein liquid chromatography
HDAC	histone deacetylase
HPLC	high performance liquid chromatography
NCoR	nuclear Co repressor
RAR	retinoic acid receptor
RARE/RXRE	retinoic acid response element/retinoid x response element
RXR	retinoid X receptor
SMRT	silencing mediator of retinoic and thyroid hormone repressor
TREp	thyroid receptor palindromic sequence

1. Introduction

1.1 Intracellular retinoic acid binding proteins

The cellular all-*trans* retinoic acid (ATRA) binding proteins (CRABP1 and CRABP2) were originally discovered as retinoid binding proteins in the 1970s (Chytil, Page, & Ong, 1975; Sani & Hill, 1974) and later cloned and more fully characterized. Both proteins belong to the large superfamily of fatty-acid binding proteins (FABP) and share a high degree of sequence identity (77%) as well as structural homology with each other (as outlined below). Consequently, they both bind ATRA with low nanomolar or sub-nanomolar dissociation constants. In addition, they can bind the 9-*cis* and 13-*cis* isomers of retinoic acid albeit with approximately one order of magnitude lower affinity compared to ATRA (Chen et al., 1995; Norris, Cheng, Giguere, Rosenberger, & Li, 1994). CRABP1 and CRABP2 are expressed at high levels in many tissues throughout the body and appear to have different primary functions. CRABP1 delivers ATRA and derivatives to catabolic enzymes enabling the Cytochrome P450 catalyzed oxidation and degradation of the compounds to form 4-hydroxy and 4-oxo derivatives, which are later glucuronidated to form soluble metabolites (Ross & Zolfaghari, 2011). The protein is therefore likely to be involved in regulating intracellular ATRA concentrations. In addition, CRABP1 is responsible for cellular activities via non-canonical pathways. Holo-CRABP1 activates the *extra-cellular regulated kinases* 1/2 (ERK1/2) pathway via the activation of the *rapidly accelerated fibrosarcoma* (Raf) kinases (Park et al., 2019; Persaud, Lin, Wu, Kagechika, & Wei, 2013). Furthermore, in certain differentiated cells such as cardiomyocytes, CRABP1 bound to ATRA has been shown to modulate the function of Ca^{2+}/calmodulin dependent protein kinase II (CaMKII) (Park et al., 2019, 2018). More detail on non-genomic retinoic acid signaling pathways can be found in "Non-canonical retinoic acid signalling" by J. Nhieu, Y.-L. Lin and L.-N. Wei.

CRABP2, in contrast, is responsible for delivering ATRA into the nucleus and to the relevant nuclear receptors—the retinoic acid receptors (RARs) (Napoli, 2017). While it was originally proposed that ATRA binding causes a significant conformational change, the crystal structures of apo- and holo-CRABP2 only showed minor changes and an overall rigidification upon ligand binding (Vaezeslami, Mathes, Vasileiou, Borhan, & Geiger, 2006). Although CRABP2 lacks a classical *nuclear localization signal* the holo-protein is recognized by importin-α enabling nuclear import (Sessler & Noy, 2005). Once inside the nucleus CRABP2 interacts directly with the RAR nuclear receptors, transferring the ligand to its target before the apo protein returns to the cytosol (Napoli, 2017). In the cytosol apo-CRABP2 binds the RNA-binding protein HuR increasing its affinity for and creating a stabilizing effect toward AU-rich elements in target mRNA. HuR in turn controls components of the nuclear import machinery responsible for the transport of holo-CRABP2 into the nucleus (Zhang, Vreeland, & Noy, 2016). It is noteworthy that in spite of the multiple functions of CRABP1 and CRABP2, mice with disrupted *crabp1* and/or *crabp2* gene(s) show no obvious morphological phenotype (Lampron et al., 1995).

Instead, studies on *crabp1* knock-out mice indicated a role in stem cell proliferation and brain activities such as memory and learning indicating more subtle and additional roles of the CRABPs that still need to be elucidated (Lin, Persaud, Nhieu, & Wei, 2017).

1.2 Retinoic acid receptors and retinoic X receptors

The first human retinoic acid receptor (RARα) was independently discovered by the groups of Evans and Chambon in 1987 (Giguere, Ong, Segui, & Evans, 1987; Petkovich, Brand, Krust, & Chambon, 1987). This was quickly followed by the identification and characterization of the closely related receptors, RARβ and RARγ (de The, Tiollais, & Dejean, 1990; Kastner et al., 1990). The sequence analyses showed that all three share the same overall sequence organization with other nuclear receptors comprising of six conserved regions designated A–F (Fig. 1). While the overall sequence identities between the RARs range from 72% to 78%, the DNA-binding region (C) with its two DNA-binding motifs is highly conserved with identities over 95%. This domain is followed by a flexible hinge region (D) leading to the ligand-binding domains (E) that share sequence identities ranging from 82% to 87%. While the function of the C-terminal F region, which is not conserved between the RARs, remains unknown the conserved A/B region has been shown to serve as secondary transcriptional activation domain (Nagpal, Friant, Nakshatri, & Chambon, 1993). However, this domain appears to be largely unstructured and no clear activation mechanisms have been suggested so far. To add to the complexity of the system, each of the three RARs exists in different isoforms with a shorter N-terminal regions due to different starting points for transcription or alternative splicing (Giguere et al., 1990; Leroy et al., 1991; Nagpal, Zelent, & Chambon, 1992; Nohno, Muto, Noji, Saito, & Taniguchi, 1991).

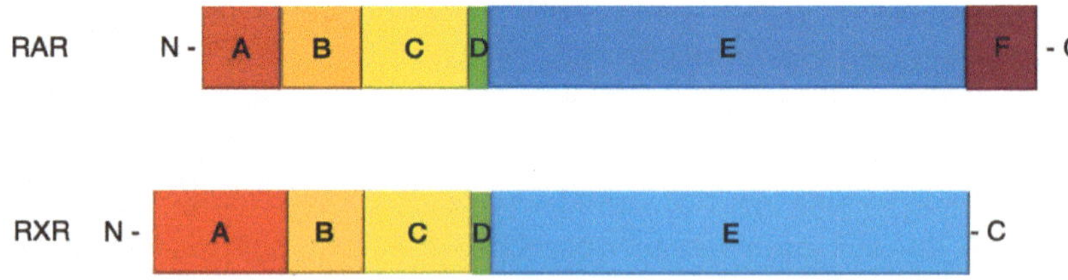

Fig. 1 Schematic gene organization and conserved region of RAR and RXR sequences. A/B: transcriptional activation, C: DNA-binding domain (DBD), D: hinge region E; Ligand-binding domain (LBD), F = unknown function, missing in RXR.

Shortly after the discovery of RARs a closely related class of nuclear receptors, designated retinoid X receptors (RXR) were cloned and characterized (Mangelsdorf, Ong, Dyck, & Evans, 1990; Yu et al., 1991). While they share slightly lower sequence identities with each other (overall 67–73%) and with members of the RAR family (31–35%) they have a similar organization of primary structures with an extended A domain but missing the F region (Fig. 1).

While it is clear that retinoids, in particular ATRA, are the physiologically relevant ligands for RAR proteins, the natural ligands of the RXRs are still being debated. It has been shown that RXRs do not bind ATRA but instead possess a high affinity for the *9-cis* isomer; however, this retinoid cannot be detected in most tissues and therefore is unlikely to be the natural RXR ligand (Heyman et al., 1992). Fatty acids including docosahexaenoic acid (DHA) and other free fatty acids have been suggested to act as natural RXR ligands. (Egea, Mitschler, & Moras, 2002; Krezel, Ruhl, & de Lera, 2019). However, although the ultimate physiologically relevant RXR ligand still needs to be established it is worth noting that RAR-RXR heterodimerization supresses the binding of known RXR ligands (Forman, Umesono, Chen, & Evans, 1995). Nevertheless, synthetic RXR ligands, shown to be modulators of RXR activities, can act as agonists or antagonists and have therefore attracted considerable attention as new therapeutics for the treatment of cancer and metabolic diseases (Dominguez, Alvarez, & de Lera, 2017; Krezel et al., 2019). RARs and RXRs form heterodimers through a tight interfacing of their ligand-binding domains. The DNA-binding domains recognize and bind specific *retinoic acid response elements* (RARE) located in the promotor region of regulated genes (Fig. 2). The classical RARE elements consist of two repeats of a hexameric motif of (A/G)-G-(G/T)-T-C-A separated by one, two or five nucleotides, also designated *direct repeats 1–5* (DR1–DR5) (Bastien & Rochette-Egly, 2004). More recently, a number of non-canonical binding sites with different spacing (DR0, DR8) and inverted repeats (IR0) have been discovered adding to the complexity of retinoic acid signaling (Moutier et al., 2012). Transcription is furthermore tightly regulated by a large number of co-repressors and co-activators. In the absence of an activating ligand such as ATRA or the presence of an *inverse agonist,* the DNA-bound RAR-RXR heterodimer is associated with co-repressors that recruit large complexes with histone deacetylase activity (Perissi, Jepsen, Glass, & Rosenfeld, 2010). Histone tail deacetylation leads to chromatin condensation and repression of gene transcription. CRABP2 is the principal

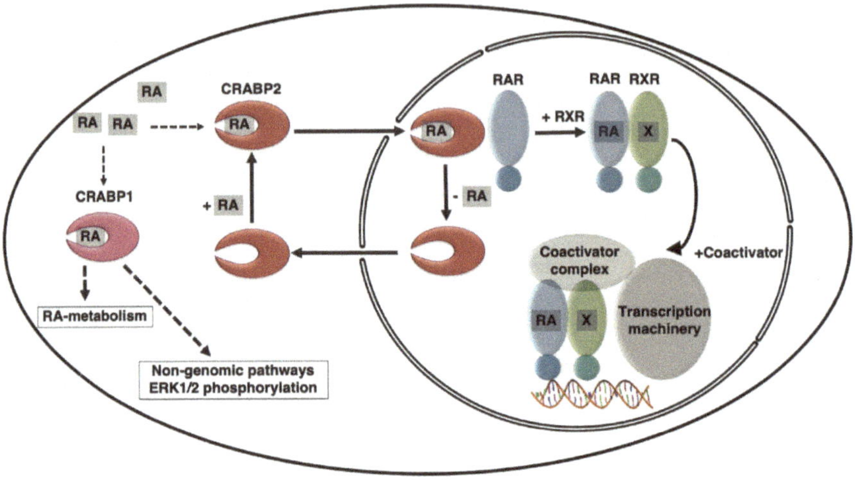

Fig. 2 Simplified schematic of the genomic pathway and principal functions of the cellular retinoic acid binding proteins, CRABP1 and CRABP2, RA = all-*trans* retinoic acid, X = possible RXR ligand.

carrier of ATRA from the cytosol into the nucleus where the protein transfers its ligand to the nuclear receptor leading to a profound conformational change, the dismissal of co-repressors and the recruitment of co-activators with histone acetyl transfer activity (Glass & Rosenfeld, 2000; Westin, Rosenfeld, & Glass, 2000). This reorganization allows the recruitment of the entire transcription machinery, which enables the activated genes to be transcribed (Dilworth & Chambon, 2001).

2. Structures and functions of cellular retinoic acid binding proteins

A large number of high-resolution crystal structure determinations of CRABP1 and CRABP2, respectively, have allowed us to investigate the protein-ligand interactions in atomic detail. Both proteins adopt the same basic structure consisting of two anti-parallel β-sheets creating an orthogonal clam-like fold (Kleywegt et al., 1994). The first two strands are connected by a long linker adopting a pair of anti-parallel α-helices closing over the ligand-binding pocket (Fig. 3A and B).

The comparison of the apo- and holo-forms shows only small conformational changes upon ligand binding. The loop linking the two helices becomes more ordered so that the second helix adopts one more turn on its N-terminus (Fig. 3C). It has been suggested that this creates the

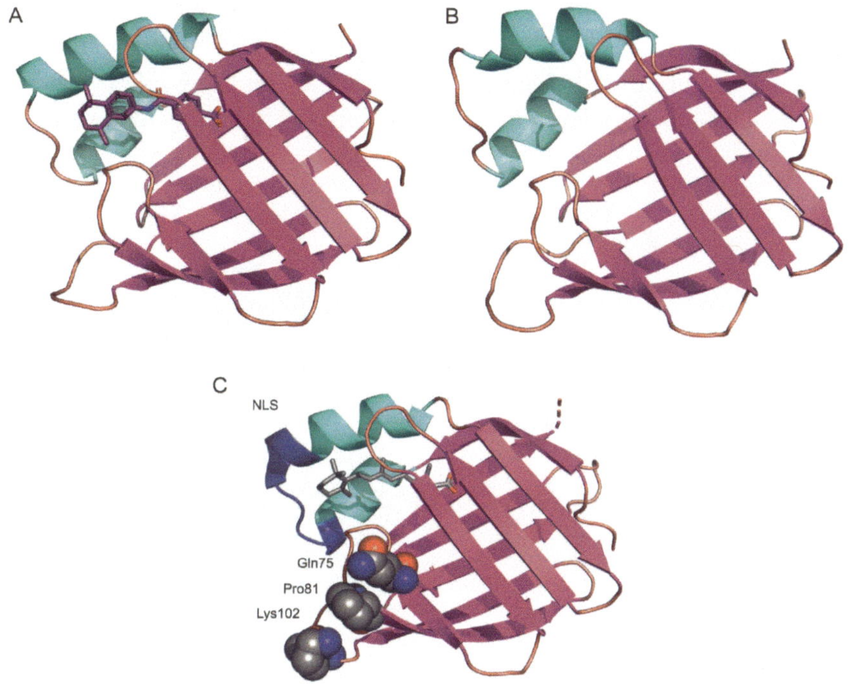

Fig. 3 Ribbon diagram of the crystal structures of cellular retinoic acid binding proteins, (A) CRABP1 in complex with retinobenzoic acid (PDB: 2CBR, Chaudhuri et al., 1999), (B) Apo-CRABP2 (R111M) (PDB: 1XCA, Chen et al., 1998), (C) CRABP2 in complex with ATRA (PDB: 1CBS, Kleywegt et al., 1994). The putative nuclear localization region (NLS) located in the loop linking the two helices are shown on darker shade. The three residues Gln75, Pro81 and Lys102 responsible for RAR binding are shown as CPK model.

three-dimensional non-canonical nuclear localization signal (Sessler & Noy, 2005). In addition, the two helices close slightly over the binding site, making the entire structure more compact and resistant against proteolysis (Napoli, 2017). In CRABP2, three residues, Gln75, Pro81 and Lys102, which are not conserved in CRABP1, located on the other side of the binding site have been shown to be responsible for CRABP2 binding to its nuclear receptor and hence ligand transfer (Budhu, Gillilan, & Noy, 2001).

3. Structural studies and mechanism of the ligand-binding domains of retinoic acid receptors

Crystallographic studies on isolated RAR domains have played a key role in deciphering the molecular mechanism of retinoid signaling.

The ligand-binding domains of all RARs and RXRs share the same overall topology with other nuclear receptors comprising of a compact fold of 12 α-helices with the N-terminal helix H12 playing a key role in signaling (Fig. 4). This helix displays a huge conformational flexibility in the different

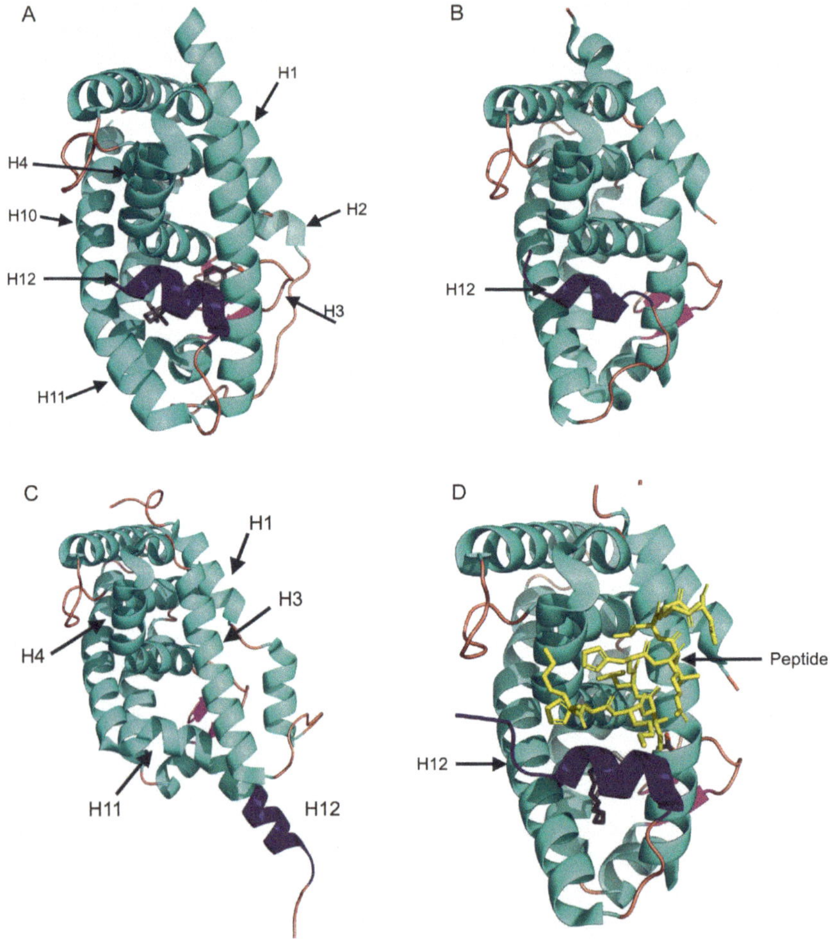

Fig. 4 (A) Ribbon diagram of one monomer of RARβ in complex with a selective retinoid shown as stick representation with the C-terminal helix H12 highlighted; PDB: 1XAP (Germain et al., 2004). (B) RXRα bound to 9-*cis*-RA with the ligand omitted for clarity PDB: 1FBY (Egea et al., 2000). (C) Structure of Apo-RXRα with its C-terminal helix in an elongated conformation PDB: 1LBD/PDB: 6HN6 (Bourguet, Ruff, Bonnier, et al., 1995; Bourguet, Ruff, Chambon, Gronemeyer, & Moras, 1995; Eberhardt, McEwen, Bourguet, Moras, & Dejaegere, 2019). (D) Structure of RXRα with docosahaexanoic acid (DHA) and a co-activator peptide.

crystal structures of homodimers, heterodimers and bound to different ligands and co-regulators. The structure of RARβ in complex with an agonist shows the distinct orientation that provides a platform for heterodimer formation and co-activator binding (Germain et al., 2004). A similar yet not identical conformation has been found for RXRα bound to 9-*cis*-RA (Fig. 4B) (Bourguet, Ruff, Chambon, et al., 1995). However, a number of biophysical studies in solution have shown that the helix adopts multiple conformations depending on the ligands bound (Cordeiro et al., 2019; Lu, Cistola, & Li, 2006; Nagy & Schwabe, 2004; Nahoum et al., 2007). High-resolution crystal structures of the different members of the nuclear receptor family also support the notion of a high degree of flexibility for the C-terminal helix. For example, in the structure of Apo-RXRα shown in Fig. 4C, H12 adopts an elongated conformation opening up the ATRA binding site and allows the ligand to enter (Eberhardt et al., 2019). While the overall fold remains the same there are additional conformational changes in particular in helices H1, H3, H4 and H11 (Fig. 4C).

Coregulators are largely intrinsically disordered proteins that provide platforms for multiple proteins to form large complexes. In the absence of a ligand, nuclear receptors recruit co-repressors with the short linear sequence Lxx**YY**xx**YY** (**Y** = hydrophobic residue). Upon ligand binding the selectivity switches toward the nuclear receptors activator box motif (LxxLL) present in co-activators with histone acetyltransferase activity (Van Roey et al., 2014). The co-crystal structure of RXRα in complex with a decapeptide containing the LxxLL motif shows how the rearrangement of helix H12 forms a hydrophobic patch for co-activator binding (depicted in a stick representation in yellow in Fig. 4D) (Egea et al., 2000).

4. RAR/RXR heterodimer formation

The importance of conformational flexibility of the C-terminal helix H12 is further illustrated by the crystal structure of the heterodimer of the ligand-binding domains of RARβ and RXRα bound to 9-*cis*-RA and the co-activator peptide (Pogenberg et al., 2005). Both proteins, RARβ (shown on the left-hand side in Fig. 5) and RXRα (right-hand side) adopt the *active* conformation upon ligand binding where the C-terminal helix H12 is folded against the core of the protein thus generating the recognition surface for the nuclear receptor peptide (Fig. 5). Based on structural data and

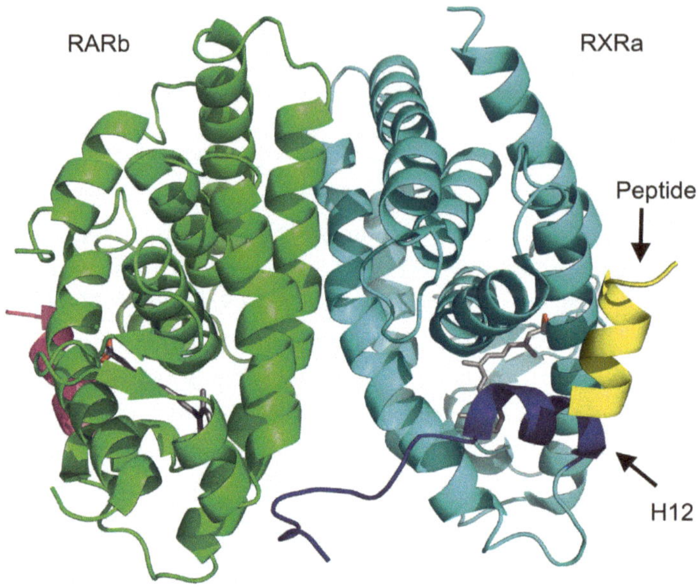

Fig. 5 Ribbon diagram of the RARβ:RXRα heterodimer (PDB: 1XDK) (Pogenberg et al., 2005). The ligands 9-*cis*-RA are shown in a stick representation, the co-activator peptide is shown in ribbon representation.

additional biophysical experiments it has been suggested that the synergist effects of RAR agonists is a result of the two contact areas for the co-repressors on each side of the heterodimer (Pogenberg et al., 2005).

5. DNA recognition and specificity of the DNA-binding domains

The DNA-binding domains (DBD) of nuclear receptors share a core sequence of approximately 67 amino acids, which form a globular domain of three α-helices with two short anti-parallel β-strands. The structure of the isolated DBD, which was first determined in solution by nuclear magnetic resonance (NMR) techniques (Holmbeck et al., 1998; Knegtel et al., 1993), is held together by two Zn-ions each coordinated by four cysteine residues. Soon afterward the crystal structure of the DNA-binding domain of RXR and RAR in complex with DNA was determined at high resolution confirming the overall fold and providing additional molecular detail on DNA recognition (Rastinejad, Wagner, Zhao, & Khorasanizadeh, 2000). The DNA used for crystallization contains the canonical response element consisting of two AGGTCA repeats with a spacing of one base pairs (DR1)

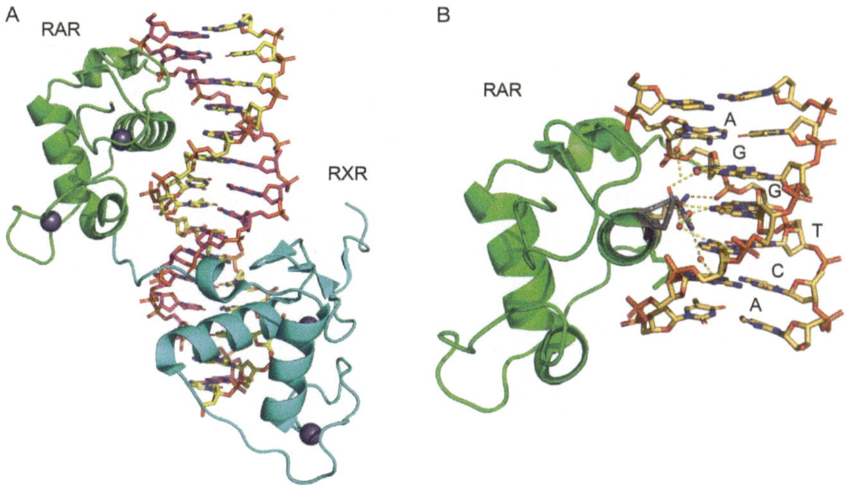

Fig. 6 (A) Ribbon diagram of the crystal structure of the DNA-binding domains of RXR-RAR in complex with the retinoic acid response element DR1, PDB: 1DSZ (Rastinejad et al., 2000). (B) Close-up of the RAR DBD with its recognition helix located in the major groove and key hydrogen bonds to the DNA shown in dashed lines. Key water molecules are shown as spheres, arginine, lysine and glutamate side chains are depicted in stick representation.

(Umesono, Murakami, Thompson, & Evans, 1991). Both DBDs adopt the same overall conformation with the first helix located in the major groove of DNA (Fig. 6A). The RAR DBD is positioned on the upstream, the RXR DBD on the downstream site of the operator. Arginine, lysine and glutamate residues located in this helix are responsible for DNA-sequence specific recognition either by direct hydrogen bonding or by water-mediated contacts (Fig. 6B). The interface is further stabilized by hydrogen bonds to the phosphate backbone (Rastinejad et al., 2000).

6. Studies toward full-length retinoic acid receptors

Structure determinations of flexible multi-domain proteins remain intrinsically challenging and to date, no structure of a full-length nuclear receptor has been determined by X-ray crystallography or cryo-electron microscopy. However, after extensive crystallization attempts, the crystal structure of the RARβ:RXRα heterodimer containing the LBD and DBD bound to DNA has been determined at a resolution of 3.5 Å (Chandra et al., 2017). In spite of the medium resolution the structure revealed a number of key findings. Both LBDs have a ligand bound, retinoic

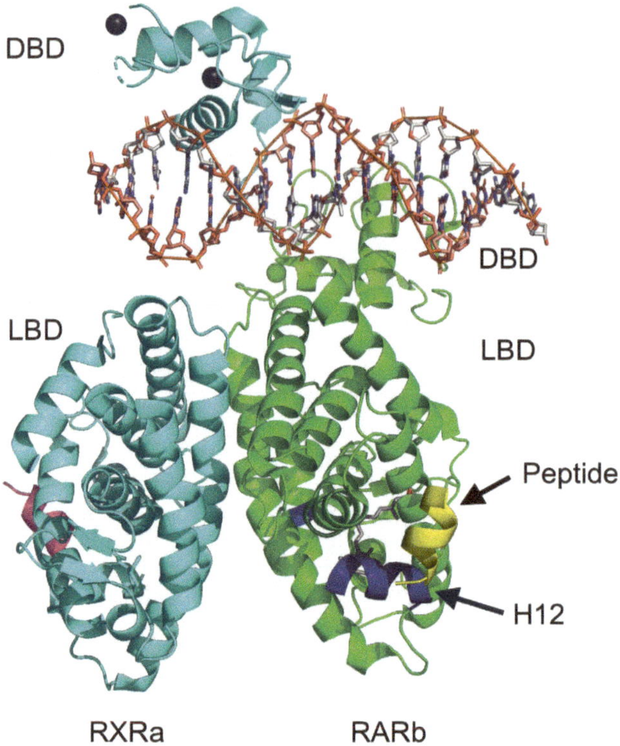

Fig. 7 Ribbon diagram of the crystal structure of the RARβ:RXRα heterodimer bound to DNA, PDB: 5UAN (Chandra et al., 2017). The ligands and the DNA are shown in stick representation. The Zn^{2+} ions of the DNA-binding domains are shown as spheres.

acid in RARβ, and 9-*cis*-retinoic acid in RXRα, leading to an overall conformation of both LBDs with helix H12 in a conformation facilitating co-activator binding (Fig. 7). Importantly, the RARβ:RXRα heterodimer on DNA adopts the same overall fold as the LBDs of the RARβ:RXRα heterodimer alone (Pogenberg et al., 2005). Furthermore, the structure reveals that each of the DBDs is bound to a co-activator peptide harboring the LxxLL motif.

The hinge region linking the DBDs and LBDs is disordered and could therefore not be traced. It is, however, important to note that this inherent flexibility is required to allow the many different heterodimers to be formed, and to recognize the different target DNA sequences and spacings (DR1–DR5, DR0, DR8 and IR0). Nevertheless, the DBD and LBD of RARβ form an intimate interface stabilized by intramolecular hydrogen bonds formed between highly conserved residues located in the loop region following helix 1. This interface may enable a direct link from ligand binding to DNA recognition. Hydrogen/deuterium exchange experiment monitored using mass spectrometric analysis suggested that the C-terminal helix 12 responsible for co-activator peptide binding undergoes conformational changes when an RA antagonist is bound to RARβ. In addition, it has been shown that different ligands lead to distinct and significant changes in DNA affinities with K_D-values ranging from 10 to 70 nM (Chandra et al., 2017). In contrast, the LBD and DBD of RXRα are not physically connected which would allow RARE sequences with different spacers to be bound in a similar way.

These findings strongly support the notion that protein-ligand studies on individual LBDs of nuclear receptors provide important information for ligand design and optimization. Structural studies complemented by a wide range of biophysical, biochemical and biological assays have greatly aided our understanding of the canonical pathways of gene regulation. However, large gaps remain, mainly due to the challenges of obtaining stable full-length proteins and assembling homogeneous complexes with co-repressor or co-activator proteins. It is likely that current advances in structural biology, in particular the on-going revolution in cryo-electron microscopy (Dubochet, 2018; Frank, 2018; Henderson, 2018) will in the near future enable full structure determinations of these multi-component complexes.

7. Nuclear co-repressor and activator complexes

So far over 300 different nuclear receptor co-regulators have been reported and are listed in the Nuclear Receptor Signalling Atlas (www.nursa.org) (McKenna et al., 2009), which from 2020 is succeeded by the Signalling Pathways Project (www.signalingpathways.org) (Ochsner et al., 2019). The possible interactions of nuclear receptor heterodimers with hundreds of co-regulators exhibiting a wide range of enzymatic functions adds a highly complex regulatory network to control gene expression.

The best studied transcriptional co-repressors SMRT/NCoR (Silencing Mediator of Retinoic and Thyroid hormone repressor/Nuclear *Co*-Repressor) are very large polypeptides with over 2400 residues (NCoR1: 2440 aa, NCoR2 2510 aa, sequence identity 38%). These proteins belong to the group of intrinsically disordered proteins with no recognizable folded domains in the sequence. Most co-repressors possess at least two receptor interaction domains (RID) with an α-helical section harboring the RAR/RXR recognition motif (Perissi et al., 1999). Structural studies have so far been restricted to complexes of RAR/RXR dimers with small peptides (Chandra et al., 2017; le Maire et al., 2010). NCoR/SMRT form large multi-component complexes with co-repressors such mSin3 (1273 residues) and histone deacetylases (HDAC) via its Swi3/Ada2/NCoR/TFIIID (SANT) sequence motifs (Heinzel et al., 1997; Nagy et al., 1997; Yu, Li, Ishizuka, Guenther, & Lazar, 2003). Deacetylation of lysines on histone tails and chromatin remodeling ultimately leads to gene silencing (Jenuwein & Allis, 2001; Schreiber & Bernstein, 2002; Xu, Glass, & Rosenfeld, 1999). The exact molecular mechanism involving a range of other components as well as post-translational modifications, however, remain to be elucidated (Mottis, Mouchiroud, & Auwerx, 2013; Stanya & Kao, 2009).

Activated nuclear receptors provide the platform to bind one of the co-activators that share the LxxLL motif located in an α-helical section of the otherwise intrinsically disordered protein. The classical co-activators include the so-called p160 subfamily of steroid receptor co-activators (SRC), such as SRC-1 (also designated NCoA1). NCoA1, which contains as many as seven LxxLL motifs within its 1440 residue polypeptide chain, possess histone acetyltransferase activity (Spencer et al., 1997). Binding of the co-activator leads to the recruitment of a range of other proteins with histone acetyl transferase, histone methyl transferase and histone demethylase activity as well as DNA-dependent ATPases (Perissi et al., 2010; Rosenfeld, Lunyak, & Glass, 2006). These multi-component complexes result in a remodeling of chromatin and the recruitment of the entire transcription machinery (Al Tanoury, Piskunov, & Rochette-Egly, 2013; Dilworth & Chambon, 2001).

While structural studies able to reveal the molecular basis of complex formation are beyond the capabilities of current techniques, the key components for ligand recognition and signaling provide a crucial starting point for the design of synthetic retinoid analogues.

8. Methods

Crystallographic studies have been the key of unraveling the molecular mechanism of ligand binding, co-activator recruitment and DNA recognition. There are many excellent reviews and text books on macromolecular crystallography and protein-ligand co-crystal structure determination that can be consulted (Danley, 2006; Muller, 2017).

Likewise, biophysical methods to investigate protein-ligand interactions such as thermal shift assays (TSA), isothermal titration calorimetry (ITC) and microscale thermophoresis (MST) have been reviewed elsewhere (see for example: Groftehauge, Hajizadeh, Swann, & Pohl, 2015; Huynh & Partch, 2015; Jelesarov & Bosshard, 1999; Jerabek-Willemsen, Wienken, Braun, Baaske, & Duhr, 2011), including in "Protein-protein interaction in the regulation of RAR-RXR heterodimer transcriptional activity" by A. le Maire, P. Germain and W. Bourguet. All molecular methods require the production of pure protein samples. In this chapter we will exemplify protein production of CRABP1 and CRABP2, respectively. Further protocols on production and characterization of proteins involved in signaling can be found on Chapters "Protein-protein interactions in the regulation of RAR–RXR heterodimers transcriptional activity" by A. le Maire, P. Germain, and W. Bourguet; "DNA recognition by retinoic acid nuclear receptors" by C. Peluso-Iltis, J. Osz, and N Rochel; and "The development of methodologies for high-throughput retinoic acid binding assays in drug discovery and beyond" by C.W.E. Tomlinson and A. Whiting, respectively.

Heterologous expression of CRABP1 in *E. coli* was used to produce high purity, homogenous protein for use in a range of biophysical experiment ranging from ligand-binding assays to macromolecular crystallography (Chisholm et al., 2019; Tomlinson, Chisholm, Valentine, Whiting, & Pohl, 2018). The methodology for protein production is outlined below.

8.1 Materials, equipment and reagents

- BL21.DE3 cells transformed with expression plasmid
- Kanamycin stock prepared at 50 mg/mL
- 1 M isopropyl β-D-1-thiogalactopyranoside (IPTG)
- LB growth media
- Orbital shaker with heating control
- High speed centrifuge for spinning down cells

8.2 Protocol

1. CRABP1 was heterologously expressed in *E. coli* BL21.DE3 cells (New England Biolabs) transformed with a pET-28a vector carrying the gene construct. Transformations were carried out under standard conditions, using a 10 s heat shock. Successful transformants were grown in LB and frozen at −80 °C in 50% glycerol.

2. Expression cultures were grown overnight from glycerol stocks by inoculation into 25 ml of LB media (Melford) with 50 µg/mL kanamycin (Sigma Aldrich). The culture was shaken overnight at 150 rpm, 37 °C.
3. The overnight cultures were transferred into 1 L expression flasks (LB/kanamycin) and grown with shaking at 37 °C, 150 rpm. Induction was carried out at an OD600 of 0.6, with 1 mL of 1 M IPTG (1 mM in culture), before shaking overnight (20h) at 25 °C.
4. The resulting cultures were spun down into pellets using an Avanti Hi-Speed centrifuge (JLA 8.1000, 4000 rpm, 25 min, 4 °C), before the supernatant was removed and the bacterial pellet transferred into a 50 mL centrifuge tubes for freezing at −80 °C

8.3 CRABP1 purification

Purification of CRABP1 was carried out using a histidine-nickel affinity method, based on the His_6-tag added to the protein. An initial purification step, followed by His_6-tag cleavage and a second pass rendered the protein pure as analyzed by SDS PAGE gel, though a further cleaning step using size exclusion could be carried out if necessary.

8.4 Materials, equipment and reagents

- Buffers (500 mM NaCl, 20 mM Tris-HCl, pH 8) with suitable imidazole concentrations for gradient elution
- Sonication equipment and high speed centrifuge
- His_6-tag nickel affinity columns, for example HisTrap High Performance columns from GE Healthcare
- Thrombin for cleavage of His_6-tag
- Peristaltic pump or FPLC suitable for loading and eluting columns

8.5 Protocol

1. CRABP1 pellet was resuspended with 15 mL loading buffer (10 mM imidazole, 500 mM NaCl, 20 mM Tris-HCl, pH 8).
2. The resulting suspension was sonicated on ice (40% power, 2 min) and centrifuged (Avanti Hi-Speed JA25.50, 20,000 rpm, 50 min, 4 °C).
3. Supernatant was loaded onto HisTrap column (1 mL, GE healthcare) and washed using 30 mL of loading buffer.
4. The column was washed with 30 mL loading buffer until the UV trace indicated all loading impurities had eluted from the column. Elution was carried out in a smooth gradient with elution buffer (500 mM imidazole,

500 mM NaCl, 20 mM Tris-HCl, pH 8). The resulting protein fractions were analyzed by SDS PAGE and concentration was estimated using Thermo Scientific Nanodrop A280.

5. CRABP1 was cleaved from the His_6-tag using Thrombin (~80 units), overnight, and dialysed into fresh buffer without imidazole (500 mM NaCl, 20 mM Tris-HCl, pH 8)
6. The results of this digest were loaded onto a HisTrap column (1 mL, GE healthcare) and cleaved protein was eluted immediately using 3 mL buffer A into 3 × 1 mL fractions.
7. Bound tag was then eluted with elution buffer and the column washed to remove any other residues.
8. The resulting protein and tag fractions analyzed by SDS PAGE (120 V, 70 min), before use in assays and crystallization experiments.

8.6 Heterologous expression of CRABP2

The routine production of pure CRABP2 protein, heterologously expressed in *E. coli* and purified as described previously (Tomlinson et al., 2018).

8.7 Materials, equipment and reagents

- BL21.DE3 cells transformed with expression plasmid
- Ampicillin prepared at 100 mg/mL
- 1 M Isopropyl β-D-1-thiogalactopyranoside (IPTG)
- LB growth media
- Orbital shaker with heating control
- High speed centrifuge for spinning down cells

8.8 Protocol

1. CRABP2 protein was expressed heterologously in BL21.DE3 (New England Biolabs) cells transformed with a pGEX 4T1 vector carrying the gene construct. Transformations were carried out under standard conditions, using a 10 s heat shock methodology. Successful transformants were grown overnight in LB and frozen for storage at −80 °C in 50% glycerol.
2. Expression cultures were grown overnight by inoculation from glycerol stocks in 25 mL of LB media (Melford) with 100 μg/mL ampicillin (Sigma Aldrich) at 37 °C, shaken at 150 rpm.

3. The overnight cultures were transferred to 1 L expression flasks (LB/ampicillin) and induced after 4 h shaking at 37 °C, 150 rpm with 1 mL of 1 M IPTG (1 mM in culture), before a 20 h overnight shake.
4. The resulting cultures were spun down into pellets using an Avanti Hi-Speed centrifuge (JLA 8.1000, 4000 rpm, 25 min, 4 °C), before the supernatant was removed and the bacterial pellet transferred into a 50 mL centrifuge tubes for freezing at −80 °C.

8.9 CRABP2 purification

Purification of the CRABP2 protein was carried out using a GST affinity chromatography procedure, based on a glutathione-s-transferase tag appended to the protein construct with a thrombin cleavage site (Redfern & Wilson, 1993). The process is outlined below.

8.10 Materials, equipment and reagents

- Lysis buffer—10 mM dithiothreitol (DTT), 100 mM $MgCl_2 \cdot 6H_2O$, in PBS pH 7.5
- Lysozyme
- Suitable sonication equipment and high speed centrifuge
- GST affinity columns—GSTrap High Performance column from GE Healthcare
- Phosphate buffered saline (PBS)
- Thrombin dissolved in PBS (~80 units/mL)
- Reduced glutathione 10 mM
- Peristaltic pump or FPLC suitable for loading and eluting columns

8.11 Protocol

1. Cell pellets were removed from the −80 °C freezer on ice and allowed to slowly defrost.
2. Cell mass was resuspended in 15 mL of cold lysis buffer (10 mM dithiothreitol (DTT), 100 mM $MgCl_2 \cdot 6H_2O$, dissolved in phosphate buffered saline (PBS) pH 7.5) and held on ice with the addition of 0.5 mg/mL lysozyme for 30 min.
3. The resulting suspension was sonicated on ice at 40% power for 2 min and centrifuged to remove cell debris (Avanti Hi-Speed, JA25.50, 20,000 rpm, 50 min, 4 °C).
4. The supernatant recovered from centrifugation was loaded at ~1 mL/min onto a GST affinity column (1 mL, GE healthcare) and washed (1 mL/min) using 30 mL PBS pH 7.5.

5. The column was then loaded with 1 mL thrombin in PBS (80 units, GE Healthcare) using a syringe adaptor to cleave the GST tag and incubated at room temperature for 16 h.
6. The cleaved protein was eluted in 3 × 1 mL fractions using PBS pH 7.5 (0.2 mL/min), and bound GST tag was eluted using 5 mL reduced glutathione (10 mM).
7. The resulting protein fractions were analyzed by SDS PAGE and liquid chromatography electron spray mass spectrometry (LC-ESI-MS). The concentration of each fraction was determined using measurements from a Thermo Scientific Nanodrop A280.
8. Optional additional purification using other separation techniques can be carried out to purify the protein to a satisfactory level—commonly size exclusion using a HiLoad Superdex 75 16/600 column (GE Healthcare).

Acknowledgments

This work was supported by an MRC Case award (MR/N0186481/1) and the MRC Confidence in Concept award (MC_PC_17157). We are grateful for additional financial support provided by HighForce Research Ltd. and by the Biophysical Sciences Institute at Durham University. We would like to thank Naomi Metting van Rijn and Ian Edwards for their excellent technical support.

References

Al Tanoury, Z., Piskunov, A., & Rochette-Egly, C. (2013). Vitamin A and retinoid signaling: Genomic and nongenomic effects. *Journal of Lipid Research, 54*, 1761–1775.

Bastien, J., & Rochette-Egly, C. (2004). Nuclear retinoid receptors and the transcription of retinoid-target genes. *Gene, 328*, 1–16.

Bourguet, W., Ruff, M., Bonnier, D., Granger, F., Boeglin, M., Chambon, P., et al. (1995). Purification, functional characterization, and crystallization of the ligand binding domain of the retinoid X receptor. *Protein Expression and Purification, 6*, 604–608.

Bourguet, W., Ruff, M., Chambon, P., Gronemeyer, H., & Moras, D. (1995). Crystal structure of the ligand-binding domain of the human nuclear receptor RXR-alpha. *Nature, 375*, 377–382.

Budhu, A., Gillilan, R., & Noy, N. (2001). Localization of the RAR interaction domain of cellular retinoic acid binding protein-II. *Journal of Molecular Biology, 305*, 939–949.

Chandra, V., Wu, D., Li, S., Potluri, N., Kim, Y., & Rastinejad, F. (2017). The quaternary architecture of RARbeta-RXRalpha heterodimer facilitates domain-domain signal transmission. *Nature Communications, 8*, 868.

Chaudhuri, B. N., Kleywegt, G. J., Broutin-L'Hermite, I., Bergfors, T., Senn, H., Le Motte, P., et al. (1999). Structures of cellular retinoic acid binding proteins I and II in complex with synthetic retinoids. *Acta Crystallographica. Section D, Biological Crystallography, 55*, 1850–1857.

Chen, X., Tordova, M., Gilliland, G. L., Wang, L., Li, Y., Yan, H., et al. (1998). Crystal structure of apo-cellular retinoic acid-binding protein type II (R111M) suggests a mechanism of ligand entry. *Journal of Molecular Biology, 278*, 641–653.

Chen, L. X., Zhang, Z. P., Scafonas, A., Cavalli, R. C., Gabriel, J. L., Soprano, K. J., et al. (1995). Arginine 132 of cellular retinoic acid-binding protein (type II) is important for binding of retinoic acid. *The Journal of Biological Chemistry, 270*, 4518–4525.

Chisholm, D. R., Tomlinson, C. W. E., Zhou, G. L., Holden, C., Affleck, V., Lamb, R., et al. (2019). Fluorescent retinoic acid analogues as probes for biochemical and intracellular characterization of retinoid signaling pathways. *ACS Chemical Biology, 14*, 369–377.

Chytil, F., Page, D. L., & Ong, D. E. (1975). Presence of cellular retinol and retinoic acid binding proteins in human uterus. *International Journal for Vitamin and Nutrition Research, 45*, 293–298.

Cordeiro, T. N., Sibille, N., Germain, P., Barthe, P., Boulahtouf, A., Allemand, F., et al. (2019). Interplay of protein disorder in retinoic acid receptor heterodimer and its corepressor regulates gene expression. *Structure, 27*(1270–1285) e6.

Danley, D. E. (2006). Crystallization to obtain protein-ligand complexes for structure-aided drug design. *Acta Crystallographica. Section D, Biological Crystallography, 62*, 569–575.

de The, H., Tiollais, P., & Dejean, A. (1990). The retinoic acid receptors. *Nouvelle Revue Française d'Hématologie, 32*, 30–32.

Dilworth, F. J., & Chambon, P. (2001). Nuclear receptors coordinate the activities of chromatin remodeling complexes and coactivators to facilitate initiation of transcription. *Oncogene, 20*, 3047–3054.

Dominguez, M., Alvarez, S., & de Lera, A. R. (2017). Natural and structure-based RXR ligand scaffolds and their functions. *Current Topics in Medicinal Chemistry, 17*, 631–662.

Dubochet, J. (2018). On the development of electron cryo-microscopy (Nobel Lecture). *Angewandte Chemie (International Ed. in English), 57*, 10842–10846.

Eberhardt, J., McEwen, A. G., Bourguet, W., Moras, D., & Dejaegere, A. (2019). A revisited version of the apo structure of the ligand-binding domain of the human nuclear receptor retinoic X receptor alpha. *Acta Crystallographica. Section F, Structural Biology Communication, 75*, 98–104.

Egea, P. F., Mitschler, A., & Moras, D. (2002). Molecular recognition of agonist ligands by RXRs. *Molecular Endocrinology, 16*, 987–997.

Egea, P. F., Mitschler, A., Rochel, N., Ruff, M., Chambon, P., & Moras, D. (2000). Crystal structure of the human RXRalpha ligand-binding domain bound to its natural ligand: 9-cis retinoic acid. *The EMBO Journal, 19*, 2592–2601.

Forman, B. M., Umesono, K., Chen, J., & Evans, R. M. (1995). Unique response pathways are established by allosteric interactions among nuclear hormone receptors. *Cell, 81*, 541–550.

Frank, J. (2018). Single-particle reconstruction of biological molecules-story in a sample (Nobel Lecture). *Angewandte Chemie (International Ed. in English), 57*, 10826–10841.

Germain, P., Kammerer, S., Perez, E., Peluso-Iltis, C., Tortolani, D., Zusi, F. C., et al. (2004). Rational design of RAR-selective ligands revealed by RARbeta crystal structure. *EMBO Reports, 5*, 877–882.

Giguere, V., Ong, E. S., Segui, P., & Evans, R. M. (1987). Identification of a receptor for the morphogen retinoic acid. *Nature, 330*, 624–629.

Giguere, V., Shago, M., Zirngibl, R., Tate, P., Rossant, J., & Varmuza, S. (1990). Identification of a novel isoform of the retinoic acid receptor gamma expressed in the mouse embryo. *Molecular and Cellular Biology, 10*, 2335–2340.

Glass, C. K., & Rosenfeld, M. G. (2000). The coregulator exchange in transcriptional functions of nuclear receptors. *Genes & Development, 14*, 121–141.

Groftehauge, M. K., Hajizadeh, N. R., Swann, M. J., & Pohl, E. (2015). Protein-ligand interactions investigated by thermal shift assays (TSA) and dual polarization interferometry (DPI). *Acta Crystallographica. Section D, Biological Crystallography, 71*, 36–44.

Heinzel, T., Lavinsky, R. M., Mullen, T. M., Soderstrom, M., Laherty, C. D., Torchia, J., et al. (1997). A complex containing N-CoR, mSin3 and histone deacetylase mediates transcriptional repression. *Nature, 387*, 43–48.

Henderson, R. (2018). From electron crystallography to single particle CryoEM (Nobel Lecture). *Angewandte Chemie (International Ed. in English), 57*, 10804–10825.

Heyman, R. A., Mangelsdorf, D. J., Dyck, J. A., Stein, R. B., Eichele, G., Evans, R. M., et al. (1992). 9-cis retinoic acid is a high affinity ligand for the retinoid X receptor. *Cell, 68*, 397–406.

Holmbeck, S. M., Foster, M. P., Casimiro, D. R., Sem, D. S., Dyson, H. J., & Wright, P. E. (1998). High-resolution solution structure of the retinoid X receptor DNA-binding domain. *Journal of Molecular Biology, 281*, 271–284.

Huynh, K., & Partch, C. L. (2015). Analysis of protein stability and ligand interactions by thermal shift assay. *Current Protocols in Protein Science, 79*, 28.9.1–28.9.14.

Jelesarov, I., & Bosshard, H. R. (1999). Isothermal titration calorimetry and differential scanning calorimetry as complementary tools to investigate the energetics of biomolecular recognition. *Journal of Molecular Recognition, 12*, 3–18.

Jenuwein, T., & Allis, C. D. (2001). Translating the histone code. *Science, 293*, 1074–1080.

Jerabek-Willemsen, M., Wienken, C. J., Braun, D., Baaske, P., & Duhr, S. (2011). Molecular interaction studies using microscale thermophoresis. *Assay and Drug Development Technologies, 9*, 342–353.

Kastner, P., Krust, A., Mendelsohn, C., Garnier, J. M., Zelent, A., Leroy, P., et al. (1990). Murine isoforms of retinoic acid receptor gamma with specific patterns of expression. *Proceedings of the National Academy of Sciences of the United States of America, 87*, 2700–2704.

Kleywegt, G. J., Bergfors, T., Senn, H., Le Motte, P., Gsell, B., Shudo, K., et al. (1994). Crystal structures of cellular retinoic acid binding proteins I and II in complex with all-trans-retinoic acid and a synthetic retinoid. *Structure, 2*, 1241–1258.

Knegtel, R. M., Katahira, M., Schilthuis, J. G., Bonvin, A. M., Boelens, R., Eib, D., et al. (1993). The solution structure of the human retinoic acid receptor-beta DNA-binding domain. *Journal of Biomolecular NMR, 3*, 1–17.

Krezel, W., Ruhl, R., & de Lera, A. R. (2019). Alternative retinoid X receptor (RXR) ligands. *Molecular and Cellular Endocrinology, 491*, 110436.

Lampron, C., Rochette-Egly, C., Gorry, P., Dolle, P., Mark, M., Lufkin, T., et al. (1995). Mice deficient in cellular retinoic acid binding protein II (CRABPII) or in both CRABPI and CRABPII are essentially normal. *Development, 121*, 539–548.

le Maire, A., Teyssier, C., Erb, C., Grimaldi, M., Alvarez, S., de Lera, A. R., et al. (2010). A unique secondary-structure switch controls constitutive gene repression by retinoic acid receptor. *Nature Structural & Molecular Biology, 17*, 801–807.

Leroy, P., Krust, A., Zelent, A., Mendelsohn, C., Garnier, J. M., Kastner, P., et al. (1991). Multiple isoforms of the mouse retinoic acid receptor alpha are generated by alternative splicing and differential induction by retinoic acid. *The EMBO Journal, 10*, 59–69.

Lin, Y. L., Persaud, S. D., Nhieu, J., & Wei, L. N. (2017). Cellular retinoic acid-binding protein 1 modulates stem cell proliferation to affect learning and memory in male mice. *Endocrinology, 158*, 3004–3014.

Lu, J., Cistola, D. P., & Li, E. (2006). Analysis of ligand binding and protein dynamics of human retinoid X receptor alpha ligand-binding domain by nuclear magnetic resonance. *Biochemistry, 45*, 1629–1639.

Mangelsdorf, D. J., Ong, E. S., Dyck, J. A., & Evans, R. M. (1990). Nuclear receptor that identifies a novel retinoic acid response pathway. *Nature, 345*, 224–229.

McKenna, N. J., Cooney, A. J., DeMayo, F. J., Downes, M., Glass, C. K., Lanz, R. B., et al. (2009). Minireview: Evolution of NURSA, the Nuclear Receptor Signaling Atlas. *Molecular Endocrinology, 23*, 740–746.

Mottis, A., Mouchiroud, L., & Auwerx, J. (2013). Emerging roles of the corepressors NCoR1 and SMRT in homeostasis. *Genes & Development*, *27*, 819–835.

Moutier, E., Ye, T., Choukrallah, M. A., Urban, S., Osz, J., Chatagnon, A., et al. (2012). Retinoic acid receptors recognize the mouse genome through binding elements with diverse spacing and topology. *The Journal of Biological Chemistry*, *287*, 26328–26341.

Muller, I. (2017). Guidelines for the successful generation of protein-ligand complex crystals. *Acta Crystallographica. Section D, Structural Biology*, *73*, 79–92.

Nagpal, S., Friant, S., Nakshatri, H., & Chambon, P. (1993). RARs and RXRs: Evidence for two autonomous transactivation functions (AF-1 and AF-2) and heterodimerization in vivo. *The EMBO Journal*, *12*, 2349–2360.

Nagpal, S., Zelent, A., & Chambon, P. (1992). RAR-beta 4, a retinoic acid receptor isoform is generated from RAR-beta 2 by alternative splicing and usage of a CUG initiator codon. *Proceedings of the National Academy of Sciences of the United States of America*, *89*, 2718–2722.

Nagy, L., Kao, H. Y., Chakravarti, D., Lin, R. J., Hassig, C. A., Ayer, D. E., et al. (1997). Nuclear receptor repression mediated by a complex containing SMRT, mSin3A, and histone deacetylase. *Cell*, *89*, 373–380.

Nagy, L., & Schwabe, J. W. (2004). Mechanism of the nuclear receptor molecular switch. *Trends in Biochemical Sciences*, *29*, 317–324.

Nahoum, V., Perez, E., Germain, P., Rodriguez-Barrios, F., Manzo, F., Kammerer, S., et al. (2007). Modulators of the structural dynamics of the retinoid X receptor to reveal receptor function. *Proceedings of the National Academy of Sciences of the United States of America*, *104*, 17323–17328.

Napoli, J. L. (2017). Cellular retinoid binding-proteins, CRBP, CRABP, FABP5: Effects on retinoid metabolism, function and related diseases. *Pharmacology & Therapeutics*, *173*, 19–33.

Nohno, T., Muto, K., Noji, S., Saito, T., & Taniguchi, S. (1991). Isoforms of retinoic acid receptor beta expressed in the chicken embryo. *Biochimica et Biophysica Acta*, *1089*, 273–275.

Norris, A. W., Cheng, L., Giguere, V., Rosenberger, M., & Li, E. (1994). Measurement of subnanomolar retinoic acid binding affinities for cellular retinoic acid binding proteins by fluorometric titration. *Biochimica et Biophysica Acta*, *1209*, 10–18.

Ochsner, S. A., Abraham, D., Martin, K., Ding, W., McOwiti, A., Kankanamge, W., et al. (2019). The signaling pathways project, an integrated 'omics knowledgebase for mammalian cellular signaling pathways. *Scientific Data*, *6*, 252.

Park, S. W., Nhieu, J., Persaud, S. D., Miller, M. C., Xia, Y., Lin, Y. W., et al. (2019). A new regulatory mechanism for Raf kinase activation, retinoic acid-bound Crabp1. *Scientific Reports*, *9*, 10929.

Park, S. W., Persaud, S. D., Ogokeh, S., Meyers, T. A., Townsend, D., & Wei, L. N. (2018). CRABP1 protects the heart from isoproterenol-induced acute and chronic remodeling. *The Journal of Endocrinology*, *236*, 151–165.

Perissi, V., Dasen, J. S., Kurokawa, R., Wang, Z., Korzus, E., Rose, D. W., et al. (1999). Factor-specific modulation of CREB-binding protein acetyltransferase activity. *Proceedings of the National Academy of Sciences of the United States of America*, *96*, 3652–3657.

Perissi, V., Jepsen, K., Glass, C. K., & Rosenfeld, M. G. (2010). Deconstructing repression: Evolving models of co-repressor action. *Nature Reviews Genetics*, *11*, 109–123.

Persaud, S. D., Lin, Y. W., Wu, C. Y., Kagechika, H., & Wei, L. N. (2013). Cellular retinoic acid binding protein I mediates rapid non-canonical activation of ERK1/2 by all-trans retinoic acid. *Cellular Signalling*, *25*, 19–25.

Petkovich, M., Brand, N. J., Krust, A., & Chambon, P. (1987). A human retinoic acid receptor which belongs to the family of nuclear receptors. *Nature*, *330*, 444–450.

Pogenberg, V., Guichou, J. F., Vivat-Hannah, V., Kammerer, S., Perez, E., Germain, P., et al. (2005). Characterization of the interaction between retinoic acid receptor/retinoid X receptor (RAR/RXR) heterodimers and transcriptional coactivators through structural and fluorescence anisotropy studies. *The Journal of Biological Chemistry, 280*, 1625–1633.

Rastinejad, F., Wagner, T., Zhao, Q., & Khorasanizadeh, S. (2000). Structure of the RXR-RAR DNA-binding complex on the retinoic acid response element DR1. *The EMBO Journal, 19*, 1045–1054.

Redfern, C. P. F., & Wilson, K. E. (1993). Ligand binding properties of human cellular retinoic acid binding protein II expressed in E. coli as a glutathione-S-transferase fusion protein. *FEBS Letters, 321*, 163–168.

Rosenfeld, M. G., Lunyak, V. V., & Glass, C. K. (2006). Sensors and signals: A coactivator/corepressor/epigenetic code for integrating signal-dependent programs of transcriptional response. *Genes & Development, 20*, 1405–1428.

Ross, A. C., & Zolfaghari, R. (2011). Cytochrome P450s in the regulation of cellular retinoic acid metabolism. *Annual Review of Nutrition, 31*, 65–87.

Sani, B. P., & Hill, D. L. (1974). Retinoic acid: A binding protein in chick embryo metatarsal skin. *Biochemical and Biophysical Research Communications, 61*, 1276–1282.

Schreiber, S. L., & Bernstein, B. E. (2002). Signaling network model of chromatin. *Cell, 111*, 771–778.

Sessler, R. J., & Noy, N. (2005). A ligand-activated nuclear localization signal in cellular retinoic acid binding protein-II. *Molecular Cell, 18*, 343–353.

Spencer, T. E., Jenster, G., Burcin, M. M., Allis, C. D., Zhou, J., Mizzen, C. A., et al. (1997). Steroid receptor coactivator-1 is a histone acetyltransferase. *Nature, 389*, 194–198.

Stanya, K. J., & Kao, H. Y. (2009). New insights into the functions and regulation of the transcriptional corepressors SMRT and N-CoR. *Cell Division, 4*, 7.

Tomlinson, C. W. E., Chisholm, D. R., Valentine, R., Whiting, A., & Pohl, E. (2018). Novel fluorescence competition assay for retinoic acid binding proteins. *ACS Medicinal Chemistry Letters, 9*, 1297–1300.

Umesono, K., Murakami, K. K., Thompson, C. C., & Evans, R. M. (1991). Direct repeats as selective response elements for the thyroid hormone, retinoic acid, and vitamin D3 receptors. *Cell, 65*, 1255–1266.

Vaezeslami, S., Mathes, E., Vasileiou, C., Borhan, B., & Geiger, J. H. (2006). The structure of Apo-wild-type cellular retinoic acid binding protein II at 1.4 A and its relationship to ligand binding and nuclear translocation. *Journal of Molecular Biology, 363*, 687–701.

Van Roey, K., Uyar, B., Weatheritt, R. J., Dinkel, H., Seiler, M., Budd, A., et al. (2014). Short linear motifs: Ubiquitous and functionally diverse protein interaction modules directing cell regulation. *Chemical Reviews, 114*, 6733–6778.

Westin, S., Rosenfeld, M. G., & Glass, C. K. (2000). Nuclear receptor coactivators. *Advances in Pharmacology, 47*, 89–112.

Xu, L., Glass, C. K., & Rosenfeld, M. G. (1999). Coactivator and corepressor complexes in nuclear receptor function. *Current Opinion in Genetics & Development, 9*, 140–147.

Yu, V. C., Delsert, C., Andersen, B., Holloway, J. M., Devary, O. V., Naar, A. M., et al. (1991). RXR beta: A coregulator that enhances binding of retinoic acid, thyroid hormone, and vitamin D receptors to their cognate response elements. *Cell, 67*, 1251–1266.

Yu, J., Li, Y., Ishizuka, T., Guenther, M. G., & Lazar, M. A. (2003). A SANT motif in the SMRT corepressor interprets the histone code and promotes histone deacetylation. *The EMBO Journal, 22*, 3403–3410.

Zhang, W., Vreeland, A. C., & Noy, N. (2016). RNA-binding protein HuR regulates nuclear import of protein. *Journal of Cell Science, 129*, 4025–4033.

CHAPTER EIGHT

Protein-protein interactions in the regulation of RAR–RXR heterodimers transcriptional activity

Albane le Maire, Pierre Germain, William Bourguet*
Center for Structural Biochemistry (CBS), Univ Montpellier, INSERM, CNRS, Montpellier, France
*Corresponding author: e-mail address: william.bourguet@cbs.cnrs.fr

Contents

Abstract

The three retinoic acid receptor subtypes (RARα, RARβ and RARγ) act as ligand-inducible transcription factors binding to DNA regulatory elements in the promoter regions of target genes by forming heterodimers with the retinoid X receptors (RXRα, RXRβ and RXRγ). They act as ligand-dependent transcription factors that regulate a large variety of genes involved in cell growth, differentiation, survival and death. The (patho)

Methods in Enzymology, Volume 637
ISSN 0076-6879
https://doi.org/10.1016/bs.mie.2020.02.007

physiological functions of RAR–RXR heterodimers rely on a dynamic sequence of protein-protein interactions, many of which being modulated by natural (retinoic acid) or synthetic ligands. Direct protein-protein interactions include heterodimerization between RARs and RXRs, recruitment (and release) of transcriptional coactivators and corepressors, cross-talk with other transcription factors, including nuclear receptors, or transient association with many enzymes involved in post-translational modifications to cite the most prominent ones. This chapter describes structural, biochemical, biophysical and cell-based assays to monitor protein-protein interactions relevant to the retinoic acid signaling pathways with a focus on those for which a structural description has been provided.

1. Introduction

All-*trans*-retinoic acid (RA) plays crucial roles in a wide variety of biological processes including embryonic morphogenesis and organogenesis, cell proliferation, differentiation and apoptosis, homeostasis, as well as in their disorders. These pleiotropic effects are mediated through retinoic acid receptors (RARs) consisting of three subtypes, α (NR1B1), β (NR1B2), and γ (NR1B3) belonging to the nuclear receptor (NR) superfamily (Germain et al., 2006a). RARs form functional heterodimers with retinoid X receptors (RXR α, β and γ, NR2B1-3) which bind to specific RA-responsive elements (RAREs) located in target gene promoters and regulate gene expression in a retinoid dependent manner (Germain et al., 2006b; Gilardi & Desvergne, 2014). Like other members of the NR family, RARs and RXRs display a modular structure with a variable N-terminal activation function (AF-1), a central conserved DNA-binding domain (DBD) and a C-terminal ligand-binding domain (LBD) (Fig. 1A). The multifunctional LBD is responsible for ligand binding or dimerization and contains a ligand-dependent activation function (AF-2), which corresponds to coregulator interaction surfaces that can be modulated by natural or pharmacological ligands (Gronemeyer, Gustafsson, & Laudet, 2004). The N-terminal domain is essentially unstructured, yet it contains phosphorylation sites which are known to modulate the interaction with Src-homology-3 (SH3) and tryptophan-tryptophan (WW) domains.

In contrast both DBD and LBD display prototypical folds. The DBD, whose core is approximately 66 residues, comprises two zinc-binding motifs with one N-terminal recognition helix inserting into the major groove of the DNA and a second perpendicular helix stabilizing the core of the DBD and the interaction (Fig. 2). The LBDs of RARs and RXRs are

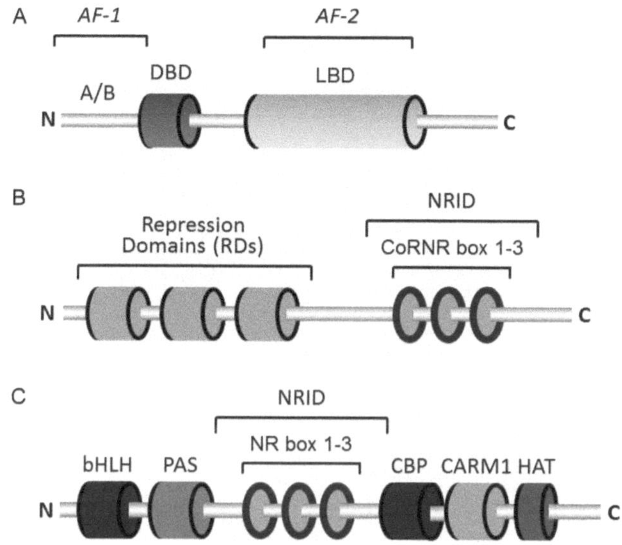

Fig. 1 Schematic illustration of the structural and functional organization of retinoid receptors and coregulators. (A) Like other NRs, RARs and RXRs contain an evolutionary well conserved DNA-binding domain (DBD), a moderately conserved ligand-binding domain (LBD) and a highly divergent N-terminal A/B region. Two transcriptional activation functions have been described in these receptors: a constitutively active AF-1 in region A/B and a ligand-inducible AF-2 in the LBD. (B) The corepressors (N-CoR and SMRT) functional domains illustrated are the C-terminal NR interaction domain (NRID) and the repression domains (RDs). (C) The functional domains of the members of the SRC-1/p160 family of coactivators illustrated are the bHLH/PAS (basic helix-loop-helix-Per/Ah receptor nuclear translocation/Sim motif), the NRID comprising three NR boxes mediating NR interactions and the CREB-binding protein (CBP), the methyltransferase (CARM1), and the histone acetyltransferase (HAT) interaction domains at the C-terminus.

organized in a primarily helical scaffold, termed "antiparallel α-helical sandwich," of 12 helices (H1 – H12) and a short β-turn (S1–S2) arranged in three layers. Helices H1–H3 form one face, H4, H5, S1–S2, H8, and H9 comprise the central layer, and H6, H7, H10 and H11 constitute the second face (Fig. 2). The C-terminal helix H12 is flexible so that its position and dynamics vary according to the nature of the bound ligand, thereby modulating the binding and release of specific transcriptional coregulators. The helical arrangement of the LBD generates a mostly hydrophobic ligand-binding pocket (LBP), a dimerization surface (H7, H9 and H10) and interaction surfaces involved in the binding of multiple coregulators (H3, H4 and H12). Historically, most of the structural information on RARs and RXRs DBDs and LBDs has been gained from studies of isolated

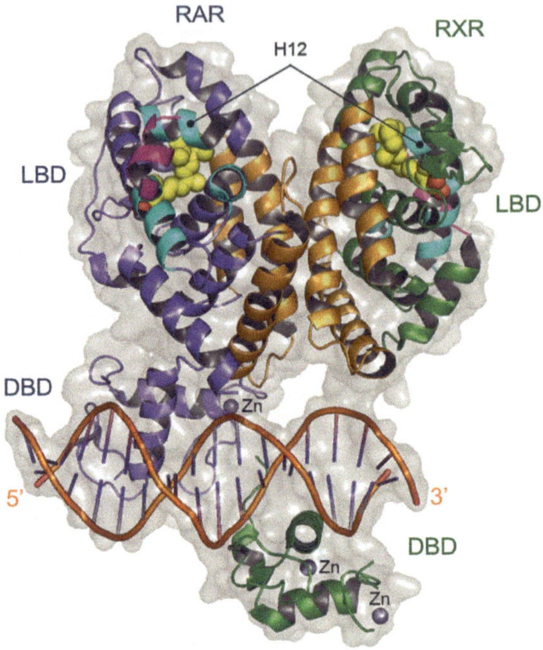

Fig. 2 Crystal structure of RARβ–RXRα heterodimer bound to DNA, ligands and coactivator peptides. Ligands are shown as yellow van der Waals spheres, coactivator peptides in magenta, zinc ions in gray, the dimerization surfaces (H7, H9, H10) in orange and the coactivator binding surfaces on RAR and RXR (H3, H4, H12) in cyan.

domains. It is only recently that the first structures of nearly full-length receptors bound to their cognate response elements have been reported based on crystallographic and solution studies, thereby revealing their overall topologies and the role of DNA in domain organization and communication (Chandra et al., 2017; Rochel et al., 2011) (Fig. 2). Interestingly, the structures of the isolated DBDs and LBDs or within the entire receptors appear very similar, indicating that no major internal domain distortion is required upon establishment of the DNA-bound multidomain heterodimer. Nevertheless, physical interconnections between the various receptor subunit domains may allow the transmission of signals between distant domains in an allosteric manner, making ligand-, DNA-, and coregulator-binding interdependent events (Khorasanizadeh & Rastinejad, 2016).

Retinoid receptors mainly function as ligand-regulated transcription factors whose activity depends on multiple interacting proteins including a diverse group of transcriptional coregulators, other transcription factors or enzymes involved in post-translational modifications, resulting in the cross-talk between different signaling pathways. Note also that a small fraction of RARs is present at the plasma membrane and exert some extra-nuclear and non-transcriptional effects through the activation of kinase signaling pathways (for a review see Piskunov, Al Tanoury, & Rochette-Egly, 2014). RXR–RAR heterodimers may act either as repressors or activators of gene transcription depending on their ligation status that in turn determines the ability of DNA-bound receptors to recruit so-called "corepressors" or "coactivators." Coactivators and corepressors correspond to large groups of proteins with a varied range of protein activities and enzymatic functions that contribute to transcription regulation by many transcription factors. Coactivator recruitment is usually ligand-dependent, whereas corepressors interact in most cases with unliganded receptors. Coactivators can be subdivided into two groups. Primary coactivators, such as those of the TIF-2/SRC-1/RAC3 (p160) family, mediate the interaction of coactivator complexes with NRs. CBP, p300, P/CAF, and some p160 coactivators themselves are reported to act as histone acetyltransferases (HATs) (Glass & Rosenfeld, 2000; Lonard & O'Malley, 2007). They are capable of acetylating specific residues in the N-terminal tails of different histones, a process that is believed to play an important role in the opening of chromatin during transcription activation (Chen, Lin, Xie, Wilpitz, & Evans, 1999; Imhof et al., 1997). Secondary coactivators represent a subgroup of molecules that are constituents of multisubunit coactivator complexes and that also contribute to the enhancement of NR-mediated transcription, but that do not directly contact the NRs. Corepressors N-CoR and SMRT have been shown to reside in, or recruit, high molecular weight complexes that display the opposite activity of coactivator complexes. While coactivator complexes acetylate histones, thereby weakening the interaction of the N-terminal histone tails with the nucleosomal DNA, corepressors recruit histone deacetylases (HDACs) that reverse this process (Heinzel et al., 1997; Nagy et al., 1997). Deacetylated histones are associated with silent regions of the genome, and it is generally accepted that histone acetylation and deacetylation shuffle nucleosomal targets between a condensed and relaxed chromatin configuration, the latter being requisite for transcriptional activation. Some histone-modifying enzymes other than HATs and HDACs, such as chromatin remodelers and histone chaperones, have been documented to serve as NR coregulators (Kato, Yokoyama, & Fujiki, 2011). Following ligand-induced nucleosome remodeling, the recruitment to the promoter region of the transcription machinery

(RNA Polymerase II, General Transcription Factors and the Mediator complex) can occur, leading to the transcription of target genes. Transcription termination could be achieved by RAR ubiquitinylation and degradation by the proteasome.

Progress in NR structure research has revolutionized our view on how NRs in general and retinoid receptors in particular act as essential regulators of fundamental cellular processes. The structural principles of the interaction of receptors with their DNA response elements, ligands and coregulators have been decoded through crystallographic and other biophysical approaches revealing the link between NR conformation, ligand-induced allosteric changes and their resulting abilities to communicate with the intracellular components.

2. Structural basis of RXR–RXR and RXR–RAR interactions

Dimerization is a general mechanism to increase binding site affinity, specificity, and diversity. In this regard, RXRs play a central role in various signal transduction pathways since they can both homodimerize and act as promiscuous heterodimerization partner for almost 15 NRs. Crystal structures of DBD and LBD homo- and heterodimers have defined the surfaces involved in dimerization. As exemplified with the RAR–RXR heterodimer (Fig. 2), two types of dimerization functions mediate homo- and heterodimerization. One involves several surface residues in the DBD that establish weak response element-specific interfaces with corresponding surfaces in the partner DBD. The second is a strong dimerization function in the LBDs of both partners that differs between homo- and heterodimers and to some extent between the partners of RXR. The structures of RXR LBD heterodimers with RAR (Bourguet et al., 2000; Pogenberg et al., 2005; Sato et al., 2010), PPAR (Gampe et al., 2000), TR (Putcha, Wright, Brunzelle, & Fernandez, 2012), LXR (Svensson et al., 2003), FXR (Wang, Zou, Xu, Zhang, & Liu, 2018), PXR (Wallace et al., 2013) and CAR (Suino et al., 2004; Xu et al., 2004) or full-length RARβ–RXRα (Chandra et al., 2017), PPARγ–RXRα (Chandra et al., 2008) and LXRβ–RXRα (Lou et al., 2014) heterodimers demonstrate a topologically conserved dimerization surface with identical structural elements generating the interface (Fig. 2). Residues from helices H7, H9, H10, as well as loops L8–9 and L9–10 of each protomer form an interface comprising a network of complementary hydrophobic and charged residues and further stabilized by neutralized basic and acidic surfaces. Amino acid variations at the surface of the various NRs determine their specific dimerization characteristics.

3. Molecular paradigms governing interactions between RXR–RAR and coregulators

3.1 Coactivators

Coactivators (Fig. 1B) interact directly with the LBD of both RXR and RAR in an agonist-dependent manner by means of short signature sequences located in their nuclear receptor interacting domains (NRIDs). These domains are typically composed of three to four repeats of LxxLL motifs, in which x is any amino acids and L is leucine, embedded in an α-helical peptide (NR box) and connected by disordered linker regions. It is now well established that these motifs are necessary and sufficient to mediate the interaction of coactivators with a cognate surface in the nuclear receptor LBD. A number of crystal structures of RAR bound to both an agonist and a coactivator fragment containing the LxxLL motif have been reported (le Maire et al., 2010; Pogenberg et al., 2005). As with other receptors, the C-terminal helix H12 seals the LBP and together with helices H3, H4, and the loop L3–4 generates a hydrophobic surface which accommodates the short LxxLL-containing helical motif. The helix is held in place *via* interactions of its leucine residues with the hydrophobic receptor groove but also through hydrogen bonds that involve two conserved residues. These amino acids are a lysine at the C-terminus of H3 and a glutamate in H12 (Fig. 3A). Both are hydrogen-bonded to a main-chain peptide bond of the LxxLL motif and together form a "charge clamp" that stabilizes the interaction and defines the precise length of the helical motif that can be accommodated by the cleft.

The fact that coactivators often contain multiple NR boxes, most of which appear to be functional in terms of NR binding *in vitro*, brought up the question of whether this multiplicity reflects redundancy or may confer some specificity to the interface. Structural studies using the entire NRID of several coactivators and various RXR heterodimers have been conducted by Small Angle X-ray Scattering (SAXS), analytical ultracentrifugation, Nuclear Magnetic Resonance (NMR), Hydrogen-Deuterium exchange coupled to mass spectrometry (HDX-MS) or Förster Resonance Energy Transfer (FRET). The data reveal that ligand and DNA binding cooperatively enhance coactivator interaction, thereby suggesting that allosteric

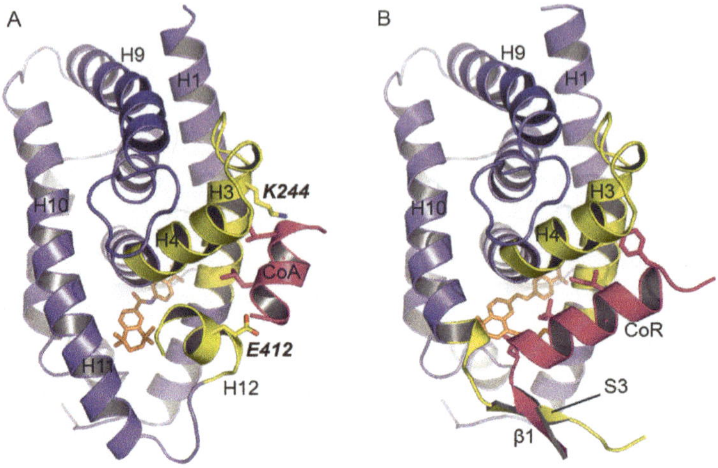

Fig. 3 Structural basis of the interaction between RAR and coregulators. (A) Overall architecture of RAR LBD bound to a synthetic agonist (AM580 shown in orange) and a coactivator-derived FxxLF peptide (magenta ribbon). Helices H3, H4 and H12 constitutive of the AF-2 surface are colored in yellow. Conserved charge-clamp residues are labeled. (B) Overall architecture of RAR LBD bound to a synthetic inverse agonist (BMS493 shown in orange) and a peptide containing the CoRNR box1 sequence of NCoR (magenta ribbon). Helices H3, H4 and S3 constitutive of the interaction surface on RAR are colored in yellow.

communication integrating signals from ligand, DNA and coactivator play a role in DNA recognition, promoter specificity and gene regulation (de Vera et al., 2017; Zhang et al., 2011). However the details of the interaction, including stoichiometry or the involvement of each protomers of the heterodimers and of the LxxLL motifs of the coactivators, remain poorly understood. Whereas ligand binding to RXR was shown to be the primary driver of coactivator recruitment to the PPARγ–RXRα heterodimer (de Vera et al., 2017), another study proposed that the coactivator only binds to the liganded partner receptor within the context of RAR–RXR (Rochel et al., 2011). In the same line, the presence of the ligands of both RXR and the partner receptor (*e.g.*, PPAR, RAR) contribute to enhance the interaction, but whether this ligand combination stabilizes the interaction of one LxxLL motif on each receptor (deck model) is not known.

3.2 Corepressors

In their unliganded form or in the presence of so-called inverse agonists, a number of NRs, including RARs, act as repressors of transcription by recruiting corepressors to their target genes. Analyses of biochemical data and protein sequences provided evidence that coactivator and corepressor recruitments share similar molecular features and involve topologically related NR surfaces, with the important difference that the helix H12 is not involved in the interaction with corepressors. Similarly to coactivators, the NRID of SMRT and N-CoR is an intrinsically disordered domain with three functional LxxI/HIxxxI/L corepressor/nuclear receptor box (CoRNR box 1–3) motifs (Fig. 1C). CoRNR boxes are not equivalent as, for example, RAR interacts preferentially with CoRNR1, and further biochemical studies revealed that residues flanking the core helical sequence determine nuclear receptor specificity (Hu, Li, & Lazar, 2001; Perissi et al., 1999; Xu et al., 2002). The structural basis of the repression function of RAR has been revealed through crystallographic studies of RARα LBD in complex with a peptide containing the CoRNR1 sequence of N-CoR (le Maire et al., 2010). The structure reveals that the interaction of RARα with corepressors utilizes the bipartite motif CoRNR1 that contains a four-turn helical motif LxxxIxxI/VIxxxF/Y docking to a hydrophobic surface formed by residues from helices H3 and H4 of the receptor and a N-terminal extended β-strand (β1) forming a specific antiparallel β-sheet with specific RAR residues (Fig. 3B). Interestingly, the RAR residues which adopt the β-strand conformation S3 in the corepressor-bound structure belong to helix H11 in the agonist-bound structure, and mutations affecting specifically the integrity of either β-strands (β1 or S3) abrogated corepressor interaction and transcriptional repression by RAR (le Maire et al., 2010). The secondary structure change from α-helix to β-strand has also been observed in the structure of the heme receptor RevErb bound to CoRNR1 of N-CoR (Phelan et al., 2010) but not with other NRs that have been shown to interact rather weakly with the shorter CoRNR2 motif of corepressors (Madauss et al., 2007; Wang et al., 2006; Xu et al., 2002). Thus, the basal repressive activity of RAR is conferred by an extended β-strand that forms an antiparallel β-sheet with specific corepressor residues. Agonist binding induces a β-strand to α-helix transition that allows helix H11 formation which in turn provokes corepressor release, stabilization of the mobile helix H12 in the so-called "active conformation," and coactivator recruitment. More recently, combination of a large set of biophysical and computational methods revealed that while the NRID of corepressors is mainly disordered, it presents transient but robust intramolecular contacts upon interaction with the heterodimer, indicating that disorder-to-order transitions of both receptors and corepressors are key events in the regulation of RXR–RAR heterodimers (Cordeiro et al., 2019). The study also showed that the NRID forms a dynamic complex with RAR–RXR with singly and doubly bound states.

4. Protocols for the study of protein–protein interactions

4.1 Two-hybrid analysis

The mammalian two-hybrid assay can be recommended to investigate interactions between nuclear receptors and transcriptional co-regulators in a cellular environment. This cell-based reporter gene assay has been adapted from the principle of yeast two-hybrid system to mammalian cells. This method was originally conceived to reveal protein–protein interaction using the GAL4 transcriptional activator of the yeast *Saccharomyces cerevisiae* (Fields & Song, 1989). The assay is based on the fact that most eukaryotic transcription factors consist of distinct functional and physical domains. The DNA-binding domain (DBD) mediates binding of these proteins to the gene promoter by sequence specific DNA recognition and the activation domain coordinates the assembly of the factors required for transcription allowing RNA polymerase II to transcribe the reporter gene downstream of the DBD. Both domains are required for the induction of gene expression, but they do not have to be present within the same protein. Thus, if the DBD of a transcription factor is fused to a bait protein and a second prey protein is fused to the activation domain of the transcription factor, efficient gene induction will be seen if the two proteins bind to each other, thereby bringing the two modules into spatial association. However, no gene expression will be observed if the two transcription factor modules are not brought in proximity to each other. Reporter genes encode for proteins that are distinguishable and easily quantified in the background of endogenous proteins. The choice of a reporter gene takes into account various parameters such as detection sensitivity, rapidity, convenience in quantitation, and the possibility of measuring gene expression in living cells. These criteria have led to the preferential use of firefly luciferase as the reporter gene in an increasing number of studies. Quantification relies on enzymatic assays of this protein. In addition to the reporter used to measure the transcriptional response, a second reporter, generally the bacterial β-galactosidase or the Renilla luciferase, serves as an internal control to normalize the data obtained from the experimental reporter. Transfection corresponds to the process of inserting foreign genetic material into a eukaryotic living cell without killing it. Virtually all mammalian cell lines can be considered. However, the chosen cell line should be relatively cheap, easy to culture, and robust enough to withstand application of the investigate methods and reagents. In this respect, COS, HeLa, and HEK293T cell lines are widely used. The following is an example of the use of a mammalian two-hybrid system to study NCoR-RARα interactions (Fig. 4).

Required materials

- COS cell line
- Luciferase-based reporter gene ((17m)5x-βGal-Luc), cytomegalovirus–β-galactosidase (CMV-βGal) vector, and BSK plasmid
- pSG5-based Gal-NCoR and RARα-VP16 chimera expression vectors
- JetPei transfectant (Ozyme, France)
- Lysis buffer: 25 mM Tris-phosphate (pH 7.8), 2 mM EDTA, 1 mM DTT, 10% glycerol, and 1% Triton X-100

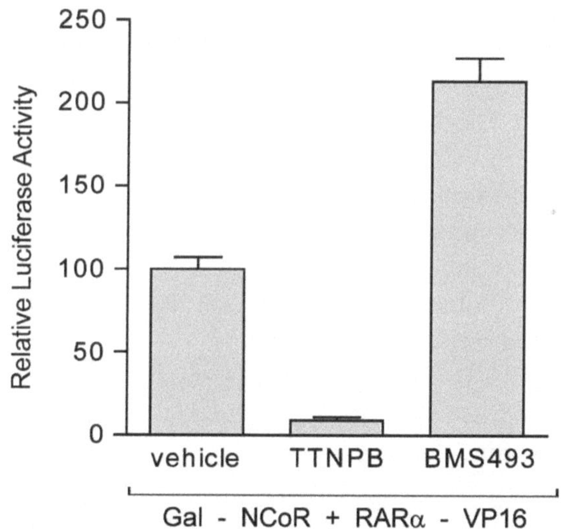

Fig. 4 The chimeric luciferase-based reporter gene ((17m)5x-βGal-Luc) was transiently transfected together with vectors expressing NCoR NRID fused to the Gal DNA-binding domain of the yeast *Saccharomyces cerevisiae* (Gal-NCoR) as bait and RARα LBD fused to the *Herpes simplex* VP16 acidic transcription activation domain as prey (RARα-VP16). As the VP16 domain confers constitutive transcription activation if it is brought close to a promoter, a specific gene induction is seen only if the two proteins bind to each other. In this assay, Gal-NCoR specifically binds to the "17m" DNA recognition site through the Gal DBD and can interact with the RARα LBD in the absence of ligand or in the presence of the inverse agonist BMS493. The RAR agonist TTNPB is able to decrease NCoR association (100%, basal transcriptional activity). Compounds were used at 1 μM in all two-hybrid assays. Error bars, s.e.m.

- Luciferin buffer: 20 mM Tris phosphate (pH 7.8), 1.07 mM $MgCl_2$, 2.67 mM $MgSO_4$, 0.1 mM EDTA, 33.3 mM DTT
- Luciferase buffer (2×): 40 mM Tris-phosphate, 2.14 mM $MgCl_2$, 5.4 $MgSO_4$, 0.2 mM EDTA, 66.6 mM DTT, pH 7.8 (store at −20 °C)
- Luciferase assay reagent: 500 μL 2× luciferase buffer, 47 μL luciferine 10 mM, 53 μL ATP 10 mM, 27 μL Coenzyme A (lithium salt) 10 mM 430 μL distilled water
- βGal buffer: 60 mM Na_2HPO_4 ($12H_2O$), 40 mM NaH_2PO_4, 10 mM KCl, 1 mM $MgCl_2$ ($6H_2O$), 50 mM β-mercaptoethanol
- ONPG (4 mg/mL)
- 24-well plates for cell culture
- Opaque white OptiPlate-96-well plates (Perkin Elmer)
- Luminometer (for instance MicroLumat LB96P, Berthold)

Protocol

1. Maintain COS cells in Dulbecco's Modified Eagle Medium (DMEM) with Glutamax and 10% (v/v) fetal calf serum (FCS)
2. Seed 10^5 cells per well in DMEM with Glutamax and 10% (v/v) FCS
3. For each well, prepare the mix of plasmids in 50 μL of 150 mM NaCl buffer (80 ng Gal-NCoR + 50 ng VP16-RARα + 150 ng (17m)5x-β Gal-Luc + 50 ng CMV-βGal + 670 ng) BSK plasmid; vortex gently and spin down briefly
4. For each well, prepare the JetPei solution (2 μL JetPei in 50 μL of 150 mM NaCl buffer); vortex gently
5. Add the JetPei solution to the DNA solution all at once; vortex gently and spin down briefly
6. Incubate for 15–30 min at room temperature
7. Add 100 μL of the mix drop-wise to the medium in each well and incubate the cells for 24 h at 37 °C in a 5% CO_2 incubator; transfect 3 wells per condition to have triplicate measurements
8. Replace the culture medium with fresh medium containing the ligand at active concentration and incubate again the cells for 24 h at 37 °C in a 5% CO_2 incubator
9. Remove the medium from the well and gently apply phosphate saline buffer (PBS) to wash the surface of the well
10. Dispense into each well 100 μL of 1× lysis buffer and place the plate on an orbital shaker with gentle rocking for 15 min
11. Distribute 30 μL of lysate and 50 μL of luciferase assay reagent per well of a 96-well white plate (luciferase assay reagent can be automatically added if a luminometer with injector is used)

12. Perform the measurement of the luciferase activity using a luminometer
13. Distribute 10 μL of lysate, 150 μL of β-Gal assay buffer and 30 μL of ONPG (4 mg/mL) in a 96-well plate
14. Incubate the plate at 37 °C and measure the time until the yellow color has appeared
15. Stop the reaction by adding 75 μL of 1 M Na_2CO_3
16. Read the absorbance immediately at 420 nM in a plate reader to measure the β-Gal activity (β-Gal unit = 100 × total volume × absorbance/(assay volume × time (h)))
17. Normalize the luciferase values with the β-Gal values.

4.2 Electrophoretic mobility shift assays

The ability of RXR-RAR heterodimers bound to specific DNA sequences to associate with transcriptional co-regulators can be assessed by electrophoretic mobility shift assays (EMSAs). EMSA is also referred as gel retardation assay, gel shift assay, or band shift assay (Hellman & Fried, 2007). In this method, solutions of proteins and nucleic acid are combined and subjected to electrophoresis under native conditions (non-denaturing TBE-polyacrylamide gels or TAE-agarose gels) to resolve protein-DNA complexes from free DNA. The adequate gel percentage is depended on the size of the used linear DNA probe as well as the size, number and charge of the interacting proteins. After electrophoresis, the distribution of species containing nucleic acid is revealed, usually by autoradiography of radiolabeled DNA target with ^{32}P-labeled by 5′ end labeling using an [γ-^{32}P]ATP and T4 polynucleotide kinase. Typically, protein-DNA complexes migrate more slowly than the corresponding free nucleic acid. This method is also recommended for investigating higher-order complexes containing several proteins, observed as a "supershift" assay. Interestingly, proteins of interest may be obtained from a crude nuclear or whole cell extract, *in vitro* transcription product or a purified preparation from various expression systems. Here a protocol is given as example of EMSA to determine the effect of various retinoid combinations on the interaction of both co-repressors (SMRT) and co-activators (TIF2) with the RXRα-RARα heterodimer bound to a specific DR5 DNA sequence (le Maire, Teyssier, Balaguer, Bourguet, & Germain, 2019) (Fig. 5).

Required materials

- The TnT® Quick Coupled Transcription/Translation System (Promega, Madison, WI 53711, USA)

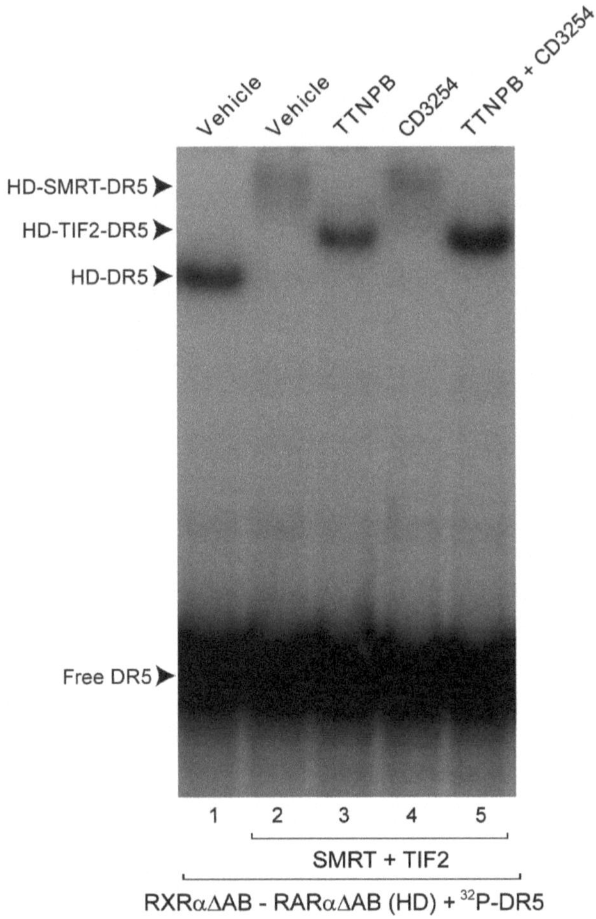

Fig. 5 EMSAs demonstrating ligand-dependent co-regulator recruitment by the RXRαΔAB-RARαΔAB heterodimer (HD). Incubation of RXRαΔAB-RARαΔAB heterodimer bound to the DR5 DNA sequence with a mix of SMRT and TIF2 produced heterodimer-SMRT complexes detectable as a super-shift. SMRT, which dissociated and was replaced by TIF2 in the presence of the RAR agonist TTNPB, remained bound to a heterodimer when exposed to the RXR agonist CD3254 alone despite the presence of TIF2, in accordance with the subordination principle. DR5 oligonucleotide, HD (0.4 μL of TNT reaction products added per assay), SMRT (1 μg per assay), and TIF2 (0.5 μg per assay) were co-incubated in the absence or in the presence of saturated amounts of TTNPB used alone or with CD3254.

- pSG5-based RXRαΔAB and RARαΔAB (receptors deleted of their AB domain, harboring both DBD and LBD) expression vectors
- Purified TIF2 (TIF2.42 residues 624–828 containing the NRID) and SMRT (SMRTct residues 982 to end containing the NRID)

- ^{32}P-DR5 (5′-TCGAGGGTAGGGGTCACCGAAAGGTCACTCG-3′; direct repeat underlined) oligonucleotide; specific radiolabeled DNA sequence recognizing RXR-RAR heterodimers
- Binding buffer: 10 mM Tris–HCl (pH 8.0), 0.1 mM EDTA, 0.4 mM DTT, 5% glycerol
- TBE (10 ×): 108 g Tris base, 55 g boric acid, 4 mL 0.5 M EDTA (pH 8.0)
- Dye (10 ×): 0.1% bromophenol blue, 0.1% xylene cyanol, 60% glycerol
- Vertical electrophoresis apparatus and electrophoresis power supply
- Autoradiography film cassette
- Kodak XAR-5 film (with film developer) or cassette and storage phosphor screen (with a phosphorimager instrument)
- Vacuum gel drier

Protocol

1. Produce RXRαΔAB and RARαΔAB receptors by *in vitro* transcription–translation using the TNT rabbit reticulocyte lysate system programmed with pSG5-based expression vectors according to the instructions provided by the manufacturer
2. Mix lysates containing RXRαΔAB and RARαΔAB on ice for 30 min (volumes of TNT reaction products containing the receptors are indicated in the legend of Fig. 5)
3. Add ligands (stock solution in ethanol) on ice at a final concentration of 1 μM and incubate for 10 min
4. Add both purified transcriptional co-regulators TIF2 and SMRT on ice for 10 min (amounts of co-regulators are indicated in the legend of Fig. 5)
5. Mix Proteins (volumes of TNT reaction products containing the receptors and concentrations of co-regulators are indicated in the legend of Fig. 5) and pre-annealed DNA probe that is labeled with ^{32}P (approximately 25,000 c.p.m.)
6. Incubate for 15 min on ice the mix proteins-DNA in a final volume of 20 μL binding buffer, containing 2 μg poly(dI-dC), as well as 150 mM KCl and 20 mg/mL BSA
7. Add 10% of 10 × dye and carefully mix by pipetting; immediately load onto the gel (use Hamilton syringe)
8. Resolve protein–DNA complexes through 6% polyacrylamide gels in 0.5 × TBE buffer for 2 h at 100 V/4 °C (pre-run the gel for 3 h at 200 V/4 °C)
9. Dry the gel for at least 45 min and expose to X-ray film for desired period of time

4.3 Protein expression and purification

The availability of homogeneous NR monomers and dimers is a prerequisite for biochemical and biophysical studies addressing interactions of NRs with partners (other NRs, coregulators, DNA). A number of protocols have been optimized for such purifications (Bourguet et al., 2000; Iyer et al., 1999). These protocols apply to both isolated LBDs and full-length receptors. The following protocol is an example given for a rapid two-step copurification procedure yielding nonaggregated and functionally homogeneous RARα-RXRα LBD heterodimers in large quantities suitable for structural and other *in vitro* studies.

Required materials

- pET-15b and pET-3a expression vectors (Novagen)
- Competent *E. coli* BL21 (DE3) cells (Novagen)
- Shaker/incubator set at 37 °C
- Sonicator
- Centrifuge
- Liquid chromatography system (AKTA purifier, GE Healthcare)
- Lysis buffer: 20 mM Tris–HCl pH 8.0, 500 mM NaCl
- Elution buffer: 20 mM Tris–HCl pH 8.0, 500 mM NaCl, 500 mM imidazole
- Gel filtration buffer: 20 mM Tris–HCl pH 7.5, 150 mM NaCl, 2 mM DTT
- HisTrap chelating columns (GE Healthcare)
- HiLoad 26/60 Superdex 75 column (GE Healthcare)
- UV–Vis spectrophotometer (NanoDrop, Thermo Scientific)

Protocol

1. Express the RARα LBD and the RXRα LBD cloned, respectively, into the pET-15b vector (as a histidine-tagged protein) or into the pET-3a vector in *E. coli* BL21 (DE3) cells. Cells are grown at 37 °C in LB supplemented with 50 μg/mL of ampicillin to an OD_{600} of 0.6–0.8, and induce T7 RNA polymerase expression by the addition of 1 mM IPTG to 1 mM. After an additional incubation overnight at 25 °C, harvest the cells by centrifugation (5000 *g* during 15 min).
2. Resuspend the pellets together (usually 3 L of RARα LBD and 2 L of RXRα LBD) in 100 mL of ice-cold lysis buffer. Sonicate the suspension three times (60% amplitude) for 2 min and centrifuge (18,000 rpm, 30 min) to obtain the crude extract.
3. Pass the crude extract over a 5 mL HisTrap chelating column previously equilibrated with 98% of lysis buffer + 2% of elution buffer, at a flow rate

of 5 mL/min. Wash the column with 20 volumes of, 98% of lysis buffer + 2% of elution buffer. Remove unspecific bounded protein and potential excess of RXRα LBD by passing 10 volumes of 90% of lysis buffer + 20% of elution buffer. Elute the heterodimer with 20 volumes of 50% of lysis buffer + 50% of elution buffer. Analyze the elution fractions by SDS–polyacrylamide gel electrophoresis, pooled, and concentrated until 5 mL.

4. Optionally, cleave the His tag from RARα LBD by adding thrombin (1 unit per mg of protein), overnight in ice.
5. Further purify the heterodimer by gel filtration using a HiLoad 26/60 Superdex 75 column preequilibrated with the gel filtration buffer. Analyze the 2 mL by SDS–polyacrylamide gel electrophoresis, pooled, and concentrated to 5 mg/mL using Centriprep 30 for subsequent *in vitro* analysis.
6. At this step, the structural homogeneity of the preparations can be evaluated by electrophoretic analysis using nondenaturating conditions (native gels), dynamic light scattering (DLS), or circular dichroism (CD).

4.4 Polarization of fluorescence

Steady-state fluorescence anisotropy relies on the differences in rotational mobility of bound vs free molecule. It is a widely used method for investigating protein–protein, protein–DNA and protein–ligand interactions as polarization measurements are done at equilibrium and can be easily and rapidly monitored. The recruitment of coregulators by RXR and RAR LBDs has been characterized by measuring the affinities (Kd) of fluorescently labeled peptides derives from CoAs and CoRs, in the absence or presence of different ligands (Cordeiro et al., 2019; le Maire et al., 2010; Pogenberg et al., 2005). The following protocol is an example given for the determination of the binding affinities of a peptide of PGC-1α corresponding to the NR box 2 (EEPSLLKKLLLAPA) for RXRα LBD in its unliganded form and in the presence of the full RXRα agonist CD3254 that reinforces the interaction (Fig. 6) (le Maire et al., 2009).

Required materials

- Safire2 microplate reader (TECAN) or CLARIOstar (BMG Labtech)
- Corning NBS 384 well low volume microplates (Corning Incorporated)
- Anisotropy buffer: 20 mM Tris–HCl (pH 7.5), 150 mM NaCl, 1 mM EDTA, 5 mM DTT, 10% glycerol
- Fluorescein-labeled PGC-1α (FITC-EEPSLLKKLLLAPA) at 20 mM in DMSO

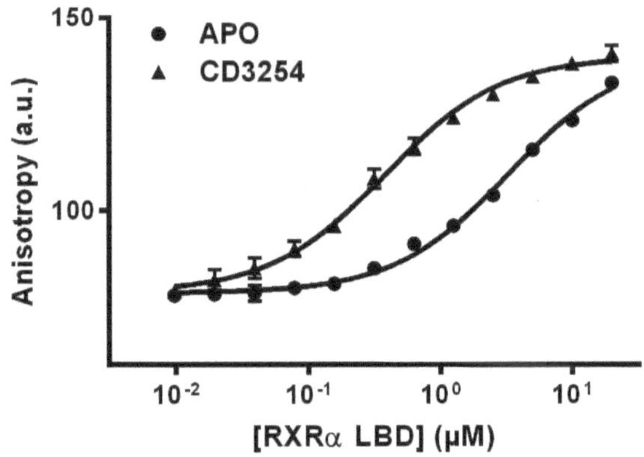

Fig. 6 Titration of fluorescein-labeled PGC-1α peptide by RXRα LBD, in the absence of ligand (APO) or in the presence of the full RXR agonist (CD3254), followed by fluorescence anisotropy. As expected, the synthetic agonist, CD3254, strongly enhances the recruitment of the coactivator peptide, since in the apo condition the Kd is around 3.3 μM against 0.4 μM in the presence of CD3254.

- Ligand CD3254 at 10 mM in DMSO
- Purified RXRα LBD at 5–10 mg/mL

Protocol

1. Prepare 60 μL of protein stock solution diluted at two times the maximal concentration in anisotropy buffer (Cmax) and add the ligand at three molar excess (3*Cmax). In fact, the protein will be diluted 1:1 with the fluorescent target (see step 5), so the final protein concentration will be half of the stock concentration. In order to saturate the binding event, the maximum protein concentration should be 10- to 20-fold above the expected Kd.
2. Dispense 30 μL in the first well. Add 30 μL of anisotropy buffer into the remaining 30 μL of the protein solution prepared in step 1, mix gently. Transfer 30 μL of this mix in the second well.
3. Repeat step 3 until the last well.
4. Prepare 500 μL of peptide solution containing 8 nM of fluorescent peptide in anisotropy buffer and dispense 30 μL in each well. Of anisotropy buffer in an additional well as a reference measurement and to ensure that the total intensity of wells containing the fluorescent peptide is at least 10 times more than ones of this reference well.
5. Determine anisotropy values of the Safire 2 or ClarioStar microplate readers with the excitation wavelength set at 470 nM and emission measured at 530 nM, for a FITC fluorophore.
6. Fit binding data and determine Kd using a sigmoidal dose-response model in GraphPad Prism (GraphPad Software, San Diego).

4.5 Electrospray ionization mass spectrometry

Since the 1990s, electrospray ionization mass spectrometry (ESI-MS) has been regularly used to study noncovalent complexes and offers new possibilities for the study of such complexes, providing direct evidence for their formation and an accurate determination of their binding stoichiometry (Fenn, Mann, Meng, Wong, & Whitehouse, 1989; Loo, 1997; Rajabi, Ashcroft, & Radford, 2015). For instance ESI-MS was previously used to monitor the ligand-mediated interactions between RXRα-RARα heterodimers and transcriptional coregulators (Sanglier et al., 2004). Hence supramolecular mass spectrometry is a powerful tool to rapidly and unambiguously NRs-protein interactions. The following protocol is an example given for the use of ESI-MS under nondenaturating conditions to monitor the interaction between the RXR–RAR LBD heterodimer and CoRNR1 of the corepressor NCoR (Fig. 7).

Required materials

- Electrospray time-of-flight mass spectrometer (LCT, Waters)
- Buffer A: 50 mM ammonium acetate pH 6.5
- Purified RXRα-RARα LBD heterodimer at 5–10 mg/mL (see above for the description of a purification method)

Protocol

1. Calibrate the instrument using the multiply charged ions produced by an injection of horse heart myoglobin diluted to 2 pmol/mL in a water/acetonitrile mixture (1:1, v/v) acidified with 1% (v/v) formic acid.
2. Prior to ESI-MS analysis, desalt the samples on Centricon PM30 microconcentrators (Amicon, Millipore) in buffer A.
3. Verify purity and homogeneity of the samples in denaturing conditions by diluting the complex solution to 5 pmol/mL in a water/acetonitrile mixture (1:1, v/v) acidified with 1% (v/v) formic acid. Record spectra in the positive ion mode on the mass range 500–2500 m/z. Verify that the measured molecular masses are in agreement with those calculated from the amino acid sequences.
4. Dilute samples to 10 pmol/mL in buffer A and infuse continuously into the ESI ion source at a flow rate of 6 mL/min through a Harvard syringe pump. The accelerating voltage (Vc) must be set to 50 V in order to preserve ternary complex formation and good mass accuracy. ESI-MS data are acquired in the positive ion mode on the mass range 1000–5000 m/z.
5. Measure the relative abundance of the different species present on ESI mass spectra from their respective peak intensities, assuming that the relative intensities displayed by the different species reflect the actual distribution of these species in solution.

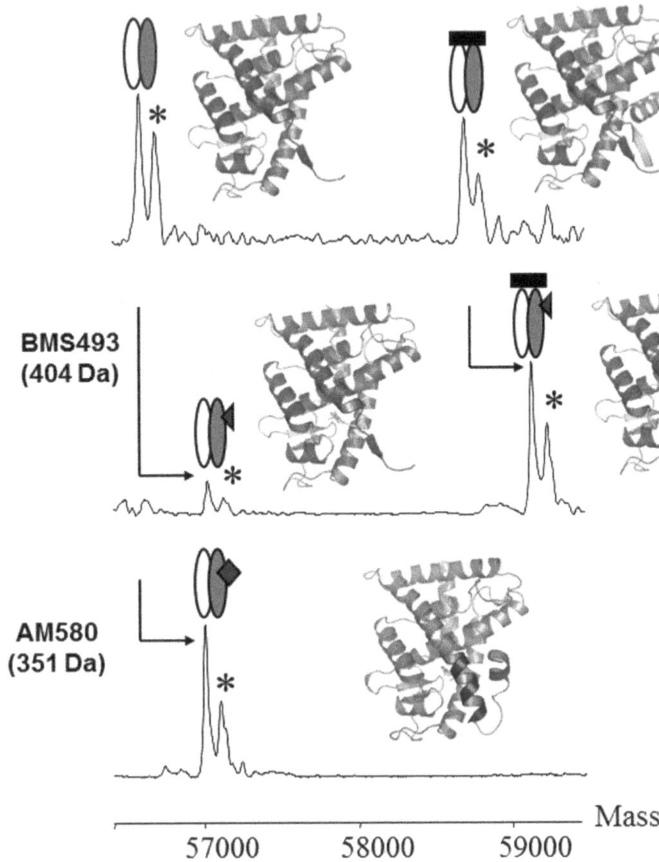

Fig. 7 Positive ESI mass spectra of RXR–RAR–CoRNR1 complexes in the absence of ligand or in the presence of the inverse agonist BMS493 or the agonist AM580. Peaks labeled with an asterisk correspond to species with an additional N-terminal methionine. Whereas the inverse agonist strongly stabilizes the interaction between the heterodimer and CoRNR1, the AM580 induces a quantitative release of the peptide.

4.6 Microscale thermophoresis (MST)

MicroScale Thermophoresis is a method to quantify biomolecular interactions and its applications range from small molecule binding events to protein–protein, protein–DNA interactions and interactions of multi-protein complexes (Jerabek-Willemsen, Wienken, Braun, Baaske, & Duhr, 2011; Zhang, Duhr, Baaske, & Laue, 2014). It measures the motion of molecules along microscopic temperature gradients and detects changes in their hydration shell, charge or size. By combining the fluorescence detections with thermophoresis, MST provides a way to accurately measure affinity constants (Kd). The following protocol is an example given for the determination of the binding affinities of the nuclear interaction domain (NRID) of the corepressor NCoR for the heterodimer RXRα-RARα LBDs in the presence of various ligands (Cordeiro et al., 2019).

Required materials

- Monolith NT.115 (NanoTemper)
- Monolith NT.115 Premium Capillaries (NanoTemper)
- MST optimized buffer: 50mM Tris–HCl pH 7.4, 150mM NaCl, 10mM $MgCl_2$, 0.05% Tween-20
- Purified N-CoR_{NID} fragment to which a fluorescent probe (Atto647N maleimide, Invitrogen) was attached to the thiol group of its sole cysteine, according to the manufacturer's instructions.
- Purified RXR-RAR LBDs heterodimer at 5–10mg/mL
- Ligands (AM580 and BMS493 from Tocris) at 10mM in DMSO

Protocol

1. Prepare 16 small reaction tubes of a volume inferior to 200μL (delivered by NanoTemper with the capillaries).
2. Prepare 20μL of protein stock solution diluted at two times the maximal concentration in MST buffer. In fact, the protein will be diluted 1:1 with the fluorescent target (step 4), so the final protein concentration will be half of the stock concentration. In order to saturate the binding event, the maximum protein concentration should be 10- to 20-fold above the expected Kd.
3. Transfer 10μL of this stock solution in tube 1. Add 10μL of MST buffer in the 10μL left of your protein stock solution and mix very well by carefully pipetting up and down several times. Transfer 10μL of this mix to tube 2. Continue until you reach tube 16 to obtain a serial dilution.
4. In each tube, add 10μL of fluorescently labeled molecule (at double the concentration of the final desired concentration, 80nM in our case) to the 10μL of the serial protein dilution and mix well by pipetting up and down several times.
5. Incubate at conditions of your choice (room temperature or on ice) before loading into the capillaries, during 10min.
6. Fill the capillaries with the mix by soaking each capillary in each tube.
7. Put the capillaries into the tray of the machine, and start an MST measurement using a LED power to obtain 200–1500 units and 40% MST power.
8. Load the results in the MO. Affinity Analysis software (NanoTemper). To derive a binding curve from the MST traces, the fluorescence ration

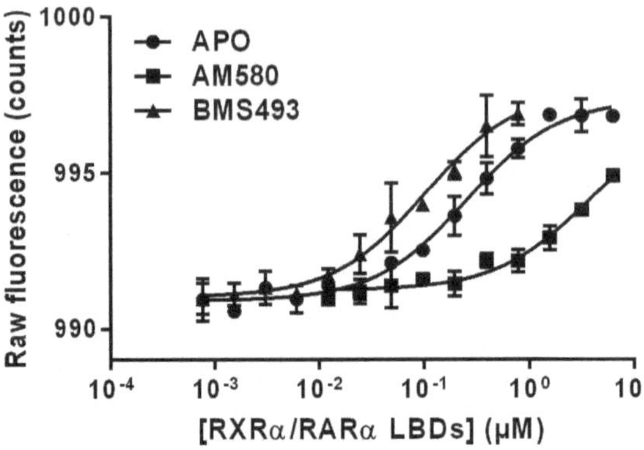

Fig. 8 Interaction study between the heterodimer RXRα-RARα LBDs and the fluorescently labeled nuclear interaction domain of NCoR ($NCoR_{NID}$) followed by MST, in the presence of a full RARα agonist (AM580) or a RARα inverse agonist (BMS493). The raw fluorescence (in counts) is plotted against the concentration of heterodimer for analysis of thermophoresis. The protein shows a significant decrease in fluorescence in the bound than in the unbound state. The Kd is fitted to 0.2 μM (apo complex), more than 5 μM (complex in the presence of AM580) and 0.1 μM (complex in the presence of BMS493) (Cordeiro et al., 2019). Error bars represent standard error of $n=3$ measurements.

before and after heating is calculated and plotted against the ligand concentration. A fitting by a sigmoidal dose-response curve allows determine the binding affinity of the interaction (Fig. 8).

4.7 Thermal shift assays (TSA)

Thermal shift assay is a powerful method for examining binding interactions in proteins (Boivin, Kozak, & Meijers, 2013). It is based on the fact that stabilities of proteins are changed (typically increased) upon addition of ligand (small molecule, protein partner…) and these changes are quantitatively linked to the affinity of the interaction. Thermal shift assay quantifies shifts in thermal denaturation temperature of a protein under varying conditions, by monitoring the thermal unfolding of the protein of interest using fluorescent dyes such as sypro orange. Data collection is rapid and straightforward using readily available real-time polymerase chain reaction instrumentation. The following protocol is an example given for the study of the interaction of RARα LBD with a CoA peptide corresponding to the NR2 motif of TIF2 (KHKILHRLLQDSS), in the presence of the agonist AM580 (Fig. 9).

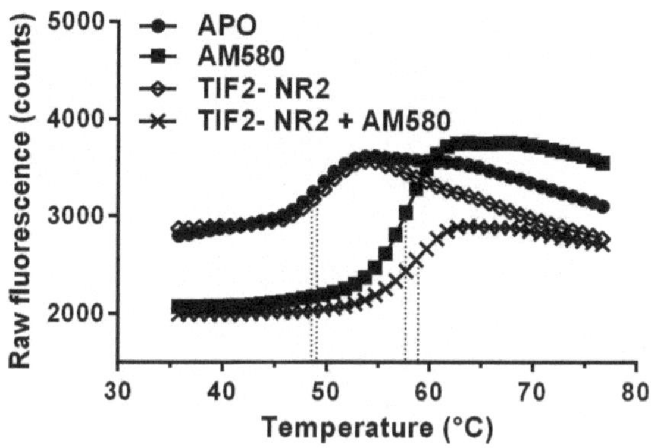

Fig. 9 Interaction study between RARα LBD and a peptide corresponding to the NR2 motif of the coactivator TIF2 (TIF2 NR2) followed by thermal shift assay, in the presence of a full RARα agonist (AM580). Thermal denaturation curves of RARα LBDs, in its unliganded form (APO), in the presence of two molar excess of AM580, in the presence of two molar excess of TIF2-NR2 and in the presence of the ligand and peptide. From these curves, fitting using a Boltzmann sigmoidal equation allows to deduce the melting temperature (Tm) of each complex (dashed lines). The RARα LBD (Tm = 48.77 °C in its unliganded form) is strongly stabilized by the addition of the ligand AM580 (Tm = 57.4 °C) and further stabilized by the concomitant addition of the peptide SCR1-NR2 (Tm = 58.2 °C), confirming the formation of the RAR-coactivator complex in the presence of an agonist ligand.

Required materials

- 7500 Real Time PCR system (Applied Biosystem)
- 96-well PCR plates and optical sealing tapes (Bio-Rad)
- Thermalshift buffer: 20 mM Tris–HCl (pH 7.5), 150 mM NaCl
- Sypro® Orange
- TIF2 NR2 peptide (KHKILHRLLQDSS)
- Ligand AM580 at 10 mM in DMSO
- Purified RARα LBD at 5–10 mg/mL

Protocol

1. Prepare 25 μL of different solutions containing 5 μM protein and eventually 10 μM of TIF2 NR2 peptide, in the presence of DMSO (condition without ligand) or in the presence of 10 μM of AM580, and 2X Sypro Orange in thermalshift buffer.
2. Transfer the solutions in the wells of a 96-well PCR plate and sealed the plates very firmly before.
3. Heat the plate in a 7500 Real Time PCR system (Applied Biosystem) from 25 to 95 °C at 1 °C intervals. Fluorescence changes in the wells

are monitored with a photomultiplier tube and the wavelengths for excitation and emission were 545 nM and 568 nM, respectively.

4. The melting temperatures (Tm) are obtained by fitting the fluorescence data with a Boltzmann model using the GraphPad Prism software

4.8 Isothermal titration calorimetry (ITC)

ITC is a label-free measurement of the binding affinity and thermodynamics of a wide variety of biomolecular interactions (Velazquez-Campoy, Leavitt, & Freire, 2004). It works by directly measuring the heat that is either released or absorbed by the calorimeter when binding occurs during gradual titration of the ligand into the sample cell containing the biomolecule of interest. The following protocol is an example given for the formation of the RXR-THR (thyroid hormone receptor) heterodimer (Fig. 10).

Required materials

- MicroCal VP-ITC system (Malvern Panalytical)
- ITC buffer: 50 mM HEPES pH 7.5, 150 mM NaCl, 5% glycerol, 1 mM TCEP
- 10 kDa molecular weight cut-off dialysis cassette (Slide-A-Lyzer 0.5 mL 10K MWCO, Thermo Scientific)
- Purified TRα LBD at 5–10 mg/mL
- Purified RXRα LBD at 5–10 mg/mL

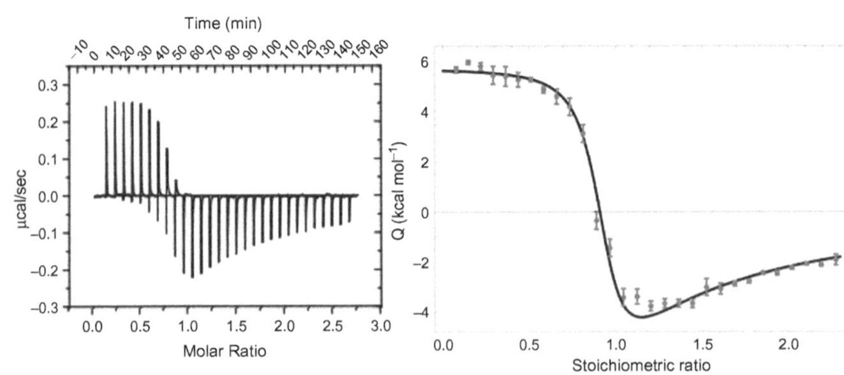

Fig. 10 Thermometric evaluation of the affinity between RXRα and THRα LBDs. ITC thermograms were obtained by titration of RXRα into THRα LBD. Isothermal calorimetric enthalpy change (left panel) and resulting binding isotherm (right panel) are shown. Two events take place: a first dissociation of the homodimer RXRα-RXRα LBDs (Kd = 29.7 µM) and a formation of the heterodimer RXRα-THRα LBDs (Kd = 65 nM) in agreement with measurement of this affinity by MST.

Protocol

1. Dialyze the two proteins (TRα and RXRα LBD) for 2 × 2 h against the same ITC buffer using dialysis cassettes.
2. Fill the reference cell of the VP-ITC with pure water.
3. Dispose 2 mL of TRα LBD at 10 μM in the sample cell.
4. Load 500 μL of RXRα LBD at 120 μM (concentration more or equal to 10 × concentration of the protein in the cell) in the syringe.
5. Set the temperature at 20 °C and monitor heat exchanges throughout titrations consisting of 30 injections (one injection of 1 μL in 2.5 s followed by 29 times 10 μL in 7.1 s, spaced by 300 s) of RXRα LBD solution into the cell containing TRα LBD.
6. Analyze data with the MicroCal ITC-ORIGIN software (Malvern Panalytical) or using Affinimeter (https://www.affinimeter.com/site/itc/) for more complex interactions.

4.9 Small angle X-ray scattering (SAXS)

Small angle X-ray scattering (SAXS) is a biophysical method to study the overall shape and structural transitions of biological macromolecules in solution (Petoukhov & Svergun, 2013; Prior, Davies, Bruce, & Pohl, 2020). It provides low resolution information on the shape, conformation and assembly state of proteins, nucleic acids and macromolecular complexes. Contrary to crystallography, this technique offers powerful means for the quantitative analysis of flexible systems, including intrinsically disordered proteins (Kachala, Valentini, & Svergun, 2015). Recent developments in SEC-SAXS have proved powerful in isolating monodisperse species from polydisperse or aggregating samples, thereby yielding structural information on transient macromolecular conformations and complexes. The following protocol is an example given for the interaction study of the nuclear interaction domain of NCoR (NCoRNID) with the heterodimer RXR/RAR LBDs (Cordeiro et al., 2019).

Required materials

- X-ray synchrotron source (ESRF, Grenoble)
- SAXS buffer: 50 mM Tris–HCl pH 7.5, 150 mM NaCl, 2 mM TCEP
- Purified complex between $NCOR_{NID}$ and RXR/RARα LBD at 5–10 mg/mL

Protocol

1. Prepare the protein complex in SAXS buffer at different concentrations ranging from 0.5 to 10 mg/mL. Keep also the buffer solution for SAXS measurement.

2. Record SAXS data by exposing the samples to a synchrotron X-ray source. Do repetitive measurements to detect and to correct for radiation damage and collect data on buffer solution before and after the measurements of each protein sample.
3. Calculate the protein SAXS curves after subtraction of the averaged buffer scattering from the protein sample patterns using Primus (ref) and merge SAXS curves of different concentrations to avoid interparticle interactions.
4. Use the ATSAS suite (Franke et al., 2017) to derive the radius of gyration, the molecular way, the maximal intraparticule distance (Dmax), the pairwise distance distribution function of the studied protein or complex. Further structural modeling is also possible using the experimental data. For that, multiple approaches are available to model dynamic conformational ensembles encoded in the scattering curve (reviewed in Brosey & Tainer, 2019) (Fig. 11).

4.10 Structural analysis by X-ray crystallography

Protein crystallization is very efficient to characterize protein–protein interactions. It is based on the preparation of crystals of the complexes followed by their analysis using X-ray beams. The determination of the high-resolution three-dimensional (3D) structures that derive from this technique provides atomic-level information on the protein–protein interface. Several crystal structures of homo- and heterodimers of NR LBDs, in complex with CoA and CoR peptides have been reported (ref). Considering that the methods used for the crystallization of protein complexes are diverse and that crystallization conditions are unique to each protein (or protein complex), we provide here a very general procedure that may apply to all kinds of NR LBDs in complex or in isolation. The following protocol is an example given for the crystallization of the RARα LBD bound to BMS493, a RARα-selective inverse agonist and in complex with CoRNR1 of NCoR (le Maire et al., 2010).

Required materials

- Crystallization robot (Mosquito)
- Crystallization screens (Molecular Dimension, Hampton Research, Qiagen, *etc.*)
- Visualization robot (Formulatrix)
- 96-Well crystallization plates (Greiner Bio-One)
- 24-Well crystallization plates (Greiner Bio-One, Molecular Dimension)

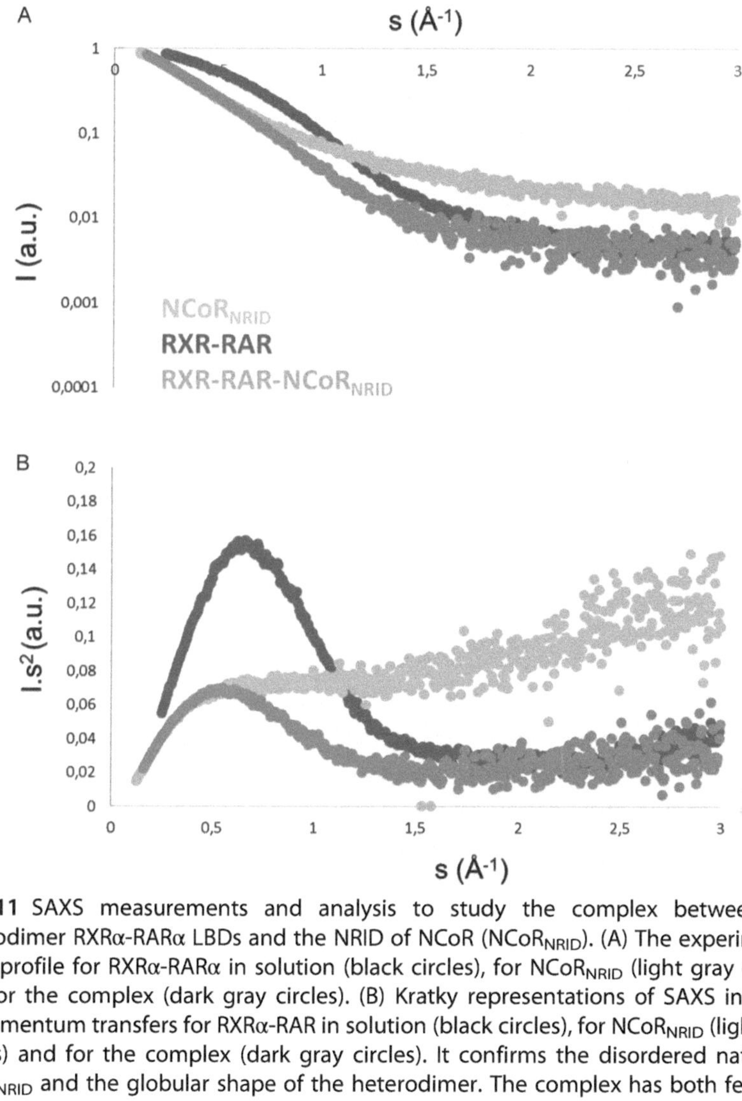

Fig. 11 SAXS measurements and analysis to study the complex between the heterodimer RXRα-RARα LBDs and the NRID of NCoR ($NCoR_{NRID}$). (A) The experimental SAXS profile for RXRα-RARα in solution (black circles), for $NCoR_{NRID}$ (light gray circles) and for the complex (dark gray circles). (B) Kratky representations of SAXS intensity vs momentum transfers for RXRα-RAR in solution (black circles), for $NCoR_{NRID}$ (light gray circles) and for the complex (dark gray circles). It confirms the disordered nature of $NCoR_{NRID}$ and the globular shape of the heterodimer. The complex has both features.

- Siliconized glass coverslips (Greiner Bio-One)
- X-ray synchrotron source (ESRF, Grenoble)
- NCoR CORNR1 peptide (EZBiolab)
- Ligand BMS493 at 10 mM in DMSO
- Purified RARα LBD at 5–10 mg/mL

Protocol

1. Mix the purified protein with a twofold molar excess of ligand and a threefold molar excess of the NCoR CoRNR1 peptide, concentrate it to 3.5 mg/mL and centrifuge for 20 min at 13,000 rpm to get rid of any aggregated material.
2. Perform an initial search for crystallization conditions by mixing the purified protein with sparse matrix screening conditions using crystallization robots, dedicated 96-well crystallization plates and the sitting-drop vapor diffusion technique. The robot automatically performs each crystallization trial by mixing 100 nL of protein with 100 nL of precipitant condition in a well suspended over a reservoir containing 40 μL of the corresponding condition.
3. Once the 96 conditions have been dispensed, seal the plates with a transparent plastic film and incubate the plates at a fixed temperature (20 °C) in the Formulatrix to take pictures of the drops at defined times. The crystals obtained at this step are generally small or not well shaped and need to be optimized through refinement of these preliminary crystallization conditions.
4. Perform the refinement by generating "home-made" grid screens where all the parameters of the initial hit are varied (pH, salt, concentration of precipitant, *etc.*). Use 24-well crystallization plates and mix 1 μL of the concentrated protein with 1 μL of each crystallization condition, the reservoir containing 500 μL of the corresponding crystallization condition. The drops are suspended from siliconized glass coverslips (hanging drop method). This scale-up of the crystallization process allows to obtain large crystals suitable for further crystallographic studies in a crystallization solution consisting of 18% (w/v) PEG3350, 0.15 M NH_4Cl.
5. Mount the crystal onto a cryoloop, transfer the crystal to a solution containing the mother liquor plus a cryoprotectant (25% glycerol) for a few seconds and flash-frozen in liquid nitrogen to preserve it before and during the X-ray diffraction experiment.
6. Collect diffraction data at a synchrotron X-ray source and solve the crystal structure following all the steps of data processing, determination of the structure by molecular replacement, model building, and refinement (see le Maire et al., 2010 for more details and programs used).

4.11 SEC-MALS

Coupling an analytical Size Exclusion Chromatography (SEC) column in-line with multi-angle light scattering (MALS), UV detector and refractometric detector provides a widespread approach and a useful tool for accurate analysis of molar mass, overall shape, aggregation, oligomerization, interactions and purity of proteins or macromolecular complexes in native solution, independent of the protein retention time analyzed by SEC (Some, Amartely, Tsadok, & Lebendiker, 2019). The following protocol is an example given for the biophysical characterization of RXRΔAB-RARΔAB-DR5 in complex with $NCOR_{NID}$ before more in-depth structural study (Fig. 12; unpublished results).

Required materials

- miniRAWN Treos—Multi-Angle Static Light Scattering + Optilab T-rex—Refractometer (Wyatt)
- 1260 Infinity II Liquid Chromatography system (Agilent)
- Superdex 200 Increase 10/300 GL (Life Sciences)

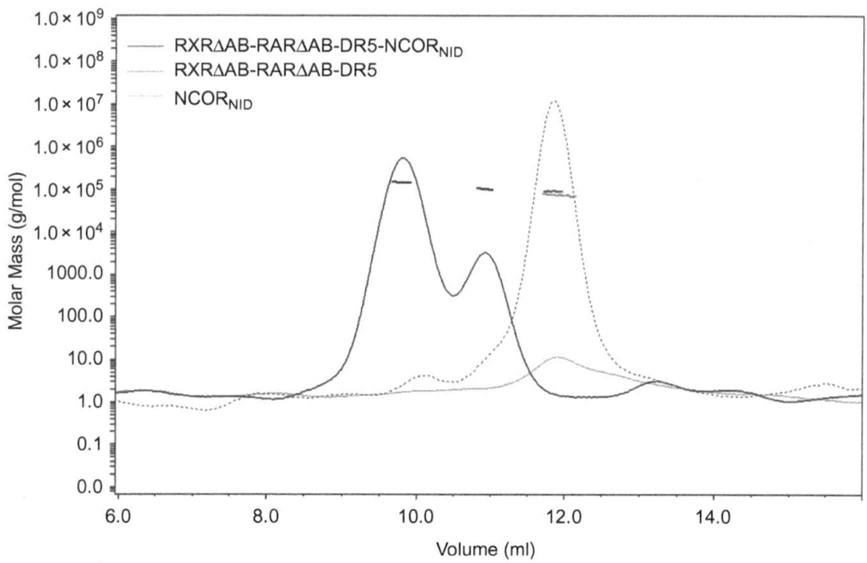

Fig. 12 SEC-MALS elution profile of RXRΔAB-RARΔAB-DR5, $NCOR_{NID}$ and the full complex of RXRΔAB-RARΔAB-DR5-$NCOR_{NID}$ after large-scale purification. The protein samples were separated on a size exclusion chromatography (SEC) column (Superdex 200 Increase 10/300 GL) equilibrated with MALS buffer. The horizontal lines crossing the SEC profile indicate molecular masses of the protein and complexes determined by connected in-line multi-angle light scattering (MALS) and refractive index (RI) detectors. The molecular weights deduced from these experiment fit well with the theoretical values: RXRΔAB-RARΔAB-DR5, $MW_{MALS} = 91$ kDa and $MW_{theo} = 82$ kDa; $NCOR_{NID}$, $MW_{MALS} = 70$ kDa and $MW_{theo} = 30$ kDa (dimer); RXRΔAB-RARΔAB-DR5-$NCOR_{NID}$, $MW_{MALS} = 140$ kDa and $MW_{theo} = 126$ kDa.

- MALS buffer: 50 mM Tris pH 7.5, 150 mM NaCl, degassed and filtrated at 0.1 μM
- $NCOR_{NID}$ purified at 2–10 mg/mL
- RXRΔAB-RARΔAB-DR5 complex purified at 2–5 mg/mL
- RXRΔAB-RARΔAB-DR5-$NCOR_{NID}$—BMS493 (RAR inverse agonist) complex purified at 2–5 mg/mL

Protocol

1. Equilibrate the system and the column in the MALS buffer. The chromatography system is controlled using the chromatography software from Agilent. Run this equilibration step until MALS detector is stable with drift lower that 10^{-4}.
2. Centrifuge the protein sample at 14,000 rpm during 20 min to remove precipitates and larger insoluble particles
3. Inject the protein sample (volume of 20–25 μL) on the column and continue the run at 0.5 mL/min (about 40 min for the 20 mL column)
4. Process the data using the ASTRA software. Set the baselines, select the peak(s) of interest, check the report, especially the Molecular Mass vs Time plot.

Acknowledgments

The CBS is a member of the France-BioImaging (FBI) and the French Infrastructure for Integrated Structural Biology (FRISBI), two national infrastructures supported by the French National Research Agency (ANR-10-INBS-04-01 and ANR-10-INBS-05, respectively).

References

Boivin, S., Kozak, S., & Meijers, R. (2013). Optimization of protein purification and characterization using Thermofluor screens. *Protein Expression and Purification, 91*(2), 192–206.

Bourguet, W., Andry, V., Iltis, C., Klaholz, B., Potier, N., Van Dorsselaer, A., et al. (2000). Heterodimeric complex of RAR and RXR nuclear receptor ligand-binding domains: Purification, crystallization, and preliminary X-ray diffraction analysis. *Protein Expression and Purification, 19*(2), 284–288.

Bourguet, W., Vivat, V., Wurtz, J. M., Chambon, P., Gronemeyer, H., & Moras, D. (2000). Crystal structure of a heterodimeric complex of RAR and RXR ligand-binding domains. *Molecular Cell, 5*(2), 289–298.

Brosey, C. A., & Tainer, J. A. (2019). Evolving SAXS versatility: Solution X-ray scattering for macromolecular architecture, functional landscapes, and integrative structural biology. *Current Opinion in Structural Biology, 58*, 197–213. https://doi.org/10.1016/j.sbi.2019.04.004.

Chandra, V., Huang, P., Hamuro, Y., Raghuram, S., Wang, Y., Burris, T. P., et al. (2008). Structure of the intact PPAR-gamma-RXR- nuclear receptor complex on DNA. *Nature, 456*(7220), 350–356.

Chandra, V., Wu, D., Li, S., Potluri, N., Kim, Y., & Rastinejad, F. (2017). The quaternary architecture of RARbeta-RXRalpha heterodimer facilitates domain-domain signal transmission. *Nature Communications*, *8*(1), 868.

Chen, H., Lin, R. J., Xie, W., Wilpitz, D., & Evans, R. M. (1999). Regulation of hormone-induced histone hyperacetylation and gene activation via acetylation of an acetylase. *Cell*, *98*(5), 675–686.

Cordeiro, T. N., Sibille, N., Germain, P., Barthe, P., Boulahtouf, A., Allemand, F., et al. (2019). Interplay of protein disorder in retinoic acid receptor heterodimer and its corepressor regulates gene expression. *Structure*, *27*(8), 1270–1285.e1276.

de Vera, I. M. S., Zheng, J., Novick, S., Shang, J., Hughes, T. S., Brust, R., et al. (2017). Synergistic regulation of coregulator/nuclear receptor interaction by ligand and DNA. *Structure*, *25*(10), 1506–1518.e1504.

Fenn, J. B., Mann, M., Meng, C. K., Wong, S. F., & Whitehouse, C. M. (1989). Electrospray ionization for mass spectrometry of large biomolecules. *Science*, *246*(4926), 64–71.

Fields, S., & Song, O. (1989). A novel genetic system to detect protein-protein interactions. *Nature*, *340*(6230), 245–246.

Franke, D., Petoukhov, M. V., Konarev, P. V., Panjkovich, A., Tuukkanen, A., Mertens, H. D. T., et al. (2017). ATSAS 2.8: A comprehensive data analysis suite for small-angle scattering from macromolecular solutions. *Journal of Applied Crystallography*, *50*(Pt. 4), 1212–1225.

Gampe, R. T., Jr., Montana, V. G., Lambert, M. H., Miller, A. B., Bledsoe, R. K., Milburn, M. V., et al. (2000). Asymmetry in the PPARgamma/RXRalpha crystal structure reveals the molecular basis of heterodimerization among nuclear receptors. *Molecular Cell*, *5*(3), 545–555.

Germain, P., Chambon, P., Eichele, G., Evans, R. M., Lazar, M. A., Leid, M., et al. (2006a). International Union of Pharmacology. LX. Retinoic acid receptors. *Pharmacological Reviews*, *58*(4), 712–725.

Germain, P., Chambon, P., Eichele, G., Evans, R. M., Lazar, M. A., Leid, M., et al. (2006b). International Union of Pharmacology. LXIII. Retinoid X receptors. *Pharmacological Reviews*, *58*(4), 760–772.

Gilardi, F., & Desvergne, B. (2014). RXRs: Collegial partners. *Sub-Cellular Biochemistry*, *70*, 75–102.

Glass, C. K., & Rosenfeld, M. G. (2000). The coregulator exchange in transcriptional functions of nuclear receptors. *Genes & Development*, *14*(2), 121–141.

Gronemeyer, H., Gustafsson, J. A., & Laudet, V. (2004). Principles for modulation of the nuclear receptor superfamily. *Nature Reviews Drug Discovery*, *3*(11), 950–964.

Heinzel, T., Lavinsky, R. M., Mullen, T. M., Soderstrom, M., Laherty, C. D., Torchia, J., et al. (1997). A complex containing N-CoR, mSin3 and histone deacetylase mediates transcriptional repression. *Nature*, *387*(6628), 43–48.

Hellman, L. M., & Fried, M. G. (2007). Electrophoretic mobility shift assay (EMSA) for detecting protein-nucleic acid interactions. *Nature Protocols*, *2*(8), 1849–1861.

Hu, X., Li, Y., & Lazar, M. A. (2001). Determinants of CoRNR-dependent repression complex assembly on nuclear hormone receptors. *Molecular and Cellular Biology*, *21*(5), 1747–1758.

Imhof, A., Yang, X. J., Ogryzko, V. V., Nakatani, Y., Wolffe, A. P., & Ge, H. (1997). Acetylation of general transcription factors by histone acetyltransferases. *Current Biology*, *7*(9), 689–692.

Iyer, J., Bonnier, D., Granger, F., Iltis, C., Andry, V., Schultz, P., et al. (1999). Versatile copurification procedure for rapid isolation of homogeneous RAR-RXR heterodimers. *Protein Expression and Purification*, *16*(2), 308–314.

Jerabek-Willemsen, M., Wienken, C. J., Braun, D., Baaske, P., & Duhr, S. (2011). Molecular interaction studies using microscale thermophoresis. *Assay and Drug Development Technologies*, *9*(4), 342–353.

Kachala, M., Valentini, E., & Svergun, D. I. (2015). Application of SAXS for the structural characterization of IDPs. *Advances in Experimental Medicine and Biology*, *870*, 261–289.
Kato, S., Yokoyama, A., & Fujiki, R. (2011). Nuclear receptor coregulators merge transcriptional coregulation with epigenetic regulation. *Trends in Biochemical Sciences*, *36*(5), 272–281.
Khorasanizadeh, S., & Rastinejad, F. (2016). Visualizing the architectures and interactions of nuclear receptors. *Endocrinology*, *157*(11), 4212–4221.
le Maire, A., Grimaldi, M., Roecklin, D., Dagnino, S., Vivat-Hannah, V., Balaguer, P., et al. (2009). Activation of RXR-PPAR heterodimers by organotin environmental endocrine disruptors. *EMBO Reports*, *10*(4), 367–373.
le Maire, A., Teyssier, C., Balaguer, P., Bourguet, W., & Germain, P. (2019). Regulation of RXR-RAR heterodimers by RXR- and RAR-specific ligands and their combinations. *Cell*, *8*(11), 1392.
le Maire, A., Teyssier, C., Erb, C., Grimaldi, M., Alvarez, S., de Lera, A. R., et al. (2010). A unique secondary-structure switch controls constitutive gene repression by retinoic acid receptor. *Nature Structural & Molecular Biology*, *17*(7), 801–807.
Lonard, D. M., & O'Malley, B. W. (2007). Nuclear receptor coregulators: Judges, juries, and executioners of cellular regulation. *Molecular Cell*, *27*(5), 691–700.
Loo, J. A. (1997). Studying noncovalent protein complexes by electrospray ionization mass spectrometry. *Mass Spectrometry Reviews*, *16*(1), 1–23.
Lou, X., Toresson, G., Benod, C., Suh, J. H., Philips, K. J., Webb, P., et al. (2014). Structure of the retinoid X receptor alpha-liver X receptor beta (RXRalpha-LXRbeta) heterodimer on DNA. *Nature Structural & Molecular Biology*, *21*(3), 277–281.
Madauss, K. P., Grygielko, E. T., Deng, S. J., Sulpizio, A. C., Stanley, T. B., Wu, C., et al. (2007). A structural and in vitro characterization of asoprisnil: A selective progesterone receptor modulator. *Molecular Endocrinology*, *21*(5), 1066–1081.
Nagy, L., Kao, H. Y., Chakravarti, D., Lin, R. J., Hassig, C. A., Ayer, D. E., et al. (1997). Nuclear receptor repression mediated by a complex containing SMRT, mSin3A, and histone deacetylase. *Cell*, *89*(3), 373–380.
Perissi, V., Staszewski, L. M., McInerney, E. M., Kurokawa, R., Krones, A., Rose, D. W., et al. (1999). Molecular determinants of nuclear receptor-corepressor interaction. *Genes & Development*, *13*(24), 3198–3208.
Petoukhov, M. V., & Svergun, D. I. (2013). Applications of small-angle X-ray scattering to biomacromolecular solutions. *The International Journal of Biochemistry & Cell Biology*, *45*(2), 429–437.
Phelan, C. A., Gampe, R. T., Jr., Lambert, M. H., Parks, D. J., Montana, V., Bynum, J., et al. (2010). Structure of Rev-erbalpha bound to N-CoR reveals a unique mechanism of nuclear receptor-co-repressor interaction. *Nature Structural & Molecular Biology*, *17*(7), 808–814.
Piskunov, A., Al Tanoury, Z., & Rochette-Egly, C. (2014). Nuclear and extra-nuclear effects of retinoid acid receptors: How they are interconnected. *Sub-Cellular Biochemistry*, *70*, 103–127.
Pogenberg, V., Guichou, J. F., Vivat-Hannah, V., Kammerer, S., Perez, E., Germain, P., et al. (2005). Characterization of the interaction between retinoic acid receptor/retinoid X receptor (RAR/RXR) heterodimers and transcriptional coactivators through structural and fluorescence anisotropy studies. *The Journal of Biological Chemistry*, *280*(2), 1625–1633.
Prior, C., Davies, O. R., Bruce, D., & Pohl, E. (2020). Obtaining tertiary protein structures by the ab initio interpretation of small angle X-ray scattering data. *Journal of Chemical Theory and Computation*. https://doi.org/10.1021/acs.jctc.9b01010.
Putcha, B. D., Wright, E., Brunzelle, J. S., & Fernandez, E. J. (2012). Structural basis for negative cooperativity within agonist-bound TR:RXR heterodimers. *Proceedings of the National Academy of Sciences of the United States of America*, *109*(16), 6084–6087.

Rajabi, K., Ashcroft, A. E., & Radford, S. E. (2015). Mass spectrometric methods to analyze the structural organization of macromolecular complexes. *Methods*, *89*, 13–21.

Rochel, N., Ciesielski, F., Godet, J., Moman, E., Roessle, M., Peluso-Iltis, C., et al. (2011). Common architecture of nuclear receptor heterodimers on DNA direct repeat elements with different spacings. *Nature Structural & Molecular Biology*, *18*(5), 564–570.

Sanglier, S., Bourguet, W., Germain, P., Chavant, V., Moras, D., Gronemeyer, H., et al. (2004). Monitoring ligand-mediated nuclear receptor-coregulator interactions by noncovalent mass spectrometry. *European Journal of Biochemistry*, *271*(23–24), 4958–4967.

Sato, Y., Ramalanjaona, N., Huet, T., Potier, N., Osz, J., Antony, P., et al. (2010). The "Phantom Effect" of the Rexinoid LG100754: Structural and functional insights. *PLoS One*, *5*(11), e15119.

Some, D., Amartely, H., Tsadok, A., & Lebendiker, M. (2019). Characterization of proteins by size-exclusion chromatography coupled to multi-angle light scattering (SEC-MALS). *Journal of Visualized Experiments*, *148*. https://doi.org/10.3791/59615.

Suino, K., Peng, L., Reynolds, R., Li, Y., Cha, J. Y., Repa, J. J., et al. (2004). The nuclear xenobiotic receptor CAR: Structural determinants of constitutive activation and heterodimerization. *Molecular Cell*, *16*(6), 893–905.

Svensson, S., Ostberg, T., Jacobsson, M., Norstrom, C., Stefansson, K., Hallen, D., et al. (2003). Crystal structure of the heterodimeric complex of LXRalpha and RXRbeta ligand-binding domains in a fully agonistic conformation. *The EMBO Journal*, *22*(18), 4625–4633.

Velazquez-Campoy, A., Leavitt, S. A., & Freire, E. (2004). Characterization of protein-protein interactions by isothermal titration calorimetry. *Methods in Molecular Biology*, *261*, 35–54.

Wallace, B. D., Betts, L., Talmage, G., Pollet, R. M., Holman, N. S., & Redinbo, M. R. (2013). Structural and functional analysis of the human nuclear xenobiotic receptor PXR in complex with RXRalpha. *Journal of Molecular Biology*, *425*(14), 2561–2577.

Wang, N., Zou, Q., Xu, J., Zhang, J., & Liu, J. (2018). Ligand binding and heterodimerization with retinoid X receptor alpha (RXRalpha) induce farnesoid X receptor (FXR) conformational changes affecting coactivator binding. *The Journal of Biological Chemistry*, *293*(47), 18180–18191.

Wang, L., Zuercher, W. J., Consler, T. G., Lambert, M. H., Miller, A. B., Orband-Miller, L. A., et al. (2006). X-ray crystal structures of the estrogen-related receptor-gamma ligand binding domain in three functional states reveal the molecular basis of small molecule regulation. *The Journal of Biological Chemistry*, *281*(49), 37773–37781.

Xu, R. X., Lambert, M. H., Wisely, B. B., Warren, E. N., Weinert, E. E., Waitt, G. M., et al. (2004). A structural basis for constitutive activity in the human CAR/RXRalpha heterodimer. *Molecular Cell*, *16*(6), 919–928.

Xu, H. E., Stanley, T. B., Montana, V. G., Lambert, M. H., Shearer, B. G., Cobb, J. E., et al. (2002). Structural basis for antagonist-mediated recruitment of nuclear co-repressors by PPARalpha. *Nature*, *415*(6873), 813–817.

Zhang, J., Chalmers, M. J., Stayrook, K. R., Burris, L. L., Wang, Y., Busby, S. A., et al. (2011). DNA binding alters coactivator interaction surfaces of the intact VDR-RXR complex. *Nature Structural & Molecular Biology*, *18*(5), 556–563.

Zhang, W., Duhr, S., Baaske, P., & Laue, E. (2014). Microscale thermophoresis for the assessment of nuclear protein-binding affinities. *Methods in Molecular Biology*, *1094*, 269–276.

CHAPTER NINE

Natural ligands of RXR receptors

Patricia García, Paula Lorenzo, Angel R. de Lera*
Departamento de Química Orgánica, Facultade de Química, CINBIO and IBIV, Universidade de Vigo, Campus As Lagoas-Marcosende, Vigo, Spain
*Corresponding author: e-mail address: qolera@uvigo.es

Contents

Abstract

Given the role of retinoid X receptors (RXRs) as promiscuous partners of heterodimeric complexes with other members of the Nuclear Receptor (NR) superfamily, RXR ligands (rexinoids) play fundamental roles in gene transcription, since upon ligand binding either transcriptionally activate the "permissive" subclass of heterodimers or synergize with partner ligands in the "non-permissive" subclass of heterodimers. The collection of natural products thus far reported to bind RXR are described, including those discovered by high-throughput screening (HTS), mere serendipity, and a combination of those. Detailed protocols for the diastereo- and enantioselective synthesis of (*R*)-9-*cis*-13,14-dihydroretinoic acid, a putative natural RXR ligand, are provided.

Methods in Enzymology, Volume 637
ISSN 0076-6879
https://doi.org/10.1016/bs.mie.2020.02.006

Abbreviations

AF	activation function
ATRA	all-*trans*-retinoic acid
CADD	computer-aided drug design
CAR	constitutive androstane receptor
ChEMBL	European Molecular Biology Laboratory Chemical Database
CoA	co-activator
Co-R	co-repressor
DHA	docosahexaenoic acid
EBI	European Bioinformatics Institute
FXR	farnesoid X receptor
GPCR	G-protein coupled receptor
GGA	geranyl geranoic acid
HETE	hydroxyeicosatetraenoic acid
LBD	ligand binding domain
LBP	ligand binding pocket
LXR	liver X receptors
NOE	nuclear Overhauser effect
NR	nuclear receptor
PPAR	peroxisome proliferator-activated receptor
PXR	pregnane X receptor
RAR	retinoic acid receptor
RXR	retinoid X receptor
RetSat	retinol saturase
SAR	structure–activity relationships
SMRT	silencing mediator for retinoid and thyroid hormone receptors
TR	thyroid hormone receptor
VDR	vitamin D receptor

1. Introduction. Structure of RXR receptors

Retinoid X receptors RXRα, β, and γ (NR2B1, NR2B2 and NR2B3) are members of the nuclear receptor (NR) superfamily of ligand-dependent transcriptional regulators that bind DNA response elements associated with target genes (Germain et al., 2006; Laudet & Gronemeyer, 2002). Similar to other NRs, RXRs feature a modular structure, which can be dissected in several domains with associated functions, most notably the N-terminal activation function (AF-1), the central DNA-binding domain (DBD) and the C-terminal ligand-binding domain (LBD).

The LBD is a single protein domain of 11–13 helices and a short β-turn arranged in three layers to form the so-called anti-parallel alpha-helical sandwich (Huang, Chandra, & Rastinejad, 2014). The multifunctional nature of this NR domain is shown not only by the control of ligand binding and receptor dimerization, but also by the interaction with the so-called co-regulators (co-repressors, CoRs, or co-activators, CoAs) through a ligand-dependent activation function (AF-2). Mechanistically, ligand binding induces the repositioning of the C-terminal helix H12 of the LBD due to alterations on the interaction surfaces with co-regulators (Huang et al., 2014; le Maire et al., 2012).

In addition, agonist ligand binding induces an exchange of CoR by CoA protein complexes and also of chromatin-remodeling enzymes. As a consequence, the condensed transcriptionally inactive state is transformed into the activated accessible state, leading ultimately to gene transcription. Other structural alterations on these complexes due to the action of ligands that act as antagonists, partial agonists and inverse agonists have also been revealed in the last few years (le Maire et al., 2012; Xu et al., 2001).

Being the common heterodimerization partner of multiple NRs, RXR occupies a central position in NR signaling (Mangelsdorf & Evans, 1995), since it can act as an active or as a silent transcription partner, and therefore the corresponding RXR-NR heterodimers are called permissive or non-permissive, respectively, if they are activated by the RXR ligand or by the partner's ligand (although synergistic effects are usually observed if both partners of the heterodimer are bound to their ligands), or only by the partner's ligand whereas ligand-bound RXR is silent or subordinated (Germain, Iyer, Zechel, & Gronemeyer, 2002).

2. Structural determinants for ligand binding to RXR receptors

As in other NRs, the CoR- or the CoA-bound states of RXR are exquisitively sensitive to ligand action (as agonists, antagonists or inverse agonists) and therefore, to their structure (Bourguet & Moras, 2015; Dominguez, Alvarez, & de Lera, 2017; Huang et al., 2014; Mendoza-Parra, Bourguet, de Lera, & Gronemeyer, 2015).

The crystal structures of RXR-ligand complexes (Bourguet & Moras, 2015; Huang et al., 2014) clearly show that the most selective RXR ligands feature a bent structure (see Fig. 1), which is enforced by a tetrahedral carbon, a 1,1-disubstituted olefin, or a twisted polyene side-chain for optimizing binding to the L-shaped RXR LBP. This sharp turn on the LBP volume is induced by the (highly conserved) conformation of a lysine residue in H11

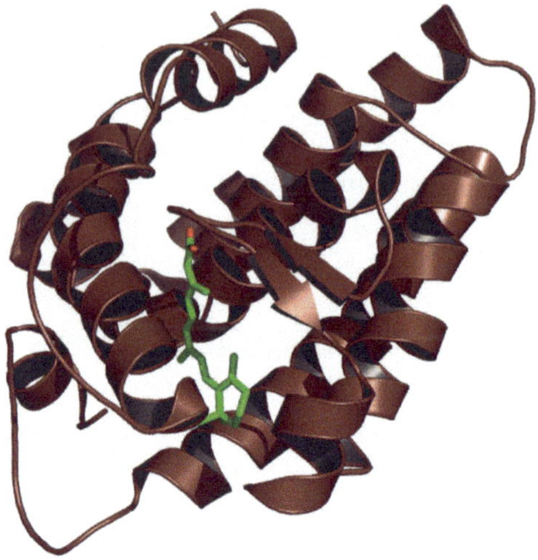

Fig. 1 RXRα–9-*cis*-retinoic acid (**31**, see Fig. 4) crystal structure (PDB code: 1k74) (Germain et al., 2009).

(residue L436 in human RXRα) (Huang et al., 2014). As a consequence, the shape of the RXR LBP is very restrictive to ligand binding, and selects those ligands that are either conformationally or configurationally twisted or sufficiently flexible to twist around single bonds. Such bending is not structurally attained by the elongated fully conjugated polyenes (although not for partially unsaturated analogs), such as all-*trans*-retinoic acid.

More than 500 RXR agonists and antagonists (with IC/IC_{50} lower than 50 μM) have been annotated in ChEMBL, the database dependent on the European Bioinformatics Institute (EBI). From the collection of RXR modulators, going from naturally occurring compounds, those discovered by high throughput screening or mere serendipity, only a few belong to the first group. The rationale for the structure-based design of the most important classes of RXR modulators has been described elsewhere (Bourguet & Moras, 2015; Dominguez et al., 2017; Hiebl, Ladurner, Latkolik, & Dirsch, 2018; Mendoza-Parra et al., 2015).

Natural ligands are compounds present in Nature, which bind their receptors with high potency (ideally, in the nanomolar range) usually leading to conformational changes in the complex structure, inducing thereby relevant physiological activities. We will discuss the biological relevance of the natural ligands reported for RXR. Needless to say, the relevant question of their presence at the right concentration on cells and tissues remains in most cases uncertain. Some of these compounds have been detected under supraphysiological conditions or nutritional intervention and their relevance for the control of biological processes dependent on RXR remains uncertain.

3. Natural RXR ligands

3.1 Dietary compounds. Natural rexinoids (Fig. 2)

A collection of flexible unsaturated natural fatty acids acquired in the diet have been reported to bind to and activate RXR (Dominguez et al., 2017; Hiebl et al., 2018). The promiscuity of RXR on binding several metabolites indicates a putative role for this receptor as an intracellular sensor of the cell metabolic status, binding either the natural ligand or fatty acids, depending upon their supply and local concentration.

Within the series of flexible unsaturated fatty acids, both the endogenous and those acquired from the diet, that bind to and activate RXR, docosahexaenoic acid (DHA) **1** (Fig. 2) isolated from mouse brain, was the first reported activator of RXR in cell-based assays (de Urquiza et al., 2000). DHA **1** is a major constituent of nutrients and is considered as a conditionally essential fatty acid. DHA **1** is highly abundant (up to 30–50% of the total fatty acid content) in mammalian brain cells, where it is found associated to membrane phospholipids, and accumulates in the retina. It is required for brain maturation in rodents and humans, and its deficiency results in abnormalities similar to those in RXRα-deficient mice. Despite these findings, it has not been demonstrated that DHA **1** is required for RXR-RAR retinoid signaling or any other RXR-NR nonretinoid signaling pathway in which RXR may be involved (Mic, Molotkov, Benbrook, & Duester, 2003; Wietrzych-Schindler et al., 2011).

DHA **1** easily adapts to the LBP of RXR, and the crystal structure of RXRα bound to DHA **1** and a co-activator peptide in a homodimeric association (Egea, Mitschler, & Moras, 2002), indicates an occupancy of about 80% of the LBP volume (compared to 72% for 9-*cis*-retinoic acid **31**, see Fig. 1). All atoms of the ligand are in direct van der Waals contact with the protein and in particular those filling the hydrophobic cavity occupied by the β-ionone ring-binding subpocket of 9-*cis*-retinoic acid (**31**, see Fig. 4) bound to RXRα. Since the endogenous levels of DHA (which range from 0.1 to 0.01 μM in animal and human serum or tissue samples) are not sufficient for transcriptional activation, which requires at best a concentration range of 5–10 μM (Lengqvist et al., 2004), its role as physiological ligand

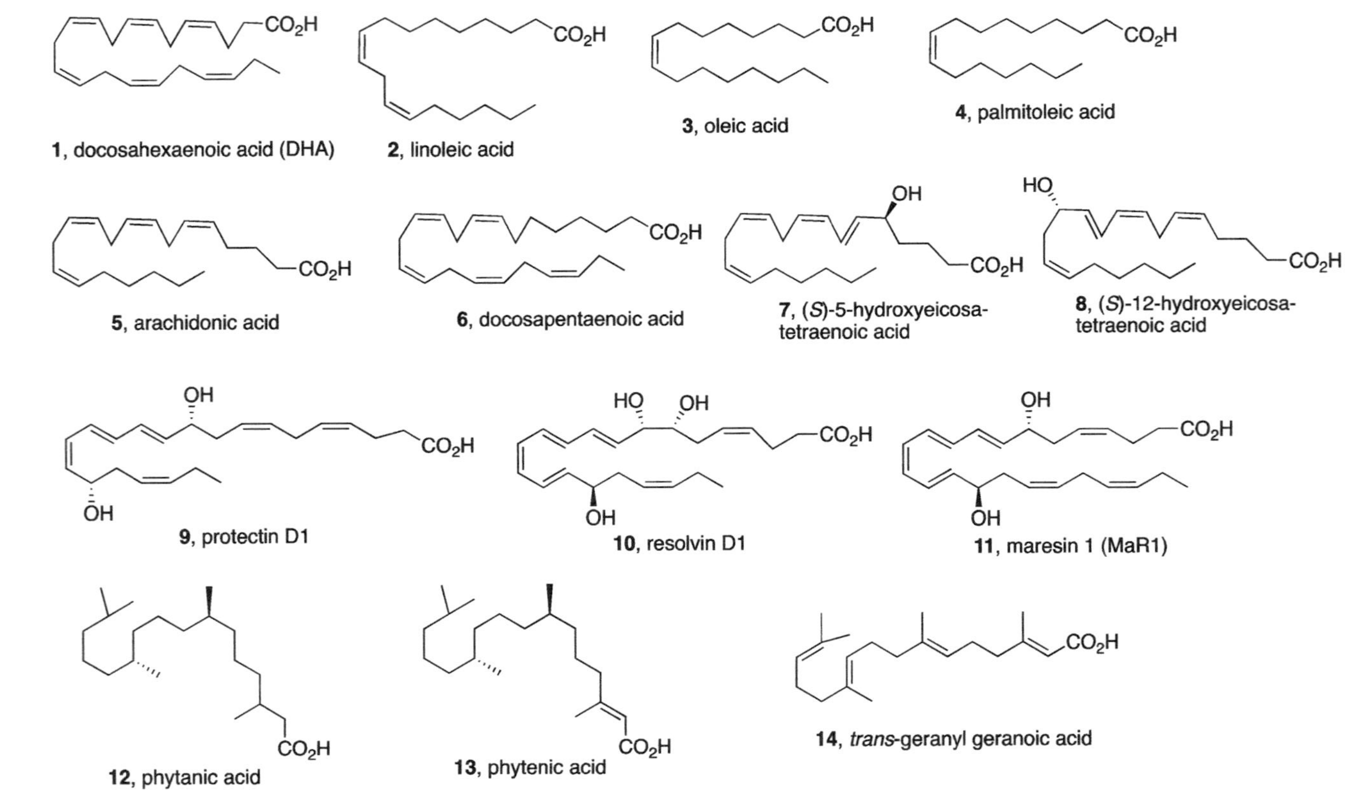

Fig. 2 Natural and dietary (unsaturated) fatty acids. Arbitrary conformations have been selected.

is questionable, despite the reported ability of vertebrates to synthesize DHA from α-linolenic acid as precursor of plant origin.

Additional fatty acids isolated from rat tissue total lipid extracts were also shown to activate RXR (Goldstein, Dobrzyn, Clagett-Dame, Pike, & DeLuca, 2003). Out of the large number of family members only the highly unsaturated fatty acids were able to bind the receptor with high affinity as shown by competition experiments with tritiated 9-*cis*-retinoic acid for binding to mRXRγ. Notably, linoleic acid **2**, which is one of the essential fatty acids, and also oleic acid **3**, palmitoleic acid **4**, arachidonic acid **5** and docosapentaenoic acid **6**, which have been isolated mainly from testis and heart, were found to activate mRXRγ at high ($EC_{50} > 200\,\mu M$) concentrations.(Lengqvist et al., 2004) Other eicosanoids, most notably hydroxyeicosatetraenoic acids such as (*S*)-5-HETE **7** and (*S*)-12-HETE **8** were reported to mimic some actions of the retinoids in cell-based assays (Eager, Brickell, Snell, & Wood, 1992).

It is not unusual to find unsaturated fatty acids bound to the LBP in the corresponding crystal structures of holo NRs (Huang et al., 2014), which are present due to the competing biosynthetic pathways of the bacterial expression host. This is the case of compound **3**, which was present in the crystal structure of the LBP of the ligand-bound RARα/F318A RXRα heterodimer complex. Oleic acid **3** conformationally adapts to the U-shape and resembles 9-*cis*-retinoic acid in the extended volume of the RXRα binding pocket due to the F318A mutation (Bourguet et al., 2000).

Several other natural products termed protectins, resolvins and maresins (**9–11**, Fig. 2), which are bis-, and tris-hydroxylated DHA metabolites (Serhan & Petasis, 2011), have also been found endogenously in vertebrates including humans (Mas, Croft, Zahra, Barden, & Mori, 2012), some of them even at higher levels than the parent DHA. They have been reported to display a large variety of biological activities through blocking or activation of various GPCRs, as well as anti-inflammatory and neuro-protective effects, similar to those regulated by RXRs (Bazan, 2018; Krishnamoorthy et al., 2010). Different mechanisms have been proposed to explain the numerous biological activities of these compounds, termed collectively Specialized Proresolving Mediators (SPMs) although binding and transactivation of RXRs has been rarely investigated (Krishnamoorthy et al., 2010; Serhan, 2014).

Phytanic acid **12** (Fig. 2), a metabolite of the phytol fragment of chlorophyll that can be obtained from dietary sources, was reported to bind RXR, but not RAR, and elicit transactivation, most likely through

RXR homodimers (Kitareewan et al., 1996; LeMotte, Keidek, & Apfel, 1996). It promoted transactivation of RXRα activities at 10 μM, and entered into competition with [^{3}H]-9-*cis*-retinoic acid for RXRα binding with a Ki of 4 μM, ca. 200-fold less efficient than 9-*cis*-retinoic acid (Zomer et al., 2000). RXR transactivation by phytanic acid **12** (at 4–10 μM) could take place under physiological levels in plasma (1–6 μM) but in mouse tissue 10–100 lower levels were measured. Therefore, physiological RXR-activation could be cell-type specific and depend on subcellular availability (including nucleus) of both this natural product and its putative metabolites. Since humans do not efficiently absorb chlorophyll, the metabolic precursor of phytanic acid, its presence in the human body is due to the digestion of meat and milk products on the diet, and therefore it cannot be considered as an endogenous ligand in humans. Thus, phytanic acid **12** should be considered as a RXR ligand candidate of nutritional origin, but its physiological role in RXR activation remains uncertain.

Its unsaturated precursor phytenic acid **13** (Fig. 2), which has been detected in bovine serum extracts, is likewise considered as RXR ligand since it was found to also displace tritiated 9-*cis*-retinoic acid **x** from the receptor in competition assays, whereas its *Z* isomer was inactive (Kitareewan et al., 1996; LeMotte et al., 1996).

Additional C20-fatty acids, like geranyl geranoic acid derivatives (GGA, **14**), which are present in diet relevant herbs have been reported as potent RXR activators (Araki, Shidoji, Yamada, Moriwaki, & Muto, 1995), although they could not be identified endogenously.

In summary, there is a strong link between DHA **1** and RXR-mediated signaling, although a clear picture of how it operates in vivo is still missing. Since the potency of these unsaturated compounds is not sufficient to make them qualify as true endogenous RXR ligands, they should instead be considered as partial RXR agonists (Bourguet et al., 2000). It appears that the promiscuity of RXR with several unsaturated metabolites reflects its fundamental role as intracellular sensor of the cell metabolic status, binding fatty acids and analogous compounds depending on their supply and their local concentrations. All together, these results underline the potential involvement of RXR in lipid homeostasis through complex feedback mechanisms in potent association with other NRs (such as PPAR or FXRs).

3.2 Rexinoids from plants (Fig. 3)

Natural product honokiol **15** (Fig. 3), isolated from *Magnolia obovata*, and positional isomer magnolol **16**, from *Magnolia officinalis*, used in traditional

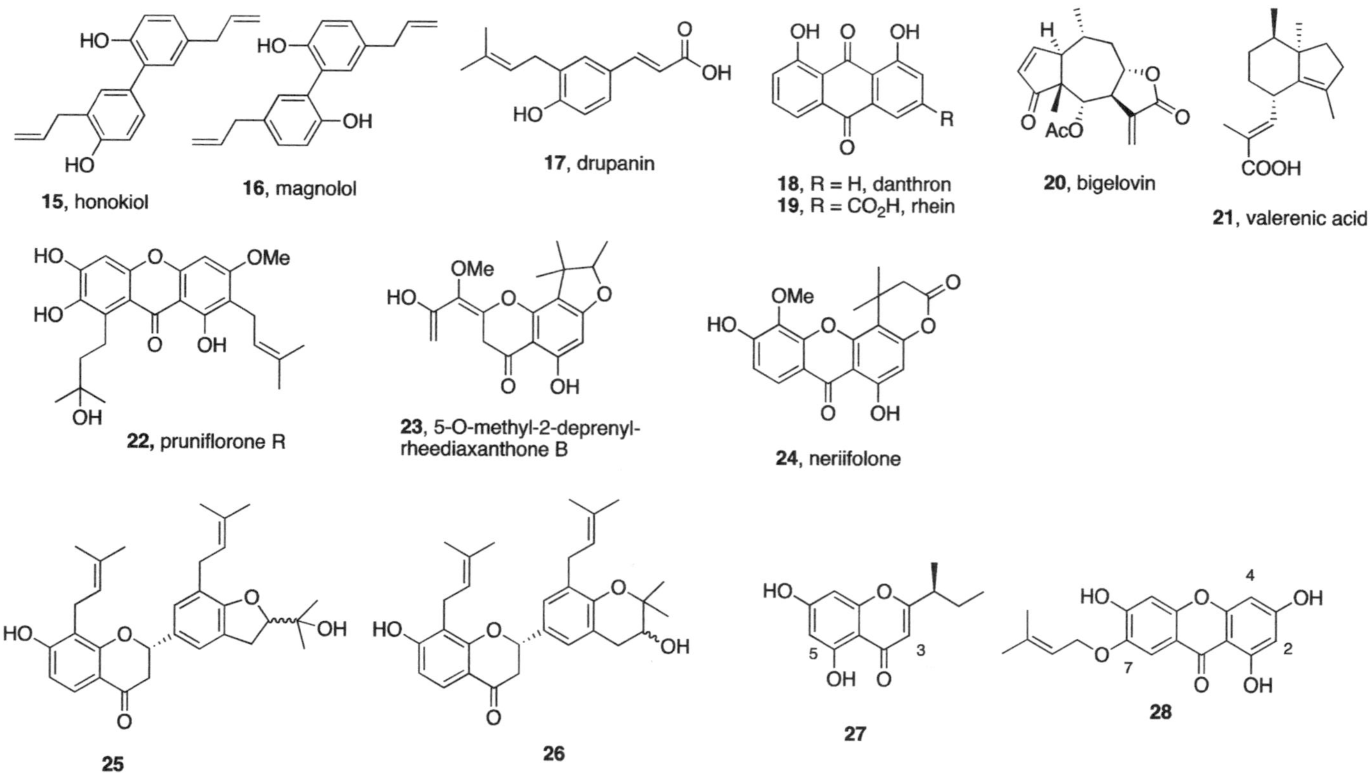

Fig. 3 Naturally occurring RXR ligands from plants with phyto-pharmacological relevance.

Chinese medicine to cure cough, diarrhea and allergic rhinitis (also metabolic syndrome) have been reported as RXR modulators (Kotani, Tanabe, Mizukami, Makishima, & Inoue, 2010; Zhang et al., 2011). Honokiol **15** was found to induce the transactivation of the PPARγ/RXRα heterodimer, but not that of the RXRα/RXRα homodimer (Kotani et al., 2010). The crystal structure of the ligand-bound RXRα revealed that the allyl phenol moieties of magnolol **16** occupy the L-shaped arms of the LBD and, since it does not contain a carboxylic acid, stabilization involves formation of a hydrogen bond with Asn306 on N-terminal helix H5 by one of the phenol groups. Ligand binding to RXRα and PPARγ was estimated as EC_{50} ~40 μM and EC_{50} ~2 μM, respectively, for **16**. The data might explain its preference for activation of RXR/PPARγ transcription at EC_{50} ~10 μM (Zhang, Xu, et al., 2011). Likewise, the reported anti-inflammatory, neuroprotective or antitumor activities of honokiol **15** in animal models of Alzheimer's or Parkinson's disease (Kotani et al., 2010) may be related to the proposed stimulation of RXRα (Wang, Dong, & Wang, 2018).

More recent Molecular Modeling and NMR spectroscopy concluded that honokiol **15** targets RXR at both sides of the interface, acting on the coactivator side of the dynamic activation function (AF2) or alternatively it switches from one to the other side of the interface (Scheepstra et al., 2014). Rational design allowed to split the dual-binding properties of the natural product honokiol **15** at the dynamic nuclear receptor interface AF2, and led to derivatives that either stabilize AF2 through binding at the LBP or destabilize AF2 through binding at the coactivator solvent-exposed side of the interface (Scheepstra et al., 2014).

Drupanin **17** (Fig. 3) isolated from green propolis (bee glue), a resinous mixture produced by honey bees, has shown selective, dual RXR/PPARγ agonistic profile in transactivation and binding assays (no binding to RARs, LXRs, VDR was observed), and also increased Ap2/FABP4 expression and accelerated lipid accumulation in differentiated 3T3-L1 cells. The dual RXR/PPARγ agonistic profile was similar to that described for magnolol **16**, although drupanin **17** can bind with higher efficacy to all RXR isotypes (EC_{50} = 2–7 μM) than to PPARγ (EC_{50} ~15 μM) (Nakashima, Murakami, Tanabe, & Inoue, 2014).

Naturally occurring anthraquinones danthron and rhein (**18**, **19**), isolated from Chinese rhubarb *Rheum palmatum* have been reported to function as antagonists of 9-*cis*-retinoic acid **31** (Fig. 4) in RXRα-LBD, Gal4-luciferase transactivation assays with IC_{50} values of 10 and 70 μM,

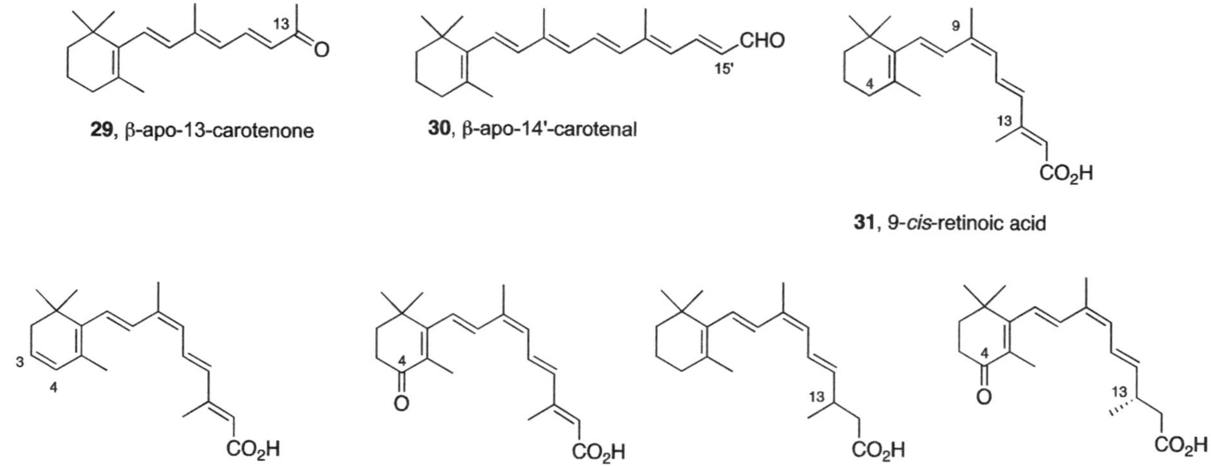

29, β-apo-13-carotenone

30, β-apo-14'-carotenal

31, 9-*cis*-retinoic acid

32, 9-*cis*-3,4-dihydroretinoic acid

33, 9-*cis*-4-oxoretinoic acid

34, 9-*cis*-13,14-dihydroretinoic acid

35, (*S*)-9-*cis*-4-oxo-13,14-dihydroretinoic acid

Fig. 4 Structures of apocarotenoids (including retinoids).

respectively (Zhang et al., 2011; Zhang, Chen, Chen, Jiang, & Shen, 2011). A crystal structure of tetrameric RXR was found to be stabilized by the anthraquinones as an inactive form with the recruitment of SMRT (silencing mediator for retinoid and thyroid hormone receptors) corepressor, and showed a strong displacement of H12 helix, making it inaccessible for formation of transcriptionally functional homo- or heterodimers. Danthron **18** and rhein **19** inhibited RXR-mediated transactivation through not only RXR-homodimers, but also through heterodimers of RXR with FXR, LXR or PPARγ.

Sesquiterpene lactone bigelovin **20** (Fig. 3), isolated from the flowers of *Inula helianthus-aquatica* or *Inula hupehensis*, which are used in folk medicine for their anti-cancer potential, binds to RXRα-LBD (no binding to LXRα, FXR or PPARγ was observed), but transactivates through the RXRα-PPARγ heterodimers (not through RXRα-FXR heterodimers) and suppresses RXR-homodimers or RXRα-LXR heterodimers on their respective response elements in the 1–10 μM concentration range (Zhang et al., 2011). The ligand-mediated remodeling of LBD observed in the crystal structure of RXR-LBD was proposed to rationalize the discriminatory activities between RXR-homodimer and heterodimer interactions, and the opposing activities of specific RXR heterodimers.

Computer-aided drug design (CADD) has also been applied to RXR receptors, leading to the discovery of rexinoids that modulate the activities of RXR in a canonical and in a non-canonical way, through binding to the receptor surface or attachment to the co-regulatory binding surface (co-activator binding site and dynamic protein interfaces).

Screening the Dictionary of Natural Products based on isomimetics of natural RXR ligands led to identification of valerenic acid **21** as new natural RXR ligand (Merk, Grisoni, Friedrich, Gelzinyte, & Schneider, 2018). This sesquiterpenoid, isolated from *Valeriana officinalis* (a flowering plant mainly present in Europe and Asia), showed agonistic activity in the micromolar range, which was reflected in the efficient induction of the RXR transcription target genes ABCA1 and ApoE. The Gal4-based reporter assays performed in HEK293T cells revealed intriguing partial selectivity for RXRβ, since valerenic acid showed better EC_{50} (5 μM) and about 10-times higher transactivation of RXRβ as compared to RXRα (EC_{50} = 27 μM), and RXRγ (EC_{50} = 43 μM). In addition, no activation of RARs, PPARs, LXRs, FXR, CAR, VDR and PXR was observed. Molecular modeling studies suggested similar binding modes of **21** to RXRα and RXRβ as classical rexinoids (Merk et al., 2018).

Among other natural products analogs isolated from *Cratoxylum formosum* ssp. pruniflorone R **22** (Fig. 3) was found to be active on RXRα, when CV-1 cells were transiently transfected with the TREpal-tk-CAT reporter (Duan et al., 2010). Xanthone **23** was the most potent of the family of analogues isolated from *Cratoxylum formosum* ssp. *pruniflorum* and assayed using the transcriptional activities of RXRα in CV-1 cells co-transfected with TREpal-tk-CAT (Duan et al., 2011). Neriifolone B **24**, isolated from the stems of *Cratoxylum cochinchinense*, was also reported as an inhibitor of the RXRα transcriptional activity (Shen et al., 2014). Similar to other derivatives and analogues, it showed concentration-dependent RXRα inhibitory activities.

RXR agonistic activity has been reported for two prenylated flavonones isolated from *Sophora tonkinensis*, which is used in traditional Chinese medicine (Inoue, Tanabe, Nakashima, Ishida, & Kotani, 2014). Both natural products (**25**, **26**) were found to bind at the nanomolar range (EC_{50} 0.77 and 0.78 μM, respectively) to the three RXR isotypes (no binding to RARα, LXRα or PPARα, β/δ or γ), and to transactivate through PPARγ/RXR, PPARδ/RXR, RXR/LXR, and RXR/RAR heterodimers, with the ability to potentiate the activity of agonists for the partner NRs. However, the concentration required for inducing a similar transcriptional activity than bexarotene (LDG1069) was 10 times higher. In addition, the prenylated flavonones were shown to induce the preferential expression of lipoprotein lipase, angiopoietin-like protein 4, and heme oxygenase-1, which are involved in prevention of heart diseases, diabetes, insulin resistance, and atherosclerosis. Therefore, the transcriptional activity of these ligands on RXR heterodimers with LXR, PPARβ/δ, PPARγ are different to those induced by bexarotene, and further studies on these prenylated flavonones are required in order to clarify their potential utility.

Xanthones **27** and **28** were reported as RXRα agonists, with K_D to RXRα-LBD of 68.3 and 14.0 μM, respectively, and showed selective cytotoxicity against HeLa cells, most likely by extrinsic pathways. Apoptosis was shown to be promoted through PARP cleavage and caspase-8 activation. A fluorescence quenching assay suggested that these two xanthones can interact with the Arg316 of the RXRα-LBD and directly bind to RXRα. Molecular docking studies taking into account the topology of these two compounds indicated their structural adaptation to occupy the L-shaped binding pocket of RXRα (Qiu et al., 2019). The proposal is in agreement with SAR studies showing that the hydroxyl group at C5 and an aliphatic chain at C2 in chromone-type compounds as well as a free

3-OH and an aliphatic chain substituted at the 7-OH in these xanthones are essential for their potent RXRα transcriptional inhibitory activities. Likewise, it is consistent with the finding that prenylated xanthones exhibited more potent effects on the RXRα transcriptional inhibitory activities than simple xanthones (Duan et al., 2010, 2011; Shen et al., 2014).

3.3 Endogenous RXR ligands (Fig. 4)

β-Apo-13-carotenone **29** (Fig. 4), has been recently demonstrated to stabilize inactive tetrameric RXR, instead of the transcriptionally active RXRα dimers (Eroglu, Hruszkewycz, Curley, & Harrison, 2010; Sun, Narayanasamy, Curley, & Harrison, 2014). Ki values for RXRα of 8 nM in a 9-*cis*-retinoic acid **31** competitive displacement assay were measured, although different modes of binding are suggested by in silico ligand docking studies with RXRα-LBD (Sun et al., 2014). In addition, this apocarotenoid appears to also act as a competitive inhibitor of RARs with a Ki of 4 nM (Eroglu et al., 2012), and to reduce ATRA-RAR-mediated activity by about 70%. Lower levels of β-apo-13-carotenone **29** (1 nM) following chronic consumption of β-carotene-enriched diets has been reported (Cooperstone et al., 2017). Since β-apo-13-carotenone **29** is generated by thermal degradation and oxidation of β-carotene during food processing (it is also found only in some fresh fruits), it cannot be produced in mammals (Harrison & Quadro, 2018). Similarly, β-apo-14′-carotenal **30** (Fig. 3) has been reported to act as a potential inhibitor of RXRα as well as of selected heterodimeric partners (PPARα, PPARβ/δ, LXRα and LXRβ, but not RARs). This apocarotenoid can be found in several melons, but its detection in human plasma was not consistent (Harrison & Quadro, 2018). It is therefore unlikely that those apocarotenoids are physiologically meaningful, despite their in vitro inhibitory activities.

Although it is traditionally considered as a natural RXR ligand, it is important to indicate that 9-*cis*-retinoic acid **31** (Fig. 4) has not been conclusively shown to be endogenously present, except in pancreatic β-cells, and recently in human serum in very low concentrations (Arnold, Amory, Walsh, & Isoherranen, 2012). Either its concentration is under the detection limit, it is present transiently or it accumulates in local cell populations (Kane, 2012).

Additional biologically potent endogenous vitamin A derivatives, among them 9-*cis*-3,4-didehydroretinoic acid **32** and 9-*cis*-4-oxo-retinoic acid **33**, have been demonstrated to bind to and activate RXR in vitro

(Mangelsdorf & Evans, 1995; Mangelsdorf et al., 1995). 9-*cis*-3,4-didehydroretinoic acid **32** could be biosynthetically derived from the precursor vitamin A_2 (all-*trans*-3,4-didehydroretinol) following similar routes to those of **31** from vitamin A (all-*trans*-retinol). 9-*cis*-4-oxo-retinoic acid **33** has been isolated from *Xenopus* embryos, where it was found to modulate axial pattern formation (Pijnappel et al., 1998), after its likely formation by isomerization of the all-*trans* isomer, which is endogenously present in the same organism (Pijnappel et al., 1993). Metabolite **33** was shown to bind RARα,β,γ and RXRα,β,γ, with IC_{50} values about 5–10 times higher than those of 9-*cis*-retinoic acid **31**, and to induce a synergistic activation of the RAR/RXR heterodimers, but not the RXR homodimers (in contrast to 9-*cis*-retinoic acid **31**) (Pijnappel et al., 1998).

The endogenous presence of 9-*cis*-13,14-dihydroretinoic acid **34** and its all-*trans* isomer, has recently been proven in several mice organs (liver, serum, brain) by HPLC-MS-MS and UV and quantified (quantitative binding affinity of K_d 90 ± 20 nM; cf. 20 ± 10 nM for 9-*cis*-retinoic acid **31** by fluorescence quenching assays) to be sufficient to maintain RXR activities (Rühl et al., 2015). Titration experiments monitored by ESI-MS showed that the affinity of RXR for the *S* enantiomer was about 65% lower than that for the *R* enantiomer. Crystal structure studies on the *R* enantiomer bound to hRXRα LBD and a 13-residue peptide comprising the nuclear receptor binding surface NR2 of NCoA2 confirmed binding to the canonical binding pocket, adopting a similar orientation than 9-*cis*-retinoic acid **31**, which is consistent with their similar transcriptional activities on several nuclear receptor-signaling pathways, possibly through the corresponding permissive heterodimers. Therefore, 9-*cis*-13,14-dihydroretinoic acid **34** was considered as the first endogenous and physiologically relevant retinoid RXR ligand in mammals (Rühl et al., 2015). This ligand and other dihydroretinoids are presumably produced by retinol saturase (RetSat) (Moise et al., 2009). The structure of 9-*cis*-13,14-dihydroretinoic acid **34** had previously been assigned to a metabolite detected in rats, together with its taurine derivative, after being fed with high doses of 9-*cis*-retinoic acid **31** (Shirley et al., 1996), but in-depth characterization of its structure and function was not pursued. 9-*cis*-retinol might be the metabolic precursor of RetSat action to generate **34** following similar biosynthetic pathways as (*R*)-all-*trans*-13,14-dihydroretinol to all-*trans*-13,14-dihydroretinal (Moise et al., 2008; Moise, Kuksa, Blaner, Baehr, & Palczewski, 2005; Moise, Kuksa, Imanishi, & Palczewski, 2004), and further metabolism to the carboxylic acid. The specific all-*trans* to 9-*cis* isomerization process is likely to occur in a fully conjugated polyene precursor, namely a carotenoid or a C9-C10-dihydrocarotenoid. Nevertheless, the precise mechanistic details and the possibility that the isomerization step could be enzymatically driven are yet unknown.

A putative metabolite of **34**, namely (*S*)-9-*cis*-4-oxo-13,14-dihydroretinoic acid **35** has been shown to be present in the liver and other tissues of both wild-type mice and rats fed a standard laboratory diet, as well as in humans (Schmidt, Volland, Hamscher, & Nau, 2002). This compound activated RAR-dependent gene transcription via RAR–RXR heterodimers, although it was less potent than all-*trans*-retinoic acid, despite its higher concentration in mice liver, but showed no transcriptional activity with either RXRα or RXRβ (Schuchardt et al., 2009).

4. Diastereo- and enantioselective synthesis of (*R*)-9-*cis*-13,14-dihydroretinoic acid 34

The preparation of (*R*)-9-*cis*-13,14-dihydroretinoic acid (*R*)-**34** was based on the Suzuki cross-coupling of enantiopure trienyliodide **40** and boronic acid **41** (Scheme 1). The synthesis of (*R*)-**3** started with (2*Z*,4*E*)-stannyldienol **35** (Domínguez, Pazos, & de Lera, 2000; Pazos & de Lera, 1999) which was transformed into the benzothiazolyl allyl sulfide **36** by Mitsunobu reaction with the corresponding thiol and subsequently oxidized to sulfone **37** with H_2O_2 and a peroxymolybdate (VI) reagent (Schultz, Freyermuth, & Buc, 1963) at −10 °C. The Julia–Kocienski olefination (Aïssa, 2009; Blakemore, 2002) was performed using a slight excess of base

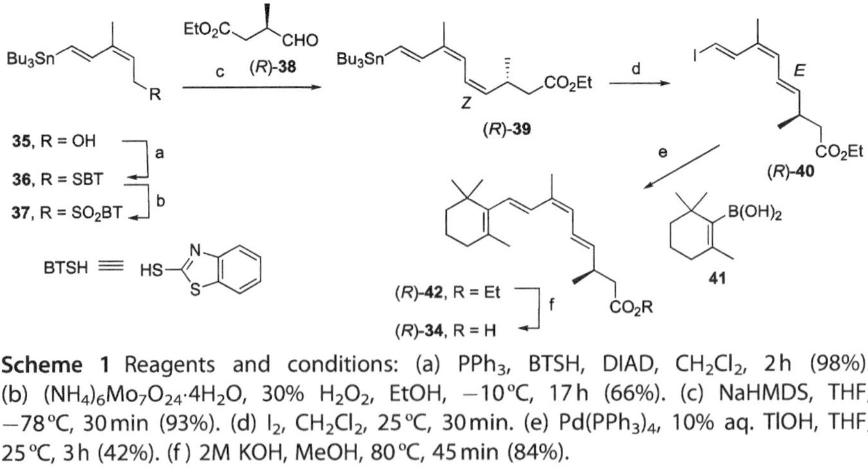

Scheme 1 Reagents and conditions: (a) PPh_3, BTSH, DIAD, CH_2Cl_2, 2 h (98%). (b) $(NH_4)_6Mo_7O_{24}\cdot 4H_2O$, 30% H_2O_2, EtOH, −10 °C, 17 h (66%). (c) NaHMDS, THF, −78 °C, 30 min (93%). (d) I_2, CH_2Cl_2, 25 °C, 30 min. (e) $Pd(PPh_3)_4$, 10% aq. TlOH, THF, 25 °C, 3 h (42%). (f) 2M KOH, MeOH, 80 °C, 45 min (84%).

(NaHMDS, 1.15 equivalents) and 1.7 equivalents of enantiopure aldehyde (*R*)-**38** (Leonard, Mohialdin, Reed, Ryan, & Swain, 1995; Moise et al., 2008). As anticipated from previous findings on the stereoselectivity of the Julia–Kocienski reactions of allylsulfones and aldehydes (Sorg & Brückner, 2005; Vaz, Alvarez, Souto, & de Lera, 2005), the geometry of the newly formed olefin of trienyl ester (*R*)-**39** is *Z* (which was confirmed by NOE experiments). Treatment of the precursor stannane with a solution of iodine in CH_2Cl_2 produced the iodide (*R*)-**40** via Sn-I exchange and iodine-promoted isomerization of the *Z,Z,E* triene to the desired *E,Z,E* geometric isomer (as confirmed by NOE experiments). The Suzuki reaction (Miyaura & Suzuki, 1995; Suzuki, 2011) of freshly prepared boronic acid **41** and trienyl iodide (*R*)-**40** using $Pd(PPh_3)_4$ as catalyst and 10% aqueous TlOH as base in THF at ambient temperature, followed by immediate work-up afforded ethyl (*R*)-9-*cis*-13,14-dihydroretinoate (*R*)-**42** in 78% yield. Saponification of (*R*)-**42** provided in 84% yield the desired carboxylic acid (*R*)-**34** (Rühl et al., 2015).

5. Materials

Silica gel column chromatography: Merck Kieselgel 60 (230–400 mesh particle size and C18-silicagel.

Aluminum plates for tlc: Merck Kieselgel 60F254.

Visualization for analytical tlc: UV irradiation (254 nm) or staining with an ethanolic solution of phosphomolybdic acid.

5.1 Reagents and solvents

2-Mercaptobenzothiazole, 97%.

Triphenylphosphine (PPh_3), ≥95%.

Tetrahydrofuran (THF), dried and distilled before use.

Acetonitrile, ACS reagent, max 0.0075% H_2O.

Diisopropyl azodicarboxylate (DIAD), 98%.

Ethanol (EtOH), reagent grade.

Ammonium heptamolybdate tetrahydrate [$(NH_4)_6Mo_7O_{24}\cdot 4H_2O$], >99%.

Hydrogen peroxide (35% in H_2O).

Diethyl ether (Et_2O), ACS reagent.

Diethyl ether (Et_2O), for synthesis, stabilized with 7 ppm BHT.

Methanol (MeOH), for HPLC ≥99.9%.

Sodium bis(trimethylsilyl)amide solution (NaHMDS, 1 M in THF).

Iodine (I_2), ≥99.8%.

Dichloromethane (CH_2Cl_2) for synthesis, stabilized with amylene.

Sodium dithionite ($Na_2S_2O_3$), ≥82%.

Sodium sulfate (Na_2SO_4), ≥99%, anhydrous powder.

Hexane, synthesis grade.

Triethylamine (Et_3N), ≥99%.

Tetrakis(triphenylphosphine)palladium [$Pd(PPh_3)_4$], 99%.

Thallium hydroxide (TlOH); a 10% aqueous solution, prepared from barium hydroxide, $Ba(OH)_2$ (≥98%) and thallium sulfate, Tl_2SO_4 (≥99.5%).

Celite®, ≥95%.

Potassium hydroxide (KOH), a 2M aqueous solution.

Hydrochloric acid (HCl), a 10% aqueous solution.

tert-Butyl methyl ether (*t*-BuOMe), ≥99.8%.

Trifluoroacetic acid (TFA), 99%.

Phosphomolybdic acid hydrate, $Na_3[P(Mo_3O_{10})_4]\cdot xH_2O$, technical grade.

6. Methods

6.1 Method for formation of BT-sulfide and BT-sulfone

6.1.1 (2Z,4E)-1-(Benzothiazol-2-yl)sulfonyl-5-(tri-n-butylstannyl)-3-methylpenta-2,4-diene (37)

1. A solution of (2*Z*,4*E*)-3-methyl-5-(tributylstannyl)penta-2,4-dien-1-ol **35** (1.0g, 2.58mmol), 2-mercaptobenzothiazole (0.65g, 3.87mmol) and PPh_3 (1.10g, 4.21mmol) in THF (14mL) in a 50mL round bottom flask equipped with a magnetic stirring bar and stirred for 5min at 0°C.
2. A solution of DIAD (0.77mL, 3.87mmol) in THF (5mL) was added dropwise to the above flask via syringe and the mixture was stirred for 30min at 25°C.
3. The solvent was removed and the residue was purified by column chromatography (C18-silica gel, CH_3CN) to afford 1.11g (78%) of (2*Z*,4*E*)-1-(benzothiazol-2-yl)sulfanyl-5-(tri-*n*-butylstannyl)-3-methylpenta-2,4-diene **36** as a colorless oil.
4. To a solution of **36** (0.48g, 0.89mmol) in EtOH (9mL), at −10°C, was added a solution of $(NH_4)_6Mo_7O_{24}\cdot 4H_2O$ (0.44g, 0.36mmol) in aqueous hydrogen peroxide (35%, 7.7mL, 89.1mmol).
5. Stirring was maintained for 17h at −10°C.
6. Work-up: the mixture was quenched with H_2O and extracted with Et_2O (3×). The combined organic layers were washed with brine (3×) and dried (Na_2SO_4), and the solvent was removed.

7. Purification: the residue was purified by chromatography (C18-silica gel, MeOH) to afford 0.33 g (66%) of a colorless oil, which was identified as 37.

6.2 Method for formation of dienylstannane

6.2.1 *Ethyl (3S,4Z,6Z,8E)-3,7-dimethyl-9-(tri-n-butylstannyl)nona-4,6,8-trienoate ((S)-39)*

1. A solution of (2*Z*,4*E*)-1-(benzothiazol-2-yl)sulfonyl-5-(tri-*n*-butylstannyl) 3-methylpenta-2,4-diene **37** (0.115 g, 0.20 mmol) in THF (9 mL) in a 50 mL round bottom flask equipped with a magnetic stirring bar was cooled down to −78 °C and then was treated with NaHMDS (0.23 mL, 1 M in THF, 0.23 mmol).
2. After stirring for 30 min at this temperature, a solution of ethyl (*S*)-3-methyl-4-oxobutanoate **38** (0.044 g, 0.30 mmol) in THF (4.5 mL) was added and the resulting mixture was stirred for 1 h at −78 °C.
3. It was then carefully allowed to reach room temperature for 3 h.
4. After lowering the temperature to 0 °C, Et_2O and water were added and the mixture was allowed to warm up to room temperature.
5. Work up: the reaction mixture was diluted with Et_2O and the layers were separated. The aqueous layer was extracted with Et_2O (3 ×), the combined organic layers were dried (Na_2SO_4) and the solvent was removed.
6. Purification: the residue was purified by column chromatography (C-18 silica gel, MeOH) to afford 0.94 g (93%) of (*S*)-**39** as pale yellow oil.

6.3 Method for stannane-iodine exchange

6.3.1 *Ethyl (3R,4E,6Z,8E)-9-iodo-3,7-dimethylnona-4,6,8-trienoate ((R)-40)*

1. To a solution of ethyl (3*R*,4*Z*,6*Z*,8*E*)-3,7-dimethyl-9-(tri-*n*-butylstannyl) nona-4,6,8-trienoate (*R*)-**39** (0.060 g, 0.121 mmol) in CH_2Cl_2 (5.3 mL) in a 25 mL round-bottomed flask equipped with a magnetic stirring bar was added dropwise a solution of iodine (0.046 g, 0.182 mmol) in CH_2Cl_2 (2.8 mL).
2. The resulting mixture was stirred for 30 min at 25 °C.
3. Work-up: a saturated $Na_2S_2O_4$ solution was then added and the reaction mixture was extracted with Et_2O (3 ×), the combined organic layers were dried (Na_2SO_4) and the solvent was removed.

4. Purification: the residue was purified by column chromatography (silica gel, 97:3 hexane/Et_3N) to afford 0.037 g (92%) of (*R*)-**40** as pale yellow oil.

6.4 Method for Suzuki cross-coupling

6.4.1 Ethyl (9Z,13R)-13,14-dihydroretinoate ((R)-42)

1. To a solution of ethyl (3*R*,4*E*,6*Z*,8*E*)-9-iodo-3,7-dimethylnona-4,6,8-trienoate (*R*)-**40** (0.036 g, 0.107 mmol) in THF (2.3 mL) in a 5 mL round-bottomed flask equipped with a magnetic stirring bar was added $Pd(PPh_3)_4$ (0.013 g, 0.011 mmol).
2. After 5 min at room temperature, 2,6,6-trimethylcyclohex-1-enylboronic acid **41** (0.027 g, 0.161 mmol) was added in one portion followed by a 10% aqueous solution of TlOH (0.75 mL, 0.407 mmol).
3. Stirring was maintained for 3 h at 25 °C.
4. Work-up: Et_2O was added and the reaction mixture was filtered through a short pad of Celite®. The filtrate was washed with $NaHCO_3$ (aqueous saturated solution), the organic layer was dried (Na_2SO_4) and the solvent was removed.
5. Purification: the residue was purified by column chromatography (silica gel, 97:3 hexane/Et_3N) to afford 0.028 g (78%) of a pale yellow oil, which was identified as ethyl (9*Z*,13*R*)-13,14-dihydroretinoate (*R*)-**42**.

6.5 Method for saponification

6.5.1 (9Z,13R)-13,14-dihydroretinoic Acid ((R)-34)

1. To a solution of ethyl (9*Z*,13*R*)-13,14-dihydroretinoate (*R*)-**42** (0.023 g, 0.069 mmol) in MeOH (4.7 mL) in a 5 mL round-bottomed flask equipped with a magnetic stirring bar was added KOH (2 M aqueous solution, 1.1 mL, 2.27 mmol).
2. The reaction mixture was allowed to stir at 80 °C for 45 min.
3. The reaction was allowed to cool down to room temperature.
4. Work-up: CH_2Cl_2 and brine were added and the layers were separated. The aqueous layer was washed with H_2O (3 ×). The combined aqueous layers were acidified with 10% HCl and extracted with CH_2Cl_2 (3 ×). The combined organic layers were dried (Na_2SO_4) and the solvent was removed.
5. Purification: the residue was purified by column chromatography (silica gel, gradient from 95:5 to 90:10 CH_2Cl_2/MeOH) to afford 0.017 g (84%) of (*R*)-**34** as pale yellow oil (Fig. 5).

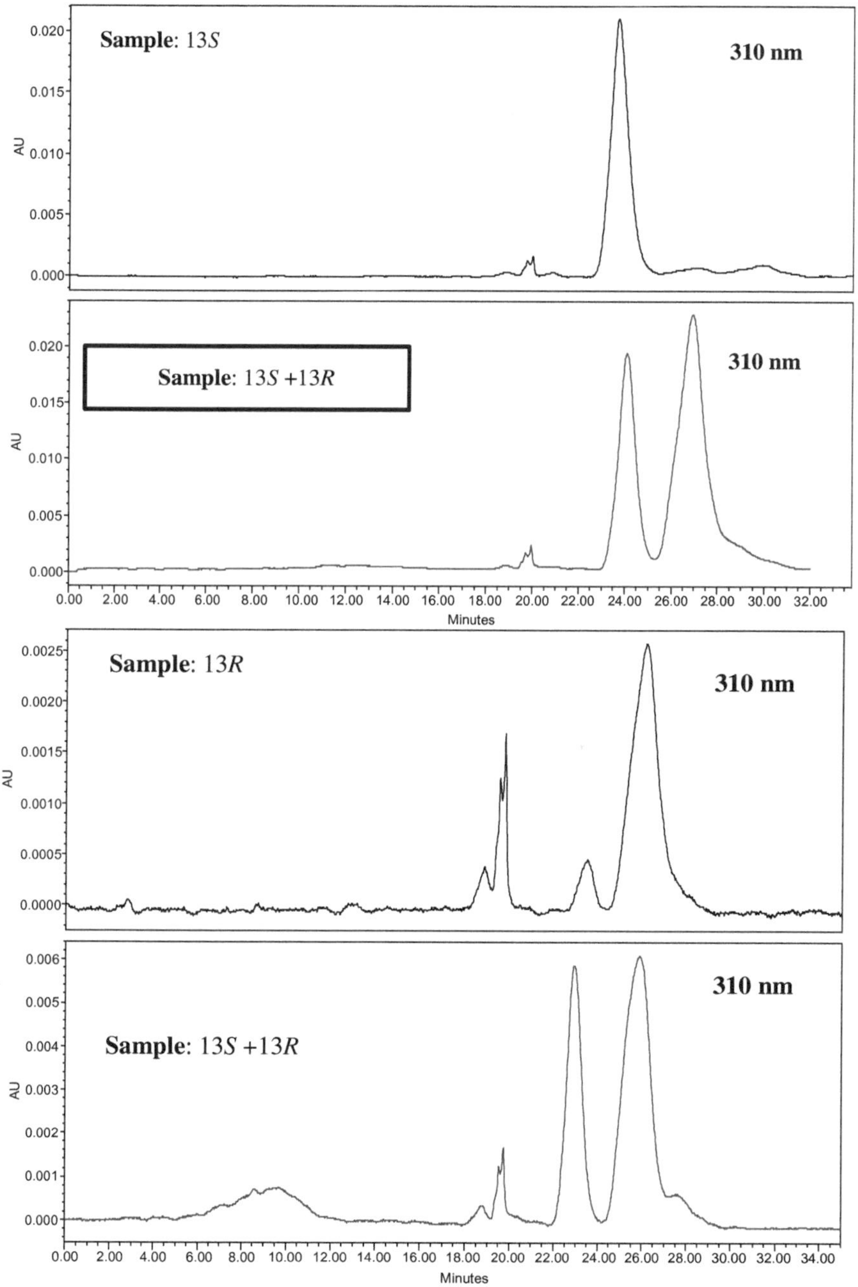

Fig. 5 Separation of 9-*cis*-13,14-dihydroretinoic acid enantiomers.

6.6 Method for enantiomeric excess determination of (9*Z*,13*R*)-13,14-dihydroretinoic acid (9*Z*,13*R*)-34

1. Stationary phase: Chiralpak IA column.
2. Eluent: 92.5:7.5 v/v solvent mixture of hexane (+0.1% TFA)/*t*-BuOMe.
3. Isocratic mode at a flow of 2 mL/min.
4. Detection by absorption at 310 nm and 254 nm.

Acknowledgments

We thank Dr. Belén Vaz and Dr. Susana Alvarez for the optimization of the enantioselective synthesis of 9-*cis*-13,14-dihydroretinoic acid, and the Spanish MINECO (SAF2016-77620-R-FEDER and BIO2016-78057-R; Xunta de Galicia (Consolidación GRC ED431C 2017/61 from DXPCTSUG; ED-431G/02-FEDER "Unha maneira de facer Europa" to CINBIO, a Galician research center 2016–19) for financial support.

References

Aïssa, C. (2009). Mechanistic manifold and new development of the Julia-Kocienski reaction. *European Journal of Organic Chemistry*, 1831–1844.

Araki, H., Shidoji, Y., Yamada, Y., Moriwaki, H., & Muto, Y. (1995). Retinoid agonist activities of synthetic geranyl geranoic acid derivatives. *Biochemical and Biophysical Research Communications*, *209*, 66–72.

Arnold, S. L. M., Amory, J. K., Walsh, T. J., & Isoherranen, N. (2012). A sensitive and specific method for measurement of multiple retinoids in human serum with UHPLC-MS/MS. *Journal of Lipid Research*, *53*, 587–598.

Bazan, N. G. (2018). Docosanoids and elovanoids from omega-3 fatty acids are pro-homeostatic modulators of inflammatory responses, cell damage and neuroprotection. *Molecular Aspects of Medicine*, *64*, 18–33.

Blakemore, P. R. (2002). The modified Julia olefination: Alkene synthesis via the condensation of metallated heteroarylalkylsulfones with carbonyl compounds. *Journal of the Chemical Society, Perkin Transactions*, *1*, 2563–2585.

Bourguet, W., & Moras, D. (2015). Retinoid receptors: Protein structure, DNA recognition and structure-function relationships. In P. Dollé & K. Niederreither (Eds.), *The retinoids: Biology, biochemistry, and disease* (pp. 131–149). Hoboken, NJ: John Wiley & Sons.

Bourguet, W., Vivat, V., Wurtz, J.-M., Chambon, P., Gronemeyer, H., & Moras, D. (2000). Crystal structure of a heterodimeric complex of RAR and RXR ligand-binding domains. *Molecular Cell*, *5*, 289–298.

Cooperstone, J. L., Riedl, K. M., Cichon, M. J., Francis, D. M., Curley, R. W., Schwartz, S. J., et al. (2017). Carotenoids and apo-carotenoids in human plasma after continued consumption of high β-carotene or high lycopene tomato juice. *The FASEB Journal*, *31*(1_supplement), 635.13-635.13.

de Urquiza, A. M., Liu, S., Sjöberg, M., Zetterström, R. H., Griffiths, W., Sjövall, J., et al. (2000). Docosahexaenoic acid, a ligand for the retinoid X receptor in mouse brain. *Science*, *290*, 2140–2144.

Dominguez, M., Alvarez, S., & de Lera, A. R. (2017). Natural and structure-based RXR ligand scaffolds and their functions. *Current Topics in Medicinal Chemistry*, *17*, 631–662.

Domínguez, B., Pazos, Y., & de Lera, A. R. (2000). Stereocontrolled synthesis of 6-s-cis and 6-s-trans-locked 9*Z*-retinoids by hydroxyl-accelerated Stille coupling of (*Z*)-tri-n-butylstannylbut-2-en-1-ol and bicyclic dienyl triflates. *The Journal of Organic Chemistry*, *65*, 5917–5925.

Duan, Y.-H., Dai, Y., Wang, G.-H., Chen, H.-F., Gao, H., Chen, J.-B., et al. (2011). Xanthone and benzophenone glycosides from the stems of *Cratoxylum formosum* ssp. *pruniflorum*. *Chemical & Pharmaceutical Bulletin*, *59*, 231–234.

Duan, Y.-H., Dai, Y., Wang, G.-H., Zhang, X., Chen, H.-F., Chen, J.-B., et al. (2010). Bioactive xanthones from the stems of *Cratoxylum formosum* ssp. pruniflorum. *Journal of Natural Products*, *73*, 1283–1287.

Eager, N. S. C., Brickell, P. M., Snell, C., & Wood, J. N. (1992). Structural and functional evidence for activation of a chick retinoid X receptor by eicosanoids. *Proceedings of the Royal Society of London Series B: Biological Sciences*, *250*, 63–69.

Egea, P. F., Mitschler, A., & Moras, D. (2002). Molecular recognition of agonist ligands by RXRs. *Molecular Endocrinology*, *16*, 987–997.

Eroglu, A., Hruszkewycz, D., Curley, R. W., & Harrison, E. H. (2010). Selected beta-apocarotenoids antagonize retinoic acid-induced transactivation of retinoic acid receptors. *The FASEB Journal*, *24*, 894.4.

Eroglu, A., Hruszkewycz, D. P., dela Sena, C., Narayanasamy, S., Riedl, K. M., Kopec, R. E., et al. (2012). Naturally occurring eccentric cleavage products of provitamin a β-carotene function as antagonists of retinoic acid receptors. *The Journal of Biological Chemistry*, *287*, 15886–15895.

Germain, P., Chambon, P., Eichele, G., Evans, R. M., Lazar, M. A., Leid, M., et al. (2006). The pharmacology and classification of the nuclear receptor superfamily. Retinoid X receptors (RXRs). *Pharmacological Reviews*, *58*, 760–772.

Germain, P., Gaudon, C., Pogenberg, V., Sanglier, S., Potier, N., Van Dorsselaer, A., et al. (2009). Differential action on coregulator interaction defines inverse retinoid agonists and neutral antagonists. *Chemistry & Biology*, *16*, 479–489.

Germain, P., Iyer, J., Zechel, C., & Gronemeyer, H. (2002). Co-regulator recruitment and the mechanism of retinoic acid receptor synergy. *Nature*, *415*, 187–192.

Goldstein, J. T., Dobrzyn, A., Clagett-Dame, M., Pike, J. W., & DeLuca, H. F. (2003). Isolation and characterization of unsaturated fatty acids as natural ligands for the retinoid-X receptor. *Archives of Biochemistry and Biophysics*, *420*, 185–193.

Harrison, E. H., & Quadro, L. (2018). Apocarotenoids: Emerging roles in mammals. *Annual Review of Nutrition*, *38*, 153–172.

Hiebl, V., Ladurner, A., Latkolik, S., & Dirsch, V. M. (2018). Natural products as modulators of the nuclear receptors and metabolic sensors LXR, FXR and RXR. *Biotechnology Advances*, *36*(6), 1657–1698.

Huang, P., Chandra, V., & Rastinejad, F. (2014). Retinoic acid actions through mammalian nuclear receptors. *Chemical Reviews*, *114*, 233–254.

Inoue, M., Tanabe, H., Nakashima, K.-i., Ishida, Y., & Kotani, H. (2014). Rexinoids isolated from *Sophora tonkinensis* with a gene expression profile distinct from the synthetic rexinoid bexarotene. *Journal of Natural Products*, 77, 1670–1677.

Kane, M. A. (2012). Analysis, occurrence, and function of 9-*cis*-retinoic acid. *Biochimica et Biophysica Acta (BBA)—Molecular and Cell Biology of Lipids*, *1821*, 10–20.

Kitareewan, S., Burka, L., Tomer, B., Parker, C. E., Deterding, L. J., Stevens, R. D., et al. (1996). Phytol metabolites are circulating dietary factors that activate the nuclear receptor RXR. *Molecular and Cellular Biology*, 7, 1153–1166.

Kotani, H., Tanabe, H., Mizukami, H., Makishima, M., & Inoue, M. (2010). Identification of a naturally occurring rexinoid, honokiol, that activates the retinoid X receptor. *Journal of Natural Products*, *73*, 1332–1336.

Krishnamoorthy, S., Recchiuti, A., Chiang, N., Yacoubian, S., Lee, C.-H., Yang, R., et al. (2010). Resolvin D1 binds human phagocytes with evidence for proresolving receptors. *Proceedings of the National Academy of Sciences of the United States of America*, *107*, 1660–1665.

Laudet, V., & Gronemeyer, H. (2002). *The nuclear receptor facts book*. San Diego: Academic Press.

le Maire, A., Alvarez, S., Shankaranarayanan, P., de Lera, A. R., Bourguet, W., & Gronemeyer, H. (2012). Retinoid receptors and therapeutic applications of RAR/RXR modulators. *Current Topics in Medicinal Chemistry*, *12*, 505–527.

LeMotte, P. K., Keidek, S., & Apfel, C. M. (1996). Phytanic acid is a retinoid X receptor ligand. *European Journal of Biochemistry*, *236*, 328–333.

Lengqvist, J., Mata de Urquiza, A., Bergman, A.-C., Willson, T. M., Sjövall, J., Perlmann, T., et al. (2004). Polyunsaturated fatty acids including docosahexaenoic and arachidonic acid bind to the retinoid X receptor α ligand-binding domain. *Molecular & Cellular Proteomics*, *3*, 692–703.

Leonard, J., Mohialdin, S., Reed, D., Ryan, G., & Swain, P. A. (1995). Stereoselective conjugate addition of organolithium and organocopper reagents to δ-oxygenated α,β-unsaturated carbonyl systems derived from glyceraldehyde acetonide. *Tetrahedron*, *51*, 12843–12858.

Mangelsdorf, D. A., & Evans, R. M. (1995). The RXR heterodimers and orphan receptors. *Cell*, *83*, 841–850.

Mangelsdorf, D. A., Thummel, C., Beato, M., Herrlich, P., Schütz, G., Umesono, K., et al. (1995). The nuclear receptor superfamily: The second decade. *Cell*, *83*, 835–839.

Mas, E., Croft, K. D., Zahra, P., Barden, A., & Mori, T. A. (2012). Resolvins D1, D2, and other mediators of self-limited resolution of inflammation in human blood following n-3 fatty acid supplementation. *Clinical Chemistry*, *58*, 1476–1484.

Mendoza-Parra, M. A., Bourguet, W., de Lera, A. R., & Gronemeyer, H. (2015). Retinod receptor-selective modulators: Chemistry, 3D structures and systems biology. In P. N. Dollé & K. Neiderreither (Eds.), *The retinoids: Biology, biochemistry, and disease* (pp. 165–192). Hoboken, N. J.: Wiley-Blackwell.

Merk, D., Grisoni, F., Friedrich, L., Gelzinyte, E., & Schneider, G. (2018). Computer-assisted discovery of retinoid X receptor modulating natural products and isofunctional mimetics. *Journal of Medicinal Chemistry*, *61*, 5442–5447.

Mic, F. A., Molotkov, A., Benbrook, D. M., & Duester, G. (2003). Retinoid activation of retinoic acid receptor but not retinoid X receptor is sufficient to rescue lethal defect in retinoic acid synthesis. *Proceedings of the National Academy of Sciences of the United States of America*, *100*, 7135–7140.

Miyaura, N., & Suzuki, A. (1995). Palladium-catalyzed cross-coupling reactions of organoboron compounds. *Chemical Reviews*, *95*, 2457–2483.

Moise, A. R., Alvarez, S., Domínguez, M., Alvarez, R., Golczak, M., Lobo, G. P., et al. (2009). Activation of retinoic acid receptors by dihydroretinoids. *Molecular Pharmacology*, *76*, 1228–1237.

Moise, A. R., Domínguez, M., Alvarez, S., Alvarez, R., Schupp, M., Cristancho, A. G., et al. (2008). Stereospecificity of retinol saturase: Absolute configuration, synthesis, and biological evaluation of dihydroretinoids. *Journal of the American Chemical Society*, *130*, 1154–1155.

Moise, A. R., Kuksa, V., Blaner, W. S., Baehr, W., & Palczewski, K. (2005). Metabolism and transactivation activity of 13,14-dihydroretinoic acid. *The Journal of Biological Chemistry*, *280*, 27815–27825.

Moise, A. R., Kuksa, V., Imanishi, Y., & Palczewski, K. (2004). Identification of all-trans-retinol: All-trans-13,14-dihydroretinol saturase. *The Journal of Biological Chemistry*, *279*, 50230–50242.

Nakashima, K.-I., Murakami, T., Tanabe, H., & Inoue, M. (2014). Identification of a naturally occurring retinoid X receptor agonist from *Brazilian green propolis*. *Biochimica et Biophysica Acta (BBA)—General Subjects*, *1840*, 3034–3041.

Pazos, Y., & de Lera, A. R. (1999). Stereoselective synthesis of 9-*cis*-retinoic acid based on stepwise or convergent Suzuki coupling reactions. *Tetrahedron Letters*, *40*, 8287–8290.

Pijnappel, W. W. M., Folkers, G. E., de Jonge, W. J., Verdegem, P. J. E., de Laat, S. W., Lugtenburg, J., et al. (1998). Metabolism to a response pathway selective retinoid ligand during axial pattern formation. *Proceedings of the National Academy of Sciences of the United States of America, 95*, 15424–15429.

Pijnappel, W. W. M., Hendriks, H. F. J., Folkers, G. E., van den Brink, C. E., Dekker, E. J., Edelenbosch, C., et al. (1993). The retinoid ligand 4-oxo-retinoic acid is a highly active modulator of positional specification. *Nature, 366*, 340–344.

Qiu, D., Zhou, M., Lin, T., Chen, J., Wang, G., Huang, Y., et al. (2019). Cytotoxic components from *Hypericum elodeoides* targeting RXRα and inducing HeLa cell apoptosis through caspase-8 activation and PARP cleavage. *Journal of Natural Products, 82*, 1072–1080.

Rühl, R., Krzyzosiak, A., Niewiadomska-Cimicka, A., Rochel, N., Szeles, L., Vaz, B., et al. (2015). 9-*cis*-13,14-dihydroretinoic acid is an endogenous retinoid acting as RXR ligand in mice. *PLoS Genetics, 11*, e1005213.

Scheepstra, M., Nieto, L., Hirsch, A. K. H., Fuchs, S., Leysen, S., Lam, C. V., et al. (2014). A natural-product switch for a dynamic protein interface. *Angewandte Chemie, International Edition, 53*, 6443–6448.

Schmidt, C. K., Volland, J., Hamscher, G., & Nau, H. (2002). Characterization of a new endogenous vitamin a metabolite. *Biochimica et Biophysica Acta (BBA)—Molecular and Cell Biology of Lipids, 1583*, 237–251.

Schuchardt, J. P., Wahlström, D., Rüegg, J., Giese, N., Stefan, M., Hopf, H., et al. (2009). The endogenous retinoid metabolite (*S*)-4-oxo-9-*cis*-13,14-dihydroretinoic acid activates retinoic acid receptor signalling both in vitro and in vivo. *The FEBS Journal, 276*, 3043–3059.

Schultz, H. S., Freyermuth, H. B., & Buc, S. R. (1963). New catalysts for the oxidation of sulfides to sulfones with hydrogen peroxide. *The Journal of Organic Chemistry, 28*, 1140–1142.

Serhan, C. N. (2014). Pro-resolving lipid mediators are leads for resolution physiology. *Nature, 510*, 92–101.

Serhan, C. N., & Petasis, N. A. (2011). Resolvins and protectins in inflammation resolution. *Chemical Reviews, 111*, 5922–5943.

Shen, Q., Dai, Y., Wang, G., Yao, F., Duan, Y., Chen, H., et al. (2014). Total synthesis and RXRα-mediated transcription studies of neriifolone B and related compounds. *Bioorganic & Medicinal Chemistry, 22*, 2671–2677.

Shirley, M. A., Bennani, Y. L., Boehm, M. F., Breau, A. P., Pathirana, C., & Ulm, E. H. (1996). Oxidative and reductive metabolism of 9-*cis*-retinoic acid in the rat. Identification of 13,14-dihydro-9-*cis*-retinoic acid and its taurine conjugate. *Drug Metabolism and Disposition, 24*, 293–302.

Sorg, A., & Brückner, R. (2005). Unexpected cis-selectivity in (Sylvestre) Julia olefinations with Bu3Sn-containing allyl benzothiazolyl sulfones: Stereoselective synthesis of 1,3-butadienyl- and 1,3,5-hexatrienylstannanes. *Synlett*, 289–293.

Sun, J., Narayanasamy, S., Curley, R. W., & Harrison, E. H. (2014). β-Apo-13-carotenone regulates retinoid X receptor transcriptional activity through tetramerization of the receptor. *The Journal of Biological Chemistry, 289*, 33118–33124.

Suzuki, A. (2011). Cross-coupling reactions of organoboranes: An easy way to construct C-C bonds. *Angewandte Chemie (International Ed in English), 50*, 6722–6737.

Vaz, B., Alvarez, R., Souto, J. A., & de Lera, A. R. (2005). γ-Allenyl allyl benzothiazole sulfonyl anions undergo cis-selective (Sylvestre) Julia Olefinations. *Synlett, 36*, 294–298.

Wang, D., Dong, X., & Wang, C. (2018). Honokiol ameliorates amyloidosis and neuroinflammation and improves cognitive impairment in Alzheimer's disease transgenic mice. *The Journal of Pharmacology and Experimental Therapeutics, 366*, 470–478.

Wietrzych-Schindler, M., Szyszka-Niagolov, M., Ohta, K., Endo, Y., Pérez, E., de Lera, A. R., et al. (2011). Retinoid X receptor gamma is implicated in docosahexaenoic acid modulation of despair Behaviors and working memory in mice. *Biological Psychiatry, 69*, 788–794.

Xu, H. E., Lambert, M. H., Montana, V. G., Plunket, K. D., Moore, L. B., Collins, J. L., et al. (2001). Structural determinants of ligand binding selectivity between the peroxisome proliferator activated receptors. *Proceedings of the National Academy of Sciences of the United States of America*, *98*, 13919–13924.

Zhang, H., Chen, L., Chen, J., Jiang, H., & Shen, X. (2011). Structural basis for retinoic X receptor repression on the tetramer. *The Journal of Biological Chemistry*, *286*, 24593–24598.

Zhang, H., Li, L., Chen, L., Hu, L., Jiang, H., & Shen, X. (2011). Structure basis of bigelovin as a selective RXR agonist with a distinct binding mode. *Journal of Molecular Biology*, *407*, 13–20.

Zhang, H., Xu, X., Chen, L., Chen, J., Hu, L., Jiang, H., et al. (2011). Molecular determinants of magnolol targeting both RXRα and PPARγ. *PLoS One*, *6*, e28253.

Zhang, H., Zhou, R., Li, L., Chen, J., Chen, L., Li, C., et al. (2011). Danthron functions as a retinoic X receptor antagonist by stabilizing tetramers of the receptor. *The Journal of Biological Chemistry*, *286*, 1868–1875.

Zomer, A. W. M., van der Burg, B., Jansen, G. A., Wanders, R. J. A., Poll-The, B. T., & van der Saag, P. T. (2000). Pristanic acid and phytanic acid: Naturally occurring ligands for the nuclear receptor peroxisome proliferator-activated receptor α. *Journal of Lipid Research*, *41*, 1801–1807.

CHAPTER TEN

DNA recognition by retinoic acid nuclear receptors

Carole Peluso-Iltis[a,b,c,d], Judit Osz[a,b,c,d], Natacha Rochel[a,b,c,d,*]

[a]Institut de Génétique et de Biologie Moléculaire et Cellulaire (IGBMC), Illkirch, France
[b]Centre National de la Recherche Scientifique, Illkirch, France
[c]Institut National de la Santé et de la Recherche Médicale, Illkirch, France
[d]Université de Strasbourg, Illkirch, France
[*]Corresponding author: e-mail address: rochel@igbmc.fr

Contents

Abstract

Retinoic acid receptors (RARs) heterodimerize with retinoid X receptors (RXRs) to regulate gene expression. The heterodimer recognizes the genome via a large and diverse repertoire of DNA response elements. Assessing the binding mode of RAR and RXR with various DNA response elements is important for understanding how they select their binding site and how DNA sequence and topology allosterically regulate RAR function. A number of complementary assays are often employed for analysis of the binding mode. To biochemically and structurally characterize RAR and RXR-DNA complexes, we describe how to express and purify RAR and RXR-DNA binding domains (DBDs) and multidomain constructs. We also describe the use of electrospray ionization mass spectrometry (ESI MS) and isothermal titration calorimetry (ITC) that give information about stoichiometry and binding affinity, as well as our approaches for co-crystallization of RAR and RXR DBDs with DNA.

Methods in Enzymology, Volume 637
ISSN 0076-6879
https://doi.org/10.1016/bs.mie.2020.03.001

1. Introduction

Retinoic acid (RA) regulates transcription by binding to its nuclear receptors (RARs). RARs consist of three subtypes, RARα (NR1B1), RARβ (NR1B2) and RARγ (NR1B3) encoded by separate genes. RARs transduce the RA signal as heterodimers with the retinoid X receptors, RXRs, which also consist of three subtypes, RXRα (NR2B1), RXRβ (NR2B2) and RXRγ (NR2B3). RARs have a well-defined domain organization and structure (Gronemeyer, Gustafsson, & Laudet, 2004), consisting mainly of a variable N-terminal domain (NTD), and two structured and highly conserved domains, a central DNA binding domain (DBD) and a C-terminal ligand binding domain (LBD; Fig. 1A). When heterodimerized with RXRs, RARs bind to specific DNA sequences or RA response elements (RAREs) located in the regulatory sequences of target genes (Fig. 1B). The classical RAREs are composed typically of two direct repeats of a core hexameric motif (A/G)G [G/T] TCA) separated by 1, 2 or 5 base pairs (referred to as DR1, DR2 and DR5) (Kurokawa et al., 1994; Mangelsdorf & Evans, 1995; Umesono, Murakami, Thompson, & Evans, 1991). In addition, RXR homodimers bind DR1 elements (Castelein, Janssen, Declercq, & Baes, 1996; Lee, Kliewer, Provencal, Wright, & Evans, 1993; Yang, Subauste, & Koenig,

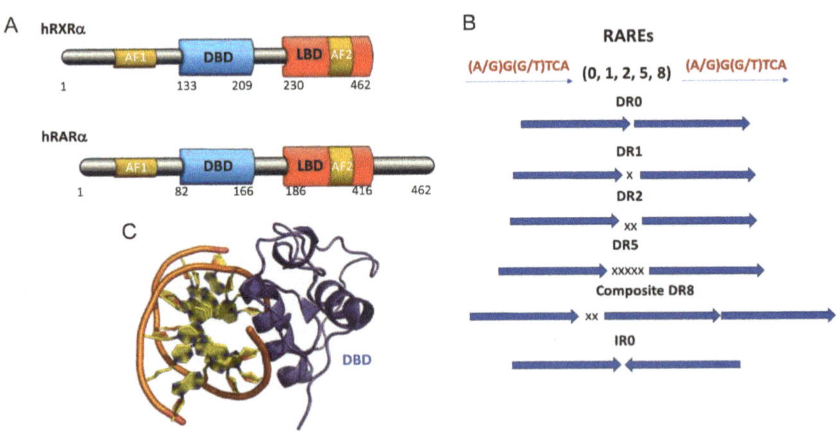

Fig. 1 (A) Schematic drawings of hRARα and hRXRα structural domains. DBD: DNA binding domain, LBD: ligand binding domain, AF: activation function. (B) Description of the RAREs with different spacing. DR: direct repeat, IR: inverted repeat. (C) 3D structure of RXR DBD bound to hexanucleotide half-site.

1995; Zechel, Shen, Chambon, & Gronemeyer, 1994; Zechel, Shen, Chen, et al., 1994). RAREs have been identified in the promoters of a number of RA-target genes implicated in a wide variety of functions including RA metabolism and cellular differentiation. RAR chromatin immunoprecipitation coupled to sequencing (ChIP-seq) studies have reported 13,000–15,000 potential RAREs (Delacroix et al., 2010; Mahony et al., 2011; Martens et al., 2010; Mendoza-Parra, Walia, Sankar, & Gronemeyer, 2011; Ross-Innes et al., 2010) and identified new RAR binding loci in the mouse genome (Chatagnon et al., 2015; Moutier et al., 2012). Indeed, RARs can occupy a larger repertoire of sites with an unexpected diversity in the spacing and topology of the DNA binding elements. These new non-canonical sites depict IR0 (inverted repeats), DR0, DR7 and DR8 spacing, with most DR8 elements comprising three half sites with DR2 and DR0 spacing. Analysis of the cistromes also revealed degenerate and non-identifiable DRs, as well as DRs overlapping with binding sites for other transcription factors, increasing the complexity of RAR-mediated gene transcription.

RARE bound dimer recruits either nuclear receptor coactivators (NCOA) or nuclear receptor corepressors (NCOR), thus directly activating or repressing transcription (Cunningham & Duester, 2015). According to the canonical model of RA-target genes transcriptional regulation, in the absence of ligand, RAR/RXR heterodimers bound to DNA are associated with repressor complexes, which maintain chromatin in a condensed and repressed state. Upon ligand binding, RARs undergo conformational changes that provoke corepressors release and initiate an ordered and coordinated recruitment of co-activator complexes involved in chromatin remodeling that pave the way for the recruitment of the transcription machinery (review in di Masi et al., 2015). However ChIP-seq experiments performed with the endogenous proteins during RA-induced differentiation of different cell types revealed that only a small set of RAR binding sites were constitutively occupied and that RA treatment induces widespread changes in RAR binding, some sets disappearing and others becoming occupied (Chatagnon et al., 2015). Such observations suggest that RA might induce rapid changes in the chromatin environment and/or rapid modifications of RARs that promote their recruitment to DNA.

Current understanding at the atomic level of the DNA recognition by RAR/RXR (Fig. 1C) is limited to the crystal structures of the RARα/RXRα DBDs (Rastinejad, Wagner, Zhao, & Khorasanizadeh, 2000) and RARβ/RXRα ΔNTD (Chandra et al., 2017) bound to an idealized DR1 with identical half sites. For RXR DBDs homodimer, the structures of DNA bound RXR complexes have been determined by X-ray crystallographic and NMR analysis using idealized and natural DR1s (Holmbeck, Dyson, & Wright, 1998; Holmbeck, Foster, et al. 1998; Lee et al., 1993; Osz et al., 2015; Osz et al., 2019; Van Tilborg et al., 1999; Zhao et al., 2000).

The availability of homogeneous dimer and dimer-DNA complex is a prerequisite for structural and biophysical studies. In the last years, we have biochemically and structurally characterized RAR and RXR-DNA complexes (Belorusova et al., 2016; Bich et al., 2010; Moutier et al., 2012; Nguyen-Huynh et al., 2016; Osz et al., 2015; Osz et al., 2019; Rochel et al., 2011). In this chapter, we describe how to produce and purify RAR and RXR DBDs or constructs truncated of their NTD (ΔNTD) for DNA interactions characterization. In addition, we describe some of the methods used for DNA interactions quantification and analysis, comprising isothermal titration calorimetry (ITC), non-covalent electrospray ionization mass spectrometry (ESI MS) and crystallization.

2. Characterization of DNA interaction of RAR and RXR DBDs

The core sequence of the nuclear receptor DBD is highly conserved in the nuclear receptor superfamily, with >40% amino acid identity over a 67-residue region. The optimal domain boundaries were determined by sequence homology and the abilities of RXR and RAR to form cooperative interactions on DNA (Rastinejad et al., 2000). The hsRARα (82–167) and hsRXRα (130 − 212) were cloned into pHXGWA (Busso, Delagoutte-Busso, & Moras, 2005) with an N-terminal thioredoxin fusion, chosen to improve protein solubility (LaVallie et al., 1993) and polyhistidine for purification. For crystallization of DNA complex, it is important to leave the smallest fusion polypeptides. Thus, after thrombin protease cleavage, the remaining N-terminal peptide is GSHM. RAR and RXR DBDs only form dimer when bound to DNA. Both proteins are thus purified separately. The purification steps are the same for both proteins (following section) and can be performed in parallel.

2.1 Expression and purification of RAR and RXR DBDs

Because the expression and the purification protocols are the same for RAR and RXR DBDs, we will use the term DBD representing both RAR and RXR DBDs. Fig. 2 shows the workflow used for the DBD expression and purification.

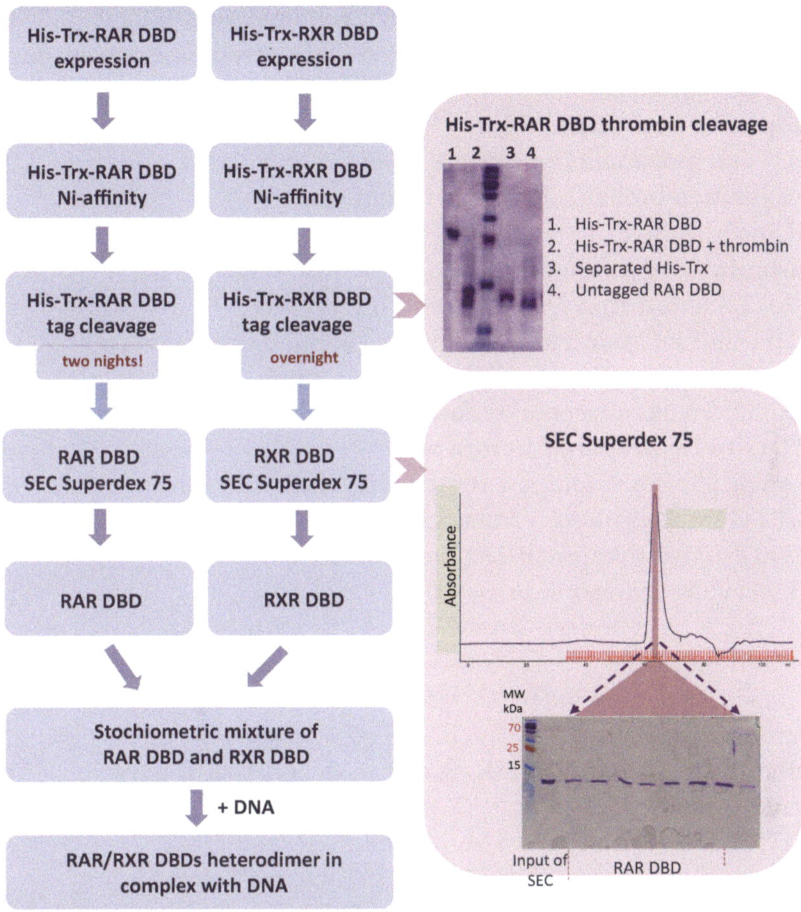

Fig. 2 Workflow of expression and purification of RAR, RXR DBDs and RAR/RXR-DNA ternary complex formation. Coomassie-stained SDS-PAGE gels showing RAR DBD at different stages of purifications.

Materials

LB Broth media (Sigma)
Ampicillin
$ZnCl_2$
IPTG (Euromedex EU0008-A)
Lysozyme (Sigma L6876)
Thrombin (MP Biomedicals)
cOmplete EDTA free Protease Inhibitor Cocktail (Roche, Ref 11873 580001)

Buffer components: Tris, NaCl, glycerol, imidazole, TCEP
Oligonucleotides containing selected DNA response elements (Sigma, HPLC purified)

Equipment and disposables

UV/vis spectrophotometer (OD600)
Sonication probe (Labsonic P, Sartorius)
Ultracentrifuge
Minisart syringe filter, 5 μm (Sartorius 17594-K)
AKTA chromatography system (GE Healthcare)
HiTrap FF Crude 5 mL Ni-affinity column (GE Healthcare)
Superdex 75 or Superdex 75 Increase column (GE Healthcare)
Amicon ultra concentration device with MWCO 10 kDa (GE Healthcare)

Buffers to be prepared before start

Ampicillin stock solution 100 mg/mL, sterile filtered
IPTG stock solution: 1 M in water
$ZnCl_2$ stock solution: 0.2 M, sterile filtered
Lysozyme: 25 mg/mL in water
Buffer A (lysis buffer): 20 mM Tris pH 8, 500 mM NaCl, 10% glycerol, 5 mM imidazole
Buffer B (elution buffer): 20 mM Tris pH 8, 500 mM NaCl, 10% glycerol, 1 M imidazole
Buffer C: 20 mM Tris pH 8, 500 mM NaCl, 10% glycerol, 40 mM imidazole
TCEP stock solution: 0.5 M, pH 8.0
Thrombin: 1 U/μL in 25 mM Tris pH 8.5, 50% glycerol
Size-exclusion chromatography (SEC) buffer: 20 mM Tris pH 8.0, 50 mM NaCl, 2 mM TCEP

2.1.1 DBDs expression

The 2 × LB Broth media is used to increase the production yield. The expression media is supplemented with $ZnCl_2$ to ensure the sufficient amount of Zn^{2+} ions for the overexpressed Zn-finger motif containing DBD.

Procedure

Day before: inoculate for each construct 200 mL LB Broth media with one colony or a glycerol freeze stock of *E.coli* BL21DE3 cells containing the expression plasmid supplemented with 100 μg/mL ampicillin. Incubate overnight at 37 °C with shaking at 180 rpm.

Day of the production: measure the OD600 of the overnight culture. Inoculate 1 L 2× LB Broth media with the overnight starter culture to achieve OD = 0.08, to start the expression culture. Supplement the media with 100 µg/mL ampicillin. Grow the culture at 37 °C with shaking at 180 rpm. When the OD600 reach 0.4–0.5, decrease the incubator temperature to 22 °C. At OD600 = 0.8, add 100 µM $ZnCl_2$, induce the protein production by adding 0.7 mM IPTG and let to grow for 3 h. After 3 h, harvest the cells by centrifugation for 20 min at 4000 *g* at 4 °C. Wash the centrifuged cell pellet with PBS + 10% glycerol and centrifuge again for 20 min at 4000 *g*. The washed cell pellet can be stored at −20 °C until purification.

Check the expression of the DBDs by 12% SDS-PAGE. The molecular mass of the expressed histidine-thioredoxin-tagged DBD is 23 kDa.

2.1.2 Cell lysis

The cell lysis and all steps of the purification should be performed at +4 °C

Procedure

(1) Resuspend 6 L DBD expression cultures in 200 mL lysis buffer supplemented with cOmplete EDTA free Protease Inhibitor Cocktail tablet.
(2) Add 1.0 mL lysozyme stock solution.
(3) Incubate 10 min on ice, 5 min at 37 °C and 10 min again on ice.
(4) Sonicate for 2 × 5 min at 60% amplitude, 0.5 s on and 0.5 s off mode.
(5) Centrifuge it at 125,000 *g* for 1 h 30 min.
(6) Filter the supernatant through 5 µm pore-size syringe filter.

Notes

(1) The lysozyme treatment is important for the effective lysis of the bacterial cell wall.
(2) Take aliquot for the SDS-PAGE at each step to follow the purification.

2.1.3 DBDs purification

The DBD purification contains four main steps: (A) Ni-affinity, (B) tag cleavage, (C) recovery of the untagged DBD and (D) size-exclusion chromatography. The average yield of the DBD purification is ~10 mg pure DBD starting from 6 L expression culture.

To succeed to crystalize the DNA complexes, it is important to remove the fusion histidine-thioredoxin and to analyze the final size-exclusion chromatography (SEC) fractions by dynamic light scattering (DLS) to conserve only the most homogeneous monodisperse ones.

Procedure

(A) Ni-affinity

(1) Equilibrate the 5 mL HisTrapp FF crude column with 50 mL of buffer A.

(2) Load the filtered supernatant on the column by the sample pump of the AKTA system with 5 mL/min flow rate.
(3) Perform the following washing steps with 5 mL/min flow rate: 75 mL buffer A, 75 mL 5% buffer B (aspecific wash), 50 mL 15% buffer B (specific elution), 50 mL 100% B buffer (final washing step).
(4) Pool together the fractions corresponding to the specific elution peak (corresponding to 150 mM imidazole).

(B) Histidine-thioredoxin tag cleavage

(5) Dilute the Ni pool with lysis buffer to decrease the imidazole concentration to 50 mM.
(6) Add TCEP at 2 mM final concentration.
(7) Measure the protein concentration and add thrombin at 4 U/mg protein concentration.
(8) Incubate for thrombin cleavage at 4 °C with rotation.
  (a) Overnight for the RXR DBD
  (b) Two nights for the RAR DBD

(C) Recover the untagged DBD and separate the histidine-thioredoxin tag

(9) Equilibrate the 5 mL HisTrapp FF crude column with 50 mL of buffer C (40 mM imidazole).
(10) Dilute the overnight incubated Ni pool + thrombin with buffer A to decrease the imidazole concentration from 50 to 40 mM.
(11) Passed the sample through the HisTrap column and collect the flow through which contains the untagged DBD.
(12) Add TCEP at 2 mM final concentration.
(13) Concentrate the untagged DBD with Amicon Ultra MWCO 10 kDa up to ≅10 mg/mL.

(D) Size-exclusion chromatography

(14) Equilibrate the Superdex 75 column with the SEC buffer.
(15) Perform the size-exclusion chromatography of the concentrated untagged DBD by the SEC buffer.
(16) Analyze the SEC fractions by 15% SDS-PAGE.
(17) Analyze the SEC fractions by DLS.
(18) Based on the results of the SDS-PAGE and the DLS measurement define the fractions suitable for crystallization.

Notes

(1) It is important to decrease the imidazole concentration of the Ni pool to keep a good quality of the protein and also to support optimal condition for cleaving the tag by thrombin.

(2) The untagged DBD recovery step could be performed either on the AKTA system or manually by using a Luer lock syringe.
(3) Take attention that the histidine-thioredoxin-tag is ≅13 kDa and after the thrombin cleavage and the RAR and RXR DBD are only 10.6 and 10.2 kDa, respectively.
(4) The concentrated DBD supplemented with 10% glycerol can be kept at −20 °C for further use.

2.1.4 *Oligonucleotides preparation*

Oligonucleotides containing selected DNA response elements were purchased from Sigma Oligos as HPLC purified quality. Each single stranded oligonucleotide was resuspended in 10 mM Hepes pH 7.0, 50 mM NaCl to 1 mM concentration. The complementary strands were mixed in equimolar ratio, heated in a boiling water bath and slowly cooled overnight at room temperature. The final DNA concentration was measured by absorbance at 260 nm.

2.2 Monitoring DNA binding by non-covalent electrospray ionization mass spectrometry

Structural mass spectrometry (MS) comprise various MS approaches that provide information about primary structure and post-translational modifications, 3D structure and dynamics as well as interactions (review in Allison & Bechara, 2019). To study the intact non-covalent complexes, non-denaturating or native MS is a useful method. Usually, electrospray ionization (ESI) or nanoelectrospray ionization (nanoESI MS) is preferred over matrix-assisted laser desorption/ionization MALDI MS for studying non-covalent complexes. ESI MS converts molecules present in solution into gas-phase and preserves weak non-covalent interactions between proteins and associated biomolecules such as DNA, cofactors and ligands. Details on the composition, stoichiometry can be obtained as well as additional information on heterogeneity or dynamic interactions. While numerous studies using ESI MS are reported on protein-protein complexes, few studies focused on DNA-protein interactions (Bich et al., 2010; Butterer et al., 2014; Chan et al., 2017; Gabelica et al., 2002; Nguyen-Huynh et al., 2016; Politis, Park, Hall, Ruotolo, & Robinson, 2013). In addition to the ability to study non-covalent complex, one advantage of the method is that it requires only a small amount of material (ng). However, special sample preparation procedures such as desalting and the use of aqueous volatile solutions at pH 7–8 as ammonium acetate, are required. Furthermore the analysis of DNA complexes are complicated by the heterogeneity of cation adducts and difficulties to achieve stable electrospray. The use of ESI MS will be illustrated to study the interaction of RXR and RAR DBDs with a specific DNA sequence, as done to evaluate the effect of DNA stoichiometry on RAR-RXR-DNA complex formation using various specific DNA sequences (Nguyen-Huynh et al., 2016).

Materials and equipment

Electrospray time-of-flight MS (MicrOTOF, BrukerDaltonic, Germany).

Buffer: 100 mM ammonium acetate pH 7.8.

Purified RXR and RAR DBS at 30–50 μM (see Section 2.1.)

Oligonucleotides containing selected DNA response (annealed, see Section 2.1.4.) (Bio-Spin, Bio-Rad)

Procedure

Before MS analysis, exchange the buffer of RXR DBD, RAR DBD and DNAs to 100 mM ammonium acetate pH 7.8 at 4 °C using Micro Bio-Spin chromatography column (exclusion limit 6 kDa). For the DNAs, the exchange should be done twice. After buffer exchange, measure the exact concentration of the proteins and the DNA.

Acquire MS data in the positive ion mode on an electrospray time-of-flight mass spectrometer (MicrOTOF, BrukerDaltonic, Germany).

First, verify the purity and sequence integrity, by performing MS acquisition in denaturing conditions measurements by diluting the proteins to 10 μM in H_2O/ACN (50/50) acidified with 1% formic acid.

Then, perform mass spectrometric analyses of the non-covalent complexes in 100 mM ammonium acetate pH 7.8. Infuse the mix of the DBDs at 32 μM and DNA at various equivalents (0.1, 0.25, 0.5, 1.2 equivalents) at 4 μL/min through a KDS100 syringe pump (KD Scientific, Holliston, MA, USA). For detection of non-covalent complexes, solution as well as instrument conditions should be carefully optimized. The source and transfer parameters of the instrument should be optimized in order to obtain the best spectrum without dissociating the complexes, especially, the capillary exit voltage (from 100 to 200 V), which accelerates the ions in the middle pressure interface. The pre-pulse time (from 30 to 40 μs), which represents a delay between transfer time and TOF pulser on for injection of ions into the TOF analyzer, should be optimized for proper transmission of high m/z ions.

In a typical experiment, RAR/RXR DBDs heterodimers were best obtained at the capillary exit of 150 V and the pre-pulse time at 40 μs (Nguyen-Huynh et al., 2016). Denaturing MS confirmed the integrity of the proteins with the RXR DBD being present as a single form corresponding to the theoretical molecular mass (10,239 Da) and RAR DBD under two forms: a major form at 10607.1 Da and a minor one at 9940.8 ± 0.5 Da which may correspond to the loss of the remaining residues of the tag GSHM plus two amino acids (proline and arginine). Fig. 3 shows the non-covalent ESI MS spectra of a mix of stoichiometric amount of RAR and RXR DBDs and increasing equivalent ration of *RARb2* DR5 (5′-AGAGGTCAGACAGAGGT CAGA-3′/3′-TCTCCAGTCTGTTCCAGTCT-5′, MW = 12,852 Da). Different oligomeric states were observed upon solution conditions with unbound RAR and RXR at 0.1 and 0.25 equivalents of DNA in concomitance with a third species at 33954.6 ± 0.4 Da corresponding to a RAR/RXR-DR5 hetero-complex bound to DNA (Fig. 3). Although no zinc was added to our ammonium acetate buffer, two Zn^{2+} remain bound to the free or bound proteins, indicating that the protein conformation has been retained in vacuum. No species revealing the existence of RAR/RAR-DR5 or RXR/RXR-DR5 homodimer complexes were furthermore observed. At higher ratio, 0.5 and 1.2 equivalents of DNA, in addition to the dimer complexes RAR-DR5 and RXR-DR5 monomeric complexes emerged in the spectra while signal due to free proteins became negligible. Because of instrument effects, the relative abundance of free protein versus protein-DNA complexes could not be extracted nor evaluated from MS spectra. However, the spectra along the DNA titration could be compared as they were acquired under the same interface parameters. In addition, under similar condition we could compare the composition of RAR/RXR complexes with various RAREs (Nguyen-Huynh et al., 2016). Thus, ESI MS method was helpful to check the stabilization of heterodimers on RARE sequences and to select the proper DNA ratio to be added.

2.3 Crystallization of DNA DBDs complexes

During crystallization of a ternary complex, it is always unknown whether the crystals contain all the components of the complex, in particular when heterogeneity of the DNA complexes are present. In addition to vary the ratio of the components, the order of the mixture of the components can help to obtain the crystal of the complex of interest. Besides the purity and homogeneity of the protein, some other aspects need to be considered for DNA complex crystallization, as the length and sequence of the DNA.

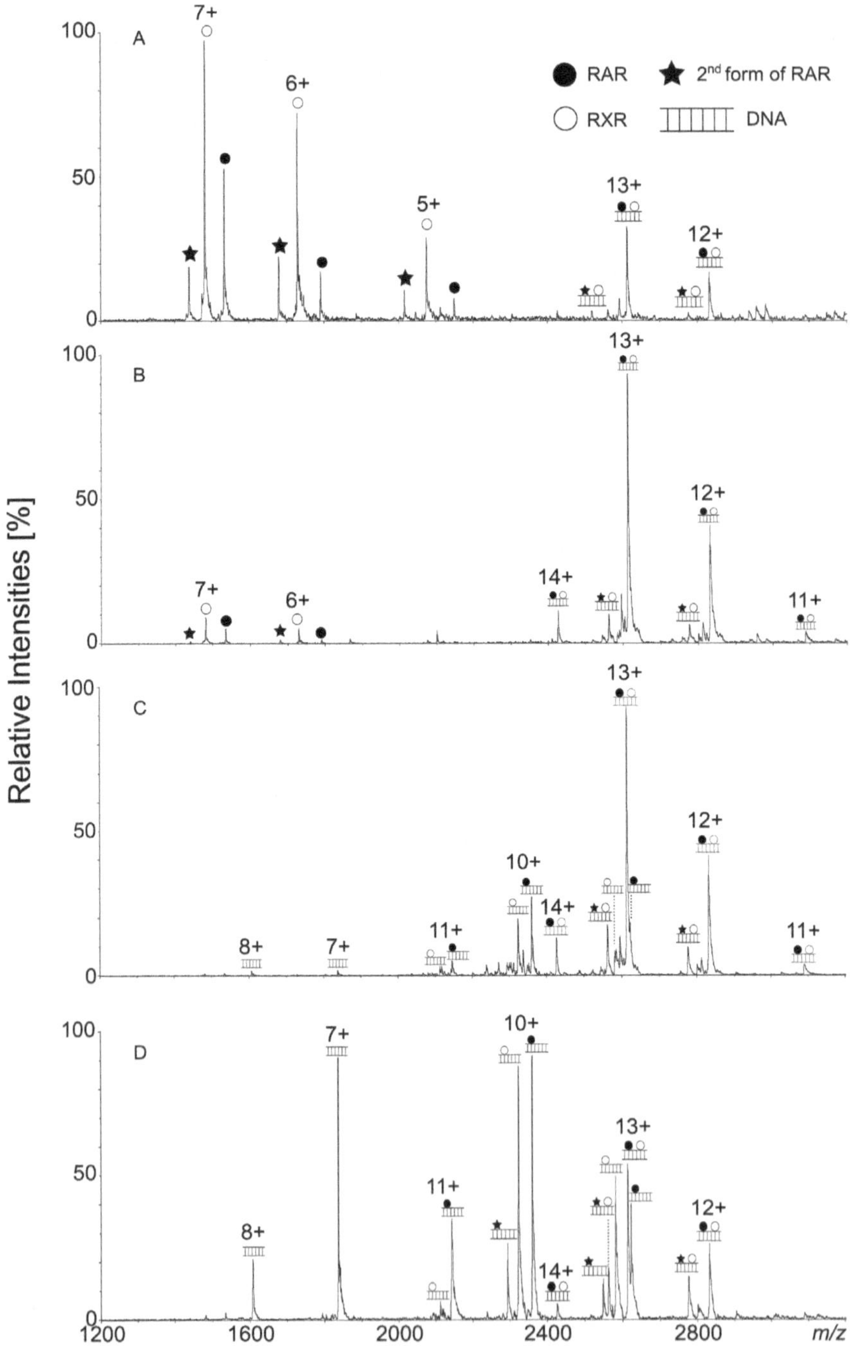

Fig. 3 Native MS showing RAR/RXR DBDs bound to DNA depending on different DNA/dimer ratio: (A) 0.1, (B) 0.25, (C) 0.5 and (D) 1.2 equivalent of *RARb2* DR5. *Adapted from Nguyen-Huynh, N. T., Osz, J., Peluso-Iltis, C., Rochel, N., Potier, N., & Leize-Wagner, E. (2016). Monitoring of the retinoic acid receptor-retinoid X receptor dimerization upon DNA binding by native mass spectrometry.* Biophysical Chemistry, 210, *2–8.*

Knowledge of the DNA sequences specifically bound by the protein is essential in order to have recognized DNA with optimal affinity and specificity. To obtain suitable diffracting crystals, both the length and the ends of the oligonucleotides must be considered (Orth, Alings, Schnappinger, Saenger, & Hinrichs, 1998; Schultz, Shields, & Steitz, 1990) (review in Hollis, 2007). To crystallize RXR and RAR/RXR DBDs in complex with natural RAREs, we used oligonucleotides containing the DR or IR elements with varying additional nucleotides at 5′ and/or 3′ ends. The ends of the DNA fragments are also a very important factor in obtaining co-crystals. We usually test various possibilities: sequences with blunt ends and sequences with cohesive ends that can be matched or not matched; the objective of the cohesive end sequences is to promote in solution an oriented and ordered association of the DNA molecules and therefore of complex. For crystallization of DNA-protein complexes, precipitant agents are often PEGs, medium saline concentrations and neutral or slightly basic pH. In addition, the effect of metallic cations (notably Mg^{2+}, Ca^{2+}, and Zn^{2+}) and polycations, especially spermine and spermidine, can be used as additives. By forming stabilizing interactions with the oxygens of the DNA phosphate group, they have been shown to be decisive in obtaining co-crystals (Schultz et al., 1990).

Materials and equipment

Purified RAR DBD and RXR DBD

Amicon ultra concentration device with MWCO 10 kDa (GE Healthcare)

Oligonucleotides containing selected DNA response (annealed, see Section 2.1.4.)

Crystallization plate for sitting drop, 96 well

Mosquito robot (TTP Labtech)

Crystallization spare matrix screens such as Natrix, PEG/pH, PEG/Ion from Hampton Research or Nucleix Acids and Morpheus from Molecular Dimensions.

DBD-DNA complex formation procedure

(1) For heterodimer, mix the RAR DBD and the RXR DBD, each of them around 1 mg/mL concentration, with 1:1 M equivalent ratio.

(2) Add $MgCl_2$ in 5 mM final concentration.

(3) Add the DNA at 0.9–1.1 M equivalent ratio and incubate 1 h on ice.

(4) Concentrate the DBDs DNA complex with Amicon Ultra MWCO 10 kDa up to 8 mg/mL for crystallization.

Crystallization procedure

Centrifuge at 20,000 *g* for 30 min the complex before crystallization trials.

Initial crystallizations are carried out using commercial spare matrix screens such as Natrix, PEG/pH, PEG/Ion from Hampton Research or Nucleix Acids and Morpheus from Molecular Dimensions.
Mix 200 nL of DNA complex with 200 nL reservoir solution in a 96 well sitting drop plates using a Mosquito robot (TTP Labtech).
After obtaining first hits, crystallization optimization can be performed.

3. Analysis of DNA complexes with RAR/RXR ΔNTD

Due to instability of the NTDs of RAR and RXR that are easily prone to degradation, we choose to work with proteins truncated of their NTDs. The optimal truncated proteins were determined by analysis of sequence homology and stability (Chen et al., 1994). The histidine-mRARα ΔNTD (84–462) was cloned in pET15b and the non-tagged-mRXRα ΔNTD (135–467) in pET3a vector. To purify the RAR/RXR ΔNTD heterodimer, we chose a co-purification strategy. The histidine-tagged RAR ΔNTD and the non-tagged RXR ΔNTD were expressed in *E. coli* separately and the two expression cultures were mixed before lysis and purification. Thus, during the lysis of the mixed expression cultures, the RAR/RXR ΔNTD dimer could be formed and be purified by the histidine-tag of the RAR ΔNTD. During the immobilized metal ion affinity chromatography, only the histidine-RAR ΔNTD monomers and the histidine-RAR/RXR ΔNTD heterodimer will be recovered and the excess of the non-tagged RXR ΔNTD will not bind to the Ni-resin. Finally, the size-exclusion chromatography step will allow to separate the stoichiometric RAR/RXR ΔNTD heterodimer form the excess of the monomeric and tetrameric forms.

The choice of the tagged protein is based on the observation that RAR is monomeric while RXR exists as a mixture of monomers, homodimers or homotetramers (Chen et al., 1998). Fig. 4 shows the workflow for the expression and co-purification of RAR/RXR ΔNTD.

3.1 Expression and purification of RAR/RXR ΔNTD

Materials

LB Broth media (Sigma)
Ampicillin
IPTG (Euromedex EU0008-A)
Lysozyme (Sigma L6876)
Buffer components: Tris, NaCl, glycerol, Chaps, imidazole, TCEP

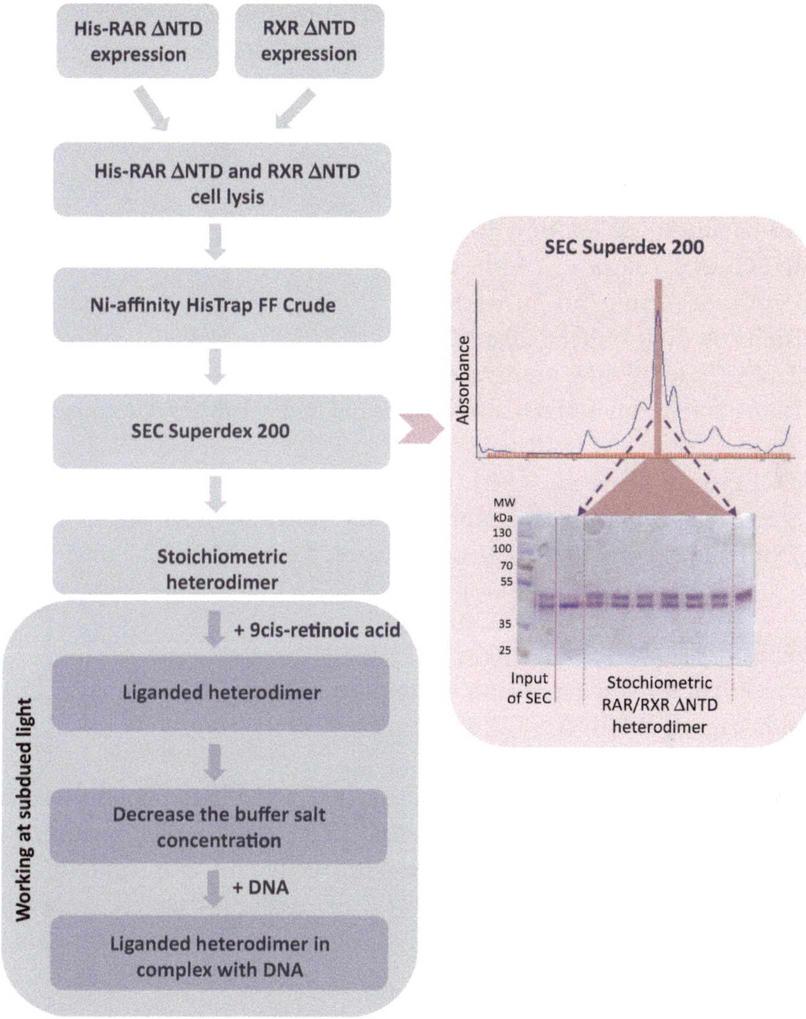

Fig. 4 Workflow of expression and co-purification of RAR/RXR ΔNTD and Coomassie-stained gel corresponding to the SEC step purification of the heterodimer.

Oligonucleotides containing selected DNA response elements (Sigma, HPLC purified)

Equipment and disposables

UV/vis spectrophotometer (OD600)

Sonication probe (Labsonic P, Sartorius)

Ultracentrifuge

Minisart syringe filter, 5 μm (Sartorius 17594-K)

AKTA chromatography system (GE Healthcare)
HiTrap FF Crude 5 mL Ni-affinity column (GE Healthcare)
Superdex 200 or Superdex 200 Increase column (GE Healthcare)
Amicon ultra concentration device with MWCO 50 kDa (GE Healthcare)

Buffers to be prepared before start

Ampicillin stock solution 100 mg/mL, sterile filtered
IPTG stock solution: 1 M in water
Lysozyme: 25 mg/mL in water
Buffer A (lysis buffer): 20 mM Tris pH 8, 500 mM NaCl, 5% glycerol, 2 mM Chaps, 5 mM imidazole
Buffer B (elution buffer): 20 mM Tris pH 8, 500 mM NaCl, 5% glycerol, 2 mM Chaps, 1 M imidazole
TCEP stock solution: 0.5 M, pH 8.0
Size-exclusion chromatography (SEC) buffer: 20 mM Hepes pH 7.5, 75 mM NaCl, 75 mM KCl, 1% glycerol, 4 mM $MgSO_4$, 1 mM TCEP

3.1.1 Expression

The protocol for the expression of the histidine-tagged RAR ΔNTD and the non-tagged RXR ΔNTD are similar.

Procedure

Day before: inoculate for each construct 200 mL LB Broth media with one colony or a glycerol freeze stock of *E. coli* BL21DE3 cells containing the expression plasmid. Supplemented with 100 μg/mL ampicillin in case of the histidine-tagged-RAR ΔNTD or with 100 μg/mL ampicillin +50 μg/mL chloramphenicol in case of the non-tagged-RXR ΔNTD plasmid. Incubate overnight at 37 °C with shaking at 180 rpm.

Day of the production: measure the OD600 of the overnight culture. Inoculate LB Broth media with the overnight starter culture to achieve OD = 0.08 to start the expression culture. Supplement the media with the antibiotic. Grow the culture at 37 °C with shaking at 180 rpm. When the OD600 reach 0.4–0.5, decrease the incubator temperature 22 °C. At OD600 = 0.8 induce the protein production by adding 1 mM IPTG and let to grow for 3 h. After 3 h, harvest the cells by centrifugation for 20 min at 4000 *g* at 4 °C. Wash the centrifuged cell pellet with PBS + 10% glycerol and centrifuge again for 20 min at 4000 *g*. The washed cell pellet can be store at −20 °C until the purification.

3.1.2 Cell lysis

It is important for the co-purification that the RAR ΔNTD and RXR ΔNTD cell suspensions are mixed before the cell lysis. The best way is to put the two frozen cell pellets in the same bottle, resuspend and homogeneize them with the lysis buffer.

The cell lysis and all steps of the purification should be performed at +4 °C.

Protocol

(1) Resuspend 4 L RAR ΔNTD and 6 L RXR ΔNTD expression cultures in 200 mL lysis buffer.
(2) Add 1.2 mL lysozyme stock solution.
(3) Incubate with the lysozyme 10 min on ice, 5 min at 37 °C and 10 min again on ice.
(4) Sonicate for 2 × 4 min at 60% amplitude, 0.5 s on and 0.5 s off mode.
(5) Centrifuge the cell lysate at 125,000*g* for 1 h 30 min.
(6) Filter the supernatant through 5 μm pore-size syringe filter.

Notes

(1) In our experience, the 2:3 ratio of the expression culture of the RAR ΔNTD and RXR ΔNTD used for the co-purification worked well. As the proper ratio could depend on the expression level of the RAR ΔNTD and RXR ΔNTD in the cultures, it may be necessary to optimize.
(2) The lysozyme treatment is important for the effective lysis of the bacterial cell wall.
(3) Take aliquot for the SDS-PAGE at each step to follow the purification.

3.1.3 Ni-affinity purification

As mentioned, in the Ni-affinity chromatography step, only histidine-RAR-ΔNTD and histidine-RAR ΔNTD/RXR ΔNTD heterodimer will be recovered and non-tagged RXR ΔNTD will not bind to the resin.

HisTrap FF crude 5 mL column is used to perform the Ni-affinity purification on the AKTA FPLC system. HisTrap FF crude is a ready-to-use column, prepacked with pre-charged Ni Sepharose 6 Fast Flow. The washing and elution done by step gradient results good resolution for the Ni-affinity purification. The 5 mM imidazole content of the lysis buffer already prevents aspecific protein binding. Further wash with 50 mM imidazole helps to remove all the non-specific binding protein. The RAR/RXR ΔNTD heterodimer is eluted from the column by 200 mM imidazole.

Procedure

(1) Equilibrate the 5mL HisTrapp FF crude column with 50mL of buffer A.

(2) Load the filtered supernatant on the column by the sample pump of the AKTA system with 3mL/min flow rate.

(3) Perform the following washing steps with 5mL/min flow rate: 75mL buffer A, 75mL 5% buffer B (non-specific wash), 50mL 20% buffer B (specific elution, 200mM imidazole), 50mL 100% B buffer (final washing step).

(4) Pool together the fractions corresponding to the specific elution peak (200mM imidazole).

(5) Dilute the specific elution fraction with the SEC buffer two times to decrease the imidazole concentration.

Notes

(1) It is important to decrease the imidazole concentration of the Ni pool to keep a good quality of the protein.

(2) At this point the purification can be interrupted. The Ni pool, after decreasing the imidazole concentration, supplemented with 2mM TCEP and 10% glycerol can be kept at −20°C for further use.

3.1.4 Size-exclusion chromatography

There are two main reasons to perform the size-exclusion chromatography (SEC) step: (1) to improve the homogeneity of the sample by removing the aggregates and (2) to separate the homogenous RAR/RXR ΔNTD heterodimer from the excess of RAR ΔNTD monomer and residual RXR ΔNTD tetramer that may result from some dissociation of the heterodimer. The Superdex 200 (or Superdex 200 Increase) column allows a separation with a good resolution. The other advantage of the size-exclusion chromatography is to decrease the buffer salt concentration.

Procedure

(1) Concentrate the Ni pool by Amicon ultra MWCO 50kDa.

(2) Perform the size-exclusion chromatography by Superdex 200 column using the SEC buffer.

(3) Analyze the gel-filtration fractions by 10% SDS-PAGE.

(4) Based on the results of the SDS-PAGE and the chromatogram, define the fractions containing the stoichiometric heterodimer (Fig. 4.)

Notes

(1) The concentration of the Ni pool is a critical step for the yield of the purification. During the centrifugation with the Amicon ultra

MWCO 50 kDa, the sample should be mixed up gently by a pipette to avoid of the very high concentration which can cause easily the precipitation of the protein.

(2) If the Ni pool yield is high, it is better to perform the SEC in two separate runs to avoid injecting very high concentration samples on the GF column (more than 7 mg/mL).

(3) For the analysis of the fractions by the SDS-PAGE, take attention to load fractions not only from the top of the main peak but also from the boundary of the peak. It is important that the fractions contained stoichiometric RAR ΔNTD and RXR ΔNTD as expected for the heterodimer. The shoulders correspond to the excess of monomer and tetramer.

(4) The pooled fractions of the purified heterodimer supplemented with 10% glycerol can be kept at −20 °C. The stability of the heterodimer is increased in the presence of the ligand. If the further use will be in holo form, it is better to freeze the heterodimer pool with ligand.

The average yield of the purification is around 12–14 mg heterodimer when starting from 4 L RAR and 6 L RXR expression culture.

3.2 RAR/RXR ΔNTD-DNA complex formation

The liganded RAR/RXR ΔNTD heterodimer forms a stable complex with the RARE response element containing oligonucleotide. Before addition of the DNA, the ligands were added to the heterodimer complex. Due to cost consideration, we used this strategy but ligands can be added from the beginning at the lysis step or at the final step. We choose 9-cis retinoic acid agonist ligand as a common ligand for both receptors. To avoid degradation and isomerization of the ligand, work should be done with no direct light.

The heterodimer is stable in 200 mM total salt condition but for efficient DNA binding, the total salt amount is decreased to 150 mM in the buffer.

Materials

9-cis retinoic acid (Sigma, R4643)

Oligonucleotide stock (0.5 mM)

Buffers to be prepared before start

9-cis-retinoic acid stock solution: 10^{-2} M in EtOH

SEC buffer without salt: Hepes 20 mM pH 7.5, glycerol 1%

Protocol

(1) Dilute the SEC-pool with the SEC buffer without NaCl and KCl to have a final salt concentration of 150 mM.

The following steps should be done in subdued light.

(2) Add 2 M equivalent ligand and incubate 10 min in ice.

(3) Add 1.2 M equivalent oligonucleotide and incubate 30 min in ice.

Notes

(1) The ligand dissolved in ethanol, it could cause the denaturation of the protein at certain concentration. As a general rule, do not exceed 2% volume of ethanol of the total volume of protein solution sample.

(2) After addition of 9-cis RA, all the further manipulation should be done in subdued light avoiding to expose the sample to direct light. The 9-cis retinoic acid stock solution or the liganded sample can be stored in amber Eppendorf tubes.

(3) For stability of the complex, it is better to add the DNA on dilute dimer and to further concentrate the DNA complex.

The heterodimer-DNA complex formation could be checked by native gel-electrophoresis. Fig. 5 shows the Coomassie-stained native gel-electrophoresis of the liganded RAR-RXR ΔNTD heterodimer alone and in complex with different molar equivalents of *RARb2* response element (5′-AGAGGTCAG ACAGAGGTCAGA-3′/3′-TCTCCAGTCTGTTCCAGTCT-5′) using Native PhastaGel Gradient 8–25% (Phasta System GE). For further structural analysis, a second size-exclusion chromatography of liganded RAR-RXR ΔNTD-DNA complex on the same SEC column as in Section 3.1.4 is necessary to remove excess of DNA and to obtain more homogenous heterodimer-DNA complex.

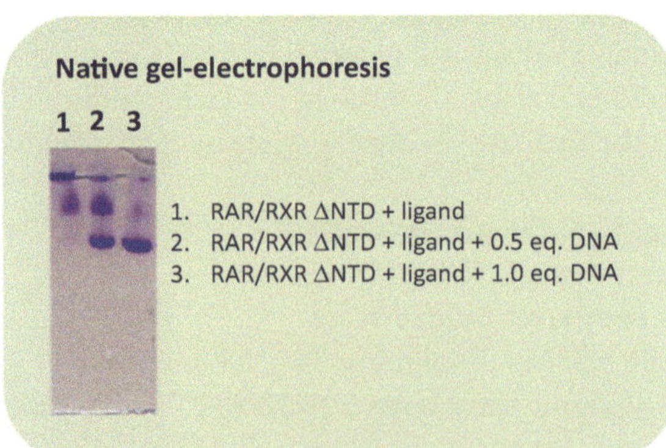

Fig. 5 Native PAGE gel of liganded RAR/RXR ΔNTD alone and in the absence and in the presence of 0.5 and 1.0 equivalent of *RARb2* DR5.

4. Monitoring DNA binding by isothermal titration calorimetry

Various biophysical methods can be used to monitor the interaction between biomolecules, in particular between protein and DNA. These methods include isothermal titration calorimetry (ITC) (Lin & Wu, 2019), fluorescence anisotropy (Zorrilla et al., 2008), microscale thermophoresis (MST) (Mueller et al., 2017), and surface plasmon resonance (SPR) (Majka & Speck, 2007). Here we describe the use of ITC to determine the binding affinity of the liganded RAR/RXR ΔNTD to RAREs. The advantage of the ITC is that it is a label free method that provides in addition to the affinity constant, the thermodynamic parameters of the DNA binding event.

An important point is the choice of a suitable buffer for the ITC experiment. Deegan, Seldeen, McDonald, Bhat, and Farooq (2010) has shown that the binding of nuclear receptor DBD to DNA is coupled to proton uptake. Therefor there is a dependency between the observed enthalpy for the binding reaction (ΔHobs) and the ionization enthalpy (ΔHion) of the reaction buffer. For that reason, the most suitable buffer is the phosphate buffer to measure DNA binding affinity. But to measure DNA binding affinity with RAR/RXR ΔNTD, it is also important to use TCEP reducing agent, a compound not stable in phosphate buffer. For this consideration, in our experiments (Moutier et al., 2012; Osz et al., 2015; Osz et al., 2019), we used Hepes buffer in which the TCEP is stable and the buffer ionization enthalpy is still acceptable for the ITC measurement (Deegan et al., 2010).

Materials

Buffer components: Hepes, NaCl, glycerol, TCEP

Dialysis membrane (GebaFlex, 6–8 K)

Purified and liganded RAR-RXR ΔNTD protein (see Section 3.1)

RARb2 oligonucleotide: (5′-AGAGGTCAGACAGAGGTCAGA-3′/ 3′-TCTCCAGTCTGTTCCAGTCT-5′) (for the preparation, see Section 2.1.4)

Equipment and disposables

UV/vis spectrophotometer (NanoDrop)

Microcal iTC200 (Malvern Pananalytical)

Buffers to be prepared before ITC experiment

ITC buffer: 20 mM Hepes pH 7.5, 100 mM NaCl, 2% glycerol, 1 mM TCEP

TCEP stock solution: 0.5 M, pH 8.0

Procedure

Wash carefully the dialysis membrane with ultra-pure water. Dialyze both the liganded RAR/RXR ΔNTD and the *RARb2* oligonucleotide in the same ITC buffer but in separate dialysis tubes for overnight at 4 °C and protected them from the light. After dialysis, measure the exact concentration both for the heterodimer and the DNA.

For the ITC measurement, fill the sample cell of the MicroCal iTC200 with the liganded RAR/RXR ΔNTD (~250–280 μL) and the syringe with the dialyzed *RARb2* oligonucleotide (80 μL). Fill also the reference cell with ultrapure water. In a typical experiment 2-μL aliquots of DNA at 80–150 μM were injected into a 10–14 μM RAR/RXR ΔNTD solution in the sample cell at 25 °C with continuous stirring 750 rpm. Perform 19 injections with 120–180 s delay between injections to permit the signal to return to base line before the next injection. The heats originated from the mixing and dilution should be subtracted from the heat of DNA binding per injection. For that, a control experiment is performed using the same ITC buffer in the sample cell titrated against the *Rarb2* oligonucleotide. The ITC titration curves were next analyzed using the software Origin 7.0 (OriginLab).

Notes

(1) The ITC experiment is performed with liganded RAR/RXR ΔNTD, therefore should be done in subdued light. The baker used during the dialysis can be covered by aluminum foil and the sample prepared for the experiment can be stored in amber Eppendorf.

(2) To determine the precise Kd, it is important to know the exact concentrations of the protein and the DNA. Because during the dialysis the concentration could change, it should be measured precisely after dialysis.

(3) As a general rule, to measure a 1 to 1 stoichiometric interaction (like the heterodimer-DNA interaction), it needs 10 times more concentrated of the injected DNA than the protein concentration in the sample cell.

(4) The volume of the sample cell and the injection syringe are 200 and 40 μL, respectively. But for the proper filling of the sample cell and injection syringe the volume requirements are around 250–280 and 80 μL, respectively.

(5) The optimal stirring speed depends on the type of the injection syringe. Use 750 rpm with the twisted paddle type injection syringe. In case of older type instrument the injection syringe can be flat type, in that case use 1000 rpm for stirring.

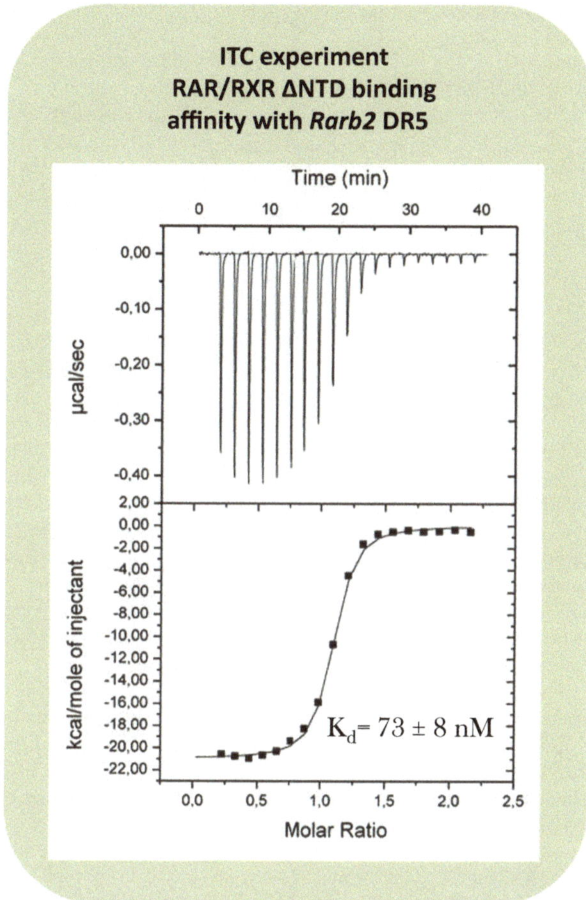

Fig. 6 Isothermal titration calorimetry titration curve for the interaction of the 9-cis retinoic acid liganded RAR/RXR ΔNTD with the *RARb2* DR5. *Adapted from Moutier, E., Ye, T., Choukrallah, M. A., Urban, S., Osz, J., Chatagnon, A., et al. (2012). Retinoic acid receptors recognize the mouse genome through binding elements with diverse spacing and topology.* The Journal of Biological Chemistry, 287*(31), 26328–26341.*

(6) Due to the extreme sensitivity of the method for the buffer composition, use the same buffer all over the assay. In general, we keep 100 mL from the dialysis buffer for further dilution of the protein or DNA and to rinse the ITC sample cell between the experiments.

Fig. 6 showing an example isothermal titration calorimetry curve for the interaction of 9-cis retinoic acid liganded RAR/RXR ΔNTD with the *Rarb2* response element (Moutier et al., 2012).

Acknowledgments

This work was supported by the Fondation ARC and the grant. This work was supported by the Centre National de la Recherche Scientifique, the Institut National de la Santé et de la Recherche Médicale and the University of Strasbourg. We acknowledge the support of the Fondation ARC, the Agence Nationale de la Recherche ANR-11-BSV8-023, and ANR-10-LABX-0030-INRT under the frame programme Investissements d'Avenir ANR-10-IDEX-0002-02. The authors acknowledge the support and the use of resources of the French Infrastructure for Integrated Structural Biology FRISBI ANR-10-INBS-05 and of Instruct-ERIC.

References

Allison, T. M., & Bechara, C. (2019). Structural mass spectrometry comes of age: New insight into protein structure, function and interactions. *Biochemical Society Transactions*, *47*(1), 317–327.

Belorusova, A., Osz, J., Petoukhov, M. V., Peluso-Iltis, C., Kieffer, B., Svergun, D. I., et al. (2016). Solution behavior of the intrinsically disordered N-terminal domain of retinoid X receptor α in the context of the full-length protein. *Biochemistry*, *55*(12), 1741–1748.

Bich, C., Bovet, C., Rochel, N., Peluso-Iltis, C., Panagiotidis, A., Nazabal, A., et al. (2010). Detection of nucleic acid-nuclear hormone receptor complexes with mass spectrometry. *Journal of the American Society for Mass Spectrometry*, *21*(4), 635–645.

Busso, D., Delagoutte-Busso, B., & Moras, D. (2005). Construction of a set gateway-based destination vectors for high-throughput cloning and expression screening in *Escherichia coli*. *Analytical Biochemistry*, *343*(2), 313–321.

Butterer, A., Pernstich, C., Smith, R. M., Sobott, F., Szczelkun, M. D., & Toth, J. (2014). Type III restriction endonucleases are heterotrimeric: Comprising one helicase-nuclease subunit and a dimeric methyltransferase that binds only one specific DNA. *Nucleic Acids Research*, *42*, 5139–5550.

Castelein, H., Janssen, A., Declercq, P. E., & Baes, M. (1996). Sequence requirements for high affinity retinoid X receptor-alpha homodimer binding. *Molecular and Cellular Endocrinology*, *119*(1), 11–20.

Chan, D. S., Seetoh, W. G., McConnell, B. N., Matak-Vinkovic, D., Thomas, S. E., Mendes, V., et al. (2017). Structural insights into the EthR–DNA interaction using native mass spectrometry. *Chemical Communications*, *53*, 3527–3530.

Chandra, V., Wu, D., Li, S., Potluri, N., Kim, Y., & Rastinejad, F. (2017). The quaternary architecture of RARβ-RXRα heterodimer facilitates domain-domain signal transmission. *Nature Communications*, *8*(1), 868.

Chatagnon, A., Veber, P., Morin, V., Bedo, J., Triqueneaux, G., Sémon, M., et al. (2015). RAR/RXR binding dynamics distinguish pluripotency from differentiation associated cis-regulatory elements. *Nucleic Acids Research*, *43*(10), 4833–4854.

Chen, Z. P., Iyer, J., Bourguet, W., Held, P., Mioskowski, C., Lebeau, L., et al. (1998). Ligand and DNA-induced dissociation of RXR tetramers. *Journal of Molecular Biology*, *275*(1), 55–65.

Chen, Z. P., Shemshedinig, L., Durand, B., Noy, N., Chambon, P., & Gronemeyer, H. (1994). Pure and functionally homogeneous recombinant retinoid X receptor. *The Journal of Biological Chemistry*, *269*(41), 25770–25776.

Cunningham, T. J., & Duester, G. (2015). Mechanisms of retinoic acid signalling and its roles in organ and limb development. *Nature Reviews Molecular Cell Biology*, *16*(2), 110–123.

Deegan, B. J., Seldeen, K. L., McDonald, C. B., Bhat, V., & Farooq, A. (2010). Binding of the ERR nuclear receptor to DNA is coupled to proton uptake. *Biochemistry*, *49*(29), 5978–5988.

Delacroix, L., Moutier, E., Altobelli, G., Legras, S., Poch, O., Choukrallah, M. A., et al. (2010). Cell-specific interaction of retinoic acid receptors with target genes in mouse embryonic fibroblasts and embryonic stem cells. *Molecular and Cellular Biology*, *30*(1), 231–244.

di Masi, A., Leboffe, L., De Marinis, E., Pagano, F., Cicconi, L., Rochette-Egly, C., et al. (2015). Retinoic acid receptors: From molecular mechanisms to cancer therapy. *Molecular Aspects of Medicine*, *41*, 1–115.

Gabelica, V., Vreuls, C., Filée, P., Duval, V., Joris, B., & Pauw, E. D. (2002). Advantages and drawbacks of nanospray for studying noncovalent protein-DNA complexes by mass spectrometry. *Rapid Communications in Mass Spectrometry*, *16*(18), 1723–1728.

Gronemeyer, H., Gustafsson, J. A., & Laudet, V. (2004). Principles for modulation of the nuclear receptor superfamily. *Nature Reviews Drug Discovery*, *3*(11), 950–964.

Hollis, T. (2007). Crystallization of protein-DNA complexes. *Methods in Molecular Biology*, *363*, 225–237.

Holmbeck, S. M., Dyson, H. J., & Wright, P. E. (1998). DNA-induced conformational changes are the basis for cooperative dimerization by the DNA binding domain of the retinoid X receptor. *Journal of Molecular Biology*, *284*(3), 533–539.

Holmbeck, S. M., Foster, M. P., Casimiro, D. R., Sem, D. S., Dyson, H. J., & Wright, P. E. (1998). High resolution solution structure of the retinoid X receptor DNA-binding domain. *Journal of Molecular Biology*, *281*(2), 271–284.

Kurokawa, R., DiRenzo, J., Boehm, M., Sugarman, J., Gloss, B., Rosenfeld, M. G., et al. (1994). Regulation of retinoid signalling by receptor polarity and allosteric control of ligand binding. *Nature*, *371*(6497), 528–531.

LaVallie, E. R., DiBlasio, E. A., Kovacic, S., Grant, K. L., Schendel, P. F., & McCoy, J. M. (1993). A thioredoxin gene fusion expression system that circumvents inclusion body formation in the *E. coli* cytoplasm. *Biotechnology (N Y)*, *11*(2), 187–193.

Lee, M. S., Kliewer, S. A., Provencal, J., Wright, P. E., & Evans, R. M. (1993). Structure of the retinoid X receptor alpha DNA binding domain: A helix required for homodimeric DNA binding. *Science*, *260*(5111), 1117–1121.

Lin, K., & Wu, G. (2019). Isothermal titration calorimetry assays to measure binding affinities in vitro. *Methods in Molecular Biology*, *1893*, 257–272.

Mahony, S., Mazzoni, E. O., McCuine, S., Young, R. A., Wichterle, H., & Gifford, D. K. (2011). Ligand-dependent dynamics of retinoic acid receptor binding during early neurogenesis. *Genome Biology*, *12*(1), R2.

Majka, J., & Speck, C. (2007). Analysis of protein-DNA interactions using surface plasmon resonance. *Advances in Biochemical Engineering/Biotechnology*, *104*, 13–36.

Mangelsdorf, D. J., & Evans, R. M. (1995). The RXR heterodimers and orphan receptors. *Cell*, *83*(6), 841–850.

Martens, J. H., Brinkman, A. B., Simmer, F., Francoijs, K. J., Nebbioso, A., Ferrara, F., et al. (2010). PML-RARalpha/RXR alters the epigenetic landscape in acute promyelocytic leukemia. *Cancer Cell*, *17*(2), 173–185.

Mendoza-Parra, M. A., Walia, M., Sankar, M., & Gronemeyer, H. (2011). Dissecting the retinoid-induced differentiation of F9 embryonal stem cells by integrative genomics. *Molecular Systems Biology*, *7*, 538.

Moutier, E., Ye, T., Choukrallah, M. A., Urban, S., Osz, J., Chatagnon, A., et al. (2012). Retinoic acid receptors recognize the mouse genome through binding elements with diverse spacing and topology. *The Journal of Biological Chemistry*, *287*(31), 26328–26341.

Mueller, A. M., Breitsprecher, D., Duhr, S., Baaske, P., Schubert, T., & Längst, G. (2017). Microscale thermophoresis: A rapid and precise method to quantify protein-nucleic acid interactions in solution. *Methods in Molecular Biology*, *1654*, 151–164.

Nguyen-Huynh, N. T., Osz, J., Peluso-Iltis, C., Rochel, N., Potier, N., & Leize-Wagner, E. (2016). Monitoring of the retinoic acid receptor-retinoid X receptor dimerization upon DNA binding by native mass spectrometry. *Biophysical Chemistry*, *210*, 2–8.

Orth, P., Alings, C., Schnappinger, D., Saenger, W., & Hinrichs, W. (1998). Crystallization and preliminary X-ray analysis of the Tet-repressor/operator complex. *Acta Crystallographica Section D, Biological Crystallography*, *54*, 99–100.

Osz, J., McEwen, A. G., Poussin-Courmontagne, P., Moutier, E., Birck, C., Davidson, I., et al. (2015). Structural basis of natural promoter recognition by the retinoid X nuclear receptor. *Scientific Reports*, *5*, 8216.

Osz, J., McEwen, A. G., Wolf, J., Poussin-Courmontagne, P., Peluso-Iltis, C., Chebaro, Y., et al. (2019). Modulation of RXR-DNA complex assembly by DNA context. *Molecular and Cellular Endocrinology*, *481*, 44–52.

Politis, A., Park, A. Y., Hall, Z., Ruotolo, B. T., & Robinson, C. V. (2013). Integrative modelling coupled with ion mobility mass spectrometry reveals structural features of the clamp loader in complex with single-stranded DNA binding protein. *Journal of Molecular Biology*, *425*, 4790–4801.

Rastinejad, F., Wagner, T., Zhao, Q., & Khorasanizadeh, S. (2000). Structure of the RXR-RAR DNA-binding complex on the retinoic acid response element DR1. *EMBO Journal*, *19*(5), 1045–1054.

Rochel, N., Ciesielski, F., Godet, J., Moman, E., Roessle, M., Peluso-Iltis, C., et al. (2011). Common architecture of nuclear receptor heterodimers on DNA direct repeat elements with different spacings. *Nature Structural & Molecular Biology*, *18*(5), 564–570.

Ross-Innes, C. S., Stark, R., Holmes, K. A., Schmidt, D., Spyrou, C., Russell, R., et al. (2010). Cooperative interaction between retinoic acid receptor-alpha and estrogen receptor in breast cancer. *Genes & Development*, *24*(2), 171–182.

Schultz, S. C., Shields, G. C., & Steitz, T. A. (1990). Crystallization of *Escherichia coli* catabolite gene activator protein with its DNA binding site. The use of modular DNA. *Journal of Molecular Biology*, *213*(1), 159–166.

Umesono, K., Murakami, K. K., Thompson, C. C., & Evans, R. M. (1991). Direct repeats as selective response elements for the thyroid hormone, retinoic acid, and vitamin D3 receptors. *Cell*, *65*(7), 1255–1266.

Van Tilborg, P. J., Mulder, F. A., de Backer, M. M., Nair, M., van Heerde, E. C., Folkers, G., et al. (1999). Millisecond to microsecond time scale dynamics of the retinoid X and retinoic acid receptor DNA-binding domains and dimeric complex formation. *Biochemistry*, *38*(7), 1951–1956.

Yang, Y. Z., Subauste, J. S., & Koenig, R. J. (1995). Retinoid X receptor alpha binds with the highest affinity to an imperfect direct repeat response element. *Endocrinology*, *136*(7), 2896–2903.

Zechel, C., Shen, X. Q., Chambon, P., & Gronemeyer, H. (1994). Dimerization interfaces formed between the DNA binding domains determine the cooperative binding of RXR/RAR and RXR/TR heterodimers to DR5 and DR4 elements. *The EMBO Journal*, *13*(6), 1414–1424.

Zechel, C., Shen, X. Q., Chen, J. Y., Chen, Z. P., Chambon, P., & Gronemeyer, H. (1994). The dimerization interfaces formed between the DNA binding domains of RXR, RAR and TR determine the binding specificity and polarity of the full-length receptors to direct repeats. *The EMBO Journal*, *13*(6), 1425–1433.

Zhao, Q., Chasse, S. A., Devarakonda, S., Sierk, M. L., Ahvazi, B., & Rastinejad, F. (2000). Structural basis of RXR-DNA interactions. *Journal of Molecular Biology*, *296*(2), 509–520.

Zorrilla, S., Lillo, M. P., Chaix, D., Margeat, E., Royer, C. A., & Declerck, N. (2008). Investigating transcriptional regulation by fluorescence spectroscopy, from traditional methods to state-of-the-art single-molecule approaches. *Annals of the New York Academy of Sciences*, *1130*, 44–51.

CHAPTER ELEVEN

Noncanonical retinoic acid signaling

Jennifer Nhieu[†], Yu-Lung Lin[†], Li-Na Wei*

Department of Pharmacology, University of Minnesota Medical School, Minneapolis, MN, United States
*Corresponding author: e-mail address: weixx009@umn.edu

Contents

Abstract

All-trans retinoic acid (atRA) is the principle active metabolite of Vitamin A. atRA is well known to act through nuclear RA receptors (RARs) to regulate gene expression involved in a wide spectrum of biological processes such as growth, differentiation, and function. Recently, novel activities of atRA, independent of the action of RARs, have been increasingly reported and referred to as noncanonical activities. We have determined cellular retinoic acid binding protein 1 (CRABP1) as the primary mediator of the noncanonical activities of atRA. At the molecular level, atRA binds CRABP1, which then immediately acts as an adaptor in the formation of specific signaling scaffolds to rapidly modulate downstream signaling pathways in a cell context-dependent manner. The first established CRABP1-atRA activity is to rapidly dampen the activation of the mitogen-activated protein kinase (MAPK) cascade in response to growth factor stimulation, thereby suppressing cell cycle progression of stem cells. The second established activity is to rapidly reduce Ca^{2+}/calmodulin dependent kinase II (CaMKII) activity in

[†] These authors contribute equally to this work.

Methods in Enzymology, Volume 637
ISSN 0076-6879
https://doi.org/10.1016/bs.mie.2020.02.012

differentiated cells such as cardiomyocyte in response to β-adrenergic stimulation. This chapter describes *in vivo* and *in vitro* experimental systems and methodologies appropriate for determining the noncanonical activities of atRA that are mediated by CRABP1 and cell context dependent.

1. Introduction

All trans retinoic acid (atRA) is a potent, active metabolite of Vitamin A. atRA affects most of the biological processes in growth, differentiation, and function. These effects are well known to be elicited by the nuclear activities of atRA through the actions of RA receptors (RARs) (Duester, 2008; Germain et al., 2006; Samarut & Rochette-Egly, 2012). Recently, certain novel activities of atRA, that are not mediated by RARs, have been increasingly recognized and referred to as noncanonical activities. The noncanonical activities of atRA exhibit three features—(a) they are independent of the action of RARs, (b) they are initiated from the cytosol, and (c) they can be detected very rapidly (within minutes) in cells exposed to atRA (Nagpal & Wei, 2019).

In searching for the direct mediator for these noncanonical activities of atRA, we have determined cellular retinoic acid binding protein 1 (CRABP1) as the direct mediator. Historically, CRABP1 has been known as a high-affinity cytosolic binding protein of atRA, which likely functions to sequester, and thereby regulate, bioavailability of atRA (D'Ambrosio, Clugston, & Blaner, 2011; Napoli, 2017; Ross, 1993). In studying stem cell proliferation, we surprisingly detected very rapid effects of atRA that could modulate their cytosolic kinase cascades (Persaud, Lin, Wu, Kagechika, & Wei, 2013), which turned out to be independent of the action of RARs. By using *crabp1* knockout (CKO) systems including *crabp1* gene targeted mice and embryonic stem cells (ESCs), we have concluded that CRABP1 serves as the direct mediator of these newly detected noncanonical activities of atRA. We were able to further define the function of CRABP1-atRA as an adaptor to molecular scaffolds that modulate different subsequent signaling cascades in a cell context-dependent manner (Lin, Persaud, Nhieu, & Wei, 2017; Park et al., 2018; Park, Nhieu, Lin, & Wei, 2019).

Currently, these activities have been validated in two types of cellular context, stem cells *versus* differentiated cells. In stem cells such as ESCs and neural stem cells (NSCs), CRABP1-atRA regulates cellular responses to growth factors and modulates cell cycle exit, thereby affecting stem cell pools (Lin et al., 2017; Persaud et al., 2013). In differentiated cells such as cardiomyocytes, CRABP1-atRA regulates cardiomyocyte function and modulates cardiac health (Park et al., 2018). In this chapter we provide a collection of *in vitro* and *in vivo* approaches and physiologically relevant experimental systems for studying these noncanonical activities of atRA.

Through rigorous studies, we have validated the physiological relevance of CRABP1-atRA signaling in whole animals, in that it can modulate the neuronal stem cell pool in the brain and the function/health of the heart. The two major experimental systems that have been carefully established for studying the noncanonical activities of atRA *in vitro* and *in vivo* are described in the following.

2. Experimental system for studying noncanonical activity of atRA in modulating stem cells

2.1 Rational

In the stem cell context, CRABP1-atRA targets rapidly accelerated fibrosarcoma (RAF) kinase, which subsequently activates its downstream kinase, mitogen-activated protein kinase kinase (MEK), and finally, its terminal kinase, extracellular signal-regulated protein kinase (ERK). This is defined as phase 1 activation of ERK by atRA and occurs minutes after atRA treatment. Phase 2 (RAR-dependent) activity typically happens at least 12h later, long after phase 1, noncanonical activity. Under growth factor stimulation, the RAF-MEK-ERK cascade is associated with mitogen-activated protein kinase (MAPK) signals for growth and proliferation (Cargnello & Roux, 2011). Interestingly, CRABP1-atRA directly interacts with RAF to compete with Ras GTPase, a molecular switch in MAPK signaling in response to oncogenic stimulus or growth factors such as epidermal growth factor (EGF). This competition dampens growth factor-stimulated ERK activation (Fig. 1). Therefore, at the molecular level, CRABP1-atRA serves as an adaptor in regulating the formation of specific signal scaffold complexes, thereby modulating the sensitivity of the RAF-MEK-ERK pathway to oncogenic or growth factor stimuli (Park, Nhieu, Lin, et al., 2019). This ultimately contributes to the control of stem cell proliferation, such as ESC culture, as well as NSC pool in the brain (Lin et al., 2017). Thus, two stem cell systems can be used to study the noncanonical activity of atRA in modulating stem cell growth—(1) ESCs established from CKO mice and (2) NSCs obtained from the hippocampus of CKO mice.

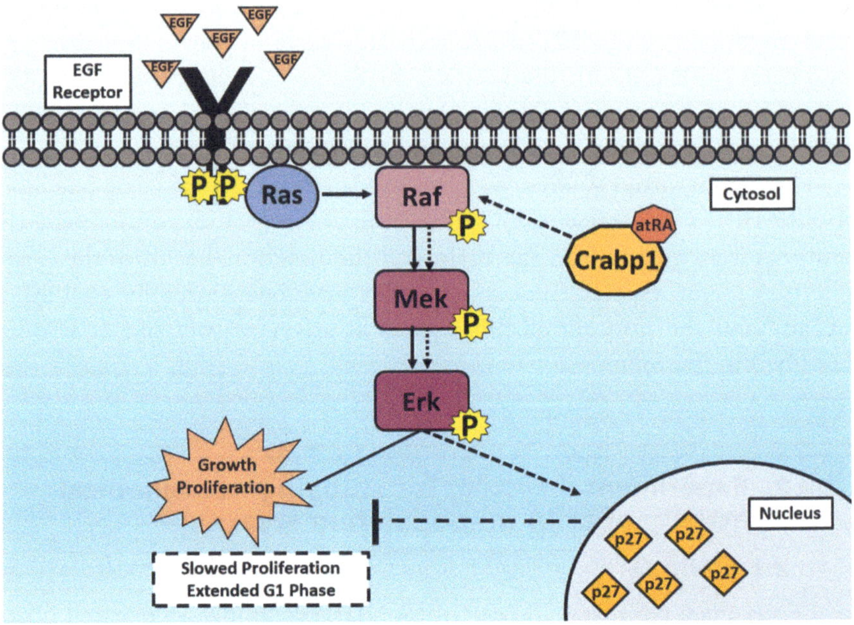

Fig. 1 Noncanonical activity of atRA in dampening growth factor-stimulated stem cell growth. Epidermal growth factor (EGF) binds to its receptor to activate Ras GTPase and the subsequent Raf-Mek-Erk kinase cascade, stimulating growth and proliferation. AtRA-CRABP1 targets Raf kinase, competing with Ras to dampen EGF-stimulated Raf-Mek-Erk kinase cascade. Altered Erk activity results in p27 accumulation in the nucleus, slowing proliferation. Growth factor stimulated signaling is depicted with solid lines and noncanonical activity of atRA is depicted with dashed lines.

2.2 Protocols

2.2.1 Embryonic stem cell (ESC) culture

Table 1 shows reagents and recipes for ESC and mouse embryonic fibroblasts (MEF) media. ESCs are routinely maintained on MEF. Once ready for experiments, ESCs should be transitioned to gelatin coated plates to prevent interference from the MEF feeder layer in experiments.

a Gelatin coated dishes

- **1** Prepare 0.2% gelatin solution
  - **(a)** Add 1 g gelatin in 500 mL double-distilled water (ddH_2O).
  - **(b)** Autoclave for 30 min and cool to room temperature.
  - **(c)** Store at room temperature.
- **2** Prepare gelatin coated dishes
  - **(a)** Add 0.2% gelatin solution into the dish (5 mL for 6 cm dish; 10 mL for 10 cm dish).

Table 1 Reagents and recipes for cell culture experiments.

Reagent	Recipe	Company (Catalog)	Comments
MEF medium	DMEM	Gibco; 11960	
	StasisTM Stem Cel FBS	Gemini; 100-125	10%
	MEM NEAA	Gibco; 11140-050	1%
	L-Glutamine	Gibco; 25030	1%
	Penicillin/Streptomycin	Gibco; 25140-122	1%
	2-Mercaptoethanol	Sigma; M3148-100 mL	5 μL
ESC medium	DMEM	Gibco; 11960	
	StasisTM Stem Cel FBS	Gemini; 100-125	20%
	MEM NEAA	Gibco; 11140-050	1%
	L-Glutamine	Gibco; 25030	1%
	Penicillin/Streptomycin	Gibco; 25140-122	1%
	2-Mercaptoethanol	Sigma; M3148-100 mL	5 μL
	ESGRO® Recombinant Mouse LIF	EMD Millipore; ESG1107	0.1%
C2C12 growth medium	DMEM	Gibco; 11965	
	FBS	Atlanta; S11150	10%
	Penicillin/Streptomycin	Gibco; 25140-122	1%

Continued

Table 1 Reagents and recipes for cell culture experiments.—cont'd

Reagent	Recipe	Company (Catalog)	Comments
C2C12 differentiation medium	DMEM	Gibco; 11965	1%
	Horse serum	Fisher Thermo; SH30074	1%
	Penicillin/Streptomycin	Gibco; 25140-122	1%
Others	Dextran-coated charcoal-treated (DCC) FBS	Atlanta; S11650	Use in place of normal FBS. DCC treated FBS has growth factors and fatty acid molecules removed
	Gelatin	Sigma, G1890	0.2%
	PMEF-CFL	Sigma, PMEF-CFL	
	Mitomycin C	Sigma, M4287-2 mg	10 μg/mL
	DPBS 1 ×	Gibco; 14190	
	Trypsin	Gibco; 25200-056	
	DMSO	Sigma, D2650-100 mL	
	atRA	Sigma, R2625	Dissolved in DMSO; storage in −20
	Mouse recombinant EGF	Stemcell; 78016.1	Dissolved in ddH_2O; storage in −20
	Duolink *in situ* PLA detection kit	Sigma-Aldrich; DUO94004	
	TUNEL assay apoptosis detection kit	Biotium; Cat#30063	

(b) Incubate at room temperature for 30 min or incubator overnight.
(c) Remove the gelatin solution and air dry for 30 min in the cell culture hood.

b MEF feeder layer

1 Thaw one vial of MEF cells by in a 37 °C water bath for 1–2 min.
2 Transfer the cell suspension to a 15 mL centrifuge tube containing 9 mL of MEF medium.
3 Collect the cells by centrifugation at 300 x*g* for 5 min at room temperature. Remove the medium by aspiration.
4 Resuspend the cells in 15 mL of fresh MEF medium by gently pipetting.
5 Transfer the cell suspension to a 15 cm tissue culture dish. Maintain cells in a cell culture incubator at 37 °C with 5% CO_2.
6 Once MEF cells reach 90% confluence, remove media and change to 20 mL of MEF medium supplemented with 10 μg/mL Mitomycin C. Incubate for 2 h.
7 Wash three times with 1 × PBS to remove Mitomycin C. Proceed to Section c "ESC culture procedure."
8 MEF Frozen Cell Stock Preparation: After dissociation with 0.5% Trypsin, the Mitomycin C-treated MEF cells can be frozen (MEF medium with 10% DMSO) and stored in liquid nitrogen. The cryovial can be used as feeder layer by plating them at 5×10^4 cells/cm^2 in gelatin-coated tissue culture dishes.

c ESC culture procedure

1 MEF feeder cells should be plated 1 day prior to adding ESCs. Change MEF feeder cells to ESC medium 1 h before adding ESCs.
2 Thaw wild type and CKO ESCs in a 37 °C water bath for 1–2 min.
3 Transfer the cell suspension to a 15 mL centrifuge tube containing 9 mL of ESC medium.
4 Collect the cells by centrifugation at 300 x*g* for 5 min at room temperature. Remove the medium by aspiration.
5 Resuspend the cells in 3 mL of fresh ESC medium by gently pipetting up and down.
6 Transfer the cell suspension to the MEF feeder layer dish. Maintain cells in an incubator at 37 °C and 5% CO_2.
7 To maintain the ESCs, change medium every day and split ESCs when they reach 70–80% confluency (Seed 2×10^5 cells/6 cm dish).

d ESC on gelatin-coated dishes

1 Prepare gelatin-coated 6 cm dishes according to manufacturer's instructions (StemCell).

2 Remove media from ESC-MEF culture. Wash dishes with 1 × PBS three times.

3 Add 1 mL 0.25% trypsin and place the dishes in 37 °C incubator until cells detach (about 3–5 min.).

4 Add 3 mL of ESC medium and disassociate the cells by pipetting.

5 Transfer medium and cells to a 15 mL tube.

6 Centrifuge cells at 300 x*g* for 5 min at room temperature. Remove the medium by aspiration.

7 Resuspend the cell pellet in 3 mL of fresh ESC medium, and transfer the cell suspension to a tissue culture dish without gelatin coating.

8 After 20 min, a majority of the MEF will attach to the dish while ESCs remain in suspension. Transfer the medium containing the ESCs to the gelatin-coated dishes. Seed 2×10^5 cells/6 cm dish.

9 Change medium every day, and split ESCs when they reach 80% confluency.

10 Repeat steps 1–9 twice. Now the ESCs are ready for experiments.

2.2.2 Detection of noncanonical atRA activity in ESC

Table 2 provides Antibody information. See Fig. 2 for examples of Western blot data.

1. Prior to the experiment, ESC media should be exchanged to dextran-coated charcoal-treated (DCC) ESC medium overnight.
2. Pretreat 0.1 μM atRA to ESC culture for a total of 0.5–1 h at 37 °C. A final concentration of 0.1 μM atRA is recommended to stay within physiological concentrations.
3. *EGF stimulation:* EGF (25 ng/mL) is treated 5 min. Prior to the end of the total time point. For example, EGF (25 ng/mL) would be treated at 55 min. After atRA pretreatment for a total of a 1-h treatment scheme.
4. Proceed to whole cell lysate isolation and Western blot as described (Persaud et al., 2013).

2.2.3 Cellular fractionation to detect noncanonical atRA activity in the cytosol

Table 3 shows buffer preparation.

1. Remove media and wash with 1–4 mL of ice cold 1 × PBS.
2. Add 0.5–1 mL of 1 × PBS.
3. Scrape and collect cells into a 1.5 mL tube. Spin at 2500 x*g* for 5 min.

Table 2 Primary and secondary antibody information and recommended concentrations.

Reagent	Recipe	Company; catalog number	Recommended concentration
Primary antibody	ErK1/2 (L34F12)	Cell Signaling; 4696	1:1000
	p-ErK1/2	Cell Signaling; 9101	1:1000
	Crabp1	Sigma-Aldrich; c1608	1:400
	cRaf (D4B3J)	Cell Signaling; 53745	1:400
	MeK1/2	Santa Cruz; SC-436	1:400
	p-MeK1/2	Cell Signaling; 9121	1:1000
	HA-antibody	Santa Cruz; sc-7392	1:1000
	pCaMKII	Santa Cruz; sc-32289	1:1000
	CaMKIIδ	Santa Cruz; sc-100362	1:1000
	pPLN T17	Santa Cruz; sc-17024	7:5000
	Total PLN	Santa Cruz; sc-21923	1:1000
	Anti-IgG-Cy3	Abcam (ab6939)	1.:200
	WGA-Alexa fluor 488	Thermofisher (W11261)	1:50
	Anti-Flag M2	Santa Cruz (F3165)	1:1000
	Anti-His	Santa Cruz (sc-8036)	1:1000
	Anti-GST	Cell Signaling (2622s)	1:1000
Secondary antibody	Goat anti-mouse IgG-HRP	Santa Cruz; SC-2005	1:2000
	Mouse anti-rabbit IgG-HRP	Santa Cruz; SC-2357	1:2000

4. Remove 1 × PBS and break cells using hypotonic buffer.
5. Spin for 10 min at 1000 x*g*. The supernatant contains the cytosolic fraction.
6. Collect the nuclear fraction by resuspending the pellet in RIPA buffer.
7. Spin resuspended pellet at 14,000 x*g* for 15 min.
8. Perform Western blot on each fraction for pERK1/2 activation.

2.2.4 ESC cell cycle studies using flow cytometry

Table 1 shows ESC medium recipe. Table 3 shows buffer preparation.

1. Induce cell cycle synchronization by exchanging ESC culture to ESC medium without FBS (Table 1) for overnight.
2. Treat cells with 0.1 μM atRA for 12 h.

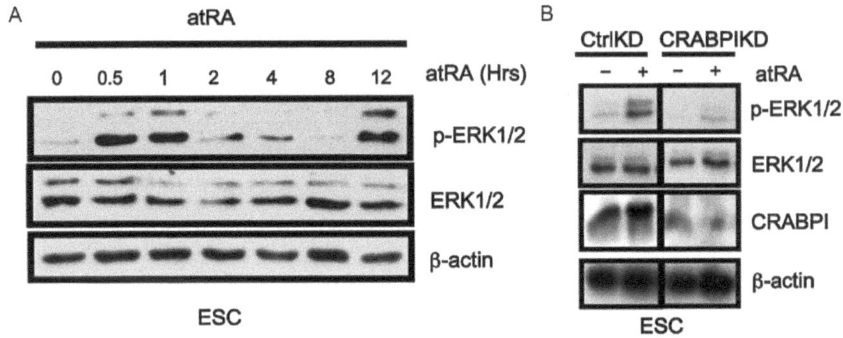

Fig. 2 Representative Western blots to demonstrate the noncanonical activity of atRA in Erk activation. (A) Without growth factor stimulation, atRA (0.1 μM) activates Erk (p-Erk1/2) in a bi-phasic manner in ESCs. Rapid, phase 1 activation occurs at 0.5–1 h while late, phase 2 (canonical, RAR-dependent) activation occurs at 12 h. (B) Phase 1 Erk1/2 activation is lost when CRABP1 expression is knocked-down, demonstrating CRABP1 as a mediator of rapid atRA-induced (0.1 μM) Erk activity. Total Erk 1/2 and β-actin were used as protein loading controls. *Modified from Persaud, S. D., Lin, Y.-W., Wu, C.-Y., Kagechika, H., & Wei, L.-N. (2013). Cellular retinoic acid binding protein I mediates rapid non-canonical activation of ERK1/2 by all-trans retinoic acid.* Cellular Signalling, 25*(1), 19–25.doi:10.1016/j.cellsig.2012.09.002).*

3. Remove media and wash with 1–4 mL of ice cold 1 × PBS.
4. Add 1 mL of 1 × PBS to the dish. Scrape and collect into a flow cytometry test tube.
5. Spin at 2500 x*g* for 5 min. Then remove 1 × PBS.
6. *Fixation:* Add 1–2 mL of 100% ice cold ethanol dropwise while vortexing.
7. Incubate for 2 h at −20°C.
8. Gently resuspend cells if visible clumps are present.
9. Spin cells at 500 x*g*. Discard supernatant.
10. Resuspend cells in staining buffer.
11. Perform fluorescence-activated cell sorting (FACS; BD Biosciences), using the FL2-A channel to detect propidium iodide signal.

Data analysis: Data analysis can be performed using FloJo software to identify ESC subpopulations at particular stages in the cell cycle as described (Pozarowski & Darzynkiewicz, 2004). Comparing wild type ESCs and CKO ESCs in these cell cycle studies validates the contribution of CRABP1 as the mediator of noncanonical atRA activity in modulating cell cycle progression of stem cells.

Table 3 Buffer recipes and reagents for *in vitro* studies.

Reagent	Recipe	Company (Catalog)	Comments
Kits	TNT® Quick Coupled Transcription/Translation System	Promega (L4610)	
Buffers	RIPA (radioimmunoprecipitation assay buffer)		150 mM sodium chloride, Triton X-100 0.5% sodium deoxycholate 0.1% SDS (sodium dodecyl sulfate) 50 mM Tris, pH 8.0
	Hypotonic buffer		10 mM HEPES, pH 7.9, 1.5 mM $MgCl_2$, 10 mM KCl, 1 mM $Na_3VO_4 \cdot 2H_2O$, 1 mM NaF, 1 mM PMSF, 1 × Halt ™Protease Inhibitor Cocktail
	FACS staining buffer		10 mL of 1 × PBS with 2 mg DNAse-free RNAse A and 0.4 mL of 500 μg/mL propidium iodine
	Kinase buffer		20 mM MOPS, pH 7.2, 0.1 mM ATP, 25 mM β-glycerophosphate, 5 mM EGTA, 1 mM sodium orthovanadate, 1 mM DTT, 18 mM $MgCl_2$, and 1 × protein inhibitor cocktail
	Coimmunoprecipitation buffer		50 mM Tris–HCl, pH 8.0, 150 mM NaCl, 0.2% (v/v) NP-40, 10% (v/v) glycerol, and 1 mM EDTA
	Triton X lysis buffer		10 mM Tris–HCl, pH 7.9, 50 mM NaCl, 0.5 M sucrose, 0.1 mM EDTA and 0.5% Triton X-100
	BSA	New England Biosciences (B9000S)	
Others	G-agarose beads	Millipore Sigma (16-266)	
	Purified Mek1 Protein	Millipore Sigma (14-429)	
	Halt™ Protease Inhibitor Cocktail, EDTA-Free	Thermofisher (78437)	Dilute to 1 ×
	Flag M2 beads	Sigma (A2220)	5 μL per 500 μL reaction

2.2.5 Pull down assay to detect direct interaction of CRABP1 with RAF

Table 2 provides antibody information. Table 3 shows buffer preparation.

1. Spin down G-agarose beads at 12,000x*g* for 30s. to remove storage solution. Beads are stored as a 50% slurry. To block nonspecific binding on the beads, resuspend in a solution of 80% 1 × PBS and 20% 10mg/mL BSA.
2. Incubate beads in blocking solution at 4°C for 1h-overnight.
3. Spin down 12,000x*g* and remove supernatant.
4. Wash three times with 1 × PBS for 30s.
5. Resuspend in 1 × PBS to create a 50% slurry.
6. Synthesize HA-cRaf using the TNT® Quick Coupled Transcription/Translation System. HA-cRaf construct was generated as described (Park, Nhieu, Lin, et al., 2019).
7. In 500μL of coimmunoprecipitation buffer (Co-IP) mix 5–10μL of HA-antibody, HA-cRAF, 4.25μg of His-CRABP1 with and without atRA. His-CRABP1 was purified as described (Park, Nhieu, Lin, et al., 2019; Park, Nhieu, Persaud, et al., 2019). Use rabbit IgG instead of anti-HA as a negative control.
8. Incubate for 2h-overnight at 4°C protected from light.
9. Spin beads down at 12,000x*g* and remove supernatant.
10. Wash three times with 500μL of Co-IP buffer for 30s.
11. Perform Western blot to detect His-CRABP1 with anti-His.

2.2.6 In situ *proximity ligation assay (PLA) for complexes between CRABP1 and RAF-MEK-ERK signaling pathway*

Table 1 provide reagent kit information. Fig. 3 depicts one set of PLA data.

1. Seed the WT and CKO ESCs on eight-well culture slides.
2. Treat the cells when they reach 70% confluency.
3. Wash with 1 × PBS and then fix by 4% (w/v) paraformaldehyde for 10min.
4. Wash with 1 × PBS for 5min three times.
5. Permeabilized using 0.2% (v/v) Triton X-100 for 5min at 4°C.
6. Wash with 1 × PBS for 5min three times.
7. *In situ* proximity ligation assay was performed using Duolink *in situ* PLA detection kit.
8. Remove PBS and add 60μL of Duolink® Blocking Solution to each well.
9. Incubate the slides in a humidity chamber for 60min at 37°C.

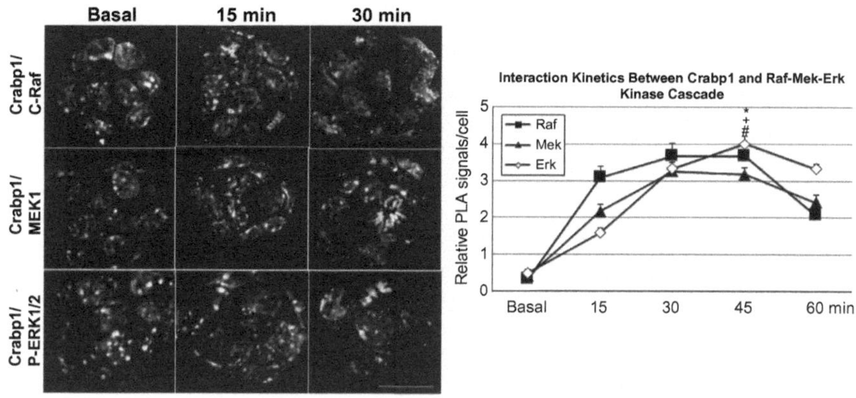

Fig. 3 PLA (Duolink) assay to demonstrate CRABP1 interaction with the MAPK cascade components-Raf, Mek, and Erk. PLA assay detects atRA-induced formation of CRABP1 complexes with the MAPK cascade components. CRABP1 first targets Raf (top panel), and subsequently recruits Mek1/2 (middle panel) and finally Erk1/2 (bottom panel). Nuclei are stained with DAPI (gray). Red puncta show complexes formed between CRABP1 and the MAPK cascade components. Scale bar is 20 μm. PLA signal (red puncta) were quantified for kinetic analysis using ImageJ (right). At 30 min PLA signal was most significant, indicating that peak complex formation occurs at this time point (Students *t*-Test: "*" Raf *P*-value ≤ 0.01, "+" Mek *P*-value ≤ 0.01, "#" Erk *P*-value ≤ 0.01; $n=4$). Error bars represent mean ± Standard Error of the Mean (SEM). (*Unpublished*.)

10. Dilute the primary antibody in the Duolink® Antibody Diluent.
11. Remove blocking solution, and add 60 μL of diluted primary antibody solution to each well and incubate the slides in a humidity chamber overnight at 4 °C.
12. Wash the slides 5 min twice with 1 × Wash Buffer A at room temperature.
13. *Duolink® PLA Probe Incubation:* Remove Wash Buffer and add 40 μL diluted PLUS and MINUS PLA probes to each well. (Mix 8 μL of PLA probe MINUS stock, 8 μL of PLA probe PLUS stock and 24 μL of Antibody Diluent.)
14. Incubate the slides in the humidity chamber for 1 h at 37 °C.
15. Wash the slides 5 min twice with 1 × Wash Buffer A at room temperature.
16. *Ligation:* Remove Wash Buffer and add 40 μL ligation solution to each well. (Mix 1 μL of Ligase to 39 μL of the 1 × ligation buffer)
17. Incubate the slides in the humidity chamber for 30 min. at 37 °C.
18. Wash the slides 5 min twice with 1 × Wash Buffer A at room temperature.

19. *Amplification:* Remove Wash Buffer and add 40 μL 1 × Amplification buffer to each well. (0.5 μL of Polymerase to 39.5 μL of the 1 × amplification buffer)
20. Incubate the slides in the humidity chamber for 100 min at 37 °C.
21. Wash the slides 5 min twice with 1 × Wash Buffer B at room temperature.
22. Wash the slides in 0.01 × Wash Buffer B for 1 min.
23. Mount the slides with a coverslip with 6 μL mounting medium with DAPI stain.

2.2.7 Semi in vitro *kinase assay for RAF activity*

Table 3 shows buffer preparation.

1. Synthesize Ha-cRaf using the TNT® Quick Coupled Transcription/Translation System.
2. Use 0.4 μg if purified Mek1 as a Raf substrate.
3. In 35 μL of Kinase buffer mix HA-cRaf, Mek1 and 1 μg of His-CRABP1 with and without 0.1 μM atRA.
4. Incubate 15 min at 30 °C protected from light.
5. Terminate reaction by adding 8.8 μL of 4 × SDS Loading buffer.
6. Proceed with Western Blot and probe for phospho-Mek1 at serine 217/221 using pMEK1/2 antibody.

3. Experimental system for studying noncanonical activity of atRA in modulating cardiomyocyte/heart function

3.1 Rational

In differentiated cardiomyocytes, CRABP1-atRA modulates the activity of Ca^{2+}/calmodulin-dependent protein kinase II (CaMKII), protecting cardiomyocytes against over-stimulation in response to physiological or pathological β-adrenergic stimulation (Park et al., 2018) (Fig. 4). CaMKII is a major mediator of Ca^{2+} signaling in cardiomyocytes. CaMKII plays important roles in multiple aspects of cardiomyocyte health/function such as maintaining excitability (contractility) and inducing apoptosis (Erickson, 2014). Normal CaMKII activity is essential for cardiomyocyte contraction, but it can also be pathologically activated. For example, over-stimulation of β-adrenergic receptors results in cardiac pathologies and heart failure. In this context, CRABP1-atRA dampens CaMKII activation, preventing devastating events caused by isoproterenol overstimulation (Mattiazzi et al., 2015). CaMKII enzymatic activity in the heart is typically detected by CaMKII

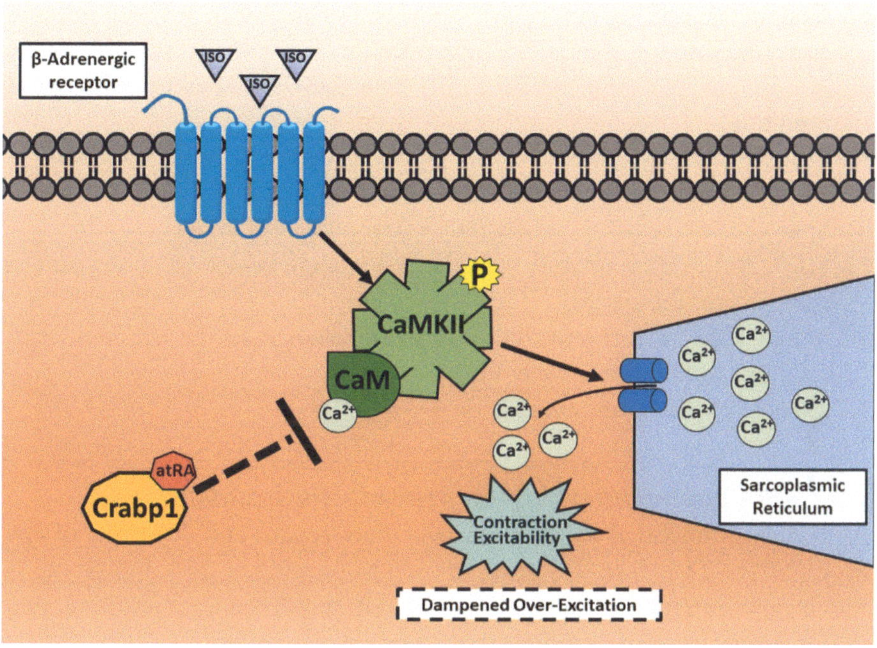

Fig. 4 Noncanonical activity of atRA in cardiomyocytes. Isoproterenol stimulates β-adrenergic receptors to activate CaMKII. Activated CaMKII targets substrates on the sarcoplasmic reticulum, causing intracellular calcium release and increasing cardiomyocyte excitability and contractility. Over-excitation of cardiomyocytes is a pathological hallmark in heart failure. AtRA-CRABP1 dampens this isoproterenol-stimulated CaMKII activity, thereby protecting against over-excitation of cardiomyocytes. Isoproterenol stimulated signaling is depicted by solid lines and noncanonical activity of atRA is shown with dashed lines.

phospho-status at threonine 287 (pCaMKII; T287) and the phosphorylation of CaMKII's substrates such as phospholamban (PLN) (Mattiazzi & Kranias, 2014). This section describes the approaches used to studying the noncanonical activity of atRA in cardiomyocytes.

3.2 Protocols

3.2.1 C2C12 cell culture and differentiation

Table 1 shows medium preparation. It is recommended to perform experiments with Day 4 differentiated C2C12 cultures to ensure that cells are responsive to stimuli as described in Park, Nhieu, Lin, et al. (2019).

1. Thaw one vial of C2C12 cells within 2 min in a 37 °C water bath.
2. Transfer the cell suspension to the 15 mL centrifuge tube containing 9 mL of DMEM medium (Table 1).

3. Collect the cells by centrifugation at 300 x*g* for 5 min at room temperature. Remove the medium by aspiration.
4. Resuspend the cells in 5 mL of fresh DMEM medium by gently pipetting.
5. Transfer the cell suspension to a 10 cm tissue culture dish with 5 mL of DMEM medium. Incubate the cells at 37 °C with 5% CO_2.
6. Split C2C12 cells 1:5, when they reach 60% confluency.
7. For C2C12 differentiation, use the cell passage below 15. The younger the passage, the better.
8. When C2C12 cells are 90% confluent, change to the differentiation medium (differentiated day 0).
9. Change the medium every 2 days.

3.2.2 AtRA pre-treatment in isoproterenol-stress cardiac model

1. It is recommended to use male mice between 16 and 20 weeks of age.
2. *atRA Pre-Treatment*: 4 h, prior to each isoproterenol injection (acute or chronic) intraperitoneally at 5 mg/kg. Since atRA is light sensitive, perform injections in low light. Use corn oil as a negative control.
3. *Acute Isoproterenol Induced Stress:* Mice were intraperitoneally injected with 160 mg/kg at two 8 h intervals. Sacrifice mice 24 h after the first ISO injection. Use saline as a negative control.
4. *Chronic Isoproterenol Induced Stress:* Mice were intraperitoneally injected with 30 mg/kg every day for 15 days or 5 weeks. Use saline as a negative control.
5. Sacrifice and collect heart tissue for protein extraction (Section 3.2.3) and histology (Section 3.2.4).

3.2.3 Heart tissue cytosolic and membrane protein extraction

Table 3 shows buffer preparation.

1. Immediately following euthanasia, excise the heart.
2. Using a razor cut the heart to separate the ventricles from the atria. Reserve the apex of the heart of histology analysis. Remove the aorta. Dab ventricles on a kim wipe to remove excess blood. Transfer to 1.5 mL tube and snap freeze in liquid nitrogen. Sample may be stored in −80 °C or immediately processed.
3. In a 5 mL Dounce homogenizer add 0.5–1 mL of Triton X-100 lysis buffer supplemented with protease inhibitors. For RNA analysis, use Trizol in place of lysis buffer.

4. Transfer frozen heart tissue to homogenizer with lysis buffer and homogenizer with at least 50 passes. Perform on ice.
5. Estimate protein concentration using Bradford Assay. Store tissue homogenates in −80 °C or immediately process for Western blot.
6. *Western Blot Procedure:* For pCaMKII T87 detection load 50 μg of protein on an 8 or 10% 1.5 mm SDS PAGE gel. For phospho-PLN (pPLN) detection load 20 μg on an 8 or 1.5 mm 10% SDS PAGE polyacrylamide gel. Perform Western as described in Park et al. (2018).

3.2.4 Preparation of heart tissue for histology

1. Extract heart as described (Section 3.2.4) and isolate the atria.
2. In a small tissue section holder add one drop of Tissue Plus O.C.T. Compound. Remove all bubbles in tissue plus submerged in O.C.T. compound to ensure high quality sections.
3. Place the heart tissue with the apex facing upwards and the atria submerged in the Tissue Plus Compound. Gently squeeze the atria with forceps to remove any trapped bubbles.
4. Fill 150 mL isopentane in 250 mL plastic beaker. Precold the isopentane by liquid nitrogen in an appropriate container.
5. Freeze sections by immersing in an isopentane slurry. The tissue block is fully frozen once the tissue plus turns completely bright white.
6. Create 7 μm sections from frozen tissue blocks by cryosection for downstream histological analysis.

3.2.5 Histological examination of cardiomyocyte size, necrosis, and apoptosis

Table 1 shows reagent kit information. Table 2 provides antibody information. See Fig. 5 for the example of histology data.

1. Rehydrate the sections with 1 × PBS for 10 min.
2. Wash with 1 × PBS for 5 min. Repeat three times.
3. Block with 5% BSA for 30 min-overnight.
4. The following primary antibodies are recommended. 1:200 of anti-IgG conjugated to Cy3. 1:50 of wheat germ agglutinin (WGA) conjugated to Alexa fluor 488.
5. Incubate for 1 h.
6. Mount slides with MM83 mounting media (CBG biotech Oakwood).
7. *Anti-IgG image acquisition:* Image whole heart sections at a 10 × magnification.

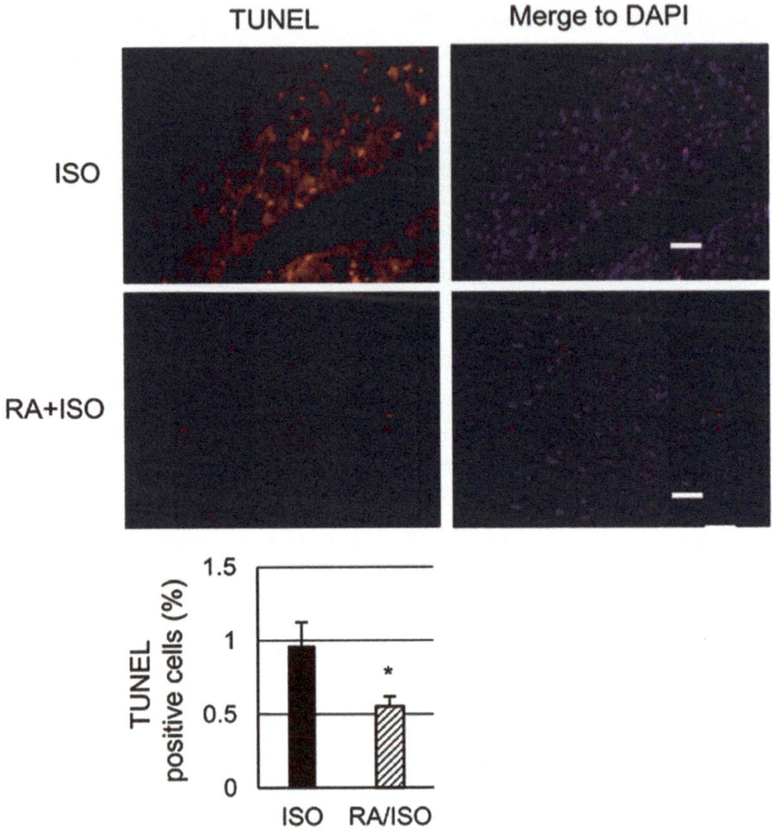

Fig. 5 Representative histological TUNEL staining of mouse heart sections. Heart tissue sections from wild-type mice treated with either isoproterenol (ISO; 160 mg/kg) or ISO+atRA (5 mg/kg) were stained with terminal deoxynucleotide transferase-mediated dUTP nick-end labeling (TUNEL) to detect apoptosis (red). Nuclei are stained with DAPI (blue). AtRA pretreatment significantly protects against iso-induced apoptosis in the wild type mice. TUNEL signal was quantified using ImageJ to analyze of apoptotic cells (bottom graph) (Student's *t*-test: "*" *P*-value $\leq$0.05; $n=3$). *Modified figure from Park, S. W., Nhieu, J., Lin, Y. W., & Wei, L. N. (2019). All-trans retinoic acid attenuates isoproterenol-induced cardiac dysfunction through Crabp1 to dampen CaMKII activation. European Journal of Pharmacology, 858, 172485. doi:10.1016/j.ejphar.2019.172485.*

8. *Apoptosis (TUNEL) assay:* Follow steps 1 and 2. Proceed with terminal deoxynucleotide transferase-mediated dUTP nick-end labeling (TUNEL) staining protocol according to manufacturer's protocol (Biotium). In addition, stain heart sections with DAPI to detect nuclei.
9. *TUNEL Image Acquisition:* Image whole heart sections at 20× magnification.

Data Analysis: Necrotic regions are represented by IgG positive areas. IgG positive areas can be calculated using ImageJ and the formula: % = Number of pixels in the thresholded IgG channel/number of pixels of myocardial tissue. Cell size analysis can be performed also using ImageJ. ~100 cardiomyocytes are recommended to determine relative cardiomyocyte size. Apoptotic cells are TUNEL positive. TUNEL signal and DAPI can be quantified using ImageJ. % apoptotic cells = TUNEL signal/DAPI signal.

3.2.6 In vitro *competition assay for CRABP1 and CaM binding to CaMKII*

Table 3 shows buffer preparation. Table 2 provides antibody information.

1. Transfer the required volume of Flag M2 beads to a 1.5 mL tube. Spin for 1 min at 500 x*g*. Discard the supernatant to remove storage solution. M2 beads are stored as 50% slurry. NOTE: Cut the tip of the pipette tip to prevent damage to M2 beads.
2. Wash beads with binding buffer three times, 30 s. For each wash. Resuspend in binding buffer to create a 50% slurry.
3. Synthesize Flag-CaMKII using the Promega TnT Kit according to kit instructions. For construct cloning see (Park et al., 2018). Flag-CaMKII serves as the prey.
4. Add 5 μL of Flag-CaMKII TnT product, 5 μL of M2 bead, and binding buffer to a final volume of 500 μL.
5. Incubate on a rocker at 4°C for 2 h. To capture Flag-CaMKII on the M2 bead. Flag-CaMKII will serve as the bait.
6. Spin at 500 x*g* and remove supernatant.
7. Combine M2-Flag-CaMKII bead, 1.5 μM His-CRABP1, GST-Calmodulin (CaM), and binding buffer in total reaction volume of 500 μL. Add CaM in increasing amounts with a recommended range of 0.1–2.0 μM.
8. *Reciprocal Assay:* Combine the above components from Step 7. Instead keep GST-CaM at a constant concentration of 0.5 μM, while increasing concentration of His-CRABP1 with a recommended range of 0.5–4.0 μM.
9. Incubate for 2 h-overnight on the rocker at 4°C.
10. Spin reaction tubes at 500 x*g* for 1 min. And discard the supernatant.
11. Wash beads by adding 500 μL binding buffer and inverting for 30 s. Then spin 500 x*g* 1 min and discard the supernatant. To avoid disturbing the beads, leave ~100 μL when removing the supernatant. Wash three times.

12 After the final wash use a pipette with a p10 tip or a 23 g needle to remove any remaining binding buffer.
13 Add 10–15 μL of 4 × SDS PAGE loading buffer.
14 Boil at 95°C for 5 min. Then spin samples at 10,000 x*g* for 2 min.
15 Load all of the sample (beads included) on to a 13.8% SDS PAGE polyacrylamide gel.
16 Proceed with Western blot.

Data Analysis: Perform densitometric analysis using software such as ImageJ or Licor StudioLite. Apparent competition is observed when increasing amounts of His-CRABP1 or GST-CaM reduce the interaction of CaMKII (bait) with the component held in a constant concentration.

4. Concluding remarks

In this chapter, we provide a collection of methods to study noncanonical activities of atRA in regulating (1) stem cell growth and (2) cardiomyocyte function. It is important to note that CRABP1 serves as the direct mediator for atRA to rapidly regulate the activities of RAF-MEK-ERK and CaMKII in a cell context-dependent manner. In addition, the mechanisms underlying these emerging CRABP1-dependent activities of atRA share the following features: (1) complete independence of RARs, (2) initiation from the cytosol, and (3) very rapid response. In studying these noncanonical activities of atRA, it is critical to take into consideration the cellular context and physiological relevance. Finally, we believe that investigating these noncanonical activities of atRA provide important new insight into the ever-increasing complexity and diversity of retinoic acid signaling in a physiological context.

Acknowledgments

This work was supported by NIH grants DK54733, DK60521, and the Dean's Commitment and the Distinguished McKnight Professorship of University of Minnesota to L.N.W.

References

Cargnello, M., & Roux, P. P. (2011). Activation and function of the MAPKs and their substrates, the MAPK-activated protein kinases. *Microbiology and Molecular Biology Reviews*, *75*(1), 50–83. https://doi.org/10.1128/MMBR.00031-10.

D'Ambrosio, D. N., Clugston, R. D., & Blaner, W. S. (2011). Vitamin A metabolism: An update. *Nutrients*, *3*, 63–103. https://doi.org/10.3390/nu3010063.

Duester, G. (2008). Retinoic acid synthesis and signaling during early organogenesis. *Cell*, *134*(6), 921–931. https://doi.org/10.1016/j.cell.2008.09.002.

Erickson, J. R. (2014). Mechanisms of CaMKII activation in the heart. *Frontiers in Pharmacology*, *5*, 59, Frontiers Media SA. https://doi.org/10.3389/fphar.2014.00059.

Germain, P., Chambon, P., Eichele, G., Evans, R. M., Lazar, M. A., Leid, M., et al. (2006). International Union of Pharmacology. LX. Retinoic acid receptors. *Pharmacological Reviews*, *58*(4), 712–725. https://doi.org/10.1124/pr.58.4.4.

Lin, Y. L., Persaud, S. D., Nhieu, J., & Wei, L. N. (2017). Cellular retinoic acid-binding protein 1 modulates stem cell proliferation to affect learning and memory in male mice. *Endocrinology*, *158*(9), 3004–3014. https://doi.org/10.1210/en.2017-00353.

Mattiazzi, A., Bassani, R. A., Escobar, A. L., Palomeque, J., Valverde, C. A., Vila Petroff, M., et al. (2015). Chasing cardiac physiology and pathology down the CaMKII cascade. *American Journal of Physiology. Heart and Circulatory Physiology*, *308*, H1177–H1191, American Physiological Society. https://doi.org/10.1152/ajpheart.00007.2015.

Mattiazzi, A., & Kranias, E. G. (2014). The role of CaMKII regulation of phospholamban activity in heart disease. *Frontiers in Pharmacology*, *5*, 5, Frontiers Media SA. https://doi.org/10.3389/fphar.2014.00005.

Nagpal, I., & Wei, L. N. (2019). All-trans retinoic acid as a versatile cytosolic signal modulator mediated by CRABP1. *International Journal of Molecular Sciences*, *20*, 3610. https://doi.org/10.3390/ijms20153610.

Napoli, J. L. (2017). Cellular retinoid binding-proteins, CRBP, CRABP, FABP5: Effects on retinoid metabolism, function and related diseases. *Pharmacology & Therapeutics*, *173*, 19–33. https://doi.org/10.1016/j.pharmthera.2017.01.004.

Park, S. W., Nhieu, J., Lin, Y. W., & Wei, L. N. (2019). All-trans retinoic acid attenuates isoproterenol-induced cardiac dysfunction through Crabp1 to dampen CaMKII activation. *European Journal of Pharmacology*, *858*, 172485. https://doi.org/10.1016/j.ejphar.2019.172485.

Park, S., Nhieu, J., Persaud, S. D., Miller, M. C., Xia, Y., Lin, Y. W., et al. (2019). A new regulatory mechanism for Raf kinase activation, retinoic acid-bound Crabp1. *Scientific Reports*, *9*(1), 10929. https://doi.org/10.1038/s41598-019-47354-7.

Park, S. W., Persaud, S. D., Ogokeh, S., Meyers, T. A., Townsend, D. W., & Wei, L. N. (2018). CRABP1 protects the heart from isoproterenol-induced acute and chronic remodeling. *Journal of Endocrinology*, *236*(3), 151–165. https://doi.org/10.1530/JOE-17-0613.

Persaud, S. D., Lin, Y.-W., Wu, C.-Y., Kagechika, H., & Wei, L.-N. (2013). Cellular retinoic acid binding protein I mediates rapid non-canonical activation of ERK1/2 by all-trans retinoic acid. *Cellular Signalling*, *25*(1), 19–25. https://doi.org/10.1016/j.cellsig.2012.09.002.

Pozarowski, P., & Darzynkiewicz, Z. (2004). Analysis of cell cycle by flow cytometry. *Methods in Molecular Biology*, *281*, 301–311. https://doi.org/10.1385/1-59259-811-0:301.

Ross, A. C. (1993). Cellular metabolism and activation of retinoids: Roles of cellular retinoid-binding proteins. *The FASEB Journal*, 7, 317–327. https://doi.org/10.1096/fasebj.7.2.8440409.

Samarut, E., & Rochette-Egly, C. (2012). Nuclear retinoic acid receptors: Conductors of the retinoic acid symphony during development. *Molecular and Cellular Endocrinology*, *348*(2), 348–360. https://doi.org/10.1016/j.mce.2011.03.025.

CHAPTER TWELVE

Retinoic acid and microRNA

Lijun Wang, Atharva Piyush Rohatgi, Yu-Jui Yvonne Wan*
Department of Pathology and Laboratory Medicine, University of California Davis Health, Sacramento, CA, United States
*Corresponding author: e-mail address: yjywan@ucdavis.edu

Contents

Abstract

Retinoic acid (RA), the biologically active metabolite of vitamin A, regulates a vast spectrum of biological processes, such as cell differentiation, proliferation, apoptosis, and morphogenesis. microRNAs (miRNAs) play a crucial role in regulating gene expression

Methods in Enzymology, Volume 637
ISSN 0076-6879
https://doi.org/10.1016/bs.mie.2020.02.009

by binding to messenger RNA (mRNA) which leads to mRNA degradation and/or translational repression. Like RA, miRNAs regulate multiple biological processes, including proliferation, differentiation, apoptosis, neurogenesis, tumorigenesis, and immunity. In fact, RA regulates the expression of many miRNAs to exert its biological functions. miRNA and RA regulatory networks have been studied in recent years. In this manuscript, we summarize literature that highlights the impact of miRNAs in RA-regulated molecular networks included in the PubMed.

1. Introduction

Retinoic acid (RA) is a metabolite of vitamin A that has a vast spectrum of biological processes, including differentiation, proliferation, apoptosis, and morphogenesis (Hu, Xu, Sun, Teng, & Xiao, 2017; Wang et al., 2012). While all-*trans*-RA is a major form, other isomers such as 13-*cis*-RA and 9-*cis*-RA are also present (Ruhl, Krezel, & de Lera, 2018; Tang & Russell, 1990). In this paper, RA is used to refer to all-*trans*-RA. Pharmacologically, retinoids including RA are used for acne and cancer treatment (Dobrotkova, Chlapek, Mazanek, Sterba, & Veselska, 2018; Leyden, Stein-Gold, & Weiss, 2017). Its effects in treating inflammation, allergy, and autoimmune diseases have also been revealed (Oliveira, Teixeira, & Sato, 2018).

The action of RA is mediated through the nuclear receptors: retinoic acid receptor (RARα, β, and γ) and retinoid x receptors (RXRα, β, and γ). Because RXR is an essential partner of many nuclear receptors, RA can regulate the function of other nuclear receptors such as peroxisome proliferator activate receptor and farnesoid x receptor. This allows RA to regulate lipid metabolism (Bushue & Wan, 2010; Chen, Wang, & Wan, 2010; Dai et al., 2003; Goto, 2019; Gyamfi, He, French, Damjanov, & Wan, 2008; Hu et al., 2017; Krezel, Ruhl, & de Lera, 2019; Liu, Wang, & Lin, 2019; Martin, Ma, & Bernard, 2019; Ruhl et al., 2018; Wan et al., 2000, 2000, 2003; Watanabe & Kakuta, 2018). Many RAR and RXR-regulated genes have been identified using genome-wide approaches such as chromatin immunoprecipitation followed by sequencing (Al Tanoury et al., 2014; Fang et al., 2013; He et al., 2013; Hu et al., 2013; Hua, Kittler, & White, 2009; Kiss et al., 2017; Liu, Ly, Hu, & Wan, 2014; Penvose, Keenan, Bray, Ramlall, & Siggers, 2019; Tang et al., 2011). In addition to transcriptional regulation, oral administration of RA alters gut microbiota

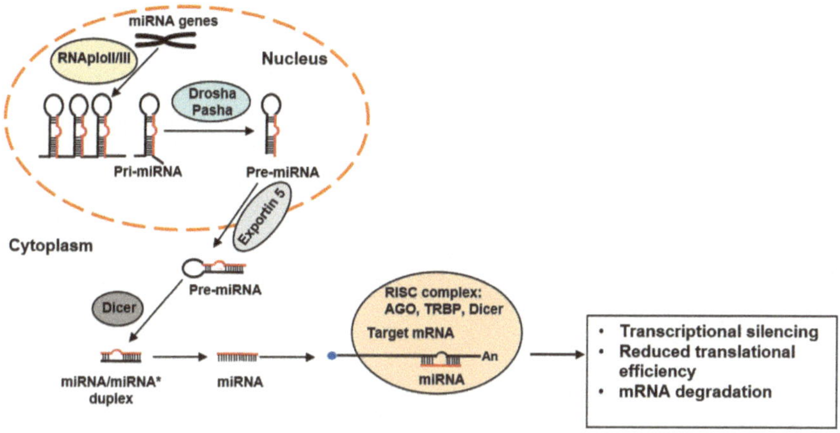

Fig. 1 miRNA Biogenesis. miRNA is transcribed into pri-mRNA, which is converted to pre-miRNA by the protein complex Drosha-Pasha. Pre-miRNA is then transported out of the nucleus via the protein exportin 5. The endoribonuclease dicer cleaves pre-miRNA to form a miRNA duplex. Then, unwinding of the two strands and asymmetric assembly of one strand into RISC (RNA-induced silencing complex) occurs. The RISC complex acts via transcriptional silencing, reduced translational efficiency, or mRNA degradation.

and has an impact on liver regeneration (Liu, Hu, & Wan, 2016). Thus, it is important to understand mechanisms by which RA exerts its effects.

miRNA is small non-coding RNA that silences messenger RNA (mRNA). Its functions via base-pairing with complementary sequences within mRNA. As a result, the targeted mRNA is reduced by either cleavage or destabilization, or has its translational efficiency reduced (Fig. 1). miRNAs regulate the expression of as much as one-third of human gene-encoded proteins (Ganapathy & Ezekiel, 2019). Like RA, miRNAs regulate multiple biological and disease processes (Ganapathy & Ezekiel, 2019; Ivey & Srivastava, 2015). miRNAs have been approved by the FDA as diagnostic tools and for treatment (Dave et al., 2019). This review paper focuses on the miRNAs that mediate or affect the effects of RA. In some cases, RA regulates the expression of miRNA through its receptors such as RARα and RARβ, which bind to the regulatory region of certain miRNAs to alter their expression levels. In other cases, miRNA may target the receptors of RA and affect the effect of RA (Fig. 2). The literature search was limited to the papers included in the PubMed using mammalian cells or animal models.

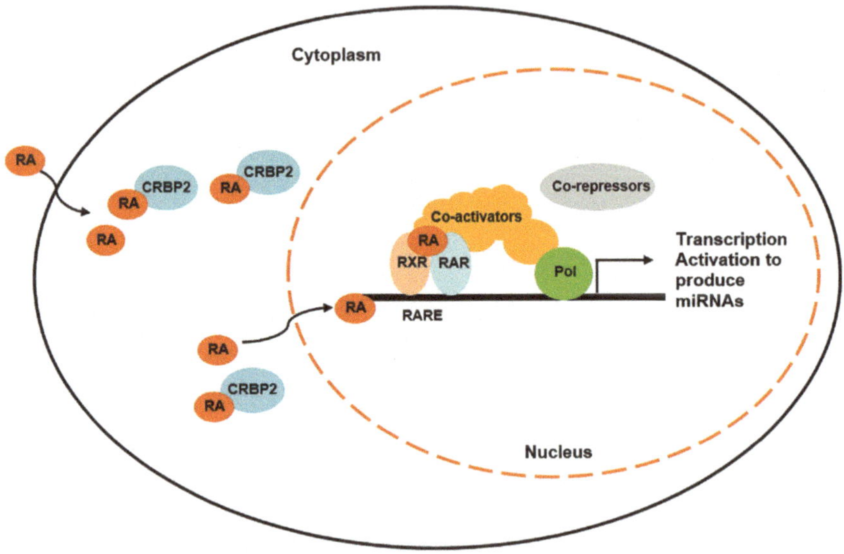

Fig. 2 Retinoic acid and its receptor-mediated gene expression. RA is transported into the cell and bind to CRABP2 (cellular RA binding protein). In the nucleus, RA binds to nuclear receptors RXR and RAR heterodimer, which recognizes the RA receptor responsive element (RARE) located on the target gene, to regulate gene expression. The transcriptional activation requires the recruitment of coactivators and the departure of co-suppressors. Once miRNA is produced, it can silence the RA receptors, the transcriptional co-regulators, or other specific targets to modulate the effects of RA.

2. miRNAs that are implicated in RA-induced differentiation

2.1 Embryonic stem cell differentiation

Dicer is an endoribonuclease that cleaves double stranded RNA critical for miRNA biogenesis. In Dicer-deficient embryonic stem cells, increased dimethylation of histone H3 at lysine 9 (H3K9me2) at over 900 CpG islands has been noted, indicating the significance of miRNAs in epigenome regulation during differentiation. Additionally, the balance between the transcriptionally favorable tri-methylation histone H3 at lysine 4 (H3K4me3) and the unfavorable tri-methylation histone H3 at lysine 27 (H3K27me3) is shifted in the Dicer-deficient epigenome. Homeobox (HOX) transcription factors play pivotal roles in many aspects of cellular physiology, embryonic development, and tissue homeostasis. In response to RA treatment, elevated H3K27me3 is found in the promoters of the caudal type homeobox 2 (*Cdx2*) and homeobox A1 (*Hoxa1*) genes of Dicer-deficient embryonic stem cells. However, forced expression of *let-7g in* Dicer-deficient embryonic cells counteracts the effect of H3K27me3 on the expression of the *Cdx2* and *Hoxa1* genes. *Let-7g* counteracts the effects by reducing the enhancer of zeste homolog 2(*Ezh2*), a histone-lysine *N*-methyltransferase enzyme, which thereby leads to the differentiated phenotype (Tennakoon, Wang, Coarfa, Cooney, & Gunaratne, 2013).

miRNAs inhibit mouse embryonic stem cell self-renewal and stabilize their differentiated states. In mouse embryonic stem cells, RA upregulates *miRNA-134*, *miR-296*, and *miR-470*, which induces differentiation by targeting the pluripotency genes *Nanog*, octamer-binding transcription factor 4 (*Oct4*), and sex determining region Y-box 2 (*Sox2*) (Tay et al., 2008; Tay, Zhang, Thomson, Lim, & Rigoutsos, 2008). *miRNA-134* inhibits *Nanog* and the liver receptor homolog 1 (*LRH1*), both of which positively regulate *Oct4* (Tay, Tam, et al., 2008). In addition, *miR-134*, *miR-296*, and *miR-470* are all involved in the lineage specific repression of mouse embryonic stem cell renewal (Zhang et al., 2015). Elevated forkhead box protein M1 (FOXM1) contributes to the maintenance of human stem cell pluripotency by directly regulating the *Oct4* promoter. *miR-134* overexpression reduces *FOXM1* and promotes the differentiation effect of RA (Chen et al., 2015). Additionally, by reducing either *miR-200b* or *miR-200c*, RA inhibits the expression of pluripotency genes and increases the ectodermal marker *Nestin*, a type VI intermediate filament protein (Zhang, Gao, et al., 2015).

2.2 Neuronal differentiation

The significance of *miR-10* in neuronal differentiation has been revealed. The *hsa-miR-10* family is substantially increased (~95-fold) during the RA-induced neuronal differentiation of human embryonic stem cells (Parsons, 2012). In addition, during RA-induced neural lineage specific progression, *hsa-miR-302* family is reduced and the expression of the *Hox* gene, *hsa-miR-10*, and *let-7* are induced (Parsons, Parsons, & Moore, 2012). Similarly, the expression of *Hoxd4* and *miR-10b* is coordinated in the RA-induced neural differentiation of P19 cells. This suggests the importance of *miR-10b* (Phua et al., 2011). Through the *let-7*-dependent mechanism, phosphorylation of a highly conserved RNA-binding protein called LIN28A fosters neural differentiation of P19 cells (Liu et al., 2017). Moreover, the differentiation of embryonic stem cells is also accompanied with increased *miR-10a-5p*, *miR-219-5p*, and *miR-219-2-3p*. The increase of *miR-219* and its mimics also promote mouse embryonic stem cell differentiation into neural cells (Wu et al., 2017). In contrast, *miR-125b-2* overexpression suppresses this differentiation (Deng, Zhang, Xu, & Ma, 2015).

Sirtuin 1 (SIRT1) plays a negative role in inducing the differentiation of induced pluripotent stem cells (iPSCs) into neural stem cells. When SIRT1 is inhibited, the differentiation continues and is marked by increased *miR-34a*, which silences *SIRT1* (Hu et al., 2014). In mouse neuroblastoma N2a cells, P19 embryonal carcinoma cells, and the mouse brain, increased *miR-124* reduces RARγ, thereby inhibiting neurite outgrowth (Wang, Yao, Lu, Li, & Ma, 2010) (Su, Gu, Zhang, Li, & Wang, 2020). There are many other miRNAs that may be implicated in neural cell differentiation (Hu et al., 2017). RA treatment of adipose-derived mesenchymal stem cells leads to changes in 76 miRNAs that have a greater than twofold difference (Hu et al., 2017).

2.3 Spermatogonia differentiation

Tyrosine-protein kinase Kit, a receptor tyrosine kinase protein, plays an essential role in cell proliferation, hematopoiesis, etc. The tyrosine-kinase receptor Kit is also essential for the maintenance of primordial germ cells in both sexes, and RA-induced differentiation of spermatogonia is accompanied by Kit induction. Recent data reveals that RA-induced *miR-26b* promotes the transition from Kit-negative to Kit-positive spermatogonia (Tu et al., 2018). *miR-26b* can silence a transcriptional factor named promyelocytic leukemia zinc finger protein (*Plzf*) in undifferentiated spermatogonia. It also downregulates the methylcytosine dioxygenase (*Tet3*) gene, thus reducing 5-hydroxymethylcytosine in spermatogonia. Together, through various mechanisms including epigenetic regulation, RA regulates spermatogonia differentiation (Tu et al., 2018).

While glial cell-derived neurotrophic factor (GDNF) is essential for the self-renewal of spermatogonia stem cells, RA is critical for the differentiation of spermatogonia. *miR-202-3p*, which is abundantly found in the testis, can be induced by GDNF but is reduced by RA. *miR-202-3p* prevents stem cells from premature differentiation by suppressing RNA binding fox-1 homolog 2 (*Rbfox2*). In addition, the inhibition of *miR-202-3p*, but not of *miR-202-5p*, induces the differentiation of spermatogonia stem cells (Chen et al., 2017).

RA-induced differentiation of spermatogonia has reduced *miR-146*, which targets the mediator complex subunit 1 (*Med1*), a co-activator of

RARs and RXRs. *miR-146* can also inhibit the expression of stimulated by retinoic acid 8 (*Stra8*) and the spermatogenesis- and oogenesis-specific basic helix-loop-helix 2 (*Sohlh2*) (Huszar & Payne, 2013). Thus, RA-reduced *miR-146* is critically important for RA-induced differentiation of spermatogonia. In addition, RA also reduces *miR-17-92* (*Mirc1*) and *miR-106b-25* (*Mirc3*), both of which function cooperatively to promote mouse spermatogonia differentiation (Tong, Mitchell, McGowan, Evanoff, & Griswold, 2012).

2.4 Pre-adipocytes and myoblast differentiation

RA inhibits the differentiation of 3T3-L1 pre-adipocytes by reducing lipid metabolism (Stoecker, Sass, Theis, Hauner, & Pfaffl, 2017). Such an action is through *miR-27a* and *miR-96* induction that leads to the rearrangement of the actin cytoskeleton as well as inhibition of the citric acid cycle (Stoecker et al., 2017).

Excess RA also inhibits myoblast differentiation and proliferation. In an embryonic tongue-derived myoblast C2C12 cell line, RA induce *miR-27b-3p*, which silences the α-dystrobrevin (*DTNA*) gene to inhibit proliferation and differentiation (Li et al., 2017). In addition, RA-induced *miR-31-5p* targets the Ca^{2+}/calmodulin-dependent protein kinase II δ gene (*CamkIIδ*) and is implicated in the RA-induced myogenic abnormalities of tongue (Liu et al., 2017). *miR-133a*, *210*, and *34a* are also involved in the RA-induced myoblasts differentiation and proliferation (Vecellio et al., 2012).

miR-10a targets the histone deacetylase 4 (*Hdac4*) gene and has a critical role in the RA-induced differentiation of mouse embryonic stem cells into smooth muscle cells (Huang et al., 2010). In addition, nuclear factor kappa B (NFκB), through binding to the *miR-10a* promoter, has a significant role in the differentiation process. Inhibiting the nuclear translocation of NFκB reduces *miR-10a* and prevents the RA-induced differentiation smooth muscle cells (Huang et al., 2010).

2.5 Hematopoietic progenitor cell differentiation

RA induces *miR-223*, which acts as a translational inhibitor to regulate lineage specification and differentiation of mouse hematopoietic stem cells (Chen, Li, Lodish, & Bartel, 2004). Transcription factors nuclear factor I-A (NFI-A) and CCAAT/enhancer-binding protein-alpha (C/EBPα) compete for binding to the *miR-223* promoter. In response to RA treatment, the induction of *miR-223* requires the release of NF1-A and the binding of C/EBPα. Taken together, the RA regulated differentiation of granulocytes is mediated by *miR-223* expression controlled by two transcriptional factors (Fazi et al., 2005).

3. miRNAs that are implicated in RA-regulated inflammatory signaling and immunity

Signal-regulatory protein α (SIRPα) is essential for modulating leukocyte inflammatory responses. The *SIRPα* gene is a commonly targeted by *miR-17*, *miR-20a*, and *miR-106a*. RA reduces these miRNAs and increases *SIRPα* expression in SIRPα-negative promyelocytic cells. However, bacterial lipopolysaccharides (LPS) increases *miR-17*, *miR-20a*, and *miR-106a* in macrophages, which leads to SIRPα reduction and macrophage activation. Thus, *miR-17*, *miR-20a*, and *miR-106a* regulate macrophage inflammatory responses by repressing SIRPα. (Zhu et al., 2013).

RA-induced *miR-10a* is abundantly found in naturally occurring regulatory T cells (Treg). *miR-10a* inhibits the development of inducible T regulatory cells (iTregs) into follicular helper T cells by suppressing both the transcriptional repressor B-cell lymphoma 6 (*Bcl6*) and the co-repressor nuclear receptor co-repressor 2 (*Ncor2*). In addition, *miR-10a* attenuates the differentiation process into the T(H)17 subset of helper T cells (Takahashi et al., 2012). In contrast to RA, transforming growth factor-β (TGF-β)-induced iTregs in vitro are unstable and do not have *miR-10a*. Thus, RA-induced *miR-10a* is selectively expressed and plays a role in stabilizing iTregs (Jeker et al., 2012).

T-cell receptor (TCR) signaling and TGF-β regulate the production of peripheral regulatory T (pTreg) cell. The expression of *miR-31* is triggered by TCR signaling but inhibited by TGF-β. *miR-31* directly targets the retinoic acid-inducible protein 3 (*Gprc5a*), a member of the type 3G protein-coupled receptor family. The deficiency of this protein leads to the impairment of pTreg-cell induction as well as an increased severity of experimental autoimmune encephalomyelitis (Zhang et al., 2015). Therefore, by targeting *Gprc5a*, *miR-31* reduces the production of pTreg (Zhang, Ke, et al., 2015).

RA induces the secretion of exosomes in gut-tropic T cells that causes an upregulation of integrin α4β7 that binds to mucosal vascular addressin cell adhesion molecule 1 (MAdCAM-1). Addressins are the ligands to the homing receptors of lymphocytes, which determine which tissue the lymphocyte will enter. miRNA profiling suggests that exosomes that originate from the gut tropic T cells contain a large quantity of miRNAs, such as *miR-132* and *miR212*, which target NKX2.3, a transcription factor important for the expression of MAdCAM-1 (Park et al., 2019).

Metabolism controls immune cell fate and function. *miR-33* not only represses genes involved in cholesterol efflux and fatty acid oxidation, but also polarization of immune cells. Macrophage-specific *miR-33* deletion promotes M2 macrophage polarization. Interestingly, inhibiting *miR-33* enhances the expression of RA-producing enzyme aldehyde dehydrogenase family 1, subfamily A2 (ALDH1A2), as well as the activity of retinal dehydrogenase in macrophages. Such findings are in consistent with the ability of RA to foster iTregs. Furthermore, treating hypercholesterolemic mice with *miR-33* inhibitors leads to accumulation of inflammation-suppressing M2 macrophages and forkhead box P3$^+$ (foxp3$^+$) Tregs in plaques. This treatment thereby inhibits the progression of atherosclerosis. Together, inhibiting *miR-33* benefits metabolism and inflammation (Ouimet et al., 2015).

4. miRNAs that are implicated in RA-induced differentiation of cancer cells

4.1 Embryonic carcinoma

RA has a profound effect in inducing the differentiation of embryonic carcinoma F9 cells into endodermal cells (Wan, Orrison, Lieberman, Lazarovici, & Ozato, 1987; Wan, Wang, & Wu, 1994; Wu, Wang, & Wan, 1992). In the F9 cell line, RA induces *miR-485*, which silences the α/β-hydrolase domain-containing protein 2 (*Abhd2*). The reduction of Abhd2 leads to ERK1/2 phosphorylation and differentiation of embryonic carcinoma cells (Yu, Zhang, Liu, Liu, & Guo, 2019).

4.2 Brain tumors

RA treatment enhances gap junctions which leads to an increased *miR-124-3p*-mediated antiproliferation in glioblastoma cells (Suzhi et al., 2015). RA also induces glioma cell death by elevating *miR-302b,* and silences E2F3, a transcriptional glioma growth regulator (Chen et al., 2014). In addition, reduced *miR-452* found in gliomas promotes their stem-like features and tumorigenesis because *miR-452* inhibits many stemness regulators like polycomb complex protein BMI-1, lymphoid enhancer binding factor 1 (LEF1), and transcription factor 4 (TCF4), (Liu et al., 2013).

4.3 Neuroblastoma

Several miRNAs have changes in their expression levels when the human neuroblastoma cell line SH-SY5Y is treated with RA and a dopamine cocktail (Das & Bhattacharyya, 2014). Among them, *miR-432* plays a key role in differentiation of human neuroblastoma cells by suppressing *NESTIN* and the repressor element-1-silencing transcription factor (REST) corepressor 1 (*RCOR1*) (Das & Bhattacharyya, 2014). RA induces neuronal differentiation via *miR-29b* and *miR-664a-5p* (Jauhari, Singh, & Yadav, 2018; Watanabe, Yamaji, & Ohtsuki, 2018). *miR-29b* reduces the P53 inhibitor P85α, which in turns increases *miR-145* to silence OCT4, SOX2, and kruppel-like factor 4 (KLF4) during SH-SY5Y differentiation (Jauhari et al., 2018). Further studies are required to understand the function of *miR-664a-5p* (Watanabe et al., 2018).

RA has anti-metastatic and anti-migratory activities via *miR-10a and miR-10b*, both of which target the SR-family splicing factor SFRS1 (*SF2/ASF*) (Meseguer, Mudduluru, Escamilla, Allgayer, & Barettino, 2011). They also induce neuronal differentiation by silencing *NCOR2* which inhibits neurite growth (Foley et al., 2011). Moreover, other RA-induced miRNAs, including *miR-9* and *miR-103*, control neuroblastoma differentiation by reducing a differentiation inhibitor molecule called Inhibitor of DNA-binding 2 (Id2) (Annibali et al., 2012). RA can also decrease *miR-17* in SH-SY5Y cells. *miR-17* is involved in the regulation of the mitogen-activated protein kinase (MAPK) signaling pathway, synaptic plasticity, and other markers of neuronal differentiation (Beveridge, Tooney, Carroll, Tran, & Cairns, 2009). RA- and brain-derived neurotrophic factor (BDNF)-induced *miR-125b* and *miR-124a* also positively regulate the differentiation of SH-SY5Y cells and neurite outgrowth of human neural progenitor ReNcell VM cells by targeting multiple genes (Le et al., 2009).

The mutant amyloid precursor protein $(APP)_{L17C}$ [the leucine (L) residue 17 is replaced by a cysteine (C) residue in amyloid β peptide (Aβ)] increases amyloid precursor protein dimerization which leads to reduced neurite outgrowth. The overexpression of a mutant amyloid precursor protein $(APP)_{L17C}$ in human neuroblastoma SH-SY5Y cells inhibits neurite outgrowth (Luu et al., 2019). *miR-34a* expression is greatly decreased in APP_{L17C} compared to APP_{WT} cells, and overexpression of *miR-34a* restores neurite outgrowth in the mutant cells (Luu et al., 2019). Thus, *miR-34a* expression has a pivotal role in APP-mediated neurite outgrowth (Luu et al., 2019).

DNA demethylation mediated by miRNAs promotes RA-induced differentiation of neuroblastoma cells. The ectopic overexpression of *miR-152* that targets the DNA methyltransferase 1 (*DNMT1*), contributes to the differentiated phenotypes, including reduced invasiveness and anchorage-independent growth in a human neuroblastoma cell line SK-N-BE (Das et al., 2010).

Other miRNAs are implicated in the differentiation of neuroblastoma cells. During the differentiation of NT2 cells, the stemness phenotype-associated *miR-302* is reduced, but *let-7*, *miR-125b*, and *miR-132* are increased (Pallocca et al., 2013). *miR-128* is also increased that in turn reduces Reelin and Doublecortin, where both are involved in the migratory potential of neural cells (Evangelisti et al., 2009). Moreover, during the RA-induced differentiation of neuroblastoma BE (2)-C cells, tumor suppressor *miR-449a* is increased. *miR-449a* targets the microfibril associated protein 4 (*MFAP4*), plakophilin 4 (*PKP4*), and TRNA splicing endonuclease subunit 15 (*TSEN15*), which are associated with poor survival of neuroblastoma patients (Zhao et al., 2015). Moreover, *miR-449a* induces cell cycle arrest by inhibiting the cyclin dependent kinase 6 (*CDK6*) and lymphoid enhancer binding factor 1 (*LEF1*) (Zhao et al., 2015).

RA-induced differentiation of the human neuroblastoma SK-N-BE cells have overexpressed *miR-34a* that directly targets the *E2F3* and lowers its protein. E2F3 is a potent transcriptional inducer for cell-cycle progression. In addition, overexpression of *miR-34a* in neuroblastoma cell lines prevents cell proliferation by inducing the caspase-dependent apoptotic pathway. Furthermore, *miR-34a* is reduced in primary neuroblastoma tumors and cell lines when compared to normal adrenal tissue (Welch, Chen, & Stallings, 2007). *miR-184* is also implicated in the apoptosis pathway as RA-induced *miR-184* promotes the apoptosis of human neuroblastoma cell lines including KELLY or SK-N-AS (Chen & Stallings, 2007).

4.4 Hematopoietic malignancies

Acute promyelocytic leukemia (APL) is caused by translocation of the promyelocytic leukemia (*PML*) gene located on chromosome 15 and the *RARα* gene on chromosome 17, which generates a fusion oncogene *PML/RARα* and differentiation arrest.

In hematopoietic malignancies, high c-Myc expression is correlated with a poor prognosis (Delgado & Leon, 2010). It has been shown that reduced PML/RARα due to loss of c-Myc is miRNA-dependent. In addition, tumor suppressor *let-7* has been identified to target both the *c-Myc* and the *PML/RARα* (Ding et al., 2016). Moreover, PML/RARα binds to the RA responsive element of the host genes such as the *let-7c* (Careccia et al., 2009; Pelosi et al., 2014). However, RA treatment induces a conformational change in the host promoter thus regulating *let-7c* (Pelosi et al., 2014). Taken together, RA by inducing *let-7c*, which silences PML/RARα, can lead to the differentiation of APL (Ding et al., 2016).

The synergism between c-Myc and *miR-17-19b*, a truncated version of the *miR-17-92* cluster, is well-established for tumor initiation (He et al., 2005; Mu et al., 2009). RA not only reduces *c-Myc* mRNA but also *miR-17-92* to induce APL cell differentiation (Yu et al., 2019). In the APL NB4 cell line, RA induces *miR-10b*, *miR-194*, *miR-195*, *miR-196a,* and others. Many of them are transcriptionally silenced by PML/RARα which leads to oncogenesis (Saumet et al., 2009). To add on, a long non-coding RNA named *HOTAIRM1* increases the RA-induced degradation of PML/RARα via the autophagy pathway. In fact, *HOTAIRM1* can function as a miRNA sponge to regulate the expression of autophagy-related genes by competing for miRNA bindings sites (Chen et al., 2017).

Other mechanisms are involved in RA-mediated differentiation of APL. Knockdown adenylate cyclase 9 (*ADCY9*), which converts ATP into cAMP, inhibits RA-induced differentiation of APL. Similarly, *miR-181a*, which directly targets *ADCY9* and reduces cAMP, has a negative role in RA-induced APL differentiation (Zhuang et al., 2014). *miR-17-5p* also inhibits RA-induced APL cell differentiation (Yu, Hu, et al., 2019). In contrast, RA induces granulocytic differentiation through *miR-382-5p* that silences the suppressor gene phosphatase and tensin homolog (*PTEN*) (Liu et al., 2019).

By regulating miRNA, RA treats APL (van Gils, Verhagen, & Smit, 2017). RA treatment causes a loss of imprinting on chromosome 14q32 that regulates the overexpression of many miRNAs (Manodoro et al., 2014). RA treatment of APL cells increases many tumor suppressor miRNAs, such as the *let-7* family (Garzon et al., 2007; van Gils et al., 2017). RA treatment of APL cells also increases *miR-107*, which silences the *NFI-A* (Fazi et al., 2005) (Garzon et al., 2007). RA also reduces many miRNAs such as *miR-17* and *miR-25* in APL cells (Garzon et al., 2007; Rossi et al., 2010). Moreover, by reducing *miR-301a-3p* to increase ubiquitin-specific protease 48 (USP48), RA is effective in APL differentiation (Li et al., 2018).

RA increases *miR-145* in myeloid leukemia cells which thereby reduces RAS-responsive element-binding protein 1 (RREB1), which prevents granulocytic differentiation of myeloid leukemia cells (Yao et al., 2019). In addition, *miR-29a*, which targets the *CDK6*, and *miR-142-3p*, which silences the TGF-β-activated kinase 1/MAP3K7 binding protein 2 (*TAB2*), contributes to RA-induced granulocytic differentiation of HL-60, THP-1, or NB4 cells. Forced expression of either *miR-29a* or *miR-142-3p* in hematopoietic stem cells induces myeloid differentiation (Wang et al., 2012). Ectopic *miR-638* overexpression, which targets the cyclin-dependent kinase 2 (*CDK2*), also increases RA-induced differentiation of leukemic cell lines and primary acute myeloid leukemia blasts (Lin et al., 2015). Moreover, RA increases the expression of tumor suppressor *miR-663* and inhibits HL-60 cell proliferation. Exotic overexpression of *miR-663* also induces HL-60 cell differentiation (Jian et al., 2011).

Autophagy regulated by miRNA has a role in the regulation of myeloid differentiation. Autophagy genes (*ATG*) are frequently repressed in primary acute myeloid leukemia patients. These reduced levels are likely due to increased *miR-106a*, which targets a transcriptional factor for several ATG genes (Jin et al., 2018). By mediating other mechanisms such as regulation of transcriptional machinery, miRNA is important for differentiation. In malignant myeloid cells, transcription factor CDX2 induces *miR-125b* expression. Increased *miR-125b* prevents the differentiation of myeloid cells and promotes leukemogenesis by inhibiting the core binding factor β (*CBFB*) translation. RA treatment decreases CDX2 activity, reduces *miR-125b*, and increases the *CBFB* during myeloid cell differentiation and in patients. Thus, *miR-125b* is important for RA-induced differentiation (Lin et al., 2011).

HOXB4, which has a function in stem cell expansion, along with *miR10a*, is highly expressed in atypical myeloproliferative neoplasms. However, overexpression of *miR-10a* has no effect on proliferation, differentiation or self-renewal of normal hematopoietic progenitors. (Dumas et al., 2018). RA reduces HOXA9 in acute leukemia cells leading to the inhibition of proliferation and promotion of differentiation of leukemia cells. The downregulation also interferes with tumor growth by modulating certain miRNAs. For example, during the leukemia cell differentiation induced by HOXA9 downregulation, *miR-663* and *miR-494* are increased and *miR-10a* and *miR-181* are decreased (Chen, Yu, Lv, & Zhang, 2017).

4.5 Breast cancer

Retinoids are promising for breast cancer prevention and treatment. In estrogen receptor (ER)-positive breast carcinoma cells (MCF-7), RA induces *miR-21* to counteract the anti-proliferative action of RA but may have a benefit in reducing cell motility. The pro-inflammatory cytokine IL1B, the adhesion molecule ICAM-1, and the tissue-type plasminogen activator (*PLAT*) are the identified *miR-21* target genes (Terao et al., 2011). In addition, *miR-21* targets the tumor suppressor gene tropomyosin 1 (*TPM1*) that is implicated in cell migration, as well as the programmed cell death 4 (*PDCD4*), and the mammary serine protease inhibitor (*Maspin*) (Terao et al., 2011). Thus, *miR-21* inhibitors may be used for breast cancer treatment.

Other miRNAs are implicated in RA-mediated anti-breast cancer effects. Nearly 70% of human breast carcinomas have reduced HOXA5. In addition, loss of HOXA5 in human breast cancer correlates with the progression to higher-grade lesions, suggesting its tumor suppressor effect (Jeannotte, Gotti, & Landry-Truchon, 2016). However, via *miR-130a* reduction, RA induces HOXA5 in breast cancer tissue (Yang, Miao, Mei, & Wu, 2013). RA also acts as a inducer for *miR-10a*, which counteracts the reduced *miR-10a* and RARβ in breast cancer cells (Khan et al., 2015). To add on, RA increases *miR-210* and *miR-23a/24-2* but decreases *miR-17/92* and *miR-424/450b* in breast cancer cells. However, estrogen can counteract the effect of RA on expression of those four miRNAs (Saumet et al., 2012).

In a specific subgroup of HER2-positive breast cancer cells, SKBR3, RA and Lapatinib treatment regulates the expression of 174 miRNAs (Fisher et al., 2015). Together, RA-regulated miRNAs have a significant impact on growth or mobility of SKBR3 cells (Fisher et al., 2015).

4.6 Digestive system cancers

RARβ is an upstream regulator and downstream effector of *miR-22*. In other words, RARβ can transcriptionally induces *miR-22*, which in turns epigenetically regulate the expression and function of RARβ and another orphan receptor named NUR77. As a result, nuclear NUR77 exports to cytosol converting its oncogenic effect into apoptotic effect thereby killing colon cancer cells (Hu et al., 2019). RA by itself or in combination with synthetic and natural HDAC inhibitors including short-chain fatty acids can induce *miR-22*. Thus, in addition to RARβ-mediated transcription, epigenetic mechanism also regulates *miR-22* expression. Additionally, it is interesting that *miR-22* targets multiple protein deacetylases including HDAC1, HDAC4, as well as SIRT1 and has HDAC inhibitory effect used to fight for cancer. Thus, RA-induced *miR-22* is a tumor suppressor for colon cancer (Hu et al., 2019). Furthermore, reduced *miR-22* is found in many other types of cancers including liver, colon, etc., and is considered as a tumor suppressor (Yang, Hu, Liu, & Wan, 2015). In consistency with the role of *miR-22* being a tumor suppressor, *miR-22* has been characterized as a metabolic silencer (Hu et al., 2020).

In metastatic pancreatic adenocarcinoma cells, *miR-10a* is overexpressed and regulates the metastatic behavior by suppressing HOXB1 and HOXB3. As mentioned above, *miR-10a* is important for RA-mediated differentiation of neuronal cells. However, in pancreatic adenocarcinoma, RAR antagonists reduces *miR-10a* and has an anti-cancer effect (Weiss et al., 2009).

Reduced *miR-452* promotes stem-like features and tumorigenesis in glioma as mentioned above (Liu et al., 2013). However, increased *miR-452* significantly promotes hepatocellular carcinoma (HCC) cell proliferation in vitro and in vivo (Zheng et al., 2016). *miR-452* is overexpressed in the chemo-resistant hepatospheres and human HCCs (Zheng et al., 2016). Thus, *miR-452* over-expression predicts poor survival for liver cancer patients (Zheng et al., 2016). Moreover, RA combined with doxorubicin reduces *miR-452* and inhibits HCC metastasis (Zheng et al., 2016). Furthermore, *miR-452* silences the *Sox7*, which suppresses the Wnt/β-catenin signaling pathway (Chan, Mak, Leung, Chan, & Ngan, 2012; Zheng et al., 2016).

4.7 Other cancers

As a tumor suppressor, reduced RARβ can be an indicator for prognosis. In papillary thyroid carcinoma, reduced RARβ is associated with increased *miR-146b-5p* (Czajka et al., 2016). Overexpression of *miR-146a-5p* and *miR-146b-5p* results in reduced *RARβ* mRNA (Czajka et al., 2016). In contrast, toll-like receptor 3 (TLR3) activation induces RARβ re-expression and tumor inhibition, which is due to increased *miR-29b, -29c, -148b,* and *-152* as well as re-expression of epigenetically silenced genes (Galli et al., 2013).

Another miRNA that targets RARβ is *miR-29b*. Tuberous sclerosis complex (TSC) is an incurable multisystem disease featured by mTORC1-hyperactive tumors. It has been found that by targeting RARβ, *miR-29b* has an oncogenic effect in TSC2-deficient cells. In addition, inhibition of *miR-29b* suppresses oncogenic property of TSC2-deficient tumors in vivo (Liu et al., 2019).

In addition to RARβ, miRNAs also affect the expression of other RA receptors. *miR-27a* overexpression, which silences RARα and RXRα, is found in aggressive rhabdomyosarcomas cells (Tombolan et al., 2015). Additionally, RA can enhance the binding of RARα to the *miR-27a* promoter, leading to the inactivation of *miR-27a* transcription, which increases glycogen synthase kinase-3β (GSK-3β) in laryngeal cancer cells. However, the mechanism by which RARα reduces *miR-27a* remains to be understood (Chen et al., 2017).

The effects of other retinoids in regulating miRNA have been studied. For example, the treatment effect of 9-*cis* RA, a ligand for RXR, is revealed in an adrenocortical xenograft mouse model. 9-*cis* RA together with mitotane reduce circulating *hsa-miR-483-5p* in the adrenocortical xenograft mouse model and can potentially be a predictor of treatment efficacy (Nagy et al., 2015). Additionally, tomato-derived bioactive antioxidants such as carotenoids and lycopene also inhibit tumor growth. In PC3 prostate cancer cells, lycopene increases *let-7f-1* to silence the *AKT2* and induce apoptosis (Li, Chen, Zhao, Hao, & An, 2016).

5. miRNAs that are implicated in other RA-regulated health issues

5.1 Liver diseases

Overexpression of *miR-34a* in human hepatoblastoma HepG2 cells reduces RXRα mRNA and protein. An inversely correlated expression pattern exists between *miR-34a* and RXRα protein in 14 liver specimens, with fibrotic livers having increased *miR-34a* but reduced RXRα. p53 can up-regulate *miR-34a* and promote liver fibrosis. Thus, by reducing RXRα, *miR-34a* has a role in liver fibrosis (Oda et al., 2014).

5.2 Atherosclerosis

miR-10a is a potential diagnostic molecule because reduced *miR-10a* in aortic endothelium and serum is linked with atherosclerosis (Lee et al., 2018). Moreover, RARα/RXRα-specific agonists induce *miR-10a* in vascular endothelial cells and inhibit atherosclerotic lesion formation. Thus, the RA-induced *miR-10a* can be a potential target for treating atherosclerosis (Lee et al., 2018).

5.3 Congenital diseases

Through hypermethylation, *miR-124a* is reduced and p38 is deactivated in RA-induced spina bifida fetuses compared with healthy rats. In addition, p38 deactivation is accompanied by increased apoptosis suggesting the role of *miR-124a* in RA-induced congenital defect (Qin et al., 2017). RA also reduces *miR-9/9**, *miR-124a*, and *miR-125b*, all of which are implicated in the development of spina bifida in rat models (Zhao et al., 2008). The up-regulation of *miR-9*, *miR-124a*, and *miR-138* as well as down-regulation of *miR-134* are found in the amniotic fluid of RA-induced spinal bifida fetuses compared with that of control fetuses (Qin et al., 2016).

5.4 Aging

miRNAs that are implicated in stress-induced premature senescence have been identified in primary cultures of human diploid fibroblasts and human trabecular meshwork cells. Senescence is associated with reduced *miR-15* and *miR-106b* as well as increased *miR-182* and *-183*. In addition, *miR-106b* and *miR-182* target the $p21^{CDKN1A}$ and the *RARG*, respectively, and are thereby implicated in senescence (Li, Luna, Qiu, Epstein, & Gonzalez, 2009).

5.5 Myopia

Mutation of the paired box gene 6 (*PAX6*) gene, which controls oculogenesis, is implicated in the development of myopia. RA induces *miR-328*, which targets the *PAX6*, in retinal pigment epithelial cells leading to increased retinal pigment epithelial cell proliferation and reduced scleral cell proliferation. Thus, the reduction of *miR-328* contributes to the prevention or treatment of myopia (Chen et al., 2012).

5.6 Alzheimer's disease

Overexpression of *miR-138* is found in the N2a/APP cells expressing human amyloid precursor proteins (APP) 695 protein (APP695, APPwt) and HEK293/tau cells (human embryonic kidney 293 cells expressing human tau protein). In HEK293/tau cells, increased *miR-138*, which directly targets the *RARα*, activates GSK-3β, and enhances tau phosphorylation. In contrast, elevated RARα inhibits GSK-3β and reduces tau phosphorylation (Wang et al., 2015).

6. Conclusion

miRNA has a significant role in mediating the functions of RA. Among the above-mentioned miRNAs, there are a few such as *let-7*, *miR-10*, and *miR-17* are implicated in multiple biological processes and disease models.

The *let-7* family has a wide variety of implications. Increased *Let-7g* leads to the differentiated phenotype of Dicer-deficient mouse ESCs. Furthermore, the *let-7* family has a variety of effects on neuronal lineage progression cells. It induces neuronal differentiation and is elevated in differentiated neuronal cultures. They can also act as tumor suppressors and lead to the differentiation of APL cells. Moreover, *let-7f-1* induced by lycopene causes apoptosis.

The *miR-10* family also has a wide variety of implications after RA induction. For example, reduced *miR-10a* is linked to atherosclerosis and can be a diagnostic tool. Furthermore, *miR-10a* promotes the differentiation of mouse embryonic stem cells into smooth muscle cells. *miR-10a* is also found in T cells and stabilizes iTregs that prevents their development into helper T cells. In addition, *miR-10a* level is low in breast cancer cells but is overexpressed in metastatic pancreatic adenocarcinoma cells.

miR-17 is generally decreased by RA. It is involved in the regulation of the MAPK signaling pathway, synaptic plasticity, and other markers of neuronal differentiation in neuroblastoma SH-SY5Y cells. It is also decreased in APL cell differentiation and in breast cancer cells. In addition, reduced levels of *miR-17* also promotes the macrophage inflammatory responses such as infiltration, phagocytosis, and pro-inflammatory cytokine secretion. *miR-17-92* also functions cooperatively with *miR-106b-25* to regulate mice spermatogonia differentiation.

Since miRNA can target many genes and pathways, it would be challenging to use them for disease treatment and prevention. Additionally, certain miRNAs such as *miR-452* has opposite effects based on the tumor types. Additional studies are required to dissect their functions and human relevance.

Funding

This study is supported by grants funded by National Institutes of Health U01CA179582 and R01CA222490.

References

Al Tanoury, Z., Piskunov, A., Andriamoratsiresy, D., Gaouar, S., Lutzing, R., Ye, T., et al. (2014). Genes involved in cell adhesion and signaling: A new repertoire of retinoic acid receptor target genes in mouse embryonic fibroblasts. *Journal of Cell Science, 127*(Pt. 3), 521–533.

Annibali, D., Gioia, U., Savino, M., Laneve, P., Caffarelli, E., & Nasi, S. (2012). A new module in neural differentiation control: Two microRNAs upregulated by retinoic acid, *miR-9* and *-103*, target the differentiation inhibitor ID2. *PLoS One*, 7(7), e40269.

Beveridge, N. J., Tooney, P. A., Carroll, A. P., Tran, N., & Cairns, M. J. (2009). Down-regulation of *miR-17* family expression in response to retinoic acid induced neuronal differentiation. *Cellular Signalling, 21*(12), 1837–1845.

Bushue, N., & Wan, Y. J. (2010). Retinoid pathway and cancer therapeutics. *Advanced Drug Delivery Reviews, 62*(13), 1285–1298.

Careccia, S., Mainardi, S., Pelosi, A., Gurtner, A., Diverio, D., Riccioni, R., et al. (2009). A restricted signature of miRNAs distinguishes APL blasts from normal promyelocytes. *Oncogene, 28*(45), 4034–4040.

Chan, D. W., Mak, C. S., Leung, T. H., Chan, K. K., & Ngan, H. Y. (2012). Down-regulation of Sox7 is associated with aberrant activation of Wnt/b-catenin signaling in endometrial cancer. *Oncotarget, 3*(12), 1546–1556.

Chen, J., Cai, T., Zheng, C., Lin, X., Wang, G., Liao, S., et al. (2017). *MicroRNA-202* maintains spermatogonial stem cells by inhibiting cell cycle regulators and RNA binding proteins. *Nucleic Acids Research, 45*(7), 4142–4157.

Chen, K. C., Hsi, E., Hu, C. Y., Chou, W. W., Liang, C. L., & Juo, S. H. (2012). *MicroRNA-328* may influence myopia development by mediating the PAX6 gene. *Investigative Ophthalmology & Visual Science, 53*(6), 2732–2739.

Chen, C. Z., Li, L., Lodish, H. F., & Bartel, D. P. (2004). MicroRNAs modulate hematopoietic lineage differentiation. *Science, 303*(5654), 83–86.

Chen, Y., Meng, L., Yu, Q., Dong, D., Tan, G., Huang, X., et al. (2015). The *miR-134* attenuates the expression of transcription factor FOXM1 during pluripotent NT2/D1 embryonal carcinoma cell differentiation. *Experimental Cell Research, 330*(2), 442–450.

Chen, P. H., Shih, C. M., Chang, W. C., Cheng, C. H., Lin, C. W., Ho, K. H., et al. (2014). *MicroRNA-302b*-inhibited E2F3 transcription factor is related to all trans retinoic acid-induced glioma cell apoptosis. *Journal of Neurochemistry, 131*(6), 731–742.

Chen, Y., & Stallings, R. L. (2007). Differential patterns of microRNA expression in neuroblastoma are correlated with prognosis, differentiation, and apoptosis. *Cancer Research, 67*(3), 976–983.

Chen, S., Sun, Y. Y., Zhang, Z. X., Li, Y. H., Xu, Z. M., & Fu, W. N. (2017). Transcriptional suppression of microRNA-27a contributes to laryngeal cancer differentiation via GSK-3beta-involved Wnt/beta-catenin pathway. *Oncotarget, 8*(9), 14708–14718.

Chen, Z. H., Wang, W. T., Huang, W., Fang, K., Sun, Y. M., Liu, S. R., et al. (2017). The lncRNA *HOTAIRM1* regulates the degradation of PML-RARA oncoprotein and myeloid cell differentiation by enhancing the autophagy pathway. *Cell Death and Differentiation, 24*(2), 212–224.

Chen, S., Wang, K., & Wan, Y. J. (2010). Retinoids activate RXR/CAR-mediated pathway and induce CYP3A. *Biochemical Pharmacology, 79*(2), 270–276.

Chen, S., Yu, J., Lv, X., & Zhang, L. (2017). HOXA9 is critical in the proliferation, differentiation, and malignancy of leukaemia cells both in vitro and in vivo. *Cell Biochemistry and Function, 35*(7), 433–440.

Czajka, A. A., Wojcicka, A., Kubiak, A., Kotlarek, M., Bakula-Zalewska, E., Koperski, L., et al. (2016). Family of *microRNA-146* regulates RARbeta in papillary thyroid carcinoma. *PLoS One, 11*(3), e0151968.

Dai, T., Wu, Y., Leng, A. S., Ao, Y., Robel, R. C., Lu, S. C., et al. (2003). RXRalpha-regulated liver SAMe and GSH levels influence susceptibility to alcohol-induced hepatotoxicity. *Experimental and Molecular Pathology*, *75*(3), 194–200.

Das, E., & Bhattacharyya, N. P. (2014). *MicroRNA-432* contributes to dopamine cocktail and retinoic acid induced differentiation of human neuroblastoma cells by targeting NESTIN and RCOR1 genes. *FEBS Letters*, *588*(9), 1706–1714.

Das, S., Foley, N., Bryan, K., Watters, K. M., Bray, I., Murphy, D. M., et al. (2010). MicroRNA mediates DNA demethylation events triggered by retinoic acid during neuroblastoma cell differentiation. *Cancer Research*, *70*(20), 7874–7881.

Dave, V. P., Ngo, T. A., Pernestig, A. K., Tilevik, D., Kant, K., Nguyen, T., et al. (2019). MicroRNA amplification and detection technologies: Opportunities and challenges for point of care diagnostics. *Laboratory Investigation*, *99*(4), 452–469.

Delgado, M. D., & Leon, J. (2010). Myc roles in hematopoiesis and leukemia. *Genes & Cancer*, *1*(6), 605–616.

Deng, S., Zhang, Y., Xu, C., & Ma, D. (2015). *MicroRNA-125b-2* overexpression represses ectodermal differentiation of mouse embryonic stem cells. *International Journal of Molecular Medicine*, *36*(2), 355–362.

Ding, Y., Wang, Z. C., Zheng, Y., Hu, Z., Li, Y., Luo, D. F., et al. (2016). C-Myc functions as a competing endogenous RNA in acute promyelocytic leukemia. *Oncotarget*, 7(35), 56422–56430.

Dobrotkova, V., Chlapek, P., Mazanek, P., Sterba, J., & Veselska, R. (2018). Traffic lights for retinoids in oncology: Molecular markers of retinoid resistance and sensitivity and their use in the management of cancer differentiation therapy. *BMC Cancer*, *18*(1), 1059.

Dumas, P. Y., Mansier, O., Prouzet-Mauleon, V., Koya, J., Villacreces, A., Brunet de la Grange, P., et al. (2018). MiR-10a and HOXB4 are overexpressed in atypical myeloproliferative neoplasms. *BMC Cancer*, *18*(1), 1098.

Evangelisti, C., Florian, M. C., Massimi, I., Dominici, C., Giannini, G., Galardi, S., et al. (2009). MiR-128 up-regulation inhibits Reelin and DCX expression and reduces neuroblastoma cell motility and invasiveness. *The FASEB Journal*, *23*(12), 4276–4287.

Fang, Y., Liu, H. X., Zhang, N., Guo, G. L., Wan, Y. J., & Fang, J. (2013). NURBS: A database of experimental and predicted nuclear receptor binding sites of mouse. *Bioinformatics*, *29*(2), 295–297.

Fazi, F., Rosa, A., Fatica, A., Gelmetti, V., De Marchis, M. L., Nervi, C., et al. (2005). A minicircuitry comprised of *microRNA-223* and transcription factors NFI-A and C/EBPalpha regulates human granulopoiesis. *Cell*, *123*(5), 819–831.

Fisher, J. N., Terao, M., Fratelli, M., Kurosaki, M., Paroni, G., Zanetti, A., et al. (2015). MicroRNA networks regulated by all-trans retinoic acid and Lapatinib control the growth, survival and motility of breast cancer cells. *Oncotarget*, *6*(15), 13176–13200.

Foley, N. H., Bray, I., Watters, K. M., Das, S., Bryan, K., Bernas, T., et al. (2011). *MicroRNAs 10a* and *10b* are potent inducers of neuroblastoma cell differentiation through targeting of nuclear receptor corepressor 2. *Cell Death and Differentiation*, *18*(7), 1089–1098.

Galli, R., Paone, A., Fabbri, M., Zanesi, N., Calore, F., Cascione, L., et al. (2013). Toll-like receptor 3 (TLR3) activation induces microRNA-dependent reexpression of functional RARbeta and tumor regression. *Proceedings of the National Academy of Sciences of the United States of America*, *110*(24), 9812–9817.

Ganapathy, A., & Ezekiel, U. (2019). Phytochemical modulation of MiRNAs in colorectal cancer. *Medicines (Basel)*, *6*(2), 48. https://doi.org/10.3390/medicines6020048.

Garzon, R., Pichiorri, F., Palumbo, T., Visentini, M., Aqeilan, R., Cimmino, A., et al. (2007). MicroRNA gene expression during retinoic acid-induced differentiation of human acute promyelocytic leukemia. *Oncogene*, *26*(28), 4148–4157.

Goto, T. (2019). A review of the studies on food-derived factors which regulate energy metabolism via the modulation of lipid-sensing nuclear receptors. *Bioscience, Biotechnology, and Biochemistry, 83*(4), 579–588.

Gyamfi, M. A., He, L., French, S. W., Damjanov, I., & Wan, Y. J. (2008). Hepatocyte retinoid X receptor alpha-dependent regulation of lipid homeostasis and inflammatory cytokine expression contributes to alcohol-induced liver injury. *The Journal of Pharmacology and Experimental Therapeutics, 324*(2), 443–453.

He, Y., Gong, L., Fang, Y., Zhan, Q., Liu, H. X., Lu, Y., et al. (2013). The role of retinoic acid in hepatic lipid homeostasis defined by genomic binding and transcriptome profiling. *BMC Genomics, 14*, 575.

He, L., Thomson, J. M., Hemann, M. T., Hernando-Monge, E., Mu, D., Goodson, S., et al. (2005). A microRNA polycistron as a potential human oncogene. *Nature, 435*(7043), 828–833.

Hu, Y., French, S. W., Chau, T., Liu, H. X., Sheng, L., Wei, F., et al. (2019). RARbeta acts as both an upstream regulator and downstream effector of *miR-22*, which epigenetically regulates NUR77 to induce apoptosis of colon cancer cells. *The FASEB Journal, 33*(2), 2314–2326.

Hu, B., Guo, Y., Chen, C., Li, Q., Niu, X., Guo, S., et al. (2014). Repression of SIRT1 promotes the differentiation of mouse induced pluripotent stem cells into neural stem cells. *Cellular and Molecular Neurobiology, 34*(6), 905–912.

Hu, Y., Liu, H. X., He, Y., Fang, Y., Fang, J., & Wan, Y. J. (2013). Transcriptome profiling and genome-wide DNA binding define the differential role of fenretinide and all-trans RA in regulating the death and survival of human hepatocellular carcinoma Huh7 cells. *Biochemical Pharmacology, 85*(7), 1007–1017.

Hu, Y., Liu, H. X., Jena, P. K., Sheng, L., Ali, M. R., & Wan, Y. J. (2020). *miR-22* inhibition reduces hepatic steatosis via FGF21 and FGFR1 induction. *JHEP Reports, 2*(2), 100093. https://doi.org/10.1016/j.jhepr.2020.100093.

Hu, F., Xu, P., Sun, B., Teng, G., & Xiao, Z. (2017). Deep sequencing reveals complex mechanisms of microRNA regulation during retinoic acid-induced neuronal differentiation of mesenchymal stem cells. *Genomics, 109*(3–4), 302–311.

Hua, S., Kittler, R., & White, K. P. (2009). Genomic antagonism between retinoic acid and estrogen signaling in breast cancer. *Cell, 137*(7), 1259–1271.

Huang, H., Xie, C., Sun, X., Ritchie, R. P., Zhang, J., & Chen, Y. E. (2010). *miR-10a* contributes to retinoid acid-induced smooth muscle cell differentiation. *The Journal of Biological Chemistry, 285*(13), 9383–9389.

Huszar, J. M., & Payne, C. J. (2013). *MicroRNA 146 (Mir146)* modulates spermatogonial differentiation by retinoic acid in mice. *Biology of Reproduction, 88*(1), 15.

Ivey, K. N., & Srivastava, D. (2015). microRNAs as developmental regulators. *Cold Spring Harbor Perspectives in Biology*, 7(7), a008144.

Jauhari, A., Singh, T., & Yadav, S. (2018). Expression of *miR-145* and its target proteins are regulated by *miR-29b* in differentiated neurons. *Molecular Neurobiology, 55*(12), 8978–8990.

Jeannotte, L., Gotti, F., & Landry-Truchon, K. (2016). Hoxa5: A key player in development and disease. *Journal of Developmental Biology, 4*(2), 13. https://doi.org/10.3390/jdb4020013.

Jeker, L. T., Zhou, X., Gershberg, K., de Kouchkovsky, D., Morar, M. M., Stadthagen, G., et al. (2012). *MicroRNA 10a* marks regulatory T cells. *PLoS One*, 7(5), e36684.

Jian, P., Li, Z. W., Fang, T. Y., Jian, W., Zhuan, Z., Mei, L. X., et al. (2011). Retinoic acid induces HL-60 cell differentiation via the upregulation of *miR-663*. *Journal of Hematology & Oncology, 4*, 20.

Jin, J., Britschgi, A., Schlafli, A. M., Humbert, M., Shan-Krauer, D., Batliner, J., et al. (2018). Low autophagy (*ATG*) gene expression is associated with an immature AML blast cell phenotype and can be restored during AML differentiation therapy. *Oxidative Medicine and Cellular Longevity*, *2018*, 1482795.

Khan, S., Wall, D., Curran, C., Newell, J., Kerin, M. J., & Dwyer, R. M. (2015). *MicroRNA-10a* is reduced in breast cancer and regulated in part through retinoic acid. *BMC Cancer*, *15*, 345.

Kiss, M., Czimmerer, Z., Nagy, G., Bieniasz-Krzywiec, P., Ehling, M., Pap, A., et al. (2017). Retinoid X receptor suppresses a metastasis-promoting transcriptional program in myeloid cells via a ligand-insensitive mechanism. *Proceedings of the National Academy of Sciences of the United States of America*, *114*(40), 10725–10730.

Krezel, W., Ruhl, R., & de Lera, A. R. (2019). Alternative retinoid X receptor (RXR) ligands. *Molecular and Cellular Endocrinology*, *491*, 110436.

Le, M. T., Xie, H., Zhou, B., Chia, P. H., Rizk, P., Um, M., et al. (2009). *MicroRNA-125b* promotes neuronal differentiation in human cells by repressing multiple targets. *Molecular and Cellular Biology*, *29*(19), 5290–5305.

Lee, D. Y., Yang, T. L., Huang, Y. H., Lee, C. I., Chen, L. J., Shih, Y. T., et al. (2018). Induction of *microRNA-10a* using retinoic acid receptor-alpha and retinoid x receptor-alpha agonists inhibits atherosclerotic lesion formation. *Atherosclerosis*, *271*, 36–44.

Leyden, J., Stein-Gold, L., & Weiss, J. (2017). Why topical retinoids are mainstay of therapy for acne. *Dermatology and Therapy*, 7(3), 293–304.

Li, D., Chen, L., Zhao, W., Hao, J., & An, R. (2016). *MicroRNA-let-7f-1* is induced by lycopene and inhibits cell proliferation and triggers apoptosis in prostate cancer. *Molecular Medicine Reports*, *13*(3), 2708–2714.

Li, G., Luna, C., Qiu, J., Epstein, D. L., & Gonzalez, P. (2009). Alterations in microRNA expression in stress-induced cellular senescence. *Mechanisms of Ageing and Development*, *130*(11–12), 731–741.

Li, N., Tang, Y., Liu, B., Cong, W., Liu, C., & Xiao, J. (2017). Retinoid acid-induced *microRNA-27b-3p* impairs C2C12 myoblast proliferation and differentiation by suppressing alpha-dystrobrevin. *Experimental Cell Research*, *350*(2), 301–311.

Li, L., Wang, Y., Zhang, X., Song, G., Guo, Q., Zhang, Z., et al. (2018). Deubiquitinase USP48 promotes ATRA-induced granulocytic differentiation of acute promyelocytic leukemia cells. *International Journal of Oncology*, *53*(2), 895–903.

Lin, Y., Li, D., Liang, Q., Liu, S., Zuo, X., Li, L., et al. (2015). *miR-638* regulates differentiation and proliferation in leukemic cells by targeting cyclin-dependent kinase 2. *The Journal of Biological Chemistry*, *290*(3), 1818–1828.

Lin, K. Y., Zhang, X. J., Feng, D. D., Zhang, H., Zeng, C. W., Han, B. W., et al. (2011). *miR-125b*, a target of *CDX2*, regulates cell differentiation through repression of the core binding factor in hematopoietic malignancies. *The Journal of Biological Chemistry*, *286*(44), 38253–38263.

Liu, X., Chen, M., Li, L., Gong, L., Zhou, H., & Gao, D. (2017). Extracellular signal-regulated kinases (ERKs) phosphorylate Lin28a protein to modulate P19 cell proliferation and differentiation. *The Journal of Biological Chemistry*, *292*(10), 3970–3976.

Liu, L., Chen, K., Wu, J., Shi, L., Hu, B., Cheng, S., et al. (2013). Downregulation of *miR-452* promotes stem-like traits and tumorigenicity of gliomas. *Clinical Cancer Research*, *19*(13), 3429–3438.

Liu, H. X., Hu, Y., & Wan, Y. J. (2016). Microbiota and bile acid profiles in retinoic acid-primed mice that exhibit accelerated liver regeneration. *Oncotarget*, 7(2), 1096–1106.

Liu, H. J., Lam, H. C., Baglini, C. V., Nijmeh, J., Cottrill, A. A., Chan, S. Y., et al. (2019). Rapamycin-upregulated *miR-29b* promotes mTORC1-hyperactive cell growth in TSC2-deficient cells by downregulating tumor suppressor retinoic acid receptor beta (RARbeta). *Oncogene*, *38*, 7367–7383.

Liu, B., Liu, C., Cong, W., Li, N., Zhou, N., Tang, Y., et al. (2017). Retinoid acid-induced *microRNA-31-5p* suppresses myogenic proliferation and differentiation by targeting CamkIIdelta. *Skeletal Muscle*, 7(1), 8.

Liu, H. X., Ly, I., Hu, Y., & Wan, Y. J. (2014). Retinoic acid regulates cell cycle genes and accelerates normal mouse liver regeneration. *Biochemical Pharmacology*, *91*(2), 256–265.

Liu, J., Wang, Y., & Lin, L. (2019). Small molecules for fat combustion: Targeting obesity. *Acta Pharmaceutica Sinica B*, *9*(2), 220–236.

Liu, D., Zhong, L., Yuan, Z., Yao, J., Zhong, P., Liu, J., et al. (2019). *miR-382-5p* modulates the ATRA-induced differentiation of acute promyelocytic leukemia by targeting tumor suppressor PTEN. *Cellular Signalling*, *54*, 1–9.

Luu, L., Ciccotosto, G. D., Vella, L. J., Cheng, L., Roisman, L. C., Multhaup, G., et al. (2019). Amyloid precursor protein dimerisation reduces neurite outgrowth. *Molecular Neurobiology*, *56*(1), 13–28.

Manodoro, F., Marzec, J., Chaplin, T., Miraki-Moud, F., Moravcsik, E., Jovanovic, J. V., et al. (2014). Loss of imprinting at the 14q32 domain is associated with microRNA overexpression in acute promyelocytic leukemia. *Blood*, *123*(13), 2066–2074.

Martin, N., Ma, X., & Bernard, D. (2019). Regulation of cellular senescence by retinoid X receptors and their partners. *Mechanisms of Ageing and Development*, *183*, 111131.

Meseguer, S., Mudduluru, G., Escamilla, J. M., Allgayer, H., & Barettino, D. (2011). *MicroRNAs-10a* and *-10b* contribute to retinoic acid-induced differentiation of neuroblastoma cells and target the alternative splicing regulatory factor SFRS1 (SF2/ASF). *The Journal of Biological Chemistry*, *286*(6), 4150–4164.

Mu, P., Han, Y. C., Betel, D., Yao, E., Squatrito, M., Ogrodowski, P., et al. (2009). Genetic dissection of the *miR-17∼92* cluster of microRNAs in Myc-induced B-cell lymphomas. *Genes & Development*, *23*(24), 2806–2811.

Nagy, Z., Baghy, K., Hunyadi-Gulyas, E., Micsik, T., Nyiro, G., Racz, G., et al. (2015). Evaluation of 9-cis retinoic acid and mitotane as antitumoral agents in an adrenocortical xenograft model. *American Journal of Cancer Research*, *5*(12), 3645–3658.

Oda, Y., Nakajima, M., Tsuneyama, K., Takamiya, M., Aoki, Y., Fukami, T., et al. (2014). Retinoid X receptor alpha in human liver is regulated by *miR-34a*. *Biochemical Pharmacology*, *90*(2), 179–187.

Oliveira, L. M., Teixeira, F. M. E., & Sato, M. N. (2018). Impact of retinoic acid on immune cells and inflammatory diseases. *Mediators of Inflammation*, *2018*, 3067126.

Ouimet, M., Ediriweera, H. N., Gundra, U. M., Sheedy, F. J., Ramkhelawon, B., Hutchison, S. B., et al. (2015). *MicroRNA-33*-dependent regulation of macrophage metabolism directs immune cell polarization in atherosclerosis. *The Journal of Clinical Investigation*, *125*(12), 4334–4348.

Pallocca, G., Fabbri, M., Sacco, M. G., Gribaldo, L., Pamies, D., Laurenza, I., et al. (2013). miRNA expression profiling in a human stem cell-based model as a tool for developmental neurotoxicity testing. *Cell Biology and Toxicology*, *29*(4), 239–257.

Park, E. J., Prajuabjinda, O., Soe, Z. Y., Darkwah, S., Appiah, M. G., Kawamoto, E., et al. (2019). Exosomal regulation of lymphocyte homing to the gut. *Blood Advances*, *3*(1), 1–11.

Parsons, X. H. (2012). MicroRNA profiling reveals distinct mechanisms governing cardiac and neural lineage-specification of pluripotent human embryonic stem cells. *Journal of Stem Cell Research & Therapy*, *2*(3), 124. https://doi.org/10.4172/2157-7633.1000124.

Parsons, X. H., Parsons, J. F., & Moore, D. A. (2012). Genome-scale mapping of microRNA signatures in human embryonic stem cell neurogenesis. *Molecular Medicine & Therapeutics*, *1*(2), 105.

Pelosi, A., Careccia, S., Sagrestani, G., Nanni, S., Manni, I., Schinzari, V., et al. (2014). Dual promoter usage as regulatory mechanism of *let-7c* expression in leukemic and solid tumors. *Molecular Cancer Research*, *12*(6), 878–889.

Penvose, A., Keenan, J. L., Bray, D., Ramlall, V., & Siggers, T. (2019). Comprehensive study of nuclear receptor DNA binding provides a revised framework for understanding receptor specificity. *Nature Communications*, *10*(1), 2514.

Phua, S. L., Sivakamasundari, V., Shao, Y., Cai, X., Zhang, L. F., Lufkin, T., et al. (2011). Nuclear accumulation of an uncapped RNA produced by Drosha cleavage of a transcript encoding miR-10b and HOXD4. *PLoS One*, *6*(10), e25689.

Qin, P., Li, L., Zhang, D., Liu, Q. L., Chen, X. R., Yang, H. Y., et al. (2016). Altered microRNA expression profiles in a rat model of spina bifida. *Neural Regeneration Research*, *11*(3), 502–507.

Qin, P., Li, L., Zhang, D., Liu, Q. L., Chen, X. R., Yang, H. Y., et al. (2017). Preliminary investigation of methylation status of microRNA-124a in spinal cords of rat fetuses with congenital spina bifida. *The Journal of Maternal-Fetal & Neonatal Medicine*, *30*(1), 23–28.

Rossi, A., D'Urso, O. F., Gatto, G., Poltronieri, P., Ferracin, M., Remondelli, P., et al. (2010). Non-coding RNAs change their expression profile after Retinoid induced differentiation of the promyelocytic cell line NB4. *BMC Research Notes*, *3*, 24.

Ruhl, R., Krezel, W., & de Lera, A. R. (2018). 9-Cis-13,14-dihydroretinoic acid, a new endogenous mammalian ligand of retinoid X receptor and the active ligand of a potential new vitamin A category: Vitamin A5. *Nutrition Reviews*, *76*(12), 929–941.

Saumet, A., Vetter, G., Bouttier, M., Antoine, E., Roubert, C., Orsetti, B., et al. (2012). Estrogen and retinoic acid antagonistically regulate several microRNA genes to control aerobic glycolysis in breast cancer cells. *Molecular BioSystems*, *8*(12), 3242–3253.

Saumet, A., Vetter, G., Bouttier, M., Portales-Casamar, E., Wasserman, W. W., Maurin, T., et al. (2009). Transcriptional repression of microRNA genes by PML-RARA increases expression of key cancer proteins in acute promyelocytic leukemia. *Blood*, *113*(2), 412–421.

Stoecker, K., Sass, S., Theis, F. J., Hauner, H., & Pfaffl, M. W. (2017). Inhibition of fat cell differentiation in 3T3-L1 pre-adipocytes by all-trans retinoic acid: Integrative analysis of transcriptomic and phenotypic data. *Biomolecular Detection and Quantification*, *11*, 31–44.

Su, X., Gu, X., Zhang, Z., Li, W., & Wang, X. (2020). Retinoic acid receptor gamma is targeted by microRNA-124 and inhibits neurite outgrowth. *Neuropharmacology*, *163*, 107657. https://doi.org/10.1016/j.neuropharm.2019.05.034.

Suzhi, Z., Liang, T., Yuexia, P., Lucy, L., Xiaoting, H., Yuan, Z., et al. (2015). Gap junctions enhance the antiproliferative effect of *microRNA-124-3p* in glioblastoma cells. *Journal of Cellular Physiology*, *230*(10), 2476–2488.

Takahashi, H., Kanno, T., Nakayamada, S., Hirahara, K., Sciume, G., Muljo, S. A., et al. (2012). TGF-beta and retinoic acid induce the microRNA *miR-10a*, which targets *Bcl-6* and constrains the plasticity of helper T cells. *Nature Immunology*, *13*(6), 587–595.

Tang, Q., Chen, Y., Meyer, C., Geistlinger, T., Lupien, M., Wang, Q., et al. (2011). A comprehensive view of nuclear receptor cancer cistromes. *Cancer Research*, *71*(22), 6940–6947.

Tang, G. W., & Russell, R. M. (1990). 13-cis-retinoic acid is an endogenous compound in human serum. *Journal of Lipid Research*, *31*(2), 175–182.

Tay, Y. M., Tam, W. L., Ang, Y. S., Gaughwin, P. M., Yang, H., Wang, W., et al. (2008). *MicroRNA-134* modulates the differentiation of mouse embryonic stem cells, where it causes post-transcriptional attenuation of Nanog and LRH1. *Stem Cells*, *26*(1), 17–29.

Tay, Y., Zhang, J., Thomson, A. M., Lim, B., & Rigoutsos, I. (2008). MicroRNAs to *Nanog*, *Oct4* and *Sox2* coding regions modulate embryonic stem cell differentiation. *Nature*, *455*(7216), 1124–1128.

Tennakoon, J. B., Wang, H., Coarfa, C., Cooney, A. J., & Gunaratne, P. H. (2013). Chromatin changes in dicer-deficient mouse embryonic stem cells in response to retinoic acid induced differentiation. *PLoS One*, *8*(9), e74556.

Terao, M., Fratelli, M., Kurosaki, M., Zanetti, A., Guarnaccia, V., Paroni, G., et al. (2011). Induction of *miR-21* by retinoic acid in estrogen receptor-positive breast carcinoma cells: Biological correlates and molecular targets. *The Journal of Biological Chemistry*, *286*(5), 4027–4042.

Tombolan, L., Zampini, M., Casara, S., Boldrin, E., Zin, A., Bisogno, G., et al. (2015). *MicroRNA-27a* contributes to rhabdomyosarcoma cell proliferation by suppressing RARA and RXRA. *PLoS One*, *10*(4), e0125171.

Tong, M. H., Mitchell, D. A., McGowan, S. D., Evanoff, R., & Griswold, M. D. (2012). Two miRNA clusters, *Mir-17-92 (Mirc1)* and *Mir-106b-25 (Mirc3)*, are involved in the regulation of spermatogonial differentiation in mice. *Biology of Reproduction*, *86*(3), 72.

Tu, J., Zhang, P., Shui Luk, A. C., Liao, J., Chan, W. Y., Qi, H., et al. (2018). *MicroRNA-26b* promotes transition from Kit(−) to Kit(+) mouse spermatogonia. *Experimental Cell Research*, *373*(1–2), 71–79.

van Gils, N., Verhagen, H., & Smit, L. (2017). Reprogramming acute myeloid leukemia into sensitivity for retinoic-acid-driven differentiation. *Experimental Hematology*, *52*, 12–23.

Vecellio, M., Meraviglia, V., Nanni, S., Barbuti, A., Scavone, A., DiFrancesco, D., et al. (2012). In vitro epigenetic reprogramming of human cardiac mesenchymal stromal cells into functionally competent cardiovascular precursors. *PLoS One*, 7(12), e51694.

Wan, Y. J., An, D., Cai, Y., Repa, J. J., Hung-Po Chen, T., Flores, M., et al. (2000). Hepatocyte-specific mutation establishes retinoid X receptor alpha as a heterodimeric integrator of multiple physiological processes in the liver. *Molecular and Cellular Biology*, *20*(12), 4436–4444.

Wan, Y. J., Cai, Y., Lungo, W., Fu, P., Locker, J., French, S., et al. (2000). Peroxisome proliferator-activated receptor alpha-mediated pathways are altered in hepatocyte-specific retinoid X receptor alpha-deficient mice. *The Journal of Biological Chemistry*, *275*(36), 28285–28290.

Wan, Y. J., Han, G., Cai, Y., Dai, T., Konishi, T., & Leng, A. S. (2003). Hepatocyte retinoid X receptor-alpha-deficient mice have reduced food intake, increased body weight, and improved glucose tolerance. *Endocrinology*, *144*(2), 605–611.

Wan, Y. J., Orrison, B. M., Lieberman, R., Lazarovici, P., & Ozato, K. (1987). Induction of major histocompatibility class I antigens by interferons in undifferentiated F9 cells. *Journal of Cellular Physiology*, *130*(2), 276–283.

Wan, Y. J., Wang, L., & Wu, T. C. (1994). The expression of retinoid X receptor genes is regulated by all-trans- and 9-cis-retinoic acid in F9 teratocarcinoma cells. *Experimental Cell Research*, *210*(1), 56–61.

Wang, X. S., Gong, J. N., Yu, J., Wang, F., Zhang, X. H., Yin, X. L., et al. (2012). *MicroRNA-29a* and *microRNA-142-3p* are regulators of myeloid differentiation and acute myeloid leukemia. *Blood*, *119*(21), 4992–5004.

Wang, X., Tan, L., Lu, Y., Peng, J., Zhu, Y., Zhang, Y., et al. (2015). *MicroRNA-138* promotes tau phosphorylation by targeting retinoic acid receptor alpha. *FEBS Letters*, *589*(6), 726–729.

Wang, S., Wang, Z., Lin, S., Zheng, W., Wang, R., Jin, S., et al. (2012). Revealing a natural marine product as a novel agonist for retinoic acid receptors with a unique binding mode and inhibitory effects on cancer cells. *The Biochemical Journal*, *446*(1), 79–87.

Wang, C., Yao, N., Lu, C. L., Li, D., & Ma, X. (2010). Mouse *microRNA-124* regulates the expression of Hes1 in P19 cells. *Frontiers in Bioscience (Elite Edition)*, *2*, 127–132.

Watanabe, M., & Kakuta, H. (2018). Retinoid X receptor antagonists. *International Journal of Molecular Sciences*, *19*(8), 2354. https://doi.org/10.3390/ijms19082354.

Watanabe, K., Yamaji, R., & Ohtsuki, T. (2018). *MicroRNA-664a-5p* promotes neuronal differentiation of SH-SY5Y cells. *Genes to Cells*, *23*(3), 225–233.

Weiss, F. U., Marques, I. J., Woltering, J. M., Vlecken, D. H., Aghdassi, A., Partecke, L. I., et al. (2009). Retinoic acid receptor antagonists inhibit miR-10a expression and block metastatic behavior of pancreatic cancer. *Gastroenterology*, *137*(6) 2136–2145.e7.

Welch, C., Chen, Y., & Stallings, R. L. (2007). *MicroRNA-34a* functions as a potential tumor suppressor by inducing apoptosis in neuroblastoma cells. *Oncogene*, *26*(34), 5017–5022.

Wu, T. C., Wang, L., & Wan, Y. J. (1992). Retinoic acid regulates gene expression of retinoic acid receptors alpha, beta and gamma in F9 mouse teratocarcinoma cells. *Differentiation*, *51*(3), 219–224.

Wu, H., Zhao, J., Fu, B., Yin, S., Song, C., Zhang, J., et al. (2017). Retinoic acid-induced upregulation of *miR-219* promotes the differentiation of embryonic stem cells into neural cells. *Cell Death & Disease*, *8*(7), e2953.

Yang, F., Hu, Y., Liu, H. X., & Wan, Y. J. (2015). *MiR-22*-silenced cyclin A expression in colon and liver cancer cells is regulated by bile acid receptor. *The Journal of Biological Chemistry*, *290*(10), 6507–6515.

Yang, F., Miao, L., Mei, Y., & Wu, M. (2013). Retinoic acid-induced *HOXA5* expression is co-regulated by *HuR* and *miR-130a*. *Cellular Signalling*, *25*(6), 1476–1485.

Yao, J., Zhong, L., Zhong, P., Liu, D., Yuan, Z., Liu, J., et al. (2019). RAS-responsive element-binding protein 1 blocks the granulocytic differentiation of myeloid leukemia cells. *Oncology Research*, *27*(7), 809–818.

Yu, X., Hu, Y., Wu, Y., Fang, C., Lai, J., Chen, S., et al. (2019). The c-Myc-regulated *miR-17-92* cluster mediates ATRA-induced APL cell differentiation. *Asia-Pacific Journal of Clinical Oncology*, *15*(6), 364–370.

Yu, M., Zhang, L., Liu, Y., Liu, D., & Guo, Z. (2019). Retinoic acid induces differentiation of mouse F9 embryonic carcinoma cell by modulating the *miR-485* targeting of *Abhd2*. *International Journal of Molecular Sciences*, *20*(9), 2071. https://doi.org/10.3390/ijms20092071.

Zhang, J., Gao, Y., Yu, M., Wu, H., Ai, Z., Wu, Y., et al. (2015). Retinoic acid induces embryonic stem cell differentiation by altering both encoding RNA and microRNA expression. *PLoS One*, *10*(7), e0132566.

Zhang, L., Ke, F., Liu, Z., Bai, J., Liu, J., Yan, S., et al. (2015). *MicroRNA-31* negatively regulates peripherally derived regulatory T-cell generation by repressing retinoic acid-inducible protein 3. *Nature Communications*, *6*, 7639.

Zhao, Z., Ma, X., Sung, D., Li, M., Kosti, A., Lin, G., et al. (2015). *microRNA-449a* functions as a tumor suppressor in neuroblastoma through inducing cell differentiation and cell cycle arrest. *RNA Biology*, *12*(5), 538–554.

Zhao, J. J., Sun, D. G., Wang, J., Liu, S. R., Zhang, C. Y., Zhu, M. X., et al. (2008). Retinoic acid downregulates microRNAs to induce abnormal development of spinal cord in spina bifida rat model. *Child's Nervous System*, *24*(4), 485–492.

Zheng, Z., Liu, J., Yang, Z., Wu, L., Xie, H., Jiang, C., et al. (2016). *MicroRNA-452* promotes stem-like cells of hepatocellular carcinoma by inhibiting *Sox7* involving Wnt/beta-catenin signaling pathway. *Oncotarget*, 7(19), 28000–28012.

Zhu, D., Pan, C., Li, L., Bian, Z., Lv, Z., Shi, L., et al. (2013). *MicroRNA-17/20a/106a* modulate macrophage inflammatory responses through targeting signal-regulatory protein alpha. *The Journal of Allergy and Clinical Immunology*, *132*(2), 426–436.e428.

Zhuang, L. K., Xu, G. P., Pan, X. R., Lou, Y. J., Zou, Q. P., Xia, D., et al. (2014). *MicroRNA-181a*-mediated downregulation of AC9 protein decreases intracellular cAMP level and inhibits ATRA-induced APL cell differentiation. *Cell Death & Disease*, *5*, e1161.

CHAPTER THIRTEEN

Analysis of vitamin A and retinoids in biological matrices

Lindsay C. Czuba, Guo Zhong, King C. Yabut, Nina Isoherranen*
Department of Pharmaceutics, School of Pharmacy, University of Washington, Seattle, WA, United States
*Corresponding author: e-mail address: ni2@uw.edu

Contents

Abstract

Vitamin A signaling pathways are predominantly driven by the cellular concentrations of *all-trans*-retinoic acid (*at*RA), as the main mechanism of retinoid signaling is *via* activation of retinoic acid receptors. *at*RA concentrations are in turn controlled by

Methods in Enzymology, Volume 637
ISSN 0076-6879
https://doi.org/10.1016/bs.mie.2020.02.010

the storage of vitamin A and enzymatic processes that synthesize and clear *at*RA. This has resulted in the need for robust and highly specific analytical methods to accurately quantify retinoids in diverse biological matrices. Tissue-specific differences in both the quantity of retinoids and background matrix interferences can confound the quantification of retinoids, and the bioanalysis requires high performance instrumentation, such as liquid chromatography mass-spectrometry (LC-MS). Successful bioanalysis of retinoids is further complicated by the innate structural instability of retinoids and their relatively high lipophilicity. Further, *in vitro* experiments with retinoids require attention to experimental design and interpretation to account for the instability of retinoids due to isomerization and degradation, sequential metabolism to numerous structurally similar metabolites, and substrate depletion during experiments. In addition, *in vitro* biological activity is often confounded by residual presence of retinoids in common biological reagents such as cell culture media. This chapter identifies common biological and analytical complexities in retinoid bioanalysis in diverse biological matrices, and in the use of retinoids in cell culture and metabolic incubations. In addition, this chapter highlights best practices for the successful detection and quantification of the vitamin A metabolome in a wide range of biological matrices.

1. Vitamin A and retinoids

Vitamin A (retinol) is a dietary fat-soluble vitamin obtained from animal sources as retinyl esters (RE) while retinaldehyde can be synthesized from provitamin A β-carotene (Fig. 1). Retinol is metabolized in the body (Fig. 1) to retinyl esters for storage and to the important active signaling molecule *all-trans*-retinoic acid (*at*RA) (Blaner et al., 2016; Harrison & Dela Seña's, 2015). *at*RA and its metabolite 4-oxo-*at*RA, are potent transcriptional activators of nuclear retinoic acid receptors (RARs) (Larange & Cheroutre, 2016; Topletz et al., 2015). *at*RA is believed to be responsible for the biological activity of vitamin A, such as regulating expression of genes important in reproduction and development (Clagett-Dame & Knutson, 2011; Duester, 2008; Hogarth, Amory, & Griswold, 2011), in energy homeostasis (Blaner, 2019; Chen & Chen, 2014; Zhao et al., 2012), in maintenance of epithelia (McCullough, Northrop-Clewes, & Thurnham, 1999), in regulation of cell cycle and differentiation (Liu, Ly, Hu, & Wan, 2014; Mezquita & Mezquita, 2019) and in immunity (Grizotte-Lake et al., 2018; Iwata et al., 2004; Larange & Cheroutre, 2016). The fact that the activation of RARs is driven by changes in ligand (*at*RA) concentrations, has resulted in a need for robust analytical methods to detect and quantify retinoids. Historically, retinoid analysis has relied on the signature UV absorbance of retinol and retinyl esters at the absorbance maximum

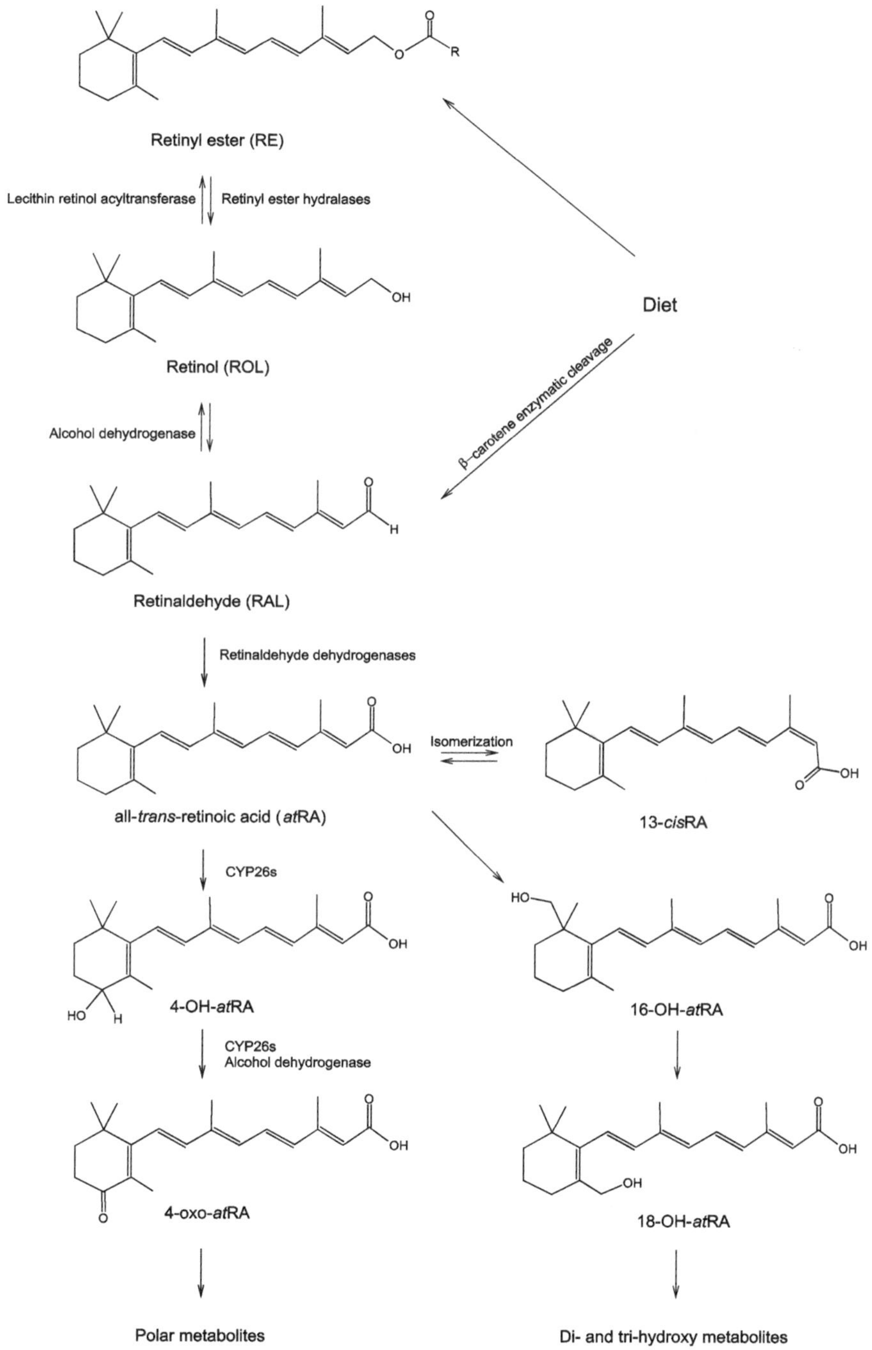

Fg. 1 Vitamin A metabolic pathway with key retinoid metabolite structures and enzymes contributing to the specific steps of *at*RA biosynthesis.

between 325 and 383 nm (Napoli, 1986), and the UV and fluorescence based measurements of retinol have been possible due to the high circulating concentrations (1–3 μM) of retinol (Gatti, Gioia, & Cavrini, 2000; Schmidt, Brouwer, & Nau, 2003). Similarly, in the liver the concentrations of retinol and retinyl esters, predominantly in the form of retinyl palmitate, range from 1.2 to 161 nmol/g liver and 12.6 nmol/g to 42.8 μmol/g liver, respectively (Zhong, Kirkwood, et al., 2019), making their UV detection feasible. In contrast, *at*RA concentrations are over 100-fold lower in circulation and tissues with measured concentrations around 1–10 nM making quantitative measurements of *at*RA challenging. Additionally, in several matrices multiple isomers of RA are present, such as 13-*cis*RA (Fig. 1), requiring robust chromatographic separation to distinguish structurally similar analytes (Arnold, Amory, Walsh, & Isoherranen, 2012). Likely due to these challenges, *at*RA and its metabolites have historically been rarely measured in animal or human tissues and plasma. The advances in liquid chromatography-mass spectrometry (LC-MS) technologies over the past 10–15 years have, however, enabled the detection and quantification of *at*RA in various tissues (Kane, Chen, Sparks, & Napoli, 2005; Kane, Folias, Wang, & Napoli, 2008). In addition, improvements in LC column quality and ultra-high-pressure liquid chromatography (UHPLC) have allowed development of methods that reproducibly separate the isomers of RA. With these advances, measurement of retinoid concentrations in tissues has become feasible for a trained analytical or bioanalytical chemist, and methods to quantify a variety of retinoids including retinol, retinyl esters (RE), *at*RA and its isomers, 4-oxo-RA and stereoisomers of 4-OH-RA have been developed (Arnold et al., 2012; Jones, Pierzchalski, Yu, & Kane, 2015; Shimshoni et al., 2012; Zhong, Kirkwood, et al., 2019). However, even with the profound analytical advancements, sample handling and preparation and sample quantity are critical determinants for successful bioanalysis of retinoids. This chapter will cover retinoid bioanalytical methods with an emphasis on best-practices to minimize inherent concerns related to analyte instability, biological matrix effects and ion suppression, matrix and/or reagent interferences, and substrate depletion.

2. Stability of retinoids in stock solutions and samples

2.1 Stability of *all-trans* retinoic acid in solutions

Light sensitivity (instability) (Napoli, 1986) and sensitivity to heat and oxidation are well documented characteristics of retinoids (Runge & Heger, 2000).

To avoid degradation, retinoids in powder form should be stored in a desiccator at −80 °C in dark amber ampules or vials under inert gas. Similarly, stock solutions of retinoids should be stored in dark amber vials at −80 °C preferably under inert gas. Since retinoids are rapidly degraded and practically insoluble in aqueous solutions, stocks should be prepared in organic solvent for maximum long-term stability. The reported solubility of *at*RA in common organic solvents is summarized in Table 1. As retinoids have an extensive conjugated double bond structure (Fig. 1), they have a signature UV absorbance peak at wavelengths 325–383 nm. This absorbance enables the use of HPLC-UV to monitor the stability of stock solutions and simultaneously detect isomerization and/or formation of auto-oxidation products (Fig. 2) (Napoli, 1986). Due to the sensitivity of retinoid stocks to handling, the stability of stock solutions should be monitored routinely before use. Fig. 2A and B show the stability of *at*RA at −80 °C in the three common solvents often used in biological analyses. While *at*RA was stable in ethanol and methanol, it degraded in DMSO over 3 weeks with degradation observed already at 1 week, as shown by the emergence of degradation products in the chromatograms. This strongly suggests that if *at*RA is used in cell culture in DMSO, the solutions should be prepared fresh for treatments. Overall, researchers should confirm the stability of their retinoid stocks if stock solutions are stored for longer than a month. The degradation of *at*RA stocks was also assessed in aqueous buffers and 5–30% degradation was observed after 30 min at 37 °C (Fig. 2C) suggesting that retinoids

Table 1 Solubility of *all-trans*-retinoic acid in common organic solvents as reported by major chemical suppliers.

	Sigma	Cayman chemicals	Santa cruz	Toronto research chemicals
DMSO	40 mg/mL	20 mg/mL	25 mg/mL	Slightly
Methanol	–	–	–	Slightly
Ethanol	2.7 mg/mL[a]	0.5 mg/mL	10 mg/mL	–
Acetone	–	–	30 mg/mL	–
Chloroform	Slightly	–	40 mg/mL	Slightly
Ethyl acetate	–	–	–	Slightly
Dimethylformamide	–	20 mg/mL	20 mg/mL	–

[a]95% Ethanol, solubility not reported.

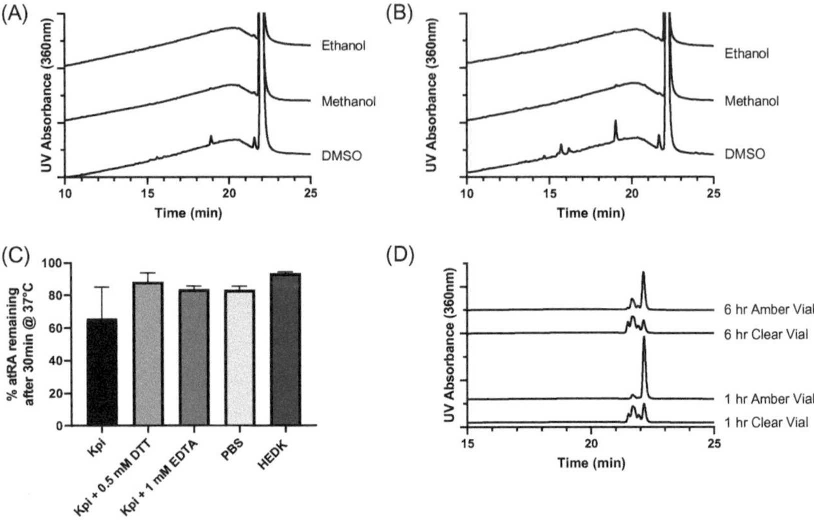

Fig. 2 Stability of *at*RA in organic solvents, aqueous buffers and with light exposure. HPLC-UV chromatograms of stock solutions of 1 mg/mL of *at*RA in ethanol, methanol or DMSO analyzed at 0 weeks (A) and after 3 weeks (B) of storage in −80 °C in amber vials are shown. Panel (C) shows the analysis of *at*RA stability in aqueous buffers. The stability of *at*RA (100 nM) was monitored over 30 min at 37 °C (common incubation temperature) in five different aqueous buffers: 100 mM potassium phosphate buffer, pH 7.4 (Kpi); Kpi with 0.5 mM dithiothreitol (DTT); Kpi with 0.1 mM ethylenediaminetetraacetic acid (EDTA); phosphate buffered saline (PBS); and 10 mM Hepes, 100 mM potassium chloride, 0.1 mM EDTA, 0.5 mM DTT, pH 8 (HEDK). The effect of light exposure on *at*RA (1 mg/mL in ethanol) stability in amber or clear glass vials is shown in (D). The solutions of *at*RA were directly exposed to ambient light (standard laboratory fluorescent lights) for an hour or 6 h and the formation of degradation products and isomerization was monitored by HPLC-UV. In panels (A), (B), and (D) the Y-axis baselines are offset for visual comparison of the appearance of degradation products.

should not be kept in aqueous solutions and buffers for prolonged periods of time. Further, as previously shown (Kane & Napoli, 2010), *at*RA degraded rapidly to various isomerization products after just 1 h of light exposure in clear vials. Isomerization was minor after 1 h of light exposure when retinoids were kept in amber vials, but significant isomerization was detected after 6 h even in amber vials (Fig. 2D). The rapid light-induced degradation of *at*RA is consistent with the observed instability in human serum when samples were not protected from light during collection (Arnold et al., 2012) and shows that work with retinoids should always be conducted under yellow lights.

Serum-free conditions are commonly used in cell culture, but consistent with the instability in aqueous solutions, retinoids have been reported to be unstable and prone to isomerization in the absence of protein-binding in culture (Sharow, Temkin, & Asson-Batres, 2012; Tsukada, Schröder, Seltmann, Orfanos, & Zouboulis, 2002). Specifically, a recent study quantified the recovery of retinoids at 24 h in cell-free incubations in serum-containing and serum-free culture medium (Sharow et al., 2012). In the absence of serum, *at*RA recovery was $<30\%$ of the starting concentration, with more than half of the total RA observed as isomers. Furthermore, retinol recovery was low ($<5\%$ recovery) in the cell-free incubations lacking serum protein due to the formation of unidentified degradation products. The recoveries of all the analytes were significantly higher in parallel cell-free incubations using serum-supplemented medium. Similarly, supplementation of serum-free medium with 6 mg/mL BSA resulted in improved recovery of all retinoids at 24 h, although *at*RA appeared to be more stable in these experiments than retinol (Sharow et al., 2012). Taken together, these experiments suggest that cell culture media should be prepared fresh prior to experiments and retinoids should not be stored in cell culture media.

2.2 Stability of retinoids in biological matrices

While the stability of retinoids in aqueous solutions is compromised by light and storage conditions, biological matrices such as serum described above may stabilize retinoids from isomerization and oxidation allowing for storage of biological samples without significant compromise in sample quality. For example, only minor ($<20\%$) light-induced retinoid depletion and isomerization was observed in human liver homogenates when monitored for up to 1 h (Zhong, Kirkwood, et al., 2019). This contrasts with human serum which is susceptible to light-induced retinoid degradation (Arnold et al., 2012). In addition, for the analysis of human or mammalian blood, serum, or plasma the initial sample exposure to ambient or clinic lighting during sample collection should be minimized. As blood draws under yellow lights are usually not feasible, it is best practice to protect samples from light by wrapping the tubes with aluminum foil and/or using covered ice containers. The initial light protection is likely more significant to retinoid recovery than the sample preparation time needed for isolating serum or plasma and the ultimate storage time of the samples. As shown in Fig. 3A, the time from the blood draw to the separation of light protected

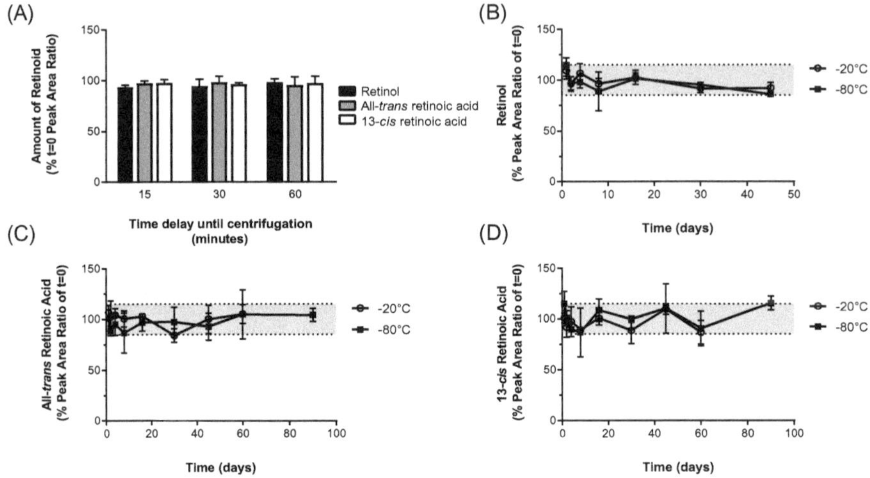

Fig. 3 The stability of retinol, *at*RA and 13-*cis*RA in blood and plasma. Panel (A) shows the effect of time from blood draw to separation of plasma from blood on retinoid stability. The blood was drawn to heparin vacutainer tubes (green top), immediately light protected with foil, and inverted. Blood samples were kept on ice for 0, 15, 30, and 60 min post-blood draw, followed by centrifugation for 10 min at $1000 \times g$ to separate plasma. Panels (B–D) show the analysis of storage conditions on retinol (B), *at*RA (C) and 13-*cis*RA (D) recovery from human plasma samples. For these analyses aliquots of plasma samples collected above were stored at −20 or −80 °C for up to 90 days. At designated time points, aliquots were thawed and analyzed by LC-MS/MS. For analysis, samples were spiked with an internal standard mixture (*at*RA-d_5, 13-*cis*RA-d_5, and ROL-d_8), protein precipitated with acetonitrile (1:1 v/v), centrifuged and analyzed as described in the protocols section. Analyte peak areas were normalized to their respective internal standard. Data is shown as means ± SD for three replicates per time point and peak area ratios are presented as % of $t = 0$ days. Shaded regions denote ±15% around $t = 0$ starting measurement to indicate range of acceptable analytical variability.

plasma by centrifugation did not impact the concentration (peak area ratio relative to the corresponding internal standard) of retinol, *at*RA, or 13*cis*RA. This finding agrees with similar studies which have demonstrated that retinol measurements from plasma are not impacted significantly by sample preparation spin-delay (Abraham et al., 2019; Craft, Brown, & Smith, 1988; Cuerq et al., 2015; Hankinson et al., 1989).

Due to the inherent lability of retinoids in aqueous solutions, the stability of retinoids in biological samples upon storage is often of concern. Yet, as shown in Fig. 3B–D, when human plasma samples were stored at −20 or −80 °C up to 3 months the storage temperature and time had minimal impact on the stability of the three analyzed retinoids. Similar data has been reported demonstrating that serum retinol was stable for 5 months when stored at −20 °C and for 28 months at −70 °C (Craft et al., 1988). If sample stability is of concern, stable-labeled retinoids, such as *at*RA-d_5, can be added to samples prior to storage to monitor for isomerization or degradation in the samples during storage (Arnold et al., 2012). However, this approach may result in challenges in quantitative analyses due to the presence of isotope labeled compounds in the samples. Notably, when we repeated an analysis of *at*RA concentrations in a set of liver samples stored in −80 °C for 1 year, the concentrations measured after a year-long storage were within analytical variability, further suggesting that retinoids in biological matrices are stable when properly stored and handled.

3. Considerations in the bioanalysis of retinoids in tissue

3.1 Analytical considerations

Quantification of endogenous tissue *at*RA concentrations is challenging, due to the low concentrations of *at*RA, poor ionization and extensive fragmentation in the MS, and need for powerful separation due to existence of retinoic acid isomers (Arnold et al., 2012; Jones et al., 2015; Zhong, Kirkwood, et al., 2019). In contrast to *at*RA, retinol and retinyl ester concentrations are relatively high in tissues, especially in the liver, making analysis and quantification feasible by a variety of methods. A major challenge in the development of bioanalytical methods for retinoid analysis is the lack of biological tissues devoid of retinoids that could be used as a blank matrix. Due to the lack of a blank matrix, the identity of the presumed retinoid peaks even in spiked samples cannot be easily confirmed and the recovery of retinoids or magnitude of matrix effects in given tissue cannot be determined. Modern LC-MS methods have improved the confidence in peak identification *via* incorporation of MS^3 analysis (MS/MS/MS) into analyte identification in addition to multiple reaction monitoring (MRM) (Arnold, Kent, Hogarth, Schlatt, et al, 2015; Jones et al., 2015; Zhong, Kirkwood, et al., 2019). This is critical as several matrix components elute close to or simultaneously with *at*RA or its isomers in LC-MS/MS runs and show identical MS/MS transitions with *at*RA. As illustrated in Fig. 4 for several tissues, other matrix components can easily be mis-analyzed as *at*RA unless careful attention is paid to peak identification and confirmation. This is less of a concern in LC-MS analysis of retinol and retinyl esters due to their much higher concentrations than *at*RA in most tissues (Fig. 5).

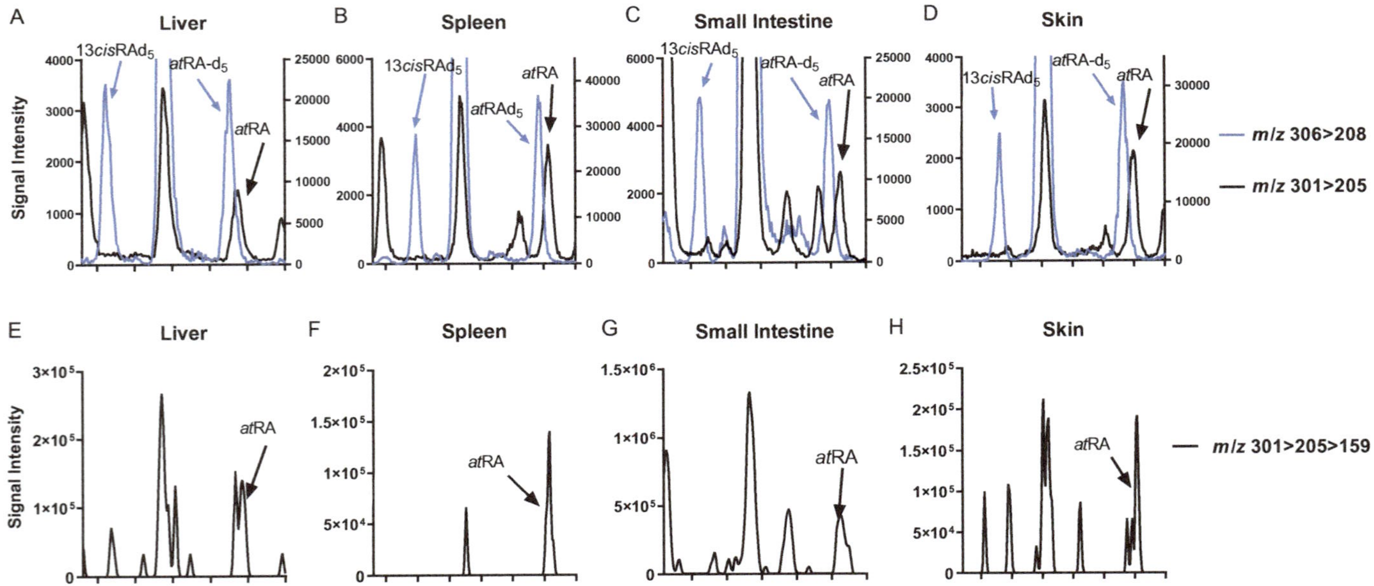

Fig. 4 Detection of *at*RA and 13-*cis*RA in various mouse tissues by different mass spectrometric methods. Panels (A–D) show the detection of endogenous *at*RA and 13-*cis*RA and their corresponding deuterium labeled internal standards (*at*RA-d_5 and 13-*cis*RA-d_5) by multiple reaction monitoring (MRM) from liver (A), spleen (B), small intestine (C) and skin (D) of mice fed normal chow. The blue lines show the internal standard trace (transition m/z 306 > 208) and black lines show the endogenous *at*RA and 13-*cis*RA trace (transition m/z 301 > 205). Panels (E–H) show the corresponding MS^3 traces for liver (E), spleen (F), small intestine (G) and skin (H) for the samples analyzed in (A–D). All samples were processed with the liquid–liquid extraction method described in Section 5. The identification and confirmation of the correct peak for *at*RA in MRM chromatograms for quantification relies on internal standard retention time together with the *at*RA peak shown in MS^3 chromatogram.

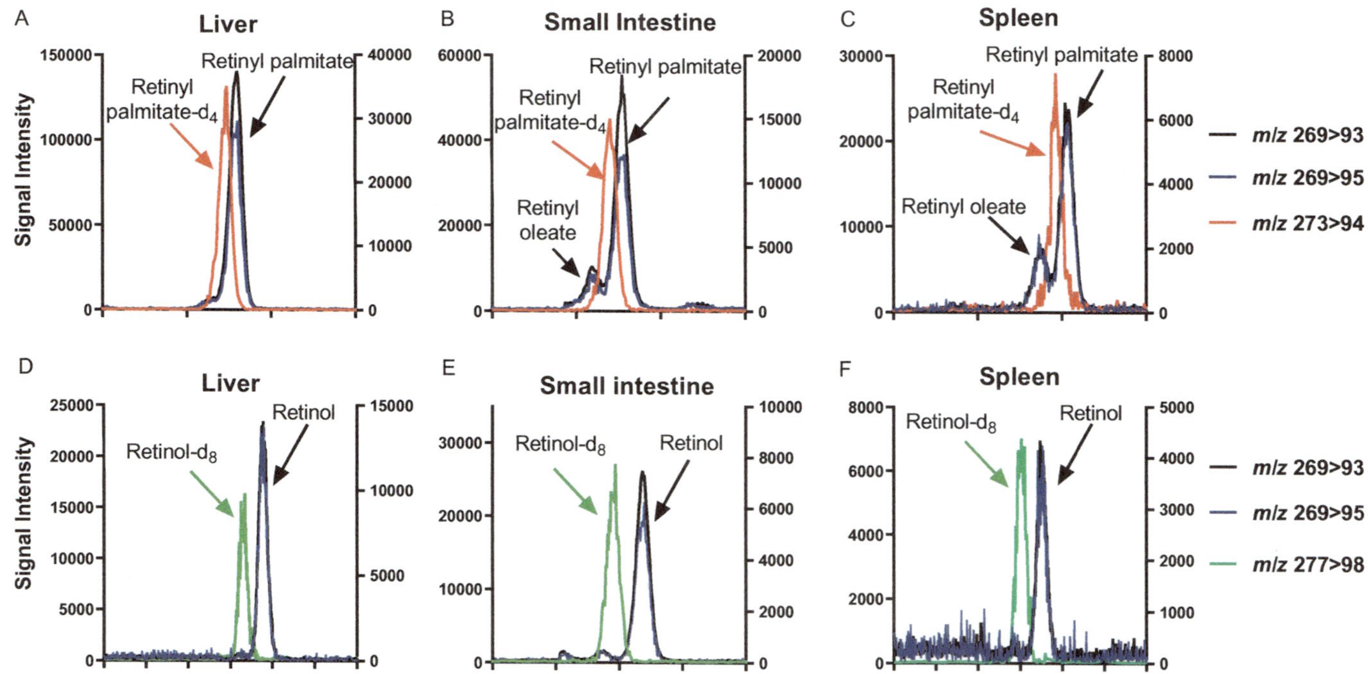

Fig. 5 Detection of retinyl palmitate and retinol in various mouse tissues by LC-MS/MS. The detection of retinyl palmitate and retinyl oleate from mouse liver (A), small intestine (B) and spleen (C) is shown with the black trace depicting MS/MS transition *m/z* 269 > 93 and the blue trace depicting transition *m/z* 269 > 95. The internal standard (retinyl palmitate-d_4) is shown as a red trace depicting MS/MS transition *m/z* 273 > 94. The detection of retinol from mouse liver (D), small intestine (E) and spleen (F) is shown with the black trace depicting MS/MS transition *m/z* 269 > 93 and the blue trace depicting transition *m/z* 269 > 95. The internal standard (retinol-d_8) is shown as a red trace depicting MS/MS transition *m/z* 277 > 98. All samples were prepared using liquid–liquid extraction as described in Section 5.

Different biological matrices and tissues may contain different endogenous compounds like hormones and lipids as is clearly evident by comparison of the interfering peaks within the chromatograms from different tissues (Fig. 4). Therefore, accurate identification and quantification of retinoids relies heavily on good separation of these interfering peaks from the target analyte peak. Even within the same organ but from different donors, the lipid content and concentrations of various endogenous compounds is expected to vary, causing matrix related differences in retinoid recovery, observable interferences in chromatograms and ion suppression occurring in the MS. For example, in three independent experiments, we compared the recovery/ion suppression of isotope-labeled internal standards, *at*RA-d_5 and 13*cis*RA-d_5, between serum, liver, skin, and colon tissues (Fig. 6). The recovery/ion suppression following extraction of *at*RA-d_5 and 13*cis*RA-d_5 was similar between liver and serum, but the overall signal for *at*RA-d_5 and 13*cis*RA-d_5 was 30–50% lower in skin and colon when compared to serum (Fig. 6). Similarly, the signal for retinyl palmitate-d_4 in the colon and small intestine was about 30% of the signal for the same concentration in serum. In contrast, the retinol-d_8 signal was similar regardless of the matrix (Fig. 7), demonstrating that the matrix effect is retinoid specific. These tissue-dependent differences in the signals are likely due to ion suppression caused by coeluting matrix components but could also be due to different absolute recovery of the retinoids in the different tissues. Hence, to correct for this variability in ion suppression/recovery, use of isotope labeled internal standards (*e.g.* retinol palmitate-d_4) and LC-MS/MS analysis is highly recommended. Chemically different internal standards such as retinyl acetate will not correct for tissue specific ion suppression due to the different retention times and likely will not have identical extraction characteristics as endogenous retinyl esters. Isotope labeled internal standards are compatible with LC-MS/MS methods, are structurally identical with the target analyte, and thus, serve as powerful tools in correcting for matrix effects. This concept was shown in a previous study demonstrating that by using an isotope-labeled internal standard, the slopes of retinoid standard curves built with charcoal treated blank human serum and mouse liver homogenates were similar (Zhong, Kirkwood, et al., 2019). In addition, retention time similarity between the target retinoid peak and the isotope labeled internal standard peak is helpful in identifying the analyte peak for quantification, especially when multiple interfering peaks are present in the chromatograms (Fig. 4).

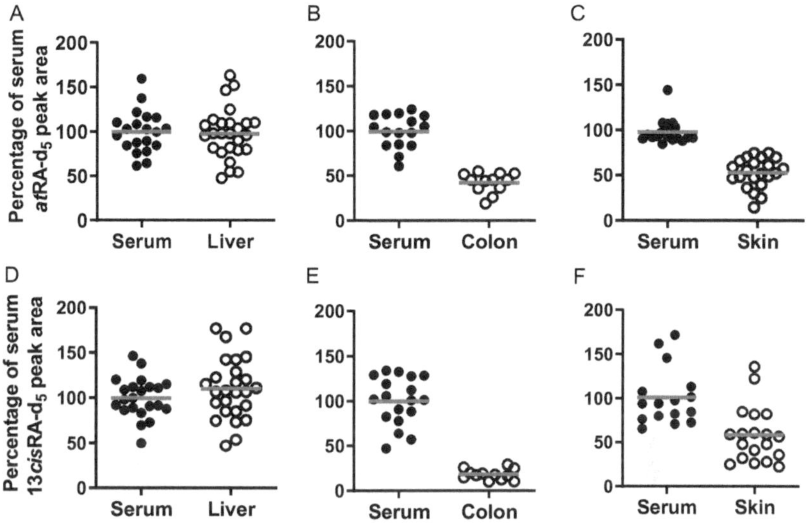

Fig. 6 The impact of tissue matrix on retinoic acid detection. Each panel represents an independent experiment in which *at*RA and 13-*cis*RA were measured in mouse tissues (liver, colon and skin). Known amount of *at*RA and 13-*cis*RA were spiked into charcoal treated serum to make standard curves and quality control (QC) samples for quantification. Each open circle represents a target tissue sample obtained from an individual mouse and each filled circle represents a RA-spiked serum sample. Within each independent measurement, the same amount of IS mixture (*at*RA-d_5 and 13*cis*RA-d_5) was added into all samples (RA-spiked serum samples and tissue samples). Samples were processed with a liquid–liquid extraction method as described in Section 5 and analyzed with LC-MS/MS methods described in Table 3. The LC-MS injection volume was the same for all samples. The average internal standard peak area for labeled retinoid spiked serum samples was calculated and the peak area ratio between the labeled analyte in a given sample and the mean internal standard peak area in serum was calculated. The data is reported as the percentage of the mean spiked serum value. In the absence of any matrix effect and with uniform recovery from each tissue the percentage value should be 100% on average. Panels (A–C) are for *at*RA-d_5 and panels (D) and (E) are for 13*cis*RA-d_5. Blue lines indicate mean values for each group.

3.2 Biological considerations

Endogenous retinoids are most commonly measured in mouse and human tissues. Mouse models are often used for retinoid-related studies and retinoids have been detected and quantified in a variety of mouse tissues including serum, liver, adipose, kidney, heart, spleen, skin, brain and testis (Arnold et al., 2012; Arnold, Kent, Hogarth, Schlatt, et al., 2015; Obrochta, Kane, & Napoli, 2014; Schmidt et al., 2003; Zhong, Hogarth, et al, 2019). Mouse liver retinyl palmitate concentration is the highest among all the retinoids

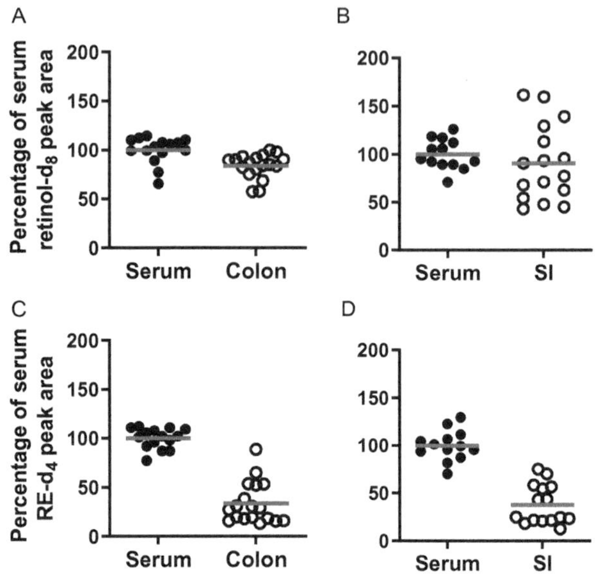

Fig. 7 The impact of tissue matrix on retinol and retinyl palmitate detection. Each panel represents an independent experiment in which retinol and retinyl palmitate were measured in mouse tissues (colon and small intestine (SI)). Known amount of retinol and retinyl palmitate was spiked into charcoal treated serum to make standard curves and quality control (QC) samples for quantification. Each open circle represents a target tissue sample obtained from an individual mouse and each filled circle represents a retinoid-spiked serum sample. For each independent measurement the same amount of internal standard mixture (retinol-d_8 and retinyl palmitate-d_4) was added into all samples (retinoid-spiked serum samples and tissue samples). Samples were processed with a liquid–liquid extraction method as described in Section 5 and analyzed with LC-MS/MS methods described in Table 3. The LC-MS injection volume was the same for all samples. The data is reported as the percentage that the internal standard peak area in each individual tissue sample is of the average internal standard peak area in the retinol and retinyl ester-spiked serum samples. In the absence of any matrix effect and with uniform recovery from each tissue the percentage value should be 100% on average. Panels (A) and (B) are for retinol-d_8 and panels (C) and (D) are for retinyl palmitate-d_4. Blue lines indicate mean values of normalized internal standard peak area.

in various tissues, at the range of 100–2000 nmol/g liver while retinyl palmitate concentrations in other mouse tissues are about 0.1–15 nmol/g tissue. Tissue retinol concentrations are in the range of 0.2–400 nmol/g and liver retinol is the highest among all tissues. *at*RA concentrations are about 1–30 pmol/g tissue while other retinoic acids and downstream metabolites, such as 13*cis*RA and 4-oxo-*at*RA, are not detected in mouse organs (Obrochta et al., 2014; Zhong, Hogarth, et al., 2019). Based on these

Table 2 Recommended sample preparation methods and sample quantities for the analysis of retinoids in tissues.

Tissue	Target retinoid	Recommended amount of tissue	Sample preparation method
Liver	Retinol and retinyl palmitate	10–40 mg (mouse and human)	Acetonitrile precipitation of proteins
Liver	Retinoic acid	100–120 mg (mouse); 50–100 mg (human)	One-round extraction with hexanes from acidified tissue homogenate
Liver	4-oxo-retinoic acid	50–100 mg (human)	Ethyl acetate extraction from acidified tissue homogenate
Serum	Retinol, retinyl palmitate, and retinoic acid	50–100 μL (mouse)	Two-round extraction with hexanes from first basic and then acidified tissue homogenate
Serum	Retinol and retinoic acid	50–100 μL (human)	Acetonitrile precipitation of proteins
Testis	Retinol, retinyl palmitate, and retinoic acid	30–60 mg (mouse)	Acetonitrile precipitation of proteins
Other organs such as heart, spleen, skin	Retinol, retinyl palmitate, and retinoic acid	120–140 mg	Two-round extraction with hexanes from first basic and then acidified tissue homogenate

concentrations, 50–100 mg of tissue is usually the minimum quantity needed for measurement of *at*RA concentrations while retinol and retinyl palmitate can often be measured from smaller quantities (Table 2).

For human tissues, retinoid concentrations have mainly been quantified in serum and plasma (Arnold et al., 2012; Jones et al., 2015; Kane et al., 2005; Kane, Folias, & Napoli, 2008; Napoli, 1986). In addition, retinoid concentrations have been reported in human liver and testis (Arnold, Kent, Hogarth, Griswold, et al., 2015; Zhong, Kirkwood, et al., 2019), while data for other human tissues is lacking. Compared to mouse liver, human liver has similar retinol and retinyl palmitate concentrations, but higher 13*cis*RA and *at*RA concentrations (Zhong, Kirkwood, et al., 2019). Similarly, human serum retinol levels are similar to those in mice but *at*RA and 13*cis*RA concentrations are higher (Arnold et al., 2012). In human serum, 9*cis*RA and 9,13-di-*cis*-RA are also detected. Furthermore, the downstream metabolites of *at*RA and 13*cis*RA, 4-oxo-*at*RA and 4-oxo-13*cis*RA, have been detected in human liver and serum and their concentrations were either similar or higher than parent RA concentrations. In contrast, neither of the 4-oxo-RA metabolites was detected in mouse tissue (Zhong, Kirkwood, et al., 2019) with the available tissue quantities.

Other than tissue and species differences, animal strain, sex and diet are also important factors that can contribute to variations in retinoid concentrations. For example, mouse strain and diet were shown to impact retinoid concentrations in mouse tissues (Obrochta et al., 2014). In another study, serum retinol and liver *at*RA concentrations were found to be significantly lower in female than male mice (Zhong, Hogarth, et al., 2019). However, sex differences in retinoid concentrations were not observed in human livers in a study including 50 healthy liver samples (Zhong, Kirkwood, et al., 2019). Taken together, when conducting retinoid analysis in biological tissues, species, tissue type, strain, sex and diet should be considered when assessing the amount of sample needed and in selection of the appropriate analytical and sample preparation methods (described in Section 6).

4. Considerations for the use of retinoids in *in vitro* biological experiments

4.1 The presence of retinoids in biological reagents

The use of retinoids in cell culture experiments is vital for understanding the basics of retinoid biology and signaling pathways, but there is little consensus for the proper procedure for their use *in vitro*. A common source of variation between studies is the degree of serum supplementation in cell culture medium varying from serum-free to 5% or 10% fetal bovine serum (FBS). While this variability in the FBS content may affect retinoid stability as described above in Section 2.1, it can also alter the overall treatment conditions. FBS is a chemically ill-defined biological matrix that includes a wide range of serum proteins, fatty acids, hormones, and cytokines (Yao & Asayama, 2017). In addition, standard FBS usually contains measurable amounts of retinol, in the range of 10–50 nM for a 10% FBS containing media (Baltes, Nau, & Lampen, 2004; Randolph & Simon, 1997;

Sharow et al., 2012). In sensitive experiments, this contamination with retinol may drive or alter biological responses. For example, one study estimated that 5% FBS-supplemented medium contained 25 nM retinol but lacked *at*RA. Yet, this media retinol content resulted in steady-state intracellular content of retinoic acid of 25–50 nM (Randolph & Simon, 1997), a concentration within normal biologically active range. Whether this concentration is sufficient to confound cell culture experiments is not well defined. In HepG2 cells with baseline treatment and culture in the presence of 5 nM *at*RA, RA was still found to dose-dependently induce CYP26A1 and RARβ (Tay, Dickmann, Dixit, & Isoherranen, 2010). One may argue that *in vivo* cells will never be in a retinoid free environment, and therefore culturing cells at some level of baseline retinoids may be advised. Yet, variability in media retinol concentrations may cause unnecessary variability in experimental results of retinoid signaling and confound dose-response experiments. Thus, contaminating retinol in reagents should be considered when interpreting the vast range of biological responses downstream of vitamin-A signaling pathways.

Fatty-acid free, or charcoal-stripped FBS is a retinoid depleted alternative for cell culture experiments that may help to overcome the lot-to-lot variation in baseline retinoid medium content. Serum alternatives, such as neonatal calf serum, may have different retinoid content and should be quantified using HPLC-UV or LC-MS methods described herein prior to use. Thus, the routine use of these products in variety of applications such as sample preparation buffers, blocking of antibody non-specific binding, flow cytometry cell sorting buffers, and sample tube coating deserves consideration to ensure that interference is minimized. Best practices may include opting for charcoal-stripped or light-exposed reagents to minimize residual retinol background signals and downstream interference such as with retinoid quantification from flow-cytometry isolated cells.

4.2 Retinoid uptake and depletion in cell culture

Retinoids are generally assumed to have high permeability across cell membranes (Noy, 1992b), and this is likely true for retinol and for *at*RA at acidic pH when *at*RA is unionized. However, the pKa of *at*RA is 4.76 (PubChem) and therefore at physiological pH *at*RA is negatively charged, likely limiting the permeability of *at*RA across membranes (Noy, 1992a). Indeed, in Caco-2 monolayers, *at*RA was shown to have relatively poor permeability (Jing et al., 2017), suggesting that *at*RA may need active transport for uptake into cells. Another process that may limit apparent cell permeability and uptake of *at*RA to cells is protein binding in the media. Apart from maintaining the stability of retinoids in solution, FBS supplementation and specifically albumin concentration, appears to be a critical determinant of both retinoid uptake into cells and the rate of metabolism/isomerization (Noy, 1992a, 1992b; Randolph & Simon, 1997; Rundhaug, Gubler, Sherman, Blaner, & Bertram, 1987; Topletz et al., 2015; Tsukada et al., 2002). Thus, when using retinoids *in vitro*, investigators should consider the potential impact of binding proteins on the data.

4.3 *In vitro* metabolic studies

The enzymes of the CYP26 family are the most important contributors to the clearance of *at*RA (Isoherranen & Zhong, 2019). However, recombinant CYP26 enzymes are notoriously difficult to express and purify *in vitro*, and at present no reports exist of successful expression of recombinant CYP26 enzymes in *E. coli* or in yeast. However, recombinant CYP26 enzymes have been successfully expressed and isolated in microsomal preparations from insect cells (Lutz et al., 2009; Topletz et al., 2012; Zhong, Ortiz, Zelter, Nath, & Isoherranen, 2018), while others have used mammalian cells for recombinant expression of CYP26s (Gomaa, Bridgens, Aboraia, et al., 2011; Gomaa, Bridgens, Veal, et al., 2011; Krivospitskaya et al., 2012). *at*RA metabolism has also been extensively studied in human liver microsomes (HLMs) and in variety of recombinant CYPs including CYP3A4, CYP3A7 and CYP2C8 (Isoherranen & Zhong, 2019; Marill, Cresteil, Lanotte, & Chabot, 2000; Thatcher, Zelter, & Isoherranen, 2010; Topletz, Zhong, & Isoherranen, 2019). The best characterized metabolite of *at*RA is 4-OH-*at*RA (Fig. 1), which is essentially always detected in CYP incubations, but detection of additional metabolites such as 18-OH-*at*RA, and 4-oxo-*at*RA is also often reported. With recombinant CYP26s, identified primary *at*RA metabolites include hydroxylation in the 4-, 16- and 18-carbons (Fig. 8). However, with CYP26 enzymes, formation of sequential metabolites from *at*RA such as 4-oxo-*at*RA, and tentatively identified 4,16- and 4,18-diOH-*at*RA is also prevalent (Lutz et al., 2009; Shimshoni et al., 2012; Topletz et al., 2015). The characterization of sequential metabolism of retinoids *in vitro* remains challenging due in part to the lack of reference standards for the sequential metabolites and assessment of the biological role of such metabolites requires "made to order" or "in house" synthesis which can be resource intensive. Furthermore, a typical incubation

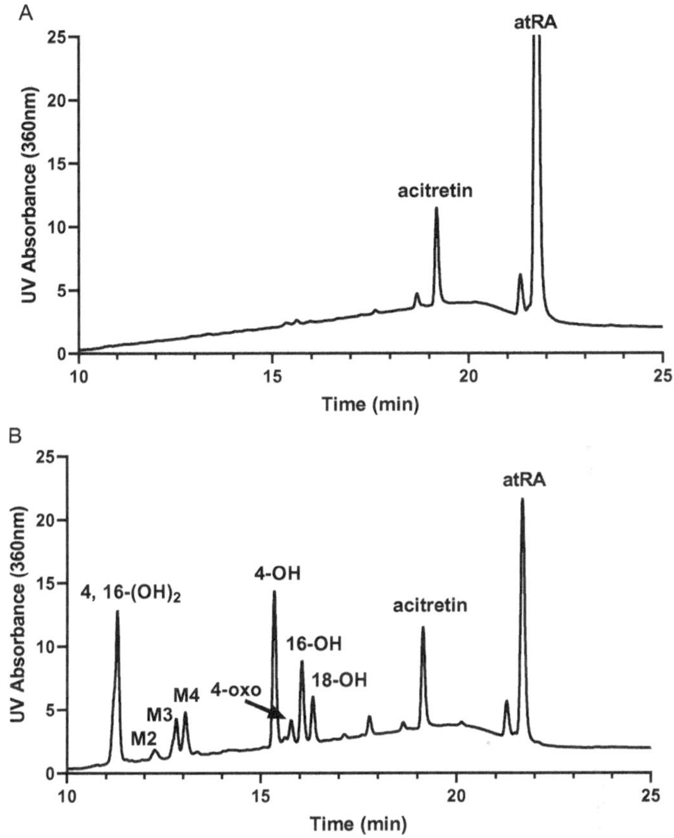

Fig. 8 Detection of the metabolites of *at*RA formed by cytochrome P450 enzymes. Panel (A) shows the HPLC-UV trace of *at*RA incubated with CYP26B1 in the absence of cofactor (NADPH). For panel (B) the metabolites were generated by incubating 500nM *at*RA with 5nM of recombinant CYP26B1 reconstituted with P450 reductase and in the presence of NADPH as described in Section 6.3. Primary metabolites were identified as 4-OH-, 16-OH-, and 18-OH-*at*RA based on previous experiments although no reference standard for 16-OH-RA exists. The di-hydroxylation sequential metabolites M2, M3, and M4 have been tentatively identified but their structures have not been confirmed *via* synthetic standards. Acitretin was used as an internal standard (Lutz et al., 2009; Thatcher et al., 2011).

reaction can generate several primary and sequential metabolites. Therefore, methods to achieve chromatographic separation of multiple structurally similar retinoid metabolites is critical for HPLC-UV analysis (Fig. 5) as well as LC-MS/MS due to the identical molecular weights of the isomeric hydroxylation products. Due to the high affinity of *at*RA to CYP26 enzymes and the high catalytic activity, substrate and primary metabolite depletion are major concerns in kinetic studies with *at*RA. The depletion of *at*RA can be >20% within 2 min when incubations are done at concentrations around K_m with just 2 nM of CYP26A1, resulting in inaccurate kinetic estimates if parameters are not corrected for substrate depletion. Adjusting protocols for lower enzyme concentrations is one approach to address substrate and metabolite depletion but the ability to detect product formation at low enzyme concentrations becomes a limitation with current quantification methods. As such, substrate depletion approach may be the most appropriate for determining accurate kinetic constants for *at*RA in *in vitro* incubations, and this approach has been used in several studies (Lutz et al., 2009; Shimshoni et al., 2012; Topletz et al., 2015). Retinoid instability and sensitivity to auto-oxidation also contribute to challenges with *in vitro* incubations.

5. Extraction methods for retinoids from biological matrices

All steps should be performed under yellow light. In general, for the majority of tissues liquid–liquid extraction is recommended, while the acetonitrile precipitation method can be used for certain analyses such as detecting retinoids in human plasma or serum (see Table 2).

5.1 Equipment

1. Workspace with yellow light
2. Borosilicate glass tubes (16 × 150 mm, 15 mL)
3. Borosilicate glass Pasteur pipettes
4. Universal glass homogenizer with glass pestle
5. Handheld electric drill
6. Bead mill homogenizer with ceramic beads
7. Analytical balance
8. Benchtop centrifuge
9. Microfuge tubes (1.7 mL)
10. Benchtop vortex
11. Nitrogen evaporator
12. UHPLC coupled to high end Tandem mass spectrometer
13. Amber Autosampler vials and caps
14. Ascentis Express RP-Amide column (2.7 μm; 150 mm × 2.1 mm; Sigma Aldrich)
15. Ascentis Express RP-Amide, Guard Cartridge (2.7 μm; 5 mm × 2.1 mm; Sigma Aldrich)

5.2 Chemicals and reagents

1. Blank human serum (DC Mass Spect Gold MSG 4000, Golden West Biologics)
2. Powdered retinoid stocks (Sigma Aldrich; purity $\geq$98%)
3. Stable-labeled retinoid internal standards (Cambridge Isotope Laboratories, Inc. or Toronto Research Chemicals)
   a. *All-trans* retinoic acid-d_5
   b. *13-cis* retinoic acid-d_5
   c. Retinyl palmitate-d_4
   d. Retinol-d_8
4. 0.25 M Potassium hydroxide (KOH) dissolved in ethanol
5. Acetonitrile (Optima LC/MS Grade)
6. Methanol (Optima LC/MS Grade)
7. Water (Optima LC/MS Grade)
8. Formic acid (Optima LC/MS Grade)
9. 4 N Hydrochloric acid (HCl)
10. Hexanes (Certified ACS)
11. Ethyl acetate (Optima)
12. Ethanol (200 proof)
13. 0.9% NaCl solution
14. Nitrogen gas

5.3 Two-round liquid–liquid extraction method for retinol, RE and RA

5.3.1. Weigh tissues; recommended tissue weights are listed in Table 3

5.3.2. Homogenize with 5 × tissue weight of saline with universal glass homogenizer on ice

5.3.3. Transfer tissue homogenate to glass tubes using glass Pasteur pipettes and keep glass tubes on ice

5.3.4. Spike in internal standard to tissue homogenates

5.3.5. Add 2 mL of 0.25 M KOH dissolved in ethanol and vortex

5.3.6. Add 10 mL of hexanes to extract

5.3.7. Centrifuge at low speed for 5 min

5.3.8. Transfer upper organic layer (retinol and RE fraction) to a new glass tube and evaporate at 25 °C under gentle N_2 flow; Dry glass tubes are kept on ice until reconstitution

5.3.9. Add 120 μL 4 M HCl to the lower aqueous layer and vortex

5.3.10. Add 10 mL hexanes to extract

Table 3 Representative LC-MS/MS parameters for retinoid analysis including target ion transitions and instrument specific ionization parameters (Topletz et al., 2015; Zhong, Kirkwood, et al., 2019).

Ionization	Analyte	MRM *m/z* (MS^3 *m/z*)	[a]CE/DP/EP (V)	[a]GS1/2	LC conditions
APCI Positive ion	Retinol	269 > 93, 95	30/35/4	70/70	*Sample injection*: 4 µL *Column*: Amide column at 40 °C, 0.4 mL/min *LC conditions:* (A) H_2O + 0.1% FA (B) ACN + 0.1% FA • 0–2 min at 60% B • Gradient 60–66% B by 9.2 min • Gradient 66–100% B by 13 min • Return to starting conditions at 23 min
	Retinol-d_8	277 > 98, 102	30/35/4	70/70	
	Retinyl palmitate	269 > 93, 95	30/35/4	70/70	
	Retinyl palmitate-d_4	273 > 94, 98	30/35/4	70/70	
	All-trans-retinoic acid and 13-*cis*-retinoic acid	301 > 205 (301 > 205 > 159)	17/80/10	80/0	*Sample injection*: 20 µL *Column*: Amide column at 25 °C, 0.4 mL/min *LC conditions:* (A) H_2O + 0.1% FA (B) ACN + 0.1% FA • 0–3 min at 70% B • Gradient 70–78% B by 12 min • Increase to 100% B for 3 min • Return to starting conditions at 15 min
	All-trans-retinoic acid-d_5 and 13-*cis*-retinoic acid-d_5	301 > 205 (306 > 208 > 162)	17/80/10	80/0	

	4-oxo-*all-trans* retinoic acid and 4-oxo-13*cis*-retinoic acid	315 > 137, 147, 241	29,26,20/50/10	60/70	*Sample injection*: 20 µL *Column*: Amide column at 25 °C, 0.5 mL/min. *LC conditions:* (A) H_2O + 0.1% FA (B) ACN + 0.1% FA • 0–2 min at 40% B • Gradient 40–67% B by 9 min • Increase to 100% B for 2 min • Return to starting conditions
	4-oxo-*all-trans* retinoic acid-d_3	318 > 137	29/50/10	60/70	
ESI Negative ion	*All-trans*-retinoic acid and 13-*cis*-retinoic acid	299 > 255	−22/−140/−7	60/60	*Sample injection*: 10 µL *Column*: C18 column at 40 °C, 0.35 mL/min. *LC conditions:* (A) H_2O + 0.1% FA (B) ACN + 0.1% FA • 0–3 min at 10% B • Gradient 50–95% B by 7.6 min • Return to starting conditions at 10.1 min
	All-trans-retinoic acid-d_5 and 13-*cis*-retinoic acid-d_5	304 > 260	−22/−140/−7	60/60	
	4-hydroxy retinoic acid	315 > 253	−28/−80/−7	60/60	
	4-hydroxy retinoic acid-d_6	321 > 259	−28/−80–7	60/60	
	4-oxo-retinoic acid	313 > 269	−22/−95/−5	60/60	
	4-oxo- retinoic acid-d_3	316 > 272	−10/−100/−5	60/60	

[a]APCI, atmospheric pressure chemical ionization; ESI, electrospray ionization; CE, collision energy; DP, declustering potential; EP, entrance potential; GS, Gas.

5.3.11. Centrifuge at low speed for 5 min

5.3.12. Transfer upper organic layer (RA fraction) to a new glass tube and evaporate at 25 °C under gentle N_2 flow; Dry glass tubes are kept on ice until reconstitution

5.3.13. Reconstitute retinol and retinyl ester fraction with 100–300 μL acetonitrile and transfer to amber MS vials for LC-MS/MS analysis

5.3.14. Reconstitute RA fraction with 60 μL acetonitrile, transfer to amber MS vials and add 40 μL water for LC-MS/MS analysis

5.4 One-round liquid–liquid extraction method for RA

Note: An ethyl acetate extraction (Section 5.6) is necessary for the extraction of RA metabolites as this one-round liquid–liquid extraction procedure will not extract the 4-oxo-RA and 4-OH-RA metabolites

5.4.1. Prepare samples the same as in steps 5.3.1–5.3.4 in method Section 5.3

5.4.2. Add 2 mL of acetonitrile + 1% formic acid to homogenates and vortex

5.4.3. Add 10 mL of hexanes to extract

5.4.4. Centrifuge at low speed for 5 min

5.4.5. Transfer upper organic layer (RA fraction) to a new glass tube and evaporate at 25 °C under gentle N_2 flow; Dry glass tubes are kept on ice until reconstitution

5.4.6. Reconstitute RA fraction with 60 μL acetonitrile, transfer to amber MS vials and add 40 μL water for LC-MS/MS analysis

5.5 Acetonitrile precipitation method

5.5.1 Weigh tissues (recommended tissue weights are listed in Table 2)

5.5.2 Homogenize with 4–5 × tissue w/v saline with universal glass homogenizer on ice or with bead mill homogenizer using ceramic beads

- **A.** *For liver tissue:* Homogenization/dilution with a saline volume equivalent to 25 × tissue weight is recommended
- **B.** *For serum:* Saline dilution is not needed

5.5.3 Transfer tissue homogenates to microfuge tubes and keep on ice

5.5.4 Add internal standard and gently vortex

5.5.5 Add ice-cold acetonitrile

- **A.** *For liver tissue*: The ratio of acetonitrile to tissue homogenates is recommended to be 15:1 to 20:1
- **B.** *For non-liver tissues*: The ratio of acetonitrile to tissue homogenates is recommended to be 1 to 1. Intensely vortex and leave on ice for 10 min

5.5.6 Centrifuge at 18,000 × *g* at 4 °C for 30 min

5.5.7 Transfer supernatant to amber MS vials for LC-MS/MS analysis. If tissue is *liver* do the following:

A. *For retinol measurement*: Transfer supernatant to amber MS vials for LC-MS/MS analysis

B. *For retinyl ester measurement*: Transfer a 10-fold diluted supernatant to amber MS vials for LC-MS/MS analysis

5.6 Ethyl acetate extraction for 4-oxo-RA

The following steps can be added after 5.3.12 or 5.4.5 to extract 4-oxo-RA

5.6.1 Add 10 mL of ethyl acetate to aqueous layer to extract

5.6.2 Add 2 mL of deionized water to facilitate phase separation

5.6.3 Centrifuge at low speed for 20 min

5.6.4 Transfer upper organic layer (4-oxo-RA fraction) to a new glass tube and evaporate at 25 °C under gentle N_2 flow; Dry glass tubes are kept on ice until reconstitution

5.6.5 Reconstitute 4-oxo-RA fraction with 60 μL acetonitrile, transfer to amber MS vials and add 40 μL water for LC-MS/MS analysis

5.7 LC-MS/MS instrumentation and conditions

While proper handling and extraction of retinoids is a critical component for optimal quantitation of retinoids, analytical sensitivity is heavily dependent on the instrumentation and separation by liquid chromatography. While many options exist for highly abundant retinoids, such as for serum retinol measurements (Gatti et al., 2000; Kane, Folias, & Napoli, 2008; Napoli, 1986; Schmidt et al., 2003), other analytes require high performance instrumentation and extraction methods (Arnold et al., 2012; Jones et al., 2015; Kane et al., 2005; Kane, Folias, Wang, et al., 2008; Kane & Napoli, 2010; Shimshoni et al., 2012; Topletz et al., 2015; Zhong, Kirkwood, et al., 2019). Currently, MS/MS using a triple-quadrupole instruments provides the highest sensitivity for detecting RA derivatives and metabolites (Kane & Napoli, 2010). Such instruments provide options for multiple-reaction monitoring, ion fragmentation energies, and ion selection. In addition, MS^3 is valuable for confirming the identity of RA and its isomers and can be set up to monitor product ions of RA from the second precursor ions (Zhong, Kirkwood, et al., 2019). Analyte separation can be obtained by coupling MS to a high-performance LC (HPLC) or ultra-high performance LC (UHPLC). The optimal analytical method is unique for each compound, and multiple LC-MS/MS methods are needed in order to quantify and separate common retinoids in a single sample. Generally, retinol, retinyl esters, *at*RA, 13-*cis*RA, and 4-oxo-RA isomers can be quantified using atmospheric pressure chemical ionization (APCI) and positive-ion mode (Zhong, Kirkwood, et al., 2019). The analytes can be separated using an amide column, such as an Ascentis Express RP Amide column (2.7 μm; 150 mm × 2.1 mm) with a corresponding pre-column and mobile phase consisting of (A) water with 0.1% formic acid (FA) and (B) ACN with 0.1% FA. However, RA isomers, retinol and retinyl esters, and 4-oxo-RA require different flow rates, mobile phase gradients, and collision energies among other parameters and thus require separate methods, some of which are summarized in Table 3. RA metabolites such as 4-OH-RA can be separated using a robust C18 column in either positive or negative electrospray ionization (Topletz et al., 2015). In addition, if using stable-labeled internal standard such as *at*RA-d_5, the corresponding transitions need to account for the impact of the deuterium label in increasing the monitored *m/z*.

6. *In vitro* characterization of *all-trans* retinoic acid metabolism

6.1 Equipment

1. Yellow light workspace
2. Borosilicate glass tubes (16 × 150 mm, 15 mL)
3. Borosilicate glass Pasteur pipettes
4. Analytical balance
5. Benchtop centrifuge
6. Microfuge tubes (1.7 mL)
7. Benchtop vortex
8. 37 °C Shaking water bath
9. Water bath nitrogen gas evaporator
10. HPLC coupled to UV or mass spectrometer
11. Amber Autosampler vials and caps
12. Zorbax extended C18 column, 2.1 × 100 mm, 3.5 μM (Agilent)
13. SecurityGuard cartridge C18, 4.0 × 3.0 mm (Phenomenex)

6.2 Chemicals and reagents

1. *All-trans* retinoic acid stock in methanol (100× of final substrate concentration)
2. 100 μM acitretin stock in methanol (internal standard for HPLC-UV)
3. 10 μM 4-oxo-d_3 (internal standard for LC-MS/MS)

4. 4-OH-*at*RA and 4-oxo-*at*RA stocks in methanol for standards
5. CYP26 microsomal fractions quantified by CO-difference spectra, human liver microsomes (HLMs) with protein content quantified, or commercial insect cell expressed recombinant CYPs
6. Purified recombinant rat cytochrome P450 reductase (CPR for recombinant enzyme incubations, not needed for human liver microsomes or commercial insect cell microsomes co-expressed with reductase)
7. Acetonitrile (Optima HPLC Grade)
8. Ethyl acetate (Analytical grade)
9. Nitrogen gas
10. 100 mM potassium phosphate buffer, pH 7.4 (Kpi)

6.3 Metabolic incubation protocol

6.3.1. Prepare microsomal solutions

- *For recombinant enzyme microsomes*: Prepare a 500 pmol/mL: 1000 pmol/mL (1:2) solution of CYP26 to CPR in Kpi; vortex and allow to sit at room temperature for 5 min before putting back on ice.
- *For HLMs*: Prepare a 20 mg/mL HLMs solution (optimized for reaction volume)
- *For commercial recombinant enzymes*: Prepare a 1000 pmol/mL supersomes or insect cell microsomes solution

6.3.2. For each incubation reaction, pipette buffer, CYP26/CPR solution, and *at*RA into a tall 15 mL borosilicate glass tubes on ice in the following volumes:

- *For recombinant enzyme microsomes*:
  - 880 μL of Kpi
  - 10 μL of CYP26/CPR solution
  - 10 μL of *at*RA stock in methanol (100 × of final incubation concentration)
- *For HLMs:*
  - 880 μL of Kpi
  - 10 μL of HLMs
  - 10 μL of *at*RA stock in methanol (100 × of final incubation concentration)
- *For commercial recombinant enzyme microsomes:*
  - 885 μL of Kpi
  - 5 μL Supersomes
  - 10 μL of *at*RA stock in methanol (100 × of final incubation concentration)

6.3.3. Vortex well and pre-incubate at 37 °C for 5 min in a shaking water bath

6.3.4. Initiate reactions with 100 μL of 10 mM NADPH (or Kpi for no cofactor control) for a final volume of 1 mL

6.3.5. Vortex and incubate reaction

- *For recombinant CYP26 enzyme microsomes*: 2 min incubation
- *For HLMs*: 30 min incubation
- *For commercial recombinant enzymes:* 10 min incubation

6.3.6. After the indicated time in 6.3.5, add 4 mL of ethyl acetate to stop the reaction, vortex well and keep on ice

6.3.7. Add 3 μL of 100 μM acitretin (LC-UV) or 2 μL of 5 μM 4-oxo-d_3 (LC-MS/MS) and vortex well

6.3.8. Centrifuge at low speed for 15 min for phase separation

6.3.9. Transfer upper organic layer to a separate borosilicate glass tube using a glass Pasteur pipette

6.3.10. Evaporate organic fraction at 25 °C under gentle nitrogen flow

6.3.11. Keep evaporated tubes on ice

6.3.12. Reconstitute in 150 μL of methanol and transfer to MS vials

6.3.13. Store samples at −20 °C until analysis *via* HPLC-UV or LC-MS/MS

7. Summary

Recent advancements in instrument sensitivity and extraction methods using stable-labeled internal standards have dramatically improved the quantitative analysis of retinoids in a wide range of human and rodent tissues. However, the biological variation of tissue and serum retinoids and variability in interfering matrix components and ion suppression poses a challenge for the analysis of retinoids from complex matrices, such as liver tissue, *via* mass spectrometry. Furthermore, *in vitro* experiments with retinoids have unique considerations that can impact experimental interpretation. For example, routine use of BSA and/or FBS can dramatically contribute to the baseline intracellular retinoid content in cells, which may be present in adequate quantities to elicit a biological response for sensitive measures. In addition, protein binding plays a critical role in the stability of retinoids in media and impacts the uptake and metabolic elimination in cell culture, which in some instances can result in retinoid depletion. Substrate depletion is also a major concern in characterizing metabolite kinetics from enzyme incubations and should be considered in the design of experiments and interpretation of results. In conclusion, the successful bioanalysis of vitamin A and retinoids requires careful handling and preparation of stocks, reagents, and samples throughout both experimental and analytical procedures.

Acknowledgments

This work was supported by grants from the National Institutes of Health: 5R01GM111772-06, 5T32DK007247-42, 2T32GM007750-41.

References

Abraham, R. A., Agrawal, P. K., Acharya, R., Sarna, A., Ramesh, S., Johnston, R., et al. (2019). Effect of temperature and time delay in centrifugation on stability of select biomarkers of nutrition and non-communicable diseases in blood samples. *Biochemia Medica, 29*(2), 359–371. https://doi.org/10.11613/BM.2019.020708.

Arnold, S. L., Amory, J. K., Walsh, T. J., & Isoherranen, N. (2012). A sensitive and specific method for measurement of multiple retinoids in human serum with UHPLC-MS/MS. *Journal of Lipid Research, 53*(3), 587–598. https://doi.org/10.1194/jlr.D019745.

Arnold, S. L., Kent, T., Hogarth, C. A., Griswold, M. D., Amory, J. K., & Isoherranen, N. (2015). Pharmacological inhibition of ALDH1A in mice decreases all-trans retinoic acid concentrations in a tissue specific manner. *Biochemical Pharmacology, 95*(3), 177–192. https://doi.org/10.1016/j.bcp.2015.03.001.

Arnold, S. L., Kent, T., Hogarth, C. A., Schlatt, S., Prasad, B., Haenisch, M., et al. (2015). Importance of ALDH1A enzymes in determining human testicular retinoic acid concentrations. *Journal of Lipid Research, 56*(2), 342–357. https://doi.org/10.1194/jlr.M054718.

Baltes, S., Nau, H., & Lampen, A. (2004). All-trans retinoic acid enhances differentiation and influences permeability of intestinal Caco-2 cells under serum-free conditions. *Development, Growth & Differentiation, 46*(6), 503–514. https://doi.org/10.1111/j.1440-169x.2004.00765.x.

Blaner, W. S. (2019). Vitamin A signaling and homeostasis in obesity, diabetes, and metabolic disorders. *Pharmacology and Therapeutics, 197*, 153–178. https://doi.org/10.1016/j.pharmthera.2019.01.006.

Blaner, W. S., Li, Y., Brun, P. J., Yuen, J. J., Lee, S. A., & Clugston, R. D. (2016). Vitamin A absorption, storage and mobilization. *Sub-Cellular Biochemistry, 81*, 95–125. https://doi.org/10.1007/978-94-024-0945-1_4.

Chen, W., & Chen, G. (2014). The roles of vitamin A in the regulation of carbohydrate, lipid, and protein metabolism. *Journal of Clinical Medicine, 3*(2), 453–479. https://doi.org/10.3390/jcm3020453.

Clagett-Dame, M., & Knutson, D. (2011). Vitamin A in reproduction and development. *Nutrients, 3*(4), 385–428. https://doi.org/10.3390/nu3040385.

Craft, N. E., Brown, E. D., & Smith, J. C. (1988). Effects of storage and handling conditions on concentrations of individual carotenoids, retinol, and tocopherol in plasma. *Clinical Chemistry, 34*(1), 44–48.

Cuerq, C., Peretti, N., Chikh, K., Mialon, A., Guillaumont, M., Drai, J., et al. (2015). Overview of the in vitro stability of commonly measured vitamins and carotenoids in whole blood. *Annals of Clinical Biochemistry, 52*(2), 259–269. https://doi.org/10.1177/0004563214542471.

Duester, G. (2008). Retinoic acid synthesis and signaling during early organogenesis. *Cell, 134*(6), 921–931. https://doi.org/10.1016/j.cell.2008.09.002.

Gatti, R., Gioia, M. G., & Cavrini, V. (2000). Analysis and stability study of retinoids in pharmaceuticals by LC with fluorescence detection. *Journal of Pharmaceutical and Biomedical Analysis*, *23*(1), 147–159. https://doi.org/10.1016/S0731-7085(00)00285-5.

Gomaa, M. S., Bridgens, C. E., Aboraia, A. S., Veal, G. J., Redfern, C. P. F., Brancale, A., et al. (2011). Small molecule inhibitors of retinoic acid 4-hydroxylase (CYP26): Synthesis and biological evaluation of imidazole methyl 3-(4-(aryl-2-ylamino)phenyl) propanoates. *Journal of Medicinal Chemistry*, *54*(8), 2778–2791. https://doi.org/10.1021/jm101583w.

Gomaa, M. S., Bridgens, C. E., Veal, G. J., Redfern, C. P. F., Brancale, A., Armstrong, J. L., et al. (2011). Synthesis and biological evaluation of 3-(1 H-imidazol- and triazol-1-yl)-2,2-dimethyl-3-[4-(naphthalen-2-ylamino)phenyl]propyl derivatives as small molecule inhibitors of retinoic acid 4-hydroxylase (CYP26). *Journal of Medicinal Chemistry*, *54*(19), 6803–6811. https://doi.org/10.1021/jm200695m.

Grizotte-Lake, M., Zhong, G., Duncan, K., Kirkwood, J., Iyer, N., Smolenski, I., et al. (2018). Commensals suppress intestinal epithelial cell retinoic acid synthesis to regulate interleukin − 22 activity and prevent microbial Dysbiosis. *Immunity*, *49*(6), 1103–1115.e6. https://doi.org/10.1016/j.immuni.2018.11.018.

Hankinson, S. E. E., London, S. J., Chute, C. G., Barbieri, R. L., Jones, L., Kaplan, L. A., et al. (1989). Effect of transport conditions on the stability of biochemical markers in blood. *Clinical Chemistry*, *35*(12), 2313–2316.

Harrison, E. H., & Dela Seña's, C. (2015). Assimilation and conversion of dietary vitamin A into bioactive retinoids. In *The retinoids: Biology, biochemistry, and disease* (pp. 35–56). Wiley Blackwell. https://doi.org/10.1002/9781118628003.ch2.

Hogarth, C. A., Amory, J. K., & Griswold, M. D. (2011). Inhibiting vitamin A metabolism as an approach to male contraception. *Trends in Endocrinology and Metabolism*, *22*(4), 136–144. https://doi.org/10.1016/j.tem.2011.01.001.

Isoherranen, N., & Zhong, G. (2019). Biochemical and physiological importance of the CYP26 retinoic acid hydroxylases. *Pharmacology & Therapeutics*, *204*, 107400. https://doi.org/10.1016/j.pharmthera.2019.107400.

Iwata, M., Hirakiyama, A., Eshima, Y., Kagechika, H., Kato, C., & Song, S. Y. (2004). Retinoic acid imprints gut-homing specificity on T cells. *Immunity*, *21*(4), 527–538. https://doi.org/10.1016/j.immuni.2004.08.011.

Jing, J., Nelson, C., Paik, J., Shirasaka, Y., Amory, J. K., & Isoherranen, N. (2017). Physiologically based pharmacokinetic model of all-trans-retinoic acid with application to cancer populations and drug interactions. *Journal of Pharmacology and Experimental Therapeutics*, *361*(2), 246–258. https://doi.org/10.1124/jpet.117.240523.

Jones, J. W., Pierzchalski, K., Yu, J., & Kane, M. A. (2015). Use of fast HPLC multiple reaction monitoring cubed for endogenous retinoic acid quantification in complex matrices. *Analytical Chemistry*, *87*(6), 3222–3230. https://doi.org/10.1021/ac504597q.

Kane, M. A., Chen, N., Sparks, S., & Napoli, J. L. (2005). Quantification of endogenous retinoic acid in limited biological samples by LC/MS/MS. *Biochemical Journal*, *388*(1), 363–369. https://doi.org/10.1042/BJ20041867.

Kane, M. A., Folias, A. E., & Napoli, J. L. (2008). HPLC/UV quantitation of retinal, retinol, and retinyl esters in serum and tissues. *Analytical Biochemistry*, *378*(1), 71–79. https://doi.org/10.1016/j.ab.2008.03.038.

Kane, M. A., Folias, A. E., Wang, C., & Napoli, J. L. (2008). Quantitative profiling of endogenous retinoic acid in vivo and in vitro by tandem mass spectrometry. *Analytical Chemistry*, *80*(5), 1702–1708. https://doi.org/10.1021/ac702030f.

Kane, M. A., & Napoli, J. L. (2010). Quantification of endogenous retinoids. *Methods in Molecular Biology*, *652*, 1–54. https://doi.org/10.1007/978-1-60327-325-1_1.

Krivospitskaya, O., Elmabsout, A. A., Sundman, E., Söderström, L. A., Ovchinnikova, O., Gidlöf, A. C., et al. (2012). A CYP26B1 polymorphism enhances retinoic acid catabolism and may aggravate atherosclerosis. *Molecular Medicine*, *18*, 712–718. https://doi.org/10.2119/molmed.2012.00094.

Larange, A., & Cheroutre, H. (2016). Retinoic acid and retinoic acid receptors as pleiotropic modulators of the immune system. *Annual Review of Immunology*, *34*(1), 369–394. https://doi.org/10.1146/annurev-immunol-041015-055427.

Liu, H. X., Ly, I., Hu, Y., & Wan, Y. J. Y. (2014). Retinoic acid regulates cell cycle genes and accelerates normal mouse liver regeneration. *Biochemical Pharmacology*, *91*(2), 256–265. https://doi.org/10.1016/j.bcp.2014.07.003.

Lutz, J. D., Dixit, V., Yeung, C. K., Dickmann, L. J., Zelter, A., Thatcher, J. E., et al. (2009). Expression and functional characterization of cytochrome P450 26A1, a retinoic acid hydroxylase. *Biochemical Pharmacology*, 77(2), 258–268. https://doi.org/10.1016/j.bcp.2008.10.012.

Marill, J., Cresteil, T., Lanotte, M., & Chabot, G. G. (2000). Identification of human cytochrome P450s involved in the formation of all-trans-retinoic acid principal metabolites. *Molecular Pharmacology*, *58*(6), 1341–1348. https://doi.org/10.1124/mol.58.6.1341.

McCullough, F. S. W., Northrop-Clewes, C. A., & Thurnham, D. I. (1999). The effect of vitamin A on epithelial integrity. *Proceedings of the Nutrition Society*, *58*(2), 289–293. https://doi.org/10.1017/S0029665199000403.

Mezquita, B., & Mezquita, C. (2019). Two opposing faces of retinoic acid: Induction of stemness or induction of differentiation depending on cell-type. *Biomolecules*, *9*(10), 567. https://doi.org/10.3390/biom9100567.

Napoli, J. L. (1986). Quantification of physiological levels of retinoic acid. *Methods in Enzymology*, *123*(C), 112–124. https://doi.org/10.1016/S0076-6879(86)23015-3.

Noy, N. (1992a). The ionization behavior of retinoic acid in aqueous environments and bound to serum albumin. *Biochimica et Biophysica Acta-Biomembranes*, *1106*(1), 151–158. https://doi.org/10.1016/0005-2736(92)90233-C.

Noy, N. (1992b). The ionization behavior of retinoic acid in lipid bilayers and in membranes. *Biochimica et Biophysica Acta-Biomembranes*, *1106*(1), 159–164. https://doi.org/10.1016/0005-2736(92)90234-D.

Obrochta, K. M., Kane, M. A., & Napoli, J. L. (2014). Effects of diet and strain on mouse serum and tissue retinoid concentrations. *PLoS One*, *9*(6), e99435. https://doi.org/10.1371/journal.pone.0099435.

Randolph, R. K., & Simon, M. (1997). Metabolism of all-trans-retinoic acid by cultured human epidermal keratinocytes. *Journal of Lipid Research*, *3*, 1374–1375.

Rundhaug, J., Gubler, M. L., Sherman, M. I., Blaner, W. S., & Bertram, J. S. (1987). Differential uptake, binding, and metabolism of retinol and retinoic acid by 10T1/2 Cells. *Cancer Research*, *47*, 5637–5643.

Runge, F. E., & Heger, R. (2000). Use of microcalorimetry in monitoring stability studies. Example: Vitamin A esters. *Journal of Agricultural and Food Chemistry*, *48*(1), 47–55. https://doi.org/10.1021/jf981163y.

Schmidt, C. K., Brouwer, A., & Nau, H. (2003). Chromatographic analysis of endogenous retinoids in tissues and serum. *Analytical Biochemistry*, *315*(1), 36–48. https://doi.org/10.1016/S0003-2697(02)00662-0.

Sharow, K. A., Temkin, B., & Asson-Batres, M. A. (2012). Retinoic acid stability in stem cell cultures. *International Journal of Developmental Biology*, *56*(4), 273–278. https://doi.org/10.1387/ijdb.113378ks.

Shimshoni, J. A., Roberts, A. G., Scian, M., Topletz, A. R., Blankert, S. A., Halpert, J. R., et al. (2012). Stereoselective formation and metabolism of 4-hydroxy-retinoic acid enantiomers by cytochrome p450 enzymes. *Journal of Biological Chemistry*, *287*(50), 42223–42232. https://doi.org/10.1074/jbc.M112.404475.

Tay, S., Dickmann, L., Dixit, V., & Isoherranen, N. (2010). A comparison of the roles of peroxisome proliferator-activated receptor and retinoic acid receptor on CYP26 regulation. *Molecular Pharmacology*, 77(2), 218–227. https://doi.org/10.1124/mol.109.059071.

Thatcher, J. E., Buttrick, B., Shaffer, S. A., Shimshoni, J. A., Goodlett, D. R., Nelson, W. L., et al. (2011). Substrate specificity and ligand interactions of CYP26A1, the human liver retinoic acid hydroxylase. *Molecular Pharmacology, 80*(2), 228–239. https://doi.org/10.1124/mol.111.072413.

Thatcher, J. E., Zelter, A., & Isoherranen, N. (2010). The relative importance of CYP26A1 in hepatic clearance of all-trans retinoic acid. *Biochemical Pharmacology, 80*(6), 903–912. https://doi.org/10.1016/j.bcp.2010.05.023.

Topletz, A. R., Thatcher, J. E., Zelter, A., Lutz, J. D., Tay, S., Nelson, W. L., et al. (2012). Comparison of the function and expression of CYP26A1 and CYP26B1, the two retinoic acid hydroxylases. *Biochemical Pharmacology, 83*(1), 149–163. https://doi.org/10.1016/j.bcp.2011.10.007.

Topletz, A. R., Tripathy, S., Foti, R. S., Shimshoni, J. A., Nelson, W. L., & Isoherranen, N. (2015). Induction of CYP26A1 by metabolites of retinoic acid: Evidence that CYP26A1 is an important enzyme in the elimination of active retinoids. *Molecular Pharmacology, 87*(3), 430–441. https://doi.org/10.1124/mol.114.096784.

Topletz, A. R., Zhong, G., & Isoherranen, N. (2019). Scaling in vitro activity of CYP3A7 suggests human fetal livers do not clear retinoic acid entering from maternal circulation. *Scientific Reports, 9*(1), 4620. https://doi.org/10.1038/s41598-019-40995-8.

Tsukada, M., Schröder, M., Seltmann, H., Orfanos, C. E., & Zouboulis, C. C. (2002). High albumin levels restrict the kinetics of 13-cis retinoic acid uptake and intracellular isomerization to all-trans retinoic acid and inhibit its anti-proliferative effect on SZ95 sebocytes. *Journal of Investigative Dermatology, 119*(1), 182–185. https://doi.org/10.1046/j.1523-1747.2002.01816.x.

Yao, T., & Asayama, Y. (2017). Animal-cell culture media: History, characteristics, and current issues. *Reproductive Medicine and Biology, 16*, 99–117. https://doi.org/10.1002/rmb2.12024.

Zhao, S., Li, R., Li, Y., Chen, W., Zhang, Y., & Chen, G. (2012). Roles of vitamin A status and retinoids in glucose and fatty acid metabolism. *Biochemistry and Cell Biology, 90*(2), 142–152. https://doi.org/10.1139/o11-079.

Zhong, G., Hogarth, C., Snyder, J. M., Palau, L., Topping, T., Huang, W., et al. (2019). The retinoic acid hydroxylase Cyp26a1 has minor effects on postnatal vitamin A homeostasis, but is required for exogenous at RA clearance. *Journal of Biological Chemistry, 294*(29), 11166–11179. https://doi.org/10.1074/jbc.ra119.009023.

Zhong, G., Kirkwood, J., Won, K. J., Tjota, N., Jeong, H., & Isoherranen, N. (2019). Characterization of vitamin A metabolome in human livers with and without nonalcoholic fatty liver disease. *Journal of Pharmacology and Experimental Therapeutics, 370*(1), 92–103. https://doi.org/10.1124/jpet.119.258517.

Zhong, G., Ortiz, D., Zelter, A., Nath, A., & Isoherranen, N. (2018). CYP26C1 is a hydroxylase of multiple active retinoids and interacts with cellular retinoic acid binding proteins. *Molecular Pharmacology, 93*(5), 189–503. https://doi.org/10.1124/mol.117.111039.

CHAPTER FOURTEEN

Effects of vitamin A and retinoic acid on mouse embryonic stem cells and their differentiating progeny

Mary Ann Asson-Batres*, Christopher W. Norwood
Tennessee State University, Nashville, TN, United States
*Corresponding author: e-mail address: massonbatres@gmail.com

Contents

Abstract

Embryonic development is controlled by retinoids, and one approach that has been used to investigate the mechanisms for retinoid actions in developmental processes has been to study the effects of adding retinoids to cultures of pluripotent embryonic stem cells (ESC). To date, most *in vitro* retinoid research has been directed at deciphering the actions of *all-trans* retinoic acid (*at*RA). *at*RA is a derivative of *all-trans* retinol (a.k.a. vitamin A, VA), which mammals can generate *via* an enzyme-catalyzed pathway. *at*RA's effects on development result from its (1) activation of receptor complexes (RARs and RXRs) in the nucleus which then bind to and activate RA response elements (RAREs) in genes and (2) its interactions with processes that are initiated in the cytoplasm. While much work has focused on the impact of *at*RA on cell differentiation, VA, itself, has been

Methods in Enzymology, Volume 637
ISSN 0076-6879
https://doi.org/10.1016/bs.mie.2020.03.006

shown to exert effects on the maintenance of ESC identity that are not dependent upon classic RA-signaling pathways. In this chapter, we present results from our laboratory and others using well-documented approaches for investigating the effects of retinoids on the differentiation of ESC *in vitro* and introduce a novel method that uses chemically-defined growth conditions. The merits of each approach are discussed.

1. Introduction

Embryonic stem cells (ESC) are derived from blastocysts of mammalian preimplantation embryos (Evans & Kaufman, 1981). Classically, they have been shown to grow and retain their nascent properties if plated on a lawn of mitotically inactivated embryonic fibroblasts in conventional cell culture medium supplemented with fetal bovine serum (FBS) and the stem cell maintenance factor, leukemia inhibitory factor (LIF), which preserves self-renewal and inhibits differentiation by activating signal transducer and activator of transcription (STAT) signaling (Nemetz & Hocke, 1998). Subsequently, as understanding of the pathways that maintain stem cells has grown, serum-free and feeder cell-free culture protocols have been developed that utilize small molecules, such as SC1, and combinations of LIF, bone morphogenetic protein (BMP), and inhibitors of glycogen synthase kinase (GSK) to promote ESC self-renewal (by relieving Ras GTPase activating protein inhibition of phosphatidylinositol 3-kinase signaling) and inhibit differentiation (by inducing inhibitor-of-differentiation (Id) proteins and blocking activation of mitogen-activated protein kinase (ERK1/2)) (Chen et al., 2006; Wei et al., 2017; Wray, Kalkan, & Smith, 2010). Under these conditions, with care and detailed attention, ESC can retain "stem" attributes and the ability to self-renew indefinitely (see Smith (2001) for a list of attributes that define mouse ESC).

The addition of *at*RA to ESC cultures can override the effects of suppressors that maintain the ESC state. ESC express enzymes that make and degrade *at*RA (ALDH1A1 and Cyp26a1) and that mediate the actions of *at*RA (retinoic acid receptors, RARα, RARγ, and RXRβ, and cellular retinoic acid binding protein, CRABP1) (Buchanan, Rochette-Egly, & Asson-Batres, 2011; Lane et al., 2008; Lu et al., 2009). RARα and RARγ are expressed in ESC and their differentiating progeny. The expression patterns are not synchronous throughout the population, suggesting that the receptive and responsive actions of individual cells are dynamic (Buchanan et al., 2011).

*at*RA affects ESC status through actions that have been described as canonical and non-canonical. Canonical programs involve the binding of *at*RA to RAR complexes in the nucleus that then bind to cognate DNA motifs where they regulate the expression of key stem cell maintenance or differentiation genes (Asson-Batres & Rochette-Egly, 2014). Non-canonical programs involve interactions of *at*RA or VA itself with extranuclear RARs and other retinoid binding proteins or effectors (Iskakova, Karbyshev, Piskunov, & Rochette-Egly, 2015; Nagpal & Wei, 2019; Persaud, Lin, Wu, Kagechika, & Wei, 2013; Piskunov, Al, & Rochette-Egly, 2014; Rochette-Egly, 2015; Rochette-Egly & Germain, 2009).

Experiments investigating the effects of retinoids on ESC have involved growing the cells as monolayer cultures or as hanging drops that are induced to form spherical, embryoid bodies (EBs) which produce early differentiated cells representing all three germ layer lineages: endoderm, ectoderm, and mesoderm (Brickman & Serup, 2017; Kim & Ong, 2012). Results have shown that *at*RA can drive the development of many types of cells using these formats (Carradori et al., 2020; Drowley et al., 2020; Dyce, Tenn, & Kidder, 2018; Guan, Chang, Rolletschek, & Wobus, 2001; Guan, Rohwedel, & Wobus, 1999; Li et al., 2019; Lu et al., 2020; Martinez-Ceballos & Gudas, 2008; Phillips, Vernochet, & Dani, 2003; Rohwedel, Guan, & Wobus, 1999).

The results of some studies suggest that neuroectodermal differentiation occurs by default (Tropepe et al., 2001; Ying, Stavridis, Griffiths, Li, & Smith, 2003), but these studies do not exclude involvement of *at*RA in the process since a B27 supplement was added to cultures, and retinol acetate is a component of B27 formulations (Chen et al., 2008). Engberg, Kahn, Petersen, Hansson, and Serup (2010) provided convincing evidence that, indeed, ESC can generate *at*RA from retinol acetate and, further, showed that spontaneous neural commitment of differentiating ESC can be inhibited if cultures are grown in media supplemented with B27 lacking retinol acetate. The potency of *at*RA in this setting was delineated by the observation that neuroectodermal differentiation could be rescued in B27-retinoid-deficient medium by adding as little as 0.1 nM *at*RA to cultures (Engberg et al., 2010). This is significant since as much as 15 μM retinol acetate may be present in B27 formulations.

Experiments focused on determining how a*t*RA affects ESC differentiation have not yielded consistent results. For example, in some protocols, nanomolar concentrations of *at*RA have been shown to promote the expression of pro-neural genes and the development of neural progenitors (Engberg et al., 2010; Lu et al., 2009), whereas in others, investigators have found that relatively low levels of *at*RA induce mesodermal cells (Rohwedel et al., 1999). More often, relatively high concentrations of *at*RA have been shown to preferentially induce neural commitment, terminal differentiation, and maturation (Clagett-Dame, McNeill, & Muley, 2006; Kim et al., 2009; Li et al., 2019; Okada, Shimazaki, Sobue, & Okano, 2004). While the intent of adding different amounts of *at*RA to ESC is to determine the optimal dose for different differentiation outcomes, what is added to the culture medium is not necessarily what the cell "experiences."

Retinoids are lipophilic compounds with low solubility in aqueous environments, but at physiological pH, a predominant fraction of acidic retinoids, such as *at*RA, are ionized, and thus, more soluble (Noy, 1992). Even so, serum or serum proteins, such as bovine serum albumin (BSA) are required to maintain *at*RA in solution. In the absence of binding proteins, *at*RA concentrations can be severely reduced in cell culture media due to the greater propensity for *at*RA to bind to lipophilic materials like plastic pipette tips and plastic cell culture vessels (Klaassen, Brakenhoff, Smeets, Snow, & Braakhuis, 1999).

Because retinoids are susceptible to oxidative damage when exposed to light, air, and high temperature (Szuts & Harosi, 1991), there was a question whether retinoids are stable under cell culture conditions. To address this question, retinoids were added to serum-supplemented and chemically-defined media and incubated with or without cells, and the concentration and integrity of *at*RA in the culture media was analyzed at 0 and 24 h post addition using high pressure liquid chromatography (HPLC). The results showed that temperature and humidity had no effect on the recovery or integrity of *at*-RA extracted from serum-supplemented media after a 24 h incubation. In contrast, there was both loss and isomerization of *at*RA in serum-free media 24 h post addition. Supplementing the chemically-defined medium with 6 mg/mL BSA, a concentration equivalent to that present in the serum-supplemented media, had a protective effect on the stability and recovery of *at*Ra after 24 h incubation. These experiments demonstrated that retinoid stability can be maintained if carrier proteins are present in the culture media and if all operations, including preparation of retinoid stocks, addition of retinoids to cultures in the cell culture hood, and viewing cells using brightfield microscopy are conducted under yellow light to prevent photodamage. In keeping with these precautions, culture dishes should be wrapped in aluminum foil and kept in the dark during incubation (Sharow, Temkin, & Asson-Batres, 2012).

Other factors that have been reported to influence how retinoids influence cell differentiation outcomes include (a) whether retinoids are added to ESC cultures that have stem cell maintenance factors present or whether they are added after withdrawal of these factors; (b) the time and duration of exposure to retinoids; (c) monolayer cell densities; (d) the size and volume of EBs; and (e) the presence or absence of extracellular proteins or feeder cells (Faherty, Kane, & Quinlan, 2005; Gassanov et al., 2008; Klaassen et al., 1999; Lu et al., 2009; Marriott, 2012; Norwood, 2011; Ohnuki & Kurosawa, 2013; Osakada & Takahashi, 2011; Walker et al., 2007; Wobus et al., 1997; Ying, Nichols, Chambers, & Smith, 2003; Zhang et al., 2011).

Much work has been done to understand how retinoids influence specific differentiation programs, but a complete understanding of how they direct ESC to differentiate into homogenous cultures of specific progenitor and mature cell types remains a research challenge. At a clinical level, discriminating how the addition of retinoids to ESC produces a physiologically-relevant outcome in the context of the whole animal is significant for tissue engineering and cell transplantation work where it will be essential to standardize conditions to maintain cells in undifferentiated or specified states of differentiation and for the development of ESC therapeutics to replace specific cells lost to injury or disease *in vivo*.

In the remaining sections of this paper, two methods for culturing ESC are presented along with sample results showing the effects of VA and *at*RA on ESC proliferation and differentiation toward neural phenotypes.

2. Direct differentiation from ESC

ESC can be induced to differentiate directly without passing through EB formation. As reported here, mESC (R1 ESC obtained from the stem cell facility at University of Toronto) were grown in Dulbecco's Minimal Essential Medium (DMEM) supplemented with LIF and 15% FBS on mitotically inactivated mouse embryonic fibroblasts (MEFs). Upon reaching 75% confluence, the cells were harvested and passed to gelatin-coated plates to eliminate the contribution of factors secreted by the MEFs. To determine the effect of VA and *at*RA on mESC number and differentiation in the presence and absence of LIF, one control (plus LIF, standard conditions) and five experimental treatments, (1) minus LIF, (2) plus VA plus LIF, (3) plus VA minus LIF, (4) plus *at*RA plus LIF, and (5) plus *at*RA minus LIF were investigated. VA and *at*RA were dissolved in dimethyl sulfoxide (DMSO) or ethanol (ETOH) and added at 100 nM concentration to the culture medium. An equal concentration of solvent was added to plus LIF and minus LIF cultures (<0.1%, final concentration, DMSO or ETOH). Initially, two experiments were conducted using DMSO as a solvent and one experiment was conducted using ETOH as a solvent. Proliferation results and morphological changes were independent of solvent used. However, based on spectrophotometric analysis, it was observed that VA and *at*RA dissolved in ETOH were more stable under the light and temperature conditions of the experiments. Thus, all subsequent experiments were carried out with retinoids dissolved in ETOH. Cells were harvested, counted, and cell viability was determined either manually or by using a Vi-CELL™ Series Cell Viability Analyzer (Beckman Coulter Inc., Fullerton, CA). OCT 3/4 expression was monitored by flow cytometry. Cells were reacted with antibodies directed against OCT 3/4, and β-tubulin III (TUJ1). Retinoid concentrations were monitored by high pressure liquid chromatography (Asson-Batres, Smith, & Clark, 2009).

2.1 Results

2.1.1 Cell growth conditions

mESC grew and retained stem cell characteristics when plated on gelatin and grown in standard R1 ESC media for up to 4 days and less than 20 passages. The plus LIF cultures with no added VA or *at*RA were used as a reference for control conditions since these conditions were most optimal for maintaining cells with stem cell properties, i.e., the ability to proliferate, self-renew, and remain pluripotent. After 72 h, cells growing under the control conditions were aggregated into well-defined cell colonies with few cells migrating away from colonies and/or differentiating into cells with different morphologies.

2.1.2 Effect of VA and atRA on ESC number, morphology, and cell differentiation

As shown in three separate experiments, similar numbers of mESC were observed in cultures exposed to R1 ESC media with or without LIF and with or without VA or *at*RA for 24 h (data not shown). Almost all cells present in cultures after 72 h were viable (97% or greater, Table 1), regardless of treatment. However, cell number and cell morphology were notably affected in the five experimental conditions.

Under control conditions (media Plus LIF with no added *at*RA or VA), there was about a 10-fold increase in the numbers of cells in cultures seeded

Table 1 Effect of VA and *at*RA on cell proliferation.

	EXPT 1: High density plating			EXPT 2/3: Low density plating		
Treatment	**Cells × 10^6**	**% viability**	**% of control**	**Mean ± (SD) cells × 10^6**	**% viability**	**% of control**
Plus LIF (Control)	4.9	98	100	2.3 (0.33)	99	100
MINUS LIF	3.8	97	78	2.4 (0.24)	99	104
Plus VA Plus LIF	4.4	98	90	1.7 (0.19)	99	74
Plus VA MINUS LIF	4.1	98	84	2.0 (0.15)	99	87
Plus *at*RA Plus LIF	3.3	98	67	1.1 (0.15)	99	48
Plus *at*RA MINUS LIF	1.8	98	37	0.6 (0.14)	99	26

R1 ESC were seeded at 0.4×10^6 cells per well (EXPT 1) or 0.1×10^6 cells per well (EXPT 2 and 3) on gelatin-coated 6-well plates and counted after 72 h growth. The mean and standard deviation is reported for data pooled from EXPT 2 and 3 ($N=4$).

at a relatively high density (EXPT 1, Table 1) and about a 20-fold increase in the number of cells seeded at a relatively lower cell density (EXPT 2/3, Table 1) after 72 h growth. The cells in the control cultures exhibited a relatively low level of differentiation and retained stem cell characteristics, as indicated by their colonial growth and positive staining for OCT 3/4, a marker for undifferentiated stem cells (Fig. 1 Plus LIF). A few cells in cell colonies grown in Plus LIF without added retinoids stained positively for TUJ1, a marker for neurons (Fig. 2).

The 72 h Minus LIF cultures (media minus LIF with no added *at*RA or VA) showed little change in cell number relative to the Plus LIF control cultures (78% of control in EXPT 1 and 104% of control in EXPT 2/3, Table 1). The cells in these cultures differentiated as indicated by their migration away from colonies, their change in morphology, and their highly reduced expression of OCT 3/4 (compare Fig. 1 Plus LIF and Minus LIF). TUJ1 expression was not detected in the minus LIF cultures (Fig. 2 Minus LIF), suggesting that factors required for differentiation along the neuron lineage were not present in these cultures.

The 72 h, Plus VA cultures showed only a slight decrease in cell number, and this effect appeared to be independent of LIF (Table 1). These cultures

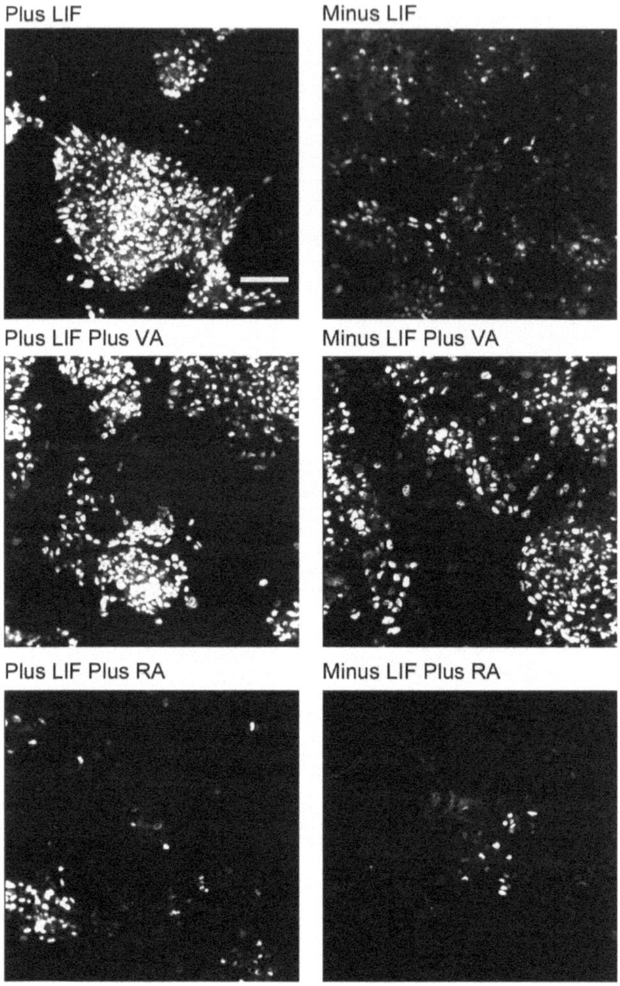

Fig. 1 Effect of VA and *at*RA on the stem and differentiation state of mESC as indicated by Oct 3/4 expression. R1 mESC were seeded on gelatin coated, 12-well plates at 0.05×10^6 cells per well and grown for 72 h. Cells were fixed with 4% paraformaldehyde and reacted with anti-OCT 3/4 antibody, a marker for undifferentiated ESC. Detection was with Cyanine 3 (CY3). Images were obtained with a Nikon TE300 inverted microscope with attached confocal imaging system using a 543 nm laser. DAPI-stained nuclei indicated that viable cells were also present in areas not stained by the OCT 3/4 antibody (data not shown). Scale bar in the Plus LIF image (upper left) represents 100 μm. All images in this and subsequent figures were formatted using Adobe Photoshop CS4 Extended software, version 11.0.2. RGB input levels were adjusted using the "levels" feature of the software.

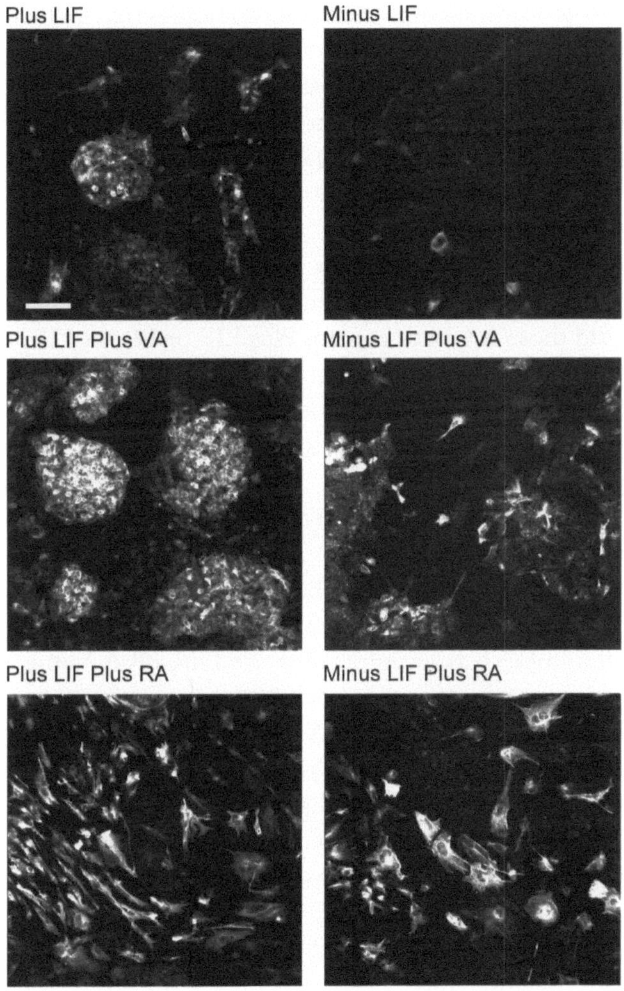

Fig. 2 Effect of VA and *at*RA on expression of TUJ1 in mESC. R1 mESC were seeded on gelatin coated, 12-well plates at 0.05×10^6 cells per well and grown for 72 h. Cells were fixed with 4% paraformaldehyde and reacted with an antibody directed against TUJ1, a marker for cells in the neuron lineage of the mammalian central nervous system. Detection was with CY3. Images were obtained as described in Fig. 1. DAPI-stained nuclei indicated that viable cells were also present in areas not stained by the OCT 3/4 antibody (data not shown). Note, the TUJ1 signal is present in cells that reside in colonies of *Plus LIF*, *Plus LIF Plus VA*, and *Minus LIF Plus VA* cultures. TUJ1 immunoreactivity in the *Plus RA* cultures is concentrated in cells that have migrated away from colonies. Positive TUJ1 immunoreactivity was not detected in *Minus LIF* cultures, although DAPI-stained cells were present. The scale bar in the *Plus LIF* image (upper left) represents 100 μm.

were made up of a striking mix of cell morphologies that included some that appeared similar to those observed in the Plus LIF control treatments (undifferentiated colonies) and some that appeared similar to those observed in the Minus LIF (without retinoids) treatments (flattened and differentiated) (Compare Fig. 1 Plus LIF, Minus LIF, and Plus and Minus LIF Plus VA). Unlike the control Minus LIF cultures, the Minus LIF Plus VA cultures retained expression of OCT 3/4 in spherical cell clusters (Fig. 1). Interestingly, some cells in these clusters also stained for TUJ1. A greater number of cells in the Plus LIF Plus VA cultures stained positively for TUJ1 than did those in either control Plus LIF or Minus LIF Plus VA cultures (Fig. 2). In each of these cases, the $TUJ1^{+}$ cells were resident in cell colonies. These results underscore the idea that VA has contrasting effects on mESC. One, cell differentiation is greater in Plus LIF Plus VA cultures than it is in control Plus LIF cultures, and two, VA preserves stem cell attributes in some cells and is most effective at doing so in cultures where LIF is not present.

*at*RA had profound effects on mESC. The Plus *at*RA Plus LIF cultures had about a 50% reduction in cell number relative to controls (Table 1). The effect on cell number was accentuated in the Plus *at*RA Minus LIF cultures where cell numbers were only about 30% of control after 72h growth (Table 1). Most cells in the Plus *at*RA treatment groups adopted a differentiated morphology whether LIF was present or not (Fig. 1). The cells migrated away from colonies and changed from a spherical shape to cells with extended processes that expressed TUJ1. Similar differentiation results were observed in the presence and absence of LIF (Fig. 2, Plus LIF Plus RA and Minus LIF Plus RA).

To quantify treatment effects on the differentiation state of mESC after a 72h exposure to either media supplemented with *at*RA, VA, or vehicle (ETOH), harvested cells were reacted with an antibody directed against OCT 3/4, and OCT 3/4 immunoreactivity was quantified by flow cytometry. Based on this assessment, differentiation occurred under all conditions (Fig. 3), but was highly accentuated by exposure of cultures to media supplemented with 500nM *at*RA (Fig. 3B and D). The effect of exposure to 500nM VA (Fig. 3C and E) was midway between that of control (Fig. 3A) and *at*RA-supplemented cultures (Fig. 3B and D). The presence of LIF had little, if any, effect on the presence of either VA or *at*RA in the cultures (compare Fig. 3B and D and C and E). These data support a conclusion that *at*RA promotes stem cell differentiation, whereas VA preempts it.

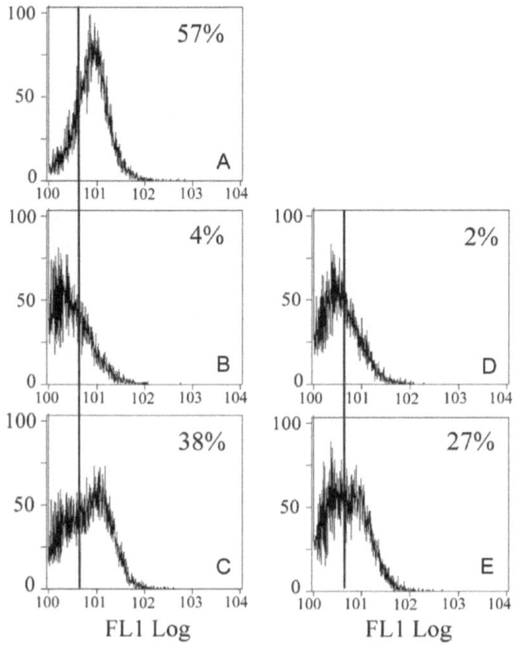

Fig. 3 Quantification of the effect of VA and *at*RA on the stem and differentiation state of mESC as indicated by Oct 3/4 expression. Images (A–E) are fluorescence intensity histograms obtained from flow cytometry analysis of R1 mESC grown in the presence (A–C) or absence (D and E) of LIF and either vehicle (ETOH, A), *at*RA (B and D) or VA (C and E) and labeled with an antibody directed against OCT 3/4, a marker for undifferentiated cells. The horizontal axis represents fluorescence intensity, and thus, peaks to the right of the reference line indicate a relatively high number of OCT $3/4^{+}$ cells in the population under study. Peaks shifted to the left of the reference line indicate a low level of fluorescence which is indicative of relatively few OCT $3/4^{+}$ cells in the sample. Percentages shown in the upper right hand corner of each histogram are estimates of the proportion of OCT $3/4^{+}$ cells in each sample, as determined by the analysis.

Four outcomes were observed in cultures of mESC grown in the presence of VA and *at*RA: (1) VA did not produce the same response in mESC as *at*RA; (2) VA maintained about 30% of mESC in an undifferentiated state, as indicated by OCT 3/4 staining after 72 h growth, whereas *at*RA induced more than 95% of the stem cells to differentiate; (3) VA plus LIF synergized to promote development of $TUJ1^{+}$ cells, but these cells remained resident in cell colonies, in contrast with *at*RA-supplemented cultures that produced $TUJ1^{+}$ cells with extended processes that migrated away from the original cell colonies; and (4) only a very few cells in cell clusters stained positively for TUJ1 in VA Minus LIF cultures, whereas the absence of LIF did not impede the expression of TUJ1 in the population of differentiated cells in the Minus LIF Plus *at*RA cultures (Figs. 1–3).

2.2 Discussion

The results presented here indicate that VA preserves stem cell attributes and mitogenic potential, whereas *at*RA reduces the mitogenic potential of mESC and induces them to differentiate. Observed changes in cell number likely reflect changes in cell proliferation; however, cell death may also contribute to the final cell number. An effect of VA on maintenance of stem cell characteristics has also been observed by Chen, Yang, Dawes, and Khillan (2007) and Chen and Khillan (2008). Their findings indicate that VA itself (not *via* mechanisms mediated by *at*RA) promotes over-expression of Nanog, a key transcription factor for stem cell self-renewal.

The TUJ1 immunoreactivity of cells in the Plus LIF control cultures was unexpected in this treatment as these conditions are optimized to maintain mESC in an undifferentiated state. This outcome may have been the consequence of unidentified differentiating factors present in the medium, including VA, which is a component of FBS. It could also reflect the fact that MEF feeder layers are needed to supply factors that preserve stem cell identity. Extracellular matrix proteins like gelatin may not be sufficient to preserve these attributes. Interestingly, the Plus VA Plus LIF treatment induced a similar, but more substantial TUJ1 expression pattern as compared with controls (Fig. 2). Clearly, the mechanisms underlying these outcomes are complex and involve synergistic and antagonistic processes that compete to promote or inhibit proliferation, cell migration, and differentiation. *Direct differentiation from ESC* methods continue to be used in the search to uncover the role that RA plays in directing the development of neuron and other cell lineages and cell types (Drowley et al., 2020; Glen, McDevitt, & Kemp, 2018).

2.3 Summary

From the standpoint of understanding how mESC develop *in vitro*, serum-supplemented, monolayer culture methods provide a useful approach for investigating how addition of VA and *at*RA (over and above the amount of retinoids already present in the standard media) affect the differentiation state of stem cells that are exposed to these retinoids. However, the use of serum-supplemented media is problematic if the goal is to identify and investigate how specific factors maintain pluripotency or promote differentiation of ESC. This is because VA, *at*RA, and other undefined factors are present in serum supplements.

3. Growth and differentiation of mESC on microscope slides in chemically-defined media

3.1 Materials, equipment, and reagents

3.1.1 Stock reagents

ICC PBS: 0.02 M KPO_4/0.15 M NaCl, pH 7.4

TNT wash buffer: 0.1 M Tris-HCl, pH 7.5/0.15 M NaCl/0.05% Tween®20

mESC growth media: ESGRO Complete Plus media + GSK3i.

Retinoids: All operations were carried out under yellow light because retinoids do not absorb light in the yellow range of the electromagnetic spectrum (570–585 nm). Concentrated stock solutions of *at*-RA and VA were prepared in ethyl acetate and stored in amber glass vials at −20 °C in the dark. ETOH stock solutions (approximately 2 mM) were prepared from the concentrated stocks, sterile filtered through a 0.2 mm syringe filter, and stored in glass vials at −20 °C in the dark. The absorbance and spectrum of each sterile retinoid stock was determined with a UV–vis spectrophotometer, and concentrations calculated using the following absorption coefficients: 44,300 L/mol × cm at 350 nm for *at*-RA in ETOH; 46,000 L/mol × cm at 325 nm for VA in ETOH.

Retinoid supplementation: mESC were seeded in growth media without retinoids and allowed to settle overnight. The next day, the culture media was replaced with fresh, retinoid-supplemented growth media, and subsequently, fresh, retinoid-supplemented media changes were made every 24 h.

3.1.2 Materials

Sources of specific reagents and equipment are provided in Table 2.

3.2 Protocols

A. Laminar Flow Hood Set-up

- **A.1.** Carry out all procedures under a BioSafety Level II Laminar Flow Hood.
- **A.2.** Leave the laminar flow hood running 24 h a day.
- **A.3.** Spray inside surfaces of the hood with 70% ETOH and wipe dry. Set up all supplies needed for operations and irradiate with ultraviolet light for 5–15 min.

Table 2 Specific reagents for growth and differentiation of mESC on microscope slides in chemically-defined media

Name of reagent	Company	Catalog number
ESGRO Complete C57/BL6 mouse ES cell line	Millipore	SF-CMTI-2
ESGRO Complete Accutase	Millipore	14190-136
ESGRO Complete PLUS Clonal Grade Medium	Millipore	SF001-500P
ESGRO Complete Supplement GSK3β Inhibitor	Millipore	SF012-250
ESGRO Complete Serum Free Basal Medium	Millipore	SF002-500
SUPERFROST® EXCELL™ slides	Thermo Scientific	4941EXC-001
Nunclon™Δ 4-well dishes	Thermo Scientific	167063
Sequenza® Tray	Thermo Scientific	733100015
Sequenza® Coverplate	Thermo Scientific	72110017
Renaissance TSA™ Fluorescence Systems Tyramide Signal Amplification Kit	Perkin Elmer Life Sciences	NEL704A Cyanine 3 with Blocking Reagent
Vectastain Elite ABC Kit	Vector Laboratories	101098-314
Neuronal Class III β-Tubulin (TUJ1) Monoclonal IgG2a Antibody	Covance	MMS-435P
Biotinylated Anti-Mouse IgG	Vector Laboratories	BA-2001
Purified Mouse Anti-OCT-3/4 Clone 40/Oct-3 (RUO)	BD Sciences	611202
FITC Rat Anti-Mouse CD90.2	BD Sciences	553004
Mouse Anti-SSEA1 (480)	Santa Cruz Biotechnology, Inc	SC-21702
Biotinylated Horse Anti-Mouse IgG Antibody, Rat Adsorbed	Vector Laboratories	BA-2001
Purified Goat Anti-GATA-4 (C-20)	Santa Cruz Biotechnology, Inc	SC-1237
Biotin-SP (long spacer) AffiniPure Donkey Anti-Goat IgG (H + L)	Jackson ImmunoResearch Laboratories, Inc	705-065-147
Hoechst 34580	Invitrogen/Molecular Probes	H21486
Retinol (VA)	Sigma-Aldrich	R7632
Retinoic Acid (*at*RA)	Sigma-Aldrich	R2625

B. Thawing mESC

B.1. If thawing mESC that were previously grown on mouse embryonic fibroblasts (MEF), grow up once on mitomycin treated or ultraviolet light (UV) irradiated MEF. These cells will need to be passed to a gelatin plate one time before transferring them to the EXCELL™ slides (step 3, below). If thawing mESC that were previously grown on gelatin, add 2–3 mL 0.1% gelatin to one 100 × 20 mm cell culture dish and incubate it at 37 °C/7.5% CO_2 for 30 min.

B.2. Add 20 mL mESC cell growth medium to the dish with MEF or the gelatin treated dish (after removing excess gelatin with a sterile pipet). Incubate for at least 15 min to pre-heat the medium and adjust the pH. (Note, if serum-free, chemically-defined medium will be used to grow the cells, the mESC need to be preadapted to grow in the medium.) Protocols for pre-adapting mESC to grow in ESGRO Complete PLUS Clonal Grade Medium are available from the manufacturer (Millipore, Billerica, MA). Alternatively, preadapted C57/BL6 mESC can be purchased (Table 2).

B.3. Transfer 10 mL mESC growth medium into a 15 mL Falcon centrifuge tube and set aside in the hood.

B.4. Add 75 mL warm tap water to a 100 mL beaker and place in the laminar flow hood. Wearing a face mask, remove one vial of mESC from storage under liquid nitrogen vapor phase (approximately 2×10^6 cells in 1 mL). Gently swirl the vial in the warm water. Keep the O-ring and cap out of the water to prevent contamination.

B.5. Remove the vial from the water bath when the contents are partially thawed (some ice still present). Spray the vial with 70% ETOH and wipe dry with a Kimwipe.

B.6. Transfer the cell suspension from the vial to the 15 mL centrifuge tube. Triturate with a 10 mL pipette 25 times to break up cell clumps and resuspend the cells.

B.7. Centrifuge at $1000 \times g$ for 3 min.

B.8. Remove the supernate. Resuspend the cell pellet in 5 mL prewarmed growth medium transferred from the pre-incubated cell culture dish. Triturate the cell suspension 25 times, and transfer back into the dish.

B.9. Swirl the dish in a circular motion gently three times to disperse the cells and let the cells settle for 5 min in the laminar flow hood.

Check under an inverted microscope to ensure even dispersal of single cells. Incubate at 37 °C/7.5% CO_2.

C. Setting up the EXCELL™ microscope slide cultures

C.1. There is no need to treat or sterilize EXCELL™ microscope slides for use in culturing mESC. They can be used directly out of the box as long as they are kept sterile. Remove cellophane from the box while in the laminar flow hood and place in a sealed Ziploc storage bag to prevent contamination after opening the box.

C.2. Place one sleeve of 4-well cell culture dishes (12.78 × 8.5 cm) in the laminar flow hood along with the EXCELL™ slides.

C.3. Remove coverplate lid from a 4-well dish, and with the use of sterile forceps, transfer one slide to each partition in the dish. Be certain that the treated side of the slide is facing up. Use the logo on the bottom of the slide to orient (Fig. 4).

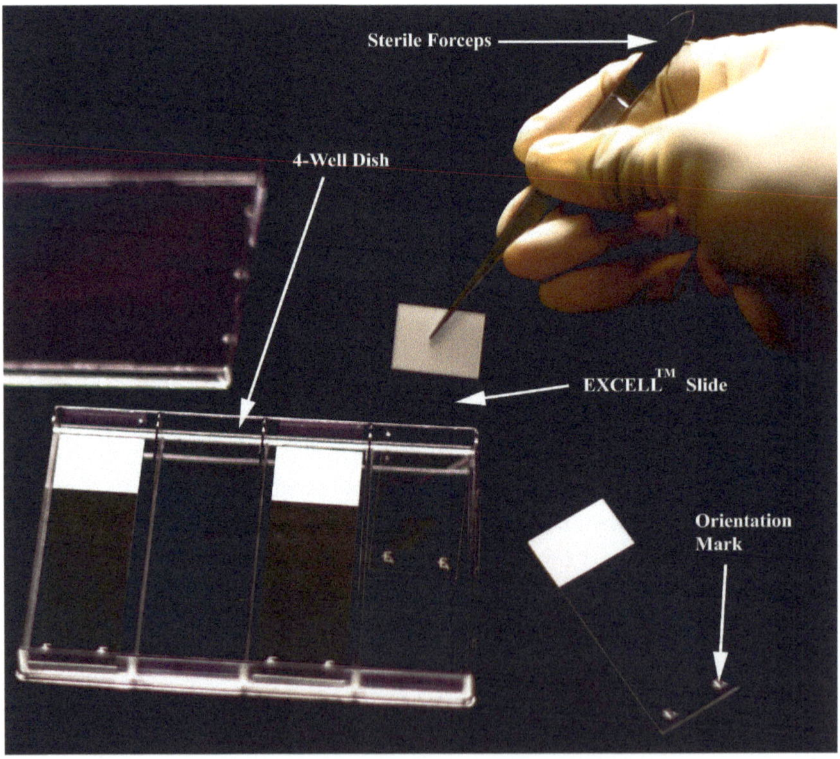

Fig. 4 SUPERFROST® EXCELL™ microscope slides and 4-well culture dishes. C57/BL6 mESC were seeded directly onto the slides.

C.4. Transfer 5 mL growth media into each well containing a slide, replace the coverplate lid, and then place the dish in the incubator to pre-heat the medium and adjust the pH.

D. Passage of mESC from dish of thawed cells grown overnight to the EXCELL™ slides

D.1. Remove mESC culture from the incubator and discard media. Wash the adherent cells in the dish with 10 mL of Dulbecco's Phosphate Buffered Saline without calcium or magnesium (DPBS) two times.

D.2. Add 3 mL enzyme dissociation solution (in our experience, Trypsin is too harsh and removes essential cell surface proteins. Thus, we recommend using Accutase™, a ready to use proprietary cell detachment solution composed of proteolytic and collagenolytic enzymes that does not require inactivation with enzyme inhibitors), and incubate for ~2 min. Check for detached cells.

D.3. Add 12 mL mESC growth medium to the detached cells and transfer cell suspension to 15 mL centrifuge tube. Gently triturate 25 times.

D.4. Centrifuge at $1000 \times g$ for 3 min.

D.5. Aspirate supernate and add 5 mL pre-incubated mESC growth media to the pellet, and gently resuspend (triturate 25 times). Count mESC.

D.6. Transfer an aliquot of the resuspension containing 0.5×10^5 mESC to each well of the 4-well dish that contains an EXCELL™ slide. Examine under an inverted microscope for even dispersal and then incubate at 37 °C/7.5% CO_2.

D.7. Cells attach to the EXCELL™ slide within several hours and the slide can then be fixed or harvested at desired time points.

E. Passing cells from EXCELL™ microscope slides or harvesting for downstream applications

E.1. Remove media from each well of the 4-well dish containing an EXCELL™ slide and wash with DBPS two times. Aim pipette at side of well to avoid knocking cells off of the slide.

E.2. Add 2 mL Accutase to each slide and incubate for ~2 min. Check for detached cells.

E.3. If passing cells, add 5 mL mESC growth medium to the detached cells. If harvesting cells, add 5 mL PBS to the detached cells.

Transfer cell suspension to a 15 mL centrifuge tube. Gently triturate 25 times.

E.4. If passing cells, follow steps D.4–D.6. If harvesting cells, pellet and proceed as specified for the downstream application.

F. Fixing mESC on slides for immunocytochemistry (ICC)

F.1. Remove 4-well culture dish with EXCELL™ slides from incubator and put on bed of ice.

F.2. Remove medium and gently wash slides in wells with 10 mL ice cold phosphate buffered saline (ICC PBS).

F.3. Incubate slides with 5 mL freshly prepared 4% paraformaldehyde/ICC PBS for 20 min at 4 °C.

F.4. Discard fixative and wash slides once with ICC PBS on ice.

F.5. Put 1–2 mL ICC PBS in wells, wrap with Parafilm®, and store at 4 °C until use.

G. Using the Sequenza® system for ICC

G.1. Place a Sequenza® system coverplate over each slide with fixed mESC. Once coverplate is in place, place the coverplate and slide into an empty compartment in the Sequenza® slide rack until they snap into proper position (Fig. 5).

G.2. Hydrate mESC with 1 mL ICC PBS. Gravitational flow moves the liquid down over the slide. Excess fluid drains through and 80 μL liquid remains in the gap between the coverplate and the slide, thus, exposing the cells to the reagent during the incubation time.

G.3. Follow preferred protocol for ICC adding a minimum of 0.2 mL reagent per step in the protocol. We have found that this volume of primary or secondary antibody works better than the 0.1 mL suggested by the manufacturer.

As an example, we used the following ICC protocol to analyze expression in mESC cultures after addition of 500 nm *at*-RA (with replacement every 24 h) for a total exposure time of 72 h. Steps a–m were carried out in the Sequenza® slide rack. Solutions were pipetted into the reservoir between the slide and the coverplate (see Fig. 5).

a. Wash with 1 ml TNT wash buffer per slide three times for 5 min at room temperature.

b. Block endogenous peroxidase activity by washing with 0.5 mL 0.3% peroxide in methanol per slide once for 20 min at room temperature.

c. Block non-specific reactive sites by adding 0.5 mL TNB (0.5% Renaissance™ TSA-Fluorescence System Kit Blocking Reagent in TNT buffer) per slide for 1 h at room temperature.

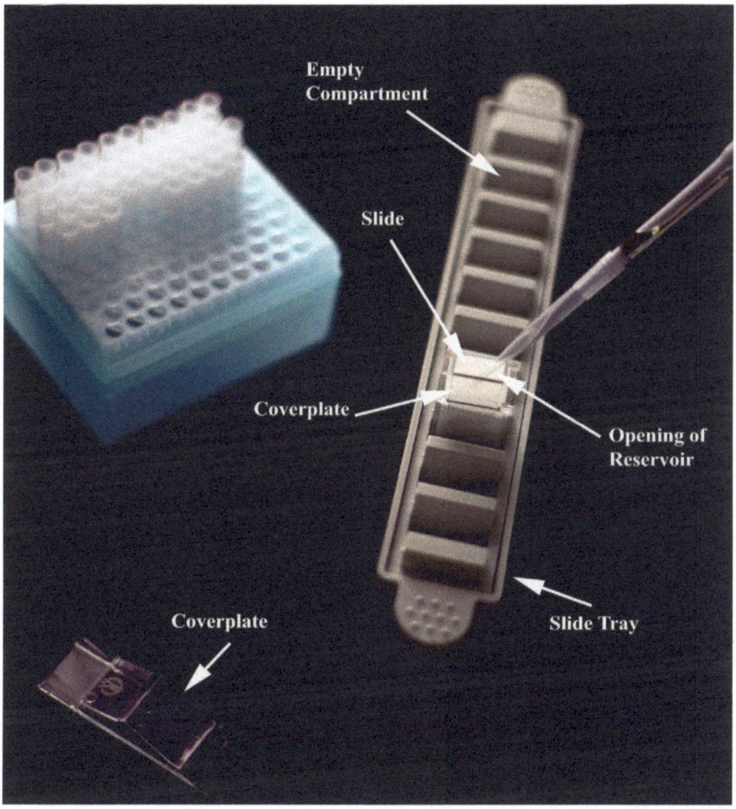

Fig. 5 Sequenza® apparatus for immunocytochemistry. Adherent cells on microscope slides were fixed with 4% paraformaldehyde. A coverplate (shown) was fitted over the microscope slide and inserted into the Sequenza® slide tray.

d. Incubate with 0.2 mL TUJ1 antibody (1:500 dilution in TNB) per slide overnight at 4 °C.

e. Repeat step a, above.

f. Incubate with 0.2 mL biotinylated anti-mouse IgG antibody (1:1000 dilution in TNB) per slide for 1 h at room temperature.

g. Repeat step a, above.

h. Incubate in 1 mL ABC reagent (prepare 2 drops Vectastain kit reagent A + 5 mL ICC PBS + 2 drops kit reagent B 30 min before use) per slide for 1 h at room temperature (this step provides additional amplification, but may be omitted in favor of using the Streptavidin-HRP reagent that is supplied with the Renaissance™ TSA-Fluorescence System Kit).

i. Repeat step a, above.

j. Incubate with 200 μL TSA-CY3 reagent (1:50 dilution in TSA-Fluorescence System Kit AMP buffer) per slide for 5 min at room temperature.
k. Repeat step a, above.
l. Stain nuclei with 1 mL Hoechst-34580 (dilute 20 μL of a 10 mg/mL stock solution in 10 mL deionized water) per slide for 5 min at room temperature.
m. Wash with 1 mL deionized water three times for 5 min at room temperature.
n. Remove slides from Sequenza® slide rack and mount coverslips with mounting medium.

3.3 Precursor techniques

Growing cells directly on the EXCELL™ microscope slides with this method solved growth problems and handling difficulties experienced when using other slide formats, including chamber slides and glass coverslips.

3.4 Discussion

The described method allows mESC to be maintained in serum-free, chemically-defined medium without a fibroblast feeder layer for up to 30 passages. The EXCELL™ slides (Fig. 4) provide a suitable substrate for optimal cell attachment without additional treatments. mESC colonies retained stem cell identity as indicated by the even distribution of colonies with typical stem cell colony morphology (i.e., tight, compact, well-defined borders) (Fig. 6A) and expression of the stem cell biomarkers, OCT 3/4 (Fig. 6B) and stage specific embryonic antigen 1 (SSEA1) (Fig. 6C). When mESC grown using this method were exposed to 500 nM *at*RA for 72 h, they lost their "stem" cell attributes and differentiated as indicated by their change in morphology and migration away from cell colony clusters (Fig. 6D). Cells in both the cardiomyocyte and neuron lineages developed in the plus *at*RA cultures as shown by the identification of cells expressing transcription factor GATA binding protein 4 (GATA-4), a key regulator of cardiac cell specification (Fig. 6E), and the neuron marker, TUJ1 (Fig. 6F).

Growth of cells on four separate slides in the 4-well dishes allows the opportunity to compare outcomes of growth in control (ESGRO Complete Plus Media) and experimental (media supplemented with different concentrations of *at*RA and VA) conditions in the same dish (Fig. 4). The EXCELL™ slides allow growth of sufficient numbers of cells for

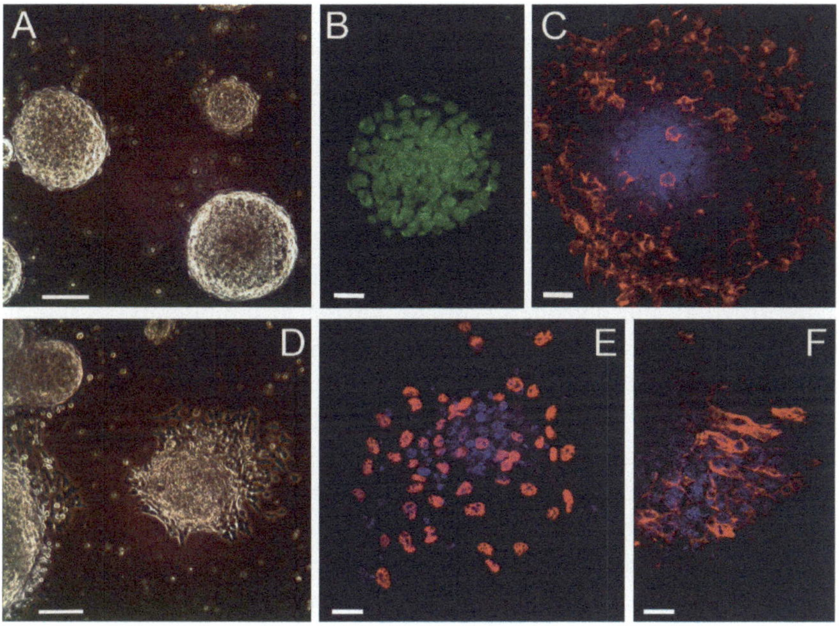

Fig. 6 Effect of *at*RA on mESC grown in chemically-defined media on EXCELL™ slides. (A–C) C57/BL6 mESC were grown on EXCELL™ slides for 72h in media supplemented with vehicle (ETOH) (A) or media without vehicle (B and C). (D–F) C57/BL6 mESC were grown on EXCELL™ slides for 72h in media supplemented with 500nM *at*RA. Images (A) and (D) are darkfield images of live cell colonies growing on a slide, scale bars represent 50μm. Cells in (B), (C), (E), and (F) were fixed directly on the slides and reacted with stem cell marker antibodies directed against OCT 3/4 (B, FITC signal) or SSEA1 (C, CY3 signal) or differentiation marker antibodies directed against transcription factor GATA binding protein 4 (GATA-4), a key regulator of cardiac cell specification (E, CY3 signal) or TUJ1, a neuron marker (CY3 signal). The image in (B) is a confocal image collected at 488nm (pseudogreen signal); the double-labeled images in (C), (E), and (F) are maximum projection, merged images of scans through the z plane collected with a confocal microscope at 543nm (pseudored signal) and 405nm to detect nuclei labeled with H-34580 (pseudoblue signal); scale bars represent 20μm.

downstream applications, including harvesting enough cells for SDS PAGE, immunoblot analyses, RNA isolation, and PCR analyses.

The slides also allow the user to carry out histochemical or ICC analyses of cells directly on the slides either using standard protocols, or, preferably, by processing the slides using a Sequenza® system (Fig. 5). Once positioned in the Sequenza® slide tray, a small gap is created between the slide and the coverplate which permits complete exposure of the cells with application of only a small volume of precious or expensive reagents (as little as 0.2mL per slide). Carrying out ICC directly on the slides allows imaging of the cells through a suitable glass coverslip that is applied to the slide at the end of the procedure. This is a distinct advantage for confocal imaging where plastic substrates are not recommended.

3.5 Summary

Ying, Nichols, et al. (2003) and Wray et al. (2010) developed a chemically defined, serum-free medium supplemented with LIF, bone morphogenetic factor (BMP4) and glycogen synthase kinase 3β inhibitor (GSK3βi) and reported it improved the growth of mESC on gelatinized culture dishes. We confirmed that C57/BL6 mESC maintain pluripotency under these growth conditions, but found that the cells detached from the dishes and often grew only around the periphery of the dish, making it difficult to obtain sufficient numbers of cells for downstream analyses. Growing ESC directly on EXCELL™ microscope slides solved problems with poor growth and also handling difficulties that arise when using other slide formats, including chamber slides or glass coverslips. The slides do not have to be sterilized and do not require pretreatment with extracellular proteins. An additional advantage of this method is that cells can be fixed directly on the microscope slides and used for histochemical analyses. Further, sufficient numbers of mESC can be grown on the slides for immunoblot and RT-PCR studies. Finally, mESC grown using this procedure can be induced to differentiate on the microscope slides, which permits investigation of retinoid effects under chemically-defined conditions.

Acknowledgments

The authors wish to thank Tamara Reid and William L. Stanford, for their assistance with some experiments and the use of facilities and protocols in laboratories at University of Toronto, Toronto, Canada. Funding for grant NIH/NIGMS/MBRS/SCORE S06 GM 008092 to M.A.A.B. was used to support these studies.

Disclosures

We have nothing to disclose.

References

Asson-Batres, M. A., & Rochette-Egly, C. (2014). *The biochemistry of retinoic acid receptors I: Structure, activation, and function at the molecular level* (2014th ed.). Springer.

Asson-Batres, M. A., Smith, W. B., & Clark, G. (2009). Retinoic acid is present in the postnatal rat olfactory organ and persists in vitamin A—Depleted neural tissue. *The Journal of Nutrition, 139*, 1067–1072.

Brickman, J. M., & Serup, P. (2017). Properties of embryoid bodies. *Wiley Interdisciplinary Reviews: Developmental Biology*, *6*, e259.

Buchanan, F. Q., Rochette-Egly, C., & Asson-Batres, M. A. (2011). Detection of variable levels of RARalpha and RARgamma proteins in pluripotent and differentiating mouse embryonal carcinoma and mouse embryonic stem cells. *Cell and Tissue Research*, *346*, 43–51.

Carradori D, Labrak Y, Miron VE, Saulnier P, Eyer J, Preat V, des RA (2020): Retinoic acid-loaded NFL-lipid nanocapsules promote oligodendrogenesis in focal white matter lesion. Biomaterials 230:119653.

Chen, S., Do, J. T., Zhang, Q., Yao, S., Yan, F., Peters, E. C., et al. (2006). Self-renewal of embryonic stem cells by a small molecule. *Proceedings of the National Academy of Sciences of the United States of America*, *103*, 17266–17271.

Chen, L., & Khillan, J. S. (2008). Promotion of feeder-independent self-renewal of embryonic stem cells by retinol (vitamin A). *Stem Cells*, *26*, 1858–1864.

Chen, Y., Stevens, B., Chang, J., Milbrandt, J., Barres, B. A., & Hell, J. W. (2008). NS21: Re-defined and modified supplement B27 for neuronal cultures. *Journal of Neuroscience Methods*, *171*, 239–247.

Chen, L., Yang, M., Dawes, J., & Khillan, J. S. (2007). Suppression of ES cell differentiation by retinol (vitamin A) via the overexpression of Nanog. *Differentiation*, *75*, 682–693.

Clagett-Dame, M., McNeill, E. M., & Muley, P. D. (2006). Role of all-trans retinoic acid in neurite outgrowth and axonal elongation. *Journal of Neurobiology*, *66*, 739–756.

Drowley, L., McPheat, J., Nordqvist, A., Peel, S., Karlsson, U., Martinsson, S., et al. (2020). Discovery of retinoic acid receptor agonists as proliferators of cardiac progenitor cells through a phenotypic screening approach. *Stem Cells Translational Medicine*, *9*, 47–60.

Dyce, P. W., Tenn, N., & Kidder, G. M. (2018). Retinoic acid enhances germ cell differentiation of mouse skin-derived stem cells. *Journal of Ovarian Research*, *11*, 19.

Engberg, N., Kahn, M., Petersen, D. R., Hansson, M., & Serup, P. (2010). Retinoic acid synthesis promotes development of neural progenitors from mouse embryonic stem cells by suppressing endogenous, Wnt-dependent nodal signaling. *Stem Cells*, *28*, 1498–1509.

Evans, M. J., & Kaufman, M. H. (1981). Establishment in culture of pluripotential cells from mouse embryos. *Nature*, *292*, 154–156.

Faherty, S., Kane, M. T., & Quinlan, L. R. (2005). Self-renewal and differentiation of mouse embryonic stem cells as measured by Oct 4 gene expression: Effects of lif, serum-free medium, retinoic acid, and dbcAMP. *In Vitro Cellular & Developmental Biology Animal*, *41*, 356–363.

Gassanov, N., Er, F., Zagidullin, N., Jankowski, M., Gutkowska, J., & Hoppe, U. C. (2008). Retinoid acid-induced effects on atrial and pacemaker cell differentiation and expression of cardiac ion channels. *Differentiation*, *76*, 971–980.

Glen, C. M., McDevitt, T. C., & Kemp, M. L. (2018). Dynamic intercellular transport modulates the spatial patterning of differentiation during early neural commitment. *Nature Communications*, *9*, 4111.

Guan, K., Chang, H., Rolletschek, A., & Wobus, A. M. (2001). Embryonic stem cell-derived neurogenesis. Retinoic acid induction and lineage selection of neuronal cells. *Cell and Tissue Research*, *305*, 171–176.

Guan, K., Rohwedel, J., & Wobus, A. M. (1999). Embryonic stem cell differentiation models: Cardiogenesis, myogenesis, neurogenesis, epithelial and vascular smooth muscle cell differentiation in vitro. *Cytotechnology*, *30*, 211–226.

Iskakova, M., Karbyshev, M., Piskunov, A., & Rochette-Egly, C. (2015). Nuclear and extranuclear effects of vitamin A. *Canadian Journal of Physiology and Pharmacology*, *93*, 1065–1075.

Kim, M., Habiba, A., Doherty, J. M., Mills, J. C., Mercer, R. W., & Huettner, J. E. (2009). Regulation of mouse embryonic stem cell neural differentiation by retinoic acid. *Developmental Biology*, *328*, 456–471.

Kim, P. T., & Ong, C. J. (2012). Differentiation of definitive endoderm from mouse embryonic stem cells. *Results and Problems in Cell Differentiation*, *55*, 303–319.

Klaassen, I., Brakenhoff, R. H., Smeets, S. J., Snow, G. B., & Braakhuis, B. J. (1999). Considerations for in vitro retinoid experiments: Importance of protein interaction. *Biochimica et Biophysica Acta*, *1427*, 265–275.

Lane, M. A., Xu, J., Wilen, E. W., Sylvester, R., Derguini, F., & Gudas, L. J. (2008). LIF removal increases CRABPI and CRABPII transcripts in embryonic stem cells cultured in retinol or 4-oxoretinol. *Molecular and Cellular Endocrinology*, *280*, 63–74.

Li, Y., Mao, X., Zhou, X., Su, Y., Zhou, X., Shi, K., et al. (2019). An optimized method for neuronal differentiation of embryonic stem cells in vitro. *Journal of Neuroscience Methods*, *330*, 108486.

Lu, J., Tan, L., Li, P., Gao, H., Fang, B., Ye, S., et al. (2009). All-trans retinoic acid promotes neural lineage entry by pluripotent embryonic stem cells via multiple pathways. *BMC Cell Biology*, *10*, 57.

Lu, Z., Xie, Y., Huang, H., Jiang, K., Zhou, B., Wang, F., et al. (2020). Hair follicle stem cells regulate retinoid metabolism to maintain the self-renewal niche for melanocyte stem cells. *eLife*, *9*, e52712.

Marriott, S. (2012). *The effects of neuregulin and retinoic acid on embryonic stem cell derived embryoid bodies*. ProQuest LLC ETD Collection for Tennessee State University, pp. 1–68, Paper AAI1519772.

Martinez-Ceballos, E., & Gudas, L. J. (2008). Hoxa1 is required for the retinoic acid-induced differentiation of embryonic stem cells into neurons. *Journal of Neuroscience Research*, *86*, 2809–2819.

Nagpal, I., & Wei, L. N. (2019). All-trans retinoic acid as a versatile cytosolic signal modulator mediated by CRABP1. *International Journal of Molecular Sciences*, *20*, 3610.

Nemetz, C., & Hocke, G. M. (1998). Transcription factor Stat5 is an early marker of differentiation of murine embryonic stem cells. *Differentiation*, *62*, 213–220.

Norwood, C. W. (2011). *Development of a reproducible in vitro method to grow and study differentiation of mouse embryonic stem cells in the presence of the vitamin A derivative, retinoic acid.* ProQuest LLC ETD Collection for Tennessee State University, pp. 1–52, Paper AAI1492284.

Noy, N. (1992). The ionization behavior of retinoic acid in aqueous environments and bound to serum albumin. *Biochimica et Biophysica Acta*, *1106*, 151–158.

Ohnuki, Y., & Kurosawa, H. (2013). Effects of hanging drop culture conditions on embryoid body formation and neuronal cell differentiation using mouse embryonic stem cells: Optimization of culture conditions for the formation of well-controlled embryoid bodies. *Journal of Bioscience and Bioengineering*, *115*, 571–574.

Okada, Y., Shimazaki, T., Sobue, G., & Okano, H. (2004). Retinoic-acid-concentration-dependent acquisition of neural cell identity during in vitro differentiation of mouse embryonic stem cells. *Developmental Biology*, *275*, 124–142.

Osakada, F., & Takahashi, M. (2011). Neural induction and patterning in mammalian pluripotent stem cells. *CNS & Neurological Disorders Drug Targets*, *10*, 419–432.

Persaud, S. D., Lin, Y. W., Wu, C. Y., Kagechika, H., & Wei, L. N. (2013). Cellular retinoic acid binding protein I mediates rapid non-canonical activation of ERK1/2 by all-trans retinoic acid. *Cellular Signalling*, *25*, 19–25.

Phillips, B. W., Vernochet, C., & Dani, C. (2003). Differentiation of embryonic stem cells for pharmacological studies on adipose cells. *Pharmacological Research*, *47*, 263–268.

Piskunov, A., Al, T. Z., & Rochette-Egly, C. (2014). Nuclear and extra-nuclear effects of retinoid acid receptors: How they are interconnected. *Sub-Cellular Biochemistry*, *70*, 103–127.

Rochette-Egly, C. (2015). Retinoic acid signaling and mouse embryonic stem cell differentiation: Cross talk between genomic and non-genomic effects of RA. *Biochimica et Biophysica Acta*, *1851*, 66–75.

Rochette-Egly, C., & Germain, P. (2009). Dynamic and combinatorial control of gene expression by nuclear retinoic acid receptors (RARs). *Nuclear Receptor Signaling*, 7, e005.
Rohwedel, J., Guan, K., & Wobus, A. M. (1999). Induction of cellular differentiation by retinoic acid in vitro. *Cells, Tissues, Organs*, *165*, 190–202.
Sharow, K. A., Temkin, B., & Asson-Batres, M. A. (2012). Retinoic acid stability in stem cell cultures. *The International Journal of Developmental Biology*, *56*, 273–278.
Smith, A. G. (2001). Embryo-derived stem cells: Of mice and men. *Annual Review of Cell and Developmental Biology*, *17*, 435–462.
Szuts, E. Z., & Harosi, F. I. (1991). Solubility of retinoids in water. *Archives of Biochemistry and Biophysics*, *287*, 297–304.
Tropepe, V., Hitoshi, S., Sirard, C., Mak, T. W., Rossant, J., & van der Kooy, D. (2001). Direct neural fate specification from embryonic stem cells: A primitive mammalian neural stem cell stage acquired through a default mechanism. *Neuron*, *30*, 65–78.
Walker, E., Ohishi, M., Davey, R. E., Zhang, W., Cassar, P. A., Tanaka, T. S., et al. (2007). Prediction and testing of novel transcriptional networks regulating embryonic stem cell self-renewal and commitment. *Cell Stem Cell*, *1*, 71–86.
Wei, Q., Liu, H., Ai, Z., Wu, Y., Liu, Y., Shi, Z., et al. (2017). SC1 promotes MiR124-3p expression to maintain the self-renewal of mouse embryonic stem cells by inhibiting the MEK/ERK pathway. *Cellular Physiology and Biochemistry*, *44*, 2057–2072.
Wobus, A. M., Kaomei, G., Shan, J., Wellner, M. C., Rohwedel, J., Ji, G., et al. (1997). Retinoic acid accelerates embryonic stem cell-derived cardiac differentiation and enhances development of ventricular cardiomyocytes. *Journal of Molecular and Cellular Cardiology*, *29*, 1525–1539.
Wray, J., Kalkan, T., & Smith, A. G. (2010). The ground state of pluripotency. *Biochemical Society Transactions*, *38*, 1027–1032.
Ying, Q. L., Nichols, J., Chambers, I., & Smith, A. (2003). BMP induction of id proteins suppresses differentiation and sustains embryonic stem cell self-renewal in collaboration with STAT3. *Cell*, *115*, 281–292.
Ying, Q. L., Stavridis, M., Griffiths, D., Li, M., & Smith, A. (2003). Conversion of embryonic stem cells into neuroectodermal precursors in adherent monoculture. *Nature Biotechnology*, *21*, 183–186.
Zhang, Q., Jiang, J., Han, P., Yuan, Q., Zhang, J., Zhang, X., et al. (2011). Direct differentiation of atrial and ventricular myocytes from human embryonic stem cells by alternating retinoid signals. *Cell Research*, *21*, 579–587.

CHAPTER FIFTEEN

Identifying vitamin A signaling by visualizing gene and protein activity, and by quantification of vitamin A metabolites

Stephen R. Shannon[a,b], Jianshi Yu[c], Amy E. Defnet[c], Danika Bongfeldt[d,e], Alexander R. Moise[d,e], Maureen A. Kane[c], Paul A. Trainor[a,b,*]

[a]Stowers Institute for Medical Research, Kansas City, MO, United States
[b]University of Kansas Medical Center, Department of Anatomy and Cell Biology, Kansas City, KS, United States
[c]University of Maryland Baltimore, School of Pharmacy, Department of Pharmaceutical Sciences, Baltimore, MD, United States
[d]Medical Sciences Division, Northern Ontario School of Medicine, Sudbury, ON, Canada
[e]Departments of Chemistry and Biochemistry, and Biology and Biomolecular Sciences Program, Laurentian University, Sudbury, ON, Canada
*Corresponding author: e-mail address: pat@stowers.org

Contents

Methods in Enzymology, Volume 637
ISSN 0076-6879
https://doi.org/10.1016/bs.mie.2020.03.011

Abstract

Vitamin A (retinol) is an essential nutrient for embryonic development and adult homeostasis. Signaling by vitamin A is carried out by its active metabolite, retinoic acid (RA), following a two-step conversion. RA is a small, lipophilic molecule that can diffuse from its site of synthesis to neighboring RA-responsive cells where it binds retinoic acid receptors within RA response elements of target genes. It is critical that both vitamin A and RA are maintained within a tight physiological range to protect against developmental disorders and disease. Therefore, a series of compensatory mechanisms exist to ensure appropriate levels of each. This strict regulation is provided by a number synthesizing and metabolizing enzymes that facilitate the precise spatiotemporal control of vitamin A metabolism, and RA synthesis and signaling. In this chapter we describe protocols that (1) biochemically isolate and quantify vitamin A and its metabolites and (2) visualize the spatiotemporal activity of genes and proteins involved in the signaling pathway.

1. Introduction

1.1 Vitamin A and retinoic acid synthesis and signaling

Vitamin A (retinol) is an essential nutrient, critical for all stages of life from embryonic development through adult homeostasis. It is found in the diet in two primary forms, provitamin A carotenoids and preformed vitamin A (Green & Fascetti, 2016). Provitamin A carotenoids are derived from plant sources whereas preformed vitamin A comes largely from animal products such as liver, milk and eggs (Green & Fascetti, 2016). Once taken into the body vitamin A is metabolized to its active metabolite retinoic acid (RA), which directly regulates gene activity and function. It is critically important to maintain appropriate levels of vitamin A and RA throughout life as both excess and deficiency are associated with birth defects and adult disorders.

Vitamin A (retinol) enters the body as either preformed vitamin A or provitamin A carotenoids. Preformed vitamin A is hydrolyzed and taken up by enterocytes of the intestinal lumen (Fig. 1). Once inside the cell, it

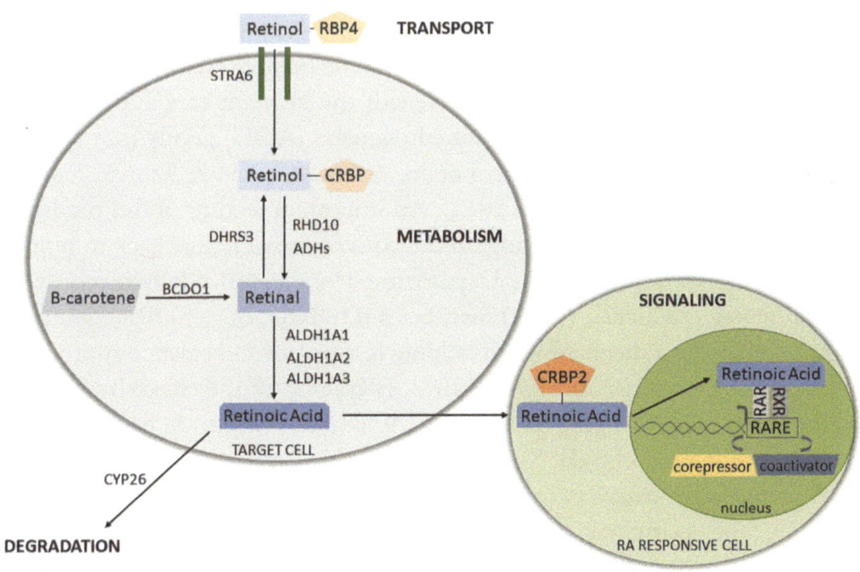

Fig. 1 Summary diagram of transport, signaling and degradation of vitamin A metabolism and RA synthesis. Retinol is transported throughout the vasculature and taken up by target cells where it is oxidized in a two-step reaction into retinoic acid (RA). β-carotene provides an alternative pathway by conversion into retinal by BCDO1 (b-carotene-15,15-dioxygenase). Once converted, RA can diffuse from the synthesizing cell and is either degraded by CYP26 enzymes or regulates cell signaling in target cells.

is esterified and packaged into chylomicrons that are secreted into circulation (Wongsiriroj et al., 2008). A large proportion of the retinol within these chylomicrons is acquired by liver hepatocytes and stored as retinyl esters in hepatic stellate cells (Moriwaki, Blaner, Piantedosi, & Goodman, 1988). Stores of retinyl esters can then be mobilized when needed via hydrolysis back to retinol, which is transported throughout the vasculature by retinol binding protein, RBP4 (Muenzner et al., 2013; Quadro et al., 2005). Target cells express the stimulated by retinoic acid 6 receptor (STRA6), which binds to circulating RBP4-retinol complexes, internalizing retinol for conversion to retinoic acid (Kawaguchi et al., 2007; Kelly, Widjaja-Adhi, Palczewski, & von Lintig, 2016).

Once inside a cell, retinol is bound by cellular retinol binding proteins (CRBP) and then undergoes two consecutive oxidation reactions to become retinoic acid (Fig. 1). In the first reaction, retinol is oxidized to all-trans-retinal (also called retinaldehyde) by microsomal retinol dehydrogenase 10 (RDH10) or other short-chain retinol dehydrogenases that recognize CRBP-bound retinol as substrate (Napoli, 2012, 2017). RDH10 is the critical enzyme necessary for the regulation of this first oxidation step during embryonic development (Sandell et al., 2007) whereas other RDH enzymes contribute postnatally (Belyaeva, Adams, Popov, & Kedishvili, 2020; Napoli, 2012). In situations of vitamin A excess, where retinol exceeds the amount of CRBP protein, members of the cytosolic alcohol dehydrogenases (ADH) family may catalyze this reaction postnatally (Deltour, Foglio, & Duester, 1999; Kumar, Sandell, Trainor, Koentgen, & Duester, 2012). An important feature of this oxidation step is that it is reversible, allowing for the conversion of retinal back to retinol. The reduction of retinal to retinol is performed by the retinaldehyde reductase, dehydrogenase/reductase (SDR) member 3 (DHRS3, or retSDR1), which in doing so provides a mechanism for helping to regulate and balance appropriate levels of retinol, retinal and RA (Adams, Belyaeva, Wu, & Kedishvili, 2014; Billings et al., 2013; Feng, Hernandez, Waxman, Yelon, & Moens, 2010; Haeseleer, Huang, Lebioda, Saari, & Palczewski, 1998).

An alternative pathway to produce retinal is via conversion of β-carotene (Fig. 1), the principal source of which comes from the provitamin A carotenoids in various fruits and vegetables (Green & Fascetti, 2016). Similar to retinol, β-carotene is transported by chylomicrons, and converted into retinal in target cells by BCDO1 (b-carotene-15,15-dioxygenase) (Harrison, 2012; von Lintig & Vogt, 2000; Wyss et al., 2000). This secondary pathway may be used for RA synthesis, as well as vitamin A storage via conversion of retinal back to retinol.

The second oxidative step in retinol metabolism converts retinal to RA (Fig. 1). This irreversible reaction is carried out by retinaldehyde dehydrogenases including, ALDH1A1, ALDH1A2, and ALDH1A3 (previously RALDH1, RALDH2 and RALDH3) (Dupe et al., 2003; Fan et al., 2003; Niederreither et al., 2001). RA is then transported from the cytosol by cellular retinoic acid binding protein (CRABPII) to the nucleus where it binds to the retinoic acid receptors (RAR) RARα, RARβ and RARγ (Germain et al., 2006a). These RARs form heterodimers with the retinoid X receptors (RXR) RXRα, RXRβ or RXRγ (Germain et al., 2006b), which then regulate gene transcription through their binding to RA response elements (RAREs) of target genes. In the absence of RA, RAR-RXRs associate with RAREs and repress gene expression through recruitment of co-repressors that include nuclear receptor corepressor (NCoR), silencing mediator of RA and thyroid hormone receptor (SMRT), as well as histone deacetylases and methyltransferases. In contrast, when RA is present, it binds as a ligand to RAR-RXR and through direct association with RAREs, initiates transcription of target genes by dis-associating the corepressors and recruiting co-activators that include steroid receptors (SRC-1, SRC-2 and SRC-3) (Al Tanoury, Piskunov, & Rochette-Egly, 2013).

RA is a small, lipophilic molecule and not only does it regulate gene activity within the cell in which it is produced, but it also acts as an autacoid acting in non-cell autonomous fashion on neighboring, RA-responsive cells. The distribution of RA creates spatial and temporal gradients that control signaling within target tissues (Shimozono, Iimura, Kitaguchi, Higashijima, & Miyawaki, 2013). However, the synthesis and diffusion of RA is constrained by catabolizing cytochrome *p*450 enzymes of the CYP26 family which degrade RA (Hernandez, Putzke, Myers, Margaretha, & Moens, 2007).

The ability to tightly regulate Vitamin A metabolism and RA synthesis and signaling is essential for proper embryogenesis and adult homeostasis. Cells and tissues therefore employ multiple feedback mechanisms to govern the levels of retinol, retinal and retinoic acid. When endogenous levels of RA are too high, Cyp26 genes are activated to help degrade excess RA. Furthermore, the enzymes responsible for oxidizing the first and second metabolic steps of vitamin A metabolism, such as *Rdh10*, and *Aldh1a1*, *Aldh1a2*, *Aldh1a3*, are downregulated. Conversely, *Dhrs3* is upregulated to further convert the intermediate molecule retinal back to retinol (Billings et al., 2013; Elizondo, Corchero, Sterneck, & Gonzalez, 2000;

Lee et al., 2012; Niederreither, McCaffery, Drager, Chambon, & Dolle, 1997; Zolfaghari, Chen, & Ross, 2012). In the opposite situation, whereby the levels of RA are too low, upregulation of *Rdh10*, *Aldh1a1*, *Aldh1a2* and *Aldh1a3,* together with downregulation of the Cyp26 degradation enzymes (Sandell, Lynn, Inman, McDowell, & Trainor, 2012) can collectively lead to an increase in RA synthesis and maintenance. In addition, many of the proteins responsible for transporting retinol and RA are encoded by genes that contain RAREs (Balmer & Blomhoff, 2002, 2005; Rhinn & Dolle, 2012). This therefore provides additional opportunities for modulating vitamin A metabolism, and RA synthesis and signaling.

1.2 The role of vitamin A and retinoic acid signaling in development and disease

Vitamin A, and RA synthesis and signaling play vital roles during embryo development and throughout adult homeostasis. Vitamin A is required for proper bone development, protection of the skin and mucosa, immune system defense, epithelial integrity, and normal development and function of the reproductive organs, hair and teeth (Clagett-Dame & DeLuca, 2002; Clagett-Dame & Knutson, 2011; Conaway, Henning, & Lerner, 2013; Endo, Mikedis, Nicholls, Page, & de Rooij, 2019; Ross, 2012; Yokota et al., 2019). However, both excess and deficiency of Vitamin A or RA causes dysfunction, hence the need to maintain appropriate levels of vitamin A and RA throughout life. The deleterious effects of excess RA were first inferred in teratogenesis studies of embryo development, whereby an excess of RA produced congenital anomalies that included defects in the formation of the brain, nervous system, heart and limbs, together with malformed eyes, jaws and palate (Cohlan, 1954; Kalter & Warkany, 1961). Retinoids are synthetic forms or derivatives of vitamin A and are typically used to treat dermatological conditions. The most well-known is isotretinoin (13-*cis*-retinoic acid), which is marketed as Accutane, and used to treat severe acne. However, vitamin A is contraindicated during pregnancy and fetal retinoid syndrome, which is characterized by craniofacial, nervous system and cardiovascular anomalies, can occur as a consequence of a mother taking retinoids during pregnancy. Conversely, congenital abnormalities and even fetal death can result from a deficiency in vitamin A or RA. Major organ systems affected by a deficiency in vitamin A, and RA synthesis and signaling include the craniofacial, cardiovascular and ocular tissues, circulatory, respiratory and urogenital systems (Antipatis et al., 1998; Moreau, Vilar, Lelievre-Pegorier, Merlet-Benichou, & Gilbert, 1998; Zile, 1998).

Appropriate levels of vitamin A are also essential for the health of the mother and adults in general, not just during embryo development. In particular, vitamin A plays an essential role in vision which relies on the 11-*cis*-retinaldehyde metabolite as the light-absorbing chromophore of opsin proteins (Dowling & Wald, 1960; Wald, 1968). Through its effects on epithelial integrity and function, vitamin A is important in the prevention of xerophthalmia, which is defined by dryness of the conjunctiva and cornea of the eye. As xerophthalmia progresses, it eventually leads to blindness, thus making vitamin A deficiency the main cause of preventable blindness (WHO, 2009). Vitamin A deficiency is therefore still considered to be a major health issue, particularly in developing countries.

Interestingly, an excess of vitamin A or RA can result in a pattern of defects similar to those caused by a deficiency of RA. This is particularly true with respect to craniofacial anomalies such as cleft palate, and cardiovascular malformations. This phenomenon is thought to occur via overcompensation to excess vitamin A or RA. Under conditions of excess, the RDH and ALDH enzymes are shut down to prevent further synthesis of RA, while Cyp26 enzymes are upregulated to catabolize pre-existing RA. However, rather than reversing these actions when RA levels return to normal, the system remains off, leading to a prolonged period of RA deficiency and consequently developmental anomalies (Lee et al., 2012).

In summary, maintaining proper levels of vitamin A and RA are essential for embryo development and adult homeostasis. This is accomplished through a series of compensatory mechanisms that have evolved to maintain an appropriate balance of vitamin A and RA to protect against developmental disorders and disease. Differential and dynamic patterns of synthesizing and metabolizing enzymes facilitate the precise spatiotemporal control of vitamin A metabolism, and RA synthesis and signaling. In this chapter we describe protocols for visualizing the spatiotemporal activity of genes and proteins involved in this process as well as biochemically isolating and quantifying vitamin A and its metabolites.

1.3 Direct quantification of RA levels and retinoid homeostasis

Numerous analytical approaches have been developed to quantify retinoids including gas chromatography-mass spectroscopy (GC/MS) (Napoli, 1990; Napoli, Pramanik, Williams, Dawson, & Hobbs, 1985), liquid chromatography with ultraviolet absorption (LC-UV) (Kane, Folias, & Napoli, 2008; Kane & Napoli, 2010; Schmidt, Brouwer, & Nau, 2003; Wyss & Bucheli, 1997), liquid chromatography-electrochemical detection (LC/ECD)

(Hagen, Washco, & Monnig, 1996; Sakhi, Gundersen, Ulven, Blomhoff, & Lundanes, 1998; Ulven et al., 2000), liquid chromatography-tandem mass spectroscopy (LC-MS/MS) (Arnold, Amory, Walsh, & Isoherranen, 2012; Kane, Chen, Sparks, & Napoli, 2005; Kane, Folias, Wang, & Napoli, 2008; Kane & Napoli, 2010), and liquid chromatography-multiple reaction monitoring cubed (LC-MRM^3 also known as liquid chromatography-multistage tandem mass spectrometry) (Jones, Pierzchalski, Yu, & Kane, 2015). The most rigorously validated quantitative methodologies for determination of endogenous RA concentrations in cell systems and tissues as well as for in vitro cell metabolism assays are liquid chromatography-tandem mass spectrometry-based techniques (Arnold et al., 2012; Jones et al., 2015; Kane et al., 2005; Kane, Folias, Wang, et al., 2008; Kane & Napoli, 2010). Absolute quantification via LC-MS/MS-based approaches rely on a unique precursor to product ion m/z transition, termed multiple reaction monitoring (MRM) or selected reaction monitoring (SRM), to yield specificity (Kane et al., 2005; Kane, Folias, Wang, et al., 2008). Tandem quadrupole instruments yield the most robust and reproducible analytical detection with typical linear ranges of 3–5 orders of magnitude and typical instrument coefficients of variation of <5% (Kane et al., 2005; Kane, Folias, Wang, et al., 2008). Quantitation is derived from calibration curves constructed from authentic standards. Stable isotope-labeled retinoids or nonendogenous retinoid derivatives are typically used as internal standards to correct for extraction efficiency and variability during data acquisition (Kane & Napoli, 2010).

Recent work has shown that a variation of LC-MS/MS-based detection that includes an additional mass transition during detection yields additional specificity for RA quantification (Jones et al., 2015). This liquid chromatography-multistage tandem mass spectrometry methodology is also interchangeably known as LC-MS^3 and LC-MRM^3. Although MRM detection in LC-MS/MS methods is one of the most selective detection modalities, abundant species in extracts of complex matrices that coelute with the analyte of interest can interfere with quantification (Jones et al., 2015; Kane, 2012). Because most extraction methods for retinoids also extract (more abundant) lipids of similar nominal mass, the potential for interferences is significant and requires additional caution and careful method validation. Current strategies typically include the use of chromatographic separation to move interfering species to retention times away from RA and/or the inclusion of additional mass transitions to impart additional selectivity (Jones et al., 2015; Kane, 2012).

RA quantification poses several additional significant analytical challenges including low endogenous concentrations, often spatially localized occurrence, and existence of endogenous geometric isomers with distinct biological functions (Kane & Napoli, 2010). Isobaric geometric isomers present an additional analytical challenge as they must be chromatographically resolved before mass spectrometric detection. Additional experimental challenges that are critical to control include the susceptibility of retinoids to light-induced isomerization and oxidation, as well as adherence to plastic surfaces. Yellow or red laboratory lights (blocking UV wavelengths $<\sim 500\,\text{nm}$) are recommended for experimental and sample preparation work in addition to efforts toward protecting samples from light by cover and/or amber vials in often non-UV blocking laboratory settings during analysis. Samples should be kept cold (on ice) during sample preparation and glass pipets, vials and containers, should be used as much as possible to prevent sample loss by adsorption to plastic (Kane & Napoli, 2010).

Typical detection limits for LC-MS/MS detection of RA are low fmol on column, which enables the determination of low nanomolar concentrations of endogenous RA in either cell systems or tissues. Typical linear ranges extend from low fmol to up to $\sim$ 1–10 pmol on column. Typical cell culture requirements range between 1×10^5 and 1×10^6 cells, depending on the cell type, level of RA present, and the lower limit of quantification (LLOQ) of the assay being used. Tissue requirements typically range from 1 to 100 mg of tissue, with the minimum amount of tissue required depending on the LLOQ for the assay being used and the levels of RA present. If multiple assays are required, then tissue requirements can be higher. Minimum amounts of tissue required and efficiency of extraction from matrices should be evaluated before quantitative experiments are undertaken. The amount of RA is typically expressed as mol of RA per g tissue (requires accurate weight of tissue), mol of RA per g protein (requires a separate protein determination), mol of RA per million cells (requires accurate cell count), or mol RA per mL of fluid, such as media or plasma (requires an accurate volume determination) (Jones et al., 2015; Kane et al., 2005; Kane, Folias, Wang, et al., 2008; Kane & Napoli, 2010).

Quantification of retinoids that serve as the substrates for RA biosynthesis as well as vitamin A storage can inform the basis of homeostatic mechanisms (Kane, Folias, & Napoli, 2008; Wang, Yoo, Obrochta, Huang, & Napoli, 2015). The concentration of different retinoid species can differ by as much as six orders of magnitude, making methods that simultaneously detect all retinoids not technically practical. For example, retinyl ester levels can reach high micromolar levels whereas endogenous RA is typically low nanomolar in concentration. Quantification of RE and retinol can be accomplished by LC-UV (Kane, Folias, & Napoli, 2008; Kane & Napoli, 2010). These more abundant retinoids (typical concentrations in micromolar range) can be feasibly detected by UV detection due to their favorable molar absorptivity (typically $\varepsilon = \sim 30{,}000–60{,}000\,M^{-1}\,cm^{-1}$), and unique absorption wavelengths, which are red-shifted from many other biological molecules (Kane, Folias, & Napoli, 2008; Kane & Napoli, 2010). LC-MS/MS methodology also exists for quantification of these more abundant RE and retinol species, which have less of a requirement for the sensitivity of LC-MS/MS-based approaches (Wang et al., 2015).

Retinal, or retinaldehyde, requires derivatization due to its reactive aldehyde group for accurate quantification. Retinaldehyde forms covalent Schiff base adducts with amine groups of proteins, phospholipids and other compounds, making it necessary to react the extract with hydroxylamine or *O*-ethylhydroxylamine to release the bound retinal and covert it to the respective *syn* and *anti*-retinaloxime adducts (Golczak, Bereta, Maeda, & Palczewski, 2010; Kane, Folias, & Napoli, 2008; Kane & Napoli, 2010; Wang et al., 2015). This is an important consideration for zebrafish, frog and avian models since retinaldehyde-vitellogenin Schiff-based adducts are the main storage form for retinoids in eggs and oocytes of oviparous animals (Costaridis, Horton, Zeitlinger, Holder, & Maden, 1996; Levi, Ziv, Admon, Levavi-Sivan, & Lubzens, 2012). Whereas, LC-UV methods have been popular, recent LC-MS/MS methodology provides more sensitive detection of retinal, which is important for samples of limited quantity that are often encountered in developmental studies (Golczak et al., 2010; Wang et al., 2015).

Additional, specialized analytical approaches for retinoids include tissues where *cis*-retinoid isomers have significant biological roles, such as the eye and pancreas. Methods for isomeric retinoid detection has been previously reviewed (Kane, 2012; Kane & Napoli, 2010). For the measurement of geometric isomers of retinoids found in the visual system, an isocratic or step-gradient normal-phase HPLC method has been developed to resolve 11-*cis* and all-trans-retinol and retinyl esters, as well as all-trans, 9-*cis*, and 11-*cis*-retinaloximes (Batten et al., 2004; Golczak et al., 2010). Chiral-HPLC can also be employed to resolve (13*R*)- and (13*S*)-enatiomers of all-*trans*-13, 14-dihydroretinol (Moise et al., 2008). For more specialized methods to measure cytotoxic lipofuscin chromophores formed in the eye such as pyridinium bisretinoids (*N*-retinyl-*N*-retinylidene ethanolamine or A2E) and retinal dimers, we refer the reader to several reviews (Bowrey et al., 2016; Crouch, Koutalos, Kono, Schey, & Ablonczy, 2015; Golczak et al., 2010).

1.4 Assessing retinoic acid signaling in cellular and animal models via reporter assays

Reporter screens can provide crucial information about the spatial resolution or extent of RA signaling within cells or tissues (reviewed in Table 1). These reporters are based on the expression of a readily assayable enzyme (beta-galactosidase or luciferase) or a fluorescent protein driven by a promoter controlled by RA signaling. Some RA signaling reporters are based on a RARE derived from the promoter of the *Rarb* gene consisting of a canonical GTTCAC direct repeat separated by 5 nucleotides (DR5) (Sucov, Murakami, & Evans, 1990). These reporters have been incorporated in cell models (Wagner, Han, & Jessell, 1992), as well as transgenic mice (Balkan et al., 1992; Rossant et al., 1991) and zebrafish (Perz-Edwards et al., 2001; Waxman & Yelon, 2011). However, a disadvantage of the use of DR5 based-reporters is that RA-signaling via other types of RAREs (Moutier et al., 2012) may not be detected. In addition, the DR5 enhancer is also recognized by other nuclear receptors such as COUP-TF (Cooney, Tsai, O'Malley, & Tsai, 1992).

Other reporters of RA signaling use a two-hybrid approach based on expression of a fusion protein of the ligand-binding domain (LBD) of RAR and the DNA-binding domain (DBD) of the GAL4 transcription factor. Expression of GAL4-nuclear receptor fusions has been a very valuable tool for high-throughput screening of compounds with agonist/antagonist activity against various nuclear receptors including RAR and RXR (Allenby et al., 1993; Grimaldi et al., 2015; Moise et al., 2009). In these systems binding of RA to the LBD of RAR induces expression of GAL4-regulated reporters (luciferase, GFP) controlled by upstream activation sequences (UAS). In comparison to DR5-based reporters, both binding of RA to the RAR-LBD and the induction of UAS-controlled reporters by GAL4-RAR chimeric receptors are very specific. Another advantage of the use of GAL4-RAR;UAS-reporter systems is that they do not rely on endogenous RAR receptors for reporter expression. However, the generation of compound GAL4-RAR;UAS-reporter expressing transgenic lines is cumbersome and time consuming since both the UAS-reporter and Gal4-RAR transgenes need to be present for activity. GAL4-RAR; UAS-reporter genes have also been introduced in mice and in zebrafish

Table 1 Summary of retinoic acid signaling reporters.

Transgenic mouse lines	Transgene construct
RARE-LacZ (Rossant, Zirngibl, Cado, Shago, & Giguere, 1991) available from Jackson Labs Stock No: 008477	Express a beta-galactosidase reporter under the control of a minimal tk promoter containing 3 RARE derived from Rarb (Balkan, Colbert, Bock, & Linney, 1992)
RARE-hsp68LacZ	Express a beta-galactosidase reporter under the control of a minimal hsp68 promoter and three copies of RARE from Rarb
Feedback-inducible nuclear-receptor-driven (FIND) expression of GAL4-RAR/UAS-hsp-lacZ (Mata De Urquiza, Solomin, & Perlmann, 1999)	Express a chimeric receptor fusion of Gal4-DBD and RAR-LBD. A second transgene UAS-hsp-lacZ expresses a lacZ reporter driven by a minimal hsp68 promoter including four UAS. The expression of the chimeric receptor Gal4-RAR is itself autoregulated by incorporating 4 UAS in a minimal hsp68 promoter

Transgenic zebrafish lines	Transgene construct
Tg(3XRARE-tk-GFP)	Express GFP under the control of a minimal promoter (tk or gata-2) including three copies of RARE (Perz-Edwards, Hardison, & Linney, 2001)
Tg(3XRARE-gata-2-GFP)	
Tg(12XRARE-ef1a:gfp)sk71	Express a GFP reporter driven by a promoter under the control of 12 concatenated RAREs (Waxman & Yelon, 2011)
Tg(β-actin:GDBD-RLBD); (UAS-reporter) or Tg(β-actin:VPBD-RLBD);(UAS:EGFP)	Express chimeric receptors based on Gal4-DBD or VP16-Gal4-DBD fused to the RARA-LBD. Ligand binding triggers receptor activation and is detected via an UAS-driven reporter (D'Aniello, Rydeen, Anderson, Mandal, & Waxman, 2013; Mandal et al., 2013)
Transgenic zebrafish expressing genetically encoded probes for RA(GEPRAs)	Created from chimeric receptors of RAR-LBD fused at the N and C-termini to CFP and YFP, respectively. Binding of RA causes a conformational change in the chimeric receptor detected via FRET (Shimozono et al., 2013)

(D'Aniello et al., 2013; Mandal et al., 2013; Mata de Urquiza & Perlmann, 2003; Mata De Urquiza et al., 1999). Constitutive overexpression of chimeric receptors based on fusions of the Gal4-DBD to the RAR-LBD leads to phenotypic effects in mice. Therefore, a feedback-inducible nuclear-receptor-driven (FIND) expression system was generated to control the expression of the GAL-RAR chimeric receptor in mice (Mata de Urquiza & Perlmann, 2003; Mata De Urquiza et al., 1999). Expression of similar chimeric receptors in zebrafish encoding the *Tg(β-actin:GDBD-RLBD)*; (*UAS-reporter*) or the *Tg(β-actin: VPBD-RLBD)*;(*UAS:EGFP*) transgenes have not been reported to cause phenotypic effects (D'Aniello et al., 2013; Mandal et al., 2013).

Whereas reporter systems offer valuable spatial information and RA signaling read-outs, caution should be exercised in the interpretation of reporter assay response in terms of concentration and temporal relationships. It is important to stress that reporter-based methods are nonquantitative and do not provide a linear response to the level of RA in tissues and thus reflect a qualitative extent of RA signaling which is correlated with RA levels. And, whereas valuable qualitative relationships can be assessed under comparable conditions, specific, absolute quantitation for RA cannot be achieved via reporter assays. Some reporter assays have cited that in addition to all-*trans*-RA, other endogenous retinoids can produce signals to varying extents including 3,4-didehydro-RA, 9cRA, 4-oxo-RA, 4-hydroxy-RA, and 4-hydroxy-retinol (McCaffery & Drager, 1994; Wagner, 1997; Wagner et al., 1992; Zetterstrom et al., 1999). Temporal relationships should also be cautiously and carefully interpreted. Since reporter assays reflect RAR activation, they reveal the longer term consequences of receptor activation with a delay in the reporter signal appearance often hours after a change in RA concentration (McCaffery & Drager, 1994; Wagner, 1997; Wagner et al., 1992; Zetterstrom et al., 1999) (Figs. 2–7). Moreover, the half-life of the RAR reporters employed thus far generally outlasts the RA input. Thus, reporter assays may not accurately correlate with RA concentration via RA signaling in real-time. Temporal delays should be considered as well as the possibility that RA may even be significantly catabolized by the time the reporter signal is visualized. Lastly, experimental caution and consideration should be paid to reports that high levels of RA can turn off reporter response in some cases (McCaffery & Drager, 1994).

The most recent addition to the RA reporter systems allows for detection of RA in transgenic zebrafish expressing genetically encoded probes for RA(GEPRAs) (Shimozono et al., 2013). These reporters express the

RAR-LBD flanked by Förster resonance energy transfer (FRET) partners CFP and YFP. Binding of RA causes a conformational change in the chimeric receptor which can be detected via FRET. Since this reporter does not require any additional substrates, it allows for the visualization of endogenous RA gradients in vivo in zebrafish embryos. This approach has great potential in studying the formation of RA in tissues in real-time, however, more studies are needed to establish the sensitivity of the GEPRAs-based reporter and its RA concentration-dependent correlation.

2. Direct quantification of retinoic acid production and endogenous retinoid levels in cells

Basal levels of RA can be determined in primary or cultured cells. If basal RA level determination is desired, skip to part 3 in Section 2.2 under "Protocols," Retinoid Extraction. However, amounts of RA can be low in cellular systems necessitating large numbers of cells for analysis (e.g., 1×10^6 cells or greater depending on the cell type). Alternatively, assessment of RA production is also a useful metric for evaluating RA homeostasis. RA production assays where the amount of RA produced from a given substrate over a defined time typically utilizes 1×10^5 cells (or greater, if available) per replicate. For RA production evaluation, similar to standard enzyme activity assays, timepoint and concentration ranges should first be determined by performing a time-course experiment and a concentration-course experiment to determine conditions within the linear range (Kane & Napoli, 2010; Wang, Kane, & Napoli, 2011). The RA production assay is amenable to a wide variety of cell types and culture conditions.

2.1 Reagents

1. All-trans-Retinol (store at −20 °C with desiccant)
2. All-trans-Retinoic acid (store at −20 °C with desiccant)
3. All trans-Retinal (store at −20 °C with desiccant)
4. Acetone (HPLC)
5. Saline (0.9% NaCl): 9 g NaCl in 1 L water
6. 0.025 M KOH in 100% ethanol
7. 50 mL 4 M HCl
8. Hexanes (Certified or highest grade available)

2.2 Protocols

Part 1. Retinol solution preparation

1. Under yellow light or red light, bring the all-trans retinol (Sigma R7632) and all-trans-Retinoic Acid (Toronto Research Chemicals Cat# R250200) to room temperature.
2. Add 10 mL ethanol to a clean borosilicate glass scintillation vial.
3. Use a clean disposable borosilicate glass Pasteur pipette to take up the desired amount of retinoid powder. The powder in the pipette tip should be about 1 mm in diameter.
4. Put the pipette in the borosilicate glass scintillation vial containing 10 mL ethanol.
5. Make sure the powder gets off the glass pipette tip.
6. Put the cap onto the vial and close the cap securely.
7. Vortex the solution for 20 s.
8. Repeat step 7 two more times.
9. Measure the retinoid absorbance spectrum using a UV/Vis spectrophotometer by scanning wavelength from 250 nm to 450 nm.
10. Use ethanol to blank the spectrophotometer.
11. If the absorbance is too high (typically absorbance should be below 1 on most spectrophotometers), dilute the retinol solution with ethanol and repeat the measurement.
12. Make at least 3 measurements.
13. Calculate the retinol concentration based on Beer's law: Absorbance $= \varepsilon Lc$ (don't forget the dilution factor if you dilute the solution).

Technical notes

- Work under yellow or red light.
- Always use borosilicate glass pipettes and containers. No plastic tubes or tips or pipettes. Retinoids stick to many plastics and variable loss of up to 40% can occur when pipetting retinoid solutions with regular (plastic) pipette tips. Loss due to surface adherence is especially pronounced at low concentrations.
- For determination of UV absorbance of retinoid solutions, cuvettes should be cleaned thoroughly before and after use with ethanol and/or acetone and then dried completely. If acetone is used, complete removal is particularly important as acetone will absorb significantly at low wavelengths. For more stringent cleaning, rinse cuvettes with concentrated nitric acid followed by water, then 100% ethanol and completely dry.
- λ_{max} of all-*trans*-retinol in ethanol is 325 nm. Molar absorptivity (ε) is wavelength (λ) and solvent dependent. Use the appropriate solvent blank

according to the solvent defining the ε value. Molar absorptivity of all-trans retinol in ethanol at 325 nm is $\varepsilon = 52{,}770\,\mathrm{M}^{-1}\,\mathrm{cm}^{-1}$. λ_{max} of all-*trans*-retinoic acid in ethanol is 350 nm. Molar absorptivity of all-trans retinoic acid at 350 nm in ethanol $\varepsilon = 45{,}300\,\mathrm{M}^{-1}\,\mathrm{cm}^{-1}$. Path length $L = 1$ cm for standard cuvette.

- Calculate concentration using the following formula $c = \frac{A}{\varepsilon L}$.
- Retinol solution preparation procedures are also applicable to other retinoids.
- Preparation of retinal can be accomplished with a similar procedure. All-trans-retinal has a λ_{max} of 338 and ε of 42,880 in ethanol or a λ_{max} of 368 and ε of 48,000 in hexane.
- The typical absorbance spectra of selected retinoids in ethanol is shown in Fig. 2.

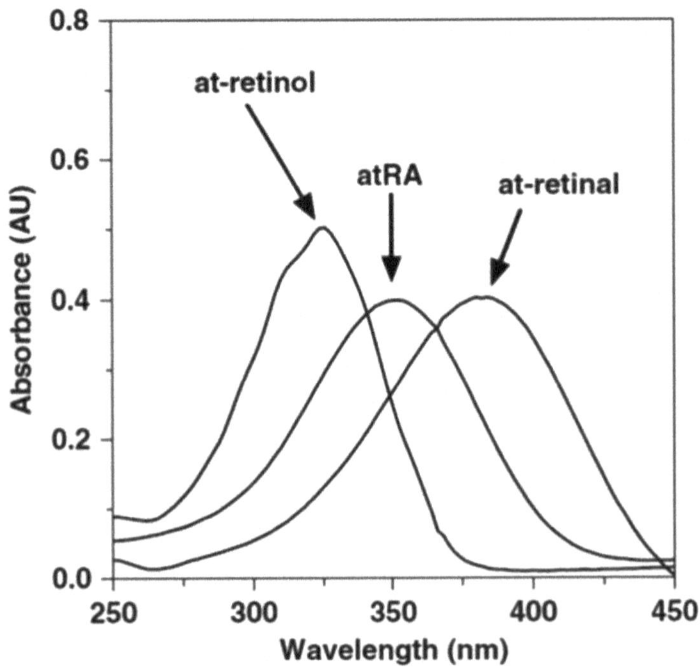

Fig. 2 Example of retinoid absorbance spectra that are typically used for determining the concentration of retinol, RA and retinal. *Reproduced from Kane, M. A., & Napoli, J. L. (2010). Quantification of endogenous retinoids.* Methods in Molecular Biology, 652, *1–54. doi:10.1007/978-1-60327-325-1_1.*

Part 2. Treating cells with retinol

1. Follow existing protocols for culturing your cells of interest. Typically, the number of cells needed is at least 0.1 million live cells per extraction. If possible, use 0.5 or 1 million live cells per replicate. Prepare a minimum of three replicates. The culture volume should be at least 1 mL. Adjust your culture vessel configuration if necessary. Typically, a culture of 0.1 million cells per well uses a 12-well plate with 1 mL culture media. Determine the live/total cell ratio by Trypan Blue staining. The ratio should be at least 90% to be consistent with high viability and experimental conditions, which yields reproducible and representative results.
2. In most cases, cell culture media should be replaced with serum-free media immediately preceding the retinol treatment period. Serum contains retinol and will provide an additional (and possibly variable) amount or retinol as substrate. Serum-free media with added retinol (or other retinoid substrate) provides a defined substrate concentration. Most cell lines and types can tolerate serum-free conditions during the treatment period, which is typically on the order of hours.
3. If you expect some cell death during your treatment or transfection, try to get a live cell estimate at the time of assay.
4. Under yellow or red light, transfer desired amount of retinoid solution (described herein as retinol) to the culture using a Hamilton syringe. Typically, the final all-trans retinol concentration should be 2 μM or another physiologically relevant concentration (typically 1–10 μM). For example, if your culture volume is 2 mL, and your fresh prepared retinol solution is 400 μM, then you need add 10 μL all trans retinol solution to the culture media. You do not have to remove 10 μL culture media before adding all-trans retinol solution, but you need to add an equal volume of ethanol (or selected vehicle) to the control wells. You do not have to use Hamilton syringe for ethanol/vehicle controls.
5. After adding retinoid solution to the culture media, mix the media gently and return the culturing cells to the incubator for the desired period of time.
6. Collect both culture media samples and cell pellet samples under yellow light or red light.
7. Put the samples in −80 °C immediately until extraction.

Technical notes

- For accurate delivery of a solution of defined retinoid concentration, proper cleaning of the Hamilton syringe used for delivery is a critical element. It is essential to clean the Hamilton syringe by flushing *before* and *after* each experiment by drawing up clean acetone 30 times. This will ensure there is no retinol carry-over. After cleaning the syringe, *dry* the syringe. Make sure both the glass barrel and the plunger are completely dry of residual acetone.

- RA production from retinal as a substrate can be performed with a similar procedure. Typical treatment concentrations range from 0.1 to 1 μM.
- Catabolism of RA can be evaluated by determining the remaining RA over time and calculating the turnover of RA via the elimination half-life.

Part 3. Retinoid extraction and quantification

1. For the RA extraction, collect 800 μL of media in a 16 × 125 mm borosilicate glass disposable culture tube as previously described (Kane & Napoli, 2010).
2. Remove the remaining media and use a cell scraper to harvest the cells for LC-MS/MS-based analysis.
3. If immediate sample preparation is not possible, add 300 μL of 0.9% Saline to each well to store the cells.
4. Cover collected cell and media samples in foil to protect from light and store at −80 °C.

RA can then be extracted from cells and media using a two-step liquid-liquid extraction, and quantified using liquid chromatography-multistage tandem mass spectrometry (Jones et al., 2015; Kane et al., 2005; Kane, Folias, Wang, et al., 2008; Kane & Napoli, 2010).

3. Direct quantification of RA in embryonic tissues

3.1 Reagents and materials

1. Acetone (HPLC)
2. Saline (0.9% NaCl)
3. 0.025 M KOH in 100% ethanol
4. 4 M HCl
5. Hexanes (Certified or highest grade available)
6. Internal standard(s) (choose according to desired analyses)
7. Glass homogenizer
8. Glass culture tubes and glass transfer pipets

3.2 Protocol

1. Tissue collection, homogenization, and extraction methodology has been described in detail previously (Kane & Napoli, 2010). Procedures and key points for analyses in developmental models are provided here.

2. Determine the minimum amount of tissue required for your RA assay by systematically testing a range of tissue quantities. Typical tissue amounts range from 1 to 100 mg, depending on the RA content of the target tissue.
3. Collect tissue under yellow (or red) lights; if using a dissecting microscope use a yellow (or red) filter.
4. Freeze tissue upon collection in liquid nitrogen or on dry ice and store at −80 °C until assay.
5. Determine an accurate tissue weight or determine the total protein content for normalization.
6. It is important to thaw tissue on ice and keep tissue on ice while homogenizing and extracting.
7. Homogenize tissue in saline (0.9% NaCl), typically 1 mL saline. (Be sure to rinse all homogenizing glassware with water followed by ethanol and/or acetone before and after use to remove retinoid residue.)
8. Extract tissues using a two-step liquid-liquid extraction that allows for extraction of both neutral retinoids (retinol, retinyl esters) and polar retinoids (RA, RA metabolites) from the same sample (Kane & Napoli, 2010).
9. If retinaldehyde quantitation is desired, an aliquot of homogenate or a separate sample is required to allow for derivatization of the aldehyde group by *O*-ethylhydroxylamine (Kane, Folias, & Napoli, 2008; Kane & Napoli, 2010; Wang et al., 2015).

Note: RA levels that have been reported in embryo and embryonic tissues are compiled in Table 2.

3.3 Analysis

3.3.1 Quantification of RA

Quantify RA using an LC-MS/MS-based assay. Typical sample numbers range from 5 to 10 biological replicates. However, the sample numbers required to demonstrate statistical significance may be more or less depending on the normal variability of RA within the target tissue and the magnitude of change in RA observed in the experimental model.

3.3.2 Considerations for RA quantification in tissue

LC-MS/MS-based quantification in tissue is applicable to any tissue and yields a direct, absolute quantification of RA. It imparts temporal accuracy

Table 2 Select RA levels in mouse embryos and tissues.

Model	Type of sample	Day of assay	Reported value for RA[a, b, c]	RA Phenotype[b]	Ref[b]
Mouse: *wild-type*	Whole embryo	E12.5	20 pmol/g tissue	Normal	Wang et al. (2018)
Mouse: *Dhrs3*$^{+/-}$	Whole embryo	E12.5	21 pmol/g tissue	Similar to normal	Wang et al. (2018)
Mouse: *Dhrs3*$^{-/-}$	Whole embryo	E12.5	27 pmol/g tissue	Excess RA	Wang et al. (2018)
Mouse: *wild-type*	Whole embryo	E14.5	5.2 pmol/g tissue	Normal	Billings et al. (2013)
Mouse: *Dhrs3*$^{+/-}$	Whole embryo	E14.5	5.2 pmol/g tissue	Similar to normal	Billings et al. (2013)
Mouse: *Dhrs3*$^{-/-}$	Whole embryo	E14.5	7.7 pmol/g tissue	Excess RA	Billings et al. (2013)
Mouse: *wild-type*	Whole embryo	E10.5	350 pmol/g protein	Normal	Ashique et al. (2012)
Mouse: *Rdh10*^{m366Asp}	Whole embryo	E10.5	275 pmol/g protein	Reduced RA	Ashique et al. (2012)
Mouse: *wild-type*	Whole embryo	E12.5	275 pmol/g protein	Normal	Ashique et al. (2012)
Mouse: *Rdh10*^{m366Asp}	Whole embryo	E12.5	100 pmol/g protein	Reduced RA	Ashique et al. (2012)
Mouse: *wild-type*	Whole embryo	E14.5	210 pmol/g protein	Normal	Ashique et al. (2012)
Mouse: *Rdh10*^{m366Asp}	Whole embryo	E14.5	110 pmol/g protein	Reduced RA	Ashique et al. (2012)
Mouse: *Rbp4*$^{-/-}$ VAS	Whole embryo	E12.5	120 pmol/g protein	Similar to normal	Fu et al. (2010)
Mouse: *Rbp4*$^{-/-}$ VAD	Whole embryo	E12.5	57 pmol/g protein	Reduced RA	Fu et al. (2010)

Mouse: *Foxc1*$^{+/}$	Meninges	E14.5	740 pmol/g protein	Normal	Siegenthaler et al. (2009)
Mouse: *Foxc1*$^{l/l}$ *(Foxc1-null)*	Meninges	E14.5	590 pmol/g protein	Reduced RA	Siegenthaler et al. (2009)
Mouse: *Foxc1*$^{+/}$	Cortex	E14.5	280 pmol/g protein	Normal	Siegenthaler et al. (2009)
Mouse: *Foxc1*$^{l/l}$ *(Foxc1-null)*	Cortex	E14.5	140 pmol/g protein	Reduced RA	Siegenthaler et al. (2009)
Mouse:wild-type	Hippocampus	E19	20 pmol/g tissue	Normal	Kane, Folias, Wang, and Napoli (2010)
Mouse: wild-type +ethanol, low BAC %	Hippocampus	E19	30 pmol/g tissue	Excess RA	Kane et al. (2010)
Mouse: wild-type +ethanol, high BAC%	hippocampus	E19	400 pmol/g tissue	Excess RA	Kane et al. (2010)
Mouse:wild-type	Cortex	E19	12 pmol/g tissue	Normal	Kane et al. (2010)
Mouse: wild-type +ethanol, low BAC %	Cortex	E19	24 pmol/g tissue	Excess RA	Kane et al. (2010)
Mouse: wild-type +ethanol, high BAC%	Cortex	E19	600 pmol/g tissue	Excess RA	Kane et al. (2010)

[a]In some cases, values have been rounded from reported data and/or estimated from graphical representations.
[b]See references indicated for specific RA data including error, statistical significance, and numbers of N as well as full description of accompanying phenotype.
[c]Note that data expressed as per g protein is typically a ~25–250 × greater value than per g tissue—depending on protein content of sample type.

and represents the level of RA at the time of collection. However, LC-MS/MS-based quantification of RA is fairly time-intensive and requires specialized instrumentation. The spatial resolution of LC-MS/MS-based analysis, which relies on homogenization of tissue regions before analysis, is limited to the ability to dissect regions of interest and does not yield any intact visualization of RA. Additionally, the typical tissue requirements for RA analysis may necessitate pooling of samples.

4. Cell-based assays of RA signaling: F9 rare-LacZ reporter

4.1 Materials and reagents

1. 0.2% gelatin coated plates
2. Dulbecco's Modified Eagle Medium: Nutrient Mixture F-12 (DMEM/F12) supplemented with 10% Heat-Inactivated Fetal Bovine Serum (FBS)
3. 1 × Penicillin-Streptomycin
4. 0.4 mg/mL G418, Geneticin

4.2 Reporter reagents (available upon request)

The RARE reporter cell line F9-RARE-lacZ (SIL15-RA) was a kind gift from Dr. Michael Wagner (SUNY Downstate Health Sciences University) and Dr. Peter McCaffery (University of Aberdeen). The RA-responsive F9 cell line expresses the *E. coli* lacZ gene under the control of a RARE element derived from the human retinoic acid receptor-β gene (RARβ) (Wagner et al., 1992).

4.3 Protocols

Culture of F9-RARE-LacZ cells

F9-RARE-LacZ cells can be cultured and maintained as previously described (Ababon, Li, Matteson, & Millonig, 2016; Wagner et al., 1992) using 0.2% gelatin coated plates and Dulbecco's Modified Eagle Medium: Nutrient Mixture F-12 (DMEM/F12) supplemented with 10% Heat-Inactivated Fetal Bovine Serum (FBS), 1 × Penicillin-Streptomycin, and 0.4 mg/mL G418, Geneticin.

F9 RARE-LacZ assays of RA-signaling activity: concentration dependent response

1. Set up each each replicate of RA-treated F9 RARE-LacZ cells in duplicate assays to allow for the measurement of RA levels by LC-MS/MS in parallel with assessment of the RA-signaling activity observed at the same time point.

2. Cells should be minimally seeded at 5×10^4 cells per well in a 0.2% Gelatin Coated 12-well cell culture plate.
3. Incubate cells with 1 mL of media supplemented with either all-trans RA (final concentration 0.5, 5 or 50 nM as in Fig. 3 or with desired concentration(s)), or unsupplemented culture media as a control.

Determination of RA levels via LC-MS/MS in F9 RARE-LacZ stable cell line

1. At the desired time points (e.g., 0, 0.5, 1, 2, 3, 6, 12, 24, 48, 72, and 96 h as in Fig. 3 or desired timepoint(s)) collect media aliquots.
2. Remove remaining culture media.
3. Use a cell scraper to harvest cells for LC-MS/MS analysis.
4. Media and cell pellets can be frozen at −80 °C until sample extraction.

F9 RARE-LacZ stable cell line sample collection and β-galactosidase enzyme reporter assay

1. For the assessment of reporter response via β-Galactosidase, remove growth medium and wash cells twice with 200 μL of PBS Buffer.
2. Add 200 μL of Reporter Lysis Buffer to the cells and rock at room temperature for 15 min.
3. Scrape off cells and transfer to a 1.5 mL centrifuge tube.
4. Vortex for 10 s, and centrifuge at 4 °C for 2 min at top speed.
5. Transfer the supernatant into a 1.5 mL DNA LoBind Tube (Eppendorf) and store at −80 °C until all required timepoints are collected.
6. A β-Galactosidase Enzyme Assay System with Reporter Lysis Buffer (RLB) (Promega) kit can be used to assay the samples according to the manufacturer's 96-well plate format. Absorbance is monitored at 420 nm with a POLARstar Omega plate reader (BMG Labtech) or comparable plate reader.

4.4 Analysis

Evaluating RA reporter response in the F9 RARE-LacZ stable cell line: cellular assay of RA-signaling activity as compared to RA levels: In Fig. 3, an example of data obtained from the F9 RARE-LacZ stable cell is shown in comparison to RA determined by LC-MS/MS-based detection. Cells were treated with varying concentrations of RA (0.5, 5, and 50 nM RA) and reporter response via β-galactosidase staining as well as RA levels were assessed over time up to 96 h. Fig. 3A, the β-galactosidase reporter assay showed an increase in signal starting at 3 h, followed by a peak in signal at 24 h, which subsequently decreased at 48 h for all atRA concentrations. No difference was seen between the 5 and 50 nM ATRA treatment. Fig. 3B and C shows the RA

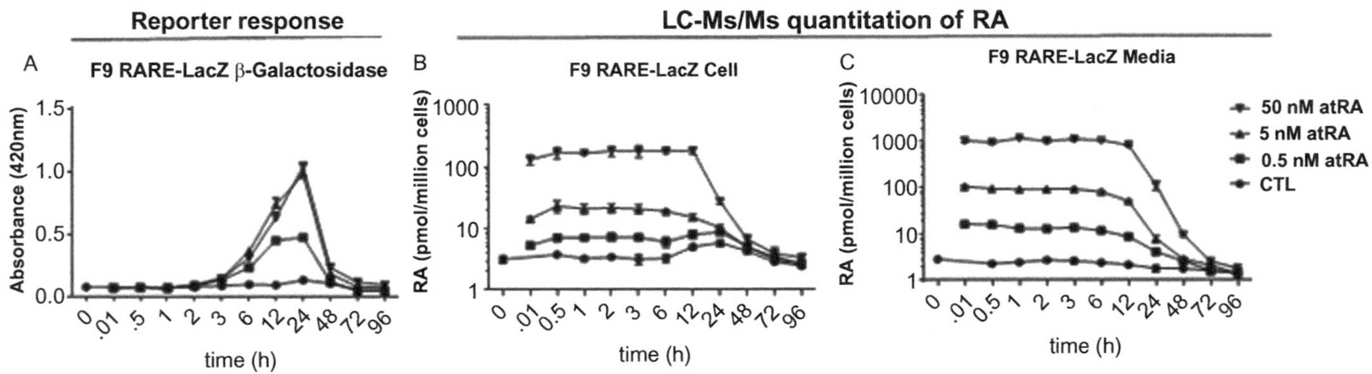

Fig. 3 Evaluating RA reporter response in the F9 RARE-LacZ stable cell line: β-Galactosidase reporter assay indicating RA-signaling activity as compared to RA levels. Cells were treated with 0–50 nM of *all-trans* retinoic acid (atRA or RA) and reporter response and RA levels were measured over time for 0–96 h. (A) The β-galactosidase reporter assay cell (B) RA in cell as determined by LC-MS/MS (C) RA in media as determined by LC-MS/MS. Cell culture for (A)–(C) and reporter assay conditions for (A) described in Section 3. LC-MS/MS quantitation for (B) and (C) as described in Section 1 and in Jones et al. Error bars represent standard deviation ($n = 3$).

levels in the cell and media, respectively. Fig. 3B and C showed a sustained, concentration dependent level of RA through 12h before decreasing, likely due to metabolism of RA.

5. Cell-based assays of RA signaling: Two-hybrid Gal4-RAR;UAS-tk-gaussia luciferase reporter

5.1 Materials and reagents

1. 0.2% gelatin coated plates
2. Dulbecco's Modified Eagle Medium:
   Nutrient Mixture F-12 (DMEM/F12)supplemented with 10% Heat-Inactivated Fetal Bovine Serum (FBS)
3. 1 × Penicillin-Streptomycin
4. 0.4mg/mL G418, Geneticin

5.2 Reporter reagents (available upon request)

Human Embryonic Kidney (HEK293) cells were obtained ATCC (Graham, Smiley, Russell, & Nairn, 1977). A construct expressing a fusion of the DNA-binding domain of the GAL4 protein (aa 1–147) to the ligand binding domain (aa 156–457) of zebrafish RARαb under the control of a CMV promoter was obtained as a kind gift from Joshua Waxman (University of Cincinnati/Cincinnati Children's Hospital) and has been previously described (D'Aniello et al., 2013; Mandal et al., 2013).

A construct expressing Gaussia luciferase under the control of the HSV thymidine kinase minimal promoter was generated by cloning 5 tandem copies of upstream activating sequences (UAS) in the Bgl2 site of the pGluc-mini-TK-2 plasmid (New England Biolabs).

5.3 Protocols

Culture HEK293 cells in Eagle's Minimum Essential Media (MEM) with 10% FBS.

Two-hybrid Gal4-RAR;UAS-tk-Gaussia luciferase reporter of RA presence reagents

1. Reverse co-transfect HEK293 cells with pGluc-mini-TK-2-UAS and pCS2p + Gal4-RAR reporter plasmids at 500ng each, per well, using Lipofectamine 2000 (ThermoFisher) in Opti-MEM (Invitrogen), and seed at 0.2 million cells per well in a 12-well cell culture plate.

2. After allowing the cells to attach, approximately 3 h, add 1 mL of Eagle's MEM to each well and incubate overnight.
3. The next day, remove cell media and replace with 1 mL of media supplemented with either *all-trans* Retinoic Acid (ATRA) (0.5, 5, 50 nM as in Fig. 4 or desired concentration(s)), holo-Retinol Binding Protein 4 (RBP4) (0.4, 2, 10 μM as in Fig. 5 or desired concentration(s)), or unsupplemented Eagle's MEM, as a control at 0 h.
4. Collect samples at 0, 0.5, 1, 2, 3, 6, 12, 24, 48, 72, and 96 h as in Figs. 4 and 5 or desired timepoint(s).

Determination of RA levels via LC-MS/MS in HEK293 cells expressing the two-hybrid Gal4-RAR;UAS-tk-Gaussia luciferase reporter

1. At the desired time points, collect media aliquots.
2. Remove the remaining culture media and use a cell scraper to harvest cells for LC-MS/MS analysis.

Gal4-RAR;UAS-tk-Gaussia luciferase expressing cells sample collection and Gaussia luciferase reporter assay

1. Collect 100 μL of media in a 1.5 mL DNA LoBind Tube (Eppendorf) for use in the reporter assay as previously performed (D'Aniello et al., 2013; Mandal et al., 2013).
2. A Gaussia Luciferase Flash Assay Kit (ThermoFisher) with a Corning 96-Well Solid White Polystyrene Microplate can then be used to determine luciferase activity according to the manufacturer's instructions.
3. Luciferase activities [luminescence (relative light units/well)] should be monitored with a POLARstar Omega plate reader (BMG Labtech) or comparable plate reader.

Technical notes

- For detachment of cells use TrypLE Express Enzyme, no Phenol Red (1 ×) (Invitrogen).
- F9 RARE-LacZ Stable Cells should be passaged frequently and should not reach more than 80–90% confluency to prevent differentiation.
- For better transfection efficiency: (1) briefly vortex the Lipofectamine before use and (2) mix the Lipofectamine 2000 and plasmids in a flat dish, rather than a microcentrifuge tube, for optimal lipid formation.
- For LC-MS/MS-based quantitation, do not use enzyme-based detachment procedures (e.g., trypsin). Remove culture media and gently use a cell scraper to detach cells.
- All experiments should be conducted under yellow light to prevent light-induced isomerization and degradation of retinoids.

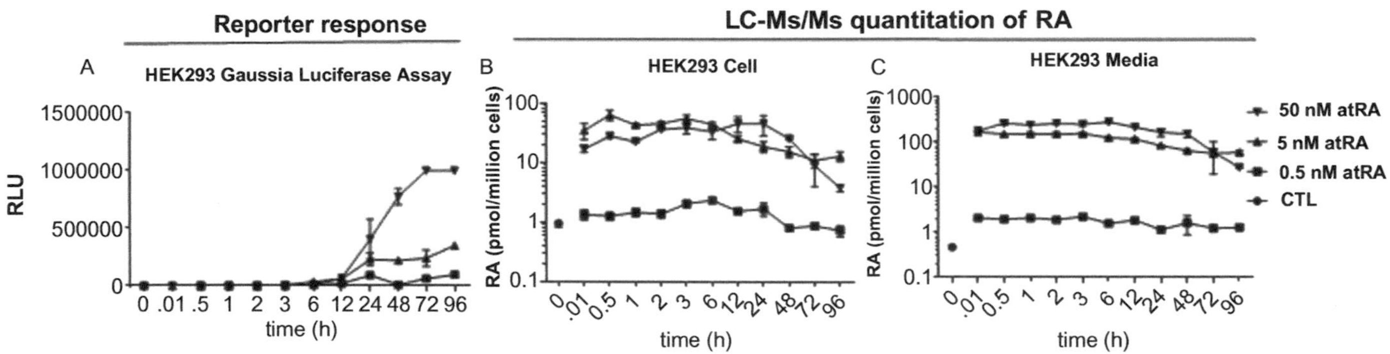

Fig. 4 Evaluating RA reporter response in the two-hybrid Gal4-RAR;UAS-tk-Gaussia luciferase reporter in HEK293 cells as compared to RA levels. Cells were treated with 0–50 nM of *all-trans* retinoic acid (atRA or RA) and reporter response and RA levels were measured over time for 0–96 h. (A) Two-hybrid Gal4-RAR;UAS-tk-Gaussia luciferase reporter (B) RA in cell as determined by LC-MS/MS (C) RA in media as determined by LC-MS/MS. Cell culture for (A)–(C) and reporter assay conditions for (A) described in Section 4. LC-MS/MS quantitation for (B) and (C) as described in Section 1 and in Jones et al. Error bars represent standard deviation ($n = 3$).

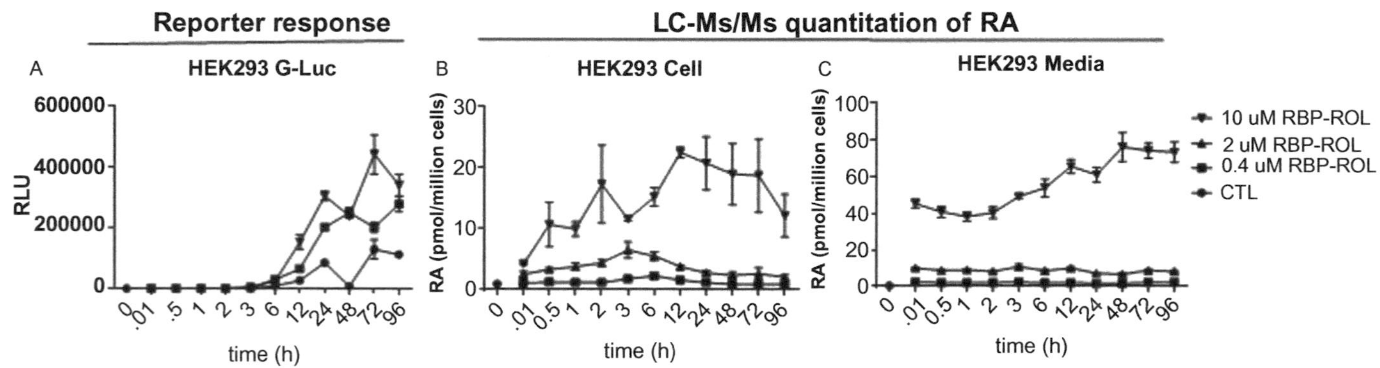

Fig. 5 Evaluating RA reporter response in the two-hybrid Gal4-RAR;UAS-tk-Gaussia luciferase reporter in HEK293 cells as compared to RA levels produced from retinol treatment. Cells were treated with 0–10 nM of holoRBP4-Retinol (RBP4-ROL) followed by quantification of reporter response and *all-trans* retinoic acid (RA) levels over time for 0–96 h. (A) Two-hybrid Gal4-RAR;UAS-tk-Gaussia luciferase reporter (B) RA in cell as determined by LC-MS/MS (C) RA in media as determined by LC-MS/MS. Cell culture for (A)–(C) and reporter assay conditions for (A) described in Section 4. Retinol treatment and LC-MS/MS quantitation for (B) and (C) as described in Section 1 and in Jones et al. Error bars represent standard deviation ($n = 3$).

- Glass pipettes and glass tubes should always be used when handling retinoids.
- Retinoid samples can be stored in DMSO or DMF at 5–10 mM at −80 °C, but long term storage at any temperature should be avoided.
- The concentration of dissolved retinoids should be confirmed by measuring the absorption of a 1000-fold dilution of the stock solution in ethanol (or hexane for more apolar carotenoids or retinyl esters) using a UV-visual spectrophotometer and deriving the concentration via the Lambert-Beer law $A_\lambda = (\varepsilon)(c)$ for a pathlength $L = 1$ cm (as described in Protocol I, Part 1). In the case of all-*trans*-RA dissolved in ethanol, the maximum absorption wavelength (λ_{max}) = 350 nm, and the molar extinction coefficient (ε_{max}) = 45,300 M^{-1} cm^{-1}. For other retinoids spectral values and absorption coefficients can be found in Kane and Napoli (2010) and references therein. In addition, LC-UV can be used to assess the purity and isomerization state of various retinoid compounds employed.

5.4 Analysis

5.4.1 Evaluating RA reporter response in the two-hybrid Gal4-RAR; UAS-tk-Gaussia luciferase reporter as compared to RA levels

Fig. 4 shows an example of data obtained from the two-hybrid Gal4-RAR; UAS-tk-Gaussia luciferase reporter in HEK293 cells in comparison to RA determined by LC-MS/MS-based detection. Cells treated with varying concentrations of RA (0.5, 5, and 50 nM RA) and reporter response via luciferase as well as RA levels were assessed over time up to 96 h. Fig. 4A, the luciferase levels in the two-hybrid Gal4-RAR;UAS-tk-Gaussia luciferase reporter showed an increase in signal starting at around 12 h with sustained signal through 96 h. Fig. 4B and C shows RA levels in the cell and media, respectively. Fig. 4B and C shows a sustained, concentration dependent level of RA through 48 h before declining, likely due to metabolism of RA.

Fig. 5 shows an example of the two-hybrid Gal4-RAR;UAS-tk-Gaussia luciferase reporter in HEK293 cells as used to evaluate reporter response from RA produced from (added) retinol substrate using the protocol described in Section 2. Similarly to Fig. 4A, the luciferase levels in the two-hybrid Gal4-RAR;UAS-tk-Gaussia luciferase reporter assay showed an increase in signal starting at around 12 h and continuing through 96 h. Fig. 5B and C shows the RA levels in the cell and media, respectively. Fig. 5B and C showed an increase in RA produced up until 12 h with sustained, concentration dependent level of RA through 72 h in both cells and media. Media maintains similar levels of RA at 96 h, whereas cellular levels of RA decline at 96 h, likely due to metabolism of RA.

5.5 Pros and cons

5.5.1 Considerations for various cellular assays

Each of the cellular assays has advantages and limitations to be considered when choosing an experimental approach as Summarized in Table 3. *RA Extraction and Quantification via LC-MS/MS* (part 3 of Section 2.2 "Protocols") offers the advantage of direct, absolute quantitation of RA that is applicable to cell lines and tissues. The quantitation of RA has temporal accuracy and represents the levels of RA that are present at the time of collection (Fig. 6). Results are very reproducible as documented in methodology publications (Jones et al., 2015; Kane et al., 2005; Kane, Folias, Wang, et al., 2008) and reproducibility representative cellular assay data are shown in Fig. 7. Limitations of this approach include more experimental effort than reporter assays and the requirement for expensive, specialized instrumentation. For tissue quantitation in embryo, LC-MS/MS approaches

Table 3 Comparison of different reporters and quantification methods for cellular systems.

F9 RARE-LacZ stable cell line β-galactosidase reporter	Two-hybrid Gal4-RAR; UAS-tk-Gaussia luciferase reporter	RA extraction and quantification via LC-MS/MS
PROS	**PROS**	**PROS**
– No transfection required – Quick results (yes/no) – Very reproducible	– Can be used in multiple cell lines – Reporter stays on once activated – Does not require cell lysis – Fastest results (yes/no) – Very reproducible	– Applicable to cell lines and tissues – Direct, absolute RA quantification – Temporal accuracy, represent levels at the time of collection – Very reproducible
CONS	**CONS**	**CONS**
– Cells can differentiate if not maintained properly – Requires cell lysis – Reporter response saturates at moderate levels of RA – Nonquantiative – Temporal delay then transient signal	– Depending on cell line, transfection can be difficult – Nonquantitative – Temporal delay	– More time intensive – Expensive, specialized instrumentation required

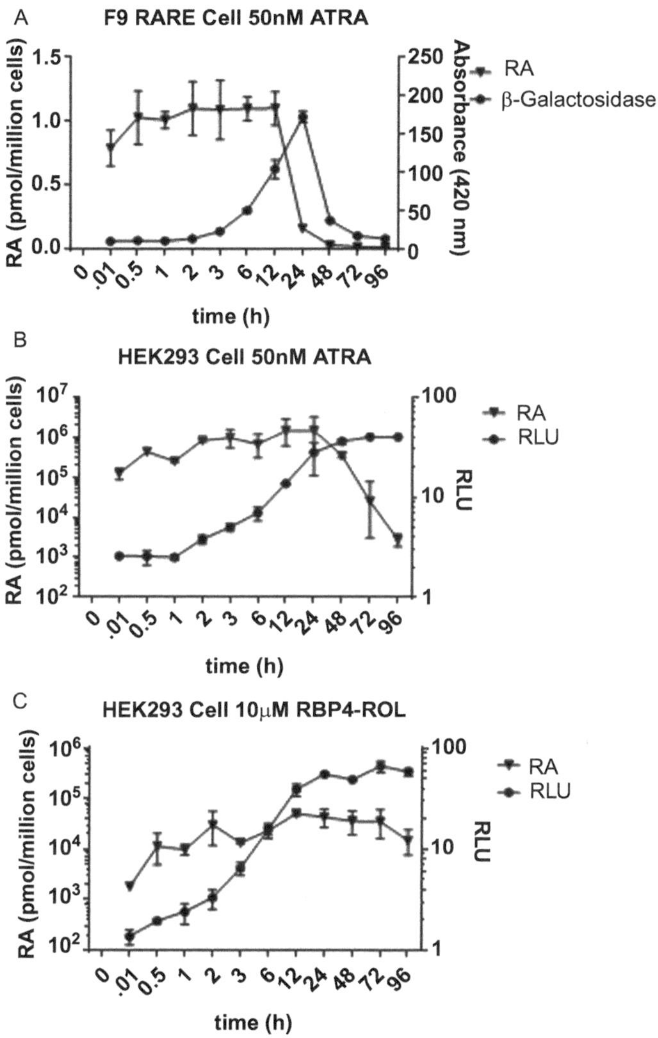

Fig. 6 Comparison of temporal responses of different cellular assays to determine RA levels or RA signaling. (A) RA reporter response after 50 nM RA treatment in the F9 RARE-LacZ stable cell line via β-galactosidase reporter assay and RA levels measured via LC-MS/MS (B) RA reporter response after 50 nM RA treatment in the two-hybrid Gal4-RAR;UAS-tk-Gaussia luciferase reporter in HEK293 cells as compared to RA levels (C) RA reporter response from RA produced after 10 μM RBP4-ROL treatment in the two-hybrid Gal4-RAR;UAS-tk-Gaussia luciferase reporter in HEK293 cells as compared to RA levels. Reporter assays for (A) as in Section 3, reporter assays for (B) and (C) as in Section 4. LC-MS/MS quantitation as described in Section 1 and in Jones et al. Error bars represent standard deviation ($n = 3$).

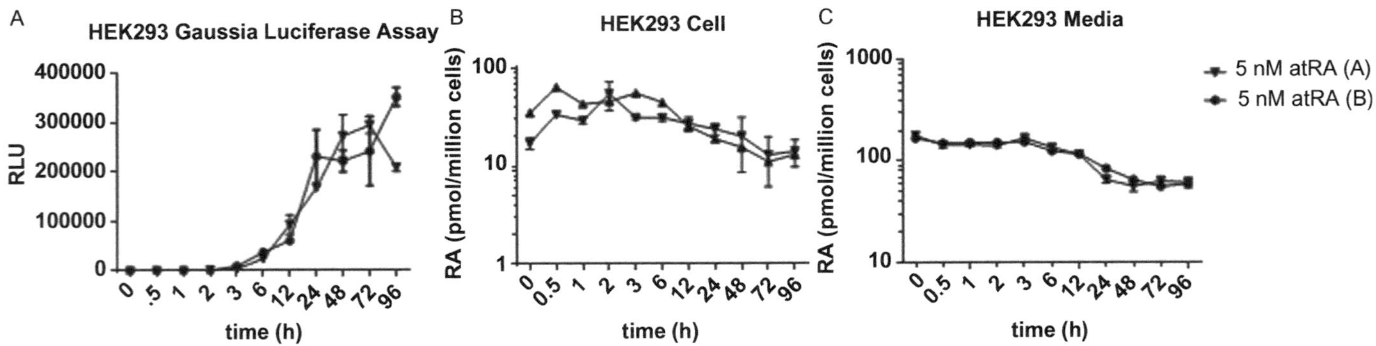

Fig. 7 Comparison of representative reproducibility between the two-hybrid Gal4-RAR;UAS-tk-Gaussia luciferase reporter in HEK293 cells as compared to RA levels. Two sets of cells were treated with 5 nM of *all-trans* retinoic acid (atRA or RA) and reporter response and RA levels were measured over time for 0–96 h. (A) Two-hybrid Gal4-RAR;UAS-tk-Gaussia luciferase reporter (B) RA in cell as determined by LC-MS/MS (C) RA in media as determined by LC-MS/MS. Cell culture for (A)–(C) and reporter assay conditions for (A) described in Section 4. LC-MS/MS quantitation for (B) and (C) as described in Section 1 and in Jones et al. Error bars represent standard deviation ($n = 3$).

do not provide intact spatial visualization, and measurement of specific spatial regions is limited by the ability to dissect those regions. Additionally, LC-MS/MS approaches have minimum tissue requirements that may necessitate pooling. The *F9 RARE-LacZ Stable Cell Line β-Galactosidase Reporter* (Section 4) offers the advantage of rapid results with no transfection required and reproducible results for comparable conditions. Limitations include the possibility that cells can differentiate if not maintained properly and the requirement for cell lysis. The F9 RARE-LacZ stable cell line β-galactosidase reporter is also limited by a response that saturates at moderate levels of RA (Fig. 3A), is nonquantitative, and displays a temporal delay post increase in RA accompanied by a transient signal response (Fig. 6). The *two-hybrid Gal4-RAR;UAS-tk-Gaussia luciferase reporter* (Section 5) has the advantage that it can be applied to any cell line amenable to transfection and does not require cell lysis. The reporter offers rapid results and remains on once activated (illustrated in Figs. 4 and 6) thereby eliminating concerns of a transient response as is encountered in the F9 RARE-LacZ cell line. Results are very reproducible as shown in Fig. 7. Limitations of the two-hybrid Gal4-RAR;UAS-tk-Gaussia luciferase reporter include transfection difficulty in some cell lines, a non-quantitative response, and a temporal delay exposure to RA (Fig. 6).

For RA production evaluation, similar to standard enzyme activity assays, the timepoint and concentration ranges for the retinoid substrate of interest should first be determined by performing a time-course experiment and a concentration-course experiment to determine conditions within the linear range (Kane & Napoli, 2010; Wang et al., 2011).

6. In situ hybridization can be used to visualize gene activity of vitamin A signaling components

The ability to accurately detect gene activity of components from the vitamin A pathway allows for the study of its role in development and disease. This section describes the use of in situ hybridization (ISH) as a sensitive method to identify spatiotemporal gene activity by localizing mRNA transcripts from a gene of interest (Fig. 8). To do this, ISH utilizes anti-sense RNA probes of varying length that are synthesized with nucleotides that contain a digoxigenin (DIG) moiety that can be bound by an enzyme linked anti-DIG antibody. Transcripts that are bound by the labeled riboprobe and antibody may be detected by a colorimetric reaction, whereby an NBT/BCIP substrate solution is metabolized to create a purple reaction

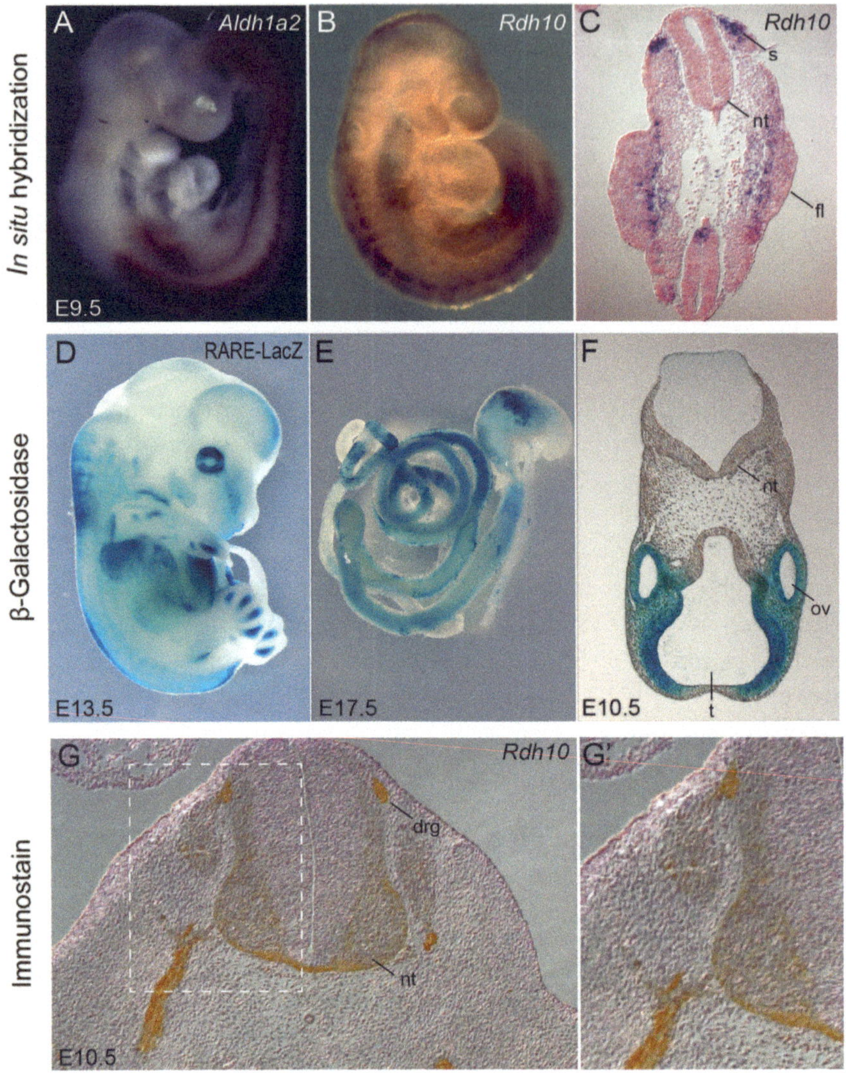

Fig. 8 Methods to identify components of vitamin A metabolism and RA synthesis. (A–C) Visualizing vitamin A metabolizing enzymes by in situ hybridization. (A) *Raldh2* is expressed in the whole embryo at E9.5 in the foregut and caudal somites. *Rdh10* (B) is expressed at E9.5 in the regions of the gut, forelimb and somites. In a transverse section (C), *Rdh10* activity can be observed within the somites and forelimb. (D–F) The *RARE-LacZ* reporter mouse contains a RA response element under the control of a minimal promoter and is used to visualize RA signaling. LacZ staining of an E13.5 embryo (D) demonstrates the spatiotemporal activity of RA signaling throughout the head and trunk including the eye, neural tube, and limbs. High levels of RA signaling are present in many organs including the gut at E17.5 (E) as well as in tissue around the optic vesicle and telencephalon as evident in a transverse section of an E10.5 embryo (F). Immunostaining for *Rdh10*, a component of vitamin A metabolism, reveals expression the dorsal root ganglia, neural tube and motor nerve projection in a transverse section (G) and inset (G') of an E10.5 embryo. Abbreviations: s, somite; nt, neural tube; fl, forelimb; ov, optic vesicle; t, telencephalon; drg, dorsal root ganglia.

product. Following the reaction, visualization of the bound mRNA transcripts can be detected in the regions of the purple color development and imaged by light microscopy.

6.1 Reagents

Note: All stock solutions should be treated with diethyl pyrocarbonate (DEPC) and autoclaved, or prepared with DEPC-H_2O.

1. 4% Paraformaldehyde in PBS
   Microwave to boiling 100 mL of DEPC-PBS and then add 4 g of paraformaldehyde powder (toxic) and stir to dissolve in a fume hood
2. Phosphate-Buffered Saline with Triton (0.1% Triton in PBS); 1 mL Triton X-100 in 1 L DEPC-PBS
3. Hybridization Buffer; 50% formamide, 5 × SSC, 0.05 M EDTA, 0.1% Triton X-100, 0.1% CHAPS, 0.05 mg/mL Heparin, 50.0 mg Yeast torula-RNA, 1.0 g Boehringer blocking powder
4. Potassium-Tris Buffer with Triton (KTBT); 0.05 M Tris/HCl (pH 7.5), 0.15 M NaCl, 0.01 M KCl, 1% Triton X-100
5. Alkaline Phosphatase Buffer; 0.01 M Tris/HCl (pH 9.5), 0.005 M NaCl, 0.005 M $MgCl_2$, 0.01% Tween20
6. Maleic Acid Buffer with Triton (MABT); 100 mM Maleic acid (pH 7.5; adjust with NaOH), 0.15 M NaCl; 1% Triton X-100

6.2 Protocols

RNA probe synthesis

1. Mix the following at room temperature:

Sterile distilled water (DEPC-H_2O)	11.5 μL
5 × transcription buffer	5.0 μL
100 mM DTT	2.5 μL
DIG 10 × nucleotide mix	2.5 μL
Linearized plasmid (1 μg/μL)	1.5 μL
Rnasin ribonuclease inhibitor	0.5 μL
Polymerase (SP6, T3 or T7)	1.5 μL
Total	25 μL

2. Incubate at 37 °C for 2 h up to 6 h.
3. Remove 1 μL aliquot and run on 1% TAE gel to check synthesis. Expect to see RNA band 10-fold more intense than plasmid band, suggesting 10–15 μg probe.
4. Add 2 μL of DNAse1 (ribonuclease free) and incubate at 37 °C for 15 min.
5. Add:

50 μL	dH_2O
25 μL	10 M ammonium acetate
200 μL	100% ethanol

6. Mix and leave on dry ice for 30 min or store at −80 °C overnight.
7. Spin in centrifuge at 4 °C for 20 min at 13,000 to 15,000 rpm.
8. Wash pellet in 50 μL of 70% ethanol and spin at 4 °C for 5 min at 13,000 to 15,000 rpm.
9. Air dry pellet for 5–10 min until obvious traces of ethanol have evaporated.
10. Redissolve pellet in 50 μL hybridization solution and store at −20 °C. If a higher concentration of probe is needed the pellet may be resuspended in 30 μL of hybridization solution.

In situ hybridization staining method

1. Dissect embryos in DEPC-PBS and fix in 4% paraformaldehyde for 2 h to overnight (use scintillation vials).
2. Wash embryos in PBT at room temperature for 10 min.
3. Dehydrate the embryos through a graded series of methanol diluted in PBT: wash in 25% methanol, 50% methanol, 75% methanol and 100% methanol for 10 min each.
4. Embryos can be stored at −20 °C up to several weeks at this stage.
5. Rehydrate the embryos through a graded series of methanol diluted in PBT: wash in 75% methanol, 50% methanol, 25% methanol for 10 min each.
6. Wash in PBT for 5 min at room temperature.
7. Incubate in 10 μg/mL proteinase K (diluted in PBT) for 5–10 min at room temperature. Incubation time will vary depending on the embryo's stage.

Incubation time by embryo stage:
E6.5: 4 min
E7.5: 4–5 min
E8.5: 6 min
E9.5: 10 min
E10.5: 15 min

8. Stop the proteinase K reaction by washing with 2 mg/mL glycine (diluted in PBT) for 5 min at room temperature (make the same day of use).
9. Wash for 5 min in PBT at room temperature.
10. Re-fix the embryos in 4% paraformaldehyde/0.25% glutaraldehyde for 20 min at 4 °C.
11. Wash for 10 min in PBT at room temperature.
12. Transfer the embryos to 2 mL Eppendorf tube with a round bottom (up to 10 embryos can be processed in a single tube).
13. Incubate embryos for at least 1.5 h in 1–2 mL of pre-warmed hybridization buffer at 62 °C to 70 °C.
14. Incubate embryos with 2 μL of RNA probe per 1 mL of hybridization buffer at 62 °C to 70 °C overnight in hybridization oven (1 mL of buffer is enough for 10 mouse embryos at E10.5).
15. Wash embryos twice in 2 mL of 2xSSC/0.1% chaps for 45 min at 62 °C.
16. Wash embryos in 2 mL of 0.2xSSC/0.1% chaps for 30 min at 62 °C.
17. Wash in KTBT for 5 min at room temperature.
18. Incubate the embryos in 2 mL of 20% goat serum (or lamb serum) in Potassium-Tris Buffer with Triton (KTBT) for at least 1 to 1.5 h at room temperature with continual rocking.
19. Replace blocking solution with fresh 20% goat serum in KTBT. Add 2 μL of DIG-alkaline phosphatase antibody and incubate at 4 °C overnight with continual rocking.
20. Rinse the embryos at least twice in KTBT.
21. Wash the embryos between 4 and 5 times in KTBT for 1 h at room temperature with continual rocking.
22. Replace with 2 mL of fresh KTBT and wash overnight at 4 °C with continual rocking.

23. In the morning, wash the embryos at least twice with KTBT for a combined time of 30 min at 4 °C.
24. Wash the embryos in 1 mL of alkaline phosphatase buffer for 10 min at room temperature with continual rocking.
25. Add 3.375 μL of NBT (100 mg/mL in DMF) and 3.5 μL of BCIP (50 mg/mL in DMF) directly to the embryos in alkaline phosphatase buffer with shaking at room temperature in the dark (wrap the Eppendorf tube in foil or place in a covered container).
26. Stop the reaction once the desired color intensity is achieved by fixing the embryos in 4% paraformaldehyde. This can take from 15 min to overnight, however 1 to 2 h is more common. Continual observation of the color development is not advisable as the background will increase.

6.3 Safety considerations and standards

Paraformaldehyde is a toxic chemical and extreme caution should be used when handling. Be sure to always use proper personal protective equipment and handle under a fume hood.

6.4 Related techniques

In addition to the technique described above, other RNA in situ hybridization assays are available that include RNAscope, ViewRNA and SABER-FISH. An advantage of these techniques is their ability to use fluorescent probes which provide a multiplexed approach to characterize the activity of many genes within the same sample. The use of fluorescence additionally provides an option to quantify gene expression (Kishi et al., 2019; Kulikova, Franken, & Bisseling, 2018; Wang et al., 2012).

6.5 Alternative methods/procedures

6.5.1 Alternative posthybridization washes in maleic acid buffer with triton (MABT)

After hybridizing the embryos overnight at 62 °C in step 14, wash the embryos in a 1:1.5 mix of hybridization buffer and MABT (described in Section 2.1 "Materials, equipment and reagents") at 65 °C for 1 h. Then wash the embryos for 30 min at 65 °C in MABT. For steps 17 through 23 replace KTBT with MABT.

6.5.2 *Alternative color development*

The color development described in step 25 of the protocol may also be carried out by using BMPurple solution in the place of NBT/BCIP. To do this, pre-warm BMPurple to 37 °C. Spin the tubes for 15–20 s in micro-centrifuge to pellet precipitate. Place supernatant onto embryos that have already incubated in alkaline phosphatase buffer. Be sure to protect from light while color develops.

6.6 Troubleshooting and optimization

See Table 4.

6.7 Summary

In situ hybridization is a sensitive method to detect gene activity by localizing mRNA transcripts within a sample. Embryos are first harvested and fixed overnight in 4% PFA (it is important all solutions are DEPC treated to prevent ribonuclease contamination). The samples are then permeabilized using proteinase K, refixed with 4% PFA/0.25% glutaraldehyde and then hybridized with riboprobe overnight between 62 °C and 70 °C. The following day, samples are processed through a series of posthybridization washes, blocked in a 20% goat serum solution and incubated at 4 °C overnight with an anti-DIG-Alkaline Phosphatase antibody. Samples then undergo a series of washes and begin a colorimetric reaction in an

Table 4 Troubleshooting for common problems when using in situ hybridization.

Problem	Solution
Riboprobes with consistently high background staining	Wash the embryos in MABT instead of SSC/SDS as described in Section 6.5 "Alternative Methods/ Procedures"
Riboprobes with consistently high background staining from probe trapping	To help alleviate probe trapping in the head in embryos at E9.5 and older, it is beneficial to poke the heads of the embryos during harvest to help clear riboprobe during washes
Poor staining quality from incomplete riboprobe penetration	To increase riboprobe penetration in whole embryos it is recommended to test increasing the incubation times with proteinase K listed in step 7
Poor staining quality from incomplete hybridization of the riboprobe	To optimize riboprobe hybridization, the temperature of hybridization in step 14 may be increased between 62 °C and 70 °C

NBT/BCIP substrate solution. Once color development is complete samples are postfixed in 4% PFA and can be imaged and/or stored at 4 °C.

7. Detection of RA signaling using a β-galactosidase assay with X-GAL

Retinoic acid signaling can be visualized in vivo using transgenic mouse lines containing a LacZ reporter. The method described in this section outlines the use of the transgenic RARE-LacZ mouse line to visualize retinoic acid (RA) activity by β-Galactosidase staining with X-gal (Fig. 8). To do this, the RARE-LacZ mouse line possesses a transgene with an RA response element (RARE) that drives expression of β-Galactosidase by RA binding to RA receptors on the RARE sequence (Rossant et al., 1991). The addition of the X-gal substrate, detailed in the below protocol, creates a purple product in the regions of RA activity, which provides an effective means to visualize the spatiotemporal activity of RA signaling. Further, other LacZ transgenic mouse lines, such as *Rdh10*$^{\beta geo}$ (Sandell et al., 2012), may be used to analyze the spatiotemporal activity of other components of the vitamin A pathway.

7.1 Reagents

1. Tissue Fixative: 0.2%glutalaldehyde/2%formalin
2. Wash Solution (to make 500 mL)
   0.1 mL NP40 in 500 mL PBS
3. Stain Solution (to make 500 mL)

5.0 mL	0.5 M K3Fe(CN)6
5.0 mL	0.5 M K4Fe(CN)6-3H_2O
1.0 mL	1.0 M $MgCl_2$
5.0 mL	1% Sodium deoxycholate
make up to 500 mL with Wash Solution	
Store in dark at 4 °C	

4. Stock Substrate
   40 mg/mL X-gal dissolved in DMF
   Store in dark at 4 °C
5. Working Stain Solution: add 1 mL of X-gal stock to 39 mL of stain solution. (store at 4 °C in the dark)

7.2 Protocol

1. Wash embryos in PBS at room temperature.
2. Fix embryos in cold 0.2%glutalaldehyde/2%formalin for 30 to 90 min at 4 °C. Fixation time depends on the embryo stage.

Fixation time by embryo stage:
E7–8: 15 min.
E9–10: 30 min.
E11–12: 1 h.
E13 and older: 1.5 h.

3. Wash embryos three times for 15 min each in wash solution at room temperature with no rocking.
4. While the samples are washing, warm the stain solution at 37 °C in the dark. Then add the X-gal substrate to the pre-warmed stain solution and return to the dark at 37 °C until ready to begin step 5.
5. Replace the wash solution with the prewarmed working stain solution and incubate in the dark at 37 °C for up to 36 h. Check the development periodically to determine when the stain is complete.
6. After staining, rinse the embryos thoroughly in PBS and store long-term in PBS/0.01% sodium azide at 4 °C.

7.3 Alternative methods/procedures

Alternatives to using X-gal in the stain solution in step 4 include Salmon-gal with tetrazolium salts, or S-gal with TNBT, which may provide a faster and more sensitive reaction in mouse embryos. It has been noted in comparative studies that Salmon-gal or S-gal were capable of detecting β-galactosidase activity with higher sensitivity than X-gal (Shen et al., 2017; Sundararajan, Wakamiya, Behringer, & Rivera-Perez, 2012).

7.4 Summary

The β-Galactosidase assay with X-gal is a straightforward method that can be used to detect gene activity of components of the vitamin A pathway by staining embryos from transgenic mouse lines, such as RARE-LacZ and *Rdh10*$^{\beta geo}$ (Sandell et al., 2012, 2007). In the six-step protocol, embryos are harvested and then fixed for a varying length of time depending on embryo stage. Once fixed, samples are washed and then incubated at 37 °C in the dark with the working stain solution containing the X-gal substrate for up to 36 h. When staining is complete the embryos can be imaged and/or stored at 4 °C with sodium azide.

8. Immunostaining can be used to visualize protein activity of vitamin A signaling components

An alternative method to study the vitamin A pathway is to detect protein activity of vitamin A signaling components. The immunostaining technique described in this section is a method to identify the spatiotemporal activity of proteins that are essential in the vitamin A pathway, such as the enzyme *Rdh10* (Fig. 8). To do this, either whole embryo or sections are fixed to preserve the protein of interest and then permeabilized to allow penetration and binding of a primary antibody to the protein. Secondary antibodies, conjugated with either an enzyme or fluorophore, are then used to bind the primary antibody. Localization of the bound protein can be achieved by either a colorimetric reaction with an enzyme conjugated secondary antibody, or by fluorescent imaging of antibodies associated with a fluorophore. Detection by either method allows for the ability to visualize the spatiotemporal activity of the protein wherever it is bound within the sample.

8.1 Reagents

Dent's Bleach; 100% MeOH: DMSO: 30% H_2O_2, 4:1:1

1. 4% Paraformaldehyde in PBS

   Microwave to boiling 100 mL of PBS (DEPC treated) and then add 4 g of paraformaldehyde powder (toxic) and stir to dissolve in a fume hood.
2. PBT (0.1% Triton in PBS): 1 mL Triton X-100 in 1 L PBS
3. BABB clearing solution; Benzyle benzoate: Benzyl alcohol, 2:1

8.2 Protocols

Section protocol

1. Section whole embryos by cryosectioning at 10 μm (slides can be stored at −20 °C).
2. Let slides defrost for a few minutes at room temperature.

3. Heat-dry with a hair dryer or slide warmer for about 15–20 s on the back of each slide.
4. Wash in PBT for 10 min.
5. Prepare a slide holder for staining. Place a piece of tissue in each slot of that container that fits perfectly and soak with water.
6. Block the sections in 3% BSA (diluted in PBT) for 30 min at room temperature (add about 200 μL of blocking solution on each slide and cover with a paraffin strip in the slide holder container).
7. Remove excess blocking solution and add about 150 μL of the primary antibody diluted in 3% BSA.
8. Cover each slide with a paraffin strip after adding the antibody solution to prevent the slides becoming dry.
9. Seal the container with a paraffin strip and store at 4 °C overnight.
10. Wash the sections three times for 5 min with PBT at room temperature.
11. Add 150 μL of the secondary antibody (typically a 1:500 dilution) and cover with a paraffin strip.
12. Incubate at room temperature for 1 h.
13. Wash the sections three times for 5 min with PBT at room temperature.
14. Add DAPI at a 1:1000 dilution and incubate for 20 min at room temperature (alternatively, DAPI can be added with the secondary antibody. Avoid coincubation if doing TUNEL or using a nuclear antibody).
15. Wash sections three times for 5 min with PBT at room temperature.
16. Use vectashield to mount the slides (you can skip adding DAPI in the previous step and use vectashield with DAPI instead).
17. Store the slides in the dark at 4 °C for long-term storage.

Whole embryo protocol

1. Incubate embryos in Dent's bleach for 2 h at room temperature (this is a critical step that promotes antibody penetration and quenches autofluorescence).
2. Remove Dent's bleach and replace with 100% methanol.
3. Equilibrate through a graded series of methanol diluted in PBT: wash in 100% methanol, 75% methanol, 50% methanol and 25% methanol for 10 min each (the tissue will become sticky during this procedure so be aware not to suck it up into pipettes etc.).
4. Wash embryos in PBT for 10 min.
5. Block in 3% BSA (diluted in PBT) for 2 h at room temperature.
6. Apply the primary antibody diluted in 3% BSA and incubate overnight at 4 °C with gentle rocking (dilution varies per antibody, determine empirically for each).

7. Wash five times for 1 h with PBT.
8. Apply the secondary antibody diluted in 3% BSA and incubate overnight at 4 °C with gentle rocking in the dark (wrap the Eppendorf tube in foil or place in a covered container).
9. Wash three times for 20 min with PBT.
10. Incubate 20 min in PBS with DAPI at a 1:1000 dilution (for small tissues 20 min DAPI incubation is enough. To completely stain a whole E10.5 embryo, incubate overnight at 4 °C with gentle rocking).
11. Rinse embryos in PBT.
12. Dehydrate the embryos through a graded series of methanol diluted in PBT: wash in 25% methanol, 50% methanol, 75% methanol and 100% methanol for 10 min each.
13. For long term storage, the tissue can be stored for weeks or months at 4 °C without loss of fluorescent intensity.

8.3 Safety considerations and standards

Paraformaldehyde is a toxic chemical and extreme caution should be used when handling. Be sure to always use proper personal protective equipment and handle under a fume hood.

If using the alternative procedure in Section 8.3 to clear whole embryos for imaging, it is important to use extra caution when handling benzyle alcohol-benzyl benzoate (BABB) as it is a powerful organic solvent that dissolves most plastics. Be sure to use and store BABB in a glass container.

8.4 Alternative methods/procedures

Prior to imaging, clearing stained embryos can help visualize internal structures in larger embryos. This can be achieved with benzyl alcohol-benzyl benzoate (BABB) by adding 1 part Benzyl alcohol to 2 parts Benzyl benzoate. Incubating the embryos for 10 min should be enough to render the tissue almost completely transparent, but also fragile. Embryos can be imaged using a glass depression slide with a ring of silicon grease to hold a coverslip in place. A Teflon ring can be used between the grease ring and coverslip if additional space is needed for the sample. After imaging, the embryos can be transferred back to 100% methanol for long term storage at 4 °C.

8.5 Troubleshooting and Optimization

See Table 5

Table 5 Troubleshooting for common problems when immunostaining.

Problem	Solution
High levels of autofluorescence	Test increasing incubation time with dents bleach to help reduce levels of autofluorescence. Additionally, the head and heart region can be poked for whole embryos at E9.5 and older to help prevent trapping by allowing the antibody to clear during washing steps
Consistently high levels of background staining	The blocking time in step 5 of the "whole embryo protocol" can be increased to help reduce high levels of background. Additionally, the number and duration of washes can be increased in step 7 after incubation with the primary antibody
Poor staining quality from incomplete penetration of the antibody	Test increasing incubation time with Dent's bleach to help increase antibody penetration in whole embryos

8.6 Summary

Immunostaining can be used to detect protein activity within the vitamin A pathway by first harvesting samples and then fixing overnight in 4% PFA. Following fixation, the samples can either be sectioned or used for whole embryo staining. If using whole embryos, incubation with Dent's bleach helps to promote antibody penetration and quench auto-fluorescence (Alanentalo et al., 2007). Both sections and whole embryos are blocked in a 3% BSA solution and incubated overnight with primary antibody followed by secondary antibody incubation the next day. Samples are then ready to be imaged and whole embryos can be cleared using benzyl alcohol-benzyl benzoate (BABB) to better visualize internal structures (Section 4.3).

Acknowledgments

We are grateful to Dr. Joshua Waxman (University of Cincinnati/Cincinnati Children's Hospital) for constructs expressing GAL4—zebrafish RARαb fusion proteins, and to Dr. Michael Wagner (SUNY Downstate Health Sciences University) and Dr. Peter McCaffery (University of Aberdeen) for the RARE reporter cell line F9-RARE-lacZ (SIL15-RA) and Janet Rossant (Hospital for Sick Children) for RARE-lacZ mice. We would also like to thank Dr. Naomi Tjaden and Dr. Lisa Sandell for the images they generously contributed to Fig. 8. The mouse animal data was collected in compliance with Stowers Institute of Medical Research (SIMR) IACUC approved protocol 2019-097. This work was supported by the Stowers Institute for Medical Research (P.A.T.), R01HD077260 (M.A.K., P.A.T., A.R.M.) and T32 GM066706 (A.E.D.) from the National Institutes of Health, and a Madison and Lila Self Graduate Fellowship from the University of Kansas (S.R.S.).

References

Ababon, M. R., Li, B. I., Matteson, P. G., & Millonig, J. H. (2016). Quantitative measurement of relative retinoic acid levels in E8.5 embryos and neurosphere cultures using the F9 RARE-Lacz cell-based reporter assay. *Journal of Visualized Experiments*, *2016*(115). e54443, https://doi.org/10.3791/54443.

Adams, M. K., Belyaeva, O. V., Wu, L., & Kedishvili, N. Y. (2014). The retinaldehyde reductase activity of DHRS3 is reciprocally activated by retinol dehydrogenase 10 to control retinoid homeostasis. *Journal of Biological Chemistry*, *289*(21), 14868–14880. https://doi.org/10.1074/jbc.M114.552257.

Al Tanoury, Z., Piskunov, A., & Rochette-Egly, C. (2013). Vitamin a and retinoid signaling: Genomic and non-genomic effects. *Journal of Lipid Research*. Jlr.R030833 [pii]. https://doi.org/10.1194/jlr.R030833.

Alanentalo, T., Asayesh, A., Morrison, H., Loren, C. E., Holmberg, D., Sharpe, J., et al. (2007). Tomographic molecular imaging and 3D quantification within adult mouse organs. *Nature Methods*, *4*(1), 31–33. https://doi.org/10.1038/nmeth985.

Allenby, G., Bocquel, M. T., Saunders, M., Kazmer, S., Speck, J., Rosenberger, M., et al. (1993). Retinoic acid receptors and retinoid X receptors: Interactions with endogenous retinoic acids. *Proceedings of the National Academy of Sciences of the United States of America*, *90*(1), 30–34.

Antipatis, C., Ashworth, C. J., Grant, G., Lea, R. G., Hay, S. M., & Rees, W. D. (1998). Effects of maternal vitamin A status on fetal heart and lung: Changes in expression of key developmental genes. *American Journal of Physiology*, *275*(6), L1184–L1191. https://doi.org/10.1152/ajplung.1998.275.6.L1184.

Arnold, S. L., Amory, J. K., Walsh, T. J., & Isoherranen, N. (2012). A sensitive and specific method for measurement of multiple retinoids in human serum with UHPLC-MS/MS. *Journal of Lipid Research*, *53*(3), 587–598. https://doi.org/10.1194/jlr.D019745.

Ashique, A. M., May, S. R., Kane, M. A., Folias, A. E., Phamluong, K., Choe, Y., et al. (2012). Morphological defects in a novel Rdh10 mutant that has reduced retinoic acid biosynthesis and signaling. *Genesis*, *50*(5), 415–423. https://doi.org/10.1002/dvg.22002.

Balkan, W., Colbert, M., Bock, C., & Linney, E. (1992). Transgenic indicator mice for studying activated retinoic acid receptors during development. *Proceedings of the National Academy of Sciences of the United States of America*, *89*(8), 3347–3351. https://doi.org/10.1073/pnas.89.8.3347.

Balmer, J. E., & Blomhoff, R. (2002). Gene expression regulation by retinoic acid. *Journal of Lipid Research*, *43*(11), 1773–1808. https://doi.org/10.1194/jlr.r100015-jlr200.

Balmer, J. E., & Blomhoff, R. (2005). A robust characterization of retinoic acid response elements based on a comparison of sites in three species. *Journal of Steroid Biochemistry and Molecular Biology*, *96*(5), 347–354. https://doi.org/10.1016/j.jsbmb.2005.05.005.

Batten, M. L., Imanishi, Y., Maeda, T., Tu, D. C., Moise, A. R., Bronson, D., et al. (2004). Lecithin-retinol acyltransferase is essential for accumulation of all-trans-retinyl esters in the eye and in the liver. *The Journal of Biological Chemistry*, *279*(11), 10422–10432. https://doi.org/10.1074/jbc.M312410200.

Belyaeva, O. V., Adams, M. K., Popov, K. M., & Kedishvili, N. Y. (2020). Generation of retinaldehyde for retinoic acid biosynthesis. *Biomolecules*, *10*(1), 5. https://doi.org/10.3390/biom10010005.

Billings, S. E., Pierzchalski, K., Butler Tjaden, N. E., Pang, X. Y., Trainor, P. A., Kane, M. A., et al. (2013). The retinaldehyde reductase DHRS3 is essential for preventing the formation of excess retinoic acid during embryonic development. *The FASEB Journal*, *27*(12), 4877–4889. https://doi.org/10.1096/fj.13-227967.

Bowrey, H. E., Anderson, D. M., Pallitto, P., Gutierrez, D. B., Fan, J., Crouch, R. K., et al. (2016). Imaging mass spectrometry of the visual system: Advancing the molecular

understanding of retina degenerations. *Proteomics Clinical Applications, 10*(4), 391–402. https://doi.org/10.1002/prca.201500103.
Clagett-Dame, M., & DeLuca, H. F. (2002). The role of vitamin A in mammalian reproduction and embryonic development. *Annual Review of Nutrition, 22*, 347–381.
Clagett-Dame, M., & Knutson, D. (2011). Vitamin A in reproduction and development. *Nutrients, 3*(4), 385–428. https://doi.org/10.3390/nu3040385.
Cohlan, S. Q. (1954). Congenital anomalies in the rat produced by excessive intake of vitamin A during pregnancy. *Pediatrics, 13*(6), 556–567.
Conaway, H. H., Henning, P., & Lerner, U. H. (2013). Vitamin a metabolism, action, and role in skeletal homeostasis. *Endocrine Reviews, 34*(6), 766–797. https://doi.org/10.1210/er.2012-1071.
Cooney, A. J., Tsai, S. Y., O'Malley, B. W., & Tsai, M. J. (1992). Chicken ovalbumin upstream promoter transcription factor (COUP-TF) dimers bind to different GGTCA response elements, allowing COUP-TF to repress hormonal induction of the vitamin D3, thyroid hormone, and retinoic acid receptors. *Molecular and Cellular Biology, 12*(9), 4153–4163.
Costaridis, P., Horton, C., Zeitlinger, J., Holder, N., & Maden, M. (1996). Endogenous retinoids in the zebrafish embryo and adult. *Developmental Dynamics, 205*(1), 41–51. https://doi.org/10.1002/(SICI)1097-0177(199601)205:1<41::AID-AJA4>3.0.CO;2-5.
Crouch, R. K., Koutalos, Y., Kono, M., Schey, K., & Ablonczy, Z. (2015). A2E and lipofuscin. *Progress in Molecular Biology and Translational Science, 134*, 449–463. https://doi.org/10.1016/bs.pmbts.2015.06.005.
D'Aniello, E., Rydeen, A. B., Anderson, J. L., Mandal, A., & Waxman, J. S. (2013). Depletion of retinoic acid receptors initiates a novel positive feedback mechanism that promotes teratogenic increases in retinoic acid. *PLoS Genetics, 9*(8), e1003689. https://doi.org/10.1371/journal.pgen.1003689.
Deltour, L., Foglio, M. H., & Duester, G. (1999). Metabolic deficiencies in alcohol dehydrogenase Adh1, Adh3, and Adh4 null mutant mice. Overlapping roles of Adh1 and Adh4 in ethanol clearance and metabolism of retinol to retinoic acid. *Journal of Biological Chemistry, 274*, 16796–16801.
Dowling, J. E., & Wald, G. (1960). The biological function of vitamin a acid. *Proceedings of the National Academy of Sciences of the United States of America, 46*(5), 587–608. https://doi.org/10.1073/pnas.46.5.587.
Dupe, V., Matt, N., Garnier, J. M., Chambon, P., Mark, M., & Ghyselinck, N. B. (2003). A newborn lethal defect due to inactivation of retinaldehyde dehydrogenase type 3 is prevented by maternal retinoic acid treatment. *Proceedings of The National Academy of Sciences of the United States of America, 100*, 14036–14041.
Elizondo, G., Corchero, J., Sterneck, E., & Gonzalez, F. J. (2000). Feedback inhibition of the Retinaldehyde dehydrogenase geneALDH1 by retinoic acid through retinoic acid receptor β± and CCAAT/enhancer-binding protein β. *Journal of Biological Chemistry, 275*(50), 39747–39753. https://doi.org/10.1074/jbc.M004987200.
Endo, T., Mikedis, M. M., Nicholls, P. K., Page, D. C., & de Rooij, D. G. (2019). Retinoic acid and germ cell development in the ovary and testis. *Biomolecules, 9*(12). https://doi.org/10.3390/biom9120775.
Fan, X., Molotkov, A., Manabe, S., Donmoyer, C. M., Deltour, L., Foglio, M. H., et al. (2003). Targeted disruption of Aldh1a1 (Raldh1) provides evidence for a complex mechanism of retinoic acid synthesis in the developing retina. *Molecular and Cellular Biology, 23*, 4637–4648.
Feng, L., Hernandez, R. E., Waxman, J. S., Yelon, D., & Moens, C. B. (2010). Dhrs3a regulates retinoic acid biosynthesis through a feedback inhibition mechanism. *Developmental Biology, 338*(1), 1–14. doi:S0012-1606(09)01288-3 [pii]. https://doi.org/10.1016/j.ydbio.2009.10.029.

Fu, M., Sato, Y., Lyons-Warren, A., Zhang, B., Kane, M. A., Napoli, J. L., et al. (2010). Vitamin A facilitates enteric nervous system precursor migration by reducing Pten accumulation. *Development, 137*(4), 631–640. https://doi.org/10.1242/dev.040550.

Germain, P., Chambon, P., Eichele, G., Evans, R. M., Lazar, M. A., Leid, M., et al. (2006a). International Union of Pharmacology. LX. Retinoic acid receptors. *Pharmacological Reviews, 58*(4), 712–725. https://doi.org/10.1124/pr.58.4.4.

Germain, P., Chambon, P., Eichele, G., Evans, R. M., Lazar, M. A., Leid, M., et al. (2006b). International Union of Pharmacology. LXIII. Retinoid X receptors. *Pharmacological Reviews, 58*(4), 760–772. https://doi.org/10.1124/pr.58.4.7.

Golczak, M., Bereta, G., Maeda, A., & Palczewski, K. (2010). Molecular biology and analytical chemistry methods used to probe the retinoid cycle. *Methods in Molecular Biology, 652*, 229–245. https://doi.org/10.1007/978-1-60327-325-1_13.

Graham, F. L., Smiley, J., Russell, W. C., & Nairn, R. (1977). Characteristics of a human cell line transformed by DNA from human adenovirus type 5. *The Journal of General Virology, 36*(1), 59–74. https://doi.org/10.1099/0022-1317-36-1-59.

Green, A. S., & Fascetti, A. J. (2016). Meeting the vitamin a requirement: The efficacy and importance of beta-carotene in animal species. *Scientific World Journal, 2016*, 7393620. https://doi.org/10.1155/2016/7393620.

Grimaldi, M., Boulahtouf, A., Delfosse, V., Thouennon, E., Bourguet, W., & Balaguer, P. (2015). Reporter cell lines for the characterization of the interactions between human nuclear receptors and endocrine disruptors. *Frontiers in Endocrinology (Lausanne), 6*, 62. https://doi.org/10.3389/fendo.2015.00062.

Haeseleer, F., Huang, J., Lebioda, L., Saari, J. C., & Palczewski, K. (1998). Molecular characterization of a novel short-chain dehydrogenase/reductase that reduces all-trans-retinal. *Journal of Biological Chemistry, 273*(34), 21790–21799. https://doi.org/10.1074/jbc.273.34.21790.

Hagen, J. J., Washco, K. A., & Monnig, C. A. (1996). Determination of retinoids by reversed-phase capillary liquid chromatography with amperometric electrochemical detection. *Journal of Chromatography B, Biomedical Applications, 677*(2), 225–231. https://doi.org/10.1016/0378-4347(95)00465-3.

Harrison, E. H. (2012). Mechanisms involved in the intestinal absorption of dietary vitamin A and provitamin a carotenoids. *Biochimica et Biophysica Acta, 1821*(1), 70–77. doi:S1388-1981(11)00084-9 [pii]. https://doi.org/10.1016/j.bbalip.2011.06.002.

Hernandez, R. E., Putzke, A. P., Myers, J. P., Margaretha, L., & Moens, C. B. (2007). Cyp26 enzymes generate the retinoic acid response pattern necessary for hindbrain development. *Development, 134*(1), 177–187. https://doi.org/10.1242/dev.02706.

Jones, J. W., Pierzchalski, K., Yu, J., & Kane, M. A. (2015). Use of fast HPLC multiple reaction monitoring cubed for endogenous retinoic acid quantification in complex matrices. *Analytical Chemistry, 87*(6), 3222–3230. https://doi.org/10.1021/ac504597q.

Kalter, H., & Warkany, J. (1961). Experimental production of congenital malformations in strains of inbred mice by maternal treatment with hypervitaminosis a. *The American Journal of Pathology, 38*(1), 1–21.

Kane, M. A. (2012). Analysis, occurrence, and function of 9-cis-retinoic acid. *Biochimica et Biophysica Acta, 1821*(1), 10–20. https://doi.org/10.1016/j.bbalip.2011.09.012.

Kane, M. A., Chen, N., Sparks, S., & Napoli, J. L. (2005). Quantification of endogenous retinoic acid in limited biological samples by LC/MS/MS. *The Biochemical Journal, 388*(Pt. 1), 363–369. https://doi.org/10.1042/BJ20041867.

Kane, M. A., Folias, A. E., & Napoli, J. L. (2008). HPLC/UV quantitation of retinal, retinol, and retinyl esters in serum and tissues. *Analytical Biochemistry, 378*(1), 71–79. https://doi.org/10.1016/j.ab.2008.03.038.

Kane, M. A., Folias, A. E., Wang, C., & Napoli, J. L. (2008). Quantitative profiling of endogenous retinoic acid in vivo and in vitro by tandem mass spectrometry. *Analytical Chemistry, 80*(5), 1702–1708. https://doi.org/10.1021/ac702030f.

Kane, M. A., Folias, A. E., Wang, C., & Napoli, J. L. (2010). Ethanol elevates physiological all-trans-retinoic acid levels in select loci through altering retinoid metabolism in multiple loci: A potential mechanism of ethanol toxicity. *The FASEB Journal*, *24*(3), 823–832. https://doi.org/10.1096/fj.09-141572.

Kane, M. A., & Napoli, J. L. (2010). Quantification of endogenous retinoids. *Methods in Molecular Biology*, *652*, 1–54. https://doi.org/10.1007/978-1-60327-325-1_1.

Kawaguchi, R., Yu, J., Honda, J., Hu, J., Whitelegge, J., Ping, P., et al. (2007). A membrane receptor for retinol binding protein mediates cellular uptake of vitamin A. *Science*, *315*(5813), 820–825.

Kelly, M., Widjaja-Adhi, M. A., Palczewski, G., & von Lintig, J. (2016). Transport of vitamin A across blood-tissue barriers is facilitated by STRA6. *FASEB Journal*, *30*(8), 2985–2995. https://doi.org/10.1096/fj.201600446R.

Kishi, J. Y., Lapan, S. W., Beliveau, B. J., West, E. R., Zhu, A., Sasaki, H. M., et al. (2019). SABER amplifies FISH: Enhanced multiplexed imaging of RNA and DNA in cells and tissues. *Nature Methods*, *16*(6), 533–544. https://doi.org/10.1038/s41592-019-0404-0.

Kulikova, O., Franken, C., & Bisseling, T. (2018). In situ hybridization method for localization of mRNA molecules in Medicago tissue sections. *Methods in Molecular Biology*, *1822*, 145–159. https://doi.org/10.1007/978-1-4939-8633-0_11.

Kumar, S., Sandell, L. L., Trainor, P. A., Koentgen, F., & Duester, G. (2012). Alcohol and aldehyde dehydrogenases: Retinoid metabolic effects in mouse knockout models. *Biochimica et Biophysica Acta*, *1821*(1), 198–205. https://doi.org/10.1016/j.bbalip.2011.04.004.

Lee, L. M., Leung, C. Y., Tang, W. W., Choi, H. L., Leung, Y. C., McCaffery, P. J., et al. (2012). A paradoxical teratogenic mechanism for retinoic acid. *Proceedings of the National Academy of Sciences of the United States of America*, *109*(34), 13668–13673. doi: 1200872109 [pii]. https://doi.org/10.1073/pnas.1200872109.

Levi, L., Ziv, T., Admon, A., Levavi-Sivan, B., & Lubzens, E. (2012). Insight into molecular pathways of retinal metabolism, associated with vitellogenesis in zebrafish. *American Journal of Physiology Endocrinology and Metabolism*, *302*(6), E626–E644. https://doi.org/10.1152/ajpendo.00310.2011.

Mandal, A., Rydeen, A., Anderson, J., Sorrell, M. R., Zygmunt, T., Torres-Vazquez, J., et al. (2013). Transgenic retinoic acid sensor lines in zebrafish indicate regions of available embryonic retinoic acid. *Developmental Dynamics*, *242*(8), 989–1000. https://doi.org/10.1002/dvdy.23987.

Mata de Urquiza, A., & Perlmann, T. (2003). In vivo and in vitro reporter systems for studying nuclear receptor and ligand activities. *Methods in Enzymology*, *364*, 463–475. Academic Press.

Mata De Urquiza, A., Solomin, L., & Perlmann, T. (1999). Feedback-inducible nuclear-receptor-driven reporter gene expression in transgenic mice. *Proceedings of the National Academy of Sciences of the United States of America*, *96*(23), 13270–13275.

McCaffery, P., & Drager, U. C. (1994). Hot spots of retinoic acid synthesis in the developing spinal cord. *Proceedings of the National Academy of Sciences of the United States of America*, *91*(15), 7194–7197. https://doi.org/10.1073/pnas.91.15.7194.

Moise, A. R., Alvarez, S., Dominguez, M., Alvarez, R., Golczak, M., Lobo, G. P., et al. (2009). Activation of retinoic acid receptors by dihydroretinoids. *Molecular Pharmacology*, *76*(6), 1228–1237. https://doi.org/10.1124/mol.109.060038.

Moise, A. R., Dominguez, M., Alvarez, S., Alvarez, R., Schupp, M., Cristancho, A. G., et al. (2008). Stereospecificity of retinol saturase: Absolute configuration, synthesis, and biological evaluation of dihydroretinoids. *Journal of the American Chemical Society*, *130*(4), 1154–1155. https://doi.org/10.1021/ja710487q.

Moreau, E., Vilar, J., Lelievre-Pegorier, M., Merlet-Benichou, C., & Gilbert, T. (1998). Regulation of c-ret expression by retinoic acid in rat metanephros: Implication in nephron mass control. *American Journal of Physiology, 275*(6), F938–F945. https://doi.org/10.1152/ajprenal.1998.275.6.F938.

Moriwaki, H., Blaner, W. S., Piantedosi, R., & Goodman, D. S. (1988). Effects of dietary retinoid and triglyceride on the lipid composition of rat liver stellate cells and stellate cell lipid droplets. *Journal of Lipid Research, 29*(11), 1523–1534.

Moutier, E., Ye, T., Choukrallah, M.-A., Urban, S., Osz, J., Chatagnon, A., et al. (2012). Retinoic acid receptors recognize the mouse genome through binding elements with diverse spacing and topology. *Journal of Biological Chemistry, 287*(31), 26328–26341.

Muenzner, M., Tuvia, N., Deutschmann, C., Witte, N., Tolkachov, A., Valai, A., et al. (2013). Retinol-binding protein 4 and its membrane receptor STRA6 control adipogenesis by regulating cellular retinoid homeostasis and retinoic acid receptor alpha activity. *Molecular and Cellular Biology, 33*(20), 4068–4082. https://doi.org/10.1128/MCB.00221-13.

Napoli, J. L. (1990). Quantification and characteristics of retinoid synthesis from retinol and beta-carotene in tissue fractions and established cell lines. *Methods in Enzymology, 189*, 470–482. https://doi.org/10.1016/0076-6879(90)89324-b.

Napoli, J. L. (2012). Physiological insights into all-trans-retinoic acid biosynthesis. *Biochimica et Biophysica Acta, 1821*(1), 152–167. https://doi.org/10.1016/j.bbalip.2011.05.004.

Napoli, J. L. (2017). Cellular retinoid binding-proteins, CRBP, CRABP, FABP5: Effects on retinoid metabolism, function and related diseases. *Pharmacology & Therapeutics, 173*, 19–33. https://doi.org/10.1016/j.pharmthera.2017.01.004.

Napoli, J. L., Pramanik, B. C., Williams, J. B., Dawson, M. I., & Hobbs, P. D. (1985). Quantification of retinoic acid by gas-liquid chromatography-mass spectrometry: Total versus all-trans-retinoic acid in human plasma. *Journal of Lipid Research, 26*(3), 387–392.

Niederreither, K., McCaffery, P., Drager, U. C., Chambon, P., & Dolle, P. (1997). Restricted expression and retinoic acid-induced downregulation of the retinaldehyde dehydrogenase type 2 (RALDH-2) gene during mouse development. *Mechanisms of Development, 62*(1), 67–78.

Niederreither, K., Vermot, J., Messaddeq, N., Schuhbaur, B., Chambon, P., & Dolle, P. (2001). Embryonic retinoic acid synthesis is essential for heart morphogenesis in the mouse. *Development, 128*, 1019–1031.

Perz-Edwards, A., Hardison, N. L., & Linney, E. (2001). Retinoic acid-mediated gene expression in transgenic reporter zebrafish. *Developmental Biology, 229*(1), 89–101. https://doi.org/10.1006/dbio.2000.9979.

Quadro, L., Hamberger, L., Gottesman, M. E., Wang, F., Colantuoni, V., Blaner, W. S., et al. (2005). Pathways of vitamin a delivery to the embryo: Insights from a new tunable model of embryonic vitamin A deficiency. *Endocrinology, 146*(10), 4479–4490.

Rhinn, M., & Dolle, P. (2012). Retinoic acid signalling during development. *Development, 139*(5), 843–858.

Ross, A. C. (2012). Vitamin A and retinoic acid in T cell-related immunity. *American Journal of Clinical Nutrition, 96*(5), 1166S–1172S. https://doi.org/10.3945/ajcn.112.034637.

Rossant, J., Zirngibl, R., Cado, D., Shago, M., & Giguere, V. (1991). Expression of a retinoic acid response element-hsp lacZ transgene defines specific domains of transcriptional activity during mouse embryogenesis. *Genes & Development, 5*, 1333–1344.

Sakhi, A. K., Gundersen, T. E., Ulven, S. M., Blomhoff, R., & Lundanes, E. (1998). Quantitative determination of endogenous retinoids in mouse embryos by high-performance liquid chromatography with on-line solid-phase extraction, column switching and electrochemical detection. *Journal of Chromatography A, 828*(1–2), 451–460. https://doi.org/10.1016/s0021-9673(98)00676-1.

Sandell, L. L., Lynn, M. L., Inman, K. E., McDowell, W., & Trainor, P. A. (2012). RDH10 oxidation of vitamin A is a critical control step in synthesis of retinoic acid during mouse embryogenesis. *PLoS One*, 7(2), e30698. https://doi.org/10.1371/journal.pone.0030698.

Sandell, L. L., Sanderson, B. W., Moiseyev, G., Johnson, T., Mushegian, A., Young, K., et al. (2007). RDH10 is essential for synthesis of embryonic retinoic acid and is required for limb, craniofacial, and organ development. *Genes and Development*, *21*(9), 1113–1124. https://doi.org/10.1101/gad.1533407.

Schmidt, C. K., Brouwer, A., & Nau, H. (2003). Chromatographic analysis of endogenous retinoids in tissues and serum. *Analytical Biochemistry*, *315*(1), 36–48. https://doi.org/10.1016/s0003-2697(02)00662-0.

Shen, X., Bao, W., Yu, W., Liang, R., Nguyen, B., & Liu, Y. (2017). An improved method with high sensitivity and low background in detecting low beta-galactosidase expression in mouse embryos. *PLoS One*, *12*(5), e0176915. https://doi.org/10.1371/journal.pone.0176915.

Shimozono, S., Iimura, T., Kitaguchi, T., Higashijima, S., & Miyawaki, A. (2013). Visualization of an endogenous retinoic acid gradient across embryonic development. *Nature*, *496*(7445), 363–366. https://doi.org/10.1038/nature12037.

Siegenthaler, J. A., Ashique, A. M., Zarbalis, K., Patterson, K. P., Hecht, J. H., Kane, M. A., et al. (2009). Retinoic acid from the meninges regulates cortical neuron generation. *Cell*, *139*(3), 597–609. https://doi.org/10.1016/j.cell.2009.10.004.

Sucov, H. M., Murakami, K. K., & Evans, R. M. (1990). Characterization of an autoregulated response element in the mouse retinoic acid receptor type beta gene. *Proceedings of the National Academy of Sciences of the United States of America*, *87*(14), 5392–5396.

Sundararajan, S., Wakamiya, M., Behringer, R. R., & Rivera-Perez, J. A. (2012). A fast and sensitive alternative for beta-galactosidase detection in mouse embryos. *Development*, *139*(23), 4484–4490. https://doi.org/10.1242/dev.078790.

Ulven, S. M., Gundersen, T. E., Weedon, M. S., Landaas, V. O., Sakhi, A. K., Fromm, S. H., et al. (2000). Identification of endogenous retinoids, enzymes, binding proteins, and receptors during early postimplantation development in mouse: Important role of retinal dehydrogenase type 2 in synthesis of all-trans-retinoic acid. *Developmental Biology*, *220*(2), 379–391. https://doi.org/10.1006/dbio.2000.9634.

von Lintig, J., & Vogt, K. (2000). Filling the gap in vitamin a research. Molecular identification of an enzyme cleaving beta-carotene to retinal. *Journal of Biological Chemistry*, *275*(16), 11915–11920.

Wagner, M. A. (1997). Use of reporter cells to study endogenous retinoid sources in embryonic tissues. *Methods in Enzymology*, *282*, 98–107. https://doi.org/10.1016/s0076-6879(97)82099-x.

Wagner, M., Han, B., & Jessell, T. M. (1992). Regional differences in retinoid release from embryonic neural tissue detected by an in vitro reporter assay. *Development*, *116*(1), 55–66.

Wald, G. (1968). Molecular basis of visual excitation. *Science*, *162*(3850), 230–239. https://doi.org/10.1126/science.162.3850.230.

Wang, F., Flanagan, J., Su, N., Wang, L. C., Bui, S., Nielson, A., et al. (2012). RNAscope: A novel in situ RNA analysis platform for formalin-fixed, paraffin-embedded tissues. *The Journal of Molecular Diagnostics*, *14*(1), 22–29. https://doi.org/10.1016/j.jmoldx.2011.08.002.

Wang, C., Kane, M. A., & Napoli, J. L. (2011). Multiple retinol and retinal dehydrogenases catalyze all-trans-retinoic acid biosynthesis in astrocytes. *Journal of Biological Chemistry*, *286*(8), 6542–6553. doi:M110.198382 [pii]. https://doi.org/10.1074/jbc.M110.198382.

Wang, J., Yoo, H. S., Obrochta, K. M., Huang, P., & Napoli, J. L. (2015). Quantitation of retinaldehyde in small biological samples using ultrahigh-performance liquid chromatography tandem mass spectrometry. *Analytical Biochemistry*, *484*, 162–168. https://doi.org/10.1016/j.ab.2015.05.016.

Wang, S., Yu, J., Jones, J. W., Pierzchalski, K., Kane, M. A., Trainor, P. A., et al. (2018). Retinoic acid signaling promotes the cytoskeletal rearrangement of embryonic epicardial cells. *FASEB Journal, 32*(7), 3765–3781, fj201701038R. doi:10.1096/fj.201701038R.

Waxman, J. S., & Yelon, D. (2011). Zebrafish retinoic acid receptors function as context-dependent transcriptional activators. *Developmental Biology, 352*(1), 128–140. https://doi.org/10.1016/j.ydbio.2011.01.022.

WHO. (2009). Global prevalence of vitamin A deficiency in populations at risk 1995–2005. WHO Global Database on Vitamin A Deficiency. Geneva: World Health Organization.

Wongsiriroj, N., Piantedosi, R., Palczewski, K., Goldberg, I. J., Johnston, T. P., Li, E., et al. (2008). The molecular basis of retinoid absorption: A genetic dissection. *Journal of Biological Chemistry, 283*(20), 13510–13519. https://doi.org/10.1074/jbc.M800777200.

Wyss, R., & Bucheli, F. (1997). Determination of endogenous levels of 13-cis-retinoic acid (isotretinoin), all-trans-retinoic acid (tretinoin) and their 4-oxo metabolites in human and animal plasma by high-performance liquid chromatography with automated column switching and ultraviolet detection. *Journal of Chromatography B, Biomedical Sciences and Applications, 700*(1–2), 31–47. https://doi.org/10.1016/s0378-4347(97)00303-4.

Wyss, A., Wirtz, G., Woggon, W., Brugger, R., Wyss, M., Friedlein, A., et al. (2000). Cloning and expression of beta,beta-carotene 15,15′-dioxygenase. *Biochemical and Biophysical Research Communications, 271*(2), 334–336.

Yokota, S., Shirahata, T., Yusa, J., Sakurai, Y., Ito, H., & Oshio, S. (2019). Long-term dietary intake of excessive vitamin a impairs spermatogenesis in mice. *Journal of Toxicological Sciences, 44*(4), 257–271. https://doi.org/10.2131/jts.44.257.

Zetterstrom, R. H., Lindqvist, E., Mata de Urquiza, A., Tomac, A., Eriksson, U., Perlmann, T., et al. (1999). Role of retinoids in the CNS: Differential expression of retinoid binding proteins and receptors and evidence for presence of retinoic acid. *The European Journal of Neuroscience, 11*(2), 407–416. https://doi.org/10.1046/j.1460-9568.1999.00444.x.

Zile, M. H. (1998). Vitamin A and embryonic development: An overview. *Journal of Nutrition, 128*(2 Suppl), 455S–458S. https://doi.org/10.1093/jn/128.2.455S.

Zolfaghari, R., Chen, Q., & Ross, A. C. (2012). DHRS3, a retinal reductase, is differentially regulated by retinoic acid and lipopolysaccharide-induced inflammation in THP-1 cells and rat liver. *American Journal of Physiology Gastrointestinal and Liver Physiology, 303*(5), G578–G588. https://doi.org/10.1152/ajpgi.00234.2012.

CHAPTER SIXTEEN

Whole mount *in situ* hybridization and immunohistochemistry for studying retinoic acid signaling in developing amphioxus

Matteo Bozzo[a], **Simona Candiani**[a,†], **Michael Schubert**[b,*,†]
[a]Laboratory of Developmental Neurobiology, Department of Earth, Environment and Life Sciences, University of Genoa, Genoa, Italy
[b]Sorbonne Université, CNRS, Laboratoire de Biologie du Développement de Villefranche-sur-Mer, Villefranche-sur-Mer, France
*Corresponding author: e-mail address: michael.schubert@obs-vlfr.fr

Contents

† These authors contributed equally to this work.

Methods in Enzymology, Volume 637
ISSN 0076-6879
https://doi.org/10.1016/bs.mie.2020.03.007

Abstract

Retinoic acid (RA) is a vitamin A-derived signaling molecule acting during development and in the adult. This chapter provides protocols to characterize the role of RA signaling during development of the invertebrate chordate amphioxus. As sister group to all other chordates and characterized by the most vertebrate-like RA signaling system of all invertebrates, amphioxus is an important model for studying the evolution of RA signaling. Focusing on the development of GABAergic neurons in the amphioxus central nervous system, we provide detailed protocols for maintaining and breeding adult animals, for performing pharmacological treatments of embryos and for analyzing the effects of these treatments by whole mount *in situ* hybridization and immunohistochemistry coupled to confocal microscopy.

1. Introduction

Retinoic acid (RA) is a signaling molecule that plays key roles in animal development and adult physiology (Laudet, Zieger, & Schubert, 2015). In the body, RA is produced from retinol (vitamin A) in a two-step oxidative process: (1) retinol is reversibly oxidized to retinaldehyde and (2) retinaldehyde is irreversibly oxidized to RA. This irreversible step is catalyzed by retinaldehyde dehydrogenase (RALDH) enzymes. Endogenous RA levels are further tightly regulated by tissue-specific degradation mediated by enzymes of the cytochrome p450 subfamily 26 (CYP26) (Ghyselinck & Duester, 2019). The cellular response to RA is generally controlled by heterodimers of two nuclear receptors, the retinoic acid receptor (RAR) and the retinoid X receptor (RXR). RA binding to RAR induces conformational changes in the heterodimer and ultimately leads to the transcriptional regulation of target genes. Although different RA isomers have been identified *in vivo*, all-*trans* retinoic acid (ATRA) is considered the natural endogenous ligand of RAR (Mic, Molotkov, Benbrook, & Duester, 2003). RA signaling controlled by RAR evolved early in animal evolution, likely at the base of bilaterians, as a low-affinity sensor system. RARs with high affinity for RA, as seen in mammals, probably evolved only later and specifically in the chordate lineage (Fonseca et al., 2019; Handberg-Thorsager et al., 2018). This marks the origin of chordates as a milestone in the evolution of RA signaling.

Cephalochordates, commonly referred to as amphioxus or lancelets, are small, filter-feeding animals that live burrowed at the bottom of temperate and tropical seas. Amphioxus are characterized by an elongated, fish-like body that, depending on the species, can reach a maximum length of between 2.5 and 8.5 cm (Poss & Boschung, 1996). According to their phylogenetic position as sister group to all other chordates (i.e., tunicates and vertebrates) (Putnam et al., 2008), amphioxus have a typical chordate body plan with a dorsal hollow nerve cord, a notochord that extends axially from rostrum to tail, metameric muscles and an anterior perforated pharynx (Bertrand & Escriva, 2011). Furthermore, in addition to a RAR with high RA affinity, amphioxus also has the most vertebrate-like RA signaling system of all invertebrates (Holland et al., 2008; Laudet et al., 2015). For example, in both amphioxus and vertebrates, RA signaling regulates anteroposterior patterning of the central nervous system by directly controlling colinear expression of *Hox* genes (Koop et al., 2010; Schubert, Holland, Laudet, & Holland, 2006). These features make amphioxus a key model for understanding the evolution of RA signaling during chordate diversification.

In this chapter, we describe how to maintain adult amphioxus in the laboratory, how to obtain gametes and how to manipulate RA signaling during embryonic development by pharmacological treatments. We also provide detailed protocols to assess the effects of these treatments on the developing neural architecture by whole mount *in situ* hybridization (ISH) and immunohistochemistry (IHC). These two approaches are complementary as the former detects mRNAs while the latter targets antigens. ISH tends to be more widely used in non-model organisms because species-specific probes can be synthesized rather easily (for details on probe synthesis see Yong, Kozmikova, & Yu, 2019). Compared to ISH, IHC is easier and faster to perform and yields higher anatomical resolution. For instance, IHC allows the detection of both cell bodies and projections of neurons, while ISH generally only stains cell bodies, a major drawback for anatomical analyses (Zieger, Candiani, Garbarino, Croce, & Schubert, 2018; Zieger, Garbarino, et al., 2018). A primary limitation of IHC is that specific antibodies are required, most of which have been developed for model organisms, such as mice and fruit flies. Although some of these antibodies have been shown to cross-react with conserved antigen epitopes in a large variety of species (Zieger, Candiani, et al., 2018; Zieger, Garbarino, et al., 2018), targeting non-conserved epitopes by IHC requires the production and subsequent validation of novel antibodies (Bozzo, Pergner, Kozmik, & Kozmikova, 2017). Thus, in addition to characterizing RA signaling functions, the protocols we provide will also be useful for the validation of novel antibodies in amphioxus, by allowing comparisons of IHC and ISH patterns.

2. Maintaining adult amphioxus in the laboratory

Sexually mature adults of the European amphioxus (*Branchiostoma lanceolatum*) are collected in Argelès-sur-Mer (France) (Fuentes et al., 2007) before the beginning of the spawning season and kept in a seawater facility in the laboratory (Carvalho, Lahaye, & Schubert, 2017; Theodosiou et al., 2011). 20–25 amphioxus adults are kept in 5L tanks containing around 3cm of autoclaved sand on a 14/10h inverted light/dark cycle (light: 9pm–11am; dark: 11am–9pm). Each tank is continuously aerated. Open systems with seawater flowing continuously through the tanks are preferable, but closed systems with filtration devices can also be used (Carvalho et al., 2017).

Animals are fed with a mixture of live microalgae (such as *Tisochrysis lutea* and *Chaetoceros gracilis*). Algae are collected in 50mL centrifuge tubes, pelleted by centrifugation at 4000 × *g* for 10min, washed twice with seawater and resuspended in fresh seawater. The resuspended algae are pooled to a final volume of 50mL and half a teaspoon of powdered green algae (such as *Spirulina*) is added to the mixture. Each tank receives 1mL of concentrated algae, corresponding to 15mL of unspun algal culture, twice a day.

3. Spawning induction, gamete collection and *in vitro* fertilization

During the spawning season (May through July), amphioxus embryos can be obtained in large numbers by *in vitro* fertilization using gametes released in response to a thermal shock (Fuentes et al., 2007).

3.1 Equipment

- 15 mL centrifuge tubes
- 5 L tank
- 6 and 15 cm plastic Petri dishes. Eggs and embryos tend to stick to plasticware. To avoid this, all Petri dishes must be rubbed with wet tissue paper and rinsed three times with artificial seawater shortly before adding eggs or embryos
- Plastic cups
- Red-light flashlight
- Room with controlled temperature and light/dark cycle
- Seawater refractometer, for instance, HANNA Instruments (HI96822)
- Stereomicroscope
- Water bath

3.2 Reagents

- Artificial seawater salt, Red Sea

3.3 Working solutions

- Artificial seawater: dissolve 400 g Red Sea seawater salt in 10 L of reverse osmosis or distilled water. Vigorously agitate using an aquarium pump for a couple of days to dissolve the salt. Measure salinity with a refractometer and adjust to 36–38 ppt by adding either seawater salt or water. Keep artificial seawater continuously agitated to prevent salt precipitation.

3.4 Protocol for obtaining and rearing amphioxus embryos

1. Select an adequate number of males and females with full gonads shortly before the end of the light period. The number of animals to be selected depends: (1) on the total number of embryos required for performing the planned experiments and (2) on the spawning efficiency of

amphioxus adults in the facility. For the experiments detailed in this chapter, four animals (two females and two males) should be sufficient. Ripe animals can be sexed using a stereomicroscope: eggs are clearly visible as small spheres inside the ovary, while testes are homogeneously filled with sperm.

2. Put the selected animals into a clean 5 L tank containing 2–3 cm of fresh seawater and no sand.
3. Place the tank into a water bath set at 23 °C with the same light/dark regime used in the animal facility. Leave overnight without food and aeration.
4. The following day, shortly before the start of the dark period, transfer each animal into an individual plastic cup with 15 mL of artificial seawater and keep them at 19 °C in the dark.
5. Check for spawning every 30 min using a red-light flashlight to avoid startling the animals. Adults typically spawn 1–4 h after the beginning of the dark period.
6. Collect eggs by gently pouring them into a 6 cm plastic Petri dish. Keep at 19 °C until fertilization.
7. Collect sperm by pouring into a 15 mL centrifuge tube and promptly move to +4°C. Sperm remains active for about 12 h, with a marked decrease in fertilization efficiency after about 2 h.
8. Fertilize eggs by adding a few drops of sperm to the Petri dish. Gently swirl the dish to homogeneously distribute the sperm, hence ensuring synchronous fertilization.
9. Verify fertilization after 2–3 min: fertilized eggs are characterized by a clearly visible fertilization envelope.
10. Immediately transfer the fertilized eggs into a new 15 cm plastic Petri dish filled with artificial seawater to prevent embryos from sticking together.
11. Let develop until the desired stage. Note that cultures need to be kept at a constant temperature to ensure development according to predefined staging systems. The culture temperature most widely used for *B. lanceolatum* is 19 °C, which was also used here.
12. After spawning, return the adults to the animal facility. Separate animals that did not spawn from those that did. While the former can be shocked again during the ongoing spawning season, the latter can be kept in the animal facility: around a third of the animals that spawned out will redevelop gonads ahead of the next spawning season (Carvalho et al., 2017).

4. Pharmacological manipulation of retinoic acid signaling in amphioxus embryos

To interfere with RA signaling, embryos were treated from the early gastrula (6 hpf, i.e., hours post-fertilization) until the late embryo stage (36 hpf) with the compounds listed in Table 1. The following protocol, presented in graphical form in Fig. 1, can be applied to the pharmacological manipulation of any signaling pathway by changing compounds and adapting treatment windows.

4.1 Equipment

- 15 mL centrifuge tubes
- 25 mL serological pipettes: wet the inside of the pipette with artificial seawater before use on embryos to avoid sticking
- 6 cm plastic Petri dish rubbed with wet tissue paper and rinsed three times with artificial seawater, as described in Section 3.1
- Filtered pipette tips
- Micropipettes
- Pipet aid
- Room with controlled temperature, set at 19 °C
- Stereomicroscope

4.2 Reagents

- Dimethyl sulfoxide (DMSO), Sigma-Aldrich (D8418)
- all-*trans* retinoic acid (ATRA), Sigma-Aldrich (R2625)
- R115866/Talarozole, Sigma-Aldrich (SML2092)

Table 1 List of chemicals used to interfere with retinoic acid signaling.

Chemical	Mechanism of action	Stock concentration	Final concentration
all-*trans*-retinoic acid (ATRA)	Endogenous RAR ligand	0.1 mM	0.1 μM
R115866 (Talarozole)	CYP26 inhibitor	0.5 mM	0.5 μM
BMS493	RAR inverse agonist	1 mM	1 μM
4-Diethyl-aminobenzaldehyde (DEAB)	RALDH inhibitor	10 mM	10 μM

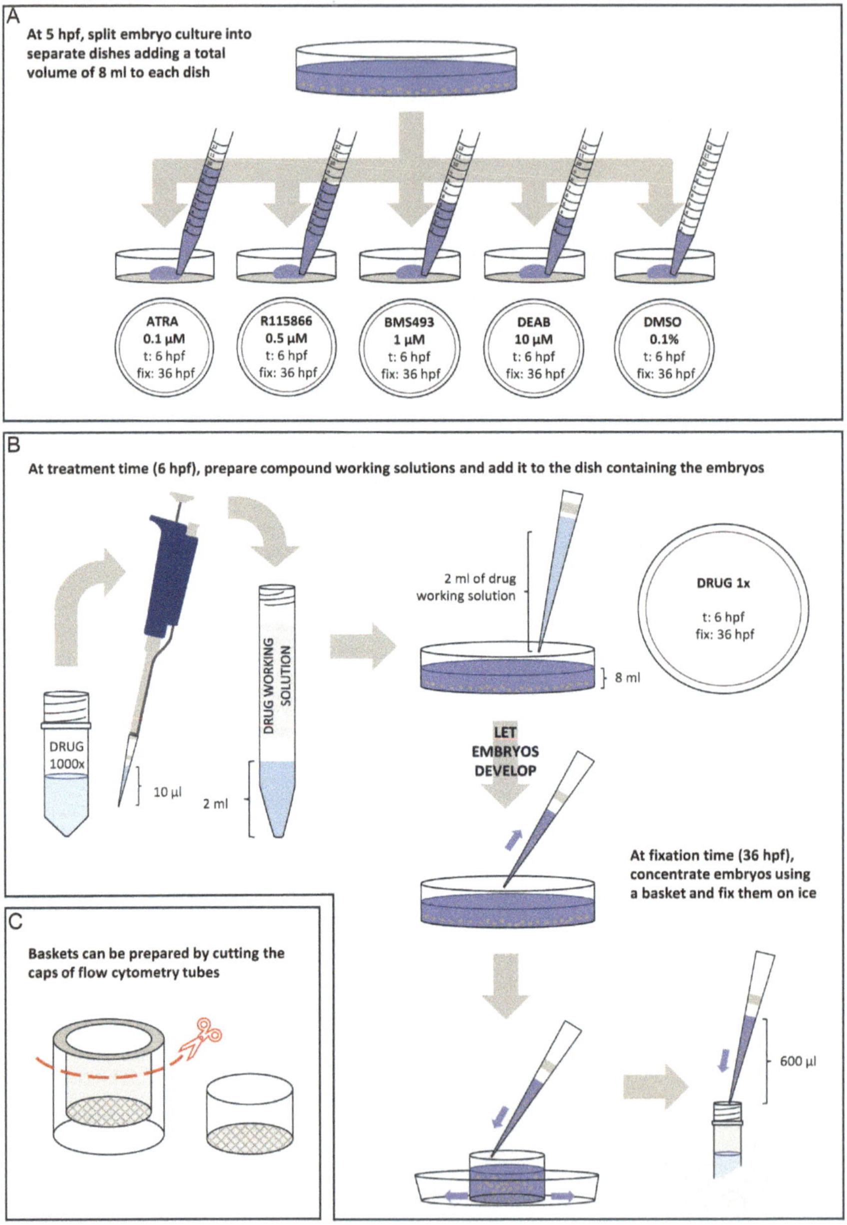

Fig. 1 Schematic protocol for the pharmacological manipulation of retinoic acid signaling in amphioxus embryos. (A) Preparation of embryo plates for treatments. (B) Treatment and fixation. (C) Preparation of embryo baskets from the caps of flow cytometry tubes.

- BMS493, Sigma-Aldrich (B6688)
- 4-Diethyl-aminobenzaldehyde (DEAB), Sigma-Aldrich (D86256)

4.3 Stock solutions

- Artificial seawater, as described in Section 3.3
- 0.1 mM ATRA in DMSO. Can be stored in aliquots in the dark at −20 °C for several months
- 0.5 mM R115866/Talarozole in DMSO. Can be stored in aliquots in the dark at −20 °C for several months
- 1 mM BMS493 in DMSO. Can be stored in aliquots in the dark at −20 °C for several months
- 10 mM DEAB in DMSO. Can be stored in aliquots in the dark at +4 °C for several weeks

4.4 Protocol for the pharmacological treatment of amphioxus embryos

1. Following fertilization, check the quality of the embryos by monitoring the first cell divisions with a stereomicroscope. Reference images of amphioxus early development can be found elsewhere (for instance, Bertrand & Escriva, 2011).
2. Ahead of the first treatment, prepare and label the required number of dishes: number of compounds plus one control for each solvent multiplied by the number of treatment windows to be analyzed. In our case, four compounds and one solvent are used with a single treatment window (6–36 hpf), so prepare and label five dishes.
3. Using a pipet aid and a 25 mL serological pipette, carefully transfer four times 2 mL of embryos from the culture plate into each dish, with each dish thus receiving a total of 8 mL. It is important to use a gentle pipetting stream to avoid damaging or removing the fertilization envelopes.
4. Verify the volume in each dish by pipetting (using the serological pipette) and, if needed, adjust to exactly 8 mL. Again, make sure to gently pipette the embryos.
5. At the desired stage (6 hpf for this experiment), add 10 μL of the appropriate 1000 × compound stock solution to 2 mL of artificial seawater in a 15 mL centrifuge tube. Mix well and add to the corresponding treatment dish. Swirl gently. Since the treatment volume is 10 mL, the resulting final treatment concentration is a 1000-fold dilution of the compound stock solution. Always remember to include control dishes containing

the corresponding concentration of the solvents used to prepare the compound stock solutions.

6. Incubate the embryos at 19 °C in the dark until the moment of fixation, 36 hpf in this study.

4.5 Expected phenotypes

If the phenotypes of the pharmacological treatments are known prior to the experiment, efficiency of the treatments can be assessed directly after fixation and prior to further analyses. For the treatments described here, it is known that upregulation of RA signaling activity in developing amphioxus shortens the pharynx, while its downregulation expands the pharynx (Escriva, Holland, Gronemeyer, Laudet, & Holland, 2002). The expected phenotypes are shown in Fig. 2.

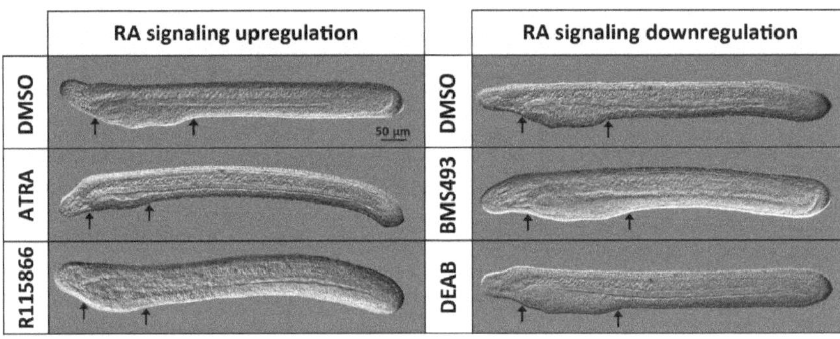

Fig. 2 Phenotype evaluation following pharmacological manipulation of retinoic acid signaling. Arrows indicate the anterior and posterior limits of the pharynx. Compound list is provided in Table 1. DMSO is solvent control. Brightfield images of embryos in lateral view with anterior to the left.

4.6 Troubleshooting and optimization

Problem	Solution
Pharmacological compound does not dissolve in solvent	Check solubility of the compound in different solvents. Note that solubility might vary greatly, and solvent must thus be chosen accordingly
Pharmacological compound precipitates in artificial seawater	Mix well after adding the compound to the artificial seawater. If necessary, vortex the stock solution tube before use. If problem persists, change stock solution aliquot. Eventually, prepare new stock solution using a different solvent

—cont'd

Problem	Solution
Solvent control embryos do not develop normally	Do not overcrowd treatment plates with embryos: a 6 cm Petri dish should hold about 150 embryos Make sure to use the lowest possible solvent concentration If abnormal development persists, verify embryo culture conditions
Treated embryos stop developing	Verify the concentration of the compound used. If you are testing a new molecule, consider lowering its concentration
Embryos do not show the expected phenotype	Verify the concentration of the compound used. If you are testing a new molecule, consider increasing its concentration For retinoids, perform treatments in the dark and make sure that stock solutions were not exposed to light for a prolonged period of time Try fresh aliquots of stock solutions. Eventually, prepare new stock solutions

5. Fixation of amphioxus embryos

Embryos for ISH are fixed with paraformaldehyde (PFA) in MOPS-EGTA buffer and stored in RNase-free 70% ethanol. Embryos for GABA immunohistochemistry are fixed in PFA/glutaraldehyde in MOPS-EGTA buffer and stored in PBS/sodium azide.

5.1 Equipment

- 4-well cell culture plates
- Baskets with a 35 μm mesh bottom, prepared by cutting the sieve out of the cap of a Corning flow cytometry tube (352235), as illustrated in Fig. 1C
- Filtered pipette tips, RNase-free
- Glass bottle, RNase-free
- Gloves
- Magnetic stirrer with hot plate and stir bar
- Micropipettes
- pH meter

- Screw-cap, low-binding microcentrifuge tubes, RNase-free
- Tweezers
- Vacuum suction system and filters

5.2 Reagents

- 3-(N-morpholino)propanesulfonic acid (MOPS), Sigma-Aldrich (69947)
- Diethyl pyrocarbonate (DEPC), Sigma-Aldrich (40718)
- Ethanol, absolute, molecular biology grade, Millipore (100983)
- Ethylene glycol-bis(2-aminoethylether)-N,N,N′,N′-tetraacetic acid (EGTA), Sigma-Aldrich (E3889)
- Glutaraldehyde solution, 50%, Sigma-Aldrich (G7651)
- Hydrochloric acid (HCl), 37%, Sigma-Aldrich (30721)
- Magnesium sulfate ($MgSO_4$), Sigma-Aldrich (M2643)
- Paraformaldehyde (PFA), Sigma-Aldrich (P6148)
- RNaseZAP, Sigma-Aldrich (R2020)
- Sodium azide, Sigma-Aldrich (71289)
- Sodium chloride (NaCl), Sigma-Aldrich (S9888)
- Sodium hydroxide solution (NaOH), 1 N, Sigma-Aldrich (79724)
- Sodium phosphate dibasic, Sigma-Aldrich (S9638)
- Sodium phosphate monobasic, Sigma-Aldrich (S3264)

5.3 Working solutions

- 4% PFA/MOPS-EGTA: weigh 4 g of PFA in an RNase-free glass bottle. Add 10 mL of 1 N NaOH and dissolve on a magnetic stirrer with hot plate by heating to 60 °C. Add 100 mL of MOPS-EGTA and leave stirring at 60 °C overnight. Bring to pH 7.4 with 37% HCl and filter. Perform the whole process under a fume hood. Prepare fresh at the beginning of the spawning season. Can be stored at +4°C for several weeks
- 70% ethanol: dilute molecular biology grade absolute ethanol to 70% (v/v) with DEPC-treated water
- Artificial seawater, as described in Section 3.3
- DEPC-treated water: add DEPC to distilled water to a final concentration of 0.1% (v/v), shake vigorously. Incubate overnight under a fume hood and autoclave. Can be stored at room temperature for several months
- MOPS-EGTA buffer: 0.1 M MOPS, 1 mM EGTA, 2 mM $MgSO_4$, 0.5 M NaCl, pH 7.5, in DEPC-treated water. Filter. Prepare fresh at the beginning of the spawning season. Can be stored at +4 °C for several weeks

- PBS/sodium azide: 2.6 mM sodium phosphate monobasic, 17.4 mM sodium phosphate dibasic, 0.9% (wt/v) NaCl, 0.1% (wt/v) sodium azide, pH 7.4. DEPC-treat and autoclave. Can be stored at room temperature for several months
- PFA/glutaraldehyde: add glutaraldehyde to 0.3% (v/v) to 4% PFA/MOPS-EGTA

5.4 Protocol for fixing amphioxus embryos

1. Perform the whole process under a fume hood.
2. Fill a 1.5 mL screw-cap microcentrifuge tube with 1 mL of ice-cold fixative, label it appropriately and place it on ice.
3. Take a dish containing treated embryos, open it, turn the lid upside down and place a basket inside.
4. To concentrate the embryos in a small volume, transfer them from the dish to the basket using a micropipette with a 1000 μL tip. Make sure that the basket does not overflow. Repeat until all the embryos have been moved from the dish into the basket. If needed, pipette the embryos up and down inside the basket to remove the fertilization envelope.
5. Using tweezers, move the basket into a well of a 4-well plate containing 600 μL of artificial seawater.
6. Using a micropipette with a 200 μL tip, transfer the embryos from the basket into the tube placed on ice and containing the fixative. Carefully add the embryos drop by drop to the tube. Do not transfer more than 600 μL to one tube.
7. Invert the tube several times and place on ice until the embryos have settled at the bottom of the tube.
8. Remove the solution and replace with fresh ice-cold fixative. Invert the tube several times. Fix at +4 °C overnight in an upright position to let the embryos settle.
9. Remove the fixative and replace with ice-cold storage solution. Invert the tube several times.
10. Place the tube on ice and wait until the embryos have settled. Remove the solution and replace with fresh ice-cold storage solution. Invert the tube several times. Leave at +4 °C overnight.
11. Remove the solution and replace with fresh ice-cold storage solution. Keep embryos at +4 °C, if stored in PBS/sodium azide, or at −20 °C, if stored in 70% ethanol. Samples stored in PBS/sodium azide should be used within weeks, while samples stored in 70% ethanol can be conserved for months.

5.5 Troubleshooting and optimization

Problem	Solution
Embryos are damaged	Minimize volume transferred from basket to fixation tube Remove as much solution as possible when replacing the fixative
Embryos are damaged, especially mid-gastrula and early neurula stages	From mid-gastrula to early neurula stages, embryos produce hatching enzymes. For these stages, fertilization envelopes must thus be removed before fixation. Make sure to remove as much liquid as possible when changing the fixative, eventually performing additional washes
Cleavage stages are dissociated	Blastomeres do not adhere very strongly and might dissociate after removal of the fertilization envelope. Fix these stages with intact fertilization envelopes and remove the envelopes right before staining, either by pipetting up and down in the well or by using insect pins

6. Whole mount *in situ* hybridization

Whole mount ISH allows the detection of mRNAs in entire embryos with cellular resolution. This technique is widely used to determine the spatiotemporal expression patterns of developmental genes. Standard protocols are based on the hybridization of the target sequence with a complementary riboprobe labeled either with digoxigenin or fluorescein. These antigens are subsequently recognized by specific antibodies that are conjugated with a reporter activity. For amphioxus embryos, protocols for both chromogenic (Yu & Holland, 2009) and fluorescent (Yong et al., 2019) ISH have already been described. In addition, double ISH approaches are available, allowing the analysis of two different mRNAs within a single embryo (Candiani, Garbarino, & Pestarino, 2015). While the protocol below uses two chromogenic substrates, NBT/BCIP and Vector® Blue, the distribution of the riboprobes within the embryo will be assessed using a confocal microscope. Compared to exclusively fluorescent ISH methods, this approach has the advantage that the progress of the staining reaction can be monitored under a stereomicroscope and stopped when the signal is fully developed. We apply this protocol to the detection of mRNA transcripts of the *glutamate decarboxylase* (*GAD*) gene that encodes an enzyme involved in the biosynthesis of the neurotransmitter GABA.

6.1 Equipment

- 200 μL and 1000 μL filtered micropipette tips, RNase-free
- 24-well cell culture plates, RNase-free
- 4-well cell culture plates, RNase-free
- Baskets, prepared as described in Section 5.1
- Gloves
- Horizontal orbital shaker
- Hybridization oven
- Micropipettes
- Paper tape
- pH meter
- Stereomicroscope
- Syringes with 0.45 μm filters
- Tweezers
- Vortex
- Water bath
- Zipper bag

6.2 Reagents

- 4-Nitro blue tetrazolium chloride (NBT), Roche (11383213001)
- 5-Bromo-4-chloro-3-indolyl phosphate 4-toluidine salt (BCIP), Roche (11383221001)
- Acetic anhydride, Sigma-Aldrich (320102)
- Anti-DIG AP-conjugated Fab fragments, Roche (11093274910)
- Bovine serum albumin (BSA), Sigma-Aldrich (A9647)
- Denhardt's solution, 50 ×, Sigma-Aldrich (D9905)
- Diethyl pyrocarbonate (DEPC), Sigma-Aldrich (D5758)
- Ethylenediaminetetraacetic acid (EDTA), Sigma-Aldrich (E6758)
- Formamide, Fisher Scientific (BP228-100)
- Glycerol, Sigma-Aldrich (G7893)
- Glycine, Sigma-Aldrich (50046)
- Heparin, Sigma-Aldrich (H4784)

- Hoechst 33258, Sigma-Aldrich (14530)
- Hydrochloric acid (HCl), 1 N, Fisher Scientific (11993123)
- Levamisole, Sigma-Aldrich (BP212)
- Magnesium chloride ($MgCl_2$), Sigma-Aldrich (M9272)
- Proteinase K, Sigma-Aldrich (P2308)
- RNase A, Sigma-Aldrich (R6513)
- RNase T1, Sigma-Aldrich (R1003)
- RNaseZAP, Sigma-Aldrich (R2020)
- Sheep serum, Sigma-Aldrich (S3772)
- Sodium acetate, 3 M, pH 5.5, Invitrogen (AM9740)
- Sodium azide, Sigma-Aldrich (71289)
- Sodium chloride (NaCl), Sigma-Aldrich (S9888)
- Sodium citrate, Sigma-Aldrich (71497)
- Sodium dodecyl sulfate (SDS), Sigma-Aldrich (L3771)
- Sodium phosphate dibasic, Sigma-Aldrich (S9638)
- Sodium phosphate monobasic, Sigma-Aldrich (S3264)
- Torula yeast RNA, Sigma-Aldrich (R3629)
- Triethanolamine, Sigma-Aldrich (90279)
- Tris-EDTA (TE) buffer, pH 7.4, Sigma-Aldrich (93302)
- Triton X-100, Sigma-Aldrich (T8787)
- TRIZMA base, Sigma-Aldrich (93362)
- Tween-20, Sigma-Aldrich (P9416)
- Vector® Blue alkaline phosphatase substrate kit, Vector Laboratories (SK-5300)

6.3 Stock solutions

- 0.1 M triethanolamine (TEA): dilute TEA to 0.1 M in DEPC-treated water. Adjust to pH 8 with 1 N HCl. Do not treat TEA with DEPC and do not autoclave. Can be stored at +4 °C for several months
- 1 M Tris-HCl, pH 8.2 or pH 9.6: dissolve RNase-free TRIZMA base in DEPC-treated water and adjust to either pH 8.2 or pH 9.6 with 1 N HCl. Do not treat with DEPC. Filter. Can be stored at +4°C for several months
- 10% BSA: dissolve to 10% (wt/v) in PBS. Can be stored in aliquots at −20 °C for several months
- 10% glycine: dissolve RNase-free glycine to 10% (wt/v) in DEPC-treated water. Can be stored in aliquots at −20 °C for several months

- 10% Tween-20: dilute to 10% (v/v) with distilled water. DEPC-treat and autoclave. Can be stored at room temperature for several months
- 10 × PBS (phosphate buffer saline): 26 mM sodium phosphate monobasic, 174 mM sodium phosphate dibasic, 9% (wt/v) NaCl, pH 7.4. DEPC-treat and autoclave. Can be stored at room temperature for several months
- 20 × saline sodium citrate (SSC): dissolve NaCl to 3 M and sodium citrate to 300 mM. Adjust to pH 7 with 1 N HCl, DEPC-treat and autoclave. Can be stored at room temperature for several months
- DEPC-treated water: prepare as described in Section 5.3
- Hoechst stock: dissolve to 10 mg/mL in distilled water. Can be stored in the dark at +4 °C for several months
- Proteinase K stock: dissolve to 10 mg/mL in DEPC-treated water. Can be stored in aliquots at −20 °C for several months
- RNase A stock: dissolve to 10 mg/mL in TE buffer. Boil for 10 min then quickly cool on ice. Can be stored in aliquots at −20 °C for several months
- RNase T1 stock: dilute RNase T1 to 10,000 U/mL in 0.1 M sodium acetate, pH 5.5. Boil for 10 min then quickly cool on ice. Can be stored in aliquots at −20 °C for several months
- Sheep serum: inactivate by heating to 55 °C for 30 min. Cool on ice and prepare 1 mL aliquots that can be stored at −20 °C for several months

6.4 Working solutions

- 0.25% or 0.5% acetic anhydride in TEA. Prepare fresh. Do not store
- 1 × PBS: Prepare diluting 10 × PBS stock with DEPC-treated water. Can be stored at room temperature for several months
- 2 mg/mL glycine: dilute 10% glycine 1:50 with PBSTw
- 80% glycerol: dilute glycerol with PBS/sodium azide. Can be stored at room temperature for several months
- Alkaline phosphatase (AP) buffer with $MgCl_2$: 100 mM NaCl, 50 mM $MgCl_2$, 100 mM Tris-HCl pH 9.6, 0.1% Tween-20, 1 mM levamisole in DEPC-treated water. Prepare fresh and filter
- Antibody solution: anti-DIG AP-conjugated antibody 1:3000, 2 mg/mL BSA, 6.5% heat-inactivated sheep serum, 1 × PBS, 0.1% TritonX-100. Can be stored in 1 mL aliquots at −20 °C for several months

- AP buffer without $MgCl_2$: 100 mM NaCl, 100 mM Tris-HCl pH 9.6, 0.1% Tween-20, 1 mM levamisole in DEPC-treated water. Prepare fresh and filter
- Blocking solution: 2 mg/mL BSA, 10% heat-inactivated sheep serum in PBSTw. Prepare fresh
- Hybridization buffer (HB): 50% (v/v) formamide, 100 μg/mL heparin, 5× SSC, 0.1% (v/v) Tween-20, 5 mM EDTA, 1 mg/mL torula yeast RNA, 1× Denhardt's solution in DEPC-treated water. Warm solution to 60 °C before adding torula yeast RNA. Can be stored in aliquots at −20 °C for several months
- PBS/sodium azide: 20 mM sodium phosphate buffer, pH 7.4, 0.9% (wt/v) NaCl, 0.1% (wt/v) sodium azide. Can be stored at room temperature for several months
- PBSTw: 20 mM sodium phosphate buffer, pH 7.4, 0.9% (wt/v) NaCl, 0.1% (v/v) Tween-20 in DEPC-treated water. Prepare fresh from 10× PBS and 10% Tween-20 stocks and filter
- Proteinase K working solution: 7.5 μg/mL in PBSTw prepared fresh from the stock. We recommend a two-step dilution as follows. First, dilute 1 μL proteinase K stock in 99 μL of PBSTw and mix vigorously. Second, add 75 μL of this solution to 925 μL of PBSTw and mix vigorously
- RNase mix: 20 μg/mL preboiled RNase A, 10 U/mL preboiled RNase T1 in WS3
- Staining solution: 2.5 μL/mL NBT, 3.5 μL/mL BCIP in AP buffer with $MgCl_2$. Mix vigorously and keep in the dark.
- Vector® Blue staining solution: add 80 μL of Reagent 1, 80 μL of Reagent 2 and 40 μL of Reagent 3 to 5 mL of 100 mM Tris-HCl pH 8.2 + 0.2% Tween-20. Mix vigorously and keep in the dark
- Wash solution 1 (WS1): 50% (v/v) formamide, 5× SSC, 1% SDS in DEPC-treated water. Mix formamide, SSC and water and heat to 60 °C before adding the SDS. Can be stored in aliquots at −20 °C for several months
- Wash solution 2 (WS2): 50% (v/v) formamide, 2× SSC, 1% SDS in DEPC-treated water. Mix formamide, SSC and water and heat to 60 °C before adding the SDS. Can be stored in aliquots at −20 °C for several months
- Wash solution 3 (WS3): 2× SSC, 0.1% Tween-20 in DEPC-treated water. Filter. Can be stored in aliquots at −20 °C for several months

- Wash solution 4 (WS4): 0.2× SSC, 0.1% Tween-20 in DEPC-treated water. Filter. Can be stored in aliquots at −20 °C for several months
- Wash solution 5 (WS5): 2 mg/mL BSA in PBSTw. Prepare fresh

6.5 Protocol for whole mount *in situ* hybridization

Before starting, clean bench, pipettes and tweezers with RNaseZAP. Always use gloves to manipulate solutions, reagents and equipment. Use only RNase-free filtered tips. If not otherwise stated, all steps of the protocol are performed in 500 μL of solution, at room temperature and rotating the embryos horizontally at 35–40 rpm. For changing solutions, use tweezers to move baskets between two wells.

6.5.1 Day 1

1. For each experimental condition, select 10–15 embryos and transfer them to a basket placed inside a well of a 24-well plate.
2. Move the basket into a new well containing PBSTw and rotate for 5 min to rehydrate the embryos. Repeat twice with fresh PBSTw.
3. Digest with freshly diluted proteinase K. Additional details for this step are provided in Section 6.6.
4. Add 10 μL of 10% glycine to each basket to stop the digestion and incubate for 1 min. Quickly transfer into 2 mg/mL glycine and incubate for 5 min.
5. Wash with 0.1 M TEA for 1 min, followed by a second wash of 5 min.
6. Transfer into freshly prepared 0.25% acetic anhydride. Incubate for 5 min. Do not rotate.
7. Transfer into freshly prepared 0.5% acetic anhydride. Incubate for 5 min. Do not rotate.
8. Wash with PBSTw for 1 min, followed by a second wash of 5 min.
9. Transfer to HB, pre-warmed to 60 °C. Incubate for 1 min.
10. Transfer into fresh pre-warmed HB. Seal the lid of the plate with paper tape then place the plate into a disposable zipper bag. Seal the bag and rock in a hybridization oven at 60 °C for at least 2 h.
11. Dilute the probe to 1 ng/μL in pre-warmed HB. Denature at 70 °C for 5 min.
12. Remove the plate from the hybridization oven, add the probe solution to a new well and transfer basket into probe solution. If more

than one basket is used, repeat the procedure, cleaning the tweezers with RNaseZAP between each basket move to prevent the transfer of probe and/or embryos between wells.

13. Re-seal the plate with paper tape, place the plate into a zipper bag and rock in the hybridization oven at 60 °C overnight.

6.5.2 Day 2

14. Pre-warm WS1 and WS2 to 60 °C. Thaw WS3 and WS4. Work quickly in steps 15–18 to keep embryos at hybridization temperature.
15. Unseal the plate and remove the probe solution, which can be stored at −20 °C for reuse.
16. Wash twice with pre-warmed WS1 at 60 °C for 5 min.
17. Wash twice with pre-warmed WS1 at 60 °C for 15 min.
18. Wash with pre-warmed WS2 at 60 °C for 5 min, then move the plate to room temperature and rotate for 10 min.
19. Wash with WS2 for 15 min.
20. Wash with WS3 for 1 min, followed by a second wash of 5 min.
21. Incubate with RNase mix at 37 °C for 20 min.
22. Wash twice with WS3 for 20 min.
23. Wash with WS4 for 20 min.
24. Wash with freshly prepared WS5 for 5 min.
25. Transfer into freshly prepared blocking solution and incubate for at least 2 h.
26. Transfer into freshly thawed antibody solution and incubate at +4 °C overnight.

6.5.3 Day 3 (for NBT/BCIP staining)

27. Remove antibody solution and store at −20 °C. Can be re-used once, if supplemented with fresh antibody solution.
28. Wash four times with fresh PBSTw for 20 min.
29. Transfer into freshly prepared AP buffer without $MgCl_2$ for a quick wash, then transfer into fresh buffer and rotate for 10 min.
30. Transfer into freshly prepared AP buffer with $MgCl_2$ and rotate for 10 min. Repeat twice with fresh buffer.
31. Take embryos out of the basket and transfer them into a four-well plate. Pipet embryos up and down in a 200 μL tip to prevent them from

sticking to the bottom of the well. Remove fibers and debris that might have accumulated in the well.

32. Change for 300 μL staining solution. Incubate at room temperature in the dark or at 37 °C with gentle rotation.
33. Monitor the progress of the staining reaction by observing the embryos with a stereomicroscope on a white background every 15–30 min. Staining time can vary significantly between probes, from 15 min to several days.
34. Every 3 h, if the staining reaction requires more time, wash briefly with AP buffer with $MgCl_2$ and change for freshly prepared staining solution. The staining reaction needs to be stopped before the appearance of conspicuous background staining. A correctly stained embryo is thus transparent or slightly pinkish, and stained cells are dark blue or purple. To optimize the signal-to-noise ratio, do not incubate in staining solution overnight. Instead, wash briefly with AP buffer with $MgCl_2$ and leave at +4 °C in fresh AP buffer with $MgCl_2$. The following day pick up the protocol at step 32.
35. When the signal is fully developed, stop the staining reaction by washing briefly with AP buffer without $MgCl_2$.
36. Change for fresh AP buffer without $MgCl_2$ and wash for 10 min.
37. Wash with PBSTw for 10 min.
38. To counterstain nuclei, incubate with freshly diluted 1:5000 Hoechst stock in PBSTw in the dark for 15 min.
39. Wash twice with PBSTw for 10 min.
40. Wash with PBS/sodium azide for 10 min.
41. Change solution for 80% glycerol. Stained embryos can be stored at +4 °C for several months.

6.5.4 Day 3 (for Vector® Blue staining)

27. Remove antibody solution and store at −20 °C. Can be re-used once, if supplemented with fresh antibody solution.
28. Wash four times with fresh PBSTw for 20 min.
29. Transfer to 100 mM Tris-HCl pH 8.2 + 0.2% Tween-20 and incubate for 10 min. Repeat twice with fresh buffer.
30. Take embryos out of the basket and transfer them into a four-well plate. Pipet embryos up and down in a 200 μL tip to prevent them from

sticking to the bottom of the well. Remove fibers and debris that might have accumulated in the well.

31. Change for 300 μL Vector® Blue staining solution. Incubate at room temperature in the dark or at 37 °C with gentle rotation.
32. Monitor the progress of the staining reaction by observing the embryos with a stereomicroscope on a white background every 15–30 min. Staining time can vary significantly between probes, from 15 min to several days.
33. Every 3 h, if the staining reaction requires more time, wash briefly with 100 mM Tris-HCl pH 8.2 + 0.2% Tween-20 and change for freshly prepared Vector® Blue staining solution. The staining reaction needs to be stopped before the appearance of conspicuous background staining. A correctly stained embryo is yellowish, and stained cells are blue. To optimize the signal-to-noise ratio, do not incubate in staining solution overnight. Instead, wash briefly with 100 mM Tris-HCl pH 8.2 + 0.2% Tween-20 and leave at +4 °C in fresh buffer. The following day pick up the protocol at step 31.
34. When the signal is fully developed, stop the staining reaction by washing briefly with 100 mM Tris-HCl pH 8.2 + 0.2% Tween-20.
35. Wash with PBSTw for 10 min.
36. To counterstain nuclei, incubate with freshly diluted 1:5000 Hoechst stock in PBSTw in the dark for 15 min.
37. Wash twice with PBSTw for 10 min.
38. Wash with PBS/sodium azide for 10 min.
39. Change solution to 80% glycerol. Stained embryos can be stored at +4 °C for several months.

6.6 Troubleshooting and optimization

Proteinase K treatment is a critical step of the ISH protocol and crucial for optimizing the signal-to-noise ratio. Digestion times depend on developmental stage, fixation quality and proteinase K batch and thus need to be adapted accordingly. The following digestion times are meant to serve as an orientation: 0–15 hpf—5 min; 15–36 hpf—10 min; over 36 hpf—15 min. The digestion should not damage the embryos, with tail and snout regions being particularly sensitive to overdigestion.

Problem	Solution
No specific signal	Make sure you work in an RNase-free environment Check riboprobe integrity on denaturing gel Verify that the riboprobe was correctly synthesized
Embryos look damaged	Reduce incubation time in proteinase K If the problem persists, consider post-fixing with 4% PFA/PBS for 1 h at room temperature immediately after the second glycine treatment
High background, especially in the tail region	Reduce incubation time in proteinase K and/or reduce probe concentration
Unspecific staining in the gut with NBT/BCIP	Make sure to include levamisole in AP buffer and staining solution
Very weak Vector® Blue signal under a stereomicroscope	Image the embryos with a confocal microscope. If the problem persists using fluorescence, try repeating color reaction with NBT/BCIP If the expression of a gene is being analyzed for the first time, we suggest using NBT/BCIP staining, as it is generally more sensitive than Vector® Blue staining

7. Whole mount immunohistochemistry

Whole mount IHC allows the detection of target antigens in entire embryos with subcellular resolution using antibodies that are either labeled (direct method) or subsequently recognized by labeled secondary antibodies (indirect method). The protocol presented below is optimized for the detection of the neurotransmitter GABA and uses a fluorochrome-conjugated secondary antibody to detect the primary anti-GABA antibody.

7.1 Equipment

- Filtered micropipette tips
- 4-well cell culture plates
- Gloves

- Horizontal orbital shaker
- Micropipettes
- pH meter
- Stereomicroscope
- Syringes with 0.45 μm filters

7.2 Reagents

- Bovine serum albumin (BSA), Sigma-Aldrich (A9647)
- Glycerol, Sigma-Aldrich (G7893)
- Goat anti-rabbit IgG Alexa Fluor 488, Invitrogen (A-11008)
- Goat serum, Sigma-Aldrich (G6767). Inactivate by heating to 55 °C for 30 min. Cool on ice and prepare 1 mL aliquots that can be stored at −20 °C for several months
- Hoechst 33258, Sigma-Aldrich (14530)
- Rabbit anti-GABA antibody, Abcam (ab8891)
- Sodium azide, Sigma-Aldrich (71289)
- Sodium chloride (NaCl), Sigma-Aldrich (S9888)
- Sodium phosphate dibasic (NaH_2PO_4), Sigma-Aldrich (S9638)
- Sodium phosphate monobasic (Na_2HPO_4), Sigma-Aldrich (S3264)
- Triton X-100, Sigma-Aldrich (T8787)

7.3 Stock solutions

- 10 × PBS (phosphate buffer saline): 26 mM sodium phosphate monobasic, 174 mM sodium phosphate dibasic, 9% (wt/v) NaCl, pH 7.4. DEPC-treat and autoclave. Can be stored at room temperature for several months
- Hoechst stock: dissolve to 10 mg/mL in distilled water. Can be stored in the dark at +4 °C for several months

7.4 Working solutions

- 1 × PBS: Prepare from 10 × PBS stock and filter
- 80% glycerol: dilute glycerol with PBS/sodium azide. Can be stored at room temperature for several months
- Blocking solution: 10% heat-inactivated goat serum, 1.5% BSA in PBSTx. Prepare fresh. Can be stored at +4 °C for a few days
- PBS/sodium azide: 1 × PBS, 0.1% (wt/v) sodium azide. Can be stored at room temperature for several months
- PBSTx: 1 × PBS, 0.3% (v/v) Triton X-100

- Secondary antibody solution: dilute Alexa-conjugated secondary antibody 1:800 and Hoechst stock 1:5000 in blocking solution. Prepare fresh. Do not store

7.5 Protocol for whole mount immunohistochemistry

If not otherwise stated, all steps of the protocol are performed in 350 μL of solution, at room temperature and rotating the embryos horizontally at 35–40 rpm. Each time you change a solution, remove fibers and debris that might have accumulated in the wells.

1. Place embryos into a four-well plate and pipet them up and down in a 200 μL tip to prevent them from sticking to the bottom of the well.
2. Wash five times with PBSTx for 5 min.
3. Change to freshly prepared blocking solution and incubate for at least 1 h.
4. Change to blocking solution containing anti-GABA antibody diluted 1:300 and incubate at room temperature overnight.
5. Wash five times with PBSTx for 20 min.
6. Change to freshly prepared secondary antibody solution and incubate at +4°C overnight with rotation. From this step onward, do not expose the plate to direct light, unless absolutely necessary. Perform washes in the dark, for instance by putting the plate inside a small cardboard box on the shaker.
7. Wash five times with PBSTx for 20 min.
8. Wash with PBS/sodium azide for 5 min.
9. Change for 80% glycerol.
10. Mount embryos on slides and image them as soon as possible. Slides and plates can be kept in the dark at +4 °C for a few weeks.

7.6 Troubleshooting and optimization

Problem	Solution
No signal is detected	If using a new primary antibody, optimize the protocol by varying antibody concentration and incubation conditions Verify whether particular fixatives and/or fixation protocols are required
Only superficial cells are stained	Antibody did not penetrate properly: incubate primary antibody at +4 °C for at least 72 h

8. Mounting embryos for microscopy

To avoid damaging the embryos, it is important to use spacers between slide and coverslip when mounting amphioxus material for microscopy. We routinely use three layers of Scotch tape.

8.1 Equipment

- 200 μL micropipette tips
- Coverslips (24 × 32 mm)
- Insect pins (size 00)
- Magic tape (3 M Scotch)
- Micropipette
- Microscope slides (26 × 76 mm)
- Transparent nail polish

8.2 Protocol for mounting amphioxus embryos for microscopy

1. On a clean microscope slide, create two spacers delimitating a central space of approximately 2 cm by placing three pieces of Scotch tape on top of each other (see Fig. 3).
2. Using a micropipette with a 200 μL tip, transfer 40–50 μL of 80% glycerol containing 5–10 embryos to the center of the slide, between the two spacers.
3. Using the tip of an insect pin, orient the embryos as desired by gently moving the glycerol around them.

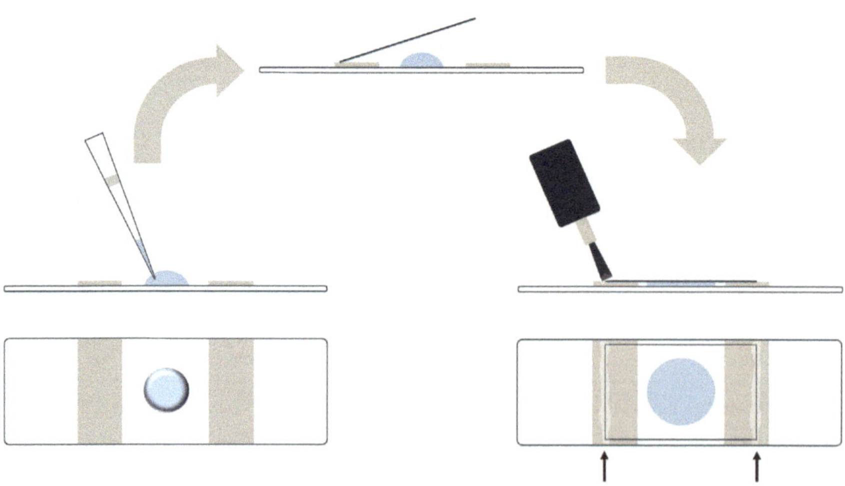

Fig. 3 Mounting whole mount amphioxus embryos for microscopy. Arrows indicate the zone of nail polish application, which corresponds to the shorter edges of the coverslip.

4. To minimize air bubble formation, place one short edge of the coverslip on one of the spacers and carefully lower the coverslip toward the other spacer.
5. For confocal microscopy, glue the coverslip to the spacers with transparent nail polish to minimize movement of the samples during stack acquisition.

8.3 Troubleshooting and optimization

Problem	Solution
Embryos are too close to the spacers	Use less mounting medium While orienting the embryos, gently push them toward the slide. Floating embryos will inevitably be moved toward the spacers
Coverslip detaches easily when using an inverted microscope	Use more mounting medium and verify its concentration Improve fixation of the coverslip using nail polish
Embryos are deformed/damaged	Verify distance between the two spacers on the slide, they might be too far apart Tape thickness might vary between brands. If you are using a tape other than 3M Scotch, a different number of tape layers might be required to appropriately mount the embryos

9. Confocal microscopy for chromogenic *in situ* hybridization and fluorescent immunohistochemistry

Chromogenic ISH and fluorescent IHC can be visualized, respectively, with a brightfield and an epifluorescence microscope. In addition, to increase the cellular resolution of the staining, the outcome of both techniques can be analyzed with a confocal microscope. While Vector® Blue, Alexa dyes and Hoechst are intended for use in fluorescence microscopy and are thus bright fluorochromes, the acquisition of NBT/BCIP fluorescence is more challenging. Two approaches are possible. Since NBT/BCIP forms crystals that reflect light, it can be visualized by setting a narrow acquisition window that includes the excitation wavelength (Jékely & Arendt, 2007). Although the signal-to-noise ratio is generally good, this method should be used with caution, as some anatomical structures (such as the amphioxus pigment spots) are also reflective and thus indistinguishable from the NBT/BCIP-derived signal. Moreover, some fluorochromes can also be reflective, limiting the use of this approach when analyzing multiple stainings in a single embryo. For these reasons, we suggest acquiring NBT/BCIP fluorescence in the far-red spectrum (Trinh et al., 2007). However, since the emission spectrum of NBT/BCIP peaks at 850 nm, which is beyond the maximum wavelength measurable by most commercial confocal microscopes, the signal tends to be weak and needs to be acquired several times, averaging the obtained values. It is thus very important to optimize the signal-to-noise ratio of the NBT/BCIP ISH procedure. Should this not be possible, NBT/BCIP can be replaced by Vector® Blue, which emits mostly in the visible spectrum. The parameters used to acquire the images presented in Figs. 4 and 5 are listed in Table 2. To avoid bleed-through artifacts, sequential scans have been performed.

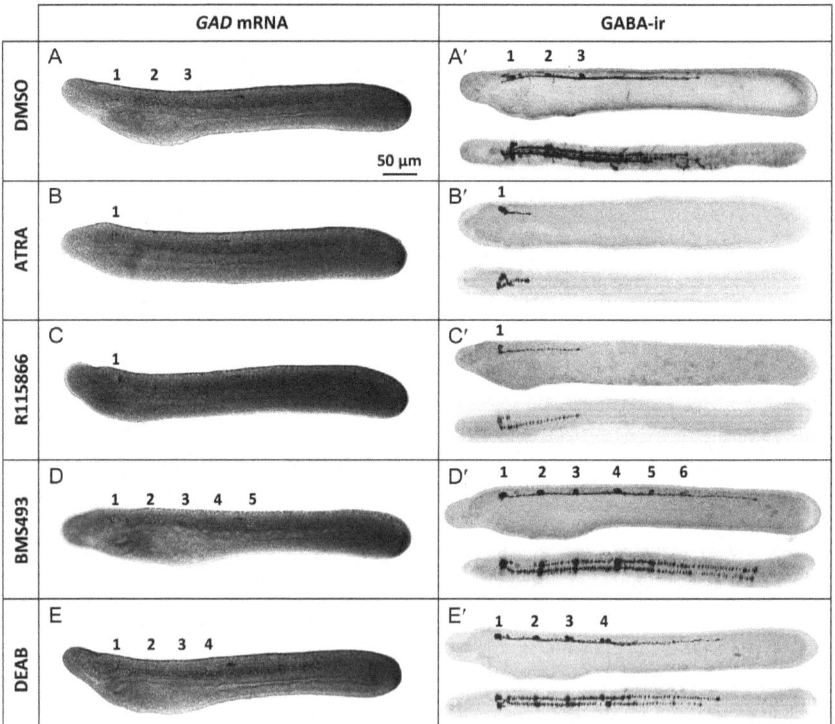

Fig. 4 Effects of pharmacological manipulation of retinoic acid signaling on the development of GABAergic neurons in the amphioxus central nervous system. Numbers indicate clusters of GABAergic neurons. Compound list is provided in Table 1. DMSO is solvent control. (A–E) Whole mount *in situ* hybridization targeting *GAD* mRNA. Confocal images of embryos in lateral view with anterior to the left. (A′–E′) Whole mount immunohistochemistry revealing GABA immunoreactivity (GABA-ir). Embryos in lateral (top) and dorsal (bottom) views with anterior to the left.

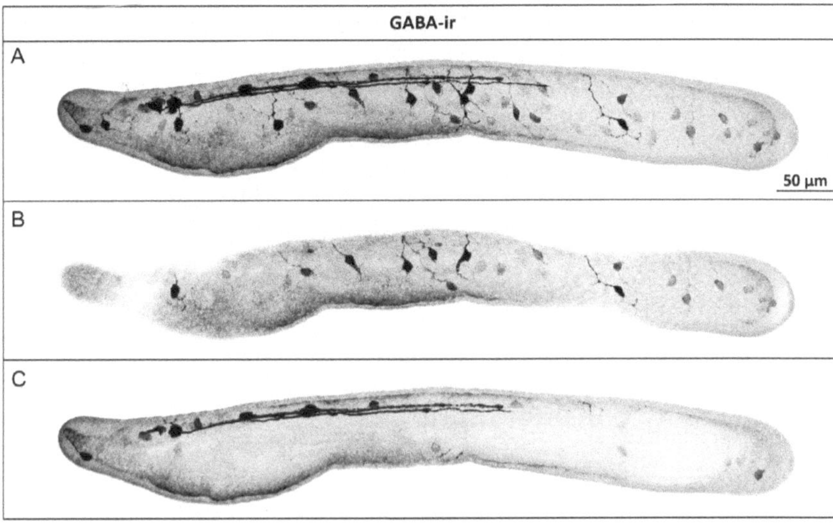

Fig. 5 Layer selection for z-projections of GABA immunoreactivity (GABA-ir). Confocal images of DMSO control embryos in lateral view with anterior to the left. (A) Projection of all focal planes. (B) Projection of layers corresponding to the position of the ectoderm on one side of the embryo. (C) Projection of layers corresponding to the position of the central nervous system.

Table 2 Parameters for fluorochrome acquisition using a confocal microscope.

Fluorochrome	Excitation	Emission	Beamsplitter
Hoechst	405 nm	420–480 nm	DD488/552
Alexa Fluor 488	488 nm	500–550 nm	TD488/553/638
NBT/BCIP	638 nm	700–800 nm	TD488/553/638
Vector® Blue	638 nm	700–800 nm	TD488/553/638

Confocal microscopy presents several advantages. In addition to increasing the cellular resolution, confocal microscopes acquire stacks of focal planes (layers) that span the whole embryo. These stacks can be used to compile 3D reconstructions of whole embryos or to project sets of a limited number of focal planes, using open-source software packages like Fiji (Schindelin et al., 2012). Whenever possible, it is advisable to acquire stacks of the entire sample during confocal imaging. Projecting the information of several focal planes into a single layer during post-production generally yields the best results. An example for the usefulness of this selection procedure is given in Section 11.

9.1 Troubleshooting and optimization

Problem	Solution
Fluorescent signal fades quickly	Make sure that the glycerol is buffered, as some fluorochromes are pH-sensitive Consider using a mounting medium containing an anti-fading agent instead of glycerol

10. Safety considerations

When handling chemicals, validated health and safety regulations need to be respected at all times. Although this protocol generally requires the manipulation of only small volumes, some substances are particularly hazardous and need to be handled with extra care. These include solutions containing fixatives, formamide and sodium azide as well as concentrated acids and bases. Furthermore, the compounds used to interfere with RA signaling are bioactive. Concerning the equipment, confocal microscopes are provided with powerful lasers that represent a potential danger to the human eye. Common laser safety protocols must thus be respected at all times.

11. Retinoic acid signaling and GABAergic neurons in amphioxus

Pharmacological treatments represent a straightforward approach for manipulating the activity of molecular signaling pathways. The ease of these treatments is particularly appealing for work on emerging animal models, for which genetic approaches remain technically challenging. Amphioxus falls into this category of emerging animal models, with several signaling pathways having already been studied using pharmacological approaches (Bertrand, Le Petillon, Somorjai, & Escriva, 2017). For RA signaling, a number of different compounds have been shown to upregulate or downregulate its activity during amphioxus development. These molecules either target the RAR/RXR heterodimer or specific components of endogenous RA metabolism, such as RALDH and CYP26 (see Table 1). Here, we made use of some of these compounds to compile protocols for the characterization of phenotypes obtained by the pharmacological modulation of RA signaling activity in amphioxus.

Previous work has demonstrated that RA signaling controls the regionalization of the amphioxus embryo along its anteroposterior axis (Escriva et al., 2002), with RA signaling being active in different developmental tissues, such as the pharyngeal endoderm, the general ectoderm and the central nervous system (Koop et al., 2010). In the central nervous system, for example, RA signaling mediates the patterning of GABAergic neurons (Zieger, Candiani, et al., 2018; Zieger, Garbarino, et al., 2018). During normal development, three clusters of GABAergic neurons are formed in the anterior central nervous system of late embryos and early larvae (Candiani, Moronti, Ramoino, Schubert, & Pestarino, 2012). Upregulation of RA signaling activity, by either ATRA or R115866, leads to the complete loss of the posterior clusters and a shortening of the remaining neurites (Fig. 4). Conversely, downregulation of RA signaling activity, by either BMS493 or DEAB, results in the formation of supernumerary clusters at more posterior levels and a concomitant posterior extension of neurites (Fig. 4). In addition to perturbing the patterning of GABAergic neurons, the treatments also severely disrupt muscular swimming, affecting both the intensity of muscular contractions and the coordination of left-right body flexing (Zieger, Candiani, et al., 2018; Zieger, Garbarino, et al., 2018). The GABAergic system might thus be required in the anterior central nervous system to coordinate sinusoidal swimming by inhibiting contralateral muscle contractions (Lacalli & Candiani, 2017).

In this chapter, we assayed the number of GABAergic neurons in the amphioxus central nervous system by ISH (using a *GAD*-specific riboprobe) and IHC (using a GABA-specific antibody). Although, for any given treatment, both approaches yielded similar results, there were marked differences in the cellular resolution. For example, *GAD* ISH only stained cell bodies, while GABA IHC marked cell bodies and neurites (Fig. 4). Thus, only the IHC results allowed us to conclude that the pharmacological treatments affected not only the number of GABAergic neurons, but also the length of their projections. Furthermore, only IHC revealed the presence of GABAergic neurons in the amphioxus ectoderm (Fig. 5). Given that this ectodermal IHC signal is very conspicuous, careful stack selection during image data analysis is required for optimizing the presentation of the obtained results.

12. Summary

RA signaling can be easily manipulated in amphioxus embryos using the protocols presented in this chapter. Prior to more refined analyses, treatment success can be verified visually by comparing pharynx lengths between control and experimental conditions. Although ISH and IHC are used routinely in a wide variety of models, the execution of both techniques presents specific challenges that were addressed in this chapter. Our results show that both ISH and IHC are powerful techniques to study the roles of RA signaling in developing amphioxus. The choice between the two depends on several factors, including the required cellular resolution and the available imaging equipment. In addition, the IHC technique is strictly dependent on the existence of specific antibodies and efforts have recently been undertaken to develop novel antibodies for IHC experiments in amphioxus (Bozzo et al., 2017). Our protocols, allowing detailed correlations of IHC and ISH patterns, will be useful for the validation of these novel antibodies.

Acknowledgments

The authors would like to thank François Lahaye for technical assistance with amphioxus spawning and embryo collection. We are grateful to the Mediterranean Culture Collection of Villefranche-sur-Mer, the Service Aquariologie of the Centre de Ressources Biologiques, and the Plateforme d'Imagerie par Microscopie of the Institut de la Mer de Villefranche-sur-Mer, all of which are supported by EMBRC-France (ANR-10-INBS-02). This work was supported by ASSEMBLE Plus grants to S.C. and M.B.

References

Bertrand, S., & Escriva, H. (2011). Evolutionary crossroads in developmental biology: Amphioxus. *Development*, *138*, 4819–4830.

Bertrand, S., Le Petillon, Y., Somorjai, I. M. L., & Escriva, H. (2017). Developmental cell-cell communication pathways in the cephalochordate amphioxus: Actors and functions. *The International Journal of Developmental Biology*, *61*, 697–722.

Bozzo, M., Pergner, J., Kozmik, Z., & Kozmikova, I. (2017). Novel polyclonal antibodies as a useful tool for expression studies in amphioxus embryos. *The International Journal of Developmental Biology*, *61*, 793–800.

Candiani, S., Garbarino, G., & Pestarino, M. (2015). Detection of mRNA and microRNA expression in basal chordates, amphioxus and ascidians. *NeuroMethods*, *99*, 279–292.

Candiani, S., Moronti, L., Ramoino, P., Schubert, M., & Pestarino, M. (2012). A neurochemical map of the developing amphioxus nervous system. *BMC Neuroscience*, *13*, 59.

Carvalho, J. E., Lahaye, F., & Schubert, M. (2017). Keeping amphioxus in the laboratory: An update on available husbandry methods. *The International Journal of Developmental Biology*, *61*, 773–783.

Escriva, H., Holland, N. D., Gronemeyer, H., Laudet, V., & Holland, L. Z. (2002). The retinoic acid signaling pathway regulates anterior/posterior patterning in the nerve cord and pharynx of amphioxus, a chordate lacking neural crest. *Development*, *129*, 2905–2916.

Fonseca, E. S. S., Hiromori, Y., Kaite, Y., Ruivo, R., Franco, J. N., Nakanishi, T., et al. (2019). An orthologue of the retinoic acid receptor (RAR) is present in the Ecdysozoa phylum Priapulida. *Genes, 10*, 985.

Fuentes, M., Benito, E., Bertrand, S., Paris, M., Mignardot, A., Godoy, L., et al. (2007). Insights into spawning behavior and development of the European amphioxus (*Branchiostoma lanceolatum*). *Journal of Experimental Zoology Part B: Molecular and Developmental Evolution, 308*, 484–493.

Ghyselinck, N. B., & Duester, G. (2019). Retinoic acid signaling pathways. *Development, 146*, dev167502.

Handberg-Thorsager, M., Gutierrez-Mazariegos, J., Arold, S. T., Nadendla, E. K., Bertucci, P. Y., Germain, P., et al. (2018). The ancestral retinoic acid receptor was a low-affinity sensor triggering neuronal differentiation. *Science Advances, 4*, eaao1261.

Holland, L. Z., Albalat, R., Azumi, K., Benito-Gutierrez, E., Blow, M. J., Bronner-Fraser, M., et al. (2008). The amphioxus genome illuminates vertebrate origins and cephalochordate biology. *Genome Research, 18*, 1100–1111.

Jékely, G., & Arendt, D. (2007). Cellular resolution expression profiling using confocal detection of NBT/BCIP precipitate by reflection microscopy. *BioTechniques, 42*, 751–755.

Koop, D., Holland, N. D., Sémon, M., Alvarez, S., de Lera, A. R., Laudet, V., et al. (2010). Retinoic acid signaling targets *Hox* genes during the amphioxus gastrula stage: Insights into early anterior-posterior patterning of the chordate body plan. *Developmental Biology, 338*, 98–106.

Lacalli, T., & Candiani, S. (2017). Locomotory control in amphioxus larvae: New insights from neurotransmitter data. *EvoDevo, 8*, 4.

Laudet, V., Zieger, E., & Schubert, M. (2015). Evolution of the retinoic acid signaling pathway. In P. Dollé & K. Niederreither (Eds.), *The retinoids: Biology, biochemistry, and disease* (pp. 75–90). Hoboken, NJ: John Wiley & Sons, Inc.

Mic, F. A., Molotkov, A., Benbrook, D. M., & Duester, G. (2003). Retinoid activation of retinoic acid receptor but not retinoid X receptor is sufficient to rescue lethal defect in retinoic acid synthesis. *Proceedings of the National Academy of Sciences of the United States of America, 100*, 7135–7140.

Poss, S. G., & Boschung, H. T. (1996). Lancelets (Cephalochordata: Branchiostomatidae): How many species are valid? *Israel Journal of Zoology, 42*, S13–S66.

Putnam, N. H., Butts, T., Ferrier, D. E. K., Furlong, R. F., Hellsten, U., Kawashima, T., et al. (2008). The amphioxus genome and the evolution of the chordate karyotype. *Nature, 453*, 1064–1071.

Schindelin, J., Arganda-Carreras, I., Frise, E., Kaynig, V., Longair, M., Pietzsch, T., et al. (2012). Fiji: An open-source platform for biological-image analysis. *Nature Methods, 9*, 676–682.

Schubert, M., Holland, N. D., Laudet, V., & Holland, L. Z. (2006). A retinoic acid-*Hox* hierarchy controls both anterior/posterior patterning and neuronal specification in the developing central nervous system of the cephalochordate amphioxus. *Developmental Biology, 296*, 190–202.

Theodosiou, M., Colin, A., Schulz, J., Laudet, V., Peyrieras, N., Nicolas, J.-F., et al. (2011). Amphioxus spawning behavior in an artificial seawater facility. *Journal of Experimental Zoology Part B: Molecular and Developmental Evolution, 316*, 263–275.

Trinh, L. A., McCutchen, M. D., Bonner-Fraser, M., Fraser, S. E., Bumm, L. A., & McCauley, D. W. (2007). Fluorescent in situ hybridization employing the conventional NBT/BCIP chromogenic stain. *BioTechniques, 42*, 756–759.

Yong, L. W., Kozmikova, I., & Yu, J. K. (2019). Using amphioxus as a basal chordate model to study BMP signaling pathway. *Methods in Molecular Biology, 1891*, 91–114.

Yu, J. K., & Holland, L. Z. (2009). Amphioxus whole-mount in situ hybridization. *Cold Spring Harbor Protocols*, *4*, pdb.prot5286.

Zieger, E., Candiani, S., Garbarino, G., Croce, J. C., & Schubert, M. (2018). Roles of retinoic acid signaling in shaping the neuronal architecture of the developing amphioxus nervous system. *Molecular Neurobiology*, *55*, 5210–5229.

Zieger, E., Garbarino, G., Robert, N. S. M., Yu, J. K., Croce, J. C., Candiani, S., et al. (2018). Retinoic acid signaling and neurogenic niche regulation in the developing peripheral nervous system of the cephalochordate amphioxus. *Cellular and Molecular Life Sciences*, *75*, 2407–2429.

CHAPTER SEVENTEEN

Design of synthetic retinoids

David R. Chisholm[a,*], Andrew Whiting[b]
[a]LightOx Ltd., Newcastle upon Tyne, United Kingdom
[b]Department of Chemistry, Durham University, Lower Mountjoy, Durham, United Kingdom
*Corresponding author: e-mail address: david.chisholm@lightox.co.uk

Contents

Abstract

This chapter has been conceived as an introductory text to aid in the understanding of the key design strategies for the development of synthetic analogs of endogenous retinoids as ligands for the retinoic acid receptors (RARs) and retinoid X receptors (RXRs). The structure and binding characteristics of the endogenous retinoids are first explained to put the main chemical design challenges in context. Existing biochemical and structural data is then used to describe the guiding principles used to develop agonists and antagonists of the RARs and RXRs. In light of the increasing proliferation of biophysical methods that employ fluorescence measurements or molecular tags, we also examine the application of retinoids as probes and the chemical principles required to develop these tools.

Methods in Enzymology, Volume 637
ISSN 0076-6879
https://doi.org/10.1016/bs.mie.2020.02.011

1. Introduction

The study of endogenous retinoids during the latter part of the twentieth century has produced an enormous wealth of biophysical and biochemical information on their wide-ranging functions and behaviors. Much of this effort coincided with significant steps forward in biophysical methods; particularly macromolecular X-ray crystallography, which has enabled the determination of the shape and structure of the retinoic acid receptors (RARs), retinoid X receptors (RXRs), cellular retinoic acid binding proteins (CRABPs) and other related proteins (Renaud et al., 1995). These models, along with structure-activity relationships identified in early efforts to develop synthetic analogs of the endogenous retinoids, have made the development of synthetic retinoids an increasingly more intuitive endeavor. In this chapter, we will discuss the principle chemical design strategies used for the development of analogs of the endogenous retinoids that exhibit increased potency, selectivity and stability.

1.1 Retinoid structure and function

Retinoids are a family of endogenous compounds derived from Vitamin A that function as potent ligands for the RARs and RXRs. These nuclear receptors regulate a range of signaling pathways by mediating the transcription of retinoic acid response elements (RAREs) (Chambon, 1996). To elicit these transcriptional responses, retinoids are shuttled from the cytoplasm to the cell nucleus via small transport proteins known as the cellular retinoic acid binding proteins (CRABPs) (Majumdar, Petrescu, Xiong, & Noys, 2011). CRABPs release retinoids for binding to the ligand-binding domain (LBD) of the RARs/RXRs, which triggers a series of conformational changes that encloses the retinoid inside the LBD. This encapsulation results in the presentation of a binding surface, known as activation-function 2 (AF-2), that binds transcription promoters and a dimerization surface that enables dimerization with another RAR/RXR. This receptor dimer is capable of binding a RARE and the resultant ternary complex acts as a stable platform for the transcription of this RARE (Chambon, 1996; Klaholz et al., 1998).

Retinoids can be described in terms of three structural regions: the hydrophobic region; linker region; and polar region (Fig. 1). The hydrophobic region is typically a large, highly lipophilic structure that fills the wide

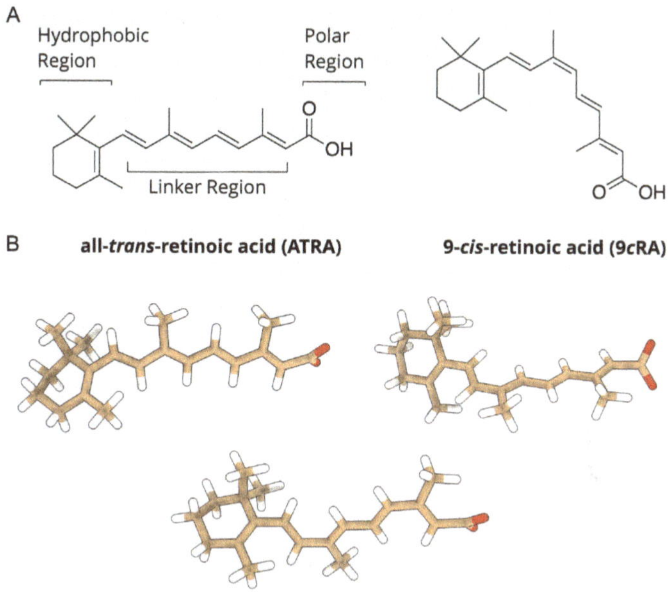

Fig. 1 (A) Chemical structures of endogenous retinoids, all-*trans*-retinoic acid (ATRA) and 9-*cis*-retinoic acid (9*c*RA). (B) Three selected low energy conformations of ATRA accessible at room temperature (Haffez et al., 2017).

hydrophobic cleft in the opening to the ligand-binding pocket (LBP) of retinoid-binding proteins. The linker region is generally a long, often conjugated, moiety that imparts length, determines the overall shape of the compound and positions the polar region for effective binding. This polar region, which is most often a carboxylic acid group, is crucial for binding to a group of charged residues at the end of the LBP of the RARs/RXRs (Klaholz et al., 1998; Renaud et al., 1995)

All-*trans*-retinoic acid (ATRA) and its corresponding isomers, such as 9-*cis*-retinoic acid (9*c*RA) shown in Fig. 1A, consist of a β-ionone-derived hydrophobic region conjugated to an extended, methylated polyene linker region, with a carboxylate polar region terminus. While ATRA is regularly drawn in an extended, all-*s*-*trans* and planar conformation, in fact, the conjugated polyene possesses a significant degree of conformational flexibility due to the connecting sigma-bonds and weak intramolecular steric effects, with the result that it fails to exhibit such a simplistic conformation—especially under physiological conditions. This is reinforced by *in silico* conformation distribution calculations (semi-empirical molecular orbital methods, AM1) which show that ATRA is capable of adopting a wide range of discrete low energy conformations at room temperature within

3–5 kcal mol^{-1} (Haffez et al., 2017). As shown in the three examples in Fig. 1B, these conformations exhibit marked differences in their 3-dimensional shape, and therefore, one must be cognizant of this behavior when designing analogs of endogenous retinoids. This conformational flexibility is also evident to varying extents in the other natural retinoic acid isomers, including 9*c*RA and 13-*cis*-retinoic acid (13*c*RA). Hence, it is not only important to bear these dynamic behaviors in mind for structural analysis, but that these effects likely have a profound effect on the diversity of biological activities and signaling pathways that retinoids elicit and mediate.

1.2 RARs and RXRs

The RAR and RXR LBDs exhibit an anti-parallel, α-helical sandwich fold that is characteristic of many nuclear receptors (Fig. 2) (Renaud et al., 1995). A ligand binding pocket (LBP) exists at the base of the assembly that is the site for retinoid binding. This LBP is a hydrophobic cavity with an opening that is adjacent to the C-terminal α-helix, H12, which acts as a trap door; closing when a retinoid enters the LBP. When H12 closes the LBP, an outer binding surface, known as AF-2, is created which is then responsible for binding coactivator (CoA) proteins that are key in the regulation of transcription processes. The maintenance of this closed-pocket form is, therefore, crucial to strong binding, efficient transcriptional activity, and ultimately, downstream biological activity (Bourguet et al., 2000; Klaholz et al., 1998; Renaud et al., 1995).

This mechanism can be observed by analyzing the structures of holo- and apo-RXRα (Bourguet et al., 2000; Bourguet, Ruff, Chambon, Gronemeyer, & Moras, 1995). Fig. 3 exemplifies the conformational

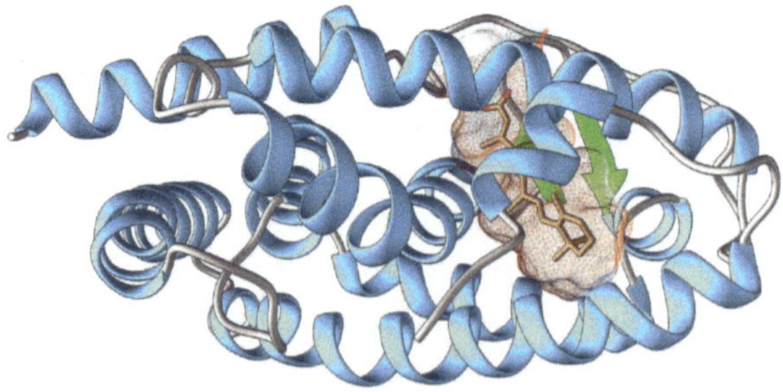

Fig. 2 X-ray crystal structure of RARγ bound to ATRA, with the hydrophobicity surface of the LBP highlighted (PDB: 2LBD) (Renaud et al., 1995).

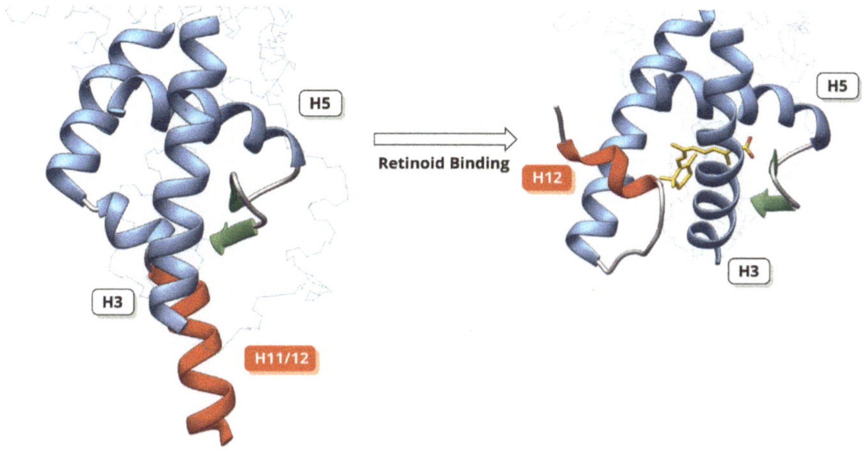

Fig. 3 Conformational changes that occur upon binding of 9*c*RA to RXRα (PDB: 1LBD/6HN6 and 1FBY) (Bourguet et al., 1995; Eberhardt, Mcewen, Bourguet, Morasa, & Dejaegerea, 2019; Egea et al., 2000).

changes that occur upon retinoid binding, where the entry of 9*c*RA into the LBP causes H11/H12 to swing up and around toward H5, causing the removal of corepressor proteins (CoR) that actively suppress transcriptional activity. This movement closes the ligand binding pocket (LBP) and stabilizes AF-2 for binding to CoAs.

Fig. 4A shows the hydrophobicity surface of the LBP of RARγ (Renaud et al., 1995) with ATRA bound inside. Both the RAR and RXRs LBP are predominantly hydrophobic, with a large cavity created between the *N*-terminal side of H5 and H7, and H11 and H12. Here, the β-ionone-derived hydrophobic region of ATRA is positioned to form favorable van der Waals interactions with a range of residues including, but not limited to, Leu268 on H5, Phe304 and Leu307 on H7, Arg396 and Leu400 on H11 and Ile412 and Met415 on H12. The center of the pocket is narrower and, likewise, mainly hydrophobic in nature, and thus, is well-suited to interaction with the lipophilic polyene linker region of ATRA through interaction with Ile275, Leu271 and Met272 on H5.

At the end of the RAR/RXR ligand binding pocket lies a cluster of polar residues (Lys236, Ser289 and Arg278 in RARγ) that are positioned to enable polar interactions with the carboxylate of ATRA (Fig. 4B). These interactions are particularly important, since this polar cluster is crucial for guiding the retinoid, carboxylate-first, into the pocket via electrostatic attraction. Furthermore, it has also been shown that although the

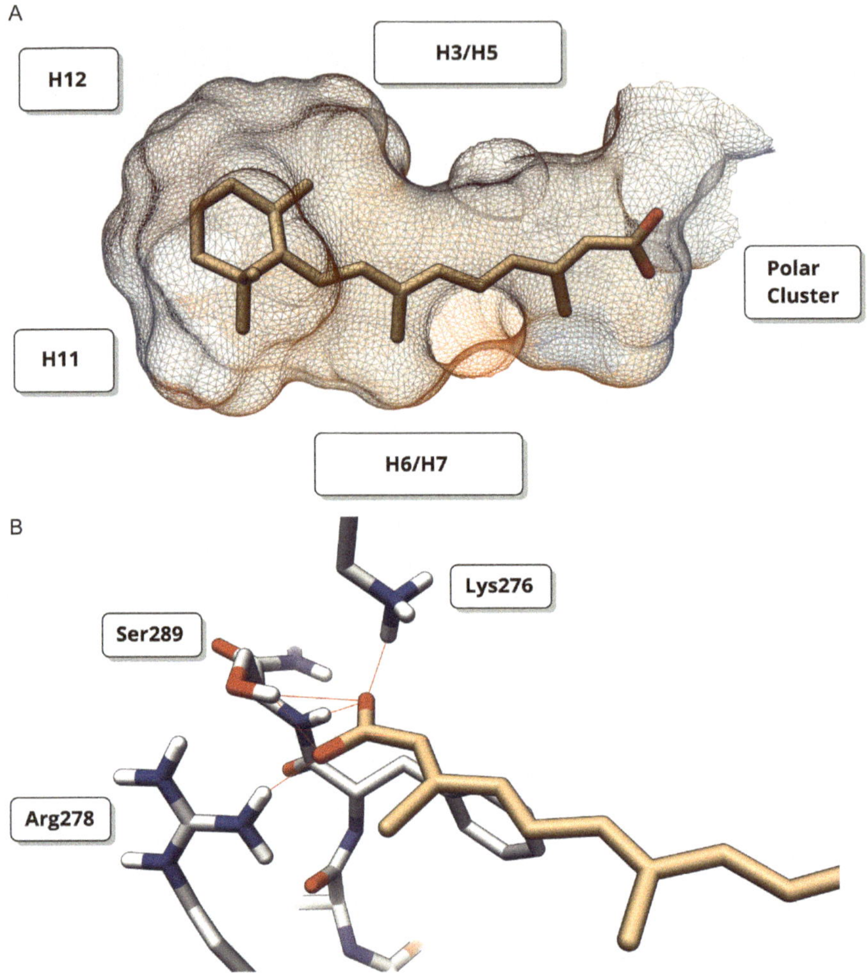

Fig. 4 (A) Hydrophobicity surface of the RARγ LBP bound to ATRA. (B) Interactions between the polar cluster at the base of the RARγ LBP and the carboxylate of ATRA (PDB: 2LBD) (Renaud et al., 1995).

majority of the strong binding affinity of a retinoid is derived from hydrophobic interactions, blocking the carboxylate (e.g., by protection as an ester) typically causes a significant reduction in transcriptional activity, even though binding affinity remains strong (Haffez et al., 2018, 2017).

The retinoid X receptor (RXR) ligand binding domain (LBD) exhibits a similar anti-parallel α-helical sandwich structure to the RAR LBD and shares much in terms of the overall structure and ligand-binding mechanism. However, the important difference lies in the shape of the LBP,

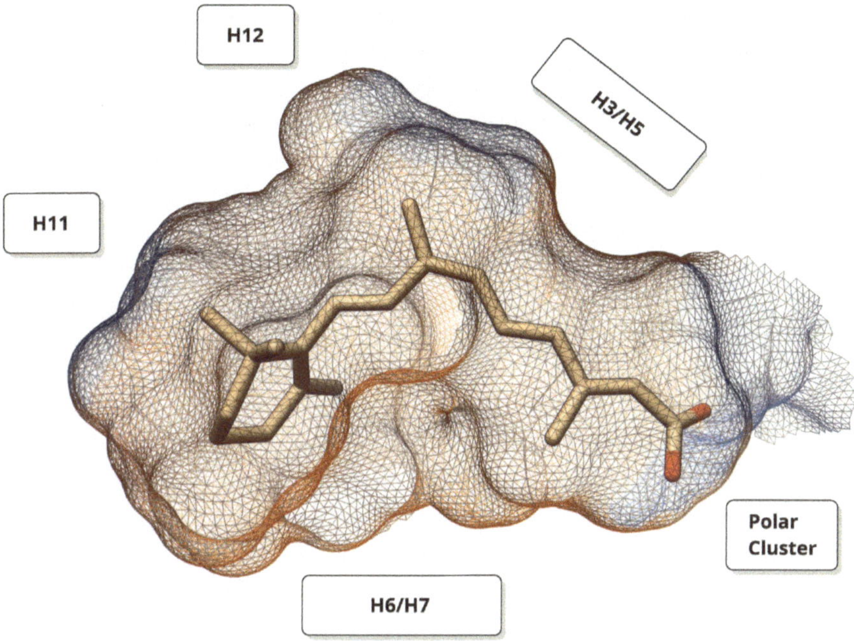

Fig. 5 Hydrophobicity surface of the RXRα LBP bound to 9*c*RA (PDB: 1FBY) (Egea et al., 2000).

Table 1 Comparison of the relative binding affinities of ATRA and 9CRA for the RARs and RXRs (NB = no binding) (Allenby et al., 1993).

	K_d/nM					
	RARα	**RARβ**	**RARγ**	**RXRα**	**RXRβ**	**RXRγ**
ATRA	0.20	0.36	0.20	NB	NB	NB
9*c*RA	0.24	0.40	0.70	15.7	18.3	14.1

i.e., the RXR LBP is L-shaped in contrast to the more linear I-shaped LBP of the RARs (Fig. 5). Consequently, the RXRs exhibit greater affinity toward retinoids with a "bent" structure or major conformation, such as 9*c*RA and 9-*cis*-13,14-dihydroretinoic acid (9*c*DHRA) (Rühl et al., 2015). Indeed, experimentally, ATRA has been shown to exhibit lower affinity for the RXRs (Allenby et al., 1993) while 9*c*RA, for example, is actually able to bind to both RARs/RXRs with high affinity, as highlighted in the comparison shown in Table 1. Both computational and experimental studies have shown that 9*c*RA is able to adopt both a conformation and a positioning that suits the I-shaped RAR LBPs, despite also exhibiting a complimentary shape and conformation more befitting of the L-shaped RXR LBP (Haffez et al., 2017; Klaholz et al., 1998).

This duality observed in the receptor-binding behavior of 9*c*RA adds further complexity to the biological signaling events that are mediated by endogenous retinoids, particularly since these compounds undergo relatively facile interconversion through both biological and biochemical stimuli. This leads one to speculate for a particular observed biological action, as to whether that phenomenon is arbitrated by a single retinoid, or perhaps by two, or more, or a mixture in a competitive manner. For example, could ATRA, its isomers and a range of other ATRA-derived metabolites we know to be produced in cells be all competing and summatively be responsible for the observed event? (Roberts, Nichols, Newton, & Sporn, 1979). This type of argument drives the idea that there is clearly a valid rationale for the design of synthetic analogs of endogenous retinoids that exhibit much more powerful biological activity through greater receptor binding selectivity resulting from stronger binding interaction. Such an approach is particularly well matched for the design of potentially novel therapeutic compounds, especially if coupled with the compelling idea of developing compounds that are also largely resistant to metabolism and isomerization. The resulting highly selective compounds, showing high levels of binding preferences as well as chemical stability, would allow one to more confidently probe and understand the rich downstream biochemical cascade that begins when a retinoid first enters its receptor target and binds.

2. Synthetic retinoid design

The chemical structures of endogenous retinoids are precisely designed in order to behave as highly complementary, adaptable and controllable ligands for their particular targets. However, from a modern medicinal chemistry point-of-view, retinoids such as ATRA are very much atypical of small molecule drugs, and indeed, exhibit a range of "structural violations" that one can anticipate alleviating, and in the process, produce more potent, more selective compounds with improved solubility and bioavailability. Chief among these is the extended polyene chain, which exhibits both high conformational flexibility and significant photosensitivity. Indeed, these properties, allied with the relatively high lipophilicity of retinoids means that endogenous retinoids are somewhat promiscuous in their binding behaviors, and exhibit measurable affinities for a range of protein targets, including the peroxisome proliferator-activated receptors (PPARs) and the fatty acid binding proteins (FABPs) (Huang, Chandra, & Rastinejad, 2014). One solution to this problem is to design compounds that exhibit reduced conformational flexibility. This would presumably provide entropic benefits to the overall ΔG of binding, but by pre-organizing the conformation of the synthetic retinoid to mimic the structure of the LBPs of the RARs/RXRs, one can also improve selectivity for this retinoid targets over other protein targets, thus minimizing off-target effects.

2.1 Pan-RAR agonists

Pan-agonists of the RARs, i.e., compounds that mediate transcriptional activity via RARα, RARβ and RARγ with roughly equal binding affinity, can be designed by mimicking the shapes and conformations that ATRA can adopt. These compounds are characterized by a relatively linear structure and an absence of hydrogen bond donor groups in the linker region. The most widely employed approach has involved the removal of the capricious polyene structure of endogenous retinoids and, instead, incorporating aromatic rings that provide planar stable structures and improved synthetic accessibility, creating the so-called arotinoids (Loeliger, Bollag, & Mayer, 1980). The substitution of the highly flexible polyene for these more rigid structures also has significant, concomitant, entropic benefits through reduction in the degrees of freedom, allowing mimicking the most effective binding conformations of ATRA in the synthetic retinoid. Fig. 6 shows an overlay of two possible binding conformations of ATRA and synthetic retinoid, EC23, derived from molecular docking simulation into RARγ, highlighting how EC23 closely mimics the length and overall shape of the endogenous compound (Haffez et al., 2017).

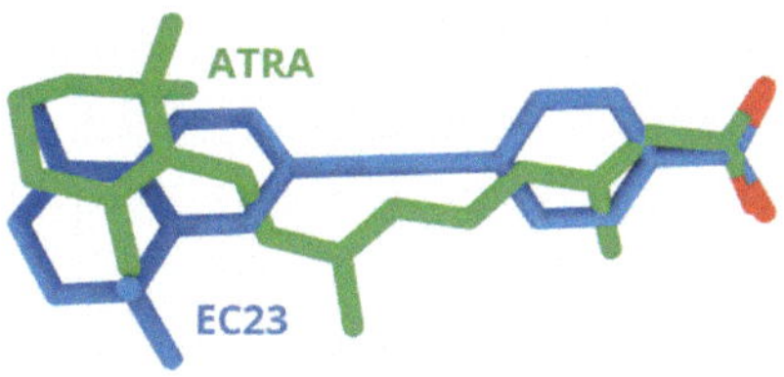

Fig. 6 Overlay of possible binding conformations of ATRA and synthetic retinoid EC23 derived from molecular docking simulations (Haffez et al., 2017).

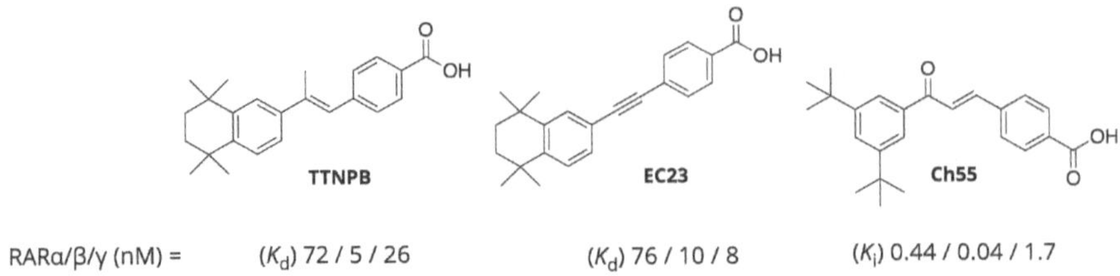

Fig. 7 Chemical structures of pan-RAR agonists TTNPB, EC23 and Ch55 (Gambone et al., 2002; Sun et al., 1997).

Fig. 7 also shows the chemical structure of EC23 along with prototypical pan-RAR agonists TTNPB and Ch55. Both TTNPB and EC23 possess a large 1,1,4,4-tetramethyl-1,2,3,4-tetrahydronaphthalene (TTN) hydrophobic region, combined with a rigid alkene or acetylene linker region with a benzoic acid polar region. These highly lipophilic, conformationally-restrained compounds exhibit strong affinity for each of the RARs and the large hydrophobic region, in particular, enhances affinity by forming more extensive interactions with the whole LBP, especially around H3 and H7 (Gambone et al., 2002; Teng, Duong, Klein, Pino, & Chandraratna, 1996). High affinity, combined with excellent chemical stability, has meant that both TTNPB and EC23 have become go-to compounds for pan-RAR agonist activity, and TTNPB, in particular, is often used as a standard in cell culture to provide robust, reproducible cellular development and differentiation (Christie et al., 2008; Murayama, Suzuki, & Matsui, 1997).

The design of Ch55 uses a similar approach, although a very large 2, 4-di-*tert*-butylphenyl hydrophobic region is employed, and the rest of the molecule must be shortened in order to accommodate this bulky hydrophobic region within the RAR LBPs. Incorporating the di-*tert*-butylphenyl motif, perhaps within a tetrahydronaphthalene-type structure, would likely be detrimental to binding since the RAR binding pockets are extremely sensitive to length—compounds that are too long or too short are transcriptionally inactive (Clemens et al., 2013). A *trans*-chalcone linker region is utilized to situate the benzoic acid polar region into the correct position for binding to the polar cluster at the end of the LBP, and the result is a compound that exhibits very strong affinity for each of the RARs (Sun et al., 1997). In summary, in order to attain strong pan-RAR agonist behavior, a synthetic retinoid should mimic the relatively linear, more extended, accessible conformations of ATRA, although with the addition of a larger hydrophobic region in conjunction with more stable aromatic groups (Fig. 8).

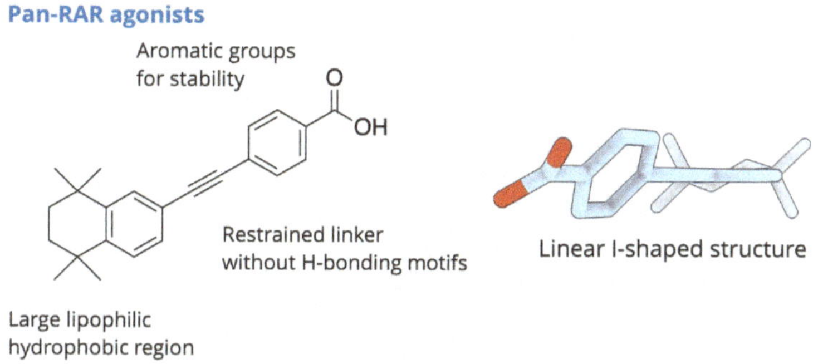

Fig. 8 Design principles for the development of pan-RAR agonists.

2.2 RAR isotype selectivity

Within both the RAR and RXR families of nuclear receptor are three isotypes: RARα, RARβ and RARγ, and RXRα, RXRβ and RXRγ. These isotypes are closely related homologs of each protein that exhibit high sequence homology within each structural region. In the RARs, the ligand-binding domain (LBD), is highly conserved between homologs (see chapter "Classical pathways of gene regulation by retinoids" by Pohl and Tomlinson) and the three, therefore, exhibit very similar structures (Chambon, 1996). The majority of these sequence differences are around the external surface of the protein, however, there are three positions of variation that have been identified in and around the LBP (Fig. 9). These differences cause subtle changes in the size and architecture of the RARα, RARβ and RARγ LBPs, to the extent that these changes can be exploited to design synthetic retinoids that exhibit isotype specificity (Ostrowski et al., 1998). In contrast, differences in the three RXR homologs are confined only to the outer structure of the LBDs, meaning that the LBPs of each of RXRα, RXRβ and RXRγ are essentially identical and, therefore, extremely difficult to differentiate between when designing selective RXR ligands.

2.2.1 RARα selectivity

An analysis of Fig. 9 shows that RARα is unique in respect of being the only RAR isotype to present a polar amino acid as a divergent, isotype-specific residue. Ser232 on H3 lies roughly in the middle of the RARα LBP, with a side chain that is oriented inwards toward the center of the pocket. Hence, this variant position could be exploited by incorporating structural motifs that act as hydrogen bond acceptors/donors into the linker region of a synthetic retinoid.

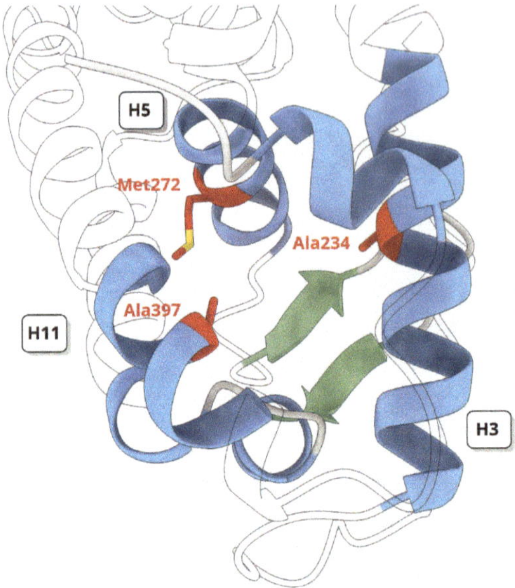

Variant Positions

	H3	H5	H11
RARα	Ser232	Ile270	Val395
RARβ	Ala225	Ile263	Val388
RARγ	Ala234	Met272	Ala397

Fig. 9 Left, structural regions that constitute the RARγ LBP (PDB:2LBD), with variant positions highlighted. Right, sequence positions within the RAR LBP that exhibit variance between RARα, RARβ and RARγ (Renaud et al., 1995).

The most straightforward modification to the linker region would be to substitute the alkene of TTNPB for an amide, as in the case of Am580. The key role of this amide is highlighted in Fig. 10, where the Am580 amide nitrogen functions as an H-bond donor, forming a hydrogen bonding interaction with the Ser232 side-chain oxygen atom (3.2 Å) (le Maire et al., 2010). In RARβ and RARγ, this interaction is not possible, and the presence of the polar amide presumably destabilizes the H3 variant position in these isotypes. As a result, Am580 exhibits good selectivity (Fig. 11) for RARα over its isotype cousins.

While the amide positioning in Am580 is clearly beneficial, it is also possible to adopt the opposite placement, i.e., with the amide nitrogen atom attached directly to the TTN hydrophobic region, e.g., Am80 in Fig. 11. Since the large TTN hydrophobic region must be oriented predominantly into the large cavity adjacent to H3, it is likely that Am80 forms a hydrogen bond interaction with Ser232 via the amide carbonyl oxygen, rather than the amide nitrogen. Binding assays have shown that this alternate positioning also imparts a degree of RARα selectivity, although Am80 generally exhibits slightly reduced overall binding affinity and transcriptional activity (two- to four-fold weaker) when compared to Am580 (Beard et al., 2002; Bernard et al., 1992).

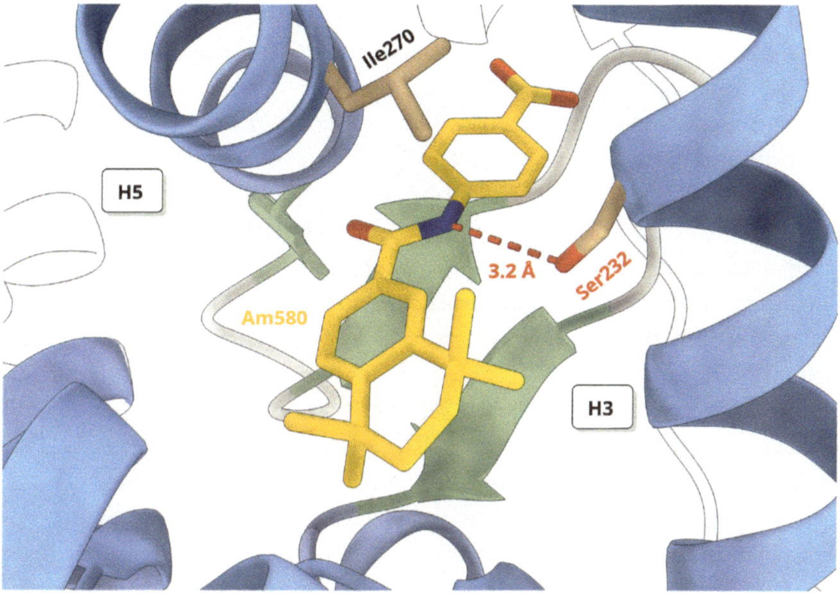

Fig. 10 Crystal structure of RARα-selective retinoid Am580 bound to RARα, with the key Ser232 interaction highlighted (PDB: 3KMR) (le Maire et al., 2010).

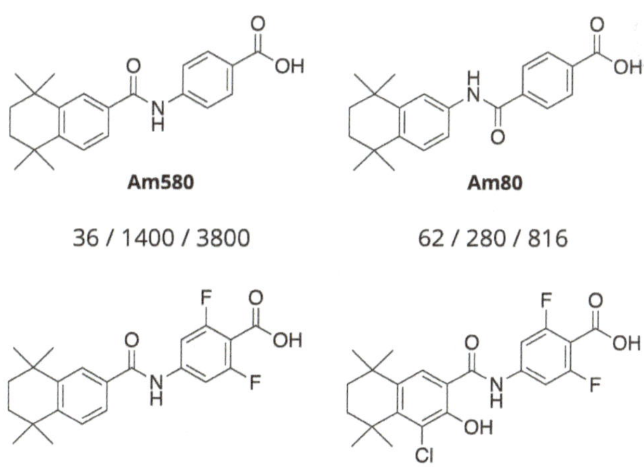

Am580 **Am80**

K_d RARα/β/γ (nM) = 36 / 1400 / 3800 62 / 280 / 816

AGN193931 **AGN195183**

7 / 1470 / 5600 3 / >10^5 / >10^5

Fig. 11 Chemical structures of RARα-selective agonists Am580, Am80, AGN193931 and AGN195183.

A 4-aminobenzoic acid polar region in synthetic retinoids has been shown to provide good RARα selectivity and strong binding affinity, but additional decoration of the structure of such systems also provides further improvements in their desired properties (Fig. 11). For example, incorporation of fluorine atoms in the 2- and 6-positions of the polar region (e.g., AGN193931) results in improved binding affinity for RARα, although with only a moderate increase in selectivity. However, incorporating a chlorine atom in the 4-position and a hydroxyl group in the 3-position of the TTN hydrophobic region (e.g., AGN195183), results in a compound that is entirely selective for RARα versus either RARβ or RARγ (Beard et al., 2002). It is not clear whether the additional hydroxyl group also interacts with Ser232, but it is most likely that the additional ring substituents cause active discrimination against both RARβ and RARγ by clashing with hydrophobic residues in these isotypes.

In summary, greater selectivity for RARα can be achieved by incorporating H-bonding motifs such as amides and pyrroles into the linker region (Fig. 12). Further selectivity can be achieved by appending small(er) halide substituents adjacent to the polar region, presumably interacting with water molecules around the polar cluster, and by incorporating further substituents at the 3- and 4-positions of the hydrophobic region, to elicit additional clashes with residues around the middle of the RARβ and RARγ LBPs.

2.2.2 *RARβ selectivity*

Analysis of the crystal structures of the three RAR homologs shows that the RARβ LBP is around 9% larger than those of RARα and RARγ, with an

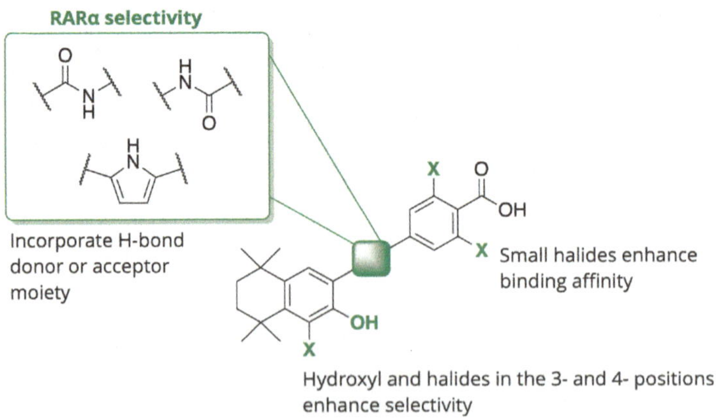

Fig. 12 Design principles for the development of RARα-selective agonists.

additional cavity near the opening of the pocket, evident between H5 and H10 as a result of the positioning of the RARβ-specific residue Ile263 on H5 (Germain et al., 2004). One can, therefore, exploit this additional space and engineer an RARβ-specific synthetic retinoid by incorporating larger groups onto the hydrophobic region that would not only block access to the narrower RARα and RARγ LBPs, but also establish beneficial interactions with the residues in the additional hydrophobic cavity.

The stilbene BMS641 (Fig. 13) adopts this approach by incorporating a bulky group onto a dihydronaphthalene hydrophobic region, thus positioning the new group outwards and away from the main section of the retinoid structure in order to fill this RARβ-specific cavity. The addition of these groups significantly increases selectivity for RARβ in binding and transcription assays (Germain et al., 2004; Nadendla et al., 2015). Furthermore, it is thought that the chlorine function on the benzoic acid polar region causes active discrimination against RARα by clashing with the RARα-specific Ser232 on H3, and the large hydrophobic region's lower positioning within the pocket causes destabilizing interactions with the RARγ-specific residue Met272 on H5 (Nadendla et al., 2015). Hence, this structural approach both augments binding with RARβ while also discouraging interaction with both RARα and RARγ.

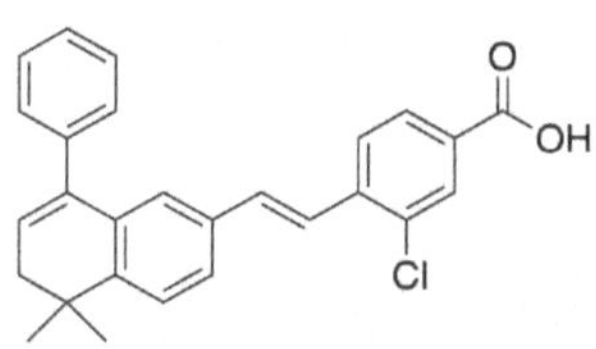

BMS641

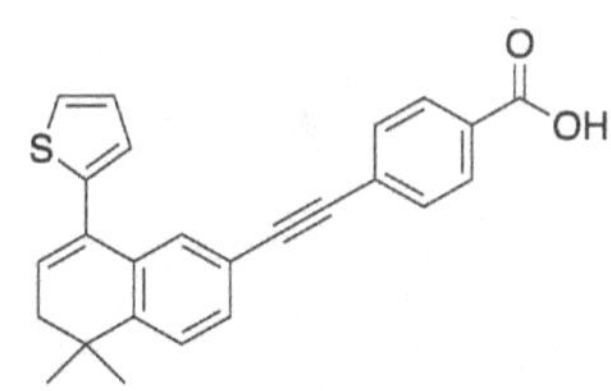

AGN193639

K_d RARα/β/γ (nM) = 225 / 2.5 / 223 129 / 20 / 104

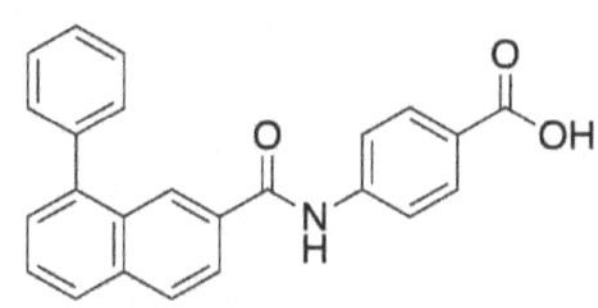

BMS948

ND / ND / ND

Fig. 13 Chemical structures of RARβ-selective agonists BMS641, AGN193639 and BMS948. ND = not determined.

The diphenylacetylene AGN193639 (Fig. 13) embodies a similar approach by incorporating a bulky thiophene group onto its dihydronaphthalene hydrophobic region. However, in contrast to the stilbene BMS641 which weakly binds to both RARα and RARγ, AGN193639 binds RARα and RARγ with reasonable affinity in comparison to RARβ, but actively silences the transcriptional activity of RARα and RARγ, presumably through an antagonistic mechanism where the thiophene blocks closure of H12 in both RARα and RARγ (Johnson, Klein, & Wang, 1996).

Perhaps the most surprising example of an RARβ-selective agonist is the amide BMS948 (Figs. 13 and 14). How an amide in this linker position generally provides RARα selectivity was discussed already, however, the amide in this system has been shown to be flipped such that the amide carbonyl moiety directs toward the RARβ-specific Ala225 on H3 at a distance of 3.1 Å, perhaps offering the ability for a favorable C—H···O interaction (Nadendla et al., 2015). Indeed, the removal of the *gem*-dimethyl group on the hydrophobic region in favor of a simple naphthalene system allows a lower positioning in the pocket that would be untenable otherwise, and

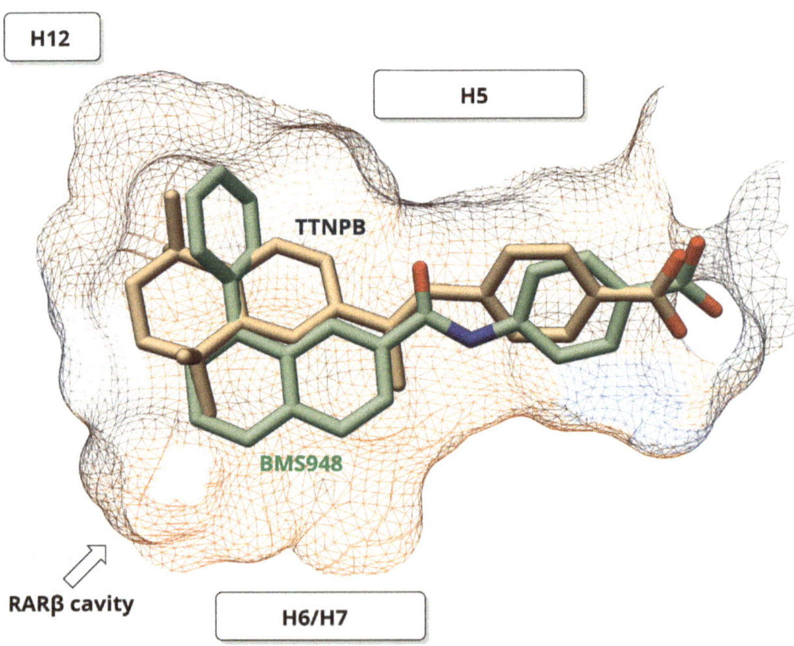

Fig. 14 Comparison between the crystal structures of pan-RAR agonist TTNPB and RARβ-selective agonist bound in the RARβ LBP (PDB: 1XAP and 4JYH) (Germain et al., 2004; Nadendla et al., 2015).

this position presumably allows this amide orientation to be favorable. As shown in Fig. 14, in comparison to pan-RAR agonist TTNPB, BMS948 fills the RARβ LBP more completely, particularly around the additional cavity unique to RARβ.

The most effective means of introducing RARβ-selectivity is, therefore, to simply add a large hydrophobic group onto the hydrophobic region such that this moiety is positioned toward the additional cavity evident in the RARβ LBP. Further discrimination can be achieved by incorporating small hydrophobic groups that are directed toward the H3 isotype variant position, but more subtle effects, for instance, the balance between actively antagonizing RARα and RARγ (or weak binding to these isotypes) can be achieved by careful modulation of the overall positioning of the compound within the pocket (Fig. 15).

2.2.3 RARγ selectivity

Selectivity for RARγ can be attained by exploiting the RARγ-specific residues on H5 and H11, i.e., Met272 and Ala397. Met272 is positioned roughly in the middle of the LBP, with the side-chain situated above the linker region of a bound retinoid, and formation of a hydrogen bond with the Met272 side-chain has been identified as a key strategy for increasing selectivity for RARγ. An amide, as used in the RARα-selective compounds, does not correctly position for a suitably strong hydrogen bond interaction, but an sp^3 hydroxyl group, or an oxime, has been found to offer a suitably strong hydrogen bond interaction such that RARγ-selectivity is improved. However, to access this key hydrogen bonding interaction, the compound must be lengthened in order to place this H-bond donor group within proximity of the Met272 side chain. Fortunately, the RARγ LBP is slightly more forgiving than the other RAR isotypes in

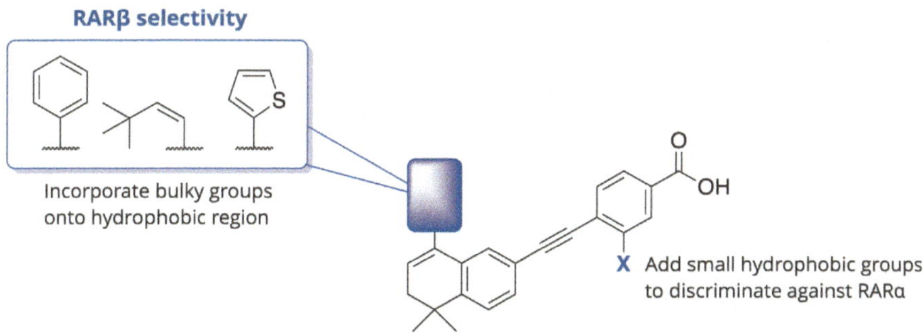

Fig. 15 Design principles for the development of RARβ-selective agonists.

regards to the accessible length of the LBP, since the variant residue on H11 is the smaller Ala397 in comparison to the larger valines of RARα and RARβ, therefore, slightly extending the LBP. In light of these differences in pocket architecture, the most effective molecular design strategy has been to incorporate a three-atom linker region, as opposed to the two atom systems used in the pan-RAR agonists or RARα and RARβ-selective compounds.

Indeed, a longer linker, e.g., ketone system CD564 (Fig. 16), reduces the affinity for RARα and enables strong binding to RARγ and also favors the large LBP of RARβ (Klaholz, Mitschler, & Moras, 2000). When the ketone is reduced to the corresponding hydroxyl, enabling the key Met272 hydrogen bonding interaction, a significant increase in selectivity for RARγ versus RARα and RARβ is observed, although with a roughly 10-fold reduction in affinity (e.g., the racemic BMS184394). However, compelling enantiomeric discrimination has also been observed in this series, where *S*-BMS184394 has been shown to be a weak agonist of RARα, RARβ and RARγ (K_d = 7500, 4095 and 816 nM respectively) but the corresponding *R*-enantiomer exhibits greater selectivity and stronger binding to RARγ (K_d = 75 nM) compared to RARα and RARβ (K_d = 16,000 and 531 nM, respectively). This is also the case in the related amide compound *R*-BMS961, which displays excellent selectivity for RARγ, although at the expense of affinity (Fig. 17) (Klaholz, Mitschler, & Moras, 2000). In this case, it appears that

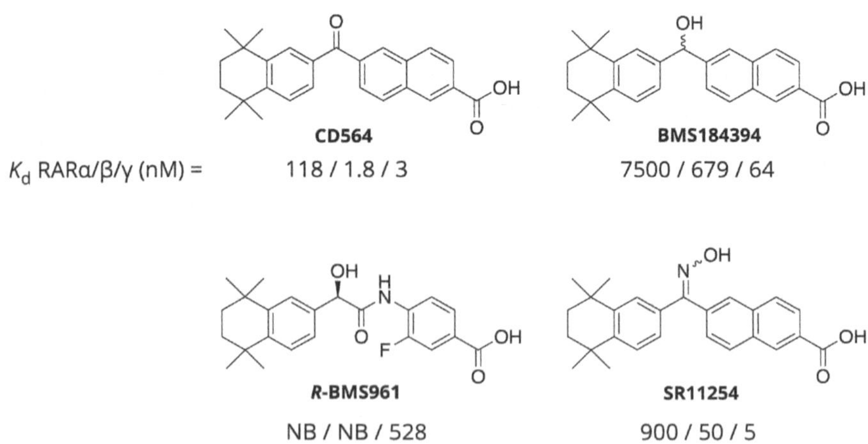

Fig. 16 Chemical structures of RARγ-selective agonists CD564, BMS184394, *R*-BMS961 and SR11254. NB = no binding.

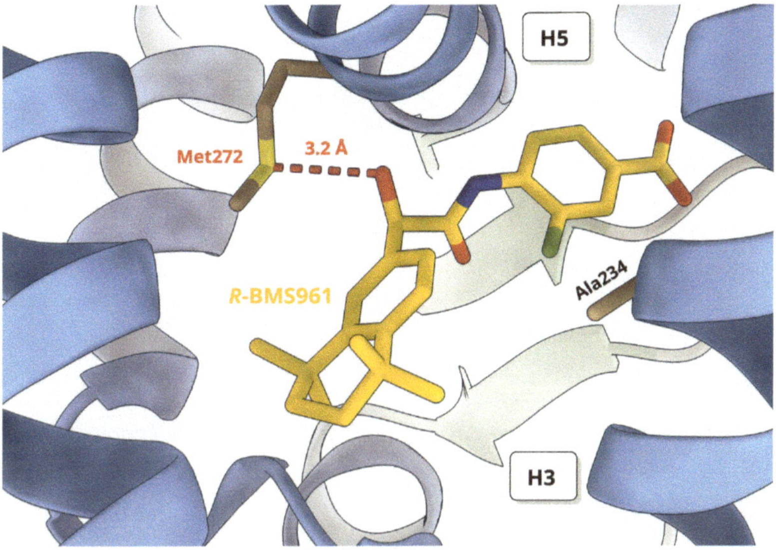

Fig. 17 Crystal structure of RARγ-selective retinoid *R*-BMS961 bound to RARα, with the key Met272 interaction highlighted (PDB: 1EXA) (Klaholz, Mitschler, Belema, Zusi, & Moras, 2000).

the discrimination against the *S*-enantiomers of these compounds is due to the adoption of an energetically unfavorable conformation and positioning within the LBP (Klaholz, Mitschler, Belema, et al., 2000). Conversion of the ketone of CD564 to an oxime (SR11254) moves the hydroxyl group closer to Met272 (2.95 Å for the *Z*-oxime isomer of SR11254 versus 3.19 Å for *R*-BMS961), also enabling it to be close enough for a favorable C—H⋯O interaction with the Met272 side-chain methyl group (2.86 Å) (Klaholz & Moras, 2002). SR11254, therefore, exhibits much stronger affinity for RARγ in comparison to the sp^3 hydroxyl systems, with similar selectivity (Yu et al., 1996).

This small series of compounds illustrates a highly effective ligand design strategy (Fig. 18) for attaining RARγ selective retinoids. However, in comparison to its RARα and RARβ cousins, the ideal marriage between selectivity and affinity has yet to be achieved in RARγ selective ligand design, and it remains a challenge that has not been fully met. Nevertheless, it is likely that incorporation of a hydroxyl group, or similar, into the linker region of a retinoid will be key to achieving this aim, particularly if derivatives of SR11254 can be elaborated such that RARβ binding is inhibited.

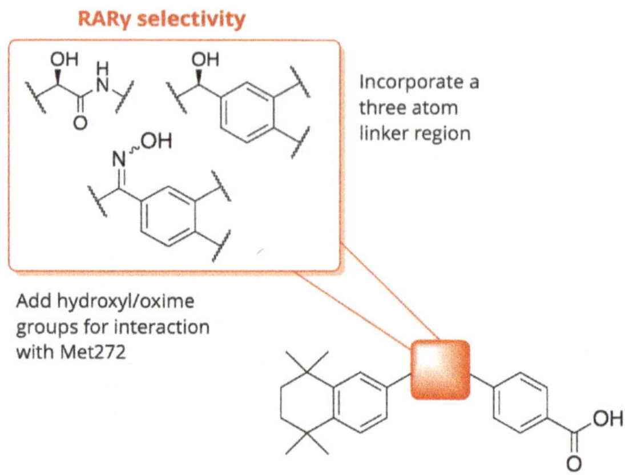

Fig. 18 Design principles for the development of RARγ-selective agonists.

2.3 RAR antagonism

Significant progress has been made in the development of retinoids that activate the RARs, but it would be equally important and useful for it to be possible to achieve selective deactivation of these powerful nuclear receptor proteins. Two molecular mechanisms can be utilized to achieve this aim: (1) incorporation of large groups that block the closing of H12, thus rendering the protein transcriptionally inactive, i.e., antagonism; or (2) improvement of the affinity of the interaction between H12 and a CoR, thus preventing removal of the CoR, and hence, inhibiting concurrent CoA binding, i.e., inverse agonism.

Selectivity for RARβ was realized by incorporating a large hydrophobic group onto the hydrophobic region, but by simply further increasing the size of this group, we can interfere with the positioning of the crucial H12 on each RAR isotype and, hence, achieve antagonistic activity. AGN194310 (Fig. 19) features an ethylbenzene-substituted thiochromene hydrophobic region designed to instigate this effect and, like the related diphenylacetylene pan-RAR agonist EC23, exhibits strong binding affinity for each of the RAR isotypes. However, the large phenyl substituent disrupts H12 repositioning on binding and, as a result, AGN194310 is a potent inhibitor of transcriptional activity elicited by RAR agonists and has been proposed as a potential antidote to retinoid-induced toxicity (Johnson, Wang, Standeven, Escobar, & Chandraratna, 1999).

The phenylacetylene-substituted stilbene BMS493 exhibits a similar structure, but crystallographic studies in RARα have shown that the phenylacetylene moiety forms particularly strong interactions with Ile396

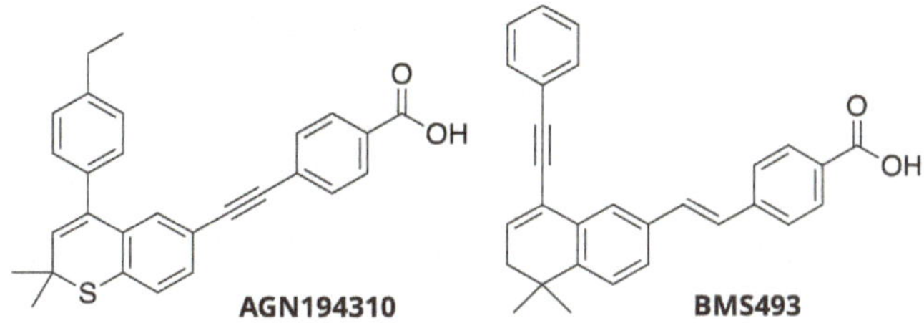

K_d RARα/β/γ (nM) = 3 / 2 / 1 4.2 / 2.9 / 98

Fig. 19 Chemical structures of pan-RAR antagonists, AGN194310 and BMS493.

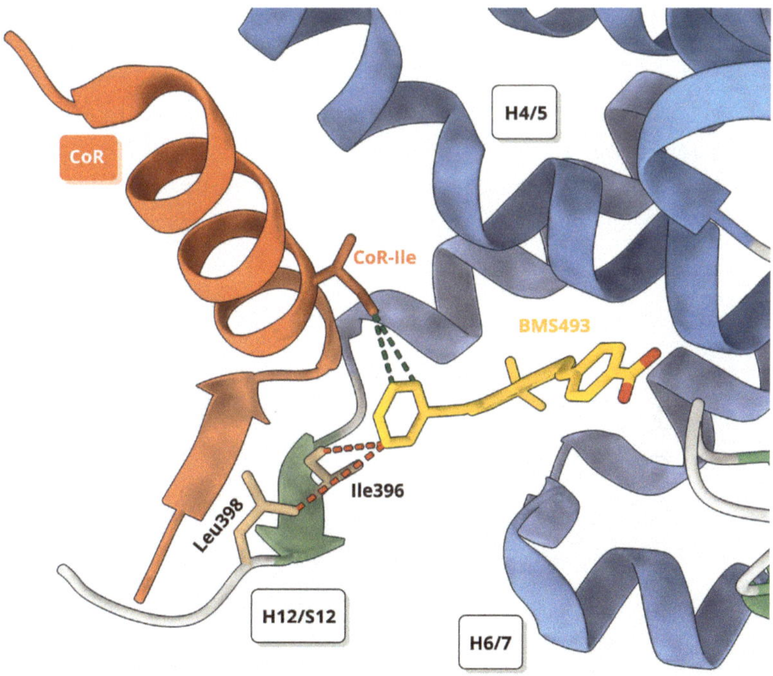

Fig. 20 Crystal structure of inverse agonist BMS493 bound to RARα, with H3 removed for clarity (PDB: 3KMZ) (le Maire et al., 2010).

and Leu398 on H12, which is thought to stabilize the association between H12 and a CoR by causing the formation of an anti-parallel β-sheet interaction as shown in Fig. 20 (Álvarez et al., 2009; Germain et al., 2009). There is clearly a subtle difference between these two inhibitory mechanisms, but it is likely that the ethylbenzene group of AGN193410 exhibits a more significant protrusion from the LBP and, hence, more directly interferes with H12, while the phenylacetylene of BMS493 interferes less obtrusively, enabling an inverse agonism mechanism (le Maire et al., 2010). Hence, by simply further increasing the size of the hydrophobic region and subtly modulating the shape and orientation of the additional group, it is possible to transition from agonist, to inverse agonist to antagonist (Fig. 21).

2.3.1 Selective RAR antagonism

Since antagonism of the RARs is achieved by disruption of H12 positioning upon retinoid binding, and not by interference with the rest of the LBP, the same design strategies can be applied that were used to attain RAR isotype selective agonists to the development of the corresponding selective antagonists. This opportunity is particularly compelling, since in terms of downstream biological activity, not activating an RAR for transcriptional regulation does not necessarily achieve the same effect as *deactivating* a particular RAR isotype, and the resulting biology from the two approaches can be completely distinct.

Fig. 22 shows three antagonists that each have both a large hydrophobic region substituted with an H12-disruption group and an isotype-selective

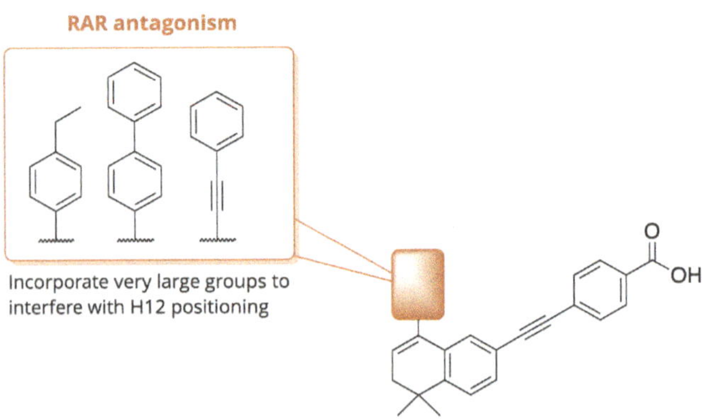

Fig. 21 Design principles for the development of pan-RAR antagonists.

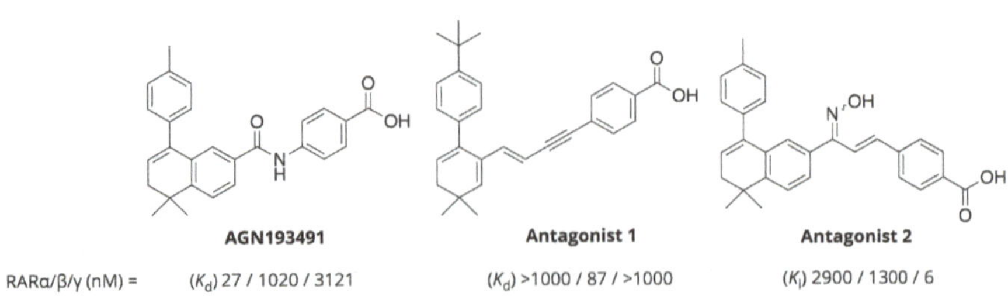

Fig. 22 Chemical structures of RAR isotype selective antagonists AGN193491, Antagonist 1 and Antagonist 2.

structural feature combined into the same structures. Both AGN193491 and Antagonist 2 utilize a tolyl-substituted hydrophobic region attached to prototypical RARα- and RARγ-selective functionalities, i.e., an amide and an oxime, respectively, the latter derived from the corresponding chalcone (Teng et al., 1997; Tsang, Sinha, Liu, Bhat, & Chandraratna, 2005). The resulting compounds exhibit excellent selectivity for binding RARα and RARγ, respectively, and are both efficient deactivators of RARα and RARγ transcriptional activity. Hence, these compounds highlight the modular nature of attaining selectivity and antagonist behavior in these three-component retinoid structures (Fig. 23). Extending the same concept to RARβ has also resulted in compounds with highly specific binding for this isotype, although in these cases many such compounds, e.g., Antagonist 1, have not tended to exhibit selective inhibition of RARβ in transactivation assays, and instead often favor RARγ (Beard, Klein, Standeven, Escobar, & Chandraratna, 2001). Nevertheless, this approach has promise and it is likely that with subtle modification to prevent modulation of RARγ, a truly selective antagonist of RARβ can be developed.

2.4 RXR agonists

The principle design strategy when developing RXR agonists is to mimic the shape and favored conformation of 9*c*RA and, hence, fix the compound in an L-shaped structure which is complementary to the L-shaped RXR-binding pocket. With a wide variety of means achieving this enforced twist, RXR agonists generally exhibit more structural variety than RAR agonists;

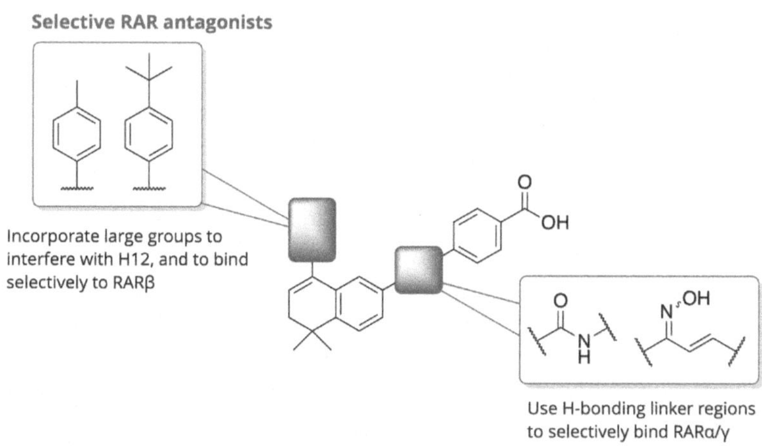

Fig. 23 Design principles for the development of isotype-selective RAR antagonists.

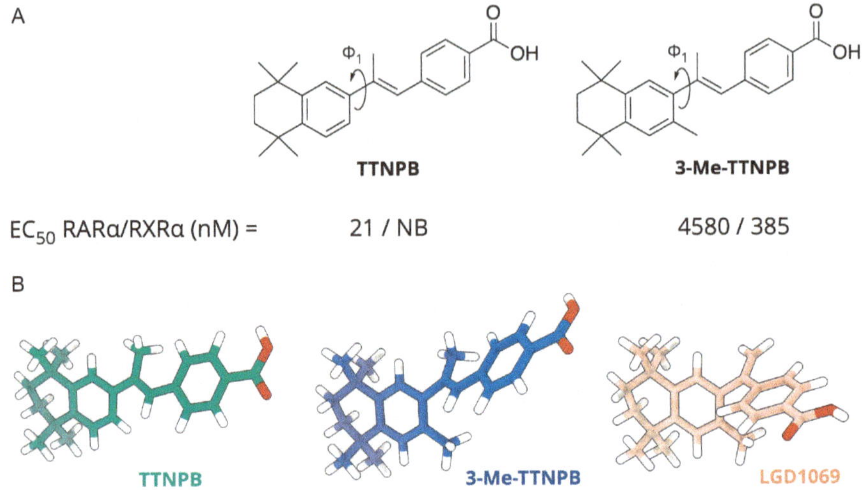

Fig. 24 (A) Chemical structures of related synthetic retinoids TTNPB and 3-Me-TTNPB. (B) Low energy conformations (AM1) of TTNPB, 3-Me-TTNPB and LGD1069, highlighting the increasingly twisted shape with increasing proximity between hydrophobic and polar region.

particularly in the linker region. One of the key breakthroughs that informed the RXR-selectivity paradigm was observed when modifying the structure of pan-RAR agonist, TTNPB (Fig. 24A).

When an additional methyl group is incorporated in the 3-position of the hydrophobic region (3-Me-TTNPB), a significant reduction in transcriptional activation through RARα was observed, but a concomitant improvement in activity mediated through RXRα was also detected (Beard et al., 1994). Conformational analysis shows that the presence of this methyl group causes a significant twist about Φ_1, to the extent that 3-Me-TTNPB exhibits a more twisted conformation in comparison to the relatively linear conformation of TTNPB, as demonstrated in Fig. 24B.

Following on from this key structural insight, by shortening the linker region to a single atom the conformational impact of the 3-Me group is augmented, thus, increasing the extent of the twist in the molecule and, hence, adopting a conformation that more precisely mimics that favored by 9*c*RA. LGD1069 (Fig. 24B and 25) incorporates a simple methylene linker region that brings the polar phenyl ring region closer to the disrupting presence of the 3-Me group, thus enforcing a greater twist about Φ_1 (Boehm et al., 1994). The result is complete discrimination against RARα, RARβ and

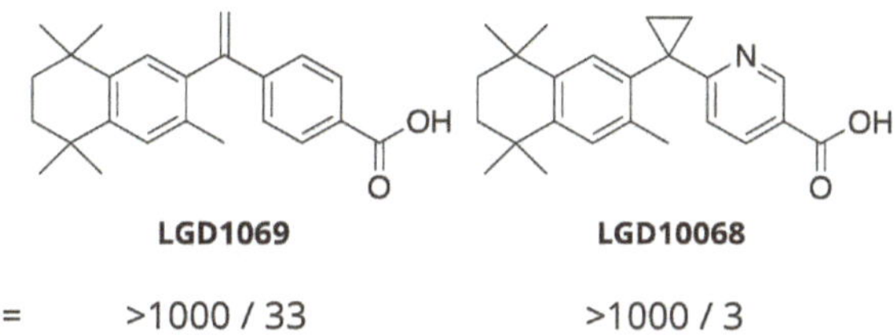

Fig. 25 Chemical structures of RXR agonists LGD1069 and LGD10068.

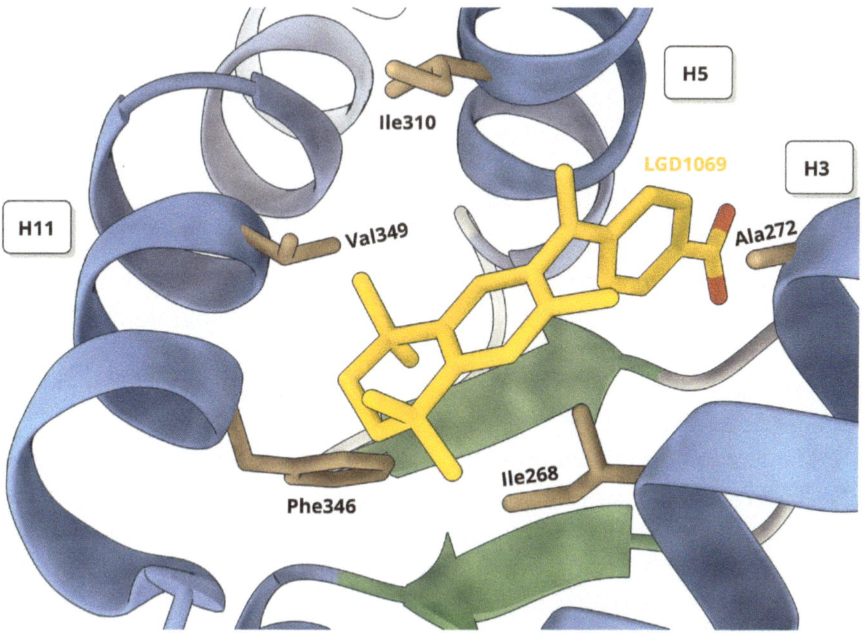

Fig. 26 Crystal structure of LGD1069 bound to RXRα, highlighting the twisted ligand conformation and the array of hydrophobic contacts made by the linker and hydrophobic regions (PDB: 4K6I) (Boerma et al., 2014).

RARγ, and enhanced binding to RXRα, RXRβ and RXRγ (Fig. 26) compared to 3-Me-TTNPB (Boehm et al., 1995).

LGD10068 elaborates on this paradigm by substituting the methylene linker region for a bulkier cyclopropane group (Fig. 25) and also incorporates a nicotinic acid polar region. LGD10068 adopts a very similar positioning to LGD1069 within the RXRα LBP, exhibiting a near identical conformation with respect to Φ_1. However, the cyclopropane analog forms

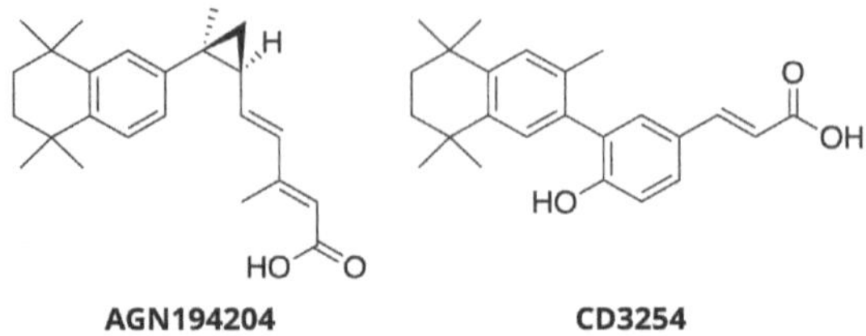

Fig. 27 Chemical structures of RXR agonists AGN194204 and CD3254.

more extensive hydrophobic interactions with residues on H10 than the corresponding methylene on LGD1069 which translates into improved binding affinity for RXRα (Love et al., 2002).

More significant departures from the TTNPB-type scaffold include AGN194204 (Fig. 27), which combines a highly lipophilic TTN hydrophobic region with a cyclopropane linker region and a dienoic acid polar region. This structure is fixed into a bent shape similar to that exhibited by 9*c*RA due to the cyclopropane function, and consequently exhibits highly potent activity in transactivation assays, without the stability issues plagued by its endogenous forebear, as a result of the increased photostability of the diene system (Apfel et al., 1995).

The design of CD3254 (Fig. 27) eschews a formal linker region in favor of a biaryl scaffold that unites a TTN hydrophobic region and a cinnamic acid polar region (Wagner et al., 2009). The angle of the plane of the TTN hydrophobic region with respect to that of cinnamic acid moiety closely mimics the twist observed in LGD1069/LGD10068 and, hence, CD3254 adopts very similar positioning and forms similar contacts with the RXRα, RXRβ and RXRγ LBPs. However, the hydroxyl group is also capable of forming a favorable hydrogen bonding interaction with Asn306 on H5 (RXRα); the modification of which has also been shown to determine agonist versus antagonist behavior (vide infra) in other compounds in the series (Santin et al., 2009).

In light of these examples, it is clear that the main determinant of RXR-agonist activity is the adoption of an L-shaped twisted ligand structure that ensures the discrimination against the more linear RAR LBP (Fig. 28). Using a highly lipophilic hydrophobic region ensures favorable interactions with the area around H12, while linker regions without hydrogen bonding motifs are generally better tolerated, with the exception of the hydroxycinnamic acid moiety used in CD3254. Less lipophilic polar regions are also

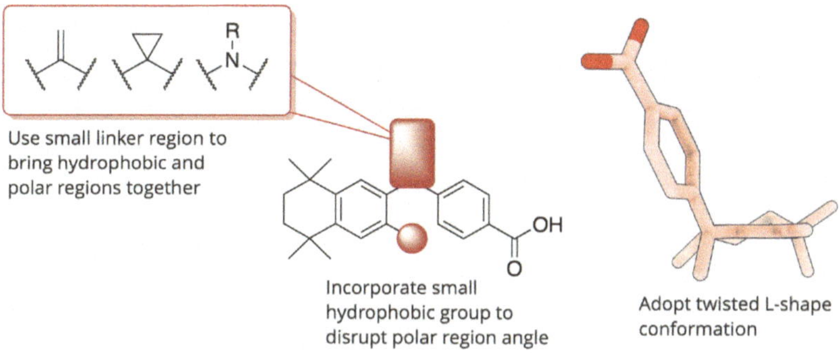

Fig. 28 Design principles for the development of pan-RXR agonists.

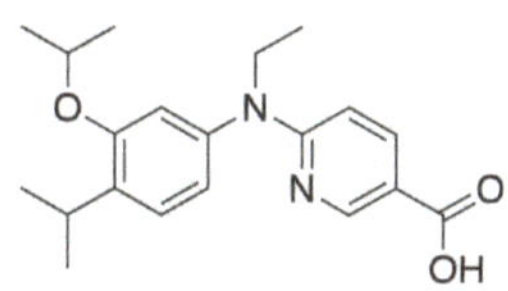

NEt-3IP

EC_{50} RXRα/β/γ (nM) = 32 / 36 / 376

Fig. 29 Chemical structure of RXRα/β-selective agonist NEt-3IP.

suitable for binding to the RXRs, e.g., pyridines and pyrimidines have been widely employed, and such compounds typically display improved ADMET/PK properties (Wagner et al., 2009).

2.5 RXR selectivity

Designing RAR isotype-selective compounds can be achieved by exploiting the subtle variances in pocket architecture caused by small sequence differences between each isotype. However, the RXR LBP is highly conserved between each of the isotypes to the extent that there are no significant differences in pocket architecture between RXRα, RXRβ and RXRγ (Takamatsu, Takano, Yakushiji, Morishita, et al., 2008). Clearly, this makes the design of compounds that are selective for each of the RXR isotypes extremely challenging. However, it has been shown in the case of a series of biarylamine compounds (Fig. 29), that selectivity for RXRα and RXRβ can be conferred by reducing overall lipophilicity (LogP), particularly around the hydrophobic region and linker regions (Takamatsu, Takano, Yakushiji, Morohashi, et al., 2008). The molecular basis of this effect is unclear, however, and the elucidation of chemical design strategies for the generation of potent RXR-selective compounds is still a challenge that must be met by the community. There are isotype-specific differences in the rest of the RXR LBD and, hence, it may be possible that an allosteric approach could modulate the structure of the LBP of specific RXR isotypes to an extent that selective binding can occur with a co-administered ligand (Bourguet et al., 1995).

2.6 RXR antagonists

While RAR isotype-specific strategies have proved unsuccessful in the RXRs, application of the molecular design rules that enabled an antagonistic effect can be directly applied to the RXRs. Here, as previously, we wish to disrupt the positioning of H12 to inhibit the closing of the LBP and, therefore, inhibit transcriptional activation through recruitment of CoA species. In the case of RXR antagonists a bulky hydrophobic group is added to the linker region, since this is the part of an RXR ligand that is oriented toward H12 due to the twisted L-shaped ligand conformation required to bind the RXRs.

This design strategy is exemplified by LG10074 (Fig. 30), which exhibits the extended polyene linker and polar region found in 9*c*RA, along with a TTN hydrophobic region substituted with a propoxy group in the 3-position. This simple appendage situated nearer to the linker region interferes directly with H12 to the extent that crystallographic studies have shown that this key helix adopts an alternative, outward positioning away from the LBP (Fig. 31) (Sato et al., 2010). It is, therefore, apparent that the protein is unable to adopt the necessary closed-pocket conformation required for activation of transcriptional processes. Indeed, in transcriptional assays, LG10074 shows potent antagonist behavior against each of the RXRs (Koch et al., 1996). UVI3003, a derivative of the pan-RXR agonist CD3254, employs a similar approach by incorporating a pentoxy

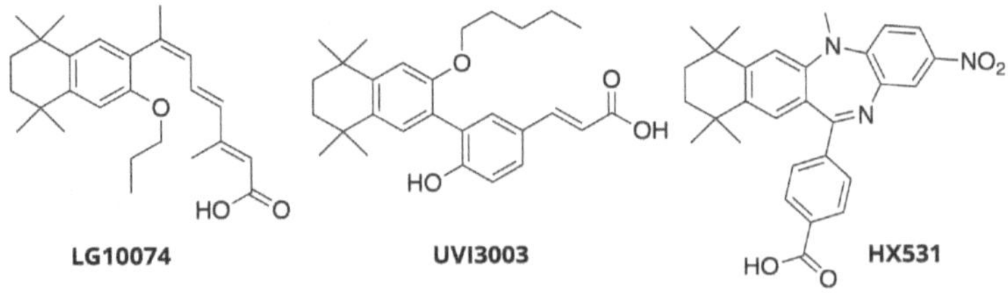

LG10074 UVI3003 HX531

IC_{50} RXRα (nM) = 16 240 18

Fig. 30 Chemical structures of pan-RXR antagonists LG10074, UVI3003 and HX531.

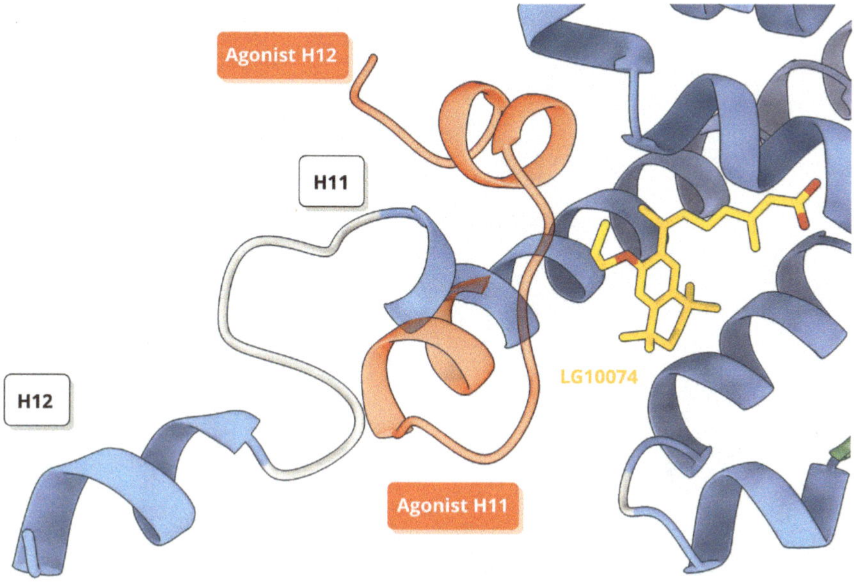

Fig. 31 Crystal structure of LG10074 bound to RXRα, with H3 hidden for clarity (PDB: 3A9E) (Sato et al., 2010). An agonist H11/H12 conformation is also shown to highlight the significant perturbation of H12 positioning due to the propoxy chain of LG10074 (PDB: 4ZSH).

group near to the linker region and it may well be the case that this group adopts very similar positioning to the propoxy group of LG10074 (Nahoum et al., 2007).

An alternative paradigm was developed, however, for the dibenzodiazepine compound, HX531 (Fig. 30) in which the TTN hydrophobic region and benzoic acid polar region are joined as a dibenzodiazepine, thus creating an extremely large linker region that fixes the compound into the requisite L-shape. However, rather than directly inhibiting H12 repositioning upon binding, HX531 is thought to clash with Trp305 on H5 and Leu438 on H10, principally through the nitro group (Ebisawa et al., 1999; Egea et al., 2000; Sato et al., 2010). This tryptophan, in particular, is understood to form stabilizing interactions with H12 when a suitable agonist is bound to the RXR and, hence, by disrupting this interaction it is possible to elicit antagonist activity in a more indirect manner. Leu438 also forms part of the LBD dimerization interface and perturbing this residue also affects the ability for the RXR to elicit transcriptional regulation (Sato et al., 2010).

It is, therefore, possible to design compounds that antagonize the RXRs in much the same manner as that used to antagonize the RARs,

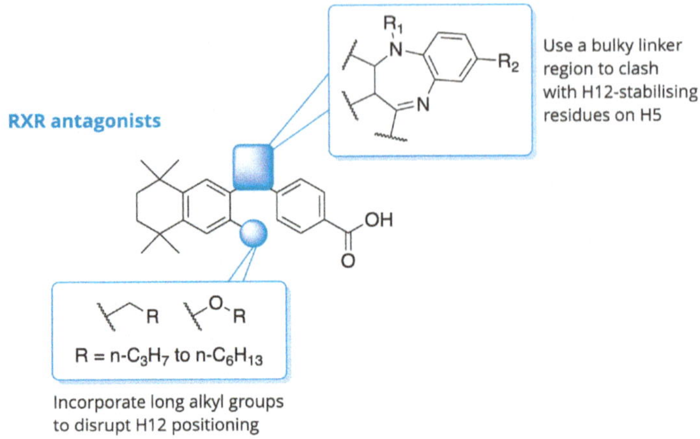

Fig. 32 Design principles for the development of pan-RXR antagonists.

i.e., by adding a bulky group that interferes with some aspect of H12 repositioning on ligand binding (Fig. 32). However, given the significant conformational differences that are required to encode RXR-binding behavior, this large group should be placed near to, or in, the linker region such that this modulatory influence is positioned toward H12 rather than the rest of the LBP.

3. Designing retinoids as probes

Much of the synthetic work in the retinoid field has been applied toward the development of highly potent, selective compounds for use in therapeutic applications. However, with the increasing availability of powerful biochemical and biophysical methods for studying biological processes in recent years, there has been rising interest in the development of retinoids as tools for understanding their activities, and using such systems as molecular probes.

3.1 Incorporation of chemical tags

For proteins with open binding pockets, the design principles for incorporation of a chemical tag for techniques such as immobilization and protein degradation are relatively straightforward; a tag can simply be appended at the site that may be easiest to access in terms of synthetic methodology and/or an area of the compound that is solvent exposed and, hence, not significantly involved in binding to a protein target. However, given the closure mechanism that operates upon binding of retinoids to both the RAR and RXR binding pockets, opportunities for the incorporation of tags are much more challenging.

One approach is to allow recognition and entrance into the binding site by retaining the carboxylate polar region, along with a suitable linker region, but to append the tag onto the hydrophobic region where the larger space around the opening of the pocket may be more accommodating (Fig. 33). This approach has been employed to append a peptide motif that activates attached cargo for proteasomal degradation via an oxime linker onto a β-ionone-derived hydrophobic region of ATRA (Itoh, Ishikawa, Naito, & Hashimoto, 2010; Okuhira et al., 2011). However, while this compound retained the majority of the structural elements required for strong binding, the modified hydrophobic region vastly reduced affinity for the RARs and the compound was only shown to be capable of binding into the open binding pocket of the CRABPs.

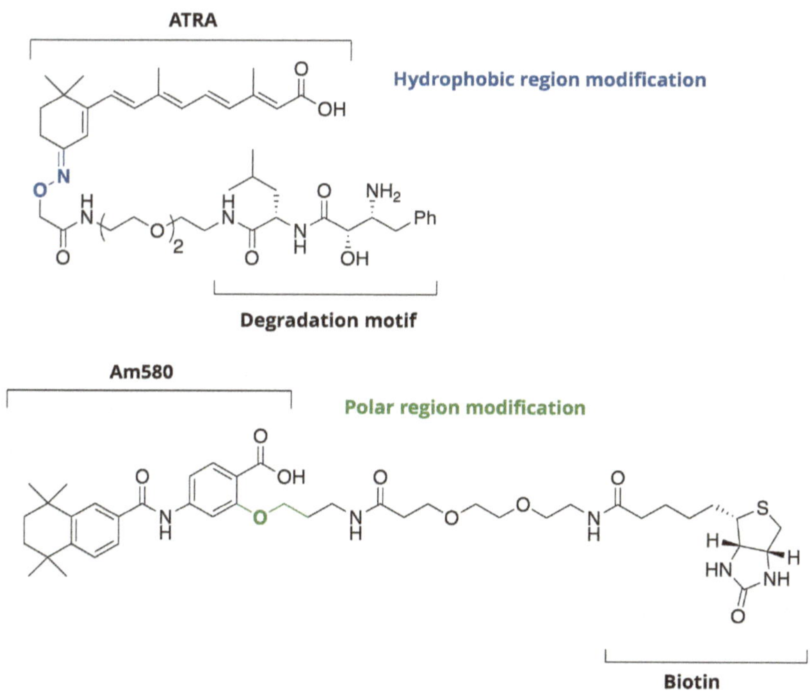

Fig. 33 Chemical structures of two retinoids that incorporate chemical tags at the hydrophobic region and at the polar region.

While modification of the hydrophobic region has seen only limited success as a tagging approach, analysis of the structure of the RARs has shown that modification of the polar region can offer much more promise. Given that we have already described (vide supra) the region around the polar cluster of both the RAR and RXR LBPs as being a closed (i.e., the highly polar region where crucial carboxylate binding takes place), this initially appears counterintuitive. However, rigorous computational analysis of the structure of RARγ has shown that, depending on the conformation of key polar residue Lys236, a small "back door" is evident at the end of the pocket, and indeed, that has been postulated as a potential unbinding, or escape, route for retinoid ligands (Kosztin, Izrailev, & Schulten, 1999). If this "back door" is large enough to accommodate an entire retinoid structure, one can also propose that it should be large enough for a chemical linker that connects the polar region of a retinoid with a suitable tag moiety that would be positioned outside of the LBP. Indeed this seems to be bourne out in practice, since both Ch55 and Am580 have been appended with a long chemical linker in the 2-position relative to the polar region carboxylate that links these retinoids to a both degradation peptide, (Itoh, Kitaguchi, Ishikawa, Naito, & Hashimoto, 2011) and a biotin motif (Fig. 33) (Fujii, Mori, Kagechika, Mendoza Parra, & Gronemeyer, 2018). Both compounds retained their ability to bind their respective RAR targets, and their tag moieties provided significant additional utility over the retinoids themselves. Crucially, however, compounds that employ polar region modification as a means to introduce a tag seem to be unable to bind the CRABPs, since a suitable "back door" does not exist in these retinoid-binding proteins. Hence, the two approaches for tagging retinoid compounds are complementary to each other, since additional selectivity for each group of retinoid-binding proteins can be introduced (Fig. 34).

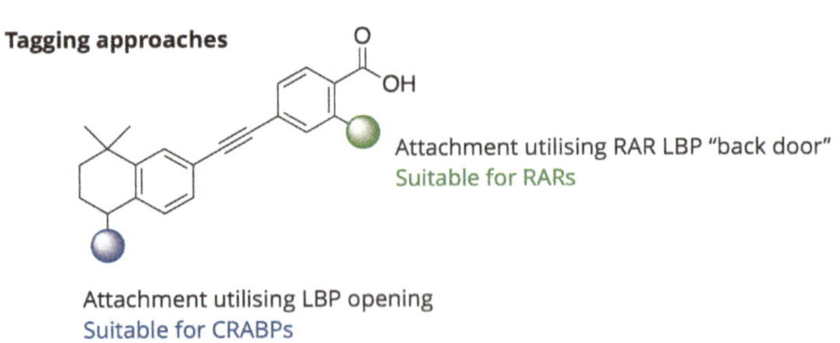

Fig. 34 Design principles for the development of retinoids with chemical tags.

3.2 Fluorescent retinoids

An orthogonal approach to the tagging of chemical recognition motifs is the incorporation of a fluorophore that can be utilized in a variety of imaging, localization or quantification studies. While it is possible to employ similar methods to the linker technology just discussed and add a fluorescent molecule at the end of a chemical linker, there would be considerable risk of changes in the behavior of this fluorescent reporter being unrepresentative of the behavior of the original retinoid at the other end of the linker due to the (often) large distance between the two. The endogenous retinoids are inherently weakly fluorescent due to their conjugated polyene structures, but photoexcitation of these compounds also results in photoisomerization and degradation, meaning utilization of their fluorescence behavior would, therefore, carry inherent unreliability which would be made even worse by low intensity signal (Murayama et al., 1997). Similarly, the majority of synthetic retinoids are also inherently highly conjugated structures and therefore, one can envisage that by careful modification of these scaffolds, it would be possible design a stable, inherently highly fluorescent retinoids that could directly report on the local environment.

Realization of this concept has been exemplified by substituting the TTN hydrophobic region of pan-RAR agonist EC23 for an electron rich dihydroquinoline, the existing donor-acceptor character of the diphenylacetylene structure is enhanced and a significant bathochromic shift in λ_{max} is observed (308–382 nm, in toluene) in the fluorescent retinoid, DC360 (Fig. 35) (Chisholm et al., 2019). The highly solvatochromatic fluorescence emission from this family of fluorescent retinoids (intense in non-polar environments such as the RAR or CRABPII LBPs but significantly quenched in polar environments) has been utilized for cellular

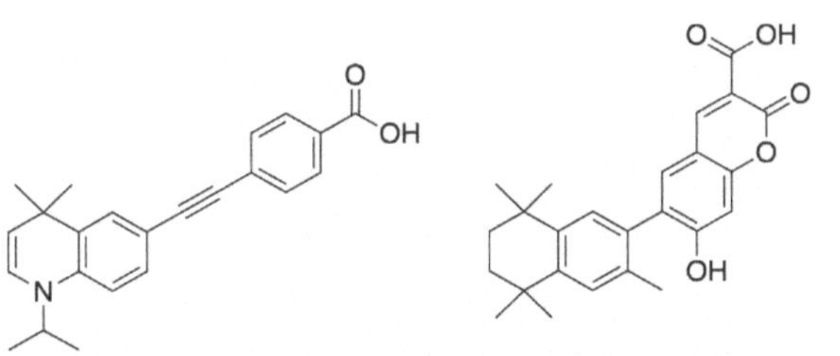

DC360
Fluorescent RAR agonist

CU-6PMN
Fluorescent RXR agonist

Fig. 35 Chemical structures of fluorescent retinoids DC360 and CU-6PMN.

imaging studies and fluorometric binding assays to quantify the binding affinity of other retinoids (Tomlinson, Chisholm, Valentine, Whiting, & Pohl, 2018).

The same molecular design strategy is less suitable when it comes to designing fluorescent RXR agonists due to the need to incorporate a twisted conformation, which is likely to reduce orbital overlap between the π-systems of the hydrophobic and linker/polar regions and, hence, adversely affect fluorescence behavior. Fortunately, however, it has been shown that the widely-utilized fluorophore, 7-hydroxycoumarin, is an ideal mimic of the linker/polar regions of a retinoid and by simply appending this structure to a 3-Me-TTN hydrophobic region a fluorescent RXR agonist, CU-6PMN, is created (Fig. 35) (Yamada et al., 2019). This compound can also be utilized in fluorometric competitive binding assays although, in contrast to DC360, the fluorescence emission of CU-6PMN is quenched when the compound enters a non-polar environment such as that of the RXR binding pocket and, hence, fluorescence enhancement is monitored to quantify binding of a competing ligand, rather than fluorescence reduction in the case of DC360. Fluorescence technology is becoming even more accessible and one can anticipate that these fluorescent retinoid probes can be applied in a wide variety of biophysical experiments that could assist in the further study of retinoid signaling pathways.

4. Summary

This chapter has been devised as a primer to explain the important milestones and models in the development of synthetic retinoids. While by no means exhaustive, it is hoped that the information presented here has underlined the key design concepts in the field. By building on these foundations, we can begin to develop the next generation of more potent, highly selective retinoids. Retinoid signaling pathways clearly exhibit immense biological influence, and the use of carefully designed synthetic retinoids is likely one of the principal means that we can further understand this fascinating area of biology.

References

Allenby, G., Bocquel, M. T., Saunders, M., Kazmer, S., Speck, J., Rosenberger, M., et al. (1993). Retinoic acid receptors and retinoid X receptors: Interactions with endogenous retinoic acids. *Proceedings of the National Academy of Sciences of the United States of America*, *90*, 30–34. Retrieved from http://www.pubmedcentral.nih.gov/articlerender.fcgi?artid=45593&tool=pmcentrez&rendertype=abstract.

Álvarez, S., Khanwalkar, H., Álvarez, R., Erb, C., Martínez, C., Rodríguez-Barrios, F., et al. (2009). C3 halogen and c8'' substituents on stilbene arotinoids modulate retinoic acid receptor subtype function. *ChemMedChem*, *4*, 1630–1640. https://doi.org/10.1002/cmdc.200900214.

Apfel, C., Kamber, M., Klaus, M., Mohr, P., Keidel, S., & LeMotte, P. (1995). Enhancement of HL-60 differentiation by a new class of retinoids with selective activity on retinoid X receptor. *The Journal of Biological Chemistry*, *270*, 30765–30772. Retrieved from http://www.jbc.org/content/270/51/30765.short.

Beard, R. L., Duong, T. T., Teng, M., Klein, E. S., Standevan, A. M., & Chandraratna, R. A. S. (2002). Synthesis and biological activity of retinoic acid receptor-alpha specific amides. *Bioorganic & Medicinal Chemistry Letters*, *12*, 3145–3148. Retrieved from http://www.ncbi.nlm.nih.gov/pubmed/12372520.

Beard, R. L., Gil, D. W., Marler, D. K., Henry, E., Colon, D. F., Gillett, S. J., et al. (1994). Structural basis for the differential RXR & RAR activity of stilbene retinoid analogs. *Bioorganic & Medicinal Chemistry Letters*, *4*, 1447–1452. Retrieved from http://www.sciencedirect.com/science/article/pii/S0960894X01805114.

Beard, R. L., Klein, E. S., Standeven, A. M., Escobar, M., & Chandraratna, R. A. (2001). Phenylcyclohexene and phenylcyclohexadiene substituted compounds having retinoid antagonist activity. *Bioorganic & Medicinal Chemistry Letters*, *11*, 765–768. Retrieved from http://www.ncbi.nlm.nih.gov/pubmed/11277515.

Bernard, B., Bernardon, J., Delescluse, C., Martin, B., Lenoir, M., Maignan, J., et al. (1992). Identification of synthetic retinoids with selectivity for human nuclear retinoic acid receptor gamma. *Biochemical and Biophysical Research Communications*, *186*, 977–983. Retrieved from http://www.sciencedirect.com/science/article/pii/0006291X92908429.

Boehm, M. F., Zhang, L., Badea, B. A., White, S. K., Mais, D. E., Berger, E., et al. (1994). Synthesis and structure-activity relationships of novel retinoid X receptor-selective retinoids. *Journal of Medicinal Chemistry*, *37*, 2930–2941. Retrieved from http://www.ncbi.nlm.nih.gov/pubmed/8071941.

Boehm, M. F., Zhang, L., Zhi, L., McClurg, M. R., Berger, E., Wagoner, M., et al. (1995). Design and synthesis of potent retinoid X receptor selective ligands that induce apoptosis in leukemia cells. *Journal of Medicinal Chemistry*, *38*, 3146–3155. Retrieved from http://www.ncbi.nlm.nih.gov/pubmed/7636877.

Boerma, L. J., Xia, G., Qui, C., Cox, B. D., Chalmers, M. J., Smith, C. D., et al. (2014). Defining the communication between agonist and coactivator binding in the retinoid x receptor α ligand binding domain. *The Journal of Biological Chemistry*, *289*, 814–826. https://doi.org/10.1074/jbc.M113.476861.

Bourguet, W., Ruff, M., Chambon, P., Gronemeyer, H., & Moras, D. (1995). Crystal structure of the ligand-binding domain of the human nuclear receptor RXR-α. *Nature*, *375*, 377–382.

Bourguet, W., Vivat, V., Wurtz, J., Chambon, P., Gronemeyer, H., & Moras, D. (2000). Crystal structure of a Heterodimeric complex of RAR and RXR ligand-binding domains. *Molecular Cell*, *5*, 289–298.

Chambon, P. (1996). A decade of molecular biology of retinoic acid receptors. *The FASEB Journal*, *10*, 940–954. Retrieved from http://www.fasebj.org/content/10/9/940.short.

Chisholm, D. R., Tomlinson, C. W. E., Zhou, G.-L., Holden, C., Affleck, V., Lamb, R., et al. (2019). Fluorescent retinoic acid analogues as probes for biochemical and intracellular characterization of retinoid signaling pathways. *ACS Chemical Biology*, *14*, 369–377. https://doi.org/10.1021/acschembio.8b00916.

Christie, V. B., Barnard, J. H., Batsanov, A. S., Bridgens, C. E., Cartmell, E. B., Collings, J. C., et al. (2008). Synthesis and evaluation of synthetic retinoid derivatives as inducers of stem cell differentiation. *Organic & Biomolecular Chemistry*, *6*, 3497–3507. https://doi.org/10.1039/b808574a.

Clemens, G., Flower, K. R., Gardner, P., Henderson, A. P., Knowles, J. P., Marder, T. B., et al. (2013). Design and biological evaluation of synthetic retinoids: Probing length vs. stability vs. activity. *Molecular BioSystems*, *9*, 3124–3134. https://doi.org/10.1039/c3mb70273a.

Eberhardt, J., Mcewen, A. G., Bourguet, W., Morasa, D., & Dejaegerea, A. (2019). A revisited version of the apo structure of the ligand-binding domain of the human nuclear receptor retinoic x receptor a. *Acta Crystallographica Section F: Structural Biology Communications*, *75*(2), 98–104. https://doi.org/10.1107/S2053230X18018022.

Ebisawa, M., Umemiya, H., Ohta, K., Fukasawa, H., Kawachi, E., Christoffel, G., et al. (1999). Retinoid X receptor-antagonistic diazepinylbenzoic acids. *Chemical & Pharmaceutical Bulletin*, *47*, 1778–1786. Retrieved from http://www.ncbi.nlm.nih.gov/pubmed/10748721.

Egea, P. F., Mitschler, A., Rochel, N., Ruff, M., Chambon, P., & Moras, D. (2000). Crystal structure of the human RXRalpha ligand-binding domain bound to its natural ligand: 9-cis retinoic acid. *The EMBO Journal*, *19*, 2592–2601. https://doi.org/10.1093/emboj/19.11.2592.

Fujii, S., Mori, S., Kagechika, H., Mendoza Parra, M. A., & Gronemeyer, H. (2018). Development of biotin-retinoid conjugates as chemical probes for analysis of retinoid function. *Bioorganic & Medicinal Chemistry Letters*, *28*(14), 2442–2445. https://doi.org/10.1016/j.bmcl.2018.06.011.

Gambone, C. J., Hutcheson, J. M., Gabriel, J. L., Beard, R. L., Chandraratna, R. A. S., Soprano, K. J., et al. (2002). Unique property of some synthetic retinoids: Activation of the aryl hydrocarbon receptor pathway. *Molecular Pharmaceutics*, *61*(2), 334–342. https://doi.org/10.1124/mol.61.2.334.

Germain, P., Gaudon, C., Pogenberg, V., Sanglier, S., Van Dorsselaer, A., Royer, C., et al. (2009). Differential action on coregulator interaction defines inverse retinoid agonists and neutral antagonists. *Chemistry & Biology*, *16*, 479–489. https://doi.org/10.1016/j.chembiol.2009.03.008.

Germain, P., Kammerer, S., Pérez, E., Peluso-Iltis, C., Tortolani, D., Zusi, F. C., et al. (2004). Rational design of RAR-selective ligands revealed by RARbeta crystal stucture. *EMBO Reports*, *5*, 877–882. https://doi.org/10.1038/sj.embor.7400235.

Haffez, H., Chisholm, D. R., Tatum, N. J., Valentine, R., Redfern, C., Pohl, E., et al. (2018). Probing biological activity through structural modelling of ligand-receptor interactions of 2,4-disubstituted thiazole retinoids. *Bioorganic & Medicinal Chemistry*, *26*, 1560–1572. https://doi.org/10.1016/j.bmc.2018.02.002.

Haffez, H., Chisholm, D. R., Valentine, R., Pohl, E., Redfern, C. P. F., & Whiting, A. (2017). The molecular basis of the interactions between synthetic retinoic acid analogues and the retinoic acid receptors. *Medicinal Chemistry Communications*, *8*, 578–592.

Huang, P., Chandra, V., & Rastinejad, F. (2014). Retinoic acid actions through mammalian nuclear receptors. *Chemical Reviews*, *114*(1), 233–254. https://doi.org/10.1021/cr400161b.

Itoh, Y., Ishikawa, M., Naito, M., & Hashimoto, Y. (2010). Protein knockdown using methyl bestatin-ligand hybrid molecules: Design and synthesis of inducers of ubiquitination-mediated degradation of cellular retinoic acid-binding proteins. *Journal of the American Chemical Society*, *132*(16), 5820–5826. https://doi.org/10.1021/ja100691p.

Itoh, Y., Kitaguchi, R., Ishikawa, M., Naito, M., & Hashimoto, Y. (2011). Design, synthesis and biological evaluation of nuclear receptor-degradation inducers. *Bioorganic & Medicinal Chemistry*, *19*(22), 6768–6778. https://doi.org/10.1016/j.bmc.2011.09.041.

Johnson, A., Klein, E., & Wang, L. (1996). Identification of retinoic acid receptor β subtype specific agonists. *Journal of Medicinal Chemistry*, *39*(26), 5027–5030. Retrieved from http://pubs.acs.org/doi/pdf/10.1021/jm960687r.

Johnson, A. T., Wang, L., Standeven, A. M., Escobar, M., & Chandraratna, R. A. S. (1999). Synthesis and biological activity of high-affinity retinoic acid receptor antagonists. *Bioorganic & Medicinal Chemistry*, 7(7), 1321–1338. Retrieved from http://www.ncbi.nlm.nih.gov/pubmed/10465407.

Klaholz, B. P., Mitschler, A., Belema, M., Zusi, C., & Moras, D. (2000). Enantiomer discrimination illustrated by high-resolution crystal structures of the human nuclear receptor hRARgamma. *Proceedings of the National Academy of Sciences of the United States of America*, *97*(12), 6322–6327. Retrieved from http://www.pubmedcentral.nih.gov/articlerender.fcgi?artid=18601&tool=pmcentrez&rendertype=abstract.

Klaholz, B. P., Mitschler, A., & Moras, D. (2000). Structural basis for isotype selectivity of the human retinoic acid nuclear receptor. *Journal of Molecular Biology*, *302*(1), 155–170. https://doi.org/10.1006/jmbi.2000.4032.

Klaholz, B., & Moras, D. (2002). C-H…O hydrogen bonds in the nuclear receptor RARgamma–a potential tool for drug selectivity. *Structure*, *10*(9), 1197–1204. Retrieved from http://www.ncbi.nlm.nih.gov/pubmed/12220491.

Klaholz, B. P., Renaud, J. P., Mitschler, A., Zusi, C., Chambon, P., Gronemeyer, H., et al. (1998). Conformational adaptation of agonists to the human nuclear receptor RAR gamma. *Nature Structural Biology*, *5*, 199–202.

Koch, S. S. C., Dardashti, L. J., Hebert, J. J., White, S. K., Croston, G. E., Flatten, K. S., et al. (1996). Identification of the first retinoid X receptor homodimer antagonist. *Journal of Medicinal Chemistry*, *39*(17), 3229–3234. https://doi.org/10.1021/jm960311d.

Kosztin, D., Izrailev, S., & Schulten, K. (1999). Unbinding of retinoic acid from its receptor studied by steered molecular dynamics. *Biophysical Journal*, *76*, 188–197. https://doi.org/10.1016/S0006-3495(99)77188-2.

le Maire, A., Teyssier, C., Erb, C., Grimaldi, M., Alvarez, S., de Lera, A. R., et al. (2010). A unique secondary-structure switch controls constitutive gene repression by retinoic acid receptor. *Nature Structural & Molecular Biology*, *17*(7), 801–807. https://doi.org/10.1038/nsmb.1855.

Loeliger, P., Bollag, W., & Mayer, H. (1980). Arotinoids: A new class of highly active retinoids. *European Journal of Medicinal Chemistry*, *15*, 9–15.

Love, J. D., Gooch, J. T., Benko, S., Li, C., Nagy, L., Chatterjee, V. K. K., et al. (2002). The structural basis for the specificity of retinoid-X receptor-selective agonists: New insights into the role of helix H12. *The Journal of Biological Chemistry*, *277*(13), 11385–11391. https://doi.org/10.1074/jbc.M110869200.

Majumdar, A., Petrescu, A. D., Xiong, Y., & Noys, N. (2011). Nuclear translocation of cellular retinoic acid-binding protein II is regulated by retinoic acid-controlled SUMOylation. *The Journal of Biological Chemistry*, *286*, 42749–42757. https://doi.org/10.1074/jbc.M111.293464.

Murayama, A., Suzuki, T., & Matsui, M. (1997). Photoisomerization of retinoic acids in ethanol under room light: A warning for cell biological study of geometrical isomers of retinoids. *Journal of Nutritional Science and Vitaminology*, *43*, 167–176. https://doi.org/10.3177/jnsv.43.167.

Nadendla, E., Teyssier, C., Delfosse, V., Vivat, V., Krishnasamy, G., Gronemeyer, H., et al. (2015). An unexpected mode of binding defines BMS948 as a full retinoic acid receptor β(RARβ, NR1B2) selective agonist. *PLoS One*, *10*, 1–20. https://doi.org/10.1371/journal.pone.0123195.

Nahoum, V., Pérez, E., Germain, P., Rodríguez-Barrios, F., Manzo, F., Kammerer, S., et al. (2007). Modulators of the structural dynamics of the retinoid X receptor to reveal receptor function. *Proceedings of the National Academy of Sciences of the United States of America*, *104*(44), 17323–17328. https://doi.org/10.1073/pnas.0705356104.

Okuhira, K., Ohoka, N., Sai, K., Nishimaki-Mogami, T., Itoh, Y., Ishikawa, M., et al. (2011). Specific degradation of CRABP-II via cIAP1-mediated ubiquitylation induced by hybrid molecules that crosslink cIAP1 and the target protein. *FEBS Letters*, *585*(8), 1147–1152. https://doi.org/10.1016/j.febslet.2011.03.019.

Ostrowski, J., Roalsvig, T., Hammer, L., Marinier, A., Starrett, J. E., Yu, K. L., et al. (1998). Serine 232 and methionine 272 define the ligand binding pocket in retinoic acid receptor subtypes. *The Journal of Biological Chemistry*, *273*(6), 3490–3495. Retrieved from http://www.ncbi.nlm.nih.gov/pubmed/9452473.

Renaud, J. P., Rochel, N., Ruff, M., Vivat, V., Chambon, P., Gronemeyer, H., et al. (1995). Crystal structure of the RARγ ligand-binding domain bound to all-trans retinoic acid. *Nature*, *378*, 681–689.

Roberts, A. B., Nichols, M. D., Newton, D. L., & Sporn, M. B. (1979). In vitro metabolism of retinoic acid in hamster intestine and liver. *The Journal of Biological Chemistry*, *254*(14), 6296–6302. Retrieved from http://www.ncbi.nlm.nih.gov/pubmed/447713.

Rühl, R., Krzyżosiak, A., Niewiadomska-Cimicka, A., Rochel, N., Szeles, L., Vaz, B., et al. (2015). 9-cis-13,14-dihydroretinoic acid is an endogenous retinoid acting as RXR ligand in mice. *PLoS Genetics*, *11*, 1–16. https://doi.org/10.1371/journal.pgen.1005213.

Santin, E. P., Germain, P., Quillard, F., Khanwalkar, H., Rodriguez-Barrios, F., Gronemeyer, H., et al. (2009). Modulating retinoid X receptor with a series of (E)-3-[4-Hydroxy-3-(3-alkoxy-5,5,8,8-tetramethyl-5,6,7,8-tetrahydronaphthalen-2-yl) phenyl]acrylic acids and their 4-Alkoxy isomers. *Journal of Medicinal Chemistry*, *17*, 3150–3158.

Sato, Y., Ramalanjaona, N., Huet, T., Potier, N., Osz, J., Antony, P., et al. (2010). The "Phantom Effect" of the Rexinoid LG100754: Structural and functional insights. *PLoS One*, *5*(11). e15119https://doi.org/10.1371/journal.pone.0015119.

Sun, S.-Y., Yue, P., Dawson, M. I., Shroot, B., Michel, S., Lamph, W. W., et al. (1997). Differential effects of synthetic nuclear retinoid receptor-selective retinoids on the growth of human non-small cell lung carcinoma cells. *Cancer Research*, *57*(21), 4931–4939.

Takamatsu, K., Takano, A., Yakushiji, N., Morishita, K., Matsuura, N., Makishima, M., et al. (2008). Reduction of lipophilicity at the lipophilic domain of RXR agonists enables production of subtype preference: RXRalpha-preferential agonist possessing a sulfonamide moiety. *ChemMedChem*, *3*, 454–460.

Takamatsu, K., Takano, A., Yakushiji, N., Morohashi, K., Morishita, K., Matsuura, N., et al. (2008). The first potent subtype-selective retinoid X receptor (RXR) agonist possessing a 3-Isopropoxy-4-isopropylphenylamino moiety, NEt-3IP (RXRα/β-dual agonist). *ChemMedChem*, *3*, 780–787.

Teng, M., Duong, T. T., Johnson, A. T., Klein, E. S., Wang, L., Khalifa, B., et al. (1997). Identification of highly potent retinoic acid receptor alpha-selective antagonists. *Journal of Medicinal Chemistry*, *40*, 2445–2451. https://doi.org/10.1021/jm9703911.

Teng, M., Duong, T. T., Klein, E. S., Pino, M. E., & Chandraratna, R. A. S. (1996). Identification of a retinoic acid receptor α subtype specific agonist. *Journal of Medicinal Chemistry*, *39*, 3035–3038. Retrieved from http://pubs.acs.org/doi/pdf/10.1021/jm9603532.

Tomlinson, C. W. E., Chisholm, D. R., Valentine, R., Whiting, A., & Pohl, E. (2018). Novel fluorescence competition assay for retinoic acid binding proteins [brief-report]. *ACS Medicinal Chemistry Letters*, *9*, 1297–1300. https://doi.org/10.1021/acsmedchemlett.8b00420.

Tsang, K., Sinha, S., Liu, X., Bhat, S., & Chandraratna, R. (2005). *Disubstituted chalcone oximes having RAR retinoid receptor antagonist activity*, Patent no. US20080097119A1.

Wagner, C. E., Jurutka, P. W., Marshall, P. A., Groy, T. L., van der Vaart, A., Ziller, J. W., et al. (2009). Modeling, synthesis and biological evaluation of potential retinoid X receptor (RXR) selective agonists: Novel analogues of 4-[1-(3,5,5,8,8-pentamethyl-5,6,7,8-tetrahydro-2-naphthyl)ethynyl]benzoic acid (bexarotene). *Journal of Medicinal Chemistry*, *52*(19), 5950–5966. https://doi.org/10.1021/jm900496b.

Yamada, S., Kawasaki, M., Fujihara, M., Watanabe, M., Takamura, Y., Takioku, M., et al. (2019). Competitive binding assay with an umbelliferone-based fluorescent Rexinoid for retinoid X receptor ligand screening [research-article]. *Journal of Medicinal Chemistry*, *62*, 8809–8818. https://doi.org/10.1021/acs.jmedchem.9b00995.

Yu, K.-L., Spinazze, P., Ostrowski, J., Currier, S. J., Pack, E. J., Hammer, L., et al. (1996). Retinoic acid receptor β,γ-selective ligands: Synthesis and biological activity of 6-substituted 2-naphthoic acid retinoids. *Journal of Medicinal Chemistry*, *39*(12), 2411–2421. https://doi.org/10.1021/jm9502293.

CHAPTER EIGHTEEN

Application of human organotypic skin raft cultures for analysis of retinoid metabolism, retinoic acid signaling, and screening of bioactive rexinoids

Alla V. Klyuyeva, Kelli R. Goggans, Natalia Y. Kedishvili, Olga V. Belyaeva*

Department of Biochemistry and Molecular Genetics, Schools of Medicine and Dentistry, University of Alabama at Birmingham, Birmingham, AL, United States

*Corresponding author: e-mail address: obeliaeva@uab.edu

Contents

Methods in Enzymology, Volume 637
ISSN 0076-6879
https://doi.org/10.1016/bs.mie.2020.02.013

Abstract

Several human enzymes of the short-chain dehydrogenase/reductase (SDR) superfamily of proteins exhibit catalytic oxidoreductive activity toward retinoid substrates *in vitro*. For some retinoid-active enzymes, their physiological significance for retinoid metabolism is supported by phenotypes linked to naturally occurring mutations in human genes or by targeted gene knockout studies of their murine homologs. However, for those enzymes that are not well conserved or display properties different from their murine counterparts, evaluation of their physiological roles can be challenging. Here, we describe the adaptation of stratified organotypic culture of human epidermis for evaluating the contribution of human putative SDR retinol dehydrogenases to biosynthesis of all-*trans*-retinoic acid in a three-dimensional cellular model highly sensitive to the levels of all-*trans*-retinol and all-*trans*-retinoic acid. In addition to providing a valuable readout of metabolic changes occurring in the presence or absence of the enzyme of interest, this model allows for evaluation of the effects of various retinoid and rexinoid therapeutic compounds on retinoic acid signaling, cell growth and differentiation.

1. Introduction

Biosynthesis of all-*trans*-retinoic acid (RA) from dietary vitamin A is enzymatically controlled. Multiple enzymes have been shown to catalyze the interconversion of retinol and retinaldehyde *in vitro* and in cell culture, while only few have been proven to affect the levels of RA *in vivo* (Belyaeva, Adams, Popov, & Kedishvili, 2019). Therefore, after establishing an enzyme's activity toward retinoid compounds *in vitro*, the necessary next step is to determine its physiological relevance for RA biosynthesis and RA signaling using *in vivo* and *ex vivo* models.

For murine enzymes, whole body targeted gene knockouts and conditional tissue specific gene knockout mouse models provided a wealth of information about their physiological roles. However, many human enzymes have properties and substrate specificities that are notably different from their mouse homologs. As a result, in some cases, gene knockouts of mouse enzymes do not faithfully reproduce the effects of loss-of-function mutations in human enzymes.

Recent years have seen an expansion of organotypic models that allow the investigation of human proteins in a physiological setting approximating the complex conditions of human tissue. We have found the organotypic culture of human epidermis to be a highly useful model in evaluating the roles of previously uncharacterized enzymes in retinoid metabolism, and specifically, in retinoic acid (RA) biosynthesis. Unlike human cell culture models, the organotypic skin raft culture allows for monitoring the effects of the enzymes' activities on the growth, differentiation and stratification of human epidermis. In addition to providing a valuable readout of metabolic changes occurring in the presence or absence of the enzyme of interest, these cultures allow for visualization of changes in the three dimensional architecture of the epidermis.

The technique for production of organotypic skin equivalents has been originally developed to study approaches to skin transplantation and problems of graft rejection (Asselineau, Bernard, Bailly, Darmon, & Pruniéras, 1986; Bell et al., 1983). Since then it has been applied more broadly to study the processes of epithelial differentiation and keratinization (Kopan, Traska, & Fuchs, 1987; Wilson, Dollard, Chow, & Broker, 1992), pathogenesis of psoriasis (Roy et al., 2018; Tjabringa et al., 2008), melanoma (Banerjee et al., 2004), skin injury and wound healing (Collawn, Banerjee, de la Torre, Vasconez, & Chow, 2012), and skin aging (Pageon et al., 2015). This technique was also used for the genetic dissection of pathogenic events and virus-host interaction associated with human papillomavirus infection (Cheng, Schmidt-Grimminger, Murant, Broker, & Chow, 1995; Chow, 2015; Chow & Broker, 1997), and for a robust recapitulation of the full infectious cycle of human papillomavirus (Wang, Broker, & Chow, 2015; Wang, Duffy, Broker, & Chow, 2009).

The approach takes advantage of the ability of primary human keratinocytes to differentiate into stratified epidermis when grown on dermal equivalents, which are later placed on metal support grids to form "rafts" at the liquid-air interface. Keratinocytes grown as submerged monolayer culture can show only limited differentiation when exposed to higher calcium or serum concentration; in organotypic cultures, however, primary keratinocytes undergo orderly stratification and differentiation to form a model of squamous epithelium.

The generation of organotypic skin substitutes requires two components: dermal equivalent and primary keratinocytes. Dermal equivalents are typically made of fibroblasts grown in rat type I collagen matrix. Both rodent and human fibroblast cells have been used to create dermal equivalents, with some reports indicating that human organotypic cultures grown on rodent fibroblast-containing collagen matrices show better resemblance to the human epithelial tissues from which keratinocytes originated (Wilson et al., 1992). Dispersed keratinocytes isolated from adult skin, neonatal foreskin and cervix have been successfully used as a source of primary epithelial cells. Alternatively, biopsies of these epithelial tissues can be directly explanted on submerged dermal equivalents. In the latter case, keratinocytes first migrate from the explant to form a cell monolayer, and then differentiate after being transferred onto supportive grids at the liquid-air interface (Chow & Broker, 1997). Established cell lines, which are generally selected for proliferative capacity, and immortalized keratinocytes exhibit impaired differentiation ability and may not be suitable for the generation of stratified epidermal structure. However, mixed organotypic cultures developed from primary foreskin keratinocytes and melanoma cell line have been used to study specificity of melanoma-targeting viruses (Banerjee et al., 2004).

While premade organotypic skin equivalents can now be procured from commercial sources, some studies require a manipulation of primary keratinocytes prior to seeding them onto dermal equivalent, or treatment of keratinocytes with various agents at different stages of stratification. For such experiments, the ability to grow organotypic skin raft cultures in-house is indispensable. Here, we describe two applications where this technique enables investigators to manipulate the expression of human genes and investigate the effects of various compounds on growth and differentiation of human epidermis.

2. Analysis of compounds modulating retinoic acid signaling

RA serves as a ligand for nuclear transcription factors, retinoic acid receptors (RARs). RARs function as heterodimers with retinoid X receptors (RXRs). Organotypic skin raft culture proved to be a useful model for analysis of RA signaling, because RA controls the proliferation and differentiation of keratinocytes and maintenance of epithelial tissue (Sorg & Saurat, 2014). Treatment of skin rafts with RAR agonists (retinoids) or RXR agonists (rexinoids) modulates the expression of RAR and RXR responsive genes (Lee, Belyaeva, Wu, & Kedishvili, 2011; Wu et al., 2016), which can be readily assessed by quantitative RT-PCR, and leads to characteristic changes in morphology, which can be detected using histological staining and immunohistochemical analysis.

Organotypic raft cultures of normal foreskin keratinocytes grown on the medium supplemented with 2 μM all-*trans*-retinol, the metabolic precursor of RA, show an upregulation of RA-inducible genes and display a proliferative phenotype (Lee et al., 2011) (Fig. 1A). Control raft cultures contain well-defined basal, spinous, granular and cornified layers, resembling the structure of normal epidermis. Cultures grown in the presence of

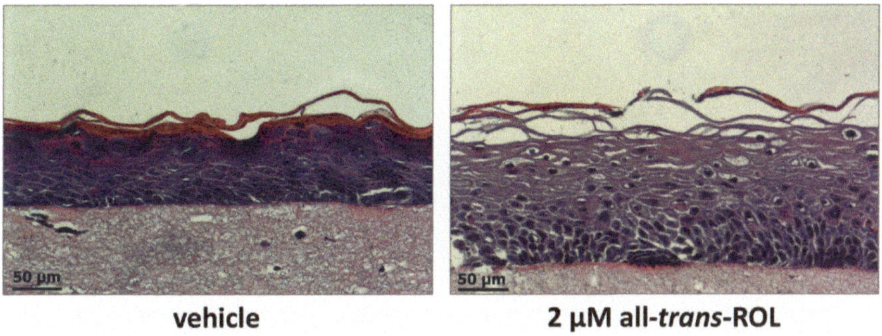

Fig. 1 Effect of all-*trans*-retinol treatment on proliferation and differentiation of organotypic skin cultures. Organotypic raft cultures developed from normal human foreskin keratinocytes were grown on medium supplemented with DMSO vehicle or 2 µM all-*trans*-retinol (ROL) continuously from the day following transfer to the air-liquid interface. Cultures were grown for 10 days, fixed, embedded in paraffin and sectioned. 6-µm sections were stained by hematoxylin and eosin (H&E). In contrast to the vehicle-treated rafts, all-*trans*-retinol—treated cultures display almost complete absence of the cornified layer or well-defined granular layer, with a thicker layer of proliferating basal keratinocytes, illustrative of the effect of retinoids on normal skin.

exogenously supplemented 2 µM retinol display an increased thickness of the basal and spinous layers, and a reduction of the granular and cornified layers (Fig. 1). This morphology is indicative of the increased proliferation and decreased differentiation, the characteristic features resulting from retinoid treatment of normal skin (Fisher & Voorhees, 1996). Increased proliferation in retinol-treated rafts can be further confirmed by incubation of skin rafts with thymidine analog 5-bromo-2′-deoxyuridine (BrdU), to label the nuclei of dividing cells prior to harvesting with subsequent immunohistochemical detection using BrdU antibody (Chien, Parker, Schmidt-Grimminger, Broker, & Chow, 2000; Lee et al., 2011). Impaired differentiation can be demonstrated by reduced immunostaining for late differentiation markers, such as filaggrin (Lee et al., 2011).

The effects of rexinoids (selective RXR-agonists) on human epidermis can be illustrated by rafts grown on the medium supplemented with 0.2 µM bexarotene (Targretin), the only RXR-agonist currently approved by FDA for clinical use (Fig. 2). Histological staining identifies a thicker basal layer in bexarotene-treated rafts in comparison to controls. Immunohistochemical staining of BrdU-treated rafts reveals an increased incorporation of the label and multiple layers of positive nuclei, in contrast to control rafts, where proliferating cells typically reside in the most basal layer.

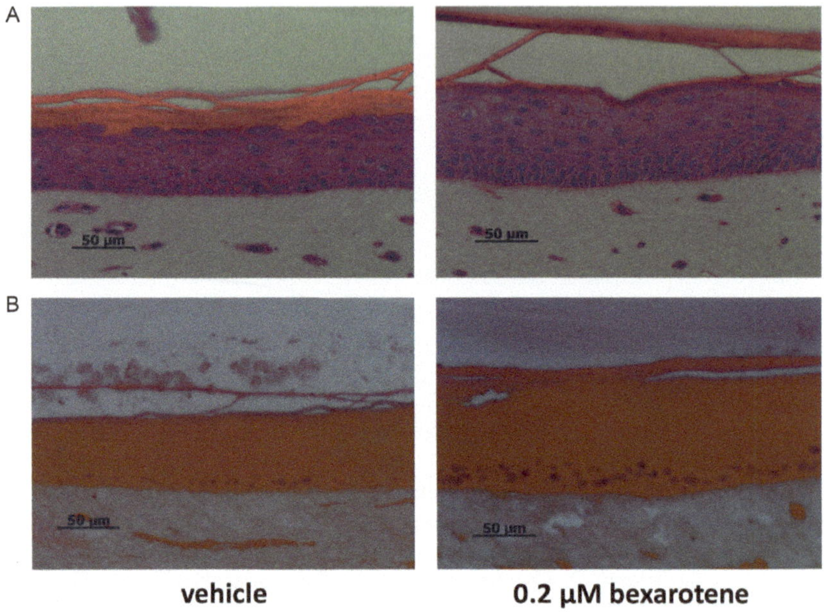

Fig. 2 Effect of RXR-agonist (bexarotene) treatment on proliferation and differentiation of organotypic skin cultures. Organotypic raft cultures developed from normal human foreskin keratinocytes were grown on medium supplemented with DMSO vehicle or 0.2 μM bexarotene continuously from the day following transfer to the air-liquid interface. Cultures were grown for 10 days; BrdU was added to the medium 12 h before harvesting to the final concentration of 50 μg/mL. Cultures were fixed, embedded in paraffin and sectioned. 6-μm sections were stained by H&E (A) or immunohistochemically for BrdU to identify cells in the S-phase (panel B). BrdU incorporation was visualized by staining with mouse anti-BrdU antibody (Invitrogen) at a 1:200 dilution followed by the incubation with 1:50 dilution of anti-rabbit, anti-mouse secondary antibody and 1:50 dilution of streptavidin-conjugated horseradish peroxidase from Super Sensitive Link-Label IHC detection kit (Biogenex, San Ramon, CA) and with Turbo 3,3′-diaminobenzidine tetrahydrochloride substrate (Innovex Biosciences, Richmond, CA).

We have applied the organotypic skin culture approach to compare the potency of a series of type I and type II rexinoids (Atigadda et al., 2015), which are being developed for use as therapeutic and cancer chemopreventive agents. For example, treatment of organotypic skin rafts with UAB30, a type II rexinoid currently undergoing clinical trials, revealed dose-dependent changes in the histology of rafts (Wu et al., 2016). Increasing concentrations of UAB30 added to the medium at the time of transfer of cultures to air-liquid interface lead to the progressive loss of cornified layer, poorly defined granular layer, and enhanced proliferation. The quantitative analysis of gene expression in rafts treated with the same concentration of UAB30 and bexarotene allowed to compare the potency of UAB30 with that of the clinically approved bexarotene, and to determine that in human epidermis UAB30 specifically targets the genes regulated by RXR/RAR heterodimers (Wu et al., 2016).

Similarly to animal tissues, the retinoid content of organotypic epithelium can be analyzed using HPLC or tandem HPLC-MS methods to corroborate the changes in retinoid metabolism inferred from the analysis of gene expression. Prior to the analysis of retinoid content samples of organotypic epidermis should be peeled off the underlying dermal equivalent, because fibroblasts in the collagen matrix also retain the ability to metabolize retinoids. Organotypic skin rafts efficiently metabolize all-*trans*-retinol present in the serum-supplemented growth medium, and, although the endogenous content of retinoids in these cultures is generally low, the retinyl esters content can be determined by HPLC in a pooled sample of four to five rafts. The same rafts can be used for detection of RA using a more sensitive LC-MS-MS method. For example, qPCR analysis indicated an upregulation of the genes responsible for the uptake and metabolism of all-*trans*-retinol to RA in UAB30-treated rafts (Wu et al., 2016). Consistent with the observed changes in gene expression, analysis of retinoid content revealed an increase in retinyl ester stores and RA in UAB30-treated rafts *versus* vehicle-treated control samples.

3. Functional analysis of proteins involved in retinoid metabolism

In addition to exogenous treatment with RAR and RXR-agonists, organotypic epidermal cultures can be used to dissect biosynthetic pathways generating endogenous agonists. To evaluate the contribution of enzymes to retinoid conversion, prior to seeding on dermal equivalents primary keratinocytes can be manipulated to overexpress the enzyme of interest, or to silence its expression. While chemical-based transfection is generally inefficient for primary keratinocytes, extensive studies by the laboratory of Dr. Louise Chow (University of Alabama at Birmingham) have generated optimized procedures for efficient transduction of neonatal foreskin keratinocytes with gene expression cassettes using recombinant amphotropic retroviruses (Banerjee, Chow, & Broker, 2005; Chow & Broker, 1997).

The coding sequences of the enzymes of interest are cloned into Moloney murine leukemia virus-derived vectors, such as pBabe, and the resulting constructs are used to transfect Bosc-23 packaging cell line. Bosc-23 cells are easily electroporated or transfected with high efficiency using lipid-based reagents, such as Lipofectamine series (Invitrogen); however, the retrovirus packaged by these cells is ecotropic and has a low titer. Therefore, the virus is next amplified using amphotropic packaging GP + envAM-12 cells, which are selected on appropriate antibiotic to generate a high-titer retrovirus capable of infecting human primary keratinocytes with high efficiency. After brief antibiotic selection, the transduced keratinocytes are expanded and seeded on dermal equivalents, which are later transferred onto metal support grids to generate organotypic skin rafts. The expression of the gene of interest in retrovirus-transduced cultures can be confirmed by Western blot analysis of total cleared lysates of organotypic epidermis.

We used this approach to evaluate the contribution of short-chain dehydrogenases/reductases (SDRs) to RA biosynthesis (Lee et al., 2011; Wu et al., 2016). Similarly to rafts generated from unmodified keratinocytes, the enzyme-expressing cultures can be analyzed for differences in histology, gene expression and products of retinoid metabolic pathways. The histology of organotypic skin rafts overexpressing human retinol dehydrogenase 10 (RDH10) (Fig. 3) mimicked that of the rafts grown on the medium supplemented with all-*trans*-retinol (Fig. 1) or RA (not shown).

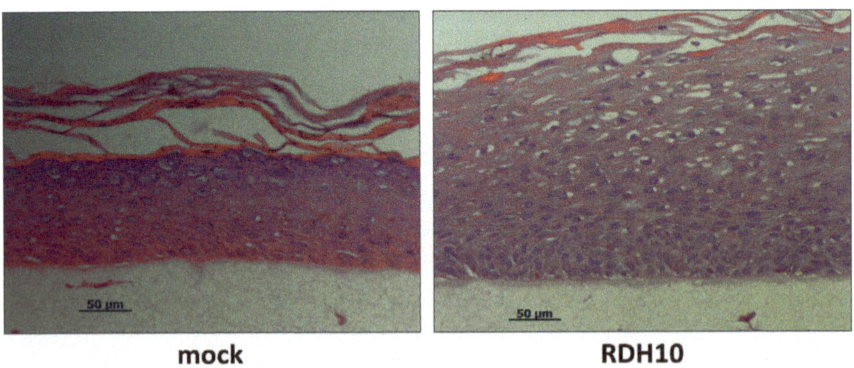

Fig. 3 Effect of overexpression of RDH10 in organotypic skin cultures. Primary human foreskin keratinocytes were infected with a retrovirus carrying the expression cassette for human RDH10 enzyme and puromycin resistance, or puromycin resistance only. Infected cells were selected by puromycin resistance and developed into organotypic skin raft cultures. After 10 days of incubation at the air-liquid interface, raft cultures were fixed, embedded in paraffin and sectioned. 6-µm sections were stained by H&E. Changes in histology of RDH10-expressing cultures closely resembled those induced by exogenously supplied all-*trans*-retinol (Fig. 1), and were consistent with the upregulation of RA signaling pathway (Lee et al., 2011).

Analysis of gene expression revealed an upregulation of RA signaling in RDH10-expressing skin rafts compared to mock-transduced skin rafts. These results provided support for the function of RDH10 as a retinol dehydrogenase, which contributes to RA biosynthesis in human tissue environment.

To improve the detection of retinoid metabolites, such as retinyl esters and RA, and to facilitate comparison between the enzyme-expressing and control cultures, the concentration of metabolites can be increased by treatment of the rafts with all-*trans*-retinol 1 day before harvesting.

4. Materials, equipment and reagents

4.1 Primary human keratinocytes

Neonatal foreskins were obtained from the Newborn Nursery of the University of Alabama at Birmingham Hospital in compliance with University of Alabama at Birmingham Institutional Review Board (IRB) regulations. As determined by the institutional IRB, the use of discarded unidentifiable foreskin tissue met the requirements for an exemption from IRB approval. Keratinocytes were isolated as described previously (Banerjee et al., 2005). Stocks prepared from early passages of keratinocytes (passage 0 or 1) were frozen in serum-free culture medium with 10% DMSO and stored in liquid nitrogen.

4.2 Cell lines

4.2.1 Bosc-23 ecotropic packaging cell line (ATCC CRL-11270)
4.2.2 GP + envAM-12 amphotropic packaging cell line (ATCC CRL-9641)
4.2.3 3T3-J2 mouse fibroblasts

4.3 Media, medium components and transfection reagents

4.3.1 Dulbecco modified Eagle medium (DMEM) (12800-017, Gibco).
4.3.2 Ham's F12 medium (21700-075, Gibco).
4.3.3 Keratinocyte serum-free medium (SFM) (DermaLife-K, complete kit, LL-0007, Lifeline Cell Technology).
4.3.4 Opti-MEM transfection medium (22600-050, Gibco).
4.3.5 Sterile Dulbecco phosphate-buffered saline (D-PBS) (21600-044, Gibco).
4.3.6 Fetal bovine serum (FBS) (S11150, Atlanta Biologicals).

4.3.7 Bovine calf serum (BCS) (16170-060 Gibco).

4.3.8 Collagen type I, from rat tail (354236, Corning, Corning NY).

4.3.9 Trypsin-EDTA solution, 0.25% (25200-056, Life Technologies).

4.3.10 Penicillin/streptomycin, 10,000 U/mL (15140-122, Life Technologies).

4.3.11 Lipofectamine 2000 (11668-019, Invitrogen).

4.3.12 Polybrene (hexadimethrine bromide) (Sigma-Aldrich). Prepare 3 mg/mL solution in PBS and sterilize through 0.2 μM syringe filter. Store at 4 °C.

4.3.13 Puromicin (P8833—10 mg, Sigma-Aldrich). Dissolve in 10 mL of water to 1 mg/mL, sterilize through 0.2 μM syringe filter and store in frozen aliquots.

4.3.14 Raft culture medium supplements:

Human epidermal growth factor (EGF) (PHG0315, 25 mcg, Gibco). Reconstitute in 5 mL of sterile water to make 5 μg/mL (10,000 ×) stock.

Cholera toxin (C8062—0.5 mg, Sigma-Aldrich). Reconstitute 0.5 mg in 1 mL of sterile water to make 5.75 mM storage solution. To make 10,000 × stock, dilute with sterile water to 1 mM. Store at 4 °C.

Insulin (I1882–100 mg, Sigma-Aldrich). Dissolve in 20 mL of sterile water acidified to pH < 2 (by the addition of glacial acetic acid) to prepare 5 mg/mL (1000 ×) stock. Store in frozen aliquots at − 20 °C.

Apotransferrin (T1147–100 mg, Sigma-Aldrich). Dissolve 100 mg in 20 mL of water to make 5 mg/mL (1000 ×) stock. Store in frozen aliquots at − 20 °C.

Hydrocortisone 21-hemisuccinate (H4881–100 mg, Sigma-Aldrich). Dissolve 100 mg in 25 mL of 100% ethanol to make 4 mg/mL (10,000 ×) stock. Store at − 20 °C.

4.3.15 Raft culture medium:

To prepare 200 mL of raft culture medium, mix 145 mL of DMEM and 35 mL of F12 medium. Add 20 mL of FBS, penicillin to 40 units/mL, streptomycin to 40 μg/mL, and add supplements from the stocks described above to the final concentrations as follows: 0.1 μM cholera toxin, 5 μg/mL apo-transferrin, 0.4 μg/mL hydrocortisone, 5 μg/mL insulin and 0.5 ng/mL EGF. Raft culture medium is stored at 4 °C and typically used within 1 week of preparation.

4.3.16 10x F12 medium:

Dissolve powdered F12 medium (21700-075, Life Technologies) in 100 mL of water instead of 1 L, filter-sterilize and store in 10-mL frozen aliquots at −20 °C. Do not adjust pH or add sodium bicarbonate.

4.3.17 10x reconstitution buffer for collagen matrix:
Dissolve 2.2 g sodium bicarbonate, 0.2 g sodium hydroxide, and 4.76 g HEPES (free acid) in 100 mL of water, filter-sterilize and store in 10-mL frozen aliquots at −20 °C.

4.4 Stocks of retinoid or rexinoid compounds in ethanol or DMSO

Stocks are prepared by dissolving crystalline compounds in 200-proof ethanol or sterile DMSO under dim red light, and stored at −80 °C.

4.5 Expression constructs

Coding DNA sequences of the enzymes of interest cloned in pBabe-puro vector upstream of puromycin expression cassette.

4.6 Equipment

4.6.1 Stainless-steel support grids, 2 × 2 × 0.4 cm (custom-made by TWP Inc., Berkeley, CA), sterilized by autoclaving.
4.6.2 Sterile spatula with angled blade for transfer of the submerged cultures on the support grids.
4.6.3 Scalpel and forceps for harvesting the samples.
4.6.4 Biosafety cabinet for cell culture work.
4.6.5 Cell culture incubator.
4.6.6 24-well and 6-well cell culture plates and 10-cm dishes.

5. Protocols

5.1 Preparation of dermal equivalents "collagen matrix"

5.1.1. Grow 3T3-J2 mouse fibroblasts to 90% confluency in DMEM medium with 10% BCS and penicillin-streptomycin (DMEM/BCS/PS) in 10-cm dishes.
5.1.2. Aspirate medium, rinse cells with D-PBS, and dissociate in 1 mL of trypsin-EDTA. Add 3 mL of DMEM/BCS/PS and pipet up and down to stop the digestion and separate cells. Take 100 μL aliquot to count the cells, transfer the rest to a 15-mL tube and centrifuge at 1000x*g* to pellet cells. Resuspend pelleted 3T3-J2 cells in FBS to 5×10^6 cells/mL and keep on ice.
5.1.3. In 15-mL tube, combine 0.8 mL of 10x reconstitution buffer, 0.8 mL of 10x F12 medium, and 6.7 mL of collagen solution. Carefully pipet up and down to mix while avoiding bubbles. All solutions must be

kept on ice prior to mixing. Addition of 10x buffer and medium neutralizes acidic collagen solution before the addition of fibroblasts. Last, add 400 μL (2×10^6) of 3T3-J2 cells in FBS, and fully mix by pipetting while still avoiding bubbles.

5.1.4. Dispense 3T3-J2 cell suspension in collagen into a 24-well plate by 0.8 mL/well. This amount of the suspension is sufficient for 10 wells and can be scaled up. However, the mixing and dispensing must be performed fast, as collagen will solidify when warmed up.

5.1.5. Transfer 24-well plate to the incubator for 30 min to let collagen matrices to solidify.

5.1.6. Add 1 mL of raft culture medium on top of collagen. Care should be taken to avoid touching the collagen matrices with pipet or shaking the plate, as this can lead to detachment and premature contraction of the collagen. Place the plate in the incubator for 5 h or longer before the seeding of keratinocytes. Dermal equivalents can be prepared up to 5 days in advance.

5.2 Preparation of "raft" cultures

5.2.1. Thaw stock of primary keratinocytes in 37 °C water bath. Transfer the suspension into a sterile centrifuge tube and centrifuge at $300 \times g$ to pellet cells. Aspirate freezing medium. Resuspend the cells in DermaLife-K SFM without antibiotics and plate on 10-cm dish.

5.2.2. When keratinocytes are nearly confluent (90%), aspirate the medium and rinse the cells with D-PBS. Aspirate D-PBS and add 2 mL of trypsin-EDTA. Keratinocyte dissociate slow, thus the plates can be placed in the 37 °C incubator for approximately 5 min. Once the complete dissociation is observed under the microscope, add 6 mL of DMEM/BCS and pipet up and down to stop the digestion and separate cells. Take 100 μL aliquot to count the cells, transfer the rest to a 15-mL tube and centrifuge at $300 \times g$ to pellet.

5.2.3. Resuspend pelleted keratinocytes in DermaLife-K SFM without antibiotics to 4×10^5 cells/mL.

5.2.4. Aspirate raft culture medium from wells containing collagen beds. Carefully dispense 0.5 mL of keratinocyte suspension (2×10^5 cells) on top of each bed. Return the plate to the incubator and incubate for 1 day. Keratinocytes will attach to the collagen bed and grow as a submerged culture (Fig. 4).

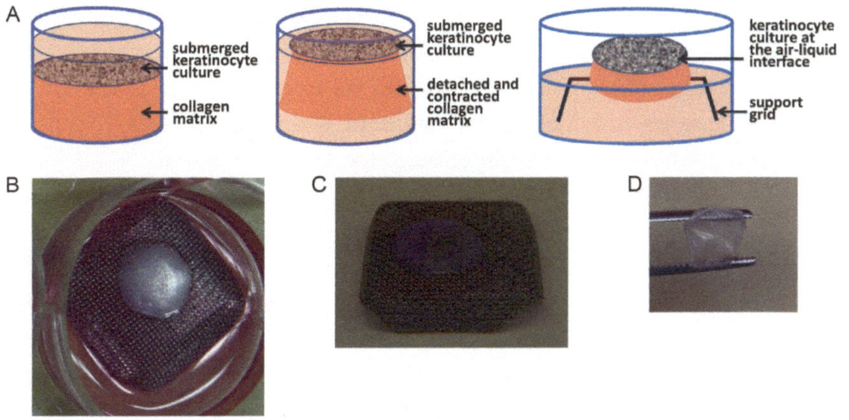

Fig. 4 Commonly used setup for the generation of organotypic skin raft cultures. (A) Diagram illustrating the preparation of organotypic skin raft cultures at the air-liquid interface. (B) Photograph of the fully developed organotypic skin culture after 10 days of incubation. (C) Support grid with an organotypic culture before harvesting. (D) The epidermis layer peeled of the dermal equivalent.

5.2.5. Next day, replace the medium in each well with 1 mL of SFM mixed with raft culture medium as 1:1. Incubate overnight.

5.2.6. Next day, aspirate the medium and run the sterile spatula around the walls of each well to detach collagen beds from the walls. Add 1 mL of raft culture medium per well and return the plate to the incubator for 2–4 h to allow collagen to contract. The top of each bed containing the monolayer of keratinocytes will visibly contract in comparison to the bottom side of the bed (Fig. 4).

5.2.7. Places sterile support grids in the wells of six-well plates and add raft culture medium to the top of the grid (diagram in Fig. 4).

5.2.8. Use sterile spatula with an angled blade to scoop collagen beds with the monolayer keratinocyte culture from the wells and place them in the middle of metal grids. The collagen bed will partially sink between the mesh wire and remain in contact with the liquid medium, while the keratinocyte monolayer will be exposed to air, thus placing the culture at the liquid-air interface.

5.2.9. Replace raft culture medium every 2 days. The stratified epidermis fully develops after 9 days of incubation at the air-liquid interface.

Note: Estimate the required number of rafts prior to the experiment. One 10-cm dish of nearly confluent keratinocytes may contain up to $3.5–4 \times 10^6$ cells, enough for seeding of up to 20 dermal equivalents. If more than one dish of confluent keratinocytes is necessary, consider splitting the cells once, or thawing additional stocks prepared from the same original culture. Multiple passages of primary keratinocytes are not recommended, as the cells may become senescent. The nearly confluent 10-cm dish of 3T3-J2 cells contains approximately 2×10^6 cells, enough for 10 collagen beds.

5.3 Treatment with retinoids or rexinoids

5.3.1. While working under dim red light, thaw the stock of the compound and make sure that it is fully dissolved, and no crystals are present. It may require warming up the stock to room temperature.

5.3.2. Dilute stock with sterile DMSO to make 500 ×–1000 × working stock, which will be added to the growth medium. For example, 2 mM all-*trans*-retinol will supply 2 μM all-*trans*-retinol and 0.1% DMSO when used as a 1000 × working stock. The final concentration of DMSO in the medium should not exceed 0.2% (v/v).

5.3.3. Next day after transferring the culture to the support grid, aspirate the medium. Prepare aliquots of raft culture medium for each treatment variant for the desired number of rafts. One well typically requires 3 mL for medium to reach the top of the support grid.

5.3.4. While working in the dark or under dim red light, add compound from the working stock to the raft culture medium aliquot, mix by pipetting, and dispense into the wells.

5.3.5. Replace medium with fresh dilution of the compound every other day until the desired time point to harvest the culture.

5.3.6. If the analysis of BrdU incorporation in the nuclei of proliferating cells is performed, 12 h before harvesting replace the medium with a fresh medium containing the diluted retinoid or rexinoid compound and BrdU added from 100 × stock to the final concentration of 50 μg/mL.

Note: Concentrated stocks of retinoids and rexinoids in ethanol or DMSO must be stored at −80 °C and handled under dim red light. Before diluting to a working stock the quality of each compound must be evaluated by comparing the UV/visual absorbance spectrum to the published reference spectrum in the same solvent, and the concentration calculated based on the published molar extinction coefficient for the corresponding solvent.

Note: To analyze the effects of compounds modulating RA signaling on tissue architecture and expression of the markers of proliferation and differentiation, treatment is conducted continuously beginning from the placement of cultures at the liquid-air interface to harvest. For analysis of retinoid metabolism by cultures overexpressing retinoid-active enzymes (part 5.4), retinol or retinaldehyde can be added to the medium for the last 24h before harvesting.

5.4 Preparation of skin raft cultures expressing retinoid-active enzymes

5.4.1. Plate Bosc-23 cells on 10-cm dishes and grow to 90% confluency in DMEM with 10% FBS and penicillin-streptomycin (DMEM/FBS/PS). One dish will be used for each of the expression constructs and a control vector. Replace growth medium with 4mL of serum-free OptiMEM without antibiotics.

5.4.2. Dilute 16μL of Lipofectamine 2000 in 0.5mL of OptiMEM, and dilute 8μg of plasmid DNA in 0.5mL of OptiMEM in a separate tube (expression construct in pBabe-puro vector, or empty pBabe vector). Mix diluted DNA and Lipofectamine 2000 and incubate for 15–20min.

5.4.3. Add DNA-Lipofectamine complexes to the dish with Bosc-23 cells, gently mix and incubate for 3–4h. Replace the transfection medium with 10mL of DMEM/FBS growth medium.

5.4.4. Next day, replace the growth medium in Bosc-23 cell dishes with 5mL of DMEM/FBS/PS. Incubate overnight.

5.4.5. On the same day, plate GP + envAM-12 cells in DMEM/FBS/PS on 10-cm dishes (1×10^6 cells per dish, or approximately 25–30% confluency).

5.4.6. Next day, aspirate the growth medium from GP + envAM-12 cells. Collect the virus-containing medium from Bosc-23 cell dishes and filter through 0.45-μM syringe filters directly into the GP + envAM-12 dishes. Add 8.3μL of polybrene per dish (to 5μg/mL), mix gently, and incubate for 4–5h.

5.4.7. Add 5mL more of DMEM/FBS/PS to each dish and incubate overnight. Allocate one dish of uninfected GP + envAM-12 cells to use a selection control.

5.4.8. Next day, begin selection of infected GP + envAM-12 cells with puromycin. Replace medium in each dish, included the uninfected control, with 10mL of fresh DMEM/FBS containing 1μg/mL of puromycin. Continue selection until control cells die, which typically takes 5 days. Replace medium with fresh puromycin-containing medium every other day.

5.4.9. When selection is completed, replace medium in each dish of infected cells with 5 mL of DMEM/FBS without antibiotics and incubate overnight.

5.4.10. On the same day, plate primary keratinocytes in DermaLife-K SFM into 6-cm dishes, preparing 2 dishes for each of the expression constructs and control vector. (To have p rimary keratinocytes ready for plating at this point, 3 days earlier thaw a stock of primary keratinocytes and plate as described above in 2.1.) Allocate one dish for use as an uninfected selection control.

5.4.11. Next day, aspirate medium from keratinocyte dishes. Collect virus-containing supernatant from infected GP + envAM-12 cells, and filter through 0.45-μM syringe filter. Dispense medium from dish infected with each of the virus variants (5 mL) to two dishes of keratinocytes in 2.5 mL—aliquots. Add 8.3 μL of polybrene per dish, mix and incubated for 90 min.

5.4.12. Add 2.5 mL of DermaLife-K SFM to each dish, and continue incubation for another 90 min.

5.4.13. Replace medium with 5 mL of fresh Dermalife-K SFM and incubate overnight.

5.4.14. Next day, begin selection of infected keratinocytes by replacing medium with SFM containing 1.5 μg/mL of puromycin. Incubate for 2 days. Confirm that uninfected keratinocytes died by the end of this time.

5.4.15. Replace medium with DermaLife-K SFM without antibiotics and allow selected keratinocytes to expand. Typically, with efficient infection they will be near confluency within 1–2 days.

5.4.16. Dissociate the transduced keratinocytes with trypsin as described above for uninfected keratinocytes, count and seed on dermal equivalents prepared as described in Section 5.1.

5.4.17. Follow the protocol in Section 5.2 to develop organotypic raft cultures.

Note: If using a different packaging cell line, retroviral transfer vector, and transfection reagent other than Lipofectamine 2000, the protocol should be optimized according to the reagent manufacturer's instructions. Electroporation can also be used to transfect Bosc-23 cells. The method suggested here allowed for efficient infection of primary keratinocytes and expression of SDR enzymes in the resulting raft cultures.

5.5 Harvesting organotypic skin raft cultures for analysis

5.5.1. To prepare organotypic raft cultures for the analysis of morphology and immunohistochemical staining, aspirate medium from the wells and rinse wells with PBS. Fill each well with 10% buffered formalin to cover the support grids and the rafts. Incubate for 45 min at room temperature. After fixation, rafts can be easily detached from the support grids together with the dermal equivalent. Rinse the detached rafts with 70% ethanol, wrap in the lens paper and place in histological cassettes for embedding.

5.5.2. To prepare tissue for the isolation of RNA, protein analysis, or analysis of metabolites, an unfixed epidermis must be separated from collagen. Remove metal grids from the culture well. While holding the grid with forceps, use scalpel to separate the culture from the grid along the perimeter of the bed, where the epidermis usually becomes attached to the wire. Once separated from the grid, use forceps to peel the epidermis from the underlying dermal equivalent (Fig. 4).

5.5.3. Place the peeled off epidermis in 1 mL of Trizol reagent for RNA isolation, or promptly freeze on dry ice or in liquid nitrogen for the analysis of metabolites and protein.

6. Safety considerations and standards

Work with primary human tissues and cells (primary human keratinocytes) requires Biosafety level 2 environment. Biosafety cabinets must be certified as BSL2 and guidelines for use of BSL2 materials must be observed.

7. Analysis

7.1 Analysis of morphology, proliferation and differentiation

7.1.1. Formalin-fixed cultures can be dehydrated and embedded in paraffin using standard embedding protocols. H&E and immunohistochemical staining can be performed as described previously (Lee et al., 2011). The presence and thickness of the different functional layers of epidermis can be compared between various treatment groups (Fig. 1, ROL treatment, Fig. 2—bexarotene treatment). The thickness of the specific layers can be measured on microscope equipped with the image-processing software.

7.1.2. In addition to the morphology of the basal layer, proliferation can be compared by assessing the localization and number of BrdU-labeled nuclei (Fig. 2—bexarotene). The localization and intensity of staining for epithelial differentiation markers can be assessed to complement the analysis of proliferation.

7.2 Analysis of gene expression

Three or more of the individual cultures in each treatment group should be allocated for the isolation of RNA and treated as biological replicates to allow for the statistical analysis of gene expression. The complete experiment must be repeated on independent set(s) of organotypic cultures developed from a different preparation of primary keratinocytes, to account for an individual variability due to different donors. RNA can be used for qRT-PCR analysis of known components of RA biosynthesis and signaling, as well as for the next-generation sequencing and discovery of additional pathways affected by the treatment (Lee et al., 2011).

7.3 Protein analysis

Crude protein homogenates prepared from individual samples of epidermis and boiled with SDS-page loading buffer are usually sufficient for the detection of overexpressed enzymes in transduced cultures by Western blot. However, several samples of epidermis can be pooled to isolate subcellular fractions for the detection of endogenous proteins.

7.4 Analysis of retinoid metabolism

The two-step protocol for the extraction of retinoids (Napoli & Horst, 1998) can be applied to extract retinol, retinyl esters stores and RA from the organotypic epidermis. As the amount of retinoids in the individual samples is marginal for the reliable quantification, it is recommended to pool several cultures for each of the replicate samples for the extraction and subsequent analysis by HPLC (all-*trans*-retinol and retinyl esters) and HPLC/MS (RA) (Wu et al., 2016).

8. Summary

Investigators interested in evaluating the roles of human retinoid-active enzymes may find the organotypic model of human epidermis highly useful as it is very sensitive to the levels of RA and can serve as a reporter model for the effects of RA signaling on cell growth and differentiation. This model is also valuable for screening various compounds that might affect the RA levels and signaling.

Acknowledgment

This work was supported by NIH P01 CA210946.

References

Asselineau, D., Bernard, B. A., Bailly, C., Darmon, M., & Pruniéras, M. (1986). Human epidermis reconstructed by culture: Is it "normal"? *Journal of Investigative Dermatology*, *86*, 181–186.

Atigadda, V. R., Xia, G., Deshpande, A., Wu, L., Kedishvili, N., Smith, C. D., et al. (2015). Conformationally defined rexinoids and their efficacy in the prevention of mammary cancers. *Journal of Medicinal Chemistry*, *58*, 7763–7774.

Banerjee, N. S., Chow, L. T., & Broker, T. R. (2005). Retrovirus-mediated gene transfer to analyze HPV gene regulation and protein functions in organotypic "raft" cultures. *Methods in Molecular Medicine*, *119*, 187–202.

Banerjee, N. S., Rivera, A. A., Wang, M., Chow, L. T., Broker, T. R., Curiel, D. T., et al. (2004). Analyses of melanoma-targeted oncolytic adenoviruses with tyrosinase enhancer/promoter-driven E1A, E4, or both in submerged cells and organotypic cultures. *Molecular Cancer Therapy*, *3*, 437–449.

Bell, E., Sher, S., Hull, B., Merrill, C., Rosen, S., Chamson, A., et al. (1983). The reconstitution of living skin. *Journal of Investigative Dermatology*, *81*, 2s–10s.

Belyaeva, O. V., Adams, M. K., Popov, K. M., & Kedishvili, N. Y. (2019). Generation of retinaldehyde for retinoic acid biosynthesis. *Biomolecules*, *10,* e5.

Cheng, S., Schmidt-Grimminger, D. C., Murant, T., Broker, T. R., & Chow, L. T. (1995). Differentiation-dependent up-regulation of the human papillomavirus E7 gene reactivates cellular DNA replication in suprabasal differentiated keratinocytes. *Genes and Development*, *9*, 2335–2349.

Chien, W. M., Parker, J. N., Schmidt-Grimminger, D. C., Broker, T. R., & Chow, L. T. (2000). Casein kinase II phosphorylation of the human papillomavirus-18 E7 protein is critical for promoting S-phase entry. *Cell Growth and Differentiation*, *11*, 425–435.

Chow, L. T. (2015). Model systems to study the life cycle of human papillomaviruses and HPV-associated cancers. *Virologica Sinica*, *30*, 92–100.

Chow, L. T., & Broker, T. R. (1997). In vitro experimental systems for HPV: Epithelial raft cultures for investigations of viral reproduction and pathogenesis and for genetic analyses of viral proteins and regulatory sequences. *Clinical Dermatology*, *15*, 217–227.

Collawn, S. S., Banerjee, N. S., de la Torre, J., Vasconez, L., & Chow, L. T. (2012). Adipose-derived stromal cells accelerate wound healing in an organotypic raft culture model. *Annals of Plastic Surgery*, *68*, 501–504.

Fisher, G. J., & Voorhees, J. J. (1996). Molecular mechanisms of retinoid actions in skin. *FASEB Journal*, *10*, 1002–1013.

Kopan, R., Traska, G., & Fuchs, E. (1987). Retinoids as important regulators of terminal differentiation: Examining keratin expression in individual epidermal cells at various stages of keratinization. *Journal of Cell Biology*, *105*, 427–440.

Lee, S. A., Belyaeva, O. V., Wu, L., & Kedishvili, N. Y. (2011). Retinol dehydrogenase 10 but not retinol/sterol dehydrogenase(s) regulates the expression of retinoic acid-responsive genes in human transgenic skin raft culture. *Journal of Biological Chemistry*, *286*, 13550–13560.

Napoli, J. L., & Horst, R. L. (1998). Quantitative analyses of naturally occurring retinoids. *Methods in Molecular Biology*, *89*, 29–40.

Pageon, H., Zucchi, H., Dai, Z., Sell, D. R., Strauch, C. M., Monnier, V. M., et al. (2015). Biological effects induced by specific advanced glycation end products in the reconstructed skin model of aging. *BioResearch Open Access*, *4*, 54–64.

Roy, B., Simard, M., Lorthois, I., Bélanger, A., Maheux, M., Duque-Fernandez, A., et al. (2018). In vitro models of psoriasis. In A. P. Marques, R. P. Pirraco, M. T. Cerqueira, & R. L. Reis (Eds.), *Skin tissue models* (pp. 103–128). Academic Press.

Sorg, O., & Saurat, J. H. (2014). Topical retinoids in skin ageing: A focused update with reference to sun-induced epidermal vitamin a deficiency. *Dermatology*, *228*, 314–325.

Tjabringa, G., Bergers, M., van Rens, D., de Boer, R., Lamme, E., & Schalkwijk, J. (2008). Development and validation of human psoriatic skin equivalents. *American Journal of Pathology*, *173*, 815–823.

Wang, H. K., Broker, T. R., & Chow, L. T. (2015). Robust HPV-18 production in organotypic cultures of primary human keratinocytes. *Methods in Molecular Biology*, *1249*, 93–109.

Wang, H. K., Duffy, A. A., Broker, T. R., & Chow, L. T. (2009). Robust production and passaging of infectious HPV in squamous epithelium of primary human keratinocytes. *Genes and Development*, *23*, 181–194.

Wilson, J. L., Dollard, S. C., Chow, L. T., & Broker, T. R. (1992). Epithelial-specific gene expression during differentiation of stratified primary human keratinocyte cultures. *Cell Growth and Differentiation*, *3*, 471–483.

Wu, L., Chaudhary, S. C., Atigadda, V. R., Belyaeva, O. V., Harville, S. R., Elmets, C. A., et al. (2016). Retinoid X receptor agonists upregulate genes responsible for the biosynthesis of all-trans-retinoic acid in human epidermis. *Public Library of Science*, *11*, e0153556.

CHAPTER NINETEEN

Tissue localization of retinoic acid receptor (RAR) active drugs

Azita Kouchmeshky[a], Timothy Goodman[b], Andrew Whiting[c], Peter McCaffery[a,*]

[a]Institute of Medical Sciences, University of Aberdeen, Aberdeen, Scotland, United Kingdom
[b]School of Pharmacy and Bioengineering, Keele University, Keele, United Kingdom
[c]Department of Chemistry, Durham University, Science Laboratories, Durham, United Kingdom
*Corresponding author: e-mail address: p.j.mccaffery@abdn.ac.uk

Contents

Methods in Enzymology, Volume 637
ISSN 0076-6879
https://doi.org/10.1016/bs.mie.2020.03.012

Abstract

The retinoic acid (RA) signaling pathway is crucial for the control of embryonic development and also regulates function of several organ systems in the adult, including the central nervous system. The retinoic acid receptors (RARs) that mediate the majority of the functions of RA can promote proliferation, differentiation, morphogenesis and cell survival. Dysregulation of this signaling pathway has been considered in the pathophysiology of various diseases including neurodegenerative disorders such Alzheimer's disease and amyotrophic lateral sclerosis. Thus, drugs targeted to the RARs have been proposed as treatments for such diseases. Understanding how these drugs distribute in the body is essential to determine their potential effectiveness. However measuring tissue levels of what are often lipophilic drugs can be difficult. Here we describe an indirect measurement of RAR ligand tissue distribution after intraperitoneal injection into rodents that uses a sensitive RA reporter cell line.

1. Introduction

1.1 Retinoic acid signaling

Retinoic acid (RA) is the active metabolite of vitamin A for many target cells where it regulates expression of a large number of genes (Mey & Mccaffery, 2004). It induces its action through the nuclear retinoic acid receptors (RARs) and retinoid X receptors (RXRs) that belong to the nuclear receptor superfamily and act as transcription factors (Mey & Mccaffery, 2004). Each of these receptors has three subtypes (α, β, and γ) and each subtype has several isoforms. Ligand binding promotes gene transcription by activating the RAR-RXR heterodimer bound to a sequence of DNA, within gene promoters and enhancers, called the RA response element (RARE); while dissociation of the ligand represses gene transcription (Mey & Mccaffery, 2004; Niederreither & Dollé, 2008). RA has also been shown to regulate extranuclear, non-transcriptional pathways which induce for instance rapid and transient kinase cascade activation essential for cellular functions, as well as controlling translation (Chen, Lau, & Sarti, 2014), and called non-genomic pathways (Al Tanoury, Piskunov, & Rochette-Egly, 2013; Rochette-Egly, 2015). RA signaling is vital during embryonic development regulating cell proliferation, differentiation, regeneration and survival and a number of such functions continue into postnatal development and adulthood. Its functions in the adult to control cell proliferation and differentiation has resulted in the use of RA as a therapeutic drug for a range of dermatological conditions and as a chemopreventative and chemotherapeutic agent in oncology (Barnard, Collings, Whiting, Przyborski, & Marder, 2009; Mey & Mccaffery, 2004). Recently, dysregulation of RA signaling in different neurodegenerative diseases has attracted scientific attention (Corcoran, So, & Maden, 2004; Esteves et al., 2015; Goodman & Pardee, 2003; Kolarcik & Bowser, 2012; Lee et al., 2009; Malaspina & Michael-Titus, 2007; Niewiadomska-Cimicka et al., 2017; Shudo, Fukasawa, Nakagomi, & Yamagata, 2009) suggesting future use of RAR ligands in the treatment of such disorders.

1.2 Detecting retinoic acid levels in tissue

The active levels of RA in tissues are regulated by synthesizing and catabolizing enzymes as well as binding proteins. In chemical nature, RA is lipophilic with a low molecular weight (~300 Da) and can diffuse readily through cell membranes (Niederreither & Dollé, 2008) and has conjugated double bonds and is easily photo-isomerized when exposed to light (Barnard et al., 2009). This photo instability, small size, and amphipathic lipid nature make it difficult to directly measure or visualize RA in tissue. Direct methods for RA measurement in tissue samples are through high performance liquid chromatography (HPLC) and UV spectrophotometry detection at 348 nm (Dzerk, Carlson, Loewen, Shirley, & Lee, 1998) or detection by mass spectrometry (Evans & McCaffery, 2010; Kane & Napoli, 2010) however sensitivity of some of these methods is not high and can require a relatively large amount of tissue or the methods require specialized equipment (Ababon, Li, Matteson, & Millonig, 2016; Sakai & Dräger, 2010). In addition, as described in Chapter 13 of this volume ("Analysis of Vitamin A and Retinoids in Biological Matrices" by Czuba, Zhong, Yabut & Isoherranen), the varied matrix components of the tissue examined also have a significant impact on the accuracy of quantification. To avoid such problems many studies instead assume the presence of RA in tissue through the expression of RA-synthesizing enzymes (and absence of RA-catabolizing enzymes). However, several very sensitive indirect methods have also been developed to determine RA levels in tissues using RA reporter systems.

1.3 Retinoic acid-reporter systems

Several RA reporter systems are based on utilizing the highly sensitive RA response element (RARE) of the *RARβ2* gene placed upstream of a reporter gene, either the *Escherichia coli lacZ* gene or the firefly *luciferase* gene, in either cell line or transgenic mice (Mendelsohn, Ruberte, Lemeur, Morriss-Kay, & Chambon, 1991; Reynolds, Mezey, & Zimmer, 1991; Rossant, Zirngibl, Cado, Shago, & Giguere, 1991; Sakai & Dräger, 2010; Stoney et al., 2016; Wagner, Han, & Jessell, 1992). Both the mouse strain generated by Rossant et al. (1991) and the F9-teratocarcinoma-derived Sil-15 reporter cell line developed by Wagner et al. (1992) use the *E. coli* β-galactosidase (*lacZ*) gene or *luciferase* gene as a reporter downstream of the RARE from human *RARβ2* (*hRARβ2*). The lacZ assay provides a histochemical detection of β-galactosidase levels in response to RA while the luciferase assay quantitatively detects the presence of RA (Wagner et al., 1992). The reporter mouse model enables visualization of the spatiotemporal distribution of RA signaling, although does not provide a quantitative measure of RA levels in tissue. The reporter cells in contrast have been effectively used to measure RA synthesis by cells in a number of different conditions (Ababon et al., 2016; Haffez et al., 2018; McCaffery & Drager, 1994; Moise, Kuksa, Blaner, Baehr, & Palczewski, 2005; Sitnik et al., 2012; Wagner et al., 1992; Yamamoto, Dräger, & McCaffery, 1998).

1.4 The F9-teratocarcinoma-derived Sil-15 reporter cell line

The F9-teratocarcinoma-derived Sil-15 reporter cell line (F9-RARE-lacZ reporter cells/Sil-15 reporter cell line) is an adherent cell derived from F9 teratocarcinoma (germ cell cancer cells). They grow very well on a gelatin-coated tissue culture-treated polystyrene surface. All three RARs (α, β, and γ) are expressed endogenously in the F9 teratocarcinoma cell line (Hu & Gudas, 1990). These cells are stably transfected with a construct containing a 64-base pair RARE from *hRARβ2* upstream of the *E. coli lacZ* gene producing β-galactosidase quantitatively in response to RA resulting in an insoluble blue precipitate in the presence of X-gal substrate (Wagner et al., 1992) (Fig. 1). The amount of blue precipitate can be measured spectrophotometrically by an ELISA reader at a maximum absorbance of 615 nm (the peak of absorption is in the range of 570–700 nm) (Levitsky, Toledo-Aral, López-Barneo, & Villadiego, 2013) to quantitatively measure RA levels (Ababon et al., 2016). The plasmid used to insert

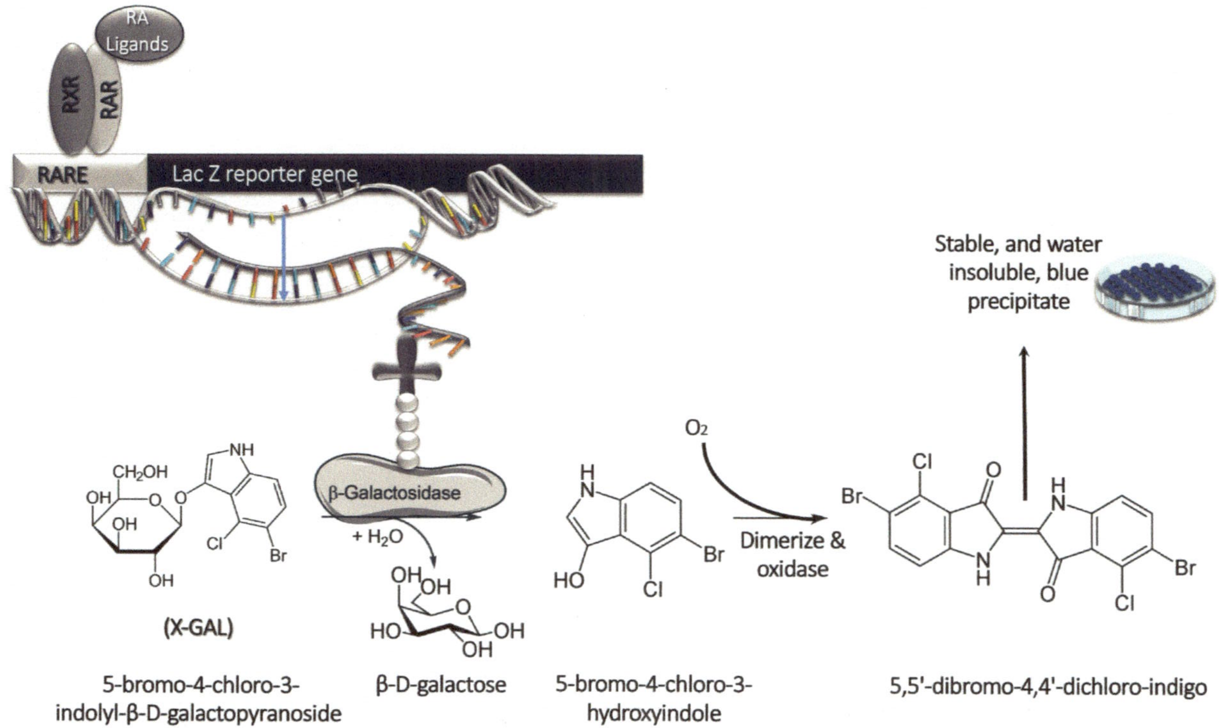

Fig. 1 Steps by which the Sil-15 reporter cells produce β-galactosidase in response to RA. The enzyme cleaves the 5-bromo-4-chloro-3-indoyl-β-D-galactosidase (X-gal), analogue of lactose, to galactose and 5-bromo-4-chloro-3-hydroxyindole (Wagner et al., 1992). Spontaneous dimerization and oxidization of the latter product make a stable and insoluble visual blue precipitate called 5,5′-dibromo-4,4′-dichloro-indigo that can be quantitatively measured by ELISA reader.

the lacZ gene includes an aminoglycoside phosphotransferase (*neo*^{R}) gene which confers resistance to the antibiotic G-418 (geneticin), providing a route to select cells for expression of the construct.

1.5 Retinoic acid receptor ligands

Synthetic RAR/RXR ligands have been designed and used for many therapeutic purposes from the treatment of acne (Tan, Schlosser, & Paller, 2018; Tyring et al., 2018) to cancer (Bushue & Wan, 2010; Siddikuzzaman, Guruvayoorappan, & Berlin Grace, 2011; Tang & Gudas, 2011). There are significant variations among synthetic RAR ligands regarding pharmacokinetic profiles, therapeutic efficacy and adverse effects (McCormick et al., 1999). Identifying synthetic RAR ligands with significant therapeutic efficacy and low side effects compatible with long-term administration is a major goal of current research. Because RARs are now proposed as potential therapeutic targets in neurodegenerative disease, the capacity of the ligands to cross the blood brain barrier is another important consideration for novel RAR ligand design. The present study provides a novel and rapid method to detect RAR ligand distribution in the body and brain after drug administration.

2. Rationale

Sil-15 cells have been used in multiple studies to detect endogenous RA levels, for instance by co-culturing reporter cells with tissue samples (Kelley, Xu, Wagner, Warchol, & Corwin, 1993; McCaffery, Lee, Wagner, Sladek, & Drager, 1992; Stoney et al., 2016; Wagner et al., 1992), adding conditioned medium to Sil-15 reporter cells from cultured embryonic spinal cord to measure RA released into the medium (McCaffery & Drager, 1994), adding zymography gel slices to reporter cells to measure RA synthetic enzyme activity (McCaffery & Dräger, 1997) or co-culturing neurospheres obtained from adult spinal cord stem cells with Sil-15 reporter cells (Ababon et al., 2016). Here we developed for the first time a bioassay using the Sil-15 reporter cell line that can map the tissue distribution of synthetic ligands to RAR injected into rodents. This new method provides a reliable protocol to track newly designed RAR targeted drugs to map their distribution and accumulation in target tissues.

3. Materials, equipment and reagents

3.1 Materials

1. C57/BL6 mice
2. F9-teratocarcinoma-derived Sil-15 reporter cell line (F9-RARE-lacZ reporter cells)

3.2 Equipment

1. Analytical balance
2. Centrifuge (refrigerated)
3. Cell counter instrument (Bio-Rad TC10TM automated cell counter instrument) or Hemocytometer
4. Cell counting slides (CAT# 145-0011, Bio-Rad)
5. Cell culture flask (T-75) (CAT# C7231, Sigma-Aldrich)
6. Centrifuge tubes 1.5 mL (CAT# Z334057, Sigma-Aldrich)
7. CO_2 chamber (for animal culling)
8. Cryotubes (CAT# 10135500, Fisher Scientific)
9. Dissection tools (forceps, tweezers, scissors)
10. Dry-Ice and Ice
11. Falcon tube 50 mL (CAT# Z618985, Sigma-Aldrich)
12. Filter (syringe filters, 0.22 μm) (CAT# 10268401, Fisher Scientific)
13. Foil
14. Freezer (−20 °C, −80 °C)
15. Freezing Container (Thermo Scientific™ Mr. Frosty™ Freezing Container, CAT# 10110051, Fisher Scientific)
16. Fridge (4 °C)
17. Homogenizing pestle (CAT# 11872913, Fisher Scientific)
18. Inverted biological microscope
19. Laminar flow hood
20. Microplate reader (ELISA plate reader in the range of 570–700 nm)
21. Multichannel pipette (10–100 μL/100–1000 μL)
22. Multichannel pipetting reservoir (CAT# 11988485, Fisher Scientific)
23. Microfuge tubes 0.6 mL (CAT# 22234-042, VWR)
24. Parafilm
25. Pipette (1–10 μL, 10–100 μL, and 100–1000 μL)
26. Pipette tips 10, 100, and 1000 μL (CAT# Z617261, Z617393, and Z617490, Sigma-Aldrich)

27. pH meter
28. Plates (96-well plate) (CAT# M0812, Sigma-Aldrich)
29. Syringe (60 mL BD Plastipak syringe) (CAT# 300866, Becton Dickinson)
30. Stripette (10 and 25 mL) (CAT# CLS4100 and CLS4250, Sigma-Aldrich)
31. Tissue culture incubator with 5% CO_2 at 37 °C (TC incubator)
32. Water bath (37 °C)

3.3 Reagents

1. RA (all-*trans* retinoic acid) (CAT# R2625, Sigma-Aldrich)
2. Butylated hydroxytoluene (CAT# W218405, Sigma-Aldrich)
3. Dulbecco's Modified Eagle Medium mixed with Ham's F-12 (1,1) supplemented with GlutaMAX (DMEM/F-12, GlutaMAX; CAT# 11514436, Fisher Scientific)
4. Dimethylformamide (DMF) (CAT# 227056, Sigma-Aldrich)
5. Dimethyl sulfoxide (DMSO) (CAT# 15303671, Fisher Scientific)
6. Ethanol for molecular biology (C_2H_5OH; CAT# 1085430250, Merck Millipore's)
7. Fetal Bovine Serum (FBS) (CAT# S1810–500, VWR)
8. G-418 (CAT# A1720, Sigma-Aldrich)
9. Gelatin (CAT# 10774751, Fisher Scientific)
10. Glutaraldehyde solution 25% (CAT# G5882, Sigma-Aldrich)
11. Isopropanol ($CH_3CHOHCH_3$; CAT# 20880.320, VWR)
12. Magnesium chloride ($MgCl_2$; CAT# M8266, Sigma-Aldrich)
13. Milli-Q deionized water
14. Penicillin/streptomycin (pen-strep) (CAT# 11548876, Fisher Scientific)
15. Potassium phosphate monobasic anhydrous ($KH_2PO_4 \cdot H_2O$; CAT# 10573181, Fisher Scientific)
16. Potassium chloride (KCl; CAT# 10355810, Fisher Scientific)
17. Potassium ferrocyanide trihydrate ($C_6FeK_4N_6 \cdot 3H_2O$; CAT# 10216460, Fisher Scientific)
18. Potassium ferricyanide ($C_6FeK_3N_6$; CAS# 10266370, Fisher Scientific)
19. RAR ligands (stock solution)
20. Sodium phosphate dibasic anhydrous (Na_2HPO_4; CAT# 10182863, Fisher Scientific)

21. Sodium chloride (NaCl; CAT# 10112640, Fisher Scientific)
22. Trypan blue dye 0.4% (CAT# 1450021, Bio-Rad)
23. Trypsin-EDTA 0.05% (CAT# 11580626, Fisher Scientific)
24. X-gal (5-bromo-4-chloro-3-indolyl-beta-D-galactopyranoside)(CAT# Bio-37035, Bioline)

4. Protocols

Prepare the following stock and working solutions:

4.1 Preparation of stock solutions

1. *Retinoic acid 0.1 M stock solution*: Dissolve 30 mg RA in 1 mL N_2-bubbled DMSO under red light in a dark room (check Section 5 for safety considerations before preparation). Aliquot into 0.6 mL microfuge tubes. Displace the air in the tubes with nitrogen. Wrap the cap of the tube with parafilm and cover the tubes with foil to protect from oxygen and light, respectively. Store in freezer at −80 degrees Celsius (°C).
2. *Retinoic acid receptor ligands 0.01 M stock solution*: Dissolve the synthetic RAR ligands in DMSO to prepare 0.01 M stock solution. Store at room temperature (RT).
3. *10× Phosphate Buffered Saline (PBS) stock solution*: Dissolve 2.4 g potassium phosphate monobasic anhydrous, 14.4 g of sodium phosphate dibasic anhydrous, 80 g of sodium chloride and 2 g potassium chloride in 800 mL Milli-Q water. The pH should be approximately pH 6.8 for the 10× stock solution and when diluted pH 7.4. If necessary, adjust the pH using hydrochloric acid or sodium hydroxide. After dissolving, add the Milli-Q water to reach a final volume of 1 L. Final concentration of 10× PBS will be: [KH_2PO_4]: 18 mM, [Na_2HPO_4]: 100 mM, [NaCl]: 1.37 M, and [KCl]: 27 mM. Keep the solution at RT. To make 1× PBS, add 100 mL of the 10× PBS stock solution to 900 mL Milli-Q water and keep at RT. Autoclave 1× PBS to sterilize for cell culture use.
4. *Magnesium chloride 1 M stock solution*: Dissolve 20.3 g magnesium chloride in 100 mL Milli-Q water to prepare the stock solution of 1 M $MgCl_2$ and keep at RT.
5. *X-gal (20 mg/mL) stock solution*: Dissolve (w/v) 200 mg of X-gal powder in 10 mL dimethylformamide. Aliquot in 1.5 mL centrifuge tubes, wrap the cap with parafilm and keep in the freezer at −20 °C.

6. *Potassium ferrocyanide 0.33 M stock solution (KFerroCN)*: Dissolve 10.9 g of potassium ferrocyanide trihydrate with 100 mL Milli-Q water. Wrap it with foil to protect from light and keep at RT.
7. *Potassium ferricyanide 0.33 M stock solution (KFerriCN)*: Dissolve 13.9 g of potassium ferricyanide with 100 mL Milli-Q water. Wrap it with foil to protect from light and keep at RT.
8. *Supplemented medium*: Add 10% FBS, with 0.8 mg/mL G-418 to the DMEM/F12, GlutaMAX medium.

4.2 Preparation of working solutions fresh on the day of the experiment

1. *0.2% Gelatin*: Add 0.50 g gelatin to 250 mL 1 × PBS. Warm in water bath at 37 °C until the gelatin dissolves completely. Filter the solution through 0.22 µm filter and store in the fridge at 4 °C.
2. *Lipid extraction solvent (LES)*: Mix (2:1) ethanol high grade: isopropanol high grade for lipid extraction (the ideal proportion of ethanol/isopropanol mixture should be optimized for type of tissue; see Section 9). As an anti-oxidant add 1 mg/mL butylated hydroxytoluene to the LES. Note: to protect retinoic acid from degradation, ketoconazole (a RA catabolizing enzyme inhibitor, CYP26 inhibitor) can be added.
3. *RA and RAR ligand serial dilutions*: Prepare 10 serial dilutions of drug for standard curve, concentration ranging from 5×10^{-4} M to 5×10^{-11} M, diluting stock solution in LES.
4. *Fixation solution (1% glutaraldehyde and 1 mM $MgCl_2$ in 1 × PBS)*: Mix 25 µL of 1 M $MgCl_2$ stock solution with 1 mL of 25% glutaraldehyde solution and 24 mL of 1 × PBS to make 25 mL fixation solution and keep at RT.
5. *X-gal development solution (20 mg/mL X-gal, 3.3 mM KFerriCN, 3.3 mM KFerroCN, and 1 mM $MgCl_2$ in 1 × PBS)*: Mix 120 µL of 0.33 M KFerriCN, 120 µL of 0.33 M KFerroCN, 12 µL of 1 mM $MgCl_2$ and 1200 µL of 20 mg/mL X-gal stock solution. Add 1 × PBS to reach a final volume of 12 mL.

4.3 Preparation and injection of retinoic acid receptor ligands or retinoic acid

To determine the tissue distribution of RAR ligands or RA after, for example, an injection of 1 mg/kg of drug, weigh animals and freshly prepare the working solutions of RA and RAR ligands in sterile 1 × PBS immediately prior to injection. Maximum 5% DMSO in 1 × PBS will be used for the vehicle control (chosen based on the percentage of DMSO used as drug vehicle to inject into the experimental mice). The tissue distribution can be examined after minutes or hours of a single ligand injection, depending on the stability of the drug. Possible side effects of such drugs mean animals should be observed for adverse effects if the drug is being used for the first time. A group of untreated animals should be set aside to provide tissue to help generate the standard curve (see Section 4.5).

4.4 Tissue collection

Weigh and label 1.5 mL centrifuge tubes for tissue collection and place on dry-ice. Centrifuge tubes should be compatible with a homogenizing pestle, with the pestle fitting snuggly into the tube for efficient homogenization. Sacrifice animals by, e.g., exposure to CO_2 at the end of the chosen drug treatment time. Rapidly remove the brain from the skull and dissect the desired subregions of the brain (e.g., frontal cortex, hippocampus, hypothalamus, striatum, and cerebellum) and other tissues such as spinal cord and eyes and directly place in the corresponding tube for immediate snap freezing either with dry-ice or liquid nitrogen. For analysis of injected RA or similar retinoids it is recommended that samples are analyzed soon after collection because of their relative sensitivity to oxygen, heat and light. For stable ligands, samples can be kept in the freezer at −80 °C until the day of experiment.

4.5 Tissue homogenization

After snap freezing of the tissues in tubes on dry-ice each is rapidly re-weighed (avoiding thawing) before homogenization which should be performed rapidly on wet ice (maximum 50 mg of tissue). Add 100 μL of LES to each tube and homogenize with pestle. Light sensitive retinoids such as RA should be homogenized under red light to prevent photo-isomerization. To generate a standard curve, homogenize the brain of non-injected mice with LES using a pestle at a μL volume of LES to mg weight of brain of 2:1. To remove insoluble material from the homogenate, centrifuge samples at a relative centrifugal force (RCF) of 10,000*g* for 5 min at 4 °C.

4.6 Assay for activation of Sil-15 reporter cells

1. *Experimental and control group*: Add 4 μL of supernatant of each sample to the wells of 96-well plates containing pre-prepared 80% confluent Sil-15 reporter cells and 196 μL supplemented medium. Treat cells for each sample in triplicate. Incubate plate overnight in a TC incubator.
2. *Generating a standard curve*: Prepare serial dilutions of RA/RAR ligand (ranging from 5×10^{-4} M to 5×10^{-11} M) for standard curve by diluting stock solutions in the supernatant of non-injected homogenized brain sample. If studying a synthetic RAR ligand ideally also prepare a RA standard curve. Triplicate blank wells containing just the supernatant without RA/RAR ligand should be included to subtract any background influence of the other lipids extracted from the brain (Fig. 2). Add 4 μL of each drug dilution to the wells in triplicate making the final concentration of serial dilution of RA/RAR ligand range from 10^{-5} M to 10^{-12} M. Incubate plate overnight in a TC incubator. For light sensitive retinoids such as RA all steps should be performed under red light. Because the cells only require overnight incubation with sample this step does not need to be performed in a sterile tissue culture hood. Use this standard curve to quantify RA level in samples.

4.7 Retinoic acid bioassay

Next day, remove the medium. Wash cells very gently twice with 1 × PBS. Fix cells by adding 100 μL of fixation solution to each well and keep for 15 min at RT. During this time prepare the X-gal development solution. Fixation solution is then removed, and wells washed twice with 1 × PBS. Add 100 μL of X-gal development solution to each well and to blank wells. Wrap the 96-well plate with foil and incubate at 37 °C. Monitor development regularly after 2 h and if the responsive is low, signal can be intensified by longer incubation up to 24 h. Read the plate with an ELISA reader at 615 nm for colorimetric detection.

4.8 Precursor techniques (set-up of RA reporter cells prior to assay)

1. *Culture and maintenance of the Sil-15 reporter cells* (*F9-RARE-lacZ reporter cells*): To start reporter cells from frozen stocks prepare a T-75 flask containing 12 mL of DMEM/F12, GlutaMAX supplemented with 10% FBS, 0.8 mg/mL G-418. Take out the cryotube from liquid nitrogen

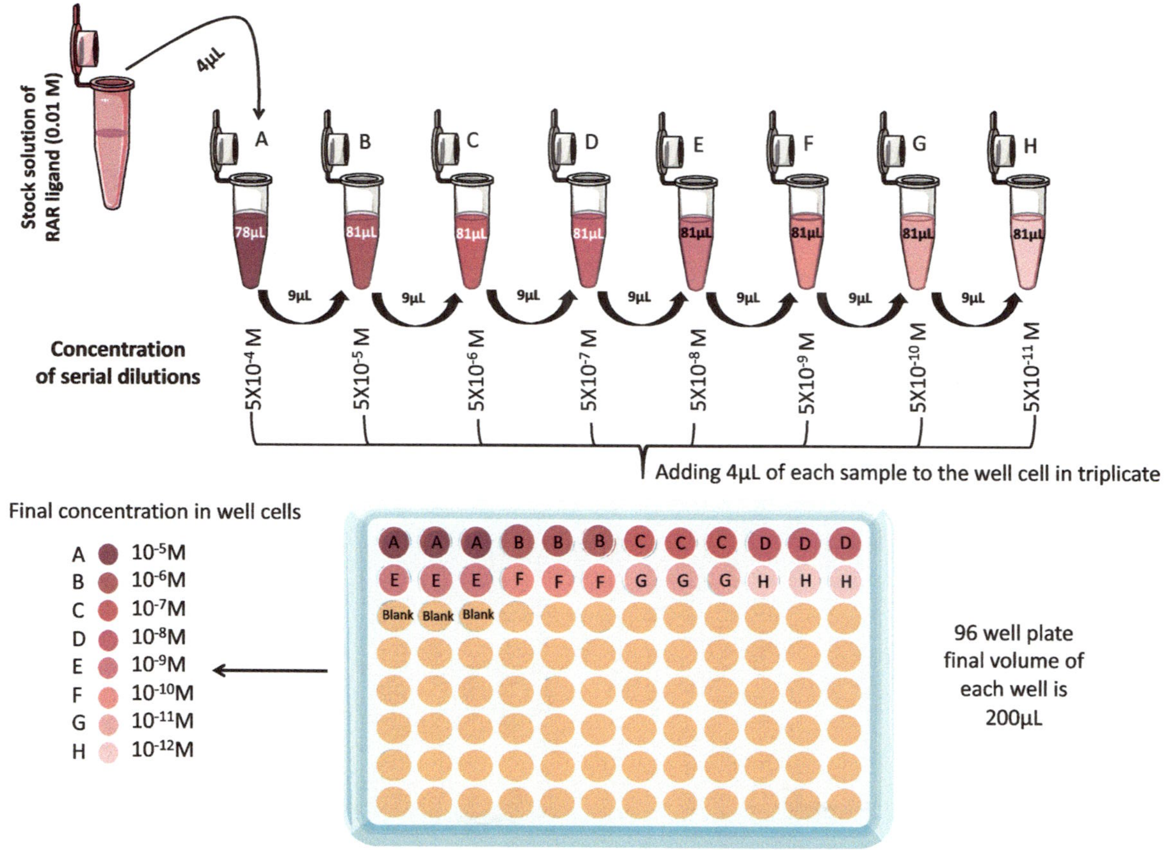

Fig. 2 Preparation of RAR ligands or RA serial dilution for standard curve generation. Prepare serial dilutions of RA/RAR ligands from 5×10^{-4} M to 5×10^{-11} M in the supernatant of non-injected homogenized brain samples. Prepare the highest concentration of 5×10^{-4} M and each subsequent concentration 10-fold diluted (i.e., 5×10^{-4} M, 5×10^{-11} M, ...) until reaching the lowest concentration of 5×10^{-11} M. Add 4 μL of each dilution to the cell in triplicate. Include three blank wells containing the supernatant of non-injected homogenized brain samples in LES to subtract any background. Incubate plate overnigh in a TC incubator.

storage and rapidly thaw the cells in a 37 °C cell culture water bath, removing immediately after the last of the ice has thawed. Transfer cells to the flask and incubate in a TC incubator. Supplemented medium must be changed following overnight culture to remove any remaining freezing medium containing DMSO. After overnight culture the cells will have adhered to the base of the flask. Replace half supplemented medium every 3–4 days with fresh medium. Passage cells once they have reached 80% confluence.

2. *Passaging reporter cells*: To detach the Sil-15 reporter cells from the flask, remove supplemented medium and wash the flask with sterile 1 × PBS. Add 2 mL trypsin to the flask and incubate for 5 min in a TC incubator. After 5 min, firmly tap the bottom of the flask to facilitate the detachment of the cells. When the cells detach completely, add a 10-fold higher volume of supplemented medium containing 10% FBS to neutralize the trypsin. Transfer the cells to a 15 mL falcon tube. Centrifuge the cells at a RCF of 226*g* for 4 min at RT to pellet the cells. Remove supernatant and replace with 1 mL supplemented medium. Count cells either by using a hemocytometer or an automated cell counter instrument by mixing 15 μL of 0.4% trypan blue with the same volume of suspended cells in a microfuge tube. Seed cells at a density of $1–2 \times 10^6$ in a T-75 flask by diluting with the appropriate volume of supplemented medium. Place the flask in a TC incubator. The medium should be changed every 3–4 days.
3. *Plating the cells in a 96-well plate*: Although the Sil-15 reporter cells are adherent to tissue culture plastic the assay requires that they stick with higher avidity and this is achieved by using 96-well plates coated with gelatin. The 96-well plates are thus coated by adding 100 μL of sterile 0.2% gelatin to each well and incubating for 2 h at RT or at 37 °C for 1 h. Remove the gelatin and rinse the plate twice with 100 μL of 1 × PBS. The gelatin must not be allowed to dry on the plate otherwise it will lose its adherent properties. Thus, after last washing the well should be left with 100 μL of 1 × PBS at 4 °C. The Sil-15 reporter cells can now be trypsinized as described in step 2 and plated at 8000–10,000 cells/well in 96-well plate. The plate can now be placed in a TC incubator, changing the medium every 3–4 days until the cells are 70–80% confluent when they will be ready for the assay. Note: Do not include G-418 in the medium for these steps.
4. *Freezing and storage of reporter cells*: Trypsinize the T-75 flask as described earlier. Centrifuge cells at a RCF of 450*g* for 7 min at 4 °C to pellet the

cells. Remove supernatant and resuspend in 2 mL of FBS containing 10% DMSO. Transfer 1 mL of the cells to each of two cryotubes. Transfer the tubes to an appropriate freezing chamber such as a "Mr. Frosty" and freeze undisturbed at −70 to −80 °C for 4 h or more before transferring to −130 °C or below for long-term storage.

5. *Assessing Sil-15 reporter cells responsiveness and sensitivity to low concentration of RA*: Because the cells vary over time in responsiveness it is advisable that they are tested first for their responsiveness by preparing a standard curve as described in Section 4.6. Serial dilution can be prepared in the culture medium instead of the supernatant of non-injected homogenized brain sample. Sensitivity should be down to at least 0.1 nM. If not, the cells can be re-selected for high response by fluorescence-activated cell sorting (FACS) using a fluorescent substrate for live cells.
6. *Fluorescence activated cell sorting (FACS) to improve Sil-15 reporter cell sensitivity*: FACS may be used to select the most sensitive Sil-15 reporter cells and improve the sensitivity of the assay. This can be achieved using, for instance, a FluoReporter™ lacZ flow cytometry kit (CAT# F1930, Fisher Scientific). In the kit fluorescein di-β-D-galactopyranoside (FDG) is used as a fluorogenic β-galactosidase substrate to quantitate enzyme activity in live cells. The competitive inhibitor, phenylethyl beta-D-thiogalactopyranoside (PETG) is used to stop the reaction. Three different concentration of RA are added to induce lacZ (10^{-8}, 10^{-10} and 10^{-12} M). The next day, a 2 mM solution of FDG is added to an equal volume of cell suspension in supplemented medium without G-418 and incubated for 1 min at 37 °C. The solution is diluted 10-fold with ice cold supplemented medium without G-418 and incubated for 30–60 min on ice. The reaction is stopped by adding PETG. FACS sorting is used to separate the top 5–10% cells using the lowest RA concentration that gives a clear separation of this group to generate a new population of highly sensitive reporter cells.

5. Safety considerations and standards

Retinoids have teratogenic effects and suitable precautions must be taken while working with them (Soprano, 1995). Glutaraldehyde solution is toxic and must be used under a fume hood. Dimethylformamide is easily absorbed through the skin and can cause liver damage and suitable steps should be taken to prevent exposure.

6. Analysis and statistics

Data were presented as mean ± SEM of independent experiments. Statistical analyses were performed using Microsoft Office Excel 2017 or GraphPad Prism 5. Student *t*-test, two-way repeated measures ANOVA and Sidak's multiple comparison test were used to determine statistically significant results.

7. Results

7.1 Optimization of LES volume and homogenate volume added to cells

Efficient homogenisation of the tissue was critical for optimal release of RA/RAR ligands into the solution. The mouse eye were used to optimize LES volume for homogenisation detecting the relatively high levels of endogenous RA in the eye (McCaffery et al., 1992). 100 μL LES buffer was optimal for detecting RA as it yielded the highest colorimetric values compared to higher LES volumes (Fig. 3). Addition of 4 μL of the

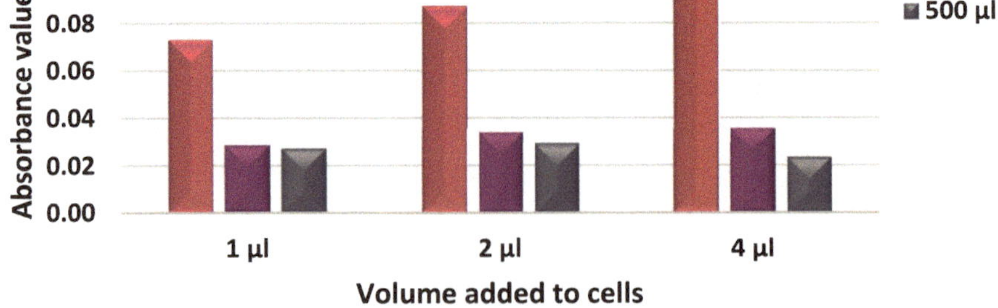

Fig. 3 Optimization of LES volume for the adult eye. Homogenization of the adult eye in 100 μL LES yielded the optimal colorimetric response in Sil-15 cells and addition of 4 μL of the homogenate to cells yielded the highest Sil-15 response.

homogenate to 196μL of media above the cells was found to yield the highest response in Sil-15 cells, although this can be modified to optimize for a partcular tissue type (see Section 9). Addition of higher volumes of homgenate to the cells was tested but 8 μL and 16 μL resulted in cell detachment due to toxicity and a corresponding decrease in colorimetric values (Fig. 4).

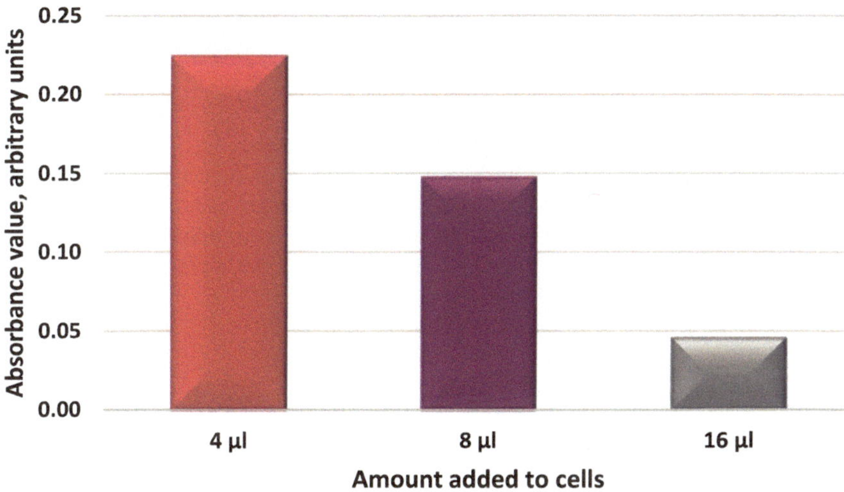

Fig. 4 Optimization of volume of homogenate added to Sil-15 cells. 4 μL in 200 μL media was the optimal volume of homogenate to add to the cells as addition of 8 and 16 μL resulted in significant detachment of cells leading to decreased colorimetric values.

7.2 Comparison between sorted and unsorted Sil-15 cell sensitivity

The sensitivity of Sil-15 cells to RA/RAR ligands could be increased by FACS. The top 5–10% of cells exposed to 10^{-11} M RA were sorted and reseeded onto a new 96-well plate. The sensitivity of this new Sil-15 population to RA was increased with a significantly higher response to 10^{-11} M RA compared to the original Sil-15 population (Fig. 5).

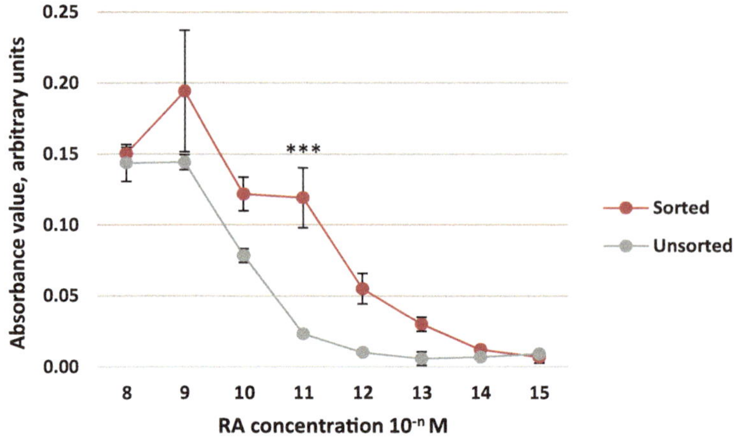

Fig. 5 Comparison of cells sorted using FACS versus control cells. Fluorescence-activated cell sorting (FACS) of Sil-15 cells increases sensitivity to RA. Sorted Sil-15 cells had significantly increased sensitivity to RA at 10^{-11} M. Significance testing was by two-way repeated measures ANOVA $P < 0.01$ and Sidak's multiple comparison test, $***P < 0.001$. (n = number on the X-axis).

7.3 Standard curve

The Sil-15 reporter cells respond well to RA standards, or RAR ligand, mixed with non-injected mouse brain tissue (Fig. 6 showing the RAR ligand DC645). This can be used to semi-quantitate DC645 in tissue after administration to an animal (Fig. 8).

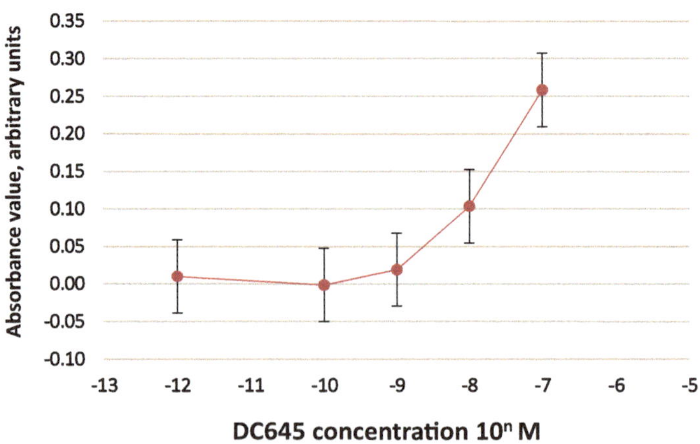

Fig. 6 Standard curve. The response of Sil-15 reporter cells to DC645 measured by an ELISA reader at 615 nm absorbance. The curve illustrates the reporter response to a range of DC645 concentrations with DC645 mixed with non-injected mouse brain tissue as described in the text. (n = number on the X-axis).

7.4 Tissue distribution of DC645 after administration into mice

Tissue distribution of DC645 in different brain regions and spinal cord 4 h after intraperitoneal injection is illustrated in Fig. 7, plotting simply against absorbance value on the y-axis and compared to the corresponding control region (DMSO injected tissue).

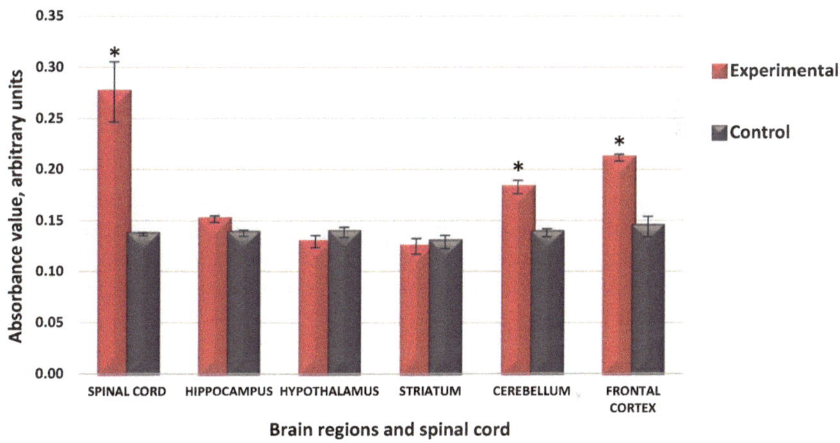

Fig. 7 Sil-15 reporter response to tissue extract from mouse injected with 1 mg/kg DC645. Absorbance values showing relative DC645 levels in different brain regions and spinal cord 4 h after single IP injection compared to DMSO injection. Significance testing was by student *t*-test $*P < 0.05$.

7.5 Relative tissue concentration of DC645 after administration into mice

To determine the relative concentration of DC645 in each tissue, the response of Sil-15 cells to extracts from brain regions and spinal cord experimental samples can be compared to the standard curve of Sil-15 cell response to known concentrations of DC645 and control (background) subtracted (Fig. 8).

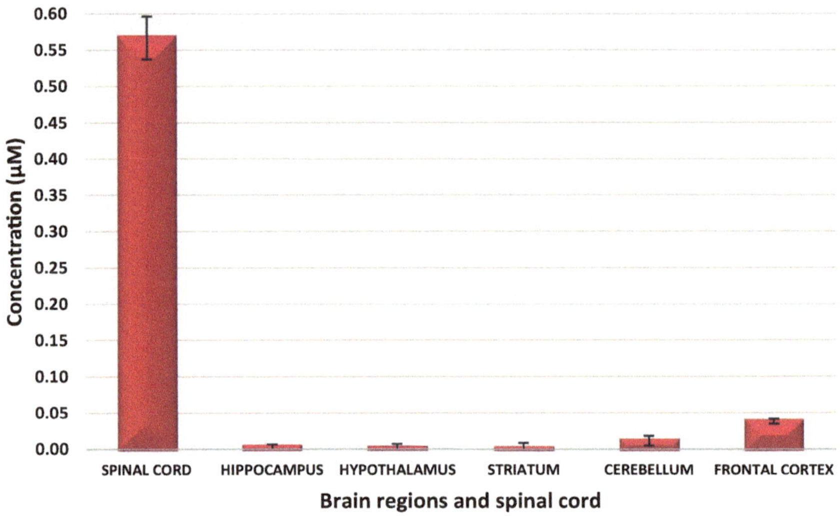

Fig. 8 Sil-15 reporter response to tissue extract from mouse injected with 1 mg/kg DC645 presented as concentration of ligand in tissue using DC645 standard curve and with control (background) subtracted. The data from Fig. 7 was replotted with the control (background) subtracted and the standard curve used to calculate actual DC645 concentrations.

8. Pros and cons of the RAR ligand quantitation bioassay

Pros	Cons
Relative quantitation of RAR ligand levels in tissue	Solvent used to extract can decrease the response of the reporter cells to RAR ligand
High sensitivity to low levels of RAR ligands (10^{-10} M RA or lower)	Occasional reduction in sensitivity of reporter cells with high passage number
Can detect RAR ligand levels in different tissues including those with high lipid content	The assay will also respond to the low levels of endogenous RA

9. Troubleshooting and optimization

Problem	Solution
RAR ligand insolubility when injecting into animals	Adjust the PBS:DMSO ratio to increase DMSO to avoid precipitation (consider though that both a high dose of DMSO or RAR ligands may be toxic for animals).
Toxicity in mice because of DMSO when high % is used to dissolve a high dose of RA or RAR ligands (>10 mg/kg)	As RA and RAR ligands are lipophilic DMSO is needed to keep them soluble, however levels of DMSO higher than 10% can have adverse effects on the animal. The solubility of these RAR ligands can be improved without high levels of DMSO by dissolving them in tissue culture medium (minimal essential medium) supplemented with 10% FBS.
Poor RAR ligand extraction	*Extraction mixture*: LES is a compromise of extraction solvents that, at the concentration used, are not toxic to the Sil-15 reporter cells. If considered that RAR ligand is not fully extracted or the LES extraction mixture is toxic for the cells, then changes in the ratio of solvents can be tested. *Volume of LES for tissue extraction*: For different tissues adjustment of the ratio of LES volume to tissue weight can further improve extraction of the drug from the tissue by LES. *Volume of tissue extraction added to cells*: The optimum volume to be added to cells from tissue extraction is 4 μL to 200 μL of media and higher volumes will reduce the colorimetric value due to killing of the Sil-15 reporter cells in the well. Lower volume however will add less extracted RAR ligand to the reporter cells. Small adjustments in the volume added to the cells may be tested to improve signal. *Exposure time*: The maximum exposure time after adding the tissue extract to the Sil-15 cells is 24 h. Increasing the exposure time can result in cell detachment due to toxicity. On the other hand, inadequate exposure time leads to a decrease in response. A suitable exposure time will be between 18 and 24 h.

Continued

—cont'd

Problem	Solution
Sil-15 reporter cell responsiveness	To maintain the health and responsiveness of the Sil-15 reporter cells it is important to consider the following: *Confluency*: During regular passaging of the cells do not allow them grow to confluency. *G-418*: Sil-15 reporter cells may reduce in their RA responsiveness over time. Antibiotic resistance selection will help maintain the response to RAR ligands. *High passages*: Later passages tend to show an inconsistent response to RAR ligand. A maximum of 5–8 passages will preserve their responsiveness. *Regular check of sensitivity of cells*: Even in the presence of G-418 and use of cells of low passage number the response of the Sil-15 reporter cells to RAR ligands may fall. It is therefore advisable to test the responsiveness of the Sil-15 reporter cells before a bioassay.
Background response to endogenous RA	Any background is subtracted from the result by normalizing to tissue from vehicle injected control animals.
Concentrations outside the linear standard curve	The dose dependent response of Sil-15 reporter cells for our results was linear between 10^{-10} and 10^{-7} M. If the sample response is above this range, then the samples require to be diluted before the assay. If below this range, then a higher amount of RAR ligand is required to be injected into the animal or larger amount of tissue extracted.
Detachment of Sil-15 reporter cells from plate	Gelatin and fixative solutions must be prepared freshly. Washing steps before fixation should be gentle. Medium should be changed regularly noting the color (pH change) of the medium as the cells release metabolite. Be aware of any contamination by microbes and that cells are healthy (cells look bright under brightfield inverted microscope). Cells should not be 100% confluent in the 96-well plate.
Variation in color development in colorimetric assay for each individual assay	It is necessary to have a standard curve for each experiment in the same 96-well plate to account for any variation in response that may occur between plates.

10. Quick summary of the protocol

1. *Preparation of Sil-15 reporter cells*: Seed the cells in a 96-well plate in 196 μL DMEM/F12, GlutaMAX supplemented with 10% FBS, and 1% pen-strep without G-418 (when using G418 do not use pen-strep since it is a competitive inhibitor of G-418). Incubate in a TC incubator until 80–90% confluency.
2. *Preparation and injection of RAR ligands or retinoic acid*: Prepare and administer suitable dose of RA/RAR ligand to animal.
3. *Tissue collection*: Cull animal at a suitable time after injection. Dissect the tissue and keep on dry-ice.
4. *Tissue homogenization*: Add 100 μL of LES (2:1) ethanol:isopropanol and homogenize by pestle. Centrifuge samples at a RCF of 10,000*g* for 5 min at 4 °C.
5. *Cell treatment*: Add 4 μL of each sample supernatant to the wells of 96-well plates in triplicate. Incubate plate overnight in a TC incubator.
6. *Standard curve*: Generate standard curve by preparing serial dilutions of the drug in the supernatant of non-injected homogenized brain sample and add to the cells.
7. *Fixation*: After 24 h, remove medium and wash cells gently twice with 1 × PBS. Add 100 μL of fixation solution to each well and incubate for 15 min at RT. Remove fixation solution and wash gently twice with 1 × PBS.
8. *Retinoic acid bioassay*: Add 100 μL of X-gal development solution to each well. Wrap the 96-well plate with foil and keep at 37 °C. Quantify the colorimetric reaction with an ELISA reader at 615 nm (absorbance range 570–700 nm) between 2 and 24 h after start of development.

11. Summary

Dysregulation of RA signaling has been considered in the onset and progression of several disorders including neurodegenerative disease. RAR ligands may be employed as future therapeutics. Determining RAR ligand tissue distribution after administration will be necessary to effectively develop these drugs and determine whether they distribute to the relevant tissues. This protocol provides a way to measure relative levels of a wide range of RAR ligand drugs in tissues.

Acknowledgments

This work was supported by the Elphinstone Scholarship University of Aberdeen, the Sir Richard Stapley Educational Trust and the "SPRINT-MND/MS PhD programme, which is funded by the Chief Scientist Office (ref MMPP/01) and the Universities of Aberdeen, Dundee, Edinburgh, Glasgow and St Andrews.

References

Ababon, M. R., Li, B. I., Matteson, P. G., & Millonig, J. H. (2016). Quantitative measurement of relative retinoic acid levels in E8.5 embryos and neurosphere cultures using the F9 RARE-Lacz cell-based reporter assay. *Journal of Visualized Experiments*, *115*, e54443 (1–9). https://doi.org/10.3791/54443.

Al Tanoury, Z., Piskunov, A., & Rochette-Egly, C. (2013). Vitamin A and retinoid signaling: Genomic and nongenomic effects. *Journal of Lipid Research*, *54*(7), 1761–1775. https://doi.org/10.1194/jlr.R030833.

Barnard, J. H., Collings, J. C., Whiting, A., Przyborski, S. A., & Marder, T. B. (2009). Synthetic retinoids: Structure-activity relationships. *Chemistry—A European Journal*, *15*(43), 11430–11442. https://doi.org/10.1002/chem.200901952.

Bushue, N., & Wan, Y.-J. Y. (2010). Retinoid pathway and cancer therapeutics. *Advanced Drug Delivery Reviews*, *62*(13), 1285–1298. https://doi.org/10.1016/j.addr.2010.07.003.

Chen, L., Lau, A. G., & Sarti, F. (2014). Synaptic retinoic acid signaling and homeostatic synaptic plasticity. *Neuropharmacology*, *78*(C), 3–12. https://doi.org/10.1016/j.neuropharm.2012.12.004.

Corcoran, J. P. T., So, P. L., & Maden, M. (2004). Disruption of the retinoid signalling pathway causes a deposition of amyloid beta in the adult rat brain. *European Journal of Neuroscience*, *20*(4), 896–902. https://doi.org/10.1111/j.1460-9568.2004.03563.x.

Dzerk, A., Carlson, A., Loewen, G., Shirley, M., & Lee, J. (1998). A HPLC method for the determination of 9-cis retinoic acid (ALRT1057) and its 4-oxo metabolite in human plasma. *Journal of Pharmaceutical and Biomedical Analysis*, *16*(6), 1013–1019. https://doi.org/10.1016/S0731-7085(97)00114-3.

Esteves, M., Cristóvão, A. C., Saraiva, T., Rocha, S. M., Baltazar, G., Ferreira, L., et al. (2015). Retinoic acid-loaded polymeric nanoparticles induce neuroprotection in a mouse model for Parkinson's disease. *Frontiers in Aging Neuroscience*, 7, 20. https://doi.org/10.3389/fnagi.2015.00020.

Evans, J. E., & McCaffery, P. (2010). *HPLC/MSN analysis of retinoids*. (pp. 149–162). https://doi.org/10.1007/978-1-60327-325-1_8.

Goodman, A. B., & Pardee, A. B. (2003). Evidence for defective retinoid transport and function in late onset Alzheimer's disease. *Proceedings of the National Academy of Sciences of the United States of America*, *100*(5), 2901–2905. https://doi.org/10.1073/pnas.0437937100.

Haffez, H., Khatib, T., McCaffery, P., Przyborski, S., Redfern, C., & Whiting, A. (2018). Neurogenesis in response to synthetic retinoids at different temporal scales. *Molecular Neurobiology*, *55*(3), 1942–1950. https://doi.org/10.1007/s12035-017-0440-7.

Hu, L., & Gudas, L. J. (1990). Cyclic AMP analogs and retinoic acid influence the expression of retinoic acid receptor alpha, beta, and gamma mRNAs in F9 teratocarcinoma cells. *Molecular and Cellular Biology*, *10*(1), 391–396. https://doi.org/10.1128/MCB.10.1.391.

Kane, M. A., & Napoli, J. L. (2010). *Quantification of endogenous retinoids. In Methods in molecular biology* (pp. 1–54). https://doi.org/10.1007/978-1-60327-325-1_1.

Kelley, M. W., Xu, X. M., Wagner, M. A., Warchol, M. E., & Corwin, J. T. (1993). The developing organ of Corti contains retinoic acid and forms supernumerary hair cells in response to exogenous retinoic acid in culture. *Development*, *119*(4), 1041–1053.

Kolarcik, C. L., & Bowser, R. (2012). Retinoid signaling alterations in amyotrophic lateral sclerosis. *American Journal of Neurodegenerative Disease*, *1*(2), 130–145. Retrieved from http://www.ncbi.nlm.nih.gov/pubmed/23383387.

Lee, H.-P., Casadesus, G., Zhu, X., Lee, H., Perry, G., Smith, M. A., et al. (2009). All-trans retinoic acid as a novel therapeutic strategy for Alzheimer's disease. *Expert Review of Neurotherapeutics*, *9*(11), 1615–1621. https://doi.org/10.1586/ern.09.86.

Levitsky, K. L., Toledo-Aral, J. J., López-Barneo, J., & Villadiego, J. (2013). Direct confocal acquisition of fluorescence from X-gal staining on thick tissue sections. *Scientific Reports*, *3*(1), 2937. https://doi.org/10.1038/srep02937.

Malaspina, A., & Michael-Titus, A. T. (2007). Is the modulation of retinoid and retinoid-associated signaling a future therapeutic strategy in neurological trauma and neurodegeneration? *Journal of Neurochemistry*, *104*(3), 584–595. https://doi.org/10.1111/j.1471-4159.2007.05071.x.

McCaffery, P., & Drager, U. C. (1994). Hot spots of retinoic acid synthesis in the developing spinal cord. *Proceedings of the National Academy of Sciences of the United States of America*, *91*(15), 7194–7197. https://doi.org/10.1073/pnas.91.15.7194.

McCaffery, P., & Dräger, U. C. (1997). A sensitive bioassay for enzymes that synthesize retinoic acid. *Brain Research Protocols*, *1*(3), 232–236. https://doi.org/10.1016/S1385-299X(96)00034-7.

McCaffery, P., Lee, M. O., Wagner, M. A., Sladek, N. E., & Drager, U. C. (1992). Asymmetrical retinoic acid synthesis in the dorsoventral axis of the retina. *Development*, *115*(2), 371–382.

McCormick, D. L., Rao, K. V. N., Steele, V. E., Lubet, R. A., Kelloff, G. J., & Bosland, M. C. (1999). Chemoprevention of rat prostate carcinogenesis by 9-cis-retinoic acid. *Cancer Research*, *59*(3), 521–524.

Mendelsohn, C., Ruberte, E., Lemeur, M., Morriss-Kay, G., & Chambon, P. (1991). Developmental analysis of the retinoic acid-inducible RAR-β2 promoter in transgenic animals. *Development*, *113*(3), 723–734.

Mey, J., & Mccaffery, P. (2004). Retinoic acid signaling in the nervous system of adult vertebrates. *The Neuroscientist*, *10*(5), 409–421. https://doi.org/10.1177/1073858404263520.

Moise, A. R., Kuksa, V., Blaner, W. S., Baehr, W., & Palczewski, K. (2005). Metabolism and transactivation activity of 13,14-dihydroretinoic acid. *Journal of Biological Chemistry*, *280*(30), 27815–27825. https://doi.org/10.1074/jbc.M503520200.

Niederreither, K., & Dollé, P. (2008). Retinoic acid in development: Towards an integrated view. *Nature Reviews Genetics*, *9*(7), 541–553. https://doi.org/10.1038/nrg2340.

Niewiadomska-Cimicka, A., Krzyżosiak, A., Ye, T., Podleśny-Drabiniok, A., Dembélé, D., Dollé, P., et al. (2017). Genome-wide analysis of RARβ transcriptional targets in mouse striatum links retinoic acid signaling with Huntington's disease and other neurodegenerative disorders. *Molecular Neurobiology*, *54*(5), 3859–3878. https://doi.org/10.1007/s12035-016-0010-4.

Reynolds, K., Mezey, E., & Zimmer, A. (1991). Activity of the β-retinoic acid receptor promoter in transgenic mice. *Mechanisms of Development*, *36*(1–2), 15–29. https://doi.org/10.1016/0925-4773(91)90068-H.

Rochette-Egly, C. (2015). Retinoic acid signaling and mouse embryonic stem cell differentiation: Cross talk between genomic and non-genomic effects of RA. *Biochimica et Biophysica Acta (BBA)—Molecular and Cell Biology of Lipids*, *1851*(1), 66–75. https://doi.org/10.1016/j.bbalip.2014.04.003.

Rossant, J., Zirngibl, R., Cado, D., Shago, M., & Giguere, V. (1991). Expression of a retinoic acid response element-hsplacZ transgene defines specific domains of transcriptional activity during mouse embryogenesis. *Genes & Development*, *5*(8), 1333–1344. https://doi.org/10.1101/gad.5.8.1333.

Sakai, Y., & Dräger, U. C. (2010). *Detection of retinoic acid catabolism with reporter systems and by in situ hybridization for CYP26 enzymes. In Methods in molecular biology* (pp. 277–294). https://doi.org/10.1007/978-1-60327-325-1_16.

Shudo, K., Fukasawa, H., Nakagomi, M., & Yamagata, N. (2009). Towards retinoid therapy for Alzheimers disease. *Current Alzheimer Research*, *6*(3), 302–311. https://doi.org/10.2174/156720509788486581.

Siddikuzzaman, Guruvayoorappan, C., & Berlin Grace, V. M. (2011). All trans retinoic acid and cancer. *Immunopharmacology and Immunotoxicology*, *33*(2), 241–249. https://doi.org/10.3109/08923973.2010.521507.

Sitnik, K. M., Kotarsky, K., White, A. J., Jenkinson, W. E., Anderson, G., & Agace, W. W. (2012). Mesenchymal cells regulate retinoic acid receptor-dependent cortical thymic epithelial cell homeostasis. *The Journal of Immunology*, *188*(10), 4801–4809. https://doi.org/10.4049/jimmunol.1200358.

Soprano, D. R. (1995). Retinoids as teratogens. *Annual Review of Nutrition*, *15*(1), 111–132. https://doi.org/10.1146/annurev.nutr.15.1.111.

Stoney, P. N., Fragoso, Y. D., Saeed, R. B., Ashton, A., Goodman, T., Simons, C., et al. (2016). Expression of the retinoic acid catabolic enzyme CYP26B1 in the human brain to maintain signaling homeostasis. *Brain Structure and Function*, *221*(6), 3315–3326. https://doi.org/10.1007/s00429-015-1102-z.

Tan, A. U., Schlosser, B. J., & Paller, A. S. (2018). A review of diagnosis and treatment of acne in adult female patients. *International Journal of Women's Dermatology*, *4*(2), 56–71. https://doi.org/10.1016/j.ijwd.2017.10.006.

Tang, X.-H., & Gudas, L. J. (2011). Retinoids, retinoic acid receptors, and cancer. *Annual Review of Pathology: Mechanisms of Disease*, *6*(1), 345–364. https://doi.org/10.1146/annurev-pathol-011110-130303.

Tyring, S. K., Kircik, L. H., Pariser, D. M., Guenin, E., Bhatt, V., & Pillai, R. (2018). Novel tretinoin 0.05% lotion for the once-daily treatment of moderate-to-severe acne vulgaris: Assessment of efficacy and safety in patients aged 9 years and older. *Journal of Drugs in Dermatology*, *17*(10), 1084–1091.

Wagner, M., Han, B., & Jessell, T. M. (1992). Regional differences in retinoid release from embryonic neural tissue detected by an in vitro reporter assay. *Development*, *116*(1), 55–66.

Yamamoto, M., Dräger, U., & McCaffery, P. (1998). A novel assay for retinoic acid catabolic enzymes shows high expression in the developing hindbrain. *Developmental Brain Research*, *107*(1), 103–111. https://doi.org/10.1016/S0165-3806(98)00003-0.

CHAPTER TWENTY

The development of methodologies for high-throughput retinoic acid binding assays in drug discovery and beyond

Charles W.E. Tomlinson, Andrew Whiting*
Department of Chemistry, Durham University, Lower Mountjoy, Durham, United Kingdom
*Corresponding author: e-mail address: andy.whiting@durham.ac.uk

Contents

Abstract

Assays for identifying the activity of retinoic acids and retinoic acid derivatives, have become increasingly accessible in the 21st century. Movement away from complex *in cellulo* expression and radiological systems toward accessible FRET or fluorescent compound led studies has increased the accessibility and high-throughput capabilities of the field. The improvement in assaying technologies will lead to both increased and faster screening of retinoids and related compounds, improving drug discovery workflows and implementation pipelines. This enabling technology will allow better understanding of the huge range of phenotypes closely linked to retinoid signaling, and through the development of improved treatments improve outlooks for diagnoses in these cases and those of closely related diseases. In this chapter we present an examination of historical and modern techniques for assaying the activity of retinoic acid derivatives against their binding proteins.

Methods in Enzymology, Volume 637
ISSN 0076-6879
https://doi.org/10.1016/bs.mie.2020.02.008

Abbreviations

13/9-*cis*-RA	13/9-*cis*-retinoic acid
ATRA	all-*trans*-retinoic acid
CAT	chloramphenicol acetyl transferase
CRABPI/II	cellular retinoic acid binding protein I/II
DHA	docosahexaenoic
EC50	effective concentration 50%
FPLC	fast protein liquid chromatography
FRET	Fourier resonance energy transfer
GST	glutathione *S*-transferase
HPLC	high performance liquid chromatography
HTRF	homogenous time resolved fluorescence
IPTG	isopropyl β-D-1-thiogalactopyranoside
PBS	phosphate buffered saline
RAR	retinoic acid receptor
RARE/RXRE	retinoic acid response element/retinoid X response element
RXR	retinoid X receptor
Tb	terbium
TREp	thyroid receptor palindromic sequence

1. Introduction

The complex and multifaceted nature of retinoid signaling has created a strong desire to understand the binding of retinoic acid and its derivatives to their cognate signaling proteins. The design, synthesis and understanding of new retinoid-like molecules has enabled the probing of these interactions at a molecular level, enlightening decision makers throughout the drug development pipeline. These technologies can in turn provide feedback to assist in more fully understanding the retinoid signaling pathway, potentially opening the field up to the creation of new and potent effectors of gene regulation. The ability to screen large libraries of retinoid and retinoid analogous drug-like molecules, as efficiently as possible, will contribute to the increasing portfolio of options going forward. To that end, accurate and high-throughput methods of assaying compounds for activity are vital, and have been the subject of research for over 30 years (Allenby et al., 1993; Heyman et al., 1992; Stafslien, Vedvik, De Rosier, & Ozers, 2007; Tomlinson, Chisholm, Valentine, Whiting, & Pohl, 2018; Yamada et al., 2019).

Naturally occurring retinoic acids including all-*trans*-retinoic-acid (ATRA), 9-*cis*-retinoic acid, and 13-*cis*-retinoic acid are prone to both photo-isomerization and oxidation under ambient conditions, reducing their biological efficacy *in vivo* and making a dedicated study of individual isomers a troublesome task, as detailed in chapter "Vitamin A and natural retinoids" by Redfern (Murayama, Suzuki, & Matsui, 1997). Additionally, these molecules are inherently weakly fluorescent due to their conjugated electronic structure, though this weakness of signal makes them difficult to isolate for direct quantitative study (Vivas, Siqueira, Silva, de Boni, & Mendonca, 2015). As a result of these undesirable physicochemical properties a wide range of indirect measurement methods have been developed to test the efficacy of both endogenous and newly designed retinoid-like molecules. Incorporating both *in vitro* and *in vivo* methodologies, these methods cover 30 years of development from cellular based assays, through radiological probing and FRET techniques to recent fluorescent displacement procedures. For more detail on the natural ligands responsible for retinoid signaling, see the introduction and chapter "Classical pathway of gene regulation by retinoids" by Tomlinson and Pohl of this book.

2. Cellular expression systems

Early methods of determining the binding of a retinoid, or retinoid-like compound, were based around the expression of relevant signaling proteins, primarily variety of RAR and RXR isoforms, as well as the retinoic acid response elements (RARE/RXRE) coupled to a reporter gene (typically luciferase) in a suitable cellular system (Heyman et al., 1992). In particular, the determination that 9-*cis*-RA may constitute the endogenous ligand for RXRα (retinoid X receptor alpha) was carried out in insect cell culture of *Drosophila melanogaster* (Schneider S2 cells). These cultures were transfected with plasmids for expression of the RXR and expression of luciferase under the control of an RXRE, to create a system by which the binding of a retinoid compound to the receptor could be measured by the activity of the luciferase reporter. Upon binding of the retinoid and presumed dimerization the RXRs would bind to the plasmid-borne response element, triggering expression of luciferase and creating a measurable emission of light (Heyman et al., 1992) (Fig. 1). This light, measured by standard spectroscopic techniques, is generated catalytically by luciferase during the oxidation of luciferin and ATP in the presence of oxygen (O_2) to form oxyluciferin, which decays to release light (Gould & Subramani, 1988;

Totter, 1981). The quantum yield for the activity of luciferase is among the highest observed for bioluminescence at 0.88 under ambient conditions (Seliger & McElroy, 1960) and makes it an ideal, high sensitivity, reporter for expression based assays. Experiments were carried out using a two phase approach; initially binding ATRA (later other test retinoids) to RXRs before lysis and purification by HPLC, and then reintroduction of fractionated lysate to fresh S2 cells, intending to determine the activity of each separated fraction, and therefore, to identify an intermediate generated in culture from ATRA (Heyman et al., 1992). Co-eluted retinoids (those bound to protein) were tested separately to those which did not bind, and a determination was made that ATRA was being isomerized into 9-*cis*-retinoic-acid (9-*cis*-RA) in culture (Fig. 1).

While this method has many advantages, including determining whether a retinoid derivative (or synthetic retinoid derivative) is suitable for activating downstream signaling, it fails to take into account the possibility of *in vivo* isomerization, and the creation of other non-native, active signaling species within the heterologous cell culture. Additionally, it is difficult, if not impossible, to measure which DNA binding heterodimers are being formed, and whether these would realistically be formed *in vivo*. While heterologous expression in an insect cell system is acknowledged as creating comparable protein modifications to those made in mammalian cells and enables ease of growth relative to human cell lines, preparing, transforming and maintaining cells can be time consuming and costly, especially when measured against bacterial cell culture (Assenberg, Wan, Geisse, & Mayr, 2013; van Oers, Pijlman, & Vlak, 2015).

In a similar vein, the initial characterization of retinoic acid receptors (RAR), first discovered through their sequence homology to other steroid and thyroid hormone receptors (see chapter "Classical pathway of gene regulation by retinoids" by Tomlinson and Pohl), was carried out using an expression system hosted in CV1 primate cells (Giguere, Ong, Segui, & Evans, 1987; Jensen, Girardi, Gilden, & Koprowski, 1964). Introduction of a series of compounds, including the retinoic acids and related derivatives, was carried out in a system with a suitable RARE DNA binding site coupled to the expression of chloramphenicol acetyl transferase (CAT) (Giguere et al., 1987). The lack of endogenous CAT activity in mammalian cells makes the CAT assay a strong reporter system for heterologous gene expression. The activity of the enzyme is measured by radiography of ^{14}C labeled chloramphenicol and the corresponding degradation products and allows for a close measurement of gene expression (Gorman, Moffat, & Howard, 1982).

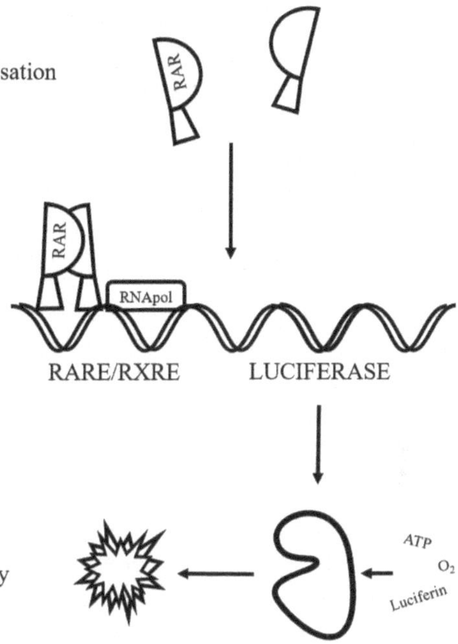

Fig. 1 Luciferase activity assays take the basic form of heterologously expressed RAR/RXR protein, alongside a luciferase reporter construct under the control of the retinoic acid response element (RARE) or retinoid X response element (RXRE) (Heyman et al., 1992). When suitable conditions exist for the dimerization of the RAR/RXR to occur (typically the binding of a ligand) the recruitment of the dimer to the promoter causes the transcription and subsequent translation of the luciferase enzyme. The active enzyme then processes ATP and the luciferin substrate, in the presence of O_2, added to the culture medium to create a light signal (Gould & Subramani, 1988; Totter, 1981). This signal is measured by standard spectroscopic techniques, and measurement can be undertaken in a high-throughput format. The luciferase reporter assay links the expression and activation of the target protein to the emission of light by the processing of luciferin, in this way a determination can be made of the activity of a particular substrate on the retinoic acid receptor or RXR.

This methodology demonstrated for the first time that the suspected novel gene product (RARα) was capable of incorporating and binding retinoic acid, based on the subsequent expression of the coupled reporter and CAT activity (Giguere et al., 1987).

The original characterization of the RXR receptor was also carried out using a similar cellular based investigation. Thyroid hormone receptor (TREp) palindromic sequence, coupled to chloramphenicol acetyl transferase activity was used as a binding target for the at-the-time unknown RXR. A mixture of retinol and retinoic acid was discovered to be the most effective modulator, as measured by CAT activity, and the receptor was shown to have greater ligand specificity than the previously discovered RARs which were activated by a wider range of metabolic derivatives of retinoic acid. This methodology relied initially on heterologous expression in S2 insect cell culture to reduce the chance of false positives generated by metabolic products from mammalian cell culture. Later experiments into specific retinoic acid derivatives were carried out in CV1 cells using an analogous approach (Mangelsdorf, Ong, Dyck, & Evans, 1990).

More contemporarily, the discovery of the RXR activating factor docosahexaenoic acid (DHA), isolated from the adult brain tissue of *mus musculus*, was carried out using heterologous expression of RXR and RAR proteins linked to a yeast transcription factor in human choriocarcinoma JEG-3 cells and a luciferase reporter construct (de Urquiza et al., 2000; Gould & Subramani, 1988). The cells were treated with extracts from both embryonic and adult mouse brain tissues, and demonstrated activation of RXR in the presence of DHA (later separated by HPLC). This was followed up by a specific investigation into DHA (and other poly unsaturated fatty acids) using a reporter system constructed in a 293T human kidney cell line and using RXRα as a binding target, which demonstrated a dose response EC_{50} for DHA of 50–100 μM (de Urquiza et al., 2000). Similar investigations carried out using luciferase reporter assays also demonstrated RXR agonist activities for other saturated and unsaturated fatty acids at micromolar affinities including palmitoleic (53 μM), oleic (82 μM) and myristic acids (>150 μM) (Goldstein, Dobrzyn, Clagett-Dame, Pike, & DeLuca, 2003).

3. Radiolabeling techniques

Radiolabeling methods have at various times been applied to retinoic acid signaling systems to detect the binding of compounds to their cognate proteins. Utilizing a tritiated derivative of ATRA (all-*trans*-[^{3}H]-RA) or 9-*cis*-retinoic acid (9-*cis*-[^{3}H]-RA), radiolabeling allows for measurement of labeled ligand free in solution *vs* that bound to the receptor, which allows for direct measurement of binding, or the measurement of the displacement affect caused by the binding of relevant test compounds (Fig. 2).

The binding of a basic suite of biologically available retinoids including ATRA, and 7-, 9-, 11- and 13-*cis* derivatives were determined using 9-*cis*-[10 −^{3}H]-RA, all-*trans*-[11,12-^{3}H]-RA and the nuclear lysate of transfected

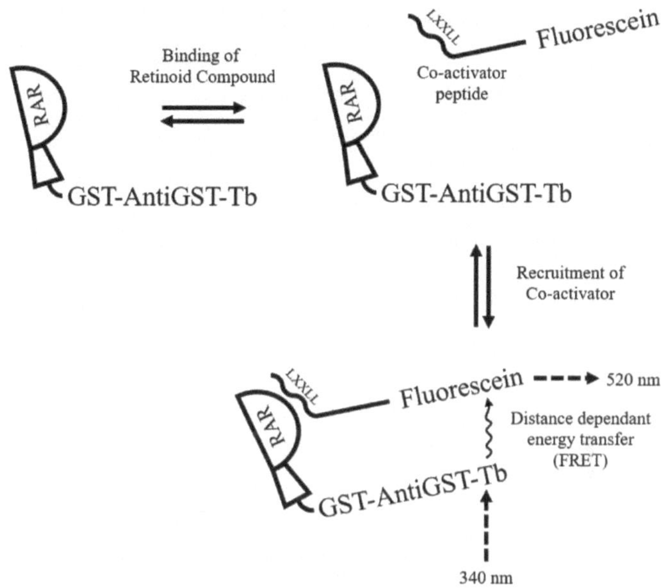

Fig. 2 The varied tritiated derivatives of ATRA and 9-*cis*-RA used in radiolabeling experiments; the synthesis of all-*trans*-[2,3-^{3}H]-RA and all-*trans*-[3,4-^{3}H]-RA are were developed to ensure high yield by including tritiation in the final steps of the process (Bennani & Boehm, 1995). All-*trans*-[11,12-^{3}H]-RA was used extensively in conjunction with 9-*cis*-[10-^{3}H]-RA to uncover the binding ability of a wide range of retinoid derivatives (both synthetic and suspected to occur naturally) (Allenby et al., 1993, 1994; Levin et al., 1992).

COS-1 primate cells—immortalized kidney cells Cercopithecus aethiops based on the CV-1 cell line (Allenby et al., 1993; Gluzman, 1981). Incubation of the cells with tritiated retinoid derivatives allowed the uptake of the ligands by heterologously expressed RARs/RXRs. After separation of the nuclear lysate a determination of binding could be carried out by titration of unlabeled ligand to displace the radiolabeled probe. This method is based on the high affinity of receptors for their cognate ligands, and the lack of these molecules in heterologous expression systems—the RARs will rapidly bind any available retinoid when expressed in a host lacking their natural ligands. Introduction of a radiolabeled probe, therefore, causes mass uptake and binding, creating a large pool of radio-ligand associated protein which can be used for competitive binding assays. Addition of a fixed concentration of competitor molecules, followed by separation by size exclusion chromatography allowed each fraction to be tested for radioactivity, determining the fraction of radio-probe still bound to protein *vs* that in the free volume

(Allenby et al., 1993). This ratio is the first step to determining the displacement effect of the previously introduced unlabeled ligand. Once ratios at a series of ligand concentrations are calculated, the rate of displacement and hence the affinity of the challenger ligand can be determined (Allenby et al., 1993; Levin et al., 1992).

Later efforts also used mouse, and human retinoic acid receptors, determining that 9-*cis*-RA and ATRA have different affinities for the different RAR isotypes (α, β and γ) (Allenby et al., 1994). This was carried out using similar heterologous expression in COS-1 cells, followed by separation by HPLC and analysis by scintillation counter. It was calculated that for each RARα, β and γ the off-rate for ATRA was similar (IC50 ~5 nM), whereas 9-*cis*-RA was displaced sixfold quicker from RARγ than RARβ, indicating a favorable interaction and ligand bias (Allenby et al., 1994). A series of experiments to determine the binding activity of all-*trans*-retinol and all-*trans*-retinaldehyde to the RARs also utilized a radio-ligand approach (Repa, Hanson, & Clagett-Dame, 1993). Both nuclear extract from human neuroblastoma and RAR proteins expressed heterologous in *E. coli* were examined by incubation with tritiated ATRA in competition with the ligands of interest, followed by separation and liquid scintillation counting. It was determined that all-*trans*-retinol was ~35-fold less effective at displacing the tritiated reagent than an ATRA control (0.09 nM *vs* 3.1 nM), and similarly, all-*trans*-retinaldehyde was ~500-fold less effective than the same (Repa et al., 1993). Given the importance of the acid head group of ATRA in binding to retinoic acid binding proteins, the contrast in binding efficiency can be attributed to the variation in oxidation states, though the nanomolar binding efficiencies attributed to each ligand still represent a significant binding ability. Alternate experimentation demonstrated that while all-*trans*-retinol might have binding affinity for the retinoic acid receptors, only 4-oxoretinol (a metabolic product produced in culture with 10–15% conversion (Leo et al., 1989; Leo & Lieber, 1985)) was capable of inducing transcription in reporter cell lines (Achkar et al., 1996).

Synthesis of tritium labeled reagents requires specialist and potentially national level facilities, and as with other non-labeled retinoids, the products require storage at low temperature and under low light to prevent isomerization (Boehm et al., 1994). While radiolabeled ATRA and *cis*-retinoids can be synthesized *de novo* it is also possible to interconvert them through isomerization and subsequent separation by HPLC (Cahnmann, 1995). The synthesis of tritiated ATRA can be carried out in a regioselective manner on the 2,3 or 3,4 positions in the cyclohexene ring, and by radiolabeling in the final steps of the reaction maintains a high radiochemical yield (Bennani & Boehm, 1995). The process is carried out in anhydrous benzene, using Wilkinson's catalyst (Meakin, Jesson, & Tolman, 1972) and allows for hydrogenation using hydrogen, deuterium or tritium—depending on application. It was determined that the 2,3-tritiation was preferable, due to the proclivity of the all-*trans*-[3,4-^{3}H]-ATRA to be oxidized in solution to form 4-oxo-retinoic acid (Bennani & Boehm, 1995).

4. Förster resonance energy transfer (FRET) techniques

Förster resonance energy transfer (FRET) is a distance dependent energy transfer between fluorescent donor and acceptor molecules, which can be harnessed to "measure" intermolecular distances and investigate interactions between molecules (Förster, 1948, 2012). A standard pairing includes a donor (which absorbs energy from an external source) and an acceptor (which absorbs the energy transferred from the donor). The energy transfer creates a drop in the emission of the donor molecule, and a corresponding increase in the fluorescence output of the acceptor. These interactions scale between 10 and 100 Å and are of increasing importance to the measurement of interactions between biomolecules, allowing calculation of contacts between substrates both *in vitro* (in cell extracts and purified proteins) and *in vivo* (using suitably tagged expression systems) (Stryer, 1978; Tsien, 1998). Since their inception, homogenous time resolved fluorescence (HTRF) assays have become widely used across the field, combining FRET measurements with a time resolved element in a single, small volume, robust format. The range of targets extends well beyond retinoid receptors, to a variety of other nuclear receptors, GPCR signaling proteins, kinases as well as general protein-protein, protein-peptide and protein-DNA interactions (Degorce et al., 2009).

This indirect method of measuring the interaction between molecules has been applied to both the RARs and RXRs to determine the binding of retinoic acids and other related molecules. In the case of RXRβ, expressed heterologously in SF9 insect cell culture, a terbium/fluorescein donor/acceptor pairing was used to determine the binding of both 9-*cis* and all-*trans*-retinoic acid as well as fatty acids phytanic acid (PA) and docosahexaenoic acid (DHA). The method relied on the recruitment of coregulatory peptides bearing a LXXLL motif which were labeled with fluorescein to the ligand binding domain of the RXR, which was itself labeled with terbium *via* a GST-tag and antibody pairing (Fig. 3).

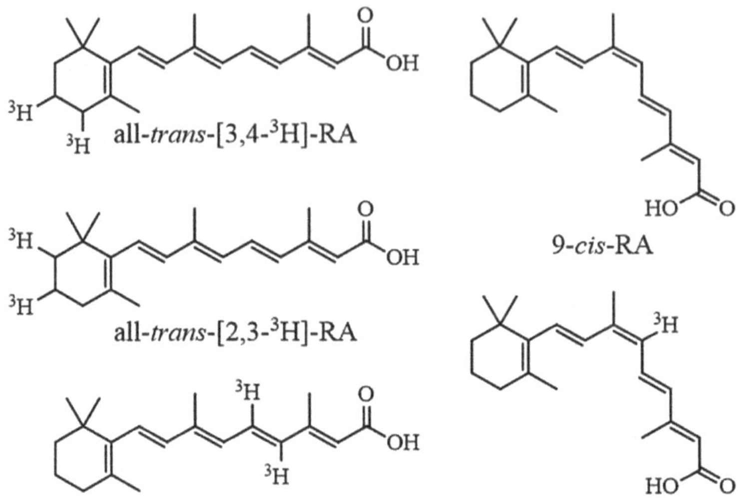

Fig. 3 Canonical FRET activity between a fluorescein (emission 520 nm) labeled co-activator peptide and terbium (excitation 320 nm) bound to RAR *via* a GST/anti-GST coupling. The addition of a test retinoid to the assay causes recruitment of the co-activator peptides bearing the fluorescein marker, and creates a FRET affect that can be measured as a facsimile for interaction between protein and ligand (Stafslien et al., 2007). Processing of this data allows for calculation of dose dependency/potency of each compound, and therefore screening for a series of ligands can be carried out as part of a drug development pipeline or binding investigation.

Recruitment of the labeled peptide to the RXR occurs upon the binding of test ligands capable of inducing a suitable conformational change and generates the FRET signal due to subsequent interaction between the terbium-fluorescein pairing. These pairs were excited at 340 nm (Tb) and had emission measured at 520 nm (fluorescein) (Stafslien et al., 2007). In this case, introduction of 9-*cis*-retinoic acid, the natural ligand of RXRα (Heyman et al., 1992), caused the greatest measurable FRET response with RXR also capable of recruiting several variations of the co-activator peptides. Phytanic acid—a dietary fatty acid found in animal products (Verhoeven, Wanders, Saudubray, & Jakobs, 1998)—was found to be similarly active, with an affinity greater than that of ATRA, with DHA following up as the lowest affinity of the tested binding compounds (Stafslien et al., 2007).

The signal resulting from a FRET interaction acts as an indicator of retinoid binding and creates an indication of the potency of each compound, based on its receptor affinity. The reported dose dependencies (EC50 values) for the ligands are suggested to be non-equivalent to the actual receptor binding affinities, as the measurement takes into account several events occurring in equilibrium including ligand recruitment, receptor conformational changes and enlistment of the co-activator peptide before the FRET signal is detectable (Stafslien et al., 2007). This makes the technique useful in a qualitative sense for indicative testing of ligands, though it may lack the thorough quantitative measurements required for modern drug development programs that can be achieved using fluorescence and other ligand focused methods.

This terbium-fluorescein pairing was also leveraged to determine the binding of retinoic acids, and EC23 derivatives to RARα/β/γ (Haffez et al., 2017). EC23 and the series of similar compounds, including 13-*cis*-RA analogue EC19, were designed to replace ATRA in cell based assays, where light and solute stability are vital (Chisholm, Zhou, Pohl, Valentine, & Whiting, 2016; Christie et al., 2008). Available as an invitrogen Lathascreen, the FRET methodology is not necessarily limited to those with established expertise, procedures and facilities, but flexibility comes at a cost. This kit-based HTRF method facilitated the determination that several synthetic retinoid derivatives (EC23 and EC19 and methyl esters of the same) had excellent RAR binding capacities, and added credence to a series of docking simulations of those molecules into the ligand binding domains of the receptor (Haffez et al., 2017). Other users of the Lanthascreen system were able to demonstrate that 4-amino-2trifluromethyl-phenyl retinate (a derivation of ATRA) preferentially binds to RXRα over RARα by nearly three orders of magnitude, and is capable of inhibiting the growth of gastric cancer cells (Hu et al., 2014). While FRET assays can be costly their use will undoubtedly continue into the future, particularly in the case of interactions between multiple proteins, DNA sections, or both, as part of the effort to better characterize retinoid signaling pathways and interactions with other signaling factors.

5. Fluorescence based assays

The development and synthesis of stable, fluorescent retinoid analogues has opened a new field of study into the binding behaviors of retinoic acid binding proteins (Chisholm et al., 2016; Clemens et al., 2013; Whiting & Marder, 2016; Yamada et al., 2010). While retinoids (including ATRA and 9-*cis*-retinoic acid) are intrinsically fluorescent, the signals corresponding to their emission are hard to measure or isolate from the background (Vivas et al., 2015). Newer, synthetic retinoid derivatives have been specifically designed to enhance this fluorescence capability, as well as to remain stable under normal laboratory light and ambient conditions (Chisholm et al., 2016; Yamada et al., 2010). These compounds additionally maintain high biological activity, and in certain cases exceed the signaling potential of ATRA (Chisholm et al., 2019; Christie et al., 2008).

One system utilizing the fluorescence of synthetic retinoids measures the interaction of retinoids with RXR (retinoid X receptors) centered on an umbelliferone-based synthetic "rexinoid" (retinoid analogue with RXR specificity). This specialist ligand was designed to maximize incorporation into the ligand binding domain of RXRα which has a structure that is distinct from that of the RARs (Yamada et al., 2010, 2019). Fluorescence was measured using spectroscopy in a plate spectrophotometer, with the probe generating a signal of greater intensity in the aqueous medium than in the ligand binding site, where its fluorescence was quenched. This was reported to be due to the deprotonation of the 7′ hydroxyl group of the ligand in an aqueous environment, creating increased charge transfer to fluorescent umbelliferone. An identical change can be induced in solution by moving from an apolar solvent (cyclohexane—low intensity fluorescence) to a polar solvent (such as NaOH—high fluorescence) (Yamada et al., 2019). The assay is highly sensitive, uses heterologously expressed RXRα and can be carried out in a standard fluorescence plate reader capable of excitation/emission 360/465 nm.

Similarly, the DC series of synthetic retinoid derivatives exhibit differential fluorescence when moved between different solvent environments (Chisholm et al., 2019). This solvatochromism manifests itself as a bathochromic shift in emission and increased fluorescence in hydrophobic environments and apolar solvents, including toluene, and decreased signal strength in aqueous solution (Chisholm et al., 2019). The effect is also detected upon movement from the hydrophobic binding pocket of a retinoic acid binding protein, such as CRABPII, to the aqueous medium and therefore allows for detection of the occupancy state of the binding site. This is reported to be due to aggregation and self-quenching in the aqueous environment. Methods to use this information have been developed, focusing on using the solvatochromic effect to detect the displacement of the fluorescent probe by other, non-fluorescent retinoid analogues in both CRAB and RAR proteins (Fig. 4) (Tomlinson et al., 2018).

This method works at concentrations as low as 100 nM (both protein and reporter), is fully developed for high-throughput work and was demonstrated initially in 96 well plates with 150 μL well contents. Fluorescence

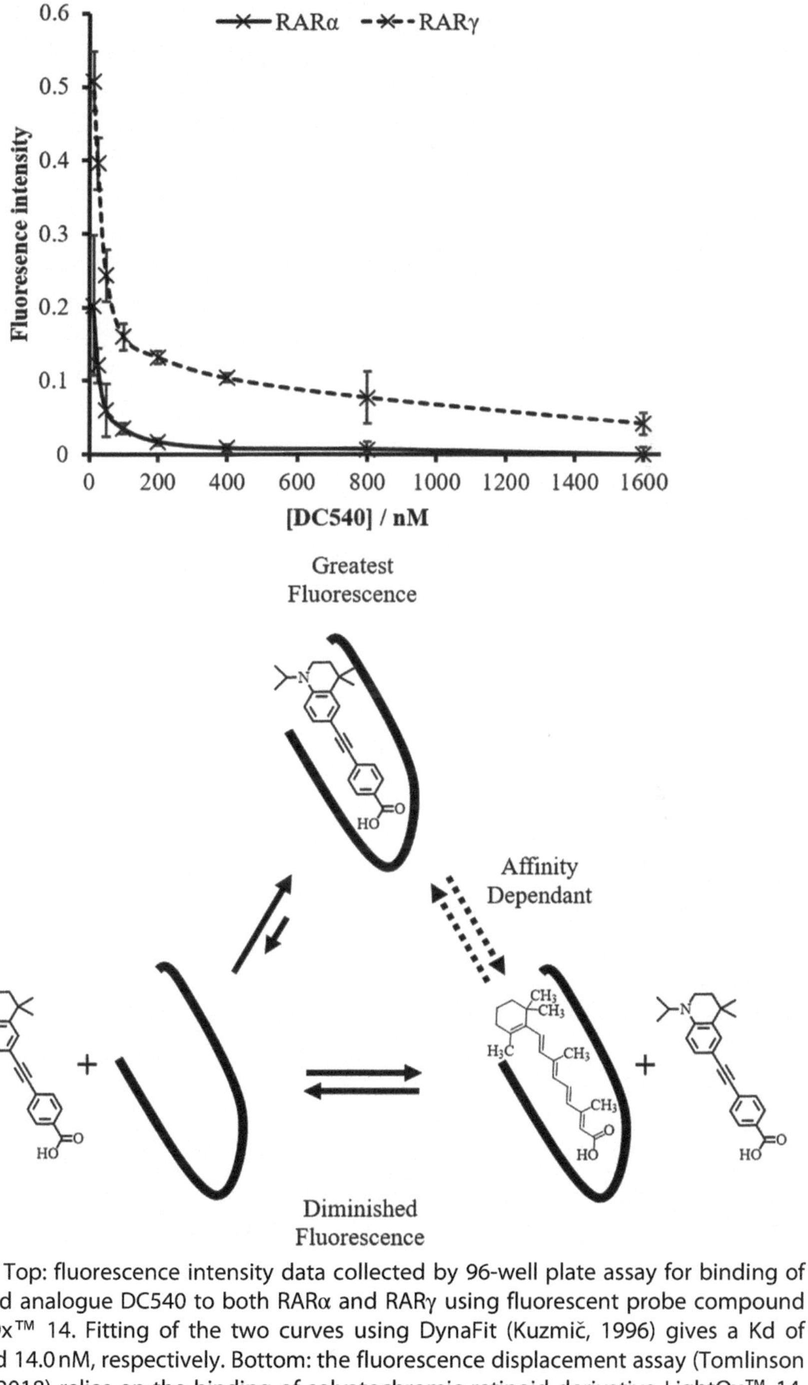

Fig. 4 Top: fluorescence intensity data collected by 96-well plate assay for binding of retinoid analogue DC540 to both RARα and RARγ using fluorescent probe compound LightOx™ 14. Fitting of the two curves using DynaFit (Kuzmič, 1996) gives a Kd of 1.3 and 14.0 nM, respectively. Bottom: the fluorescence displacement assay (Tomlinson et al., 2018) relies on the binding of solvatochromic retinoid derivative LightOx™ 14, which once displaced from the pocket by an alternative ligand has its fluorescence diminished. Measurement and interpretation of this change in fluorescence allows for calculation of binding affinities for the compounds under scrutiny.

is generated and detected at an excitation/emission of 355/460nm so the only requirements are the compounds (solvated in a dilute solution of ethanol) and the binding protein to which ligand affinity is to be measured, as well as access to a plate reading fluorescence spectrophotometer with suitable filters (Tomlinson et al., 2018). Unlike other methods, fluorescence displacement directly measures the occupancy state of the ligand binding site and does not require separation of unbound probe as is the case with radiolabeling techniques. Using this technique it has been possible to measure the affinity of a host of non-fluorescent, synthetic retinoid derivatives, to CRABPII as well as both RARα and RARγ (Fig. 4) (Tomlinson, 2020). The level of precision achievable makes it possible to differentiate between the binding affinities of the RAR isoforms, and will pave the way for development of further ligands capable of specific binding and modulation of one isoform in preference to any other.

Both systems are viable alternatives to those previously mentioned and exhibit multiple benefits over other methodologies. While the culturing of human or primate cells is not beyond the laboratory capabilities of most, it is undeniably more complex than heterologous expression in *E. coli*. These heterologously expressed proteins can also be frozen under cryogenic conditions and are, therefore, readily available when necessary for assaying. There is additionally no need to separate bound from unbound compounds, as with radiolabeling, as the intrinsic fluorescence properties report on the occupancy of the binding site. The methods utilized in the production of heterologously expressed retinoid binding proteins, and their use in binding assays, is outlined below.

6. Methods

Binding assays using fluorescent retinoids have been carried out on CRABPII, the cellular retinoic acid binding protein, and details of the expression for the protein can be found in chapter "Classical pathway of gene regulation by retinoids" by Tomlinson and Pohl (Tomlinson et al., 2018). This methodology was adapted for use with the retinoic acid receptors to better determine how further development of molecules for drug discovery can be undertaken, and the expression of that protein is detailed below.

6.1 CRABPII competitive binding assay

Competitive binding assays were carried out as outlined below, with the intention of determining the binding efficiency of several retinoid derivative compounds to CRABPII, using the intrinsic fluorescence of the LightOx™ 14 ligand as an *in situ* indicator (Tomlinson et al., 2018).

6.1.1 Materials, equipment and reagents

- LightOx™ 14 probe retinoid (available from Merck/Sigma Aldrich)
- CRABPII protein from heterologous expression
- 96 well black fluorescence plates with non-binding treatment
- EC23, ATRA, or other known binder for displacement controls
- Dilution series of test compounds
- Plate reader capable of excitation and emission 355/460 nm

6.1.2 Protocol

1. LightOx™ 14 (150 nM) solvated in 1% EtOH after dilution and CRABPII (150 nM in 10 mM HEPES, 150 mM NaCl, pH 7.5, 0.5 mM EDTA +0.5 mM TCEP) solutions were prepared and mixed before use to create an equimolar solution.
2. A Corning, black non-binding surface (NBS), 96 well plate was cleaned using compressed air and maximum response, median response (half displacement by known ligand) and all test compound wells were loaded with 100 μL CRABPII/LightOx™ 14 solution.
3. Minimum response wells were loaded with 100 μL control solution of LightOx™ 14 mixed with assay buffer to replicate the effect of the experiment in the absence of protein when all probe molecules are a minimum fluorescence.
4. Median response wells were then loaded with 50 μL EC23 (600 nM in 1% EtOH), calculated to displace half of the probe molecules and create a suitable median signal for calibration.
5. Maximum and minimum response wells were loaded with 50 μL of a 1% EtOH solution, to create a control for the effect of the competitor ligand solution.
6. Test compounds were prepared as dilution series' (suitable to create concentrations of 1600–12.5 nM on the plate) and applied to the plate as 50 μL aliquots.
7. Plates were spun at 1700 rpm, 2 min to ensure mixing.
8. Plates were read using a Synergy H4 plate reader, at ex/em 355/460 nm. The total well volume is 150 μL, the on-plate concentration of CRABPII and LightOx™ 14 is 50 nM.

9. Curve fitting was carried out using Dynafit (Kuzmič, 1996) after determination of the Z' factor for each plate, typically scoring greater than 0.6 (Zhang, Chung, & Oldenburg, 1999).

6.2 RARα/γ expression

Heterologous expression of RARα and RARγ has allowed us access to high quality, high concentration protein for both assaying and crystallography. In all cases the ligand binding domain of the relevant RAR was expressed for testing in fluorescence assays. The methodology developed is outlined below.

6.2.1 Materials, equipment and reagents

- BL21.DE3 cells transformed with expression plasmid
- Ampicillin prepared at 100 mg/mL
- 1 M Isopropyl β-D-1-thiogalactopyranoside (IPTG)
- LB growth media
- Orbital shaker with heating control
- High speed centrifuge for spinning down cells

6.2.2 Protocol

1. RARα/γ protein was heterologously expressed in BL21.DE3 cells (New England Biolabs) transformed with a popinS3C vector carrying the gene construct (Bird, 2011). Transformations were carried out under standard conditions, using a 10 s heat shock. Successful transformants were grown overnight in LB and frozen for storage at −80 °C in 50% glycerol.
2. Expression cultures were grown overnight from glycerol stocks by inoculation into 25 mL of LB media (Melford) with 100 μg/mL ampicillin (Sigma Aldrich). The culture was shaken overnight at 150 rpm, 37 °C.
3. The overnight cultures were transferred into 1 L expression flasks (LB/ampicillin) and grown shaking at 37 °C, 150 rpm. Induction was carried out at an OD600 of 0.6, with 1 mL of 1 M IPTG (1 mM in culture), before shaking overnight (20 h) at 25 °C.
4. The resulting cultures were spun down into pellets using an Avanti Hi-Speed centrifuge (JLA 8.1000, 4000 rpm, 25 min, 4 °C), before the supernatant was removed and the bacterial pellet transferred into a 50 mL centrifuge tubes for freezing at −80 °C

6.3 RARα/γ purification

Purification of the RAR proteins was carried out using a histidine-nickel affinity method, based on the 6-his tag appended to the protein by the popin vector (Bird, 2011). A first pass purification, followed by HIS tag cleavage and a second pass rendered the protein pure as analyzed by SDS PAGE gel, though a second polishing step using size exclusion or other chromatography step could be carried out if necessary.

6.3.1 Materials, equipment and reagents

- Buffers (500 mM NaCl, 20 mM Tris-HCl, pH 8) with suitable imidazole concentrations for gradient elution
- Sonication equipment and high-speed centrifuge
- HIS trap nickel affinity columns—HisTrap HP from GE Healthcare
- HRV3C protease for cleavage of HIS tag
- Peristaltic pump or FPLC suitable for loading and eluting columns

6.3.2 Protocol

1. RARα (popinS3C, BL21) frozen as a pellet was resuspended with 15 mL buffer A (5 mM imidazole, 500 mM NaCl, 20 mM Tris-HCl, pH 8) and lysed with addition of 0.5 mg/mL lysozyme for 30 min.
2. The resulting suspension was sonicated on ice (40% power, 2 min) and centrifuged (Avanti Hi-Speed JA25.50, 20,000 rpm, 50 min, 4 °C).
3. Supernatant was loaded onto HIS trap column (1 mL, GE Healthcare) and washed using 30 mL buffer A (5 mM imidazole, 500 mM NaCl, 20 mM Tris-HCl, pH 8).
4. The column was washed with 30 mL buffer B (50 mM imidazole, 500 mM NaCl, 20 mM Tris-HCl, pH 8) before elution in gradient with buffer C (150 mM imidazole, 500 mM NaCl, 20 mM Tris-HCl, pH 8). The resulting protein fractions were analyzed by SDS PAGE and concentration was estimated using Thermo Scientific Nanodrop A280.
5. RAR proteins were cleaved from 3C-SUMO-HIS tag using 1 unit/100 μg protein HRV3C protease, based on Nanodrop A280 calculated protein concentration, overnight, and dialyzed into fresh buffer without imidazole (500 mM NaCl, 20 mM Tris-HCl, pH 8)
6. The results of this digest were loaded onto a HIS trap column (1 mL, GE Healthcare) and cleaved protein was eluted immediately using 3 mL buffer A (5 mM imidazole, 500 mM NaCl, 20 mM Tris-HCl, pH 8) into 3 × 1 mL fractions.

7. Bound tag was then eluted with buffer C (150 mM imidazole, 500 mM NaCl, 20 mM Tris-HCl, pH 8) and the column washed to remove any other residues.
8. The resulting protein and tag fractions analyzed by SDS PAGE (120 V, 70 min), before use in assays and crystallization experiments.

6.4 RARα/γ binding assays

Binding efficiency of synthetic retinoids to RARα and RARγ was determined using a competitive binding assay protocol, adapted to take into account the natural binding affinity of the RAR proteins (Tomlinson, 2020). Comparison of binding effects between the different RAR isoforms will form the basis for development of specialized ligands, with the ultimate aim of being able to target one isoform preferentially *in vivo*, and hence influence retinoid signaling in cells.

6.4.1 Materials, equipment and reagents

- LightOx™ 14 probe retinoid—available from Merck/Sigma Aldrich
- RARα/γ protein from heterologous expression
- 96 well black fluorescence plates with non-binding treatment
- EC23, ATRA, or other known binder for displacement controls
- Dilution series of test compounds
- Plate reader capable of excitation and emission 350/440 nm

6.4.2 Protocol

1. Purified RAR-α and -γ protein (500 mM NaCl, 20 mM Tris-HCl, pH 8) was combined with LightOx™ 14 (diluted in ~1% ethanol) in an equimolar ratio (300 nM).
2. 100 μL was aliquoted into each well of a 96 well, low binding black fluorescence plate (Fisher Brand), cleaned of any storage dust with compressed air.
3. Maximum, minimum and median controls were established using protein/fluorophore, protein/fluorophore/known competitor and buffer/fluorophore respectively to mimic the conditions in the test wells.
4. An eight factor dilution series of each test compound (1600–12.5 nM effective concentration) was added on top, in triplicate, and the plates were spun at 1500 rpm for 2 min.
5. Each plate was read for fluorescence emission (ex350/em440) with a Synergy H4 plate reader.

6. Fitting of the resulting binding curves was carried out using the Dynafit software (Kuzmič, 1996) and a competitive binding instruction set. Dynafit operates using a least squares regression to fit a suitable curve to the provided data, based on an equation generated from a reaction scheme provided by the user. In this case the two retinoids are treated as competitors for the protein binding site, with LightOx™ 14 having a known affinity to the RAR and the test compound affinity being the variable for determination. Each plate was also treated to a Z' factor calculation to assess robustness, typically scoring greater than 0.6 (Zhang et al., 1999).

   N.B—it may be necessary to determine the emission peak of the compound incorporated in the binding site, especially in the case of testing non RAR proteins or mutants, to ensure thorough results.

7. Evaluation

The demand for a rapid, cost effective and high-throughput screening methodology for interpreting the interactions of retinoid-like molecules with the retinoic acid (and retinoid X) receptors has led to the development of a multitude of evolving methods. The shift from live cell techniques (Heyman et al., 1992), using reporter constructs, and radiolabeling methods (Allenby et al., 1993) toward *in vitro* FRET (Stafslien et al., 2007) and fluorescence interrogation (Chisholm et al., 2019; Tomlinson et al., 2018; Yamada et al., 2019) using heterologous expression techniques has created a slew of improvements in both cost/time efficiency and accessibility. The additional ability to combine assays, to both determine *in vitro* and putative *in vivo* activity of a ligand, is being used in the determination of binding for new and upcoming retinoid and non-retinoid-like RAR/RXR modulators (Busby et al., 2011). There is a wide scope to incorporate a series of assays into a single development program, using fluorescence assays to determine compound biding in tandem with promoter activation assays to ensure that compounds still retain efficacy on target (or do not if that is beneficial). The advantage of FRET and fluorescence over older radiolabeling-based detection is found in the absence of radiological materials to source, handle and dispose of, as well as the cessation of the need to separate bound from unbound fractions. While FRET and HTRF relies on costly terbium, or other exotic metal labeling, as well as antibody sourcing, fluorescence detection methods tend toward small molecules that often have extensive shelf-lives, as well as protein expressed heterologously. Direct fluorescence measurement additionally has the advantage of reducing the "degrees of separation" between measurement and biological effect, in stark contrast to the second-hand measurements associated with the recruitment of FRET donor and acceptors to a signaling protein. By using a surrogate ligand buried directly in the binding site, and measuring its displacement through intrinsic fluorescence, the incorporation of a new (non-fluorescent) ligand can be directly measured.

The power of high-throughput assaying of library targets for drug development is widely known, and a variety of suitable assays that can be utilized improve the opportunities for progress in the field. Modern fluorescence assays in particular will enable high-throughput, low cost evaluation of ligands in development pipelines, leading to improved progress against disease targets, as well as further understanding of the intrinsic signaling pathways that are modulated by retinoic and its derivatives.

References

Achkar, C. C., Derguini, F., Blumberg, B., Langston, A., Levin, A. A., Speck, J., et al. (1996). 4-Oxoretinol, a new natural ligand and transactivator of the retinoic acid receptors. *Proceedings of the National Academy of Sciences of the United States of America*, *93*(10), 4879–4884.

Allenby, G., Bocquel, M.-T., Saunders, M., Kazmer, S., Speck, J., Rosenberger, M., et al. (1993). Retinoic acid receptors and retinoid X receptors: Interactions with endogenous retinoic acids. *Proceedings of the National Academy of Sciences of the United States of America*, *90*(1), 30–34.

Allenby, G., Janocha, R., Kazmer, S., Speck, J., Grippo, J. F., & Levin, A. A. (1994). Binding of 9-cis-retinoic acid and all-trans-retinoic acid to retinoic acid receptors alpha, beta, and gamma. Retinoic acid receptor gamma binds all-trans-retinoic acid preferentially over 9-cis-retinoic acid. *Journal of Biological Chemistry*, *269*(24), 16689–16695.

Assenberg, R., Wan, P. T., Geisse, S., & Mayr, L. M. (2013). Advances in recombinant protein expression for use in pharmaceutical research. *Current Opinion in Structural Biology*, *23*(3), 393–402.

Bennani, Y. L., & Boehm, M. F. (1995). Syntheses of high specific activity 2, 3-and 3, 4-[3H] 2-9-cis-retinoic acid. *The Journal of Organic Chemistry*, *60*(5), 1195–1200.

Bird, L. E. (2011). High throughput construction and small scale expression screening of multi-tag vectors in Escherichia coli. *Methods*, *55*(1), 29–37.

Boehm, M. F., McClurg, M. R., Pathirana, C., Mangelsdorf, D., White, S. K., Hebert, J., et al. (1994). Synthesis of high specific activity tritium-labeled [3H]-9-cis-retinoic acid and its application for identifying retinoids with unusual binding properties. *Journal of Medicinal Chemistry*, *37*(3), 408–414.

Busby, S. A., Kumar, N., Kuruvilla, D. S., Istrate, M. A., Conkright, J. J., Wang, Y., et al. (2011). Identification of a novel non-retinoid pan inverse agonist of the retinoic acid receptors. *ACS Chemical Biology*, *6*(6), 618–627.

Cahnmann, H. (1995). A fast photoisomerization method for the preparation of tritium-labeled 9-cis-retinoic acid of high specific activity. *Analytical Biochemistry*, *227*(1), 49–53.

Chisholm, D. R., Tomlinson, C. W., Zhou, G.-L., Holden, C., Affleck, V., Lamb, R., et al. (2019). Fluorescent retinoic acid analogues as probes for biochemical and intracellular characterization of retinoid signaling pathways. *ACS Chemical Biology*, *14*(3), 369–377.

Chisholm, D. R., Zhou, G.-L., Pohl, E., Valentine, R., & Whiting, A. (2016). Practical synthetic strategies towards lipophilic 6-iodotetrahydroquinolines and-dihydroquinolines. *Beilstein Journal of Organic Chemistry*, *12*, 1851.

Christie, V. B., Barnard, J. H., Batsanov, A. S., Bridgens, C. E., Cartmell, E. B., Collings, J. C., et al. (2008). Synthesis and evaluation of synthetic retinoid derivatives as inducers of stem cell differentiation. *Organic & Biomolecular Chemistry*, *6*(19), 3497–3507.

Clemens, G., Flower, K. R., Gardner, P., Henderson, A. P., Knowles, J. P., Marder, T. B., et al. (2013). Design and biological evaluation of synthetic retinoids: Probing length vs. stability vs. activity. *Molecular BioSystems*, *9*(12), 3124–3134.

de Urquiza, A. M., Liu, S., Sjöberg, M., Zetterström, R. H., Griffiths, W., Sjövall, J., et al. (2000). Docosahexaenoic acid, a ligand for the retinoid X receptor in mouse brain. *Science*, *290*(5499), 2140–2144.

Degorce, F., Card, A., Soh, S., Trinquet, E., Knapik, G. P., & Xie, B. (2009). HTRF: A technology tailored for drug discovery—A review of theoretical aspects and recent applications. *Current Chemical Genomics*, *3*, 22.

Förster, T. (1948). Zwischenmolekulare energiewanderung und fluoreszenz. *Annalen der Physik*, *437*(1–2), 55–75.

Förster, T. (2012). Energy migration and fluorescence. *Journal of Biomedical Optics*, *17*(1), 011002.

Giguere, V., Ong, E. S., Segui, P., & Evans, R. M. (1987). Identification of a receptor for the morphogen retinoic acid. *Nature*, *330*(6149), 624.

Gluzman, Y. (1981). SV40-transformed simian cells support the replication of early SV40 mutants. *Cell*, *23*(1), 175–182.

Goldstein, J. T., Dobrzyn, A., Clagett-Dame, M., Pike, J. W., & DeLuca, H. F. (2003). Isolation and characterization of unsaturated fatty acids as natural ligands for the retinoid-X receptor. *Archives of Biochemistry and Biophysics*, *420*(1), 185–193.

Gorman, C. M., Moffat, L. F., & Howard, B. H. (1982). Recombinant genomes which express chloramphenicol acetyltransferase in mammalian cells. *Molecular and Cellular Biology*, *2*(9), 1044–1051.

Gould, S. J., & Subramani, S. (1988). Firefly luciferase as a tool in molecular and cell biology. *Analytical Biochemistry*, *175*(1), 5–13.

Haffez, H., Chisholm, D. R., Valentine, R., Pohl, E., Redfern, C., & Whiting, A. (2017). The molecular basis of the interactions between synthetic retinoic acid analogues and the retinoic acid receptors. *Medicinal Chemistry Communications*, *8*(3), 578–592.

Heyman, R. A., Mangelsdorf, D. J., Dyck, J. A., Stein, R. B., Eichele, G., Evans, R. M., et al. (1992). 9-cis retinoic acid is a high affinity ligand for the retinoid X receptor. *Cell*, *68*(2), 397–406.

Hu, K.-W., Pan, X.-H., Chen, F.-H., Qin, R., Wu, L.-M., Zhu, H.-G., et al. (2014). A novel retinoic acid analog, 4-amino-2-trifluoromethyl-phenyl retinate, inhibits gastric cancer cell growth. *International Journal of Molecular Medicine*, *33*(2), 415–422.

Jensen, F. C., Girardi, A. J., Gilden, R. V., & Koprowski, H. (1964). Infection of human and simian tissue cultures with Rous sarcoma virus. *Proceedings of the National Academy of Sciences of the United States of America*, *52*(1), 53.

Kuzmič, P. (1996). Program DYNAFIT for the analysis of enzyme kinetic data: Application to HIV proteinase. *Analytical Biochemistry*, *237*(2), 260–273.

Leo, M. A., Lasker, J. M., Raucy, J. L., Kim, C.-I., Black, M., & Lieber, C. S. (1989). Metabolism of retinol and retinoic acid by human liver cytochrome P450IIC8. *Archives of Biochemistry and Biophysics*, *269*(1), 305–312.

Leo, M. A., & Lieber, C. S. (1985). New pathway for retinol metabolism in liver microsomes. *Journal of Biological Chemistry*, *260*(9), 5228–5231.

Levin, A. A., Sturzenbecker, L. J., Kazmer, S., Bosakowski, T., Huselton, C., Allenby, G., et al. (1992). 9-cis retinoic acid stereoisomer binds and activates the nuclear receptor RXRα. *Nature*, *355*(6358), 359.

Mangelsdorf, D. J., Ong, E. S., Dyck, J. A., & Evans, R. M. (1990). Nuclear receptor that identifies a novel retinoic acid response pathway. *Nature*, *345*(6272), 224.

Meakin, P., Jesson, J., & Tolman, C. (1972). Nature of chlorotris (triphenylphosphine) rhodium in solution and its reaction with hydrogen. *Journal of the American Chemical Society*, *94*(9), 3240–3242.

Murayama, A., Suzuki, T., & Matsui, M. (1997). Photoisomerization of retinoic acids in ethanol under room light: A warning for cell biological study of geometrical isomers of retinoids. *Journal of Nutritional Science and Vitaminology*, *43*(2), 167–176.

Repa, J. J., Hanson, K. K., & Clagett-Dame, M. (1993). All-trans-retinol is a ligand for the retinoic acid receptors. *Proceedings of the National Academy of Sciences of the United States of America*, *90*(15), 7293–7297.

Seliger, H. H., & McElroy, W. D. (1960). Spectral emission and quantum yield of firefly bioluminescence. *Archives of Biochemistry and Biophysics*, *88*(1), 136–141.

Stafslien, D. K., Vedvik, K. L., De Rosier, T., & Ozers, M. S. (2007). Analysis of ligand-dependent recruitment of coactivator peptides to RXRβ in a time-resolved fluorescence resonance energy transfer assay. *Molecular and Cellular Endocrinology*, *264*(1–2), 82–89.

Stryer, L. (1978). Fluorescence energy transfer as a spectroscopic ruler. *Annual Review of Biochemistry*, *47*(1), 819–846.

Tomlinson, C. (2020). *Unpublished results*.

Tomlinson, C. W. E., Chisholm, D. R., Valentine, R., Whiting, A., & Pohl, E. (2018). Novel fluorescence competition assay for retinoic acid binding proteins. *ACS Medicinal Chemistry Letters*, *9*(12), 1297–1300. https://doi.org/10.1021/acsmedchemlett.8b00420.

Totter, J. R. (1981). In M. A. DeLuca & W. D. McElroy (Eds.), *Bioluminescence and chemiluminescence: basic chemistry and analytical applications* (pp. 179–180). New York/London: Academic Press.

Tsien, R. Y. (1998). The green fluorescent protein. In *Annual reviews 4139 El Camino way*. CA, United States: Palo Alto.

van Oers, M. M., Pijlman, G. P., & Vlak, J. M. (2015). Thirty years of baculovirus–insect cell protein expression: From dark horse to mainstream technology. *Journal of General Virology*, *96*(1), 6–23.

Verhoeven, N., Wanders, R., Saudubray, J.-M., & Jakobs, C. (1998). The metabolism of phytanic acid and pristanic acid in man: A review. *Journal of Inherited Metabolic Disease*, *21*(7), 697–728.

Vivas, M., Siqueira, J., Silva, D., de Boni, L., & Mendonca, C. (2015). Investigation of the nonlinear absorption spectrum of all-trans retinoic acid by using the steady and transient two-photon absorption spectroscopy. *RSC Advances*, *5*(91), 74531–74538.

Whiting, A., & Marder, T. (2016). *Fluorescent synthetic retinoids*, WO 2016/055800 A2.

Yamada, S., Kawasaki, M., Fujihara, M., Watanabe, M., Takamura, Y., Takioku, M., et al. (2019). Competitive binding assay with an umbelliferone-based fluorescent rexinoid for retinoid X receptor ligand screening. *Journal of Medicinal Chemistry*, *62*(19), 8809–8818.

Yamada, S., Ohsawa, F., Fujii, S., Shinozaki, R., Makishima, M., Naitou, H., et al. (2010). Fluorescent retinoid X receptor ligands for fluorescence polarization assay. *Bioorganic & Medicinal Chemistry Letters*, *20*(17), 5143–5146.

Zhang, J.-H., Chung, T. D., & Oldenburg, K. R. (1999). A simple statistical parameter for use in evaluation and validation of high throughput screening assays. *Journal of Biomolecular Screening*, *4*(2), 67–73.

CHAPTER TWENTY-ONE

Using the human CYP26A1 gene promoter as a suitable tool for the determination of RAR-mediated retinoid activity

Reza Zolfaghari[a,*], Floyd J. Mattie[a], Cheng-Hsin Wei[a], David R. Chisholm[b], Andrew Whiting[b], A. Catharine Ross[a,c]

[a]Department of Nutritional Sciences, Pennsylvania State University, University Park, PA, United States
[b]Centre for Sustainable Chemical Processes, Department of Chemistry, Science Laboratories, Durham University, Durham, United Kingdom
[c]Center for Molecular Immunology and Infectious Disease, Pennsylvania State University, University Park, PA, United States
*Corresponding author: e-mail address: rxz7@psu.edu

Contents

Methods in Enzymology, Volume 637
ISSN 0076-6879
https://doi.org/10.1016/bs.mie.2020.03.013

Abstract

We have used a shortened construct form of the CYP26A1 gene promoter, in a promoter-less vector with either luciferase (known as E4) or a red fluorescent protein, RFP (known as E4.2) as the reporter gene and examined their responses to retinoids in transfected HepG2 and HEK293T cells. The promoter responded linearly to a wide concentration range of at-RA in cells cotransfected with retinoic acid receptors (RAR). The promoter also responded quantitatively to retinol and various other retinoids. An isolated clonal line of HEK293T cells that was permanently transfected with the promoter driving the expression of RFP responded to both at-RA and retinol, and the responses could be measured by fluorescence microscopy and flow cytometry. The promoter was also used to assess the retinoid activity of 3 novel synthetic retinoid analogues. Among them, EC23 was shown to be more potent than at-RA at lower concentrations and also more stable than at-RA. The promoter was also used to estimate the retinoid activities of intact rat serum samples as well as extracts of rat liver and lung, using retinol and at-RA as the reference standards. The retinoid activities could be measured in control rat serum samples and were increased in the serum of at-RA-treated rats. The total retinol and at-RA levels in the rat liver and lung samples determined by this promoter-based assay were compared with total retinol levels determined by the UPLC as the conventional methods. This system should offer a biologically-based alternative to mass-based retinoid analysis.

1. Introduction

Retinoids refer to a class of chemicals, which are natural and synthetic derivatives of vitamin A. As a natural retinoid, all-trans retinoic acid (at-RA), the major active metabolite of vitamin A, functions as an endogenous ligand for three family members of the nuclear receptors, the retinoic acid receptors RARα, β, and γ, which dimerize with retinoid X receptors (RXRα, β, γ) and bind to specific DNA sites, known as retinoic acid response elements (RARE), typically located in the promoters of genes that are transcriptionally regulated by at-RA (Carrier & Rochette-Egly, 2015; Mendoza-Parra, Bourguet, de Lera, & Gronemeyer, 2015; Urban, Ye, & Davidson, 2015). These nuclear receptors are the key regulatory molecules, which are usually targeted by synthetic retinoids for combating the diseases such as cancers, immune disorders, and skin abnormalities (Altucci, Leibowitz, Ogilvie, de Lera, & Gronemeyer, 2007; Chien, 2018; de Thé & Fenaux, 2015; Spinella, Freemantle, & Dmitrovsky, 2015).

Of the various procedures for vitamin A analysis, almost all use chemical determination, principally using HPLC for separation through either reverse or normal phase columns, coupled to UV or electrochemical detection and/or mass spectrometry for quantification (Kane & Napoli, 2010). However, a few genetic tools using synthetically designed promoters containing RARE/RXRE elements have also been designed to analyze retinoid activity (Jurutka & Wagner, 2019; Lee, Tam, McCaffery, & Shum, 2019). Among numerous mammalian genes regulated by vitamin A, CYP26A1 has been shown to be the most highly inducible one that is regulated by at-RA as well as other retinoids (Isoherranen & Zhong, 2019; Ross & Zolfaghari, 2011). In this chapter, we describe how the native promoter of this human gene, modified for use in a reporter construct, is potentially useful in human cultured cells to determine the retinoid activities of novel synthetic retinoids. This tool may also be used as an alternative to estimate vitamin A activity of tissues including, in our examples, rat serum, liver and lung.

2. The promoter of CYP26A1 gene is highly responsive to retinoids in cultured cells

CYP26A1, a member of the CYP26 subfamily of cytochromeP450 enzymes, specifically catalyzes the oxidative inactivation of *all-trans* retinoic acid (at-RA) by converting it to more oxidized products. The CYP26A1 gene is highly responsive to the presence of at-RA as shown by the elevated expression of CYP26A1 mRNA during embryonic development as well as in retinoid-exposed adult tissues and cultured cells (Pennimpede et al., 2010; Ross & Zolfaghari, 2011). This induction is contributed mainly by the presence of four functional RARE elements contained in the full-length (FL) promoter, including one RARE in the region proximal to the transcription start site, and a cluster of 3 elements located in the distal region of the promoter further away from the proximate region of the transcription start site (Loudig, Maclean, Dore, Luu, & Petkovich, 2005; Zhang, Zolfaghari, & Ross, 2010) (see Fig. 1). By elimination of the intermediate region between these elements to bring the distal region closer to the proximate region, we created a construct made in a *p*GL3-basic-*luc* vector (a promoter-less vector with luciferase as the reporter gene), named as E4 (Fig. 1) (Zhang et al., 2010). This construct which was shown to have a higher response to at-RA as compared to the FL promoter when transiently

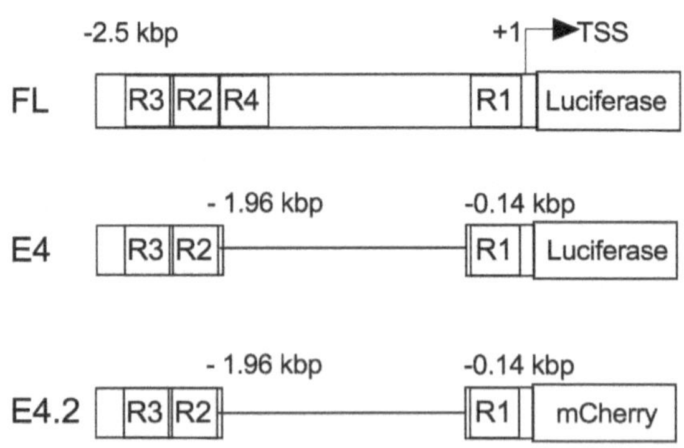

Fig. 1 Schematic representation of the full-length (FL) and deleted forms of human CYP26A1 gene promoter in *p*GL3-basic vector containing luciferase (E4) and mCherry (E4.2) as the reported genes. R1, R2, R3 and R4 represent retinoic acid response elements (RARE). TSS is the transcription start site.

cotransfected with RARα/RXRα into HepG2 cells, a human hepatoma cell line (Zolfaghari et al., 2019; Zolfaghari & Ross, 2014). We have tested this construct transiently in HepG2 cells for: (1) kinetics and dose response to at-RA; (2) response to retinol and RAR and RXR specific ligands; and (3) determination of the retinoid activity of novel synthetic retinoids.

2.1 Equipment

All equipment and reagents are specified by manufacturer/vendor as in the authors' laboratory, but other comparable items may suffice.

1. Nano-Drop instrument (Thermo)
2. Step One Plus Real-Time Thermo-Cycler System (Applied Biosystem)
3. Luminometer TD-20/20 (Promega)
4. CO_2-air incubator

2.2 Material

1. Ethanol
2. Dimethyl sulfoxide (DMSO) (Sigma-Aldrich)
3. All-trans-retinol (1 and 10 mM in ethanol) (Sigma-Aldrich)
4. All-trans-retinoic acid (at-RA) (1 and 10 mM in ethanol) (Sigma-Aldrich)
5. 9-Cis-retinoic acid (1 mM in ethanol) (Sigma-Aldrich)
6. LG-268, an RXR specific ligand (1 mM in ethanol) (Sigma-Aldrich), Am580, an RARα specific ligand (1 mM in ethanol) kindly provided to us by Dr. K. Shudo and Y. Kagechika (Tokyo Medical and Dental University, Tokyo, Japan)
7. Synthetic retinoids including EC23, DC360, and DC324 (Fig. 2) were synthesized as reported (Chisholm et al., 2019) and prepared in at 1 mM solution in DMSO
8. Human hepatoma HepG2 cells (ATCC)
9. Human embryonic kidney HEK293T cells (ATCC)
10. Eagle's minimum essential medium (EMEM) (ATCC)
11. Dulbecco's modified Eagle's medium (DMEM) (GIBCO)

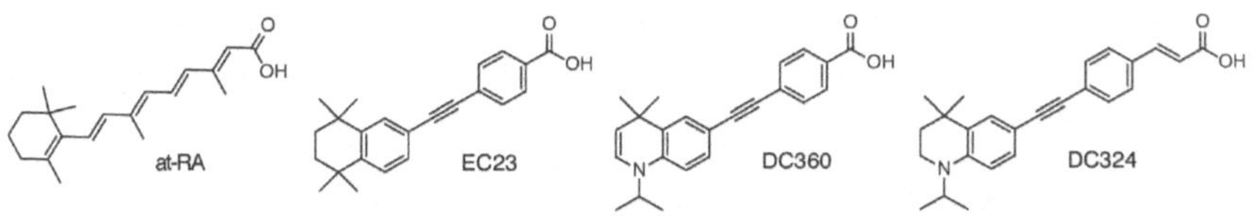

Fig. 2 Structure of synthetic retinoids. EC23 and DC360 are synthetic analogues of at-RA, and DC324 is a non-active analogue that exhibits poor binding affinity for RARs due to the extended structure of the compound.

12. Heat inactivated fetal bovine serum (FBS) (GIBCO)
13. Antibiotics (100 × penicillin-streptomycin) (ATCC)
14. Phosphate-buffered saline (PBS) (Corning)
15. Trypsin 0.25%-EDTA 2.2 mM (Corning)
16. pGL3-basic-luc (Promega)
17. pRLTK vector (Promega)
18. pcDNA3.1(+) (Invitrogen)
19. Restriction enzymes including *Stu*I, *Sma*I, and *Bst*BI (NE Biolab)
20. T4-DNA ligase (Promega)
21. Trizol reagent (Invitrogen)
22. Diethyl pyrocarbonate (DEPC), (Sigma)
23. MMLV-Reverse transcriptase (Promega)
24. OligodT nucleotide (Promega)
25. dNTP (Promega)
26. Real time PCR kit (Applied Biosystem)
27. Lipofectamine 2000 (Invitrogen)
28. DRL luciferase assay system including lysis buffer (Promega)

3. Methods

3.1 Cultured cells

HepG2 cells were grown in EMEM and HEK293T cells, human embryonic kidney cells, were maintained in DMEM, each supplemented with 10% FBS and 0.5% penicillin-streptomycin at 37 °C in a 5% CO_2-air incubator. The cells were plated and used at 60–70% confluency.

3.2 RNA isolation and analysis

Total RNA was extracted from the cells using Trizol reagent and then dissolved in DEPC-treated autoclaved water for analysis (Zolfaghari et al., 2019). The first stranded cDNA was synthesized from total RNA and then analyzed by real-time PCR (Zolfaghari et al., 2019) for CYP26A1 mRNA using primer pair of sense, 5'-GCTGCCTCTCTAACCTGCAC-3′, and antisense 5′-TGCTTTAGTGCCTGCATGTC-3′ and for 18S ribosomal RNA as an internal control using the primer pair of sense 5′-CGCGGTTCTATTTTGTTGGT-3′ and antisense, 5′-AGTCGGCAT-CGTTTATGGTC-3′. The PCR program was set to run first at 94 °C for 10 min for activation of the polymerase and then 40 cycles of 20 s at 94 °C, 30 s at 60 °C, 30 s at 72 °C (Zolfaghari et al., 2019) using Step One Plus Real-Time PCR System.

3.3 Plasmid clones

The FL and E4 promoters of the CYP26A1 gene cloned in pGL3-basic-luciferase vector are reported previously (Zhang et al., 2010). Open reading frames of human RARα, RARβ, RARγ, and RXRα were synthesized from poly A^+ RNA from human liver tissue sample and each was cloned into pcDNA3.1 as described (Zolfaghari et al., 2019). For construction of pcDNA3.1-hRARα-hRXRα expression vector, the pcDNA3.1-RAR was first double digested with *Stu*I or *Sma*I/*Bst*BI to remove the DNA reading frame for neomycin and replaced with ORF for RXRα as described (Zolfaghari et al., 2019). Clones created in this section have been deposited to the Addgene repository (Table 1).

3.4 Transient transfection and luciferase assay protocol

1. Grow the cells in full growth medium with 10% FBS in T-75 flask at 37 °C in 5% CO_2-air incubator.
2. Trypsinize and transfer the cells into 12- or 24-well plates 1–2 day before transfection when the cells are about 60–70% confluent.
3. Aspirate out the medium and replace it with the same medium containing 3% FBS with no antibiotics about 1 h before transfection.
4. Prepare DNA samples as follow: For each well of the 24-well plates mix 0.7 μg of FL or E4 promoter construct of the human CYP26A1 gene and pRLTK (7:1 ratio by weight), 0.1 μg of pcDNA3.1-hRARα. hRXRα in 50 μL EMEM or DMEM without FBS and antibiotics and for each well of the 12-well plate double the amount for each in

Table 1 Relevant plasmid clones deposited with Addgene.org.

Addgene ID	Clone name
135397	pcDNA3.1-hRARα
135400	pcDNA3.1-hRXRα
135411	pcDNA3.1-hRARα.hRXRα
135415	pcDNA3.1-hRARβ.hRXRα
135416	pcDNA3.1-hRARγ.hRXRα
135478	pGL3-Basic-hCYP26AlP-E4.2-mCherry
135566	pGL3-Basic-hCYP26A1-FL-luciferase
135592	pGL3-Basic-hCYP26A1P-E4-luciferase
135910	pcDNA3.1-hRXRα

100 μL of EMEM or DMEM without FBS and antibiotics. For cells without expression vector add pcDNA3.1 empty vector DNA.

5. Mix 2 μL of Lipofectamine-2000 in 50 μL of EMEM or DMEM without FBS and antibiotics for each well of 24-well plate or 4 μL of Lipofectamine-2000 in 100 μL of EMEM or DMEM without FBS and antibiotics for each well of 12-well plate and incubate at room temperature for 5 min.
6. Mix DNA solution (Section 4) with diluted Lipofectamine-2000 (Section 5) and incubate at room temperature for 20 min before adding to the cells.
7. Add the solution prepared in the Section 6 to the cells, mix by swirling and incubate overnight at 37 °C.
8. Following transfection aspirate the medium and incubate the cells with the full-growth medium containing 10% FBS and 0–1 μM at-RA, or retinol in ethanol as the vehicle or the synthetic retinoids in DMSO as the vehicle. The final concentration of either vehicle in the growth media was 0.001%.
9. After incubation at 37 °C for up to 24 h (incubation time was chosen for convenience) wash the cells with PBS and then lyse to assay for Firefly- and Renilla-luciferase activities using the DRL luciferase assay system from Promega in Luminometer following the protocol from manufacturer.
10. Promoter activity is defined as the ratio of Firefly- to Renilla-luciferase activity. Each reported activity is an average of 3 wells with standard error of the mean.

4. Results

4.1 RXRα enhances the effectiveness of RARα on the promoter activity of CYP26A1 gene

We have shown here that RXRα alone had no significant effect on the promoter activity of E4 promoter in HepG2 cells in response to at-RA; however, it increased the effectiveness of RARα by about twofold in response to at-RA (Fig. 3A). Based on this result, an expression vector containing the open reading frames for both RARα and RXRα (*p*cDNA3.1-RARα.RXRα) was constructed (Zolfaghari et al., 2019) and used for cotransfection with the promoter of CYP26A1 gene into HepG2 cells for a better response to at-RA treatment.

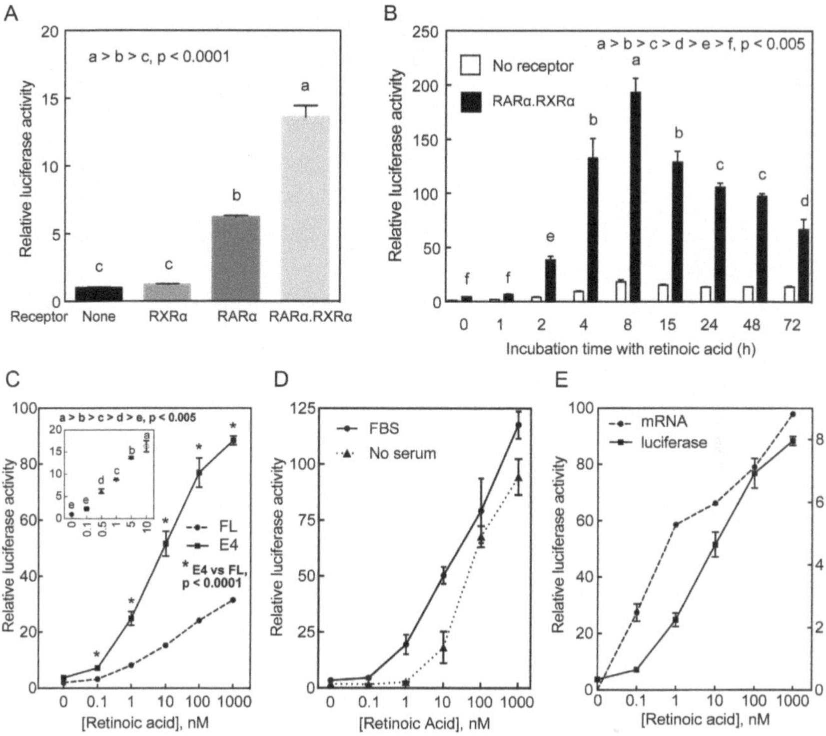

Fig. 3 CYP26A1 gene promoter is highly responsive to at-RA in HepG2 cells. (A) HepG2 cells cotransfected with E4 construct together with an expression vector containing both RARα and RXRα were more responsive to at-RA (1 nM) than with the expression vector containing only RARα (Zolfaghari et al., 2019). (B) Kinetic response to at-RA (1 μM) of HepG2 cells cotransfected with the E4 promoter construct of hCYP26A1 gene without or with the RARα.RXRα expression vector. A similar pattern of kinetic with maximum response at 8 h was observed when a lower concentration of at-RA, 100 nM, was used. (C) Full-length (FL) or E4 promoter constructs of hCYP26A1 gene responded dose dependently to at-RA in HepG2 cells cotransfected with RARα.RXRα expression vector. (D) At all doses of at-RA, the E4 promoter responses were higher in the presence than the absence of FBS. (E) The response of the E4 promoter to a range of concentrations of at-RA was in parallel with the expression levels of endogenous CYP26A1 mRNA in HepG2 cells.

4.2 CYP26A1 gene E4 promoter response to at-RA is rapid and linear

In a kinetic experiment, the E4 promoter construct responded to at-RA within 2 h (Fig. 3B), reached to maximum at 8 h, and then decreased probably due to the at-RA induction of the expression of endogenous CYP26A1 which inactivates the at-RA untaken by the cells. Similar kinetic behavior was observed when lower at-RA concentration (100 nM) was used (data not shown). The E4 promoter response was much higher than that of the full-length over different concentrations of the at-RA ranging from 10^{-10} to 10^{3} M (Fig. 3C). In the logarithmic dose-response study, which encompassed up to 1 μM of at-RA concentration, the E4 promoter response was nearly linear in the range of 0.1–10 nM (Insert of Fig. 3C).

In a separate experiment we also did at-RA dose responses in the absence and presence of FBS. At zero concentration of at-RA the average promoter response was 3.44 ± 0.02 in the presence of FBS as compared to 1.889 ± 0.013 in the absence of FBS. At all doses of at-RA, the promoter responses were significantly higher in the presence than the absence of FBS (Fig. 3D).

The promoter response to at-RA might be higher if an effective inhibitor for CYP26A1 were used during incubation of the cells. However, we did not make any attempt to use any inhibitor because there is not any specific inhibitor for CYP26A1 available in the market and, in addition, we did not want to complicate the system by using a non-specific inhibitor.

The E4 promoter responses to different concentrations of at-RA were shown to be in parallel with the expression levels of endogenous CYP26A1 mRNA in HepG2 cells (Fig. 3E). Among three RAR subtypes, RARα and RARβ, compared to RARγ, were shown to have greater effect on the E4 promoter in response to at-RA in HepG2 cells as well as in HEK 293T cells (Zolfaghari et al., 2019).

4.3 CYP26A1 promoter responds to retinol (ROH) and to various retinoid analogues

The E4 promoter of CYP26A1 gene did not respond significantly to retinol treatment at or below a concentration of 10 nM (Fig. 4A), however, at 100 nM retinol the relative luciferase activity increased significantly to a level similar to that obtained by 1 nM at-RA. It is not known whether retinol acts as a potential ligand for retinoic acid receptors, or if it needs to be oxidized first for formation of at-RA, although it has been reported that all-trans-retinol may act as a functional ligand for RARs (Repa, Hanson, & Clagett-Dame, 1993).

The promoter responded comparably to Am580, an RARα specific ligand (Fig. 4B) as well as 9-cis-RA (Zolfaghari et al., 2019), which binds to both RAR and RXR receptors. However, the promoter did not respond to LG-268, an RXR specific ligand, unless it was used in combination with at-RA or RAR specific ligands (Fig. 4B). Nuclear receptors such as RARs, TRs, and VDR are considered non-permissive receptors, which are

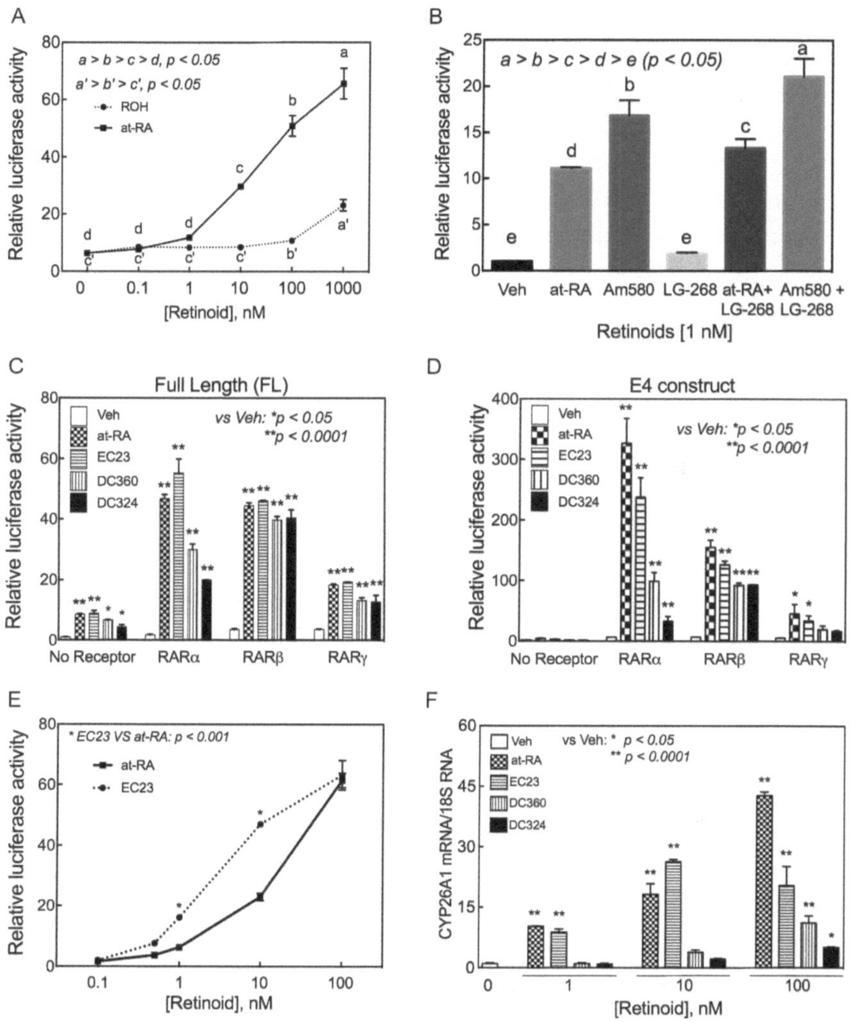

Fig. 4 CYP26A1 promoter responds to retinol and to different retinoid analogues in HepG2 cells. (A) HepG2 cells cotransfected with E4 promoter construct together with hRARα.hRXRα expression vector were treated with either at-RA or retinol at different concentrations for 24 h and then lysed for luciferase assay. The promoter responded to retinol but at higher concentrations than that of at-RA. (B) E4 promoter did not respond to RXR specific ligand alone unless it was used in combination with RAR specific ligands. (C) HepG2 cells grown in 24-well plates were cotransfected with either FL or (D) E4 promoter of human CYP26A1 promoter without or with retinoid receptors and then treated with vehicle or retinoid compounds including at-RA at 0.1 μM in the medium for 24 h after which the cells were collected for luciferase assay. (E) HepG2 cells cotransfected with E4 promoter together with hRARα.hRXRα were incubated with either at-RA or EC23 at different concentrations for 24 h after which the cells were collected for luciferase activity assay. (F) HepG2 cells grown in 12-well plates and treated with either vehicle or retinoid compounds including at-RA at 3 different concentrations in triplicate for 4 h and then collected for RNA analysis by RT-PCR. In the bar graphs, the relative expression of CYP26A1/18S ribosomal RNA level was set as 1.0 in HepG2 cells treated with vehicle only.

enhanced by their own agonists but are not affected by RXR ligands (Gilardi & Desvergne, 2014). This is in contrast to the permissive action of RXR heterodimers, which can be affected by either RXR ligands and/or the ligands of the nuclear receptors such as PPARs, LXR, PXR, FXR and CAR (Gilardi & Desvergne, 2014).

The CYP26A1 FL (Fig. 4C) and E4 (Fig. 4D) promoters were both used to determine the retinoid activities of three novel synthetic retinoids, namely, EC23, DC360 and DC324. EC23 was shown to have higher activity than the other two retinoids, DC360 and DC324 (Fig. 4C and D). EC23 is also more effective than at-RA at low concentrations based on the data from promoter activity (Fig. 4E) as well as based on CYP26A1 gene expression in at-RA-treated HepG2 cells (Fig. 4F). Similar to the results obtained using at-RA (Fig. 3E), the promoter responses to the synthetic retinoids were shown to vary in parallel with the expression levels of endogenous CYP26A1 mRNA in HepG2 cells (Fig. 4C–F).

5. mCHERRY RFP may be used as a reporter for the promoter of the CYP26A1 gene

The E4 promoter response to at-RA levels as measured by luciferase activity assay in cells is relatively specific and sensitive, and, in addition, it is reproducible in repeated experiments for measurement of at-RA levels. However, for luciferase activity assay, the cells have to be lysed and processed using an expensive assay kit. To overcome this problem, we first substituted the luciferase open reading frame in our E4 promoter construct with mCherry cDNA, as the reporter gene (Fig. 1), which upon expression produces a red fluorescent protein that can be observed by fluorescence microscopy in live cells. We tested this construct, which we refer to as E4.2 (Fig. 1), to determine how it responds to at-RA in a transient transfection mode in HepG2 cells as well as in HEK293T cells.

5.1 Instruments

1. Olympus IX70 inverted fluorescence microscope equipped with an Olympus DP72 camera and CellSens standard 1.14 image acquisition software, or a similar microscope
2. NIH ImageJ 1.49 software for analysis

5.2 Material

1. *p*CMV-YFP

6. Methods

6.1 Construction of E4.2 vector

A DNA fragment of the mCherry ORF was first amplified by PCR from a commercial plasmid vector using aaccatggTGAGCAAGGGCGAGG as sense and aatctagaTTACTTGTACAGCTCGTCCATG as antisense primer pairs, and then the fragment was cloned into pGEM-T Easy vector by TA-cloning as described (Zolfaghari et al., 2019). An isolated clone was subjected to DNA sequencing for confirmation and then double digested with *Nco*I/*Xba*I to obtain the mCherry fragment for subcloning into pGL3-Basic plasmid vector. For construction of pGL3-Basic-hCYP26A1P-E4-mCherry (E4.2 vector) clone containing the human CYP26A1 promoter, the pGL3-Basic-hCYP26A1P-E4-luciferase was first double digested with NcoI/XbaI to remove luciferase DNA fragment and then replaced with the mCherry ORF as described (Zolfaghari et al., 2019). The pGL3-Basic-hCYP26A1P-E4.2-mCherry clone created in this section has been deposited to the Addgene repository (Table 1).

6.2 Transient transfection and RFP analysis

1. Grow monolayer cells with the full growth medium containing 10% FBS in 12-well plates 1–2 day before transfection.
2. When the cells are about 60–70% confluent aspirate out the medium and replace it with the same medium containing 3% fetal bovine serum with no antibiotics about 1 h before transfection.
3. Prepare DNA samples as follow: For each well of the 12-well plates mix 1.2 μg of E4.2 and 0.2 μg pYFP, 0.2 μg of pcDNA3.1-hRARα.hRXRα in 100 μL EMEM or DMEM without FBS and antibiotics. For the cells without pcDNA3.1-hRARα.hRXRα add pcDNA3.1 empty vector DNA.
4. Mix 4 μL of Lipofectamine-2000 in 100 μL of EMEM or DMEM without FBS with no antibiotics for each well and incubate at room temperature for 5 min.
5. Mix DNA solution (Section 3) with diluted Lipofectamine-2000 (Section 4) and incubate at room temperature for 20 min before adding to the cells.
6. Add the solution prepared in the Section 5 to the cells, mix by swirling and incubate overnight at 37 °C.

7. Following transfection aspirate the medium and incubate the cells with the full-growth medium containing 10% fetal bovine serum and 0–1 μM at-RA.
8. Observe and evaluate the cells under fluorescence microscopy for expression of RFP in response to at-RA treatment and compare to the expression of YFP as an internal control for transfection efficiency.
9. At least three regions of the monolayer cells in each well were evaluated and recorded. The RFP data were analyzed and quantified using NIH ImageJ 1.49d.

7. Results

7.1 Expression results of RFP in response to at-RA are comparable to those of luciferase activities in HepG2 cells

Expression of RFP was almost none (undetectable) in HepG2 cells treated with vehicle, however, treatment of these cells with as low as 0.1 nM at-RA resulted in quantifiable expression of RPF in the cells as visualized by microscopy (Fig. 5). A gradual increase in the expression of RFP was observed in the cells treated with up to 10 nM at-RA (Fig. 5), similar to that of luciferase used as the reporter gene (Fig. 3C). We also observed that both RARα and RARβ had stronger effects on RFP expression than RARγ at all the at-RA concentrations examined in HepG2 cells and in HEK293T cells, similar to those results obtained by luciferase assay (Zolfaghari et al., 2019).

8. Permanent transfection of E4.2-CYP26A1 vector in HEK293T cells results in production of a suitable cell line for determination of retinoid activity

Based on the quantitative responses of the CYP26A1 promoter to at-RA in transient transfection mode, we decided to produce a permanent cell line possessing the CYP26A1 promoter with RFP, as a reporter gene (E4.2 Vector), which can be directly measured in live cells treated with retinoids without repeated transfection. For this, HEK293T cells were transfected with E4.2 vector and screened for selection of an appropriate cell line, which had to have low background for the expression of RFP in the absence of at-RA, while still responding to at-RA significantly upon treatment with receptor ligand. We selected HEK293T cells over HepG2 cells for several reasons: (1) these cells are easier to clone; (2) they grow

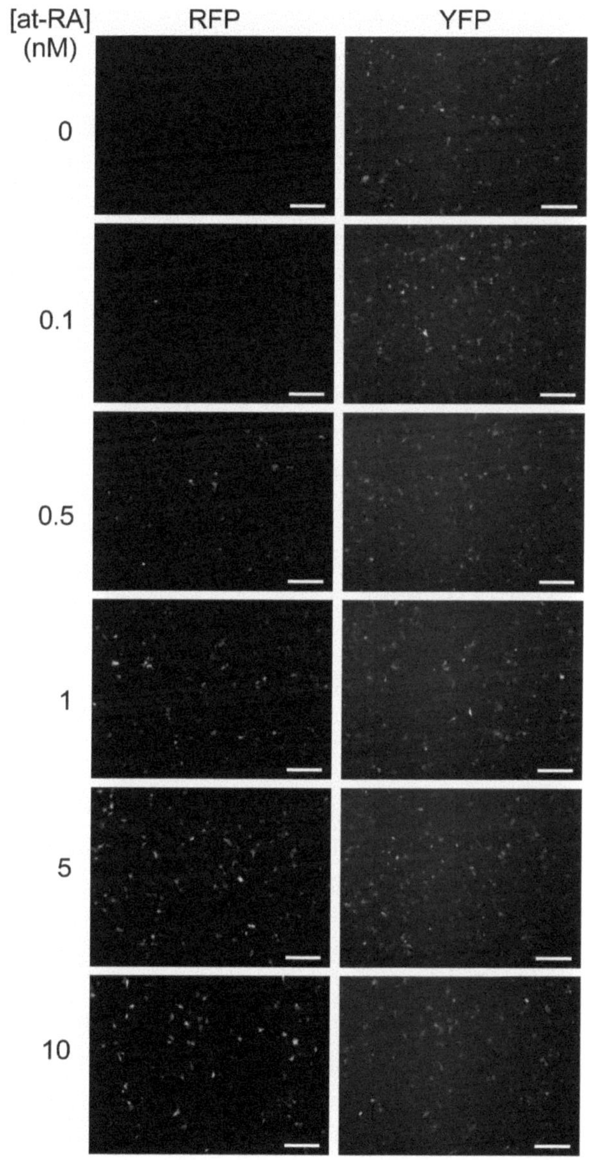

Fig. 5 E4.2 promoter of the CYP26A1 gene responds to at-RA dose-dependently for expression of RFP in HepG2 cells. Monolayer cells grown in 12-well plates were cotransfected with E4.2 promoter construct together with hRARα.hRXRα expression vector and *p*CMV-YFP as the internal control for transfection efficiency and then incubated with at-RA at different concentrations at 37 °C for 6 h. The live cells in three sections of each well were observed by microscopy for relative expression of RFP and YFP. Scale bar represents 200 μm.

rapidly and are easily trypsinized into individual cells for flow cytometry; (3) they have an apparently higher copy number of the reporter gene fused in the right places in the genome, and therefore a higher signal in response to retinoids; and (4) they tolerate high level of retinoids including at-RA. A monoclonal cell line, namely HEK293T 1A1 cells, was isolated and used to estimate the retinoid activities of novel synthetic retinoids as well as of extracts prepared from rat tissue samples, using both retinol and at-RA as the standard references for the assay.

8.1 Instruments

1. Beckman Coulter MoFlo Astrios EQ Cell Sorter, or an equivalent sorter
2. Accuri C6 flow cytometer (Accuri, BD Biosciences, San Jose, CA), or equivalent analytical equipment

8.2 Material

1. Geneticin, G418 (Thermo)

9. Methods

9.1 Permanent transfection of HEK293T cells with E4.2 vector

Monolayers of HEK293T cells grown in 12-well plates were cotransfected with E4.2 vector and expression vectors for hRARα and hRXRα (see Section 6.2). After 2 days, the cells were transferred and grown in a new 12-well plate with full growth medium containing 300 μg/mL G418 as the selection agent. Following 2 weeks, the live cells were transferred into a new plate and incubated with 1 μM at-RA for 24 h after which the cells expressing RFP were first cloned by limiting dilution, then sorted into individual cells using a Beckman Coulter MoFlo Astrios EQ Cell Sorter, with discrimination based on at-RA-induced RFP expression. The individual cells were grown in 96-well plates for 2 weeks and then transferred into 12-well plates with full growth medium containing G418 (300 μg/mL as above). The resulting monoclonal lines were assessed for expression of RFP after treatment with at-RA under a fluorescence microscope and a cell line, named HEK293T 1A1, was selected for further processing and analysis (Zolfaghari et al., 2019).

9.2 Treatment of HEK293T 1A1 cells with at-RA for RNA extraction

1. Grow HEK293T 1A1 cells with full growth medium containing 10% FBS and 300 μg/mL G418 in12-well plates for 1–2 days
2. When the cells are 60–70% confluent aspirate the medium and replace with the fresh full growth medium with at-RA at different concentrations and incubate at 37 °C for 24 h.
3. After observing the cells under fluorescence microscope for expression of RFP aspirate out the medium and wash the cells with PBS.
4. Extract total RNA from the cells with Trizol reagent and analyze for expression of mCherry mRNA by RT-PCR as described on Section 3.2 using primer pairs for mCherry (see Section 6.1) and for18S ribosomal RNA as the control (see Section 3.2).
5. Analyze the data using Step One Plus Real Time PCR System (see Section 3.2).

9.3 Treatment of HEK293T 1A1 cells with at-RA treated conditioned media from HepG2 cells for RFP expression analysis by flow Cytometry

1. Treat HepG2 cells with at-RA at different concentrations in a 12-well plate for 24 h and collect the medium and store frozen at −80 °C.
2. Grow monolayer HEK293T 1A1 cells in full growth medium containing 10% FBS and 300 μg/mL G418 in 12-well plates and treat them with either the conditioned media from HepG2 cells (from Section 1) or at-RA as the reference standard at different concentrations for 24 h.
3. Aspirate out the medium and wash the cells with 1 mL of PBS, add 150 μL of trypsin-EDTA, and incubate at 37 °C for 5 min.
4. Add 1 mL of the full growth medium and pipet up and down to break the cells into individual ones.
5. Transfer the cells into each 5-mL tube and wash the wells with 1 mL more PBS and transfer to the tubes.
6. Centrifuge the tubes at 1000 × g for 15 min, aspirate out the medium and then resuspend the cells into 150 μL PBS. Pipet up and down to disperse the cells.
7. Analyze the cells with flow cytometer using wild type HEK293T cells treated with at-RA for 24 h as the base.

10. Results

10.1 mCherry mRNA was expressed in HEK293T 1A1Cells proportionally in response to retinoic acid levels

HEK293T 1A1 cells were treated with at-RA at different concentrations for 24 h and then collected for RNA extraction and analysis. Similar to the RFP signal, the mCherry mRNA level was almost undetectable when the cells were treated with vehicle, however, it was increased proportionally in response to at-RA concentrations (Fig. 6A).

10.2 Expression of RFP in HEK293T 1A1 cells could be measured by flow Cytometry

Monolayer cells grown in 12-well plates were incubated for 24 h with either at-RA-treated conditioned media from HepG2 cells or an at-RA reference standard at concentrations from 1 to 500 nM and then prepared for flow cytometry. RFP expression levels were linearly correlated in response to at-RA treatment ($R^2 = 0.96$, $P < 0.0001$) (Fig. 6B). The conditioned media was as effective as the at-RA reference standard at the lower concentrations, however, they were more effective than the reference standards at the higher concentrations of at-RA. For example, RFP expression induced in the cells by the conditioned medium was about twice that induced by the at-RA reference standard alone at 500 nM concentration (Fig. 6B). It should be mentioned that we found no particular by-product of at-RA in the conditioned medium as analyzed by Mass-Spectroscopy (data not shown).

10.3 HEK293T 1A1 cells could be used to measure the retinoid activity of synthetic retinoid analogues

Monolayer HEK293T 1A1 cells in 12-well plates were incubated with either synthetic retinoid analogues or at-RA, each at three different concentrations of 10, 100, and 1000 nM for 24 h and then analyzed by fluorescence microscopy for expression of RFP. The RFP expressing cells were counted and the relative fluorescence intensity levels were calculated based relative to the expressed RFP in cells treated with vehicle (Fig. 7A). The cells were significantly more responsive in term of RFP expression to EC23 than to at-RA at either 10 or 100 nM (Fig. 7B and C). However, at 1000 nM, EC23 was less effective than at-RA, probably indicating saturation of EC23 as a potent ligand for retinoic acid receptors. In fact, the cells

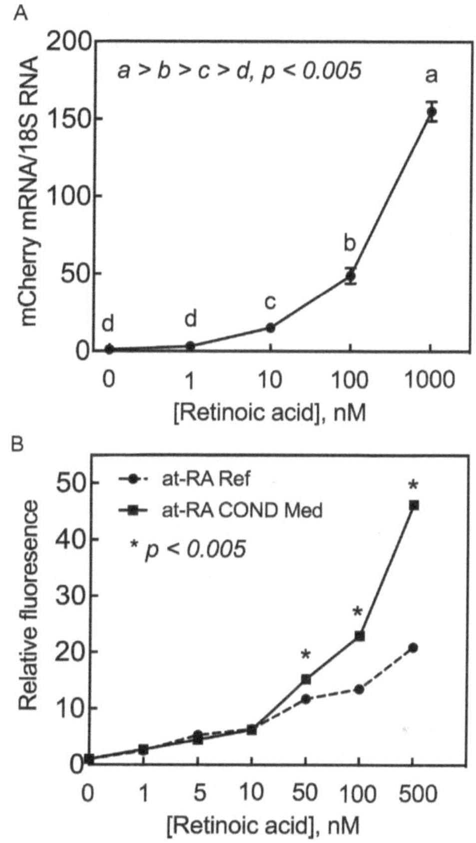

Fig. 6 Red fluorescent protein (RFP) is expressed quantitatively in HEK293T 1A1 cells in response to at-RA. (A) HEK293T 1A1 cells grown in 12-well plates were incubated with at-RA at different concentrations for 24h after which the cells were washed and collected for total RNA isolation. The RNA samples were then subjected to RT-PCR analysis for expression of mCherry RNA in triplicate using 18S ribosomal RNA expression as the internal control. (B) HEK293T 1A1 cells grown in 12-well plates were incubated with either at-RA at different concentrations or 24-h at-RA-treated conditioned media from HepG2 cells. For conditioned media preparation, HepG2 cells were incubated with at-RA at different concentrations in full growth media for 24h. The media were collected and then incubated at 50% with HEK293T 1A1 cells without addition of any at-RA. For reference standard, the HEK293T 1A1 cells were incubated with 50% HepG2 cell 24-h vehicle-treated conditioned media with added at-RA at different concentrations for 24h. The HEK293T 1A1 cells in duplicate wells were then trypsinized and collected for flow cytometry to quantify the cells expressing RFP.

responded to at-RA by the expression of RFP dose-dependently. Two other retinoid analogues, DC360 and DC324, were significantly less effective than at-RA in inducing expression of RFP in these cells (Fig. 7B and C).

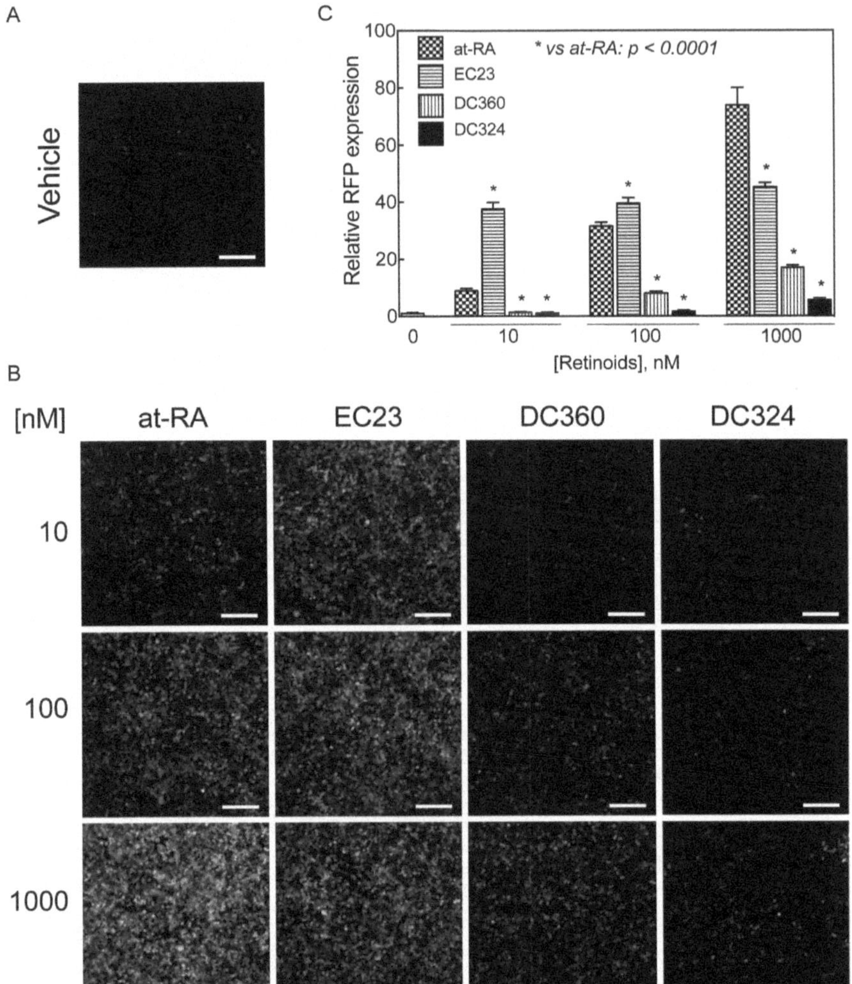

Fig. 7 HEK293T 1A1 cells responds to the synthesized retinoid analogues. (A) and (B) Monolayer HEK293T 1A1 cells in 12-well plates were incubated with either at-RA or retinoid analogues, each at different concentrations for 24 h and then observed under a microscope for expression of RFP. (C) The RFP expressing cells were counted using the NIH ImageJ program and the relative fluorescence intensity levels were calculated based on the expression of RFP in vehicle-treated cells. Values represent the mean of 3 different regions of the well ± SD. Scale bar represents 200 μm.

In a separate experiment, monolayer HEK293T 1A1 cells in 12-well plates were treated with either EC23 or at-RA at concentrations from 0 to 100 nM for 24 h, after which the cells were analyzed by fluorescence microscopy for RFP. EC23 was shown to be more effective than at-RA in inducing RFP expression in these cells at concentrations from 0.5 to 100 nM (Fig. 8A and B).

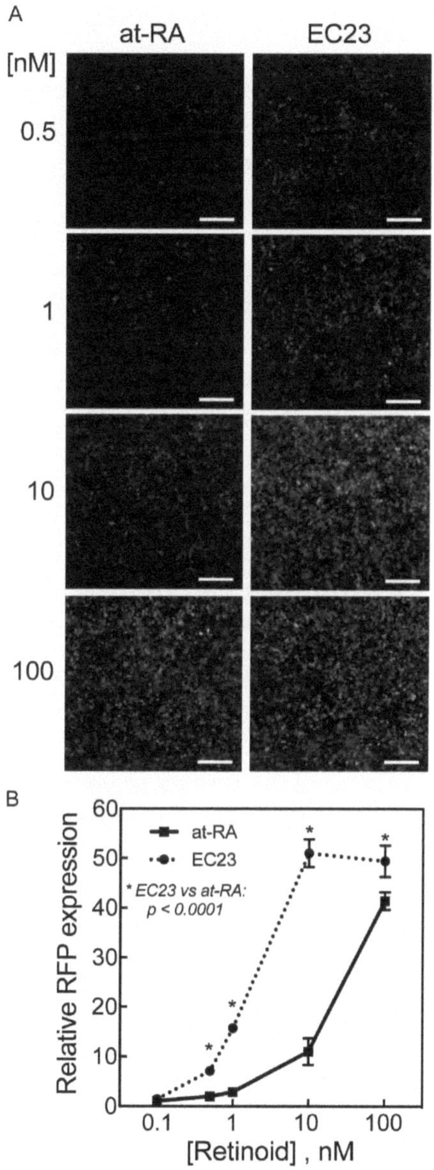

Fig. 8 EC23 was shown to have more retinoid activity than at-RA at lower concentrations. (A) Monolayer HEK293T 1A1 cells in 12-well plates were incubated with either EC23 or at-RA, each at different concentrations for 24 h and then observed under microscope for RFP expression. (B) The RFP expressing cells were counted using NIH ImageJ program and the relative fluorescence intensity levels were calculated based on the expressed RFP in cells treated with vehicle. The values represent the mean of 3 different regions of the well ± SD. Scale bar represents 200 μm.

These results were comparable to those obtained in HepG2 cells transfected with CYP26A1 E4 promoter with luciferase as the reporter gene (Fig. 4E).

Compared to at-RA, these synthetic retinoid analogues were shown to exhibit a longer-lasting effect on RFP expression in HEK293T 1A1 cells, measured after treatment with different concentrations for either 1 or 4 days. The RFP signals in the cells treated with EC23, compared to at-RA, were higher in the 4-day treatment than in the 24-h treatment (Fig. 9A, B and E). Likewise, RFP signals were doubled in cells treated with either DC360 or DC324 for 4 days as compared to those in the cells treated for 24 h (Fig. 9C–E).

at-RA is not only the inducer of CYP26A1 but also it is its specific substrate whereas EC23 and other synthetic retinoids used in this study are apparently considered as the inducers of CYP26A1 expression but are not the apparent substrates for CYP26A1.

11. CYP26A1 promoter can be used to estimate the retinoid activity levels in tissue samples

As an alternative to conventional methods we used CYP26A1 promoter to estimate the retinoid activity in tissue samples, using rat tissues as tests. For this, we treated rats orally with either vehicle as the control or at-RA and then euthanized the animals to collect blood, liver, and lung samples. Total retinoid activity was measured in serum and tissue extracts using the E4 plasmid construct in HepG2 cells for luciferase activity determination and HEK293T 1A1 cells for RFP analysis. As the reference standards, the cells were treated with either retinol or at-RA at different concentrations. For comparison to conventional method total retinol levels in tissue samples were measured by UPLC.

11.1 Instruments

1. CO_2 Tank
2. N2 Tank
3. Water bath
4. Surgery tools
5. Table top centrifuge
6. Homogenizer
7. Ultra Performance Liquid Chromatography (UPLC) (Waters), or a comparable instrument system

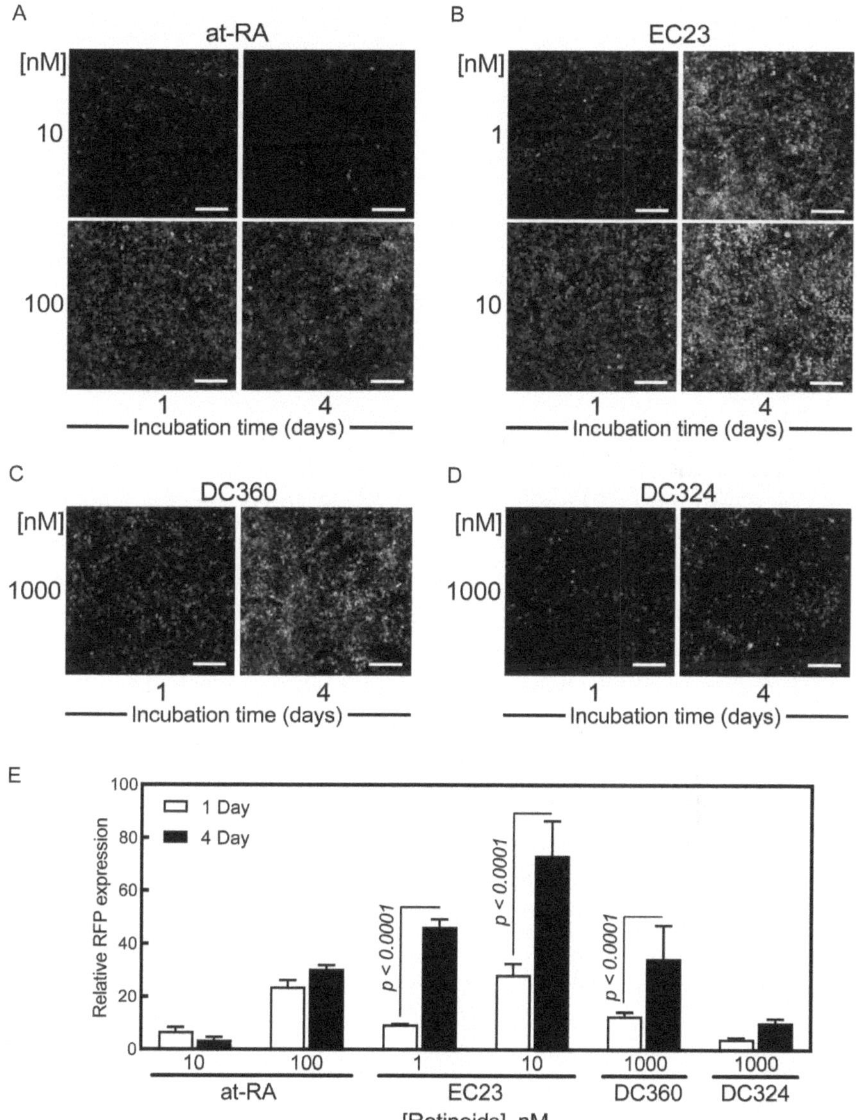

Fig. 9 Synthetic retinoid analogues have apparently more extended retinoid activity than at-RA in expression of RFP in HEK293T 1A1 cells. Monolayer HEK293T 1A1 cells grown in 12-well plates were incubated with either at-RA at 10 and 100 nM (A), EC23 at 1 and 10 nM concentrations (B), DC360 (C) or DC324 (D), each at 1000 nM for either 1 or 4 days and RFP expression was observed by fluorescence microscopy, and, (E) measured using NIH ImageJ. The values represent the mean of 3 different regions of the well ± SD. Scale bar represents 200 μm.

11.2 Material

1. Potassium hydroxide (Sigma)
2. Pyrogallol (Sigma)
3. Hexane
4. Trimethylmethoxyphenyl (TMMP)-retinol (Sigma)
5. Methanol
6. Acetonitrile
7. Vegetable oil

12. Methods

12.1 Animal experiment

The protocol for animal use in this study was approved by The Animal Care and Use Committee of The Pennsylvania State University. Adult female rats were orally treated once with either vegetable oil or at-RA at 1 mg, 2 mg, or 5 mg/kg body weight for 6 h and then euthanized by CO_2 inhalation. Blood was withdrawn and liver and lung tissue samples were snap frozen in liquid nitrogen immediately. Serum samples were prepared by centrifugation and stored frozen at −20 °C while liver and lung samples were stored frozen at −80 °C until use.

12.2 Retinoid extraction

Individual serum samples (0.1 mL) were subjected to ethanol and then hexane extraction for total retinoid partition by UPLC (Zolfaghari & Ross, 2002). Part of the serum samples was used directly to incubate with the cells for retinoid activity determination. Liver and lung samples (about 0.5 g) were homogenized in ethanol (1:10, w:v) and then saponified after which the samples were mixed with 12 mL of hexane to partition the total free retinoids (Zolfaghari & Ross, 2000). A portion of the ethanol and hexane extract, 1 mL from each, was evaporated to dryness under a light flow of N_2 gas and immediately dissolved in 1 mL of ethanol before dilution for incubation with cells for retinoid determination. Liver (>2000 times diluted) and lung (>80 times diluted) extracts were separately diluted before addition to the cell media. Retinol and at-RA at different concentrations were used as the reference standards. The hexane extract of the tissue samples was also subjected to UPLC for total free retinol determinations (Zolfaghari & Ross, 2000). We used this method because we assume most of the retinoids, if not entirely, in the tissues consists of retinol and retinyl esters.

13. Results

13.1 HEK293T 1A1 cells were used to estimate the retinoid activity levels in rat serum

HEK293T 1A1 cells were directly incubated with rat serum at a 20% dilution in DMEM media without FBS for 24 h and then observed under fluorescence microscope for RFP expression; either exogenous retinol or at-RA was included in separate wells as reference standards, added in DMEM media with 20% FBS. As in the previous experiments, there was no significant expression of RFP in 1A1 cells treated with vehicle (Fig. 10A). Whereas HEK293T 1A1 cells responded effectively to at-RA in the expression of RFP at a concentration as low as 1 nM, retinol did not have a significant effect on the expression of RFP at a concentration of 100 nM (Fig. 10B and C). As compared to the at-RA reference standards, the control serum sample from the vehicle-treated rat was equivalent to ~2.5 nM at the 20% serum dilution used (Fig. 10B and C). Then, serum samples from rat treated with 1, 2 and 5 mg of at-RA/kg BW were tested. Serum (after 1 mg at-RA/kg) increased the RFP expression slightly (Fig. 10B and C), however, serum from rats treated with 2 and 5 mg at-RA/kg BW elevated the at-RA concentrations in the serum to about 30 and 180 nM, respectively, as determined by RFP expression in HEK293T 1A1 cells (Fig. 10B and C). Total retinol concentrations of serum as determined by UPLC were all in the normal ranges of 0.98 for the control rat and 0.77, 0.74, and 1.74 μM for rats treated with at-RA at 1, 2, and 5 mg/kg body weight, respectively.

13.2 Retinoid activity levels were estimated in tissue extract of rat liver and lung

We also estimated the retinoid activity levels in rat liver and lung samples using luciferase activity assay in HepG2 cells and RFP expression in HEK293T 1A1 cells. For this, total retinoids were extracted from liver and lung samples and then incubated with either HepG2 cells cotransfected with E4 plasmid vector and RARα.RXRα expression vector for luciferase expression or HEK293T 1A1 cells for RFP expression. Based on the retinol reference standards the total retinoid activity levels of liver and lung samples as measured by luciferase expression are comparable to those retinol levels determined by UPLC as the conventional method (Fig. 11). However, the estimated values as determined by RFP in HEK293T 1A1 cells were somehow higher than those determined by UPLC.

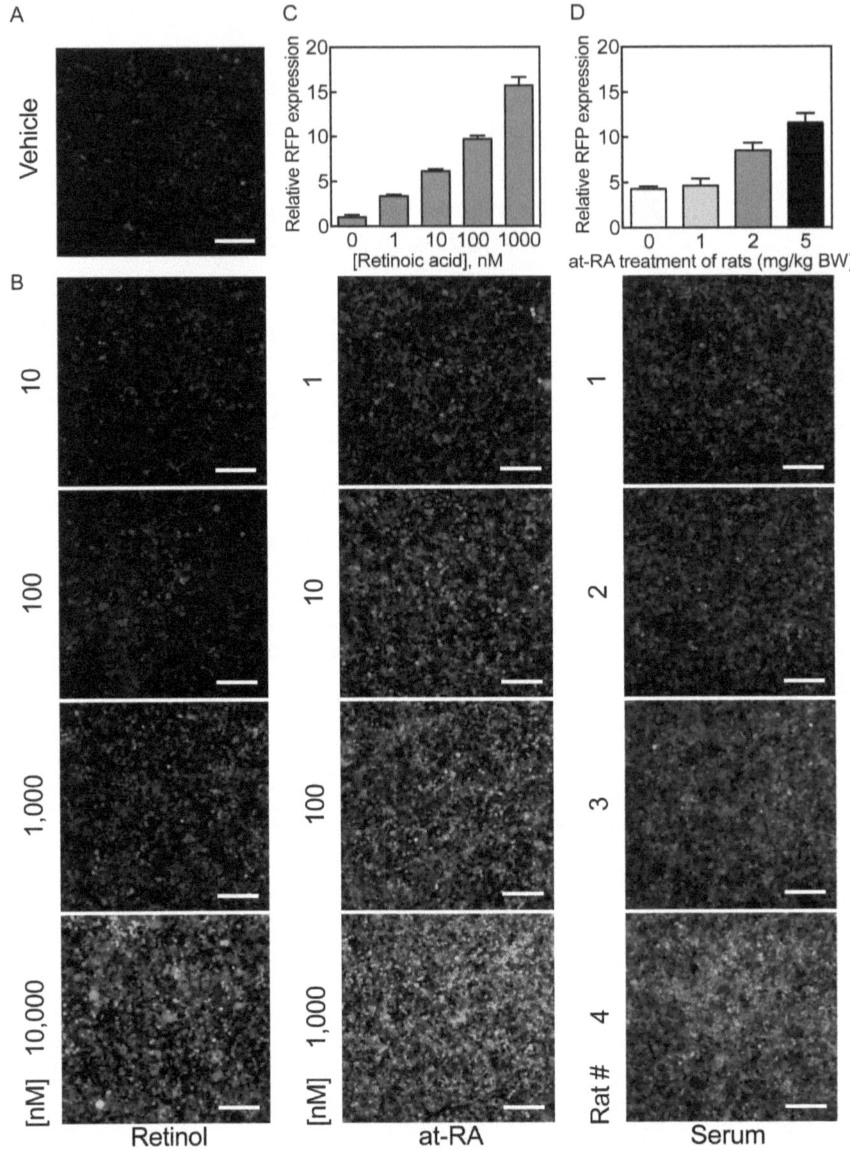

Fig. 10 Total retinoid activity may be estimated in intact rat serum by HEK293T 1A1 cells. Adult female rats were treated orally with either oil as the vehicle control or at-RA at 1, 2, and 5 mg/kg body weight for 6 h and then sacrificed for blood, liver and lung tissue collection. Serum samples were prepared by centrifugation and stored at −20 °C until analysis. Total serum retinol concentrations as determined by UPLC were all in the normal ranges of 0.98 for the control rat and 0.77, 0.74, and 1.74 μM for rats treated with at-RA at 1, 2, and 5 mg/kg body weight, respectively. (A) and (B) Cells were incubated with intact rat serum samples at 20% in DMEM media without added FBS for 24 h, and then observed by fluorescence microscopy for expression of RFP using retinol and at-RA at different concentrations in DMEM media with added 20% FBS as the standards. (C) and (D) RFP expressing cells were counted using NIH ImageJ program and the relative fluorescence intensity levels were calculated based on the expressed RFP in cells treated with vehicle. The retinoid activity levels of the serum samples tested were calculated based on the at-RA standard. The values represent the mean of three different regions of the well ± SD. Scale bar represents 200 μm.

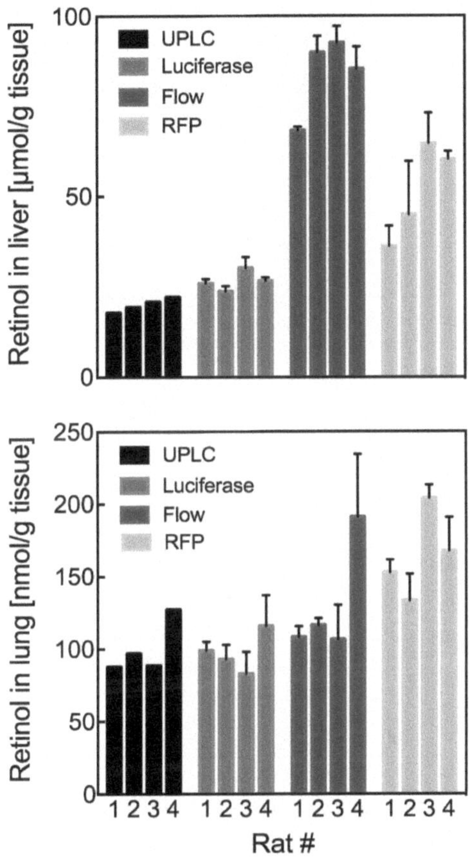

Fig. 11 HEK293T cells 1A1 cells were used to estimate total retinol based on the retinoid activity in the rat liver and lung. Liver and lung samples (about 0.5 g) from rats treated with either oil or at-RA for 6 h (see Sections 12.1 and 12.2 and the legend of Fig. 10) were homogenized in ethanol (1:10 w:v) and saponified before addition of hexane (12 mL) for partitioning. The hexane extract samples (2 mL) were dried and dissolved in 1 mL of ethanol which was then diluted (2000 times for liver extract and 80 times for lung extract) before incubation with HepG2 cells for luciferase assay and HEK293T 1A1 cells for RFP expression using retinol and at-RA at different concentrations as the reference standards. Part of the hexane extract samples was used for determination of total retinol by UPLC.

Although most, if not all, of the vitamin A contents in tissues including liver and lung are in the form of retinol and retinyl esters, we also determined the total retinoid activity levels based on at-RA as the reference standard in the liver and lung samples using luciferase activity in HepG2 cells (Table 2).

Table 2 Retinoic acid activity levels in the liver and lung of rats.[a]

Rat no.	Treatment of rat with at-RA (mg/Kg body weight)	at-RA levels (nmol/g tissue)α	
		Liver	Lung
1	0	189 ± 18	2.62 ± 0.11
2	1	109 ± 07	2.91 ± 0.23
3	2	148 ± 07	2.68 ± 0.30
4	5	197 ± 35	3.62 ± 0.53

[a]The hexane extract from the rat liver and lung samples (see Section 12.2) were dried and dissolved in ethanol before incubation with HepG2 cells for luciferase assay using at-RA at different concentrations as the reference standards (Mean ± SD).

Although both retinol and at-RA may be used as the standard references, we should be cautious that retinol is much less stable than retinoic acid. Whereas at-RA in either oil or ethanol is stable even after 2 years, retinol is not stable even after a few months when it is stored at −20 °C. This is probably the reason that TMMP-retinol or retinyl acetate is used as the reference standard in UPLC (Zolfaghari & Ross, 2000). We have not tried either of these compounds in this culture cell system.

14. Summary

A short version of the promoter of the CYP26A1 gene driving the expression of either luciferase (E4) or RFP in a promoter-less vector (E4.2) was used to test its response to at-RA in HepG2 cells and HEK293T cells cotransfected with RARs. The promoter responded nearly linearly to a wide range of the log-10 concentrations of at-RA and was sensitive enough to detect at-RA activity as low as 0.5 nM. The promoter responded better to at-RA when the cells were cotransfected with RARα.RXRα as compared to the other RARs. It responded to retinol and was used to estimate the retinoid activities of three synthetic retinoids. A stable cell line of HEK293T cells was established to express RFP under the CYP26A1 E4.2 promoter to respond to at-RA as well as to retinol. This cell line could be observed live for expression of RFP using fluorescence microscopy or trypsinized easily into individual cells for quantification of RFP expression using flow cytometry. This system was used to estimate the retinoid activities of synthetic retinoids as well as biological specimens such as serum and the retinoid extracts obtained from liver and lung.

References

Altucci, L., Leibowitz, M. D., Ogilvie, K. M., de Lera, A. R., & Gronemeyer, H. (2007). RAR and RXR modulation in Cancer and metabolic disease. *Nature Reviews Drug Discovery*, *6*(10), 793–810. https://doi.org/10.1038/nrd2397.

Carrier, M., & Rochette-Egly, C. (2015). Control of gene expression by nuclear retinoic acid receptors. In P. Dollé & K. Neiderreither (Eds.), *The Retinoids* (pp. 91–116). John Wiley & Sons, Inc.

Chien, A. L. (2018). Retinoids in acne management: Review of current understanding, future considerations, and focus on topical treatments. *Journal of Drugs in Dermatology*, *17*(12 Suppl), s51–s55.

Chisholm, D. R., Tomlinson, C. W. E., Zhou, G.-L., Holden, C., Affleck, V., Lamb, R., et al. (2019). Fluorescent retinoic acid analogues as probes for biochemical and intracellular characterization of retinoid Signaling pathways. *ACS Chemical Biology*, *14*(3), 369–377. https://doi.org/10.1021/acschembio.8b00916.

de Thé, H., & Fenaux, P. (2015). Retinoic acid in acute myeloid Leukemia's. *The Retinoids*, 543–555. https://doi.org/10.1002/9781118628003.ch25.

Gilardi, F., & Desvergne, B. (2014). RXRs: Collegial Partners. In M. A. Asson-Batres & C. Rochette-Egly (Eds.), *The biochemistry of retinoic acid receptors I: Structure, activation, and function at the molecular level* (pp. 75–102). Dordrecht: Springer Netherlands.

Isoherranen, N., & Zhong, G. (2019). Biochemical and physiological importance of the CYP26 retinoic acid hydroxylases. *Pharmacology & Therapeutics*, 107400. https://doi.org/10.1016/j.pharmthera.2019.107400.

Jurutka, P. W., & Wagner, C. E. (2019). Methods to assess activity and potency of Rexinoids using rapid luciferase-based assays: A case study with NEt-TMN. In S. K. Ray (Ed.), *Retinoid and rexinoid signaling: Methods and protocols* (pp. 95–108). New York, NY: Springer New York.

Kane, M. A., & Napoli, J. L. (2010). Quantification of endogenous retinoids. *Methods in Molecular Biology*, *652*, 1–54.

Lee, L. M. Y., Tam, S. T. K., McCaffery, P. J., & Shum, A. S. W. (2019). Highly sensitive quantitative determination of retinoic acid levels, retinoic acid synthesis, and catabolism in embryonic tissue using a reporter cell-based method. In S. K. Ray (Ed.), *Retinoid and rexinoid Signaling: Methods and protocols* (pp. 181–192). New York, NY: Springer New York.

Loudig, O., Maclean, G. A., Dore, N. L., Luu, L., & Petkovich, M. (2005). Transcriptional co-operativity between distant retinoic acid response elements in regulation of Cyp26A1 Inducibility. *Biochemical Journal*, *392*(Pt 1), 241–248.

Mendoza-Parra, M.-A., Bourguet, W., de Lera, A. R., & Gronemeyer, H. (2015). Retinoid receptor-selective modulators *The Retinoids* (pp. 165-192): John Wiley & Sons, Inc.

Pennimpede, T., Cameron, D. A., MacLean, G. A., Li, H., Abu-Abed, S., & Petkovich, M. (2010). The role of CYP26 enzymes in defining appropriate retinoic acid exposure during embryogenesis. *Birth Defects Research Part A, Clinical and Molecular Teratology*, *88*(10), 883–894.

Repa, J. J., Hanson, K. K., & Clagett-Dame, M. (1993). All-trans-retinol is a ligand for the retinoic acid receptors. *Proceedings of the National Academy of Sciences*, *90*(15), 7293–7297.

Ross, A. C., & Zolfaghari, R. (2011). Cytochrome P450s in the regulation of cellular retinoic acid metabolism. *Annual Review of Nutrition*, *31*(1), 65–87. https://doi.org/10.1146/annurev-nutr-072610-145127.

Spinella, M. J., Freemantle, S. J., & Dmitrovsky, E. (2015). Advances in the use of retinoids in Cancer therapy and prevention. *The Retinoids*, 557–574. https://doi.org/10.1002/9781118628003.ch26.

Urban, S., Ye, T., & Davidson, I. (2015). How the RAR–RXR heterodimer recognizes the genome. In P. Dollé & K. Neiderreither (Eds.), *The retinoids* (pp. 151–163). John Wiley & Sons, Inc.

Zhang, Y., Zolfaghari, R., & Ross, A. C. (2010). Multiple retinoic acid response elements cooperate to enhance the inducibility of CYP26A1 gene expression in liver. *Gene*, *464*(1–2), 32–43.

Zolfaghari, R., Mattie, F. J., Wei, C. H., Chisholm, D. R., Whiting, A., & Ross, A. C. (2019). CYP26A1 gene promoter is a useful tool for reporting RAR-mediated retinoid activity. *Analytical Biochemistry*, *577*, 98–109. https://doi.org/10.1016/j.ab.2019.04.022.

Zolfaghari, R., & Ross, A. C. (2000). Lecithin:retinol acyltransferase from mouse and rat liver. cDNA cloning and liver-specific regulation by dietary vitamin a and retinoic acid. *Journal of Lipid Research*, *41*(12), 2024–2034.

Zolfaghari, R., & Ross, A. C. (2002). Lecithin:Retinol Acyltransferase expression is regulated by dietary vitamin a and exogenous retinoic acid in the lung of adult rats. *The Journal of Nutrition*, *132*(6), 1160–1164.

Zolfaghari, R., & Ross, A. C. (2014). Hepatocyte nuclear factor 4α (HNF4α) in coordination with retinoic acid receptors increases all-trans-retinoic acid-dependent CYP26A1 gene expression in HepG2 human hepatocytes. *Journal of Cellular Biochemistry*, *115*(10), 1740–1751. https://doi.org/10.1002/jcb.24839.

FSC
www.fsc.org
MIX
Papier | Fördert
gute Waldnutzung
FSC® C083411

Kolibri 360 GmbH
Ferdinand-Jühlke-Straße 7
99095 Erfurt, Deutschland
produktsicherheit@kolibri360.de